Introduction to Information Optics

Introduction to Information Optics

Edited by

FRANCIS T. S. YU

The Pennsylvania State University

SUGANDA JUTAMULIA

Blue Sky Research

SHIZHUO YIN

The Pennsylvania State University

ACADEMIC PRESS

A Harcourt Science and Technology Company

San Diego San Francisco New York Boston
London Sydney Tokyo

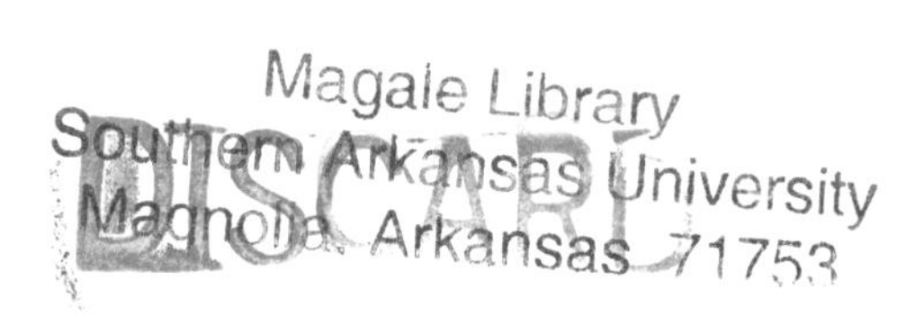

This book is printed on acid-free paper. ∞

ACADEMIC PRESS
A Harcourt Science and Technology Company
525 B Street Suite 1900, San Diego, California 92101-4495, USA
http://www.academicpress.com

ACADEMIC PRESS
24-28 Oval Road, London NW1 7DX, UK

Library of Congress Cataloging Number: 2001089409

ISBN: 0-12-774811-3

PRINTED IN THE UNITED STATES OF AMERICA
01 02 03 04 05 SB 9 8 7 6 5 4 3 2 1

Contents

Preface

We live in a world bathed in light. Light is not only the source of energy necessary to live—plants grow up by drawing energy from sunlight; light is also the source of energy for information - our vision is based on light detected by our eyes (but we do not grow up by drawing energy from light to our body through our eyes). Furthermore, applications of optics to information technology are not limited to vision and can be found almost everywhere. The deployment rate of optical technology is extraordinary. For example, optical fiber for telecommunication is being installed about one kilometer every second worldwide. Thousands of optical disk players, computer monitors, and television displays are produced daily.

The book summarizes and reviews the state of the art in information optics, which is optical science in information technology. The book consists of 12 chapters written by active researchers in the field. Chapter 1 provides the theoretical relation between optics and information theory. Chapter 2 reviews the basis of optical signal processing. Chapter 3 describes the principle of fiber optic communication. Chapter 4 discusses optical switches used for communication and parallel processing. Chapter 5 discusses integral transforms, which can be performed optically in parallel. Chapter 6 presents the physics of optical interconnects used for computing and switching. Chapter 7 reviews pattern recognition including neural networks based on optical Fourier transform and other optical techniques. Chapter 8 discusses optical storage including holographic memory, 3D memory, and near-field optics. Chapter 9 reviews digital optical computing, which takes advantage of parallelism of optics. Chapter 10 describes the principles of optical fiber sensors. Chapter 11 introduces advanced displays including 3D holographic display. Chapter 12 presents an overview of fiber optical networks.

The book is not intended to be encyclopedic and exhaustive. Rather, it merely reflects current selected interests in optical applications to information technology. In view of the great number of contributions in this area, we have not been able to include all of them in this book.

Chapter 1 Entropy Information and Optics

Francis T.S. Yu
THE PENNSYLVANIA STATE UNIVERSITY

Light is not only the mainstream of energy that supports life; it also provides us with an important source of information. One can easily imagine that without light, our present civilization would never have emerged. Furthermore, humans are equipped exceptionally good (although not perfect) eyes, along with an intelligent brain. Humans were therefore able to advance themselves above the rest of the animals on this planet Earth. It is undoubtedly true that if humans did not have eyes, they would not have evolved into their present form. In the presence of light, humans are able to search for the food they need and the art they enjoy, and to explore the unknown. Thus *light*, or rather *optics*, provide us with a very valuable source of information, the application of which can range from very abstract artistic interpretations to very efficient scientific usages.

This chapter discusses the relationship between entropy information and optics. Our intention is not to provide a detailed discussion, however, but to cover the basic fundamentals that are easily applied to optics. We note that *entropy information* was not originated by optical scientists, but rather by a group of mathematically oriented electrical engineers whose original interest was centered on electrical communication. Nevertheless, from the very beginning of the discovery of entropy information, interest in its application has never totally been absent from the optical standpoint. As a result of the recent development of optical communication, signal processing, and computing, among other discoveries, the relationship between optics and entropy information has grown more profound than ever.

1.1. INFORMATION TRANSMISSION

Although we seem to know the meaning of the word *information*, fundamentally that may not be the case. In reality, information may be defined as related to usage. From the viewpoint of mathematic formalism, entropy information is basically a *probabilistic* concept. In other words, without probability theory there would be no entropy information.

An information transmission system can be represented by a block diagram, as shown in Fig. 1.1. For example, a message represents an information source which is to be sent by means of a set of written characters that represent a code. If the set of written characters is recorded on a piece of paper, the information still cannot be transmitted until the paper is illuminated by a visible light (the transmitter), which obviously acts as an information carrier. When light reflected from the written characters arrives at your eyes (the receiver), a proper decoding (translating) process takes place; that is, character recognition (decoding) by the user (your mind). This simple example illustrates that we can see that a suitable encoding process may not be adequate unless a suitable decoding process also takes place. For instance, if I show you a foreign newspaper you might not be able to decode the language, even though the optical channel is assumed to be perfect (i.e., noiseless). This is because a suitable decoding process requires *a priori* knowledge of the encoding scheme; for example, the knowledge of the foreign characters. Thus the decoding process is also known as *recognition process*. Information transmission can be in fact represented by *spatial* and *temporal* information. The preceding example of transmitting written characters obviously represents a spatial information transmission. On the other hand, if the written language is transmitted by coded light pulses, then this

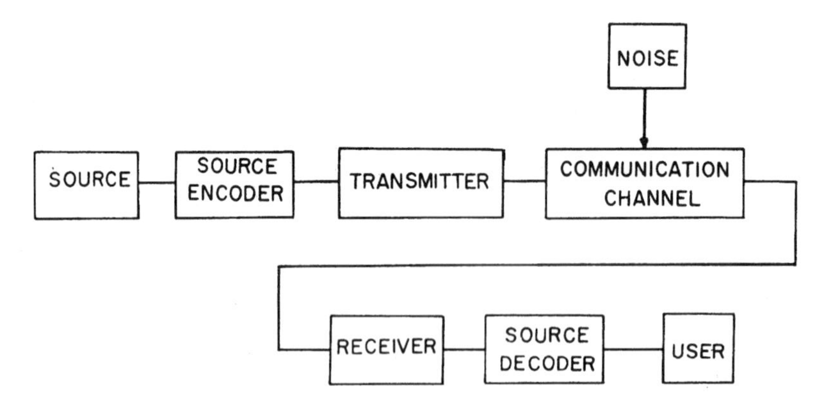

Fig. 1.1. Block diagram of a communication system.

language should be properly (temporally) encoded; for instance, as transmitted through an optical fiber, which represents a *temporal communication channel.* Needless to say, at this receiving end a temporally decoding process is required before the temporal coded language is sent to the user. Viewing a television show, for example, represents a one-way spatial–temporal transmission. It is interesting to note that temporal and spatial information can be *traded* for information transmission. For instance, television signal transmission is a typical example of exploiting the temporal information transmission for spatial information transmission. On the other hand, a movie sound track is an example of exploiting spatial information transmission for temporal information transmission.

Information transmission has two basic disciplines: one developed by Wiener [1.1, 1.2], and the other by Shannon [1.3, 1.4]. Although both Wiener and Shannon share a common interest, there is a basic distinction between their ideas. The significance of Wiener's work is that, if a signal (information) is corrupted by some physical means (e.g., noise, nonlinear distortion), it may be possible to recover the signal from the corrupted one. It is for this purpose that Wiener advocates correlation detection, optimum prediction, and other ideas. However, Shannon carries his work a step further. He shows that the signal can be optimally transferred provided it is properly encoded; that is, the signal to be transferred can be processed before and after transmission. In other words, it is possible to combat the disturbances in a communication channel by properly encoding the signal. This is precisely the reason that Shannon advocates the information measure of the signal, communication channel capacity, and signal coding processes. In fact, the main objective of Shannon's theory is the efficient utilization of the information channel.

A fundamental theorem proposed by Shannon may be the most surprising result of his work. The theorem can be stated approximately as follows: Given a stationary finite-memory information channel having a channel capacity C, if the binary information transmission rate R (which can be either spatial or temporal) of the signal is smaller than C, there exists a channel encoding and decoding process for which the probability of error per digit of information transfer can be made arbitrarily small. Conversely, if the formation transmission rate R is larger than C, there exists no encoding and decoding processes with this property; that is, the probability of error in information transfer cannot be made arbitrarily small. In other words, the presence of random disturbances in an information channel does not, by itself, limit transmission accuracy. Rather, it limits the transmission rate for which arbitrarily high transmission accuracy can be accomplished.

To conclude this section, we note that the distinction between the two communication disciplines are that Wiener assumes, in effect, that the signal in question can be processed after it has been corrupted by noise. Shannon suggests that the signal can be processed both before and after its transmission through the communication channel. However, both Wiener and Shannon share the same basic objective; namely, faithful reproduction of the original signal.

1.2. ENTROPY INFORMATION

Let us now define the information measure, which is one of the vitally important aspects in the development of Shannon's information theory. For simplicity, we consider discrete input and output message ensembles $A = \{a_i\}$ *and* $B = \{b_j\}$, respectively, as applied to a communication channel, as shown in Fig. 1.2. If a_i is an input event as applied to the information channel and b_j is the corresponding transmitted output event, then the information measure about the received event b_j specifies a_i, can be written as

$$I(a_i; b_j) \triangleq \log_2 \frac{P(a_i/b_j)}{P(a_i)} \text{ bits.} \tag{1.1}$$

Where $P(a_i/b_j)$ is the conditional probability of input event a_i depends on the output event b_j, $P(a_i)$ is the *a priori* probability of input event a_i, $i = 1, 2, \ldots, M$ *and* $j = 1, 2, \ldots, N$.

By the symmetric property of the joint probability, we show that

$$I(a_i; b_j) = I(b_j; a_i). \tag{1.2}$$

In other words, the amount of information transferred by the output event b_j from a_i is the same amount as provided by the input event a_i that specified b_j. It is clear that, if the input and output events are *statistically independent*; that is, if $P(a_i, b_j) = P(a_i)P(b_j)$, then $I(a_i; b_j) = 0$

Furthermore, if $I(a_i; b_j) > 0$, then $P(a_i, b_j) > P(a_i)P(b_j)$, there is a higher joint probability of a_i and b_j. However, if $I(a_i; b_j) < 0$, then $P(a_i, b_j) < P(a_i)P(b_j)$, there is lower joint probability of a_i *and* b_j.

Fig. 1.2. An input–output communication channel.

As a result of the conditional probabilities, $P(a_i/b_j) \leqslant 1$ *and* $P(b_j/a_i) \leqslant 1$, we see that

$$I(a_i; b_j) \leqslant I(a_i), \tag{1.3}$$

and

$$I(a_i; b_j) \leqslant I(b_j), \tag{1.4}$$

where

$$I(a_i) \triangleq -\log_2 P(a_i),$$
$$I(b_j) \triangleq -\log_2 P(b_j),$$

$I(a_i)$ *and* $I(b_j)$ are defined as the respective *input* and *output self-information* of event a_i and event b_j. In other words, $I(a_i)$ *and* $I(b_j)$ represent the amount of information provided at the input and output of the information channel of event a_i and event b_j, respectively. It follows that the mutual information of event a_i and event b_j is equal to the self-information of event a_i if and only if $P(a_i/b_j) = 1$, then

$$I(a_i; b_j) = I(a_i). \tag{1.5}$$

It is noted that, if Eq. (1.5) is true for all i; that is, the input ensemble, then the communication channel is *noiseless*. However, if $P(b_j/a_i) = 1$, then

$$I(a_i; b_j) = I(b_j). \tag{1.6}$$

If Eq. (1.6) is true for all the output ensemble, then the information channel is *deterministic*.

To conclude this section, we note that the information measure as defined in the preceding can be easily extended to higher product spaces, such as

$$I(a_i; b_j/c_k) \triangleq \log_2 \frac{P(a_i/b_j c_k)}{P(a_i/c_k)}. \tag{1.7}$$

Since the measure of information can be characterized by ensemble average, the average amount of information provided at the input end can be written as

$$I(A) \triangleq -\sum_A P(a) \log_2 P(a) \triangleq H(A), \tag{1.8}$$

where the summation is over the input ensemble A.

Similarly, the average amount of self-information provided at the output end can be written as

$$I(B) \triangleq -\sum_{B} P(b) \log_2 P(a) \triangleq H(B). \tag{1.9}$$

These two equations are essentially the same form as the *entropy equation* in statistical thermodynamics, for which the notations $H(A)$ and $H(B)$ are frequently used to describe *information entropy*. As we will see later, indeed $H(A)$ and $H(B)$ provide a profound relationship between the information entropy and the physical entropy. It is noted that entropy H, from the communication theory point of view, is a measure of *uncertainty*. However, from the statistical thermodynamic standpoint, H is a measure of disorder.

In addition, we see that

$$H(A) \geqslant 0, \tag{1.10}$$

where $P(a)$ is always a positive quantity. The equality of Eq. (1.10) holds if $P(a) = 1$ or $P(a) = 0$. Thus, we can conclude that

$$H(A) \leqslant \log_2 M, \tag{1.11}$$

where M is the number of input events. The equality holds for *equiprobability* input events; that is, $P(a) = 1/M$. It's trivial to extend the ensemble average to the conditional entropy:

$$I(B/A) \triangleq -\sum_{B}\sum_{A} p(a, b) \log_2 p(b/a) \triangleq H(B/a). \tag{1.12}$$

And the entropy of the product ensemble AB can also be written as

$$H(AB) = -\sum_{A}\sum_{B} p(a, b) \log_2 p(b/a). \tag{1.13}$$

Thus, we see that

$$H(AB) = H(A) + H(B/A). \tag{1.14}$$

$$H(A/B) = H(B) + H(A/B). \tag{1.15}$$

In view of $\ln u \leqslant u - 1$, we also see that

$$H(B/a) \leqslant H(B), \tag{1.16}$$

$$H(A/B) \leqslant H(A), \tag{1.17}$$

where the equalities hold if and only if a and b are statistically independent.

Let us now turn to the definition of *average mutual information*. We consider first the conditional average mutual information:

$$I(A;\ b) \triangleq \sum_{A} p(a/b) I(a;\ b). \tag{1.18}$$

Although the mutual information between input event and output event can be negative, $I(a; b) < 0$, the average conditional mutual information can never be negative:

$$I(A; b) \geqslant 0, \tag{1.19}$$

with the equality holding if and only if events A are statistically independent of b; that is, $p(a/b) = p(a)$, for all a.

By taking the ensemble average of Eq. (1.19), the average mutual information can be defined as

$$I(a; B) \triangleq \sum_{B} p(b) I(A'b). \tag{1.20}$$

Equation (1.20) can be written as

$$\begin{aligned} I(A;\ B) &\triangleq \sum_{B}\sum_{A} p(a; b) \log_2 \frac{p(a/b)}{p(a)} \\ &= \sum_{A}\sum_{B} p(a; b) I(a; b). \end{aligned} \tag{1.21}$$

Again we see that

$$I(A; B) \geqslant 0. \tag{1.22}$$

The equality holds if and only if a and b are statistically independent. Moreover, from the symmetric property of $I(a; b)$, we see that

$$I(A; B) = I(B; A). \tag{1.23}$$

In view of Eqs. (1.3) and (1.4), we also see that

$$I(A;\ B) \leqslant H(A) = I(A), \tag{1.24}$$

$$I(A;\ B) \leqslant H(B) = I(B), \tag{1.25}$$

which implies that the mutual information (the amount of information transfer) cannot be greater than the entropy information (the amount of information provided) at the input or the output ends, whichever comes first. We note that if the equality holds for Eq. (1.24) then the channel is *noiseless*; on the other hand, if the equality holds for Eq. (1.25), then the channel is *deterministic*.

Since Eq. (1.13) can be written as

$$H(AB) = H(A) + H(B) - I(A; B), \tag{1.26}$$

we see that,

$$I(A; B) = H(A) - H(A/B), \tag{1.27}$$

$$I(A; B) = H(B) - H(B/A), \tag{1.28}$$

where $H(A/B)$ represents the amount of *information loss* (e.g., due to noise) or the *equivocation* of the channel, which is the average amount of information needed to specify the noise disturbance in the channel. And $H(B/A)$ is referred to as the *noise entropy* of the channel.

To conclude this section, we note that the entropy information can be easily extended to continuous product space, such as

$$H(A) \triangleq -\int_{-\infty}^{\infty} p(a) \log_2 p(a) da, \tag{1.29}$$

$$H(B) \triangleq -\int_{-\infty}^{\infty} p(b) \log_2 p(b) db, \tag{1.30}$$

$$H(B/A) \triangleq -\int_{-\infty}^{\infty} \int_{-\infty}^{\infty} p(a, b) \log_2 p(b/a) da\, db, \tag{1.31}$$

$$H(A/B) \triangleq -\int_{-\infty}^{\infty} \int_{-\infty}^{\infty} p(a, b) \log_2 p(a/b) da\, db, \tag{1.32}$$

and

$$H(AB) \triangleq -\int_{-\infty}^{\infty} \int_{-\infty}^{\infty} p(a, b) \log_2 p(a, b) da\, db, \tag{1.33}$$

where the p's are the probability density distributions.

1.3. COMMUNICATION CHANNEL

An optical communication channel can be represented by an input–output block diagram, for which an input event a can be transferred into an output event b, as described by a transitional probability $p(b/a)$. Thus, the input–output transitional probability $P(B/A)$ describes the random noise disturbances in the channel.

Information channels are usually described according to the type of input–output ensemble and are considered *discrete* or *continuous*. If both the input and output of the channel are discrete events (discrete spaces), then the channel is called a *discrete channel*. But if both the input and output of the channel are represented by continuous events, the channel is called a *continuous channel*. However, a channel can have a discrete input and a continuous output, or vice versa. Accordingly, the channel is then called a *discrete–continuous* or *continuous–discrete* channel.

We note again that the terminology of discrete and continuous communication channels can also be extended to spatial and temporal domains. This concept is of particular importance for optical communication channels.

A communication channel can, in fact, have multiple inputs and multiple outputs. If the channel possesses only a single input terminal and a single output terminal, it is a *one-way* channel. However, if the channel possesses two input terminals and two output terminals, it is a *two-way* channel. It is trivial. One can have a channel with n input and m output terminals.

Since a communication channel is characterized by the input–output transitional probability distribution $P(B/A)$, if the transitional probability distribution remains the same for all successive input and output events, then the channel is a *memoryless channel*. However, if the transitional probability distribution changes with the preceding events, whether at the input or the output, then the channel is a *memory channel*. Thus, if the memory is finite; that is, if the transitional probability depends on a finite number of preceding events, the channel is a *finite-memory* channel. Furthermore, if the transitional probability distribution depends on stochastic processes and the stochastic processes are assumed to be nonstationary, then the channel is a *nonstationary channel*. Similarly, if the stochastic processes the transitional probability depends on are *stationary*, then the channel is a *stationary channel*. In short, a communication channel can be fully described by the characteristics of its transitional probability distribution; for example, a *discrete nonstationary memory* channel.

Since a detailed discussion of various communication channels is beyond the scope of this chapter, we will evaluate two of the simplest, yet important, channels in the following subsection.

1.3.1. MEMORYLESS DISCRETE CHANNEL

For simplicity, we let an input message to be transmitted to the channel be

$$\alpha^n = \alpha_1 \alpha_2 \cdots \alpha_n,$$

and the corresponding output message be

$$\beta^n = \beta_1 \beta_2 \cdots \beta_n,$$

where α_i and β_j are any one of the input and output events of A and B, respectively.

Since the transitional probabilities for a memoryless channel do not depend on the preceding events, the composite transitional probability can be written as

$$P(\beta^n/\alpha^n) = P(\beta_1/\alpha_1)P(\beta_2/\alpha_2) \cdots P(\beta_n/\alpha_n).$$

Thus, the joint probability of the output message β^n is

$$P(\beta^n) = \sum_{A^n} P(\alpha^n)P(\beta^n/a^n),$$

where the summation is over the A^n product space.

In view of entropy information measure, the average mutual information between the input and output messages (sequences) of α^n and β^n can be written as

$$I(A^n; B^n) = H(B^n) - H(B^n/A^n), \tag{1.34}$$

where B^n is the output product space, for which $H(B^n)$ can be written as

$$H(B^n) = -\sum_{B^n} P(B^n) \log_2 P(B^n).$$

The conditional entropy $H(B^n/A^n)$ is

$$H(B^n/A^n) = -\sum_{A^n}\sum_{B^n} P(\alpha^n)P(\beta^n/\alpha^n) \log_2 P(\beta^n/\alpha^n). \tag{1.35}$$

Since $I(A^n; B^n)$ represents the amount of information provided by then n output events about the given n input events, $I(A^n; B^n)/n$ is the amount of mutual information per event. If the channel is assumed memoryless, $I(A^n; B^n)/n$ is only a function of $P(\alpha^n)$ and n. Therefore, the capacity of the channel would be the maximum value of $I(A^n; B^n)/n$, for a possible probability distribution $P(\alpha^n)$ and length n; that is,

$$C \triangleq \max_{P(\alpha_n), n} \frac{I(A^n;\ B^n)}{n} \qquad \text{bits/event.} \tag{1.36}$$

It is also noted that, if the input events are *statistically independent* (i.e., from a memoryless information source), then the channel capacity of Eq. (1.36) can be written as

$$C \triangleq \max_{P(\alpha_n)} I(A; B) \qquad \text{bits/event.} \tag{1.37}$$

We emphasized that evaluation of the channel capacity is by no means simple, and it can be quite involved.

1.3.2. CONTINUOUS CHANNEL

A channel is said to be continuous if and only if the input and output ensembles are represented by continuous Euclidean spaces. For simplicity, we restrict our discussion to only the one-dimensional case, although it can be easily generalized for a higher dimension.

Again, we denote by A and B the input and output ensembles, but this time A and B are continuous random variables. It is also noted that a continuous channel can be either *time discrete* or *time continuous.* We first discuss time-discrete channels and then consider time-continuous channels.

Like a discrete channel, a continuous channel is said to be memoryless if and only if its transitional probability density $p(b/a)$ remains the same for all successive pairs of input and output events. A memoryless continuous channel is said to be disturbed by an *additive noise* if and only if the transitional probability density $p(b/a)$ on the difference between the output and input random variables, $b - a$:

$$p(b/a) = p(c), \tag{1.38}$$

where $c = b - a$.

Thus, for additive channel noise the conditional entropy $H(B/a)$ can be shown:

$$\begin{aligned} H(B/a) &= -\int_{-\infty}^{\infty} p(b/a) \log_2 p(b/a) db \\ &= -\int_{-\infty}^{\infty} p(c) \log_2 p(c) dc. \end{aligned} \tag{1.39}$$

We see that $H(B/a)$ is independent of a, which is similar to the fact that the channel is uniform from input. The average conditional entropy is

$$H(B/a) = \int_{-\infty}^{\infty} p(a)H(B/a)da$$

$$= -\int_{-\infty}^{\infty} p(c) \log_2 p(c)dc = H(B/a). \tag{1.40}$$

In the evaluation of the channel capacity, we would first evaluate the average mutual information $I(A; B)$ and then maximize the $I(A; B)$ under the constraint of $p(a)$.

In view of $I(A; B) = H(B) - H(B/A)$, we see that if one maximizes $H(B)$, then $I(A; B)$ is maximized. However, $H(B)$ cannot be made infinitely large, since $H(B)$ is always restricted by certain physical constraints; namely, the available power. This power constraint corresponds to the mean-square fluctuation of the input signal:

$$\sigma_a^2 = \int_{-\infty}^{\infty} a^2(a)da.$$

Without loss of generality, we assume that the average value of the additive noise is zero:

$$\bar{c} = \int_{-\infty}^{\infty} cp(c)dc = 0.$$

Then the mean-square fluctuation of the output signal can be written as

$$\sigma_b^2 = \int_{-\infty}^{\infty} b^2 p(b)db.$$

Since $b = a + c$ (i.e., signal plus noise), one can show that

$$\sigma_b^2 = \sigma_a^2 + \sigma_c^2, \tag{1.41}$$

where

$$\sigma_c^2 = \int_{-\infty}^{\infty} c^2 p(c)dc.$$

From the preceding equation, we see that setting an upper limit to the mean-square fluctuation of the input signal is equivalent to setting an upper limit to the mean-square fluctuation of the output signal. Thus, for a given mean-square value of σ_b^2, one can show that for the corresponding entropy, derived from $p(b)$, there exists an upper bound:

$$H(B) \leqslant \tfrac{1}{2}\log_2(2\pi e\sigma_c^2), \tag{1.42}$$

where the equality holds if and only if the probability density $p(b)$ is Gaussianly distributed with zero mean and the variance equals to σ_b^2.

From the additivity property of the channel noise $H(B/A)$, which is dependent solely on $p(c)$, we see that

$$H(B/A) \leqslant \tfrac{1}{2}\log_2(2\pi e\sigma_c^2), \tag{1.43}$$

where the equality holds when $p(c)$ is Gaussianly distributed, with zero mean and the variance equal to σ_c^2.

Thus, if the additive noise in a memoryless continuous channel has a Gaussian distribution, with zero mean and the variance equal to N, where N is the average noise power, then the average mutual information is bounded by the following inequality:

$$I(A;\, B) \leqslant \tfrac{1}{2}\log_2(2\pi e\sigma_b^2) - \tfrac{1}{2}\log_2(2\pi eN) \leqslant \tfrac{1}{2}\log_2\frac{\sigma_b^2}{N}. \tag{1.44}$$

Since the input signal and the channel noise are assumed to be statistically independent,

$$\sigma_b^2 = \sigma_a^2 + \sigma_c^2 = \sigma_a^2 + N, \tag{1.45}$$

we have

$$I(a;\, b) \leqslant \tfrac{1}{2}\log_2\frac{\sigma_a^2 + N}{N}. \tag{1.46}$$

By maximizing the preceding equation, we have the channel capacity as written by

$$C = \tfrac{1}{2}\log_2\left(1 + \frac{S}{N}\right), \tag{1.47}$$

where the input signal is Gaussianly distributed, with zero mean and a variance equal to S.

Let's now evaluate one of the *most useful* channels; namely, a memoryless, time-continuous, band-limited, continuous channel. The channel is assumed to be disturbed by an additive white Gaussian noise, and a band-limited time-continuous signal, with an average power not to exceed a given value S, is applied at the input end of the channel.

Note that, if a random process is said to be a stationary Gaussian process, then the corresponding joint probability density distribution, assumed by a time function at any finite time interval, is independent of the time origin, and it has a Gaussian distribution. If a stationary Gaussian process is said to be *white*, then the power spectral density must be uniform (constant) over the entire range of the frequency variable.

Let us denote $c(t)$ be a white Gaussian noise. By the Karhunen-Loeve expansion theorem [1.5, 1.6], $c(t)$ can be written over a time interval $-T/2 \leqslant t < T/2$:

$$c(t) = \sum_{i=-\infty}^{\infty} c_i \phi_i(t), \tag{1.48}$$

where the $\phi_i(t)$'s are *orthonormal functions* that can be represented by

$$\int_{-T/2}^{T/2} \phi_i(\mathrm{t})\phi_j(\mathrm{t})\mathrm{dt} = \begin{cases} 1, & i = j, \\ 0, & i \neq j, \end{cases} \tag{1.49}$$

and c_i are real coefficients commonly known as *orthogonal expansion coefficients*. Furthermore, the c_i's are statistically independent, and the individual probability densities have a stationary Gaussian distribution, with zero mean and the variance equal to $N_o/2T$, where N_o is the corresponding power spectral density.

Now we consider an input time function $a(t)$ as applied to the communication channel, where the frequency spectrum is limited by the channel bandwidth $\Delta\nu$. Since the channel noise is assumed an additive white Gaussian noise, the output response of the channel is

$$b(t) = a(t) + c(t). \tag{1.50}$$

Such a channel is known as a *band-limited channel* with additive white Gaussian noise.

Again by the Karhunen-Loeve expansion theorem, the input and output time functions can be expanded:

$$a(t) = \sum_{i=-\infty}^{\infty} a_i \phi_i(t),$$

and

$$b(t) = \sum_{i=-\infty}^{\infty} b_i \phi_i(t).$$

Thus, we see that

$$b_i = a_i + c_i. \tag{1.51}$$

Since the input function $a(t)$ is band-limited by Δv, only $2T\Delta v$ coefficients a_i, $i = 1, 2, \ldots, 2T\Delta v$, within the passband are considered. In other words, the input signal ensemble can be represented by a $2T\Delta v$–order product ensemble over a; that is, $A^{2T\Delta v}$. This is similarly true for the output ensemble over b; that is, $B^{2T\Delta v}$. Thus, the average mutual information between the input and output ensembles is

$$I(A^{2T\Delta v};\ B^{2T\Delta v}) = H(B^{2T\Delta v}) - H(B^{2T\Delta v}/A^{2T\Delta v}). \tag{1.52}$$

It is also clear that a, b, and c each form a $2T\Delta v$–dimensional *vector space*, for which we write

$$\mathbf{b} = \mathbf{a} + \mathbf{c}. \tag{1.53}$$

If we let $p(\mathbf{a})$ and $p(\mathbf{c})$ be the probability density distribution of $\mathbf{a}$ and $\mathbf{c}$ respectively, then the transitional probability density of $p(\mathbf{b}/\mathbf{a})$ is

$$p(\mathbf{b}) = p(\mathbf{b} - \mathbf{a}) = p(\mathbf{c}),$$

where $\mathbf{a}$ and $\mathbf{c}$ are statistically independent. For simplicity, we let $X \triangleq A^{2T\Delta v}$ be the vector space (the product space) of $\mathbf{a}$. The probability density distribution of $\boldsymbol{b}$ can be determined by

$$p(\mathbf{b}) = \int_X p(\mathbf{a})p(\mathbf{c})dX,$$

where the integral is over the entire vector space X.

Similarly, $Y \triangleq B^{2T\Delta v}$ and $Z \triangleq C^{2T\Delta v}$ represent the vector space of $\mathbf{b}$ and $\mathbf{c}$, respectively. The average mutual information can therefore be written by

$$I(X;\ Y) = H(Y) - H(Z), \tag{1.54}$$

where

$$H(Y) = -\int_Y p(\text{b}) \log_2 p(\text{b}) dY,$$

and

$$H(Z) = H(Y/X) = -\int_Z p(\mathrm{c}) \log_2 p(\mathrm{c}) dZ.$$

The channel capacity can be determined by maximizing the $I(X; Y)$; that is,

$$C \triangleq \max_{T, p(\mathbf{a})} \frac{I(X; Y)}{T}, \qquad \text{bits/time.} \tag{1.55}$$

Under the constraint of the signal mean-square fluctuation that cannot exceed a specified value S,

$$\int_X |\mathbf{a}|^2 p(\mathbf{a}) dX \leqslant S. \tag{1.56}$$

Since each of the vectors **a**, **b**, and **c** are represented by $2T\Delta\nu$ continuous variables, and each c_i is statistically independent Gaussianly distributed with zero mean, and has a variance equal to $N_o/2T$, we see that

$$I(X; Y) = I(A^{2T\Delta\nu}; B^{2T\Delta\nu}) = \sum_{i=1}^{2T\Delta\nu} I(A_i; B_i). \tag{1.57}$$

Thus, from Eq. (1.43), we have

$$H(Z) = 2T\Delta\nu H(C_i), \tag{1.58}$$

where

$$H(C_i) = \tfrac{1}{2} \log_2(2\pi e \sigma_{c_i}^2).$$

If we let $N = \sigma_{c_i}^2 = N_o \Delta\nu$, then $H(Z)$ can be written as

$$H(Z) = T\Delta\nu \log_2 \left(\frac{\pi e N_o}{T} \right). \tag{1.59}$$

In view of Eq. (1.42), we see that

$$H(B_i) \leqslant \log_2(2\pi e \sigma_{b_i}^2), \tag{1.60}$$

where the equality holds if and only if b_i is Gaussianly distributed, with zero mean and a variance equal to $\sigma_{b_i}^2$. Since $\mathbf{b} = \mathbf{a} + \mathbf{c}$, $p(\mathbf{b})$ is Gaussianly distributed if and only if $p(\mathbf{a})$ is also Gaussianly distributed with zero mean. The average mutual information can therefore be written as

$$I(X;\ Y) = \sum_{i=1}^{2T\Delta v} H(B_i) - H(Z)$$

$$= \tfrac{1}{2}\log_2\left[\prod_{i=1}^{2T\Delta v}(2\pi e \sigma_{b_i}^2)\right] - T\Delta v \log_2\left(\frac{\pi e N_o}{T}\right), \tag{1.61}$$

where Π denotes the product ensemble. Since a_i *and* c_i are statistically independent the variance of $\boldsymbol{p}(\mathbf{b})$ is given by

$$\sigma_{b_i}^2 = \sigma_{a_i}^2 + \sigma_{c_i}^2 = \sigma_{a_i}^2 + \frac{N_o}{2T}.$$

In view of Eq. (1.56), we see that

$$\sum_{i=1}^{2T\Delta v} \sigma_{b_i}^2 = \sum_{i=1}^{2T\Delta v} \sigma_{a_i}^2 + N_o \Delta v \leqslant S + N, \tag{1.62}$$

where $N = N_o \Delta v$. The equality holds for Eq. (1.62) when the input probability density distribution $\boldsymbol{p}(\mathbf{a})$ is also Gaussianly distributed with zero mean and a variance equal to S. Furthermore, from Eq. (1.62), we can write

$$\prod_{i=1}^{2T\Delta v} \sigma_{b_i}^2 \leqslant \left(\frac{S+N}{2T\Delta v}\right)^{2T\Delta v},$$

where the equality holds if and only if the $\sigma_{b_i}^2$ are all equal and $\boldsymbol{p}(\mathbf{a})$ is Gaussianly distributed with zero mean and a variance equal to S.

Therefore, the corresponding channel capacity can be written as

$$C = \max_{T, p(\mathbf{a})} \frac{I(X;\ Y)}{T} = \Delta v \log_2\left(1 + \frac{S}{N}\right) \quad \text{bits/sec} \tag{1.63}$$

where S/N is the signal-to-noise ratio. We note that the preceding result is one of the most popular equations as derived by Shannon [1.3] and independently by Wiener [1.5], for a memoryless additive Gaussian channel. Because of its conceptual and mathematical simplicity, this equation has been widely used in practice and has also been occasionally misused. We note that this channel capacity is derived under the assumption of additive white Gaussian noise regime, and the average input signal power cannot exceed a specified value of S. We further stress that the channel capacity equation is obtained under the assumption that input signal is also Gaussianly distributed with zero mean.

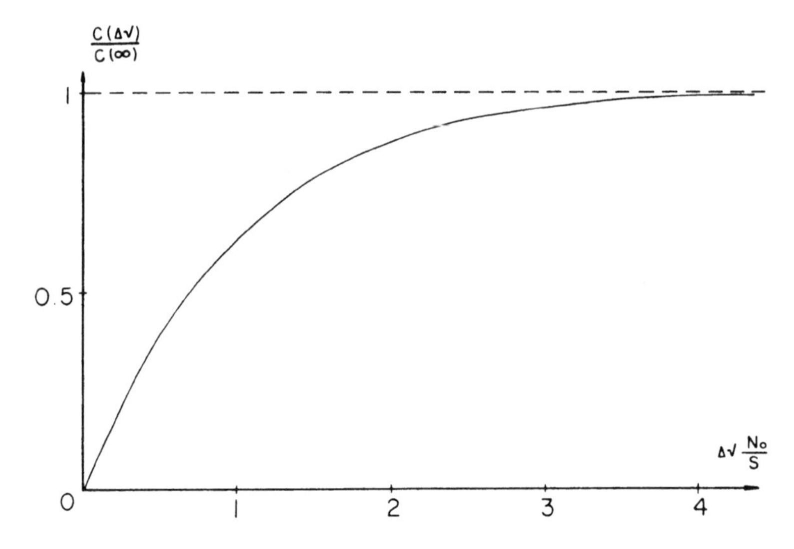

Fig. 1.3. The capacity of an additive white Gaussian channel as a function of bandwidth $\Delta\nu$.

Since the average noise power within the channel bandwidth is $N_o \Delta\nu$, as the bandwidth increases to infinitely large the capacity channel approaches a definite value, such as

$$C(\infty) = \lim_{\Delta\nu \to \infty} C(\Delta\nu) = \frac{S}{N_o} \log_2 e. \tag{1.64}$$

This result provides an important physical significance: the measurement or observation of any physical quantity is practically always limited by *thermal noise agitation*. This thermal agitation is usually considered as an additive white Gaussian noise within the channel bandwidth. The noise spectral density N_o is related to the thermal temperature T as given by

$$N_o = kT, \tag{1.65}$$

where k is Boltzmann's constant, and T is in degree Kelvins. It follows that the signal energy transmitted through a physical communication channel must be at least kT per nat of information. In other words, it takes at least kT energy for a nat of information to be properly transmitted through the channel. A plot of the channel capacity as a function of bandwidth is shown in Fig. 1.3, in which the capacity started rapidly increases and then asymptotically approaches to $C(\infty)$ as $\Delta\nu$ becomes very large.

1.4. BAND-LIMITED ANALYSIS

In practice, all communication channels (spatial and temporal as well) are band limited. An information channel can be a *low-pass*, *bandpass*, or *discrete bandpass* channel. But a strictly high-pass channel can never happen in practice. A low-pass channel represents nonzero transfer values from zero frequency to a definite frequency limit ν_m. Thus, the bandwidth of a low-pass channel can be written as

$$\Delta\nu = \nu_m.$$

On the other hand, if the channel possesses nonzero transfer values from a lower frequency limit ν_1 to a higher frequency limit ν_2, then it is a bandpass channel and the bandwidth can be written as

$$\Delta\nu = \nu_2 - \nu_1. \tag{1.66}$$

It is trivial to note the analysis of a bandpass channel can be easily reduced to an equivalent low-pass channel. Prior to our discussion a basic question may be raised: What sort of response would we expect from a band-limited channel? In other words, from the frequency domain standpoint, what would happen if the signal spectrum is extended beyond the passband?

To answer this basic question, we present a very uncomplicated but significant example. For simplicity, the transfer function of an ideal low-pass channel, shown in Fig. 1.4, is given by

$$H(\nu) = \begin{cases} 1, & |\nu| \leqslant \nu_m \\ 0, & |\nu| > \nu_m. \end{cases} \tag{1.67}$$

If the applied signal to this low-pass channel is a finite duration Δt signal, then to have good output reproduction of the input signal the channel bandwidth $\Delta\nu$ must be greater than or at least equal to $1/\Delta t$ the signal bandwidth; that is,

$$\Delta\nu \geqslant \frac{1}{\Delta t}, \tag{1.68}$$

which can also be written as

$$\Delta t\,\Delta\nu \geqslant 1, \tag{1.69}$$

where $\Delta\nu = 2\nu_m$ is the channel bandwidth. The preceding equations show an interesting duration-bandwidth product relationship. The significance is that if the signal spectrum is more or less concentrated in the passband of the channel;

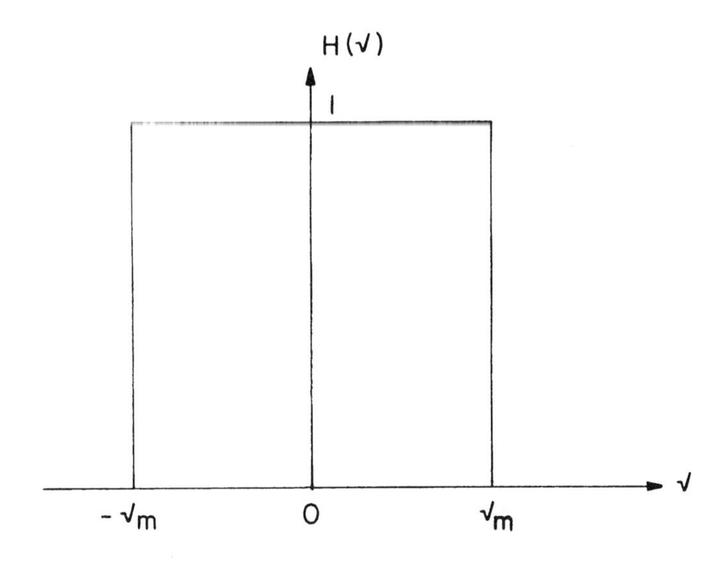

Fig. 1.4. An ideal low-pass channel.

that is, $1/\Delta t \leqslant \nu_m$, then the output response will quite faithfully reproduce the input signal. However, if the signal spectrum spreads beyond the passband of the channel, the output response will be severely distorted, which fails to reproduce the input signal. We note that this duration-bandwidth product provides a profound relationship with the *Heisenberg's uncertainty principle* in quantum mechanics [1.7], such as

$$\Delta x\, \Delta p \geqslant h, \tag{1.70}$$

where Δx and Δp are the position and momentum errors, respectively, and h is Plank's constant. In other words, the uncertainty principle states that the position variable x and its momentum variable p cannot be observed or measured simultaneously with arbitrary accuracy.

In fact, the uncertainty relation can also be written in the form of energy and time variables:

$$\Delta E\, \Delta t \geqslant h, \tag{1.71}$$

where ΔE and Δt are the corresponding energy and time errors. Since $E = h\nu$, hence $\Delta E = h\Delta\nu$, we see that

$$\Delta E\, \Delta t = h\Delta\nu\, \Delta t \geqslant h. \tag{1.72}$$

Thus, we see that $\Delta\nu\, \Delta t \geqslant 1$ is the *Heisenberg uncertainty relation.*

Let us provide an example to show that the uncertainty relation is indeed relevant for information transmission through a communication channel. We assume a short symmetric pulse of duration of Δt, as applied to the channel; that is,

$$f(t) = f(t) = \begin{cases} f(t), & |t| \leqslant \dfrac{\Delta t}{2} \\ 0, & |t| > \dfrac{\Delta t}{2} \end{cases}, \tag{1.73}$$

where

$$f(0) \geqslant f(t).$$

The corresponding signal spectrum can be obtained as given by

$$F(v) = \mathscr{F}[f(t)],$$

where $\mathscr{F}$ denotes the Fourier transformation.

Let us define a *nominal signal duration* Δt and a *nominal signal bandwidth* which are equivalent to the duration of a rectangular pulse of amplitude $f(0)$ and a rectangular pulse spectrum $F(0)$, as given by

$$\Delta t\, f(0) \triangleq \int_{-\infty}^{\infty} f(t)dt, \tag{1.74}$$

and

$$\Delta v F(0) \triangleq \int_{-\infty}^{\infty} F(v)dv. \tag{1.75}$$

From the definition of the Fourier transformation, the nominal quantities can be written as

$$\Delta t = \frac{F(0)}{f(0)}, \tag{1.76}$$

$$\Delta v = \frac{f(0)}{F(0)}, \tag{1.77}$$

for which we have the lower bound condition of the uncertainty relations; that is,

$$\Delta t\, \Delta v = 1. \tag{1.78}$$

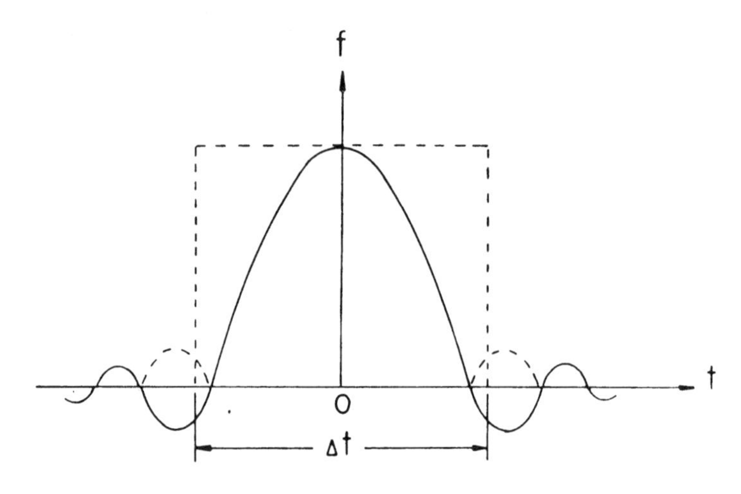

Fig. 1.5. Symmetric pulse function.

It should be noted that if the symmetric pulse $f(t)$ contains *negative* values, as shown in Fig. 1.5, then the definitions of nominal duration and bandwidth must be modified:

$$\Delta t|f(0)| \triangleq \int_{-\infty}^{\infty} |f(t)|dt \geqslant \left|\int_{-\infty}^{\infty} f(t)dt\right| = |F(0)|, \tag{1.79}$$

and

$$\Delta v|F(0)| \triangleq \int_{-\infty}^{\infty} |F(v)|dv \geqslant \left|\int_{-\infty}^{\infty} F(v)dv\right| = |f(0)|. \tag{1.80}$$

These definitions give rise to the uncertainty relation

$$\Delta t\, \Delta v \geqslant 1,$$

which is essentially the condition of Eq. (1.69). In view of Fig. 1.5, we see that the nominal duration is determined by equating the area under the rectangular pulse function to the area under the curve of $|f(t)|$. It is evident that the nominal duration Δt is *wider* under this new definition, provided $f(t)$ contains negative values. Similarly, the nominal bandwidth Δv can be interpreted in the same manner.

1.4.1. DEGREES OF FREEDOM

Let $f(t)$ be a band-limited signal the spectrum of which extends from zero frequency to a definite limit of ν_m. Assume that $f(t)$ extends over a time interval of T, where $\nu_m T \gg 1$. Strictly speaking, a band-limited signal cannot be time limited or vice versa. Now a question arises: How many sampling points (i.e., degrees of freedom) are required to describe the function $f(t)$, over T, uniquely? To answer this fundamental question we present an example.

First, we let $f(t)$ repeat itself at every time interval of T; that is,

$$f(t) = f(t + T).$$

Thus, the function $f(t)$ over the period T, can be expanded in a Fourier series:

$$f(t) = \sum_{n=-M}^{M} C_n \exp(i2\pi n \nu_o t), \tag{1.81}$$

where $\nu_o = 1/T$, *and* $M = \nu_m T$.

From this Fourier expansion, we see that $f(t)$ contains a finite number of terms:

$$N = 2M + 1 = 2\nu_m T + 1,$$

which includes the zero-frequency Fourier coefficient C_o. If the duration T is sufficiently large, we see that the number of degree of freedom reduces to

$$N \approx 2\nu_m T. \tag{1.82}$$

In other words, it requires a total of N equidistant sampling points of $f(t)$, over T, to describe the function

$$t_s = \frac{T}{N} = \frac{1}{2\nu_m}, \tag{1.83}$$

where t_s is known as the *Nyquist sampling interval* and the corresponding *sampling frequency* is

$$f_s = \frac{1}{t_s} = 2\nu_m, \tag{1.84}$$

which is known as the *Nyquist sampling rate* or sampling frequency. Thus, we see that

$$f_s \geqslant 2\nu_m. \tag{1.85}$$

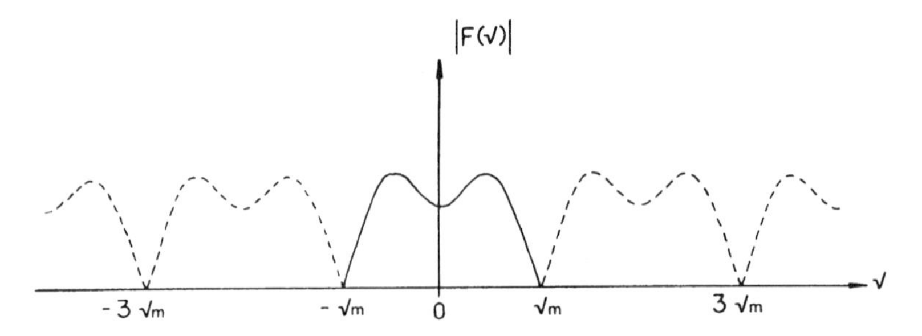

Fig. 1.6. Periodic representation of the Fourier spectrum.

In words, the sampling frequency has to be at least equal or twice the highest frequency limit of $f(t)$.

By taking the Fourier transform of $f(t)$, we have

$$F(\nu) = \int_{-\infty}^{\infty} f(t)\exp(-i2\pi\nu t)dt.$$

Since $f(t)$ is assumed band limited, $F(\nu)$ may be repeated over the frequency domain as shown in Fig. 1.6. Thus, the corresponding Fourier expansion can be written as

$$F(\nu) = \begin{cases} \sum_{n=-\infty}^{\infty} K_n \exp\left(\frac{i\pi n\nu}{\nu_m}\right), & |\nu| \leqslant \nu_m, \\ 0, & |\nu| > \nu_m, \end{cases} \tag{1.86}$$

where

$$K_n = \int_{-\nu_m}^{\nu_m} F(\nu)\exp\left(-i\,\frac{\pi n\nu}{\nu_m}\right) d\nu.$$

$F(\nu)$ is the Fourier transform of $f(t)$, which can be written as

$$f(t) = \int_{-\nu_m}^{\nu_m} F(\nu)\exp(i2\pi\nu t)d\nu. \tag{1.87}$$

In particular, at sampling points $t = -n/2\nu_m$, we have

$$f\left(-\frac{n}{2\nu_m}\right) = \int_{-\nu_m}^{\nu_m} F(\nu)\exp\left(-i\,\frac{\pi n\nu}{\nu_m}\right) d\nu = K_n. \tag{1.88}$$

Thus, we see that if $f(t)$ is given at various Nyquist intervals ($t = n/v_m$), then the corresponding Fourier coefficient K_n can be obtained. From Eq. (1.86), however, we see that $F(v)$ can in turn be determined, and from Eq. (1.87) that the knowledge of $F(v)$ implies a knowledge of $f(t)$. Therefore, if we substitute Eq. (1.86) into Eq. (1.87) we have

$$f(t) = \int_{-v_m}^{v_m} \sum_{n=-\infty}^{\infty} K_n \exp\left(\frac{i\pi n v}{v_m}\right) \exp(i2\pi v t) dv.$$

By interchanging the integration and summation in the preceding equation, we obtained

$$\begin{aligned} f(t) &= \sum_{n=-\infty}^{\infty} K_n \frac{\sin 2\pi v_m(t + n/2v_m)}{2\pi v_m(t + n/2v_m)} \\ &= \sum_{n=-\infty}^{\infty} f\left(\frac{n}{2v_m}\right) \frac{\sin 2\pi v_m(t - n/2v_m)}{2\pi v_m(t - n/2v_m)}, \end{aligned} \tag{1.89}$$

in which the weighting factor $[(\sin x)/x]$ is known as the *sampling function*. This is, in fact, the output response of an ideal low-pass channel having a cutoff frequency at v_m, when the samples $f(n/2v_m)$ are applied at the input end of the channel.

1.4.2. GABOR'S INFORMATION CELL

Let us consider the frequency versus the time coordinate shown in Fig. 1.7, in which v_m denotes the maximum frequency limit and T the finite time sample of the signal $f(t)$. This frequency–time space can be subdivided into elementary information cells that Gabor called *logons*, such as

$$\Delta v \, \Delta t = 1. \tag{1.90}$$

We quickly recognize that Eq. (1.90) is essentially the lower bound of the uncertainty relation of Eq. (1.69). However, note that the signal in each of the information cells has two possible *elementary signals* (*symmetric and antisymmetric*) having the same bandwidth Δv and the same duration Δt. Notice that the amplitudes of these signals should be given so that the signal function $f(t)$ can be uniquely described. In view of Fig. 1.7, we see that over the (v_m, T) space there are a total number of information cells; that is,

$$N_1 = v_m T. \tag{1.91}$$

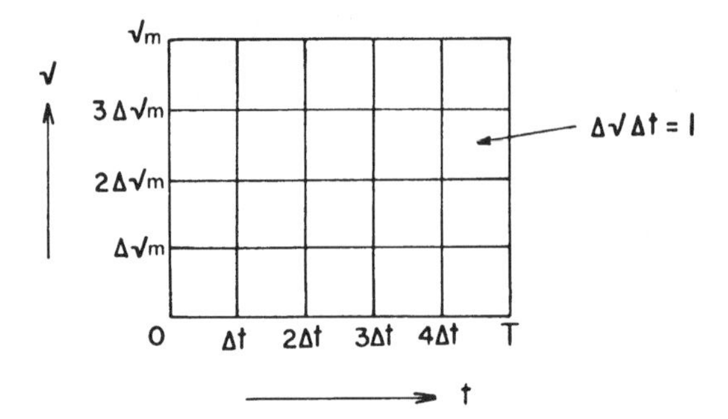

Fig. 1.7. Gabor's information cell.

Since each of the information cells is capable of having two elementary signals, the total number of elementary signals within the (v_m, T) domain is

$$N = 2N_1 = 2v_m T, \tag{1.92}$$

which is essentially the Nyquist sampling rate.

Notice that the *shapes* of the information cells are not particularly critical, but their unit area is, $\Delta v\, \Delta t = 1$. The sampling function as we see it takes place along the time coordinate, whereas the Fourier analysis is displayed along the vertical axis of the frequency coordinate. We further note that the elementary signals as Gabor [1.7, 1.8] suggested the use of *Gaussine cosine* and *Gaussian sine* signals, are also known as *wavelets*, as shown in Fig. 1.8.

Equation (1.91) shows that the elementary information cell suggested by Gabor is, in fact, the lower bound of the Heisenberg uncertainty principle in quantum mechanics. We further emphasize that the band-limited signal must be a very special type, for which the function must be well behaved. The function contains no discontinuity and sharp angles, as illustrated in Fig. 1.9. In other words, the signal must be an *analytic function* over T.

1.5. SIGNAL ANALYSIS

A problem of considerable importance in signal processing is the extraction of signal from random noise and distortion. There are, however, two major approaches to this issue; the extraction of signals that have been lost in random noise as examples, called *signal detection*, and the reconstruction of unknown signals that have been distorted as examples, called *signal recovery* or *restoration*. As we see, the optimum signal detection can be achieved by maximizing

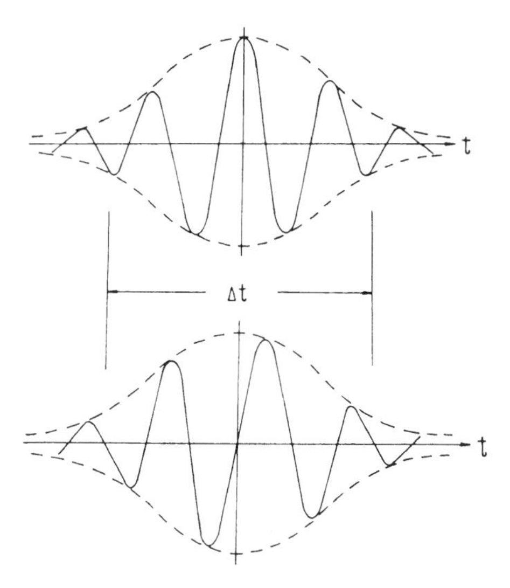

Fig. 1.8. The cosine and sine elementary signals, with Gaussian envelope.

the output signal-to-noise ratio, and will eventually lead to *matched filtering*, while optimum signal recovery can be obtained by minimizing the mean-square error, which will lead to the *Wiener-Hopf solution*. We discuss some of these issues in the following subsection.

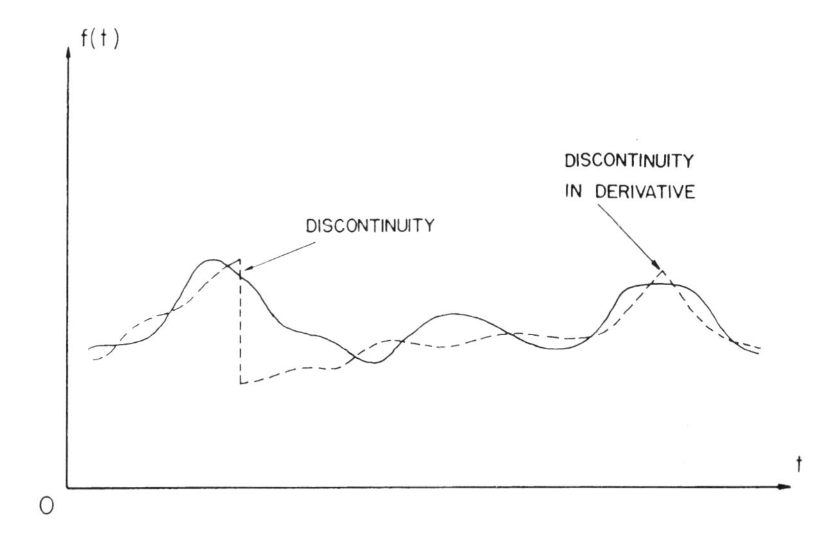

Fig. 1.9. Typical example of a band-limited signal.

1.5.1. SIGNAL DETECTION

It is well known that the signal-to-noise ratio at the output end of a correlator can be greatly improved by matched filtering. Let us consider the input excitation to a linear filtering system to be an additive mixture signal $s(t)$ and $n(t)$; that is,

$$f(t) = s(t) + n(t), \tag{1.93}$$

in which we assume that the noise is a stationary process. If we denote $s_o(t)$ to be the corresponding output signal and $n_o(t)$ to be the output noise, then the output signal-to-noise ratio, at $t = 0$, would be

$$\frac{S}{N} \triangleq \frac{|s_o(o)|^2}{\sigma^2} = \frac{\left|\int_{-\infty}^{\infty} H(v)S(v)dv\right|^2}{\int_{-\infty}^{\infty} |H(v)|^2 N(v)dv}, \tag{1.94}$$

where σ^2 is the mean-square value of the output noise and $H(v)$ and $N(v)$ are the filter transfer function and input noise power spectral density, respectively. In view of *Schwarz's inequality*, the preceding equation can be shown as

$$\frac{S}{N} \leqslant \int_{-\infty}^{\infty} \frac{|S(v)|^2}{N(v)} dv, \tag{1.95}$$

in which the equality holds if and only if the filter function is

$$H(v) = K \frac{S^*(v)}{N(v)}, \tag{1.96}$$

where K is a proportional constant and the superasterisk represents the complex conjugation. We note that, if the noise is *stationary* and *white*, then the optimum filter function is

$$H(v) = KS^*(v), \tag{1.97}$$

which is proportional to the conjugate of the signal spectrum. This optimum filter is also known as *matched filter*, since the transfer function $H(v)$ matches $S^*(v)$.

1.5.2. STATISTICAL SIGNAL DETECTION

In the preceding section we illustrated that signal detection can be achieved by improving the output signal-to-noise ratio. In fact, the increase in signal-to-noise ratio is purposely to minimize the probability of error in signal detection. However, in certain signal detections, an increase in the signal-to-noise ratio does not necessarily guarantee minimizing the detection error. Nevertheless, minimizing the detection error can *always* be obtained by using a *decision process.*

Let us consider the detection of *binary signals.* We establish a *Bayes's decision rule,* in which the conditional probabilities are given by

$$P(a/b) = \frac{P(a)P(b/a)}{P(b)}. \tag{1.98}$$

One can write

$$\frac{P(a = 0/b)}{P(a = 1/b)} = \frac{P(a = 0)P(b/a = 0)}{P(a = 1)p(b/a = 1)}, \tag{1.99}$$

where $P(a)$ is the *a priori* probability of a; that is, $a = 1$ corresponds to the signal presented, and $a = 0$ corresponds to no signal.

A logical decision rule is that, if $P(a = 0/b) > P(a = 1/b)$, we decide that there is no signal ($a = 0$) for a given b. However, if $P(a = 0/b) < P(a = 1/b)$, then we decide that there is a signal ($a = 1$) for a given b. Thus, the Bayes's decision rule can be written as:

Accept $a = 0$ if

$$\frac{P(b/a = 0)}{P(b/a = 1)} > \frac{P(a = 1)}{P(a = 0)}. \tag{1.100}$$

Accept $a = 1$ if

$$\frac{P(b/a = 0)}{P(b/a = 1)} < \frac{P(a = 1)}{P(a = 0)}. \tag{1.101}$$

Two possible errors can occur: if we accept that the received event b contains no signal ($a = 0$), but the signal does in fact exist ($a = 1$); and vice versa. In other words, the error of accepting $a = 0$ when $a = 1$ has actually occurred is a *miss,* and the error of accepting $a = 1$ when $a = 1$ has actually not occurred is a *false alarm.*

If we further assign the cost values C_{00}, C_{01}, C_{10}, C_{11}, respectively, to the following cases: (1) $a = 0$ is actually true, and the decision is to accept it; (2) $a = 0$ is actually true, and the decision is to reject it; (3) $a = 1$ is actually true, and the decision is to reject it; and (4) $a = 1$ is actually true, and the decision is to accept it, then the overall *average cost* is

$$\bar{C} = \sum_{i=0}^{1} \sum_{j=0}^{1} C_{ij} P(a_i) P(B_j/a_i), \tag{1.102}$$

where $a_o = 0$, $a_1 = 1$, $P(a_i)$ is the *a priori* probability of a_i, *and* $P(B_j/a_i)$ is the conditional probability that b falls in B_j *if* a_i is actually true, as shown in Fig. 1.10.

To minimize the average cost, it is desirable to select a region B_o, where $B_1 = B - B_o$, in such a manner that the average cost is minimum; in other words, to place certain restrictions on the cost values C_{ij} so that $\bar{C}$ will be minimum for a desirable B_o. For example, a miss or a false alarm may be costlier than correct decisions:

$$C_{10} > C_{11}, \tag{1.103}$$

and

$$C_{01} > C_{00}. \tag{1.104}$$

In view of the cost function, we have

$$\bar{C} = C_{01}P(a = 0) + C_{11}P(a = 1) + \int_{B_o} P(a = 1)(C_{10} - C_{11})P(b/a = 1)db$$
$$- \int_{B_o} P(a = 0)(C_{01} - C_{00})P(b/a = 0)db, \tag{1.105}$$

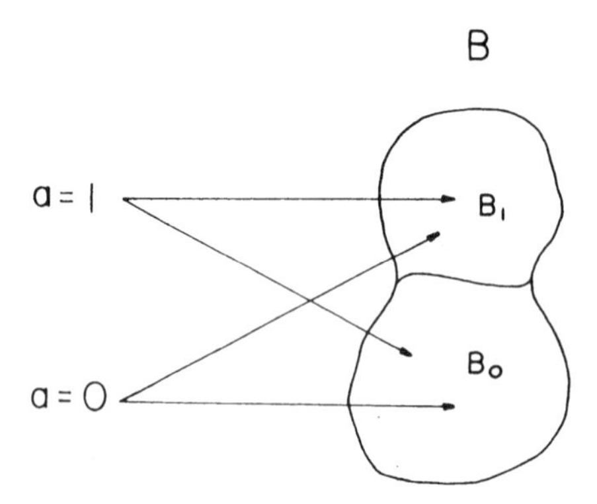

Fig. 1.10. Hypothesis of the received events for two possible transmitted events.

where

$$P(B_j/a_i) = \int_{B_j} P(b/a_i)db, \qquad i, j = 0, 1,$$

$$P(B_1/a_i) = 1 - P(B_o/a_i), \qquad i = 0, 1.$$

To minimize the $\bar{C}$, an optimum region B_o can be selected. In view of the cost in imposition of Eqs. (1.103) and (1.104), it is sufficient to select region B_o such that the second integral of Eq. 1.105 is larger than the first integral, for which we conclude

$$\frac{P(b/a = 0)}{P(b/a = 1)} > \frac{P(a = 1)(C_{10} - C_{11})}{P(a = 0)(C_{01} - C_{00})}. \tag{1.106}$$

Let us write

$$\alpha \triangleq \frac{P(b/a = 0)}{P(b/a = 1)}, \tag{1.107}$$

which is the *likelihood* ratio, and

$$\beta \triangleq \frac{P(a = 1)(C_{10} - C_{11})}{P(a = 0)(C_{01} - C_{00})} \tag{1.108}$$

which is simply a constant incorporated with the *a priori* probabilities and the error costs. The decision rule is to select the hypothesis for which the signal is actually absent, if $\alpha > \beta$.

If the inequality of Eq. 1.106 is reversed ($\alpha < \beta$) then one chooses B_1 instead. In other words, Bayes' decision rule (Eq. 1.106) ensures a *minimum average cost* for the decision making.

Furthermore, if the costs of the errors are equal, $C_{10} = C_{01}$, then the decision rule reduces to Eqs. 1.101 and 1.102. If the decision making has sufficient information on the error costs, then one uses Bayes's decision rule of Eq. 1.106 to begin with. However, if information on the error costs is not provided, then one uses the decision rule as given by Eqs. 1.101 and 1.102.

We also noted that the Bayesian decision process depends on the *a priori* probabilities $P(a = 0)$ and $P(a = 1)$. However, if the *a priori* probabilities are not provided but one wishes to proceed with decision making alone, then the likelihood ratio test can be applied. That is, if

$$\frac{P(b/a = 0)}{P(b/a = 1)} > 1, \tag{1.109}$$

then one accepts $a = 0$ for the received event of b. But if

$$\frac{P(b/a = 0)}{P(b/a = 1)} < 1, \tag{1.110}$$

then one accepts $a = 1$.

From Eq. 1.106, we see that Eq. 1.109 implies that

$$P(a = 1)(C_{10} - C_{11}) = P(a = 0)(C_{01} - C_{00}).$$

Thus, if $C_{10} - C_{11} = C_{01} - C_{00}$, then the *a priori* probabilities of $P(a)$ are equal. Note that applications of the likelihood ratio test are limited, since the decision making takes place without knowledge of the *a priori* probabilities. Thus, the results frequently are quite different from those from the minimum-error criterion.

Although there are times when there is no satisfactory way to assign appropriate *a priori* probabilities and error costs, there is a way of establishing an optimum decision criterion; namely, the *Neyman-Pearson criterion* [1.9]. One allows a fixed false-alarm probability and then makes the decision in such a way as to minimize the miss probability.

1.5.3. SIGNAL RECOVERING

An interesting problem in signal processing must be signal recovering (i.e., restoration). The type of optimum filters is the solution based on minimization of the *mean-square error*, as given by

$$\min \langle \overline{\varepsilon^2(t)} \rangle = \min \left\langle \lim_{T \to \infty} \frac{1}{2T} \int_{-T}^{T} [f_o(t) - f_d(t)]^2 \, dt \right\rangle, \tag{1.111}$$

where $f_o(t)$ *and* $f_d(t)$ are the actual output and the desired output response of the filter. A linear filter that is synthesized by using minimum–mean-square error criterion is known as the *optimum linear filter*.

Let us look at the mean-square-error performance of a linear filter function as given by

$$\overline{\varepsilon^2(t)} = \lim_{T \to \infty} \frac{1}{2T} \int_{-T}^{T} \left[\int_{-\infty}^{\infty} h(\tau) f_i(t - \tau) d\tau - f_d(t) \right]^2 dt, \tag{1.112}$$

where $h(t)$ is the impulse response of the filter and $f_i(t)$ is the input signal function. By expanding the preceding equation, it can be shown that the mean-square error can be written as

$$\overline{\varepsilon^2(t)} = \int_{-\infty}^{\infty} h(\tau)d\tau \int_{-\infty}^{\infty} h(\sigma)d\sigma R_{ii}(\tau - \sigma) - 2\int_{-\infty}^{\infty} h(\tau)d\tau R_{id}(\tau) + R_{dd}(0), \tag{1.113}$$

where $R_{ii}(\tau)$ is the autocorrelation function of the input signal, $R_{id}(\tau)$ is the cross-correlation of the input with the desired-output function, and $R_{dd}(0)$ is the mean-square value of the desired-output function, as given by

$$R_{ii}(\tau) = \int_{-\infty}^{\infty} f_i(t) f_i(t + \tau)dt,$$

$$R_{id}(\tau) = \int_{-\infty}^{\infty} f_i(t) f_d(t + \tau)dt,$$

$$R_{dd}(\tau) = \int_{-\infty}^{\infty} f_d(t) f_d(t + \tau)dt,$$

$$R_{dd}(0) = \int_{-\infty}^{\infty} [f_d^2(t)]\, dt. \tag{1.114}$$

It is trivial to see that the mean-square error is a function of **h(t)**, which is the central problem to determine **h(t′)** by minimizing the error. The minimization of function is subject to *calculus of variation*, which leads to the *Wiener-Hopf* integral equation, as given by

$$\int_{-\infty}^{\infty} h(\sigma)R_{ii}(\tau - \sigma)d\sigma - R_{id}(\tau) = 0. \tag{1.115}$$

We note that by following the realizable constraint of a physical filter, strictly speaking the preceding result should be written as

$$\int_{-\infty}^{\infty} h(\sigma)R_{ii}(\tau - \sigma)d\sigma - R_{id}(\tau) \begin{cases} \neq 0, & \text{for } \tau < 0 \\ = 0, & \text{for } \tau \geqslant 0 \end{cases}. \tag{1.116}$$

In view of the Wiener-Hopf equation (1.115) as written by,

$$R_{id}(\tau) = \int_{-\infty}^{\infty} h_{opt}(\tau)R_{ii}(\tau - \sigma)d\sigma, \tag{1.117}$$

by which the optimum filter $h_{opt}(\tau)$ is needed to solve.

If the desired output could be obtained without the physical realizable constraint, the input–output cross-correlation of the optimum filter is, in fact, the input and the desired-output cross-correlation itself, which holds not only for $\tau \geqslant 0$ but also $\tau < 0$. Now the problem becomes a trivial one. By taking the Fourier transform of Eq. 1.117, we have

$$R_{id}(v) = H_{opt}(v) R_{ii}(v), \tag{1.118}$$

where $H_{opt}(v)$ is the transform function of the optimum filter, written by

$$H_{opt}(v) = \frac{R_{id}(v)}{R_{ii}(v)}, \tag{1.119}$$

which is, in fact, the solution of our problem.

We do not exactly know the desired output response, but we have been demanding the filter more than our expectation; namely, by minimizing the mean-square error, which gives rise to the *best* approximation. In other words, this method attempts to get the best result we can, similar to bargaining for an automobile with a used-car salesperson. As we demanded more than a filter can be synthesized, the input-output crosscorrelation *cannot* be equal to the crosscorrelation of the input and the desired output. According to the Wiener-Hopf equation, the filter design should be based on the minimum error criterion and the impulse response should be compelled to the inequality of $\tau \geqslant 0$, but allows it to vary somewhat for $\tau < 0$, such as

$$q(\tau) = R_{id}(\tau) - \int_{-\infty}^{\infty} h_{opt}(\sigma) R_{ii}(\tau - \sigma) d\sigma, \tag{1.120}$$

where $q(\tau) \neq 0$ for $\tau < 0$ but vanishes for $\tau \geqslant 0$.

1.5.4. SIGNAL AMBIGUITY

The choice of a transmitted waveform for radar seems more trivial than it is for communication. However, it is actually not so fundamental. For example, it is the reflected (or echo) wave from the target that the temporal and doppler shift (i.e., frequency) provide the range and radial velocity information of the target. Whenever time and frequency are mentioned, we anticipate relating the quantities to the uncertainty relationship. In other words, no wave form can occupy a very short time while having a very narrow bandwidth. Thus, there is a limitation imposed on the combined resolution in time and frequency.

To simplify the analysis, we set scalar wave field $u(t)$ be constrained by the following equation:

$$\int |u(t)|^2 \, dt = \int |U(v)|^2 \, dv = 1. \tag{1.121}$$

In view of the Wiener-Khinchin and Parseval's Theorems, we see that in time domain, we have

$$R(\tau) = \int u(t)u^*(t + \tau)dt = \int |U(v)|^2 \, e^{i2\pi v\tau} \, dv, \tag{1.122}$$

and

$$T = \int |R(\tau)|^2 \, d\tau = \int |U(v)|^4 \, dv,$$

where the superasterisk denotes the complex conjugate. Similarly, in frequency domain we have

$$K(v) = \int U^*(v')U(v' + v)dv' = \int |u(t)|^2 \, e^{i2\pi vt} \, dt, \tag{1.123}$$

and

$$F = \int |K(v)|^2 \, dv = \int |u(t)|^4 \, dt,$$

where F is defined as the frequency resolution cell, which provides the well-known uncertainty relationship as given by

$$\text{(frequency resolution cell)} \times \text{(time span)} = 1. \tag{1.124}$$

Notice that Eq. (1.122) applies frequently to the resolution of detecting stationary targets at different ranges, while Eq. (1.123) applies to the resolution of detecting moving targets at different radial velocities. However, when the targets are both at different ranges, and moving with different radial velocities, (neither quantity being known in advance), the separation of time (e.g., range) and frequency (i.e., radial velocity) resolution do not always provide the actual resolving power of the target. Thus, we need a more general description of time and frequency shift to interpret the resolving power of the signal, which is known as the *ambiguity function*, as given by

$$\chi(\tau, \nu) = \int u(t)u^*(t + \tau)e^{-i2\pi\nu t}\,dt = \int U^*(\nu')U(\nu' + \nu)e^{i2\pi\nu'\tau}\,d\nu'. \quad (1.125)$$

The significance of $\chi(\tau, \nu)$ is that the range and (radial) velocity of a target cannot be resolved at $(\tau_0 + \tau, \nu_0 + \nu)$, where τ_0 and ν_0 are the mean range and velocity of the target. By virtue of normalization, we let

$$\iint |\chi(\tau, \nu)|^2\,d\tau\,d\nu = 1, \quad (1.126)$$

which is called the *area of ambiguity* in the time–frequency domain.

One of the simplest examples illustrating the ambiguity of a signal resulting from a single Gaussian pulse is given by

$$u(t) = \sqrt[4]{2}e^{-\pi t^2}.$$

By substituting into Eq. (1.125), we have

$$\chi(\tau, \nu) = \exp(-\tfrac{1}{2}\pi\tau^2)\exp(-\tfrac{1}{2}\pi\nu^2)\exp(i\pi\nu\tau).$$

From the preceding equation we see that the magnitude of ambiguity distribution, as given by

$$|\chi(\tau, \nu)|^2 = e^{-\pi(\tau^2 + \nu^2)},$$

describes a circular pattern in the (τ, ν) domain, as can be seen in Fig. 1.11.

If the pulse is shortened, the circular ambiguity pattern would become elliptical, as illustrated in Fig. 1.12.

A less trivial example is that $\mathbf{u(t)}$ represents a train of Gaussian pulses, modulated by a broad Gaussian envelope, as shown in Fig. 1.13(a). The corresponding ambiguity $|\chi(\tau, \nu)|^2$ is sketched in Fig. 1.13(b), in which the ambiguity in frequency resolution is split up into narrow bands, whereas in time it repeats itself. Notice that the shaded areas of the redistributed ambiguity are in fact equal to the area of the single pulse ambiguity; i.e.,

$$\iint |\chi(\tau, \nu)|^2\,d\tau\,d\nu = 1.$$

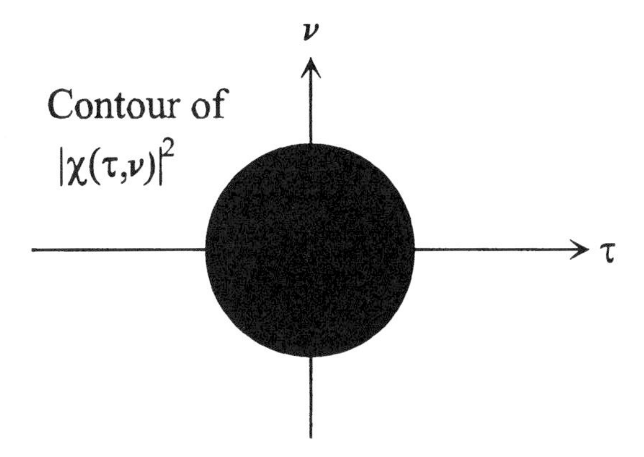

Fig. 1.11. Ambiguity of a Gaussian pulse.

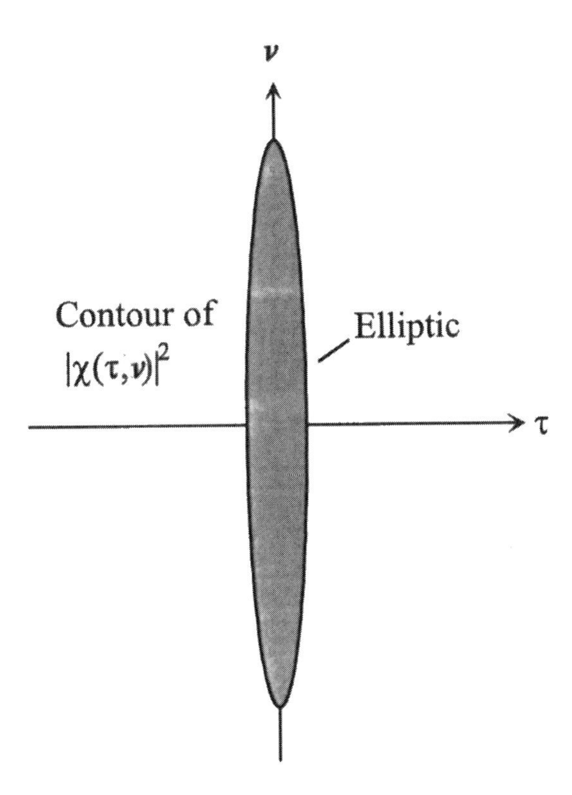

Fig. 1.12. Ambiguity of a very short Gaussian pulse.

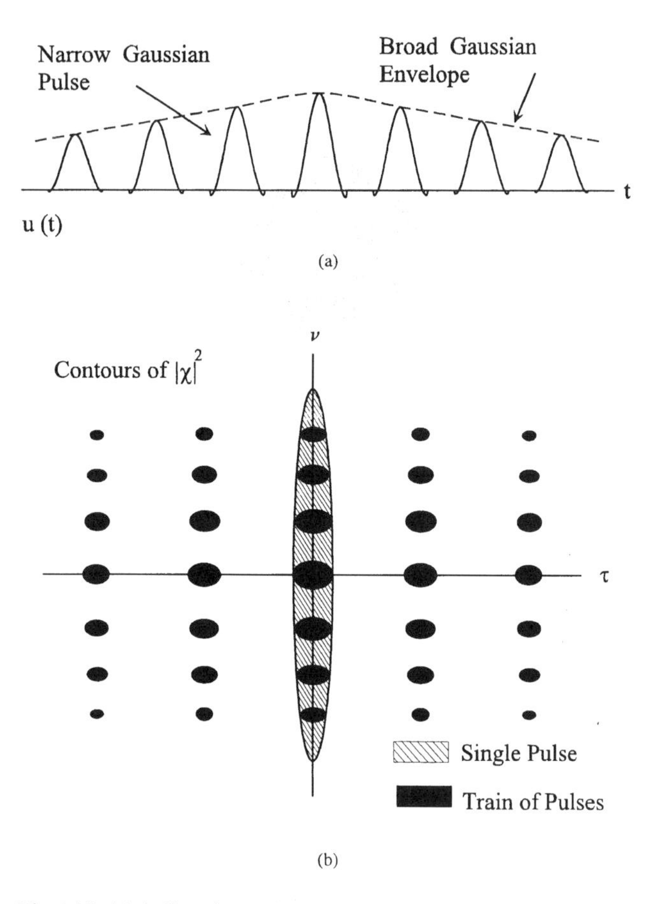

Fig. 1.13. (a) A Gaussian pulse-train, (b) Ambiguity diagram of (a).

Note that in actual application, we are not interested in resolving all the ranges of the target, for which the redundant parts of the ambiguity diagram remote from the origin have no practical usage. For example, a target 10 miles away should not be confused with one 1010 miles away. In practice, we may draw a rectangular region around the origin of the (τ, ν) coordinate plane to discriminate the ambiguity outside the region. Therefore, we see that the pulse train example will be a more acceptable waveform for range and (radial) velocity discrimination, as compared with the ambiguity distribution obtained from a single pulse. However, when interpreting the ambiguity diagram, one should remember that τ and ν do not represent the actual range and (radial) velocity, but rather the difference between the ranges and velocities of any two targets that need to be resolved.

1.5.5. WIGNER DISTRIBUTION

There is another form of time–frequency signal representation, similar to ambiguity function, defined by

$$W(\tau, \nu) = \int u(t)u^*(\tau - t)e^{-i2\pi\nu t}\,dt = \int U^*(\nu')U(\nu - \nu')e^{i2\pi\nu' t}\,d\nu', \quad (1.127)$$

which is known as the *Wigner distribution function.* Instead of using the correlation operator [i.e., $u(t)u^*(t + \tau)$], Wigner used the convolution operator [i.e., $u(t)u^*(\tau - t)$] for his transformation. The physical properties of the Wigner distribution function (WDF) can be shown as

$$\int W(\tau, \nu)\,d\nu = |u(t)|^2, \quad (1.128)$$

$$\int W(\tau, \nu)\,d\tau = |U(\nu)|^2, \quad (1.129)$$

$$\int W(t, \nu)\,dt\,d\nu = 1, \quad (1.130)$$

in which the WDF has been normalized to unity for simplicity.

One of the interesting features of the WDF is the inversion. Apart from the association with a constant phase factor, the transformation is unique, as can be shown,

$$\int W(t, \nu)e^{-i4\pi\nu t}\,dt = U(2\nu)U^*(0), \quad (1.131)$$

$$\int W(\tau, \nu)e^{i4\pi\nu t}\,d\nu = u(2\tau)u^*(0). \quad (1.132)$$

If the constant phase factors $\mathbf{u}^*(\mathbf{0})$ or $\mathbf{U}^*(\mathbf{0})$ happen to be very small or even zero, then we can reformulate the transformations by using the maximum value of $|u(\tau)|^2$ or $|U(\tau)|^2$, which occurs at $\tau_{\max}$, or $\nu_{\max}$, respectively, as given by,

$$\int W(\tau, \nu)\exp[-4\pi(\nu - \nu_{\max})\tau]d\tau = U(2\nu - \nu_{\max})U^*(\nu_{\max}), \quad (1.133)$$

$$\int W(\tau, \nu)\exp[i4\pi\nu(\tau - \tau_{\max})]\,d\nu = u(2\tau - \tau_{\max})u^*(\tau_{\max}). \quad (1.134)$$

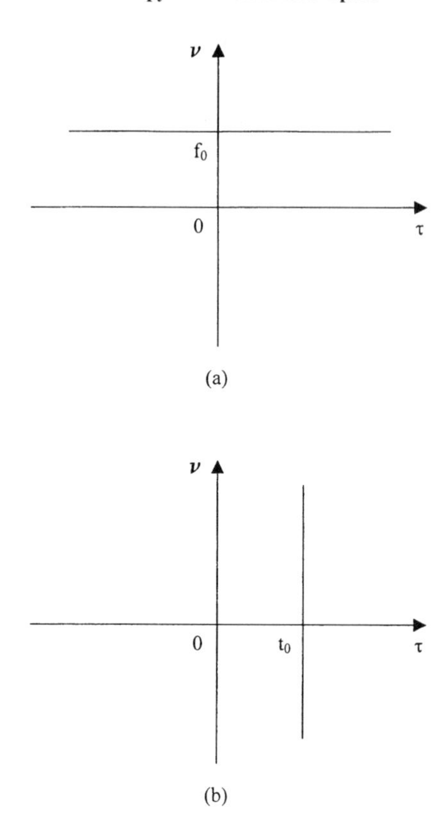

Fig. 1.14. Wigner distributions, (a) For a single tone signal, (b) For a pulse signal.

In other words, WDF does provide the frequency content of a time-varying signal, instead of providing the time–frequency resolution as the ambiguity function does.

In the following, we provide a couple of simple examples to illustrate of the WDF. First, let us consider a single tone signal, as given by

$$u(t) = e^{i2\pi\nu_0 t}.$$

By substitution into Eq. (1.127), we have

$$W(\tau, \nu) = \delta(\nu - \nu_o),$$

which is shown in Fig. 1.14a.

On the other hand, if the signal is a single pulse function, such as

$$u(t) = \delta(t - t_0),$$

then the corresponding WDF would be

$$W(\tau, v) = \delta(\tau - t_0),$$

which is plotted in Fig. 1.14b.

1.6. TRADING INFORMATION WITH ENTROPY

Let us consider a *nonisolated* system, in which the structure complexity has been established *a priori* in N equiprobable status; the entropy of the system can be written as [1.10]

$$S_o = k \ln N, \tag{1.135}$$

where k is Boltzmann's constant. If the system structure is reduced by outside intervention to M state, $(M < N)$, then its entropy would be

$$S_1 = k \ln M. \tag{1.136}$$

Since $S_o > S_1$, the decrease in entropy in the system is obviously related to the entropy information I that can be acquired from external sources:

$$\Delta S = S_1 - S_o = -kI \ln 2, \tag{1.137}$$

where $I = \log_2 N/M$. Thus, we see that the amount of information required for this reduction should be proportional to amount of entropy as ΔS decreases in the system. One of the most intriguing laws in thermodynamics must be the *second law* [1.11], in which it stated that for an *isolate* system its entropy can only be increased or remain constant; that is,

$$\Delta S_1 = \Delta(S_o - kI \ln 2) \geqslant 0. \tag{1.138}$$

In other words, any further increase in entropy ΔS_1 can be due to ΔS_o or ΔI, or both. Although in principle it is possible to distinguish the changes in ΔS_o and ΔI separately, in some cases the separation of the changes due to ΔS_o and ΔI may be difficult to discern.

It is interesting to note that, if the initial entropy S_o of the system corresponds to some complexity of a structure but not the *maximum*, and if S_o remains constant ($\Delta S_o = 0$), then after a certain free evolution without the influence of external sources, from Eq. (1.138), we will have

$$\Delta I \leqslant 0, \tag{1.139}$$

since $\Delta S_o = 0$.

In view of the preceding equation, we see that the changes in information ΔI are negative, or decreasing. The interpretation is that, when we have no *a priori* knowledge of the system complexity, the entropy S_o is assumed to be maximum (i.e., the equiprobable case). The information provided by the system structure is maximum, therefore, $\Delta I \leqslant 0$ is due to the fact that the system entropy has to increase $\Delta S_o > 0$. In other words, a decrease of information is needed in the system for its entropy to increase. This means that information can be provided or transmitted (a souce of *negentropy*) only by increasing the entropy of the system. However, if the initial entropy S_o is at a maximum state, then $\Delta I = 0$ which implies that the system *cannot* be used as a source of *negentropy* (or information). Thus, we see that information and (physical) entropy can be simply traded,

$$\Delta I \rightleftarrows \Delta S. \tag{1.140}$$

1.6.1. DEMON EXORCIST

One of the most interesting applications of the entropy theory of information must be the *Maxwell's Sorting Demon*, for which we consider a *thermally isolated* chamber as shown in Fig. 1.15. The chamber is divided into two chambers by a partition wall equipped with a small trapdoor for the demon to

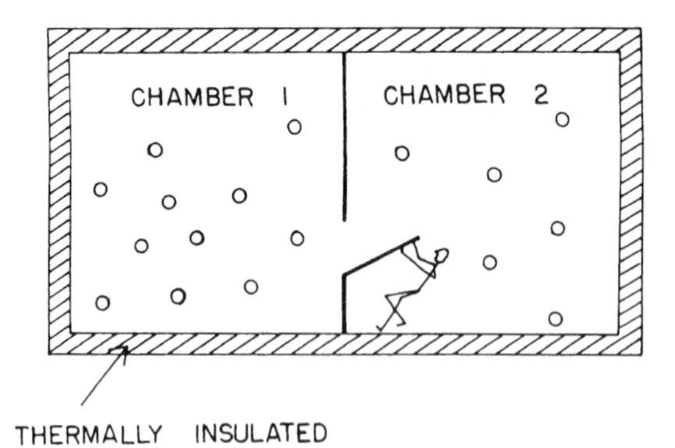

Fig. 1.15. Operation of Maxwell's demon.

operate. The demon is able, we assume, to see every individual molecule, so he can allow a molecule to pass from one chamber to the other. Now the demon decides to let the fast molecules in chamber 1 pass into chamber 2, and the slower molecules in chamber 2 to pass into chamber 1. In this manner we eventually are able to collect all the fast molecules in one chamber and all the slower molecules in other chambers. Thus, without any expenditure of work, the demon is able to raise the temperature in one chamber and lower the temperature in the other. Hence, work can be performed by the difference in temperature between the two chambers. However, in a thermally isolated chamber, this radiation within the chamber is only *black body* radiation, and it is impossible for the demon to see anything inside the chamber. Raising the temperature of the chamber would not help, since the demon would be able to see only the thermal radiation and its random fluctuations, but not the molecules. Thus, under these isolated conditions, the demon is unable to operate the trapdoor.

In order for the demon to see the *individual* molecules, we can provide the demon with a fresh flashlight (i.e., the source of negentropy or information). Then, by using the amount of information provided by the flashlight, the demon is able to operate the trapdoor. In this manner he can decrease the entropy of the chamber to a lower state. However, the entropy increase in the flashlight can be shown to be larger than, or at least equal to, the entropy decrease in the chamber. The lowest entropy increase by the flashlight can be seen as

$$\Delta S_d = \frac{h\nu}{T_o}, \tag{1.141}$$

where h is Planck's constant ν is the frequency of the flashlight radiation, and T_o is the temperature of the chamber in degrees Kelvin. The corresponding amount of information provided for the demon is

$$I_d = \frac{\Delta S_d}{k \ln 2}, \tag{1.142}$$

(we assume that the demon is capable of absorbing this information). By using this data the demon is able to reduce the chamber entropy from S_o to S, and then the amount entropy decreased per down operation at best would be

$$\Delta S_1 = S_1 - S_o = k \ln\left(1 - \frac{\Delta N}{N_o}\right)$$
$$\approx -k\frac{\Delta N}{N}, \quad \text{for } \frac{\Delta N}{N} \ll 1, \tag{1.143}$$

where N_o and ΔN denote the initial and the net change in complexity of the chamber.

However, from the second law of thermodynamics, the net entropy increase in the isolated chamber should be

$$\Delta S = \Delta S_d + \Delta S_1 = k\left[\frac{k\nu}{kT_o} - \frac{\Delta N}{N_o}\right] > 0. \tag{1.144}$$

So we see that the amount of entropy provided by the flashlight is higher than the amount of the chamber entropy that the demon can reduce.

Since we are in the computer era, we further assume that the demon is a *diffraction-limited demon*. Even though we assume that the demon can see the molecules under constant temperature conditions, he is so diffraction limited that he cannot distinguish an individual molecule from a group of molecules that are approaching the trapdoor. In order for him to do so, we equip the demon with the most powerful computer ever developed. To perceive the arriving molecules, the demon turns on the flashlight, which is required to emit at least a quanta of light to make the observation, i.e.,

$$h\nu = kT. \tag{1.145}$$

We also assume that the quanta of light reflected by the approaching molecules is totally absorbed by the diffractive-limited eye of the demon, which corresponds to an increase of entropy in the demon; that is,

$$\Delta S_d = \frac{h\nu}{T}, \tag{1.146}$$

or equivalent to the amount of information provided to the demon; i.e.,

$$I_d = \frac{\Delta S_d}{k \ln 2}. \tag{1.147}$$

Because of the diffraction-limited eye, the demon needs to process the absorbed quanta to a higher resolution, so that he is able to resolve the molecules and to allow the passages of the fast or slower molecules through the trapdoor. The amount of information gain, through the processing by the equipped computer, constitutes an equivalent amount of entropy increased, as given by

$$\Delta S_p = k \Delta I_d \ln 2, \tag{1.148}$$

where ΔI_d is the incremental amount of information provided by the computer.

With this amount of information gain, the demon is able to reduce the entropy of the chamber to a lower state. Again we can show that the overall net entropy changed in the chamber, per trapdoor operation, by the demon, would be

$$\Delta S = \Delta S_d + \Delta S_p + \Delta S_1 = k \left| \frac{h\nu}{kT} + \Delta I_d \ln 2 - \frac{\Delta N}{N_0} \right| > 0, \qquad (1.149)$$

in which we see that the diffraction-limited demon's exorcist is still within the limit of the second law of thermodynamics.

1.6.2. MINIMUM COST OF ENTROPY

One question still unanswered is, What would be the minimum cost of entropy required for the demon to operate the trapdoor? Let the arrival molecules at the trapdoor at an instant be one, two, or more molecules. Then the computer is required to provide the demon with a "yes" or a "no" information. For example, if a single molecule is approaching the trapdoor (say, a high-velocity one), the demon will open the trapdoor to allow the molecule to go through. Otherwise, he will stand still.

For simplicity, let us assume that the probability of one molecule arriving at the trapdoor is a 50% chance; then the demon needs one additional bit of information from the computer for him to open the trapdoor. This additional bit of information corresponds to the amount of entropy increased provided by the computer; i.e.,

$$\Delta S_p = k \ln 2 \approx 0.7k, \qquad (1.150)$$

which is the *minimum cost of entropy* required for the demon to operate the trapdoor. Thus, the overall net entropy increased in the chamber is

$$\Delta S = k \left[\frac{h\nu}{kT} + 0.7 - \frac{\Delta N}{N_0} \right] > 0. \qquad (1.151)$$

If one takes into account the other bit of "no" information provided by the computer, then the average net entropy increased in the chamber per operation would be

$$\Delta S_{ave} = k \left[\frac{2h\nu}{kT} + 1.4 - \frac{\Delta N}{N_0} \right] \gg 0, \qquad (1.152)$$

in which *two quantas* of light radiation are required. It is trivial, if one includes the possibility of a slower molecule approaching the trapdoor; the average cost of entropy per operation is even higher.

Even though we omit the two quantas of light in the calculation, the overall net entropy change in the chamber is still increasing; i.e.,

$$\Delta S_{ave} = k\left[1.4 - \frac{\Delta N}{N_0}\right] > 0, \tag{1.153}$$

In other words, the entropy compensated by the computer is still higher than the entropy reduced by the demon. With this argument, we note that the computer provided for the demon is also operated within the second law of thermodynamics.

We should now discuss the cost of entropy required to increase the resolution beyond the diffraction limit. Let a classical imaging system have the following resolution limit,

$$r_m = \frac{1.22\lambda f}{D}, \tag{1.154}$$

where r_m is the minimum separation, λ is the wavelength of the light source, and f and D are the focal length and the diameter of the imaging aperture. To increase the resolution of the system, one could either reduce the wavelength, enlarge the aperture of the system, or both. However, if the wavelength reduction and the aperture enlargement are not the subject to be considered, the observed image can be processed by a computer beyond the resolution limit. However, the amount of resolution gain would be compensated by the amount of entropy increased. Since the minimum cost of entropy to resolve a point object is $0.7k$, for n object points resolution, the minimum cost of entropy will be $0.7nk$. Nevertheless, in practice, the actual cost of entropy is very excessive.

For example, let us denote A as the field of view of an imaging system and ΔA_0 as the observation error, as shown in Fig. 1.16. The amount of information obtained by this observation would be

$$I = \log_2 \frac{A}{\Delta A_0}, \tag{1.155}$$

where $\Delta A_0 = \pi(r_m)^2$, and r_m is the minimum resolvable separation of object points by the optical system. If the observed image is then processed to a higher resolution, a net amount of information gain is anticipated. In other words, the observation error of the imaging system can be reduced, by processing the observed image, from ΔA_0 and ΔA_1. The net information gain provided by the

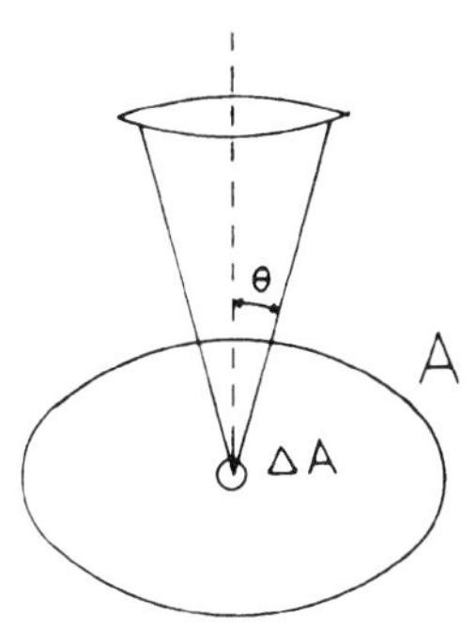

Fig. 1.16. A high-accuracy observation.

computer is

$$\Delta I = \log_2 \frac{\Delta A_0}{\Delta A_1}, \qquad \Delta A_1 \ll \Delta A_0. \tag{1.156}$$

Thus, we see that the minimum cost of entropy is

$$\Delta S = \Delta I k \ln 2 = k \ln 2 \log_2 \frac{\Delta A_0}{\Delta A_1}, \tag{1.157}$$

where ΔA_0 and ΔA_1 are the initial and final observation errors.

In short, we have achieved a very fundamental conclusion: one cannot get something from nothing—there is always a price to pay. Resolution beyond the resolution limit is possible, but only through the increase of entropy from a certain physical system (in our example, the computer). However, in practice, the cost of entropy is very excessive. In fact, the additional information gain at best can only approach the equivalent amount of entropy traded off:

$$\Delta I \leqslant \frac{\Delta S}{k \ln 2}. \tag{1.158}$$

1.7. ACCURACY AND RELIABILITY OBSERVATION

We discuss problems of observation within a space-and-time domain. This involves observing a position on a spatial plane, at an instant in time, with a certain observation error. Note that the spatial domain of observation must be bounded; otherwise, the information cannot be defined, since the information provided would lead to an infinite amount.

Let us assume a spatial domain A, which corresponds to the total field of view of an optical system. The spatial domain A is then subdivided into small subareas ΔA, which are limited by the resolvable power of the optical system. Of course, light is necessary for the observation, in which a particle or object is assumed to be wandering within the spatial domain A. In practice, we look at each subarea ΔA until we locate the particle; then the *accuracy of observation* can be written as

$$\mathscr{A} \triangleq \frac{A}{\Delta A} = \alpha \tag{1.159}$$

where α is the total number of ΔA's within A. To look for the particle we simply illuminate each ΔA by a beam of light, and each ΔA is assumed to be equipped with a photodetector able to detect scattered light, if any, from the particle. We further assume that each of the photodetectors is maintained at a constant Kelvin temperature T. Let us now investigate each of the succeeding photodetectors until a positive reading is obtained; say, from the $qth\,\Delta A$, where $q \leqslant \alpha$. The reading may be caused by thermal fluctuation in the detector, or it could be a positive reading for which the particle has been found in one of the ΔA's out of the q possibilities. Hence the amount of information obtained by this *sequential observation* is

$$I = \log_2 q \text{ bits.} \tag{1.160}$$

Since the positive reading was obtained from the absorption of scattered light by the qth photodetector, the accompanying entropy increase in the qth detector is

$$\Delta S \geqslant -k \ln[1 - (\tfrac{1}{2})^{1/q}], \qquad q \geqslant 1. \tag{1.161}$$

For a large value of q, the right-hand side of the preceding equation can be written as

$$\Delta S \geqslant k(\ln q + 0.367) > k \ln 2. \tag{1.162}$$

Thus

$$\Delta S - Ik \ln 2 \geqslant 0.367\text{k} > 0, \tag{1.163}$$

which is a positive quantity.

For the case of *simultaneous observations*, we have simultaneously observed γ positive readings from the α photodetectors. The amount of information obtained is therefore

$$I = \log_2 \frac{\alpha}{\gamma} \text{ bits}, \tag{1.164}$$

where $\gamma \leqslant \alpha$. Again, any reading could be due to thermal fluctuation. Since there are γ detectors absorbing the scattered light and the observations are made on all the α photodetectors, the overall amount of entropy increase in the γ photodetectors is

$$\Delta S \geqslant -\gamma k \ln[1 - \tfrac{1}{2}^{1/\alpha}]. \tag{1.165}$$

For $\alpha \gg 1$, the preceding equation can be approximated by

$$\Delta S \geqslant \gamma k(\ln \alpha + 0.367), \tag{1.166}$$

which increases with respect to γ and α. Thus

$$\Delta S - Ik \ln 2 \geqslant k[\ln \gamma + (\gamma - 1)\ln \alpha + 0.367\gamma] > 0. \tag{1.167}$$

It is interesting to note that, if it takes only one of the α photodetectors to provide a positive reading ($\gamma = 1$), then Eq. (1.167) is essentially identical to Eq. (1.163). However, if $\gamma \gg 1$, then the amount of information obtained from the simultaneous observations of Eq. (1.164) is somewhat less than that obtained from the sequential observations of Eq. (1.160), for $q \gg 1$, and the amount of entropy increase for the simultaneous observations is also greater.

Since it is assumed that only one particle is wandering in the spatial domain A, for $\gamma = 1$, Eq. (1.167) yields the *smallest trade-off* of entropy and information. At the same time, for a large number of photodetectors ($\alpha \gg 1$), Eq. (1.166) is the asymptotic approximation used in *high-accuracy observation* ($h\nu \gg kT$). Remember that any other arrangement of the photodetectors may result in higher entropy. For example, if all the photodetectors are arranged to receive light directly, rather than from scattered light, it can be seen that a much higher cost of entropy will be paid.

We now turn our attention to *high-accuracy observation*. Note that, if ΔA becomes very small, then higher-frequency illumination (a shorter wavelength) is necessary for the observation. As illustrated in Fig. 1.16, this observation cannot be efficient unless the wavelength of the light source is shorter than $1.64d \text{ in } \theta$ [1.12]:

$$\lambda \leqslant 1.64d \sin \theta, \tag{1.168}$$

where d is the diameter of ΔA, and θ is the subtended half-angle of lens aperture. By referring to the well-known resolving power of the microscope, we have

$$d = \frac{1.22\lambda}{2\sin\theta}, \tag{1.169}$$

where $2\sin\theta$ is the *numerical aperture*. The frequency required for the observation must satisfy the inequality

$$\nu = \frac{c}{\lambda} \geqslant \frac{c}{1.64d\sin\theta}, \tag{1.170}$$

where c is the speed of light.

By using the lower bound of the preceding equation, the characteristic diameter (or distance) of the detector can be defined. We assume the detector maintains at a constant temperature T:

$$\frac{h\nu}{kt} = \frac{hc}{1.64kTd\sin\theta} = \frac{d_o}{1.64d\sin\theta}. \tag{1.171}$$

Then the *characteristic diameter* d_o can be shown as

$$d_o = \frac{hc}{kT} \cong \frac{1.44}{T}. \tag{1.172}$$

Thus for *high-frequency observation*, $(h\nu \gg kT)$, we see that the resolving distance d is smaller than the characteristic distance d_o; that is,

$$d \ll d_o. \tag{1.173}$$

However, for *low-frequency observation*, $(h\nu \ll kT)$, we have

$$d \gg d_o. \tag{1.174}$$

We stress that d_o possesses no physical significance except that, at a given temperature T, for which d_o provides a distance boundary for low- and high-frequency observations.

Let us recall the high-frequency observation for which we have

$$h\nu > E_o = -kT\ln[1 - (\tfrac{1}{2})^{1/\alpha}], \tag{1.175}$$

where E_o is the *threshold energy level* for the photodetector.

For $\alpha \gg 1$ the preceding equation can be written as

$$h\nu > kT(\ln \alpha + 0.367). \quad (1.176)$$

Since the absorption of *one quantam* of light is adequate for a positive response, the corresponding entropy increase is

$$\Delta S = \frac{h\nu}{T} > k(\ln \alpha + 0.367). \quad (1.177)$$

The amount of information obtained would be

$$I = \log_2 \alpha \text{ bits}. \quad (1.178)$$

Thus we see that

$$\Delta S - Ik \ln 2 > 0.367k > 0. \quad (1.179)$$

Except for the equality, this ΔS is *identical* to that of the *low-frequency observation* of Eq. (1.162). However, the entropy increase is much higher, since ν is very high. Although *fine observation* can be obtained by using higher frequency, there is a price to be paid; namely, higher cost of entropy.

We now come to the *reliable observation.* One must distinguish the basic difference between accuracy and reliability in observations. A reliable observation is dependent on the chosen decision threshold level E_o; that is, the higher the threshold level, the higher the reliability. However, accuracy in observation is inversely related to the spread of the detected signal; the narrower the spread, the higher the accuracy. These two issues are illustrated in Fig. 1.17. It is evident that the higher threshold energy level E_o chosen would have higher the reliability. However, higher reliability also produces higher probability of *misses.* On the other hand, if E_o is set at a lower level, a less reliable observation is expected. In other words, high probability of error (false alarms) may produce, for example, due thermal noise fluctuation.

1.7.1. UNCERTAINTY OBSERVATION

All physical systems are ultimately restricted by some limitations. When quantum conditions are in use, all limitations are essentially imposed by the basic Heisenberg uncertainty principle:

$$\Delta E \, \Delta t \geqslant h, \quad (1.180)$$

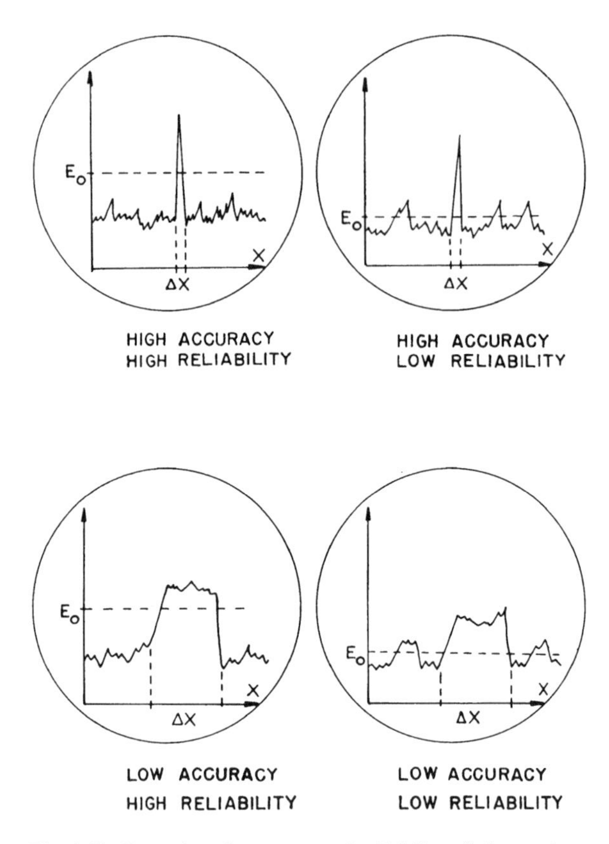

Fig. 1.17. Examples of accuracy and reliability of observation.

where ΔE denotes the energy perturbation and Δt is the time-interval observation.

By reference to the preceding observation made by radiation, one can compare the required energy ΔE with the mean-square thermal fluctuation of the photodetector γkT, where γ is the number of degrees of freedom, which is essentially the number of low-frequency vibrations ($h\nu \ll kT$). Thus, if $\Delta E < \gamma kT$, we have

$$\Delta t \gg \frac{h}{\gamma kT}. \tag{1.181}$$

From this inequality we see that a larger time resolution Δt is required for low-frequency observation. Since ΔE is small, the perturbation within Δt is very small and can by comparison be ignored.

However, if the radiation frequency ν becomes higher, such as $\Delta E = h\nu > \gamma kT$, then we have

$$\Delta t \leqslant \frac{h}{\gamma kT}. \tag{1.182}$$

We see that, as the radiant energy required for the observation increases, the more accurate time resolution can be made. But the perturbation of the observation is also higher. Thus, the time resolution Δt obtained by the observation may not even be correct, since ΔE is large.

In the classical theory of light, observation has been assumed nonperturbable. This assumption is generally true for many particles problems, for which a large number of quanta is used in observation. In other words, accuracy of observations is not expected to be too high in classical theory of light, since its imposed condition is limited far away from the uncertainty principle; that is

$$\Delta E\,\Delta t \gg h,$$

or equivalently,

$$\Delta p\,\Delta x \gg h.$$

However, as quantum conditions occur, a nonperturbing system simply does not exist. When a higher-quantum $h\nu$ is used, a certain perturbation within the system is bound to occur; hence, high-accuracy observation is limited by the uncertainty principle.

Let us look at the problem of observing extremely small distance Δx between two particles. One must use a light source having a wavelength λ that satisfies the condition

$$\lambda \leqslant 2\Delta x \tag{1.183}$$

Since Δx is assumed to be extremely small, a high-frequency light source is required for the observation, which corresponds to a higher momentum:

$$p = \frac{h}{\lambda} \geqslant \frac{h}{2\Delta x}. \tag{1.184}$$

In turn, this high-frequency source of radiation corresponds to a higher-quantum $h\nu$, in which it interacts with the observed specimen (as well as with the observing equipment), which causes the changes of momentum from $-p$ to p. Thus, we have

$$\Delta p = 2p \gg \frac{h}{\Delta x}. \tag{1.185}$$

The radiant energy provided can be written as

$$\Delta E = h\nu = \frac{hc}{\lambda} \geqslant \frac{hc}{2\Delta x}, \tag{1.186}$$

for which we have

$$\Delta E\,\Delta x \geqslant \frac{hc}{2}. \tag{1.187}$$

Theoretically speaking, there is no lower limit to Δx, as long as ΔE is able to increase. However, in reality, as ΔE increases, the perturbation of the observation cannot be ignored. Therefore in practice, when ΔE reaches a certain quantity, the precise observation of Δx is obstructed and the observation of smaller and smaller particles presents ever-increasing difficulty.

Finally, let us emphasize that Heisenberg's principle of uncertainty observation is restricted to the ensemble point of view; that is, for a special observation, the uncertainty may be violated. However, we have never been able to predict when this observation will occur. Therefore, a meaningful answer to the Heisenberg uncertainty principle, is only true under the statistical ensemble.

1.8. QUANTUM MECHANICAL CHANNEL

In the preceding sections we have presented an information channel from a *many particles* point of view. Intuitively, the formulation of the channel as we assumed is quite correct. However, when we deal with a communication channel in quantum mechanical regime, the results we have evaluated may lead to erroneous consequences. For instance, the capacity of a continuous additive Gaussian channel is given by

$$C = \Delta\nu \log_2\left(1 + \frac{S}{N}\right) \text{ bits/sec,}$$

in which we see that, if the average noise power N approaches zero, the channel capacity approaches infinity. This is obviously contradictory to the basic physical constraints. Therefore, as the information transmission moves to high-frequency regime, where the quantum effect takes place, the communication channel naturally leads to a discrete model. This is where the quantum theory of radiation replaces the classical wave theory.

We consider a quantum mechanical channel, for which the information source represents an optical signal (e.g., temporal signal). The signal propagated through the quantum mechanical channel is assumed perturbed by an additive thermal noise. For simplicity, we assume that the signal is one dimensional (i.e., the photon fluctuation is restricted to only one polarized state), in which the propagation occurs in the same direction as the wave vectors. Note that the corresponding occupation number of the quantum levels can be fully described. Obviously, these occupation quantum levels correspond to the *microsignal* structure of the information source. We can therefore assume that these occupation numbers can be uniquely determined for those representing an input signal ensemble. We further assume that an ideal receiver (an ideal photon counter) is used at the output end of the channel; that is, the receiver is capable of detecting the photon signal. We stress that the interaction of the photon signal and the detector are assumed statistical; that is, a certain amount of information loss is expected at the output end of the receiver. The idealized model of the receiver we have proposed mainly simplifies the method by which the quantum effect on the channel can be calculated. Let us now assume that the input photon signal can be quantized as represented by a set of microstate signals $\{a_i\}$, and each a_i is capable of channeling to the state of b_j, where the set $\{b_j\}$, $j = 1 = 2, \ldots, n$, represents the *macroscopic* states of the channel. Note that each macroscopic state b_j is an ensemble of various *microscopic* states within the channel. Thus by denoting $P(b_j/a_i)$ corresponding transitional probability, for each applied signal a_i the corresponding conditional entropy $H(B/a_i)$ can be written as

$$H(B/a_i) = -\sum_{j=1}^{n} P(b_j/a_i) \log_2 P(b_j/a_i).$$

Thus, the entropy equivocation of the channel is

$$H(B/A) = \sum_{i=1}^{n} P(a_i) H(B/a_i). \tag{1.188}$$

Since the (output) entropy can be written as

$$H(B) = \sum_{j=1}^{n} P(b_j) \log P(b_j). \tag{1.189}$$

The mutual information provided by the quantum mechanical channel can be determined by

$$I(A;\, B) = H(B) - H(B/A), \tag{1.190}$$

which corresponds to physical entropy of

$$\Delta S = I(A;\ B)k \ln 2. \tag{1.191}$$

This result can be interpreted as follows: the microstate of the channel acts as the transmission of information through the channel. It is noted that $I(A; B)$ is derived under a strictly stationary *ergodic* assumption. In practice, however, the output signal should be *time limited*, and the channel should have some memory. In view of the second law of thermodynamics, the physical entropy of the channel should be higher than that of the actual information transfer:

$$\Delta S > I(A;\ B)k \ln 2. \tag{1.192}$$

1.8.1. CAPACITY OF A PHOTON CHANNEL

Let us denote the mean quantum number of the photon signal by $\bar{m}(\nu)$, and the mean quantum number of a noise by $\bar{n}(\nu)$. We have assumed an additive channel; thus, the signal plus noise is

$$\bar{f}(\nu) = \bar{m}(\nu) + \bar{n}(\nu).$$

Since the phonton density (i.e., the mean number of photons per unit time per frequency) is the mean quantum number, the signal energy density per unit time can be written as

$$E_s(\nu) = \bar{m}(\nu)h\nu,$$

where h is Planck's constant. Similarly, the noise energy density per unit time is

$$E_N(\nu) = \bar{n}(\nu)h\nu.$$

Due to the fact that the mean quantum noise (blackbody radiation at temperature T) follows Planck's distribution (also known as the Bose-Einstein distribution),

$$E_N(\nu) = \frac{h\nu}{\exp(h\nu/kT) - 1}, \tag{1.193}$$

the noise energy per unit time (noise power) can be calculated by

$$N = \int_{\varepsilon}^{\infty} E_N(\nu)d\nu = \int_{\varepsilon}^{\infty} \frac{h\nu}{\exp(h\nu/kT) - 1} d\nu = \frac{(\pi kT)^2}{6h}, \tag{1.194}$$

where ε is an arbitrarily small positive constant.

Thus, the minimum required entropy for the signal radiation is

$$\Delta S = \int_{o}^{N} \frac{dE_s(T')}{T'} = \int_{o}^{T} \frac{dE_s(T')}{dT'} \frac{dT'}{T'}, \tag{1.195}$$

where $E(T)$ is signal energy density per unit time as a function of temperature T', and T is the temperature of the blackbody radiation. Thus, in the presence of a signal the output radiation energy per unit time (the power) can be written

$$P = S + N,$$

where S and N are the signal and the noise power, respectively. Since the signal is assumed to be deterministic (i.e., the microstate signal), the signal entropy can be considered zero. Remember that the validity of this assumption is mainly based on the independent statistical nature between the signal and the noise, for which the photon statistics follow Bose-Einstein distribution. However, the Bose-Einstein distribution cannot be used for the case of *fermions*, because, owing to the *Pauli exclusion principle*, the microstates of the noise are restricted by the occupational states of the signal, or vice versa. In other words, in the case of fermions, the signal and the noise can *never* be assumed to be statistically independent. For the case of Bose-Einstein statistics, we see that the amount of entropy transfer by radiation remains unchanged:

$$H(B/A) = \frac{\Delta S}{k \ln 2}. \tag{1.196}$$

Since the mutual information is $I(A; B) = H(B) - H(B/A)$, we see that $I(A; B)$ reaches its maximum when $H(B)$ is maximum. Thus, for maximum information transfer, the photon signal should be chosen randomly. But the maximum value of entropy $H(B)$ occurs when the ensemble of the microstates of the total radiation (the ensemble B) corresponds to Gibbs's distribution, which reaches to the thermal equilibrium. Thus, the corresponding mean occupational quantum number of the total radiation also follows Bose-Einstein distribution at a given temperature $T_e \geqslant T$:

$$\bar{f}(\nu)) = \frac{h\nu}{\exp(h\nu/kT_e) - 1}, \tag{1.197}$$

where T_e is defined as the *effective temperature.*

Since the output entropy can be determined by

$$H(B) = \frac{1}{k \ln 2} \int_o^{T_e} \frac{dE_N(T')}{dT'} \frac{dT'}{T'}, \tag{1.198}$$

the quantum mechanical channel capacity can be evaluated as given by

$$C = H(B) - H(B/A) = \frac{1}{k \ln 2} \int_T^{T_e} \frac{dE_N(T')}{dT'} \frac{dT'}{T'}, \tag{1.199}$$

where

$$E_N(T') = \frac{(\pi k T')^2}{6h}.$$

In view of the total output power

$$P = S + N = \frac{(\pi k T_e)^2}{6h} = S + \frac{(\pi k T)^2}{6h}, \tag{1.200}$$

the effective temperature T_e can be written as

$$T_e = \left[\frac{6hS}{(\pi k)^2} + T^2\right]^{1/2}. \tag{1.201}$$

Thus, the capacity of a photon channel can be shown to be

$$C = \frac{\pi^2 kT}{3h \ln 2}\left[\left(1 + \frac{6hS}{(\pi k T)^2}\right)^{1/2} - 1\right]. \tag{1.201}$$

We would further note that high signal-to-noise ratio corresponds to high-frequency transmission ($h\nu \gg kT$), for which we have

$$\frac{6hS}{(\pi k T)^2} \gg 1.$$

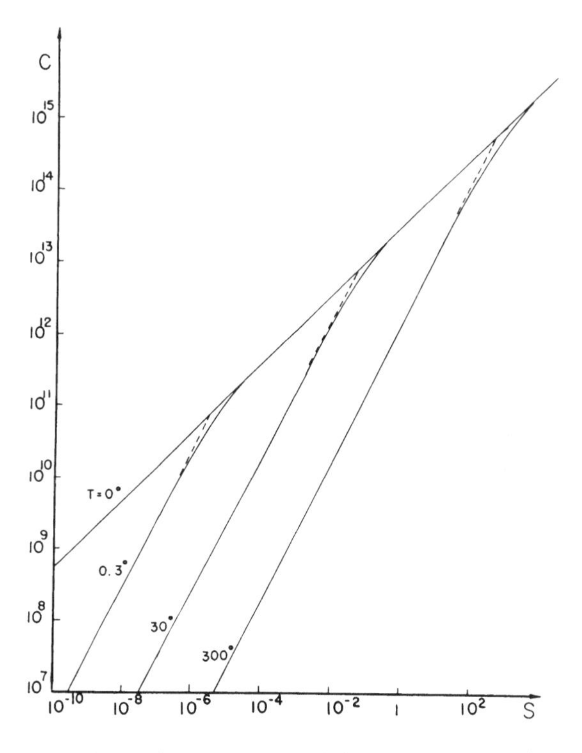

Fig. 1.18. Photon channel capacity as a function of signal powers, for various values of thermal noise temperature *T*. Dashed lines represent the classical asymptotes of Eq. (1.203).

Then the photon channel capacity is limited by the quantum statistic; that is,

$$C_{\text{quant}} = \frac{\pi}{\ln 2}\left(\frac{2S}{3h}\right)^{1/2}. \tag{1.203}$$

However, if the signal-to-noise ratio is low ($h\nu \ll kT$), the photon channel capacity reduces to the classical limit:

$$C_{\text{class}} = \frac{S}{kT \ln 2}. \tag{1.204}$$

Figure 1.18 shows a plot of the photon channel capacity as a function of signal energy. We see that at the low signal-to-noise ratio ($h\nu \ll kT$) regime, the channel capacity approaches this *classical limit*. However, for high signal-to-noise ratio ($h\nu \gg kT$), the channel capacity approaches the *quantum limit* of Eq (1.203), which offers a much higher transmission rate.

REFERENCES

1.1 N. Wiener, *Cybernetics*, MIT Press, Cambridge, Mass., 1948.
1.2 N. Wiener, *Extrapolation, Interpolation, and Smoothing of Stationary Time Series.*
1.3 C. E. Shannon, 1948, "A Mathematical Theory of Communication," *Bell Syst. Tech. J.*, vol. 27, 379–423, 623–656.
1.4 C. E. Shannon and W. Weaver, *The Mathematical Theory of Communication*, University of Illinois Press, Urbana, 1949.
1.5 W. Davenport and W. Root, *Random Signals and Noise*, McGraw-Hill, New York, 1958.
1.6 M. Loeve, *Probability Theory*, 3rd ed., Van Nostrand, Princeton, N.J., 1963.
1.7 D. Gabor, 1946, "Theory of Communication," *J. Inst. Electr. Eng.*, vol. 93, 429.
1.8 D. Gabor, 1950, "Communication Theory and Physics", *Phi., Mag.*, vol. 41, no. 7, 1161.
1.9 J. Neyman and E.S. Pearson, 1928, "On the Use and Interpretation of Certain Test Criteria for Purposes of Statistical Inference", *Biometrika*, vol. 20A, 175, 263.
1.10 L. Brillouin, *Science and Information Theory*, 2nd ed., Academic, New York, 1962.
1.11 F. W. Sear, *Thermodynamics, the Kinetic Theory of Gases, and Statistical Mechanics*, Addison-Wesley, Reading, Mass., 1953.
1.12 F. T. S. Yu and X. Yang, *Introduction to Optical Engineering*, Cambridge University Press, Cambridge, UK, 1997.

EXERCISES

1.1 A picture is indeed worth more than a thousand words. For simplicity, we assume that an old-fashioned monochrome Internet screen has a 500×600 pixel-array and each pixel element is capable of providing eight distinguishable brightness levels. If we assume the selection of each information pixel-element is equiprobable, then calculate the amount of information that can be provided by the Internet screen. On the other hand, a broadcaster has a pool of 10,000 words; if he randomly picked 1000 words from this pool, calculate the amount of information provided by these words.

1.2 Given an n-array symmetric channel, its transition probability matrix is written by

$$[P] = \begin{bmatrix} 1-p & \frac{p}{n-1} & \frac{p}{n-1} & \cdots & \frac{p}{n-1} \\ \frac{p}{n-1} & 1-p & \frac{p}{n-1} & \cdots & \frac{p}{n-1} \\ \vdots & \vdots & \vdots & \cdots & \vdots \\ \frac{p}{n-1} & \frac{p}{n-1} & \frac{p}{n-1} & \cdots & 1-p \end{bmatrix}$$

(a) Evaluate the channel capacity.
(b) Repeat part (a) for a binary channel.

1.3 Show that an information source will provide the maximum amount of information, if and only if the probability distribution of the ensemble information is equiprobable.

1.4 Let an input ensemble to a discrete memoryless channel be $A = \{a_1, a_2, a_3\}$ with the probability of occurrence $p(a_1) = \frac{1}{2}$, $p(a_2) = \frac{1}{4}$, $p(a_3) = \frac{1}{4}$, and let $B = \{b_1, b_2, b_3\}$ be a set of the output ensemble. If the transition matrix of the channel is given by

$$\begin{bmatrix} \frac{1}{3} & \frac{1}{3} & \frac{1}{3} \\ \frac{1}{2} & \frac{1}{4} & \frac{1}{4} \\ \frac{1}{4} & \frac{1}{4} & \frac{1}{2} \end{bmatrix},$$

a. Calculate the output entropy $H(B)$.
b. Compute the conditional entropy $H(A/B)$.

1.5 Let us consider a memoryless channel with input ensemble $A = a_1, a_2, \ldots, a_r$ and output ensemble $B = b_1, b_2, \ldots, b_s$, and channel matrix $[p(b_j/a_i)]$. A random decision rule may be formalized by assuming that if the channel output is b_j for every $i = 1, 2, \ldots, s$, the decoder will select a_i with probability $q(a_i/b_j)$, for every $i = 1, 2, \ldots, r$. Show that for a given input distribution there is no random decision rule that will provide a lower probability of error than the ideal observer.

1.6 The product of two discrete memoryless channels C_1 and C_2 is a channel the inputs of which are ordered pairs (a_i, a'_j) and the outputs of which are ordered pairs (b_k, b'_l) where the first coordinates belong to the alphabet of C_1 and the second coordinates to the alphabet of C_2. If the transition probability of the product channel is

$$P(b_k, b'_l/a_i, a'_j) = P(b_k/a_i)\, p(b'_l/a'_j),$$

determine the capacity of the product channel.

1.7 Develop a maximum likelihood decision and determine the probability of errors for a discrete memoryless channel, as given by,

$$P = \begin{bmatrix} \frac{1}{2} & \frac{1}{3} & \frac{1}{6} \\ \frac{1}{6} & \frac{1}{2} & \frac{1}{3} \\ \frac{1}{3} & \frac{1}{6} & \frac{1}{2} \end{bmatrix},$$

where this input ensembles are

$$P(a_1) = \tfrac{1}{2}, \quad p(a_2) = \tfrac{1}{4}, \quad p(a_3) = \tfrac{1}{4}.$$

1.8 *A memoryless continuous channel is perturbed by an additive Gaussian noise with zero mean and variance equal to N*. The output entropy and the noise entropy can be written as

$$H(B) \leqslant \tfrac{1}{2}\log_2(2\pi e\sigma_b^2),$$

$$H(B/A) \leqslant \tfrac{1}{2}\log_2(2\pi e\sigma_c^2),$$

where $b = a + c$ (i.e., signal plus noises), and $\sigma_b^2 = \sigma_a^2 + \sigma_c^2 = S + N$. Calculate the capacity of the channel.

1.9 The input to an ideal low-pass *filter* is a narrow rectangular pulse signal as given by

$$H(v) = \begin{cases} 1, & |v| < \dfrac{\Delta v}{2}, \\ 0, & \text{otherwise}^{\cdot} \end{cases}$$

a. Determine the output response $g(t)$.
b. Sketch $g(t)$ when $\Delta t\,\Delta v < 1$, $\Delta t\,\Delta v = 1$ and $\Delta t\,\Delta v > 1$.
c. If

$$H(v) = \begin{cases} e^{-i2\pi v t_o,} & |v| \leqslant \dfrac{\Delta v}{2}, \\ 0, & \text{otherwise}, \end{cases}$$

show how this linear phase low-pass filter affects the answers in part (b).

1.10 Consider an $m \times n$ array spatial channel in which we spatially encode a set of coded images. What would be the spatial channel capacity?

1.11 A complex Gaussian pulse is given by

$$u(t) = K\exp(-At^2),$$

where M is a constant and

$$A = a + ib.$$

Determine and sketch the ambiguity function, and discuss the result as it applies to radar detection.

1.12 Given a chirp signal, as given by

$$u(t) = \exp[i2\pi v_1 t^2 + v_0 t],$$

calculate and sketch the corresponding Wigner distribution.

1.13 A sound spectrograph can display speech signals in the form of a frequency versus time plot which bears the name of logon (spectrogram). The analyzer is equipped with a narrow band $\Delta\nu = 45\,\text{Hz}$ and a wide band $\Delta\nu = 300\,\text{Hz}$.
a. Show that it is impossible to resolve the frequency and time information simultaneously by using only one of these filters.
b. For a high-pitched voice that varies from 300–400 Hz, show that it is not possible to resolve the time resolution, although it is possible to resolve the fine-frequency content.
c. On the other hand, for a low-pitched voice that varies from 20–45 Hz, show that it is possible to resolve the time resolution but not the fine-frequency resolution.

1.14 We equip the Szillard's demon with a beam of light to illuminate the chamber, in which it has only one molecule wandering in the chamber, as shown in Fig. 1.19. Calculate
a. the net entropy change per cycle of operation,
b. the amount of information required by the demon,
c. the amount of entropy change of the demon, and
d. show that with the intervention of the demon, the system is still operating within the limit of the second law of thermodynamics.

1.15 A rectangular photosensitive paper may be divided into an $M \times N$ array of information cells. Each cell is capable of resolving K distinct gray levels. By a certain recording (i.e., encoding) process, we reduce the $M \times N$ to a $P \times Q$ array of cells with $P < M$ and $Q < N$
a. What is the amount of entropy decrease in reducing the number of cells?
b. Let an image be recorded on this photosensitive paper. The amount of information provided by the image is I_i bits, which is assumed to be smaller than the information capacity of the paper. If I_j (i.e., $I_j < I_i$)

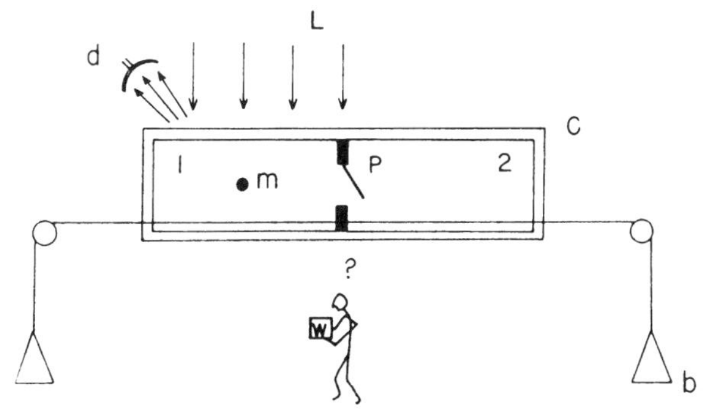

Fig. 1.19. Szilard's machine by the intervention of the demon. D, photodetector; L, light beams; C, transparent cylinder; P, piston; m, molecule.

bits of the recorded information are considered to be in error, determine the entropy required for restoring the image.

1.16 For low-frequency observation, if the quantum stage $g = m$ is selected as the decision threshold, we will have an error probability of 50% per observation. If we choose $g = 5m$ calculate the probability of error per observation.

1.17. In high-frequency observation, show that the cost of entropy per observation is greater. What is the minimum amount of entropy required per observation? Compare this result with the low-frequency case and show that the high-frequency case is more reliable.

1.18. For the problem of many simultaneous observations, if we assume that an observation gives a positive (correct) result (i.e., any one of the α photodetectors give rise to an energy level above $E_0 = gh\nu$) and a 25% chance of observation error is imposed, then calculate

a. the threshold energy level E_0 as a function of α and

b. the corresponding amount of entropy increase in the photodetectors.

1.19 In low-frequency simultaneous observations, if we obtain γ simultaneous correct observations out of α photodetectors ($\gamma < \alpha$), determine the minimum cost of entropy required. Show that for the high-frequency case, the minimum entropy required is even greater.

1.20 Given an imaging device; for example, a telescope, if the field of view at a distant corner is denoted by A and the resolution of this imaging system is limited to a small area ΔA over A, calculate the accuracy of this imaging system and the amount of information provided by the observation.

1.21 With reference to the preceding problem,

a. Show that a high-accuracy observation requires a light source with a shorter wavelength, and the reliability of this observation is inversely proportional to the wavelength employed.

b. Determine the observation efficiency for higher-frequency observation.

1.22 The goal of a certain experiment is to measure the distance between two reflecting walls. Assume that a plane monochromatic wave is to be reflected back and forth between the two walls. The number of interference fringes is determined to be 10 and the wavelength of the light employed is 600nm. Calculate

a. the separation between the walls,

b. the amount of entropy increase in the photodetector ($T = 300°K$),

c. the amount of information required,

d. the corresponding energy threshold level of the photodetector if a reliability $\mathscr{R} = 4$ is required, where $\mathscr{R} = 1/\text{probability of error}$, and

e. the efficiency of this observation.

1.23 Repeat the preceding problem for the higher-frequency observation.

1.24 With reference to the problem of observation under a microscope, if a square (*sides* $= r_0$) instead of a circular wavelength is used, calculate the minimum amount of entropy increase in order to overcome the thermal background fluctuations. Show that the cost of entropy under a microscope observation is even greater than $kI \ln 2$.

1.25 Show that the spatial information capacity of an optical channel under coherent illumination is generally higher than under incoherent illumination.

1.26 Let us consider a band-limited periodic signal of bandwidth $\Delta\nu_m$, where ν_m is the maximum frequency content of the signal. If the signal is passed through an ideal low-pass filter of bandwidth $\Delta\nu_f$ where the cutoff frequency ν_f is lower than that of ν_m, estimate the minimum cost of entropy required to restore the signal.

1.27 Refer to the photon channel previously evaluated. What would be the minimum amount of energy required to transmit a bit of information through the channel?

1.28 High and low signal-to-noise ratio is directly related to the frequency transmission through the channel. By referring to the photon channels that we have obtained, under what condition would the classical limit and quantum statistic meet? For a low transmission rate, what would be the minimum energy required to transmit a bit of information?

1.29 It is well known that under certain conditions a band-limited photon channel capacity can be written as

$$C \approx \Delta\nu \left[\log_2 \left(1 + \frac{S}{h\nu} \right) + \frac{S}{h\nu} \log_2 \left(1 + \frac{h\nu}{S} \right) \right].$$

The capacity of the channel increases as the mean occupation number $\bar{m} = S/h\nu$ increases. Remember, however, that this is valid only under the condition $\Delta\nu/\nu \ll 1$, for a narrow-band channel. It is incorrect to assume that the capacity becomes infinitely large as $\nu \to 0$. Strictly speaking, the capacity will never exceed the quantum limit as given in Eq. (1.202). Show that the narrow-band photon channel is in fact the Shannon's continuous channel under high signal-to-noise ratio transmission.

Chapter 2 Signal Processing with Optics

Francis T. S. Yu
PENNSYLVANIA STATE UNIVERSITY

Optical processing can perform a myriad of processing operations. This is primarily due to its complex amplitude processing capability. Optical signal processors can perform one- or two-dimensional spatial functions using single linear operators, such as conventional linear systems. However, all those inherent processing merits of optical processing cannot happen without the support of good coherence property of light. For this reason, we shall begin our discussion with the fundamental coherence theory of light.

2.1. COHERENCE THEORY OF LIGHT

When radiation from two sources maintains a fixed-phase relation between them, they are said to be *mutually coherent*. Therefore, an extended source is coherent if all points of the source have fixed-phase differences among them. We first must understand the basic theory of coherent light.

In the classic theory of electromagnetic radiation, it is usually assumed that the electric and magnetic fields are always measurable quantities at any position. In this situation there is no need for the coherence theory to interpret the property of light. There are scenarios, however, in which this assumption cannot be made; for these it is essential to apply coherence theory. For example, if we want to determine the diffraction pattern caused by radiation

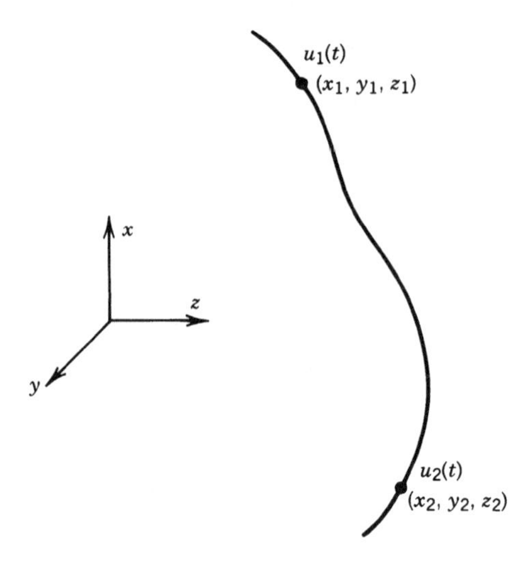

Fig. 2.1. A wavefront propagates in space.

from several sources, we cannot obtain an exact result unless the degrees of coherence among the separate sources are taken into account. In such a situation, it is desirable to obtain a *statistical* ensemble average for the most likely result; for example, from any combination of sources. It is therefore more useful to provide a statistical description than to follow the dynamic behavior of a wave field in detail.

Let us assume an electromagnetic wave field propagating in space, as depicted in Fig. 2.1, where $u_1(t)$ and $u_2(t)$ denote the instantaneous wave disturbances at positions 1 and 2, respectively. The *mutual coherence function* (i.e., *cross-correlation function*) between these two disturbances can be written as

$$\Gamma_{12}(\tau) \triangleq \langle u_1(t+\tau)u_2^*(t)\rangle = \lim_{T\to\infty} \frac{1}{T}\int_0^T u_1(t+\tau)u_2^*(t)\,dt, \tag{2.1}$$

where the superasterisk denotes the complex conjugate, the $\langle\,\rangle$ represents the time ensemble average, and τ is a time delay.

The *complex degree of coherence* between $u_1(t)$ and $u_2(\tau)$ can be defined as

$$\gamma_{12}(\tau) \triangleq \frac{\Gamma_{12}(\tau)}{[\Gamma_{11}(0)\Gamma_{22}(0)]^{1/2}}, \tag{2.2}$$

where $\Gamma_{11}(\tau)$ and $\Gamma_{22}(\tau)$ are the *self-coherence functions* of $u_1(t)$ and $u_2(t)$, respectively.

Needless to say, the *degree of coherent* is bounded by $0 \leqslant |r_{12}(\tau)| \leqslant 1$, in which $|r_{12}| = 1$ represents strictly coherent, and $|r_{12}| = 0$ represents strictly noncoherent. We note that, in high-frequency regime, it is not easy or impossible to directly evaluate the degree of coherence. However, there exists a practical fringe visibility relationship for which the degree of coherence $|r_{12}|$ can be directly measured, by referring to the Young's experiment in Fig. 2.2, in which Σ represents a monochromatic extended source. A diffraction screen is located at a distance t_{10} from the source, with two small pinholes in this screen, Q_1 and Q_2, separated at a distance d. On the observing screen located r_{20} away from the diffracting screen, we observe an interference pattern in which the maximum and minimum intensities $I_{\max}$ and $I_{\min}$ of the fringes are measured. The *Michelson visibility* can then be defined as

$$V \triangleq \frac{I_{\max} - I_{\min}}{I_{\max} + I_{\min}}. \tag{2.3}$$

We shall now show that under the equal intensity condition (i.e., $I_1 = I_2$), the visibility measure is equal to the degree of coherence. The electromagnetic wave disturbances $u_1(t)$ and $u_2(\tau)$ at Q_1 and Q_2 can be determined by the wave equation, such as

$$\nabla^2 u = \frac{\partial^2 u}{c^2 \partial t^2},$$

where c is the velocity of light. The disturbance at point P, on the observing screen, can be written as

$$u_p(t) = c_1 u_1\left(t - \frac{r_1}{c}\right) + c_2 u_2\left(t - \frac{r_2}{c}\right),$$

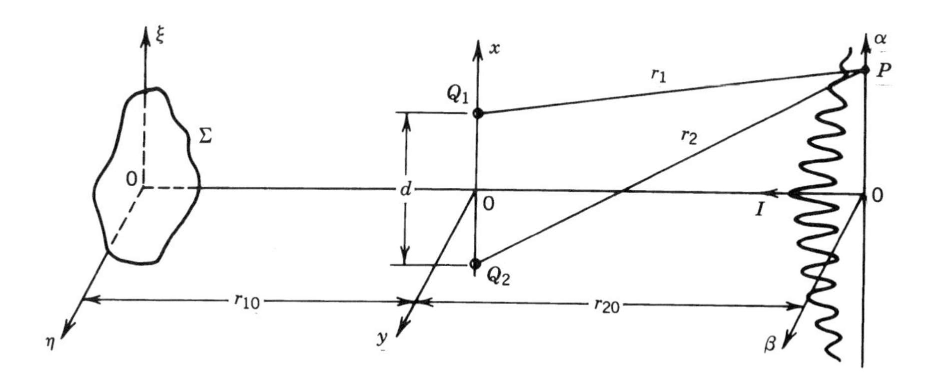

Fig. 2.2. Young's experiment.

where c_1 and c_2 are the appropriate complex constants. The corresponding irradiance at P is written by

$$I_p = \langle u_p(t)u_p^*(t)\rangle = I_1 + I_2 + 2\,\mathrm{Re}\left\langle c_1 u_1\left(t - \frac{r_1}{c}\right) c_2^* u_2^*\left(t - \frac{r_2}{c}\right)\right\rangle,$$

where I_1 and I_2 are proportional to the squares of the magnitudes of $u_1(t)$ and $u_2(t)$. By letting

$$t_1 = \frac{r_1}{c}, \quad t_2 = \frac{r_2}{c} \quad \text{and} \quad \tau = t_2 - t_2,$$

the preceding equation can be written as

$$I_p = I_1 + I_2 + 2c_1c_2^*\,\mathrm{Re}\langle u_1(t + \tau)u_2^*(t)\rangle.$$

In view of the mutual coherence and self-coherence function, we show that

$$I_p = I_1 + I_2 + 2(I_1I_2)^{1/2}\,\mathrm{Re}[\gamma_{12}(\tau)].$$

Thus, we see that for $I = I_1 = I_2$, the preceding equation reduces to

$$I_p = 2I[1 + |\gamma_{12}(\tau)| \cos \phi_{12}(\tau)],$$

in which we see that

$$V = |\gamma_{12}(\tau)|, \tag{2.4}$$

the visibility measure is equal to the degree of coherence.

Let us now proceed with the Young's experiment further, by letting d increase. We see that the visibility drops rapidly to zero, then reappears, and so on, as shown in Fig. 2.3. There is also a variation in fringe frequency as d varies. In other words, as d increases the *spatial frequency* of the fringes also increases. Since the visibility is equal to the degree of coherence, it is, in fact, the degree of *spatial coherence* between points Q_1 and Q_2. If we let the source size Σ deduce to very small, as illustrated in Fig. 2.3, we see that the degree of (spatial) coherence becomes unity (100%), over the diffraction screen. In this point of view, we see that the degree of spatial coherence is, in fact, governed by the source size.

As we investigate further, when the observation point P moves away from the center of the observing screen, visibility decreases as the path difference $\Delta r = r_2 - r_1$ increases, until it eventually becomes zero. The effect also depends

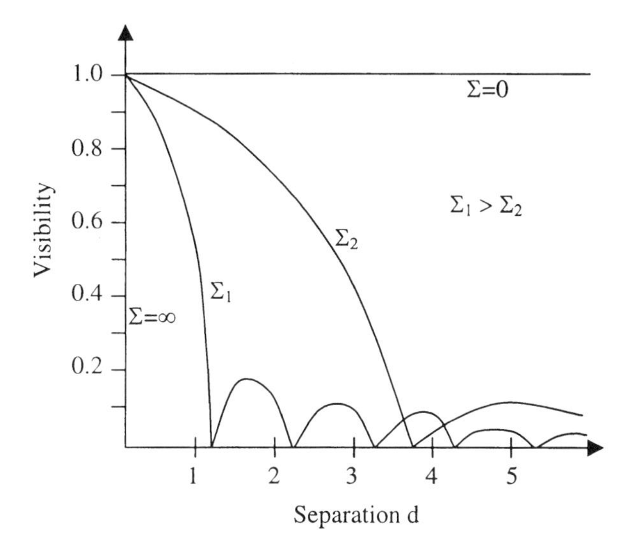

Fig. 2.3. Vibility as a function of separation.

on how nearly monochromatic the source is. The visibility as affected by the path difference can be written as

$$\Delta r \approx \frac{c}{\Delta v}, \tag{2.5}$$

where c is the velocity of light and Δv is the spectral bandwidth of the source. The preceding equation is also used to define the *coherence length* (or *temporal coherence*) of the source, which is the distance at which the light beam is longitudinally coherent.

In view of the preceding discussion, one sees that spatial coherence is primarily governed by the source size and temporal coherence is governed by the spectral bandwidth of the source. In other words, a monochromatic point source is a *strictly coherent source*, while a monochromatic source is a *temporal coherent source* and a point source is a *spatial coherence source*. Nevertheless, it is not necessary to have a completely coherent light to produce an interference pattern. Under certain conditions, an interference pattern may be produced from an incoherent source. This effect is called *partial coherence*. It is worthwhile to point out that the degree of temporal coherence from a source can be obtained by using the *Michelson interferometer*, as shown in Fig. 2.4. In short, by varying one of the minors, an interference fringe pattern can be viewed at the observation plane. The path difference, after the light beam is

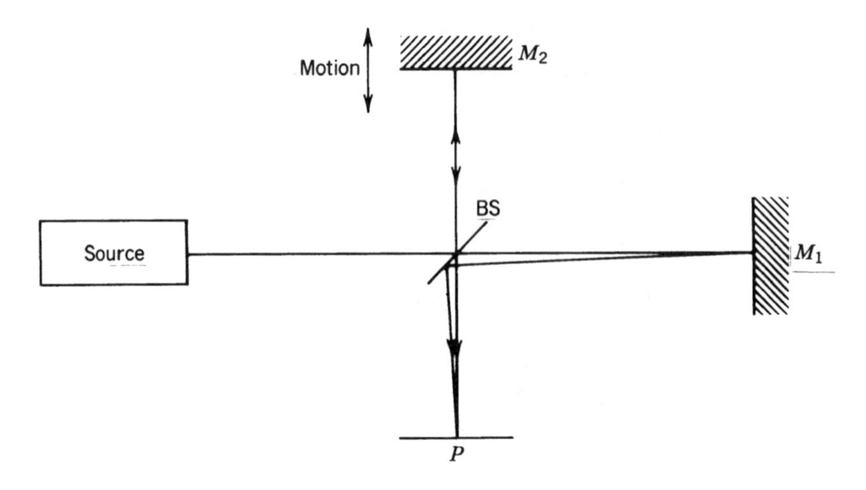

Fig. 2.4. The Michelson interferometer. BS; beam splitter; M, mirrors; P, observation screen.

split, is given by

$$\Delta r = \Delta tc \approx \frac{c}{\Delta v}, \tag{2.6}$$

where the *temporal coherent length* of the source is

$$\Delta t \approx \frac{1}{\Delta v}. \tag{2.7}$$

In fact, the coherent length of the source can also be shown as

$$\Delta r \approx \frac{\lambda^2}{\Delta \lambda}, \tag{2.8}$$

where λ is the center wavelength, and $\Delta\lambda$ is the spectral bandwidth of the light source.

2.2. PROCESSING UNDER COHERENT AND INCOHERENT ILLUMINATION

Let a hypothetical optical processing system be shown in Fig. 2.5. Assume that the light emitted by the source Σ is monochromatic, and let $u(x, y)$ be the complex light distribution at the input signal plane due to an incremental source $d\Sigma$. If the complex amplitude transmittance of the input plane is $f(x, y)$,

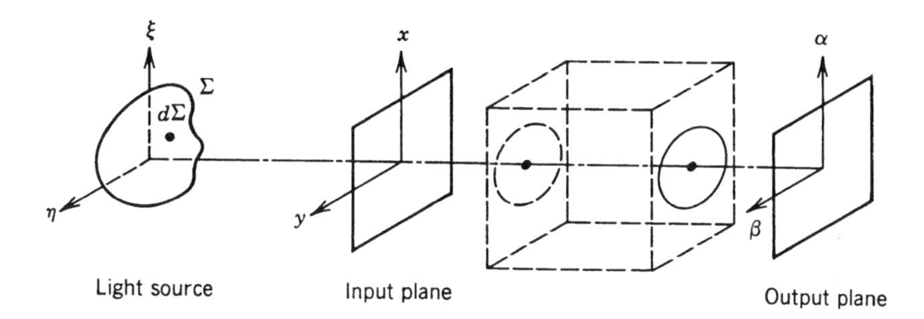

Fig. 2.5. A hypothetical optical processing system.

the complex light field immediately behind the signal plane would be $u(x, y)f(x, y)$. We assume the optical system (i.e., block box) is linearly spatially invariant with a spatial impulse response of $h(x, y)$; the output complex light field, due to $d\Sigma$, can be calculated by

$$g(\alpha, \beta) = [u(x, y)f(x, y)] * h(x, y),$$

which can be written as

$$dI(\alpha, \beta) = g(\alpha, \beta)g^*(\alpha, \beta)\, d\Sigma,$$

where the asterisk represents the convolution operation and the superasterisk denotes the complex conjugation. The overall output intensity distribution is therefore

$$I(\alpha, \beta) = \iint |g(\alpha, \beta)|^2\, d\Sigma,$$

which can be written in thc following convolution integral:

$$I(\alpha, \beta) = \iiiint \Gamma(x, y; x', y')h(\alpha - x, \beta - y)h^*(\alpha - x', \beta - y') \cdot f(x, y)f^*(x', y')\, dx\, dy\, dx'\, dy' \tag{2.9}$$

where

$$\Gamma(x, y: x', y') = \iint_\Sigma u(x, y)u^*(x'y')\, d\Sigma$$

is the *spatial coherence function*, also known as *mutual intensity function*, at the input plane (x, y).

By choosing two arbitrary points Q_1 and Q_2 at the input plane, and if r_1 and r_2 are the respective distances from Q_1 and Q_2 to $d\Sigma$, the complex light disturbances at Q_1 and Q_2 due to $d\Sigma$ can be written as

$$u_1(x, y) = \frac{[I(\xi, \eta)]^{-1/2}}{r_1} \exp(ikr_1)$$

and

$$u_2(x', y') = \frac{[I(\xi, \eta)]^{-1/2}}{r_1} \exp(ikr_2)$$

where $I(\xi, \eta)$ is the intensity distribution of the light source. By substituting preceding equations in Eq. (2.9), we have

$$\Gamma(x, y; x'y') = \iint_{\Sigma} \frac{I(\xi, \eta)}{r_1 r_2} \exp[ik(r_1 - r_2)]\, d\Sigma. \tag{2.10}$$

In the paraxial case, $r_1 - r_2$ may be approximated by

$$r_1 - r_2 \approx \frac{1}{r}[\xi(x - x') + \eta(y - y')],$$

where r is the separation between the source plane and the signal plane. Then Eq. (2.10) can be reduced to

$$\Gamma(x, y; x' y') = \frac{1}{r^2} \iint_{\Sigma} I(\xi, \eta) \exp\left\{i \frac{k}{r}[\xi(x - x') + \eta(y - y')]\right\} d\xi\, d\eta \tag{2.11}$$

which is known as the *Van Cittert-Zernike theorem*. Notice that Eq. (2.11) forms an inverse Fourier transformation of the source intensity distribution.

One of the two extreme cases is by letting the light source become infinitely large; for example, $I(\xi, \eta) \approx K$; then Eq. (2.11) becomes

$$\Gamma(x, y; x'y') = K_1 \delta(x - x', y - y'), \tag{2.12}$$

which describes a completely *incoherent illumination*, where K_1 is an appropriate constant.

On the other hand, if the light source is vanishingly small; i.e., $I(\xi, \eta) \approx K, \delta(\xi, \eta)$, Eq. (2.11) becomes

$$\Gamma(x, y; x'y') = K_2, \tag{2.13}$$

which describes a completely *coherent illumination*, where K_2 is a proportionality constant. In other words, a monochromatic point source describes a strictly coherent processing regime, while an extended source describes a strictly incoherent system. Furthermore, an extended monochromatic source is also known as a *spatially incoherent* source.

By referring to the completely incoherent illumination, we have

$$\Gamma(x, y; x'y') = K_1 \delta(x - x', y - y'),$$

and the intensity distribution at the output plane can be shown as

$$I(\alpha, \beta) = \iint |h(\alpha - x, \beta - y)|^2 |f(x, y)|^2 \, dx \, dy, \tag{2.14}$$

in which we see that the output intensity distribution is the convolution of the input signal intensity with respect to the intensity impulse response. In other words, for the *completely incoherent illumination*, the optical signal processing system is linear in *intensity*; that is,

$$I(\alpha, \beta) = |h(x, y)|^2 * |f(x, y)|^2 \tag{2.15}$$

where the asterisk denotes the convolution operation. On the other hand, for the completely coherent illumination; i.e., $\Gamma(x, y; x'y') = K_2$, the output intensity distribution can be shown as

$$I(\alpha, \beta) = g(\alpha, \beta) g^*(\alpha, \beta) = \iint h(\alpha - x, \beta - y) f(x, y) \, dx \, dy \cdot \iint h^*(a - x', \beta - y') f^*(x', y') \, dx' \, dy' \tag{2.16}$$

when

$$g(\alpha, \beta) = \iint h(a - x, \beta - y) f(x, y) \, dx \, dy, \tag{2.17}$$

for which we can see that the optical signal processing system is linear in *complex amplitude*. In other words, a coherent optical processor is capable of processing the information in complex amplitudes.

2.3. FRESNEL-KIRCHHOFF AND FOURIER TRANSFORMATION

2.3.1. FREE SPACE IMPULSE RESPONSE

To understand the basic concept of optical Fourier transformation, we begin our discussion with the development of the Fresnel-Kirchhoff integral. Let us start from the *Huygens principle*, in which the complex amplitude observed at the point p' of a coordinate system $\sigma(\alpha, \beta)$, due to a monochromatic light field located in another coordinate system $\rho(x, y)$, as shown in Fig. 2.6, can be calculated by assuming that each point of light source is an infinitesimal spherical radiator. Thus, the complex light amplitude $h_l(\alpha, \beta; k)$ contributed by a point p in the (x, y) coordinate system can be considered to be that from an unpolarized monochromatic point source, such as

$$h_l = -\frac{i}{\lambda r}\exp[i(kr - \omega t)], \tag{2.18}$$

where λ, k, and ω are the wavelengths, wave number, and angular frequency of the point source, respectively, and r is the distance between the point source and the point observation.

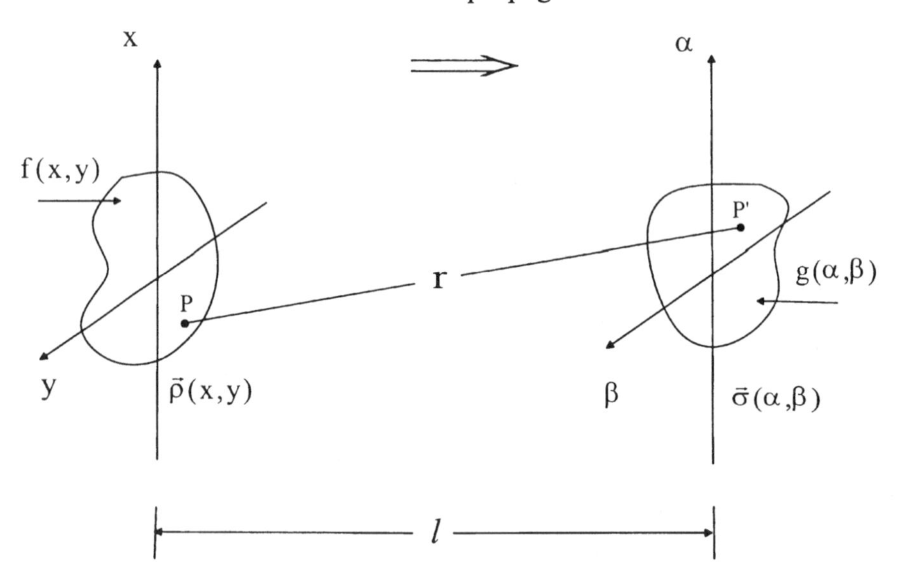

Fig. 2.6. Fresnel-Kirchhoff theory.

If the separation l of the two coordinate systems is assumed to be large compared to the regions of interest in the (x, y) and (α, β) coordinate systems, r can be approximated by

$$r = l + \frac{(\alpha - x)^2}{2l} + \frac{(\beta - y)^2}{2l}, \tag{2.19}$$

which is known as *paraxial approximation.* By substituting into Eq. (2.18) we have

$$h_l(\sigma - \rho; k) = -\frac{i}{\lambda l} \exp\left\{ik\left[l + \frac{(\alpha - x)^2}{2l} + \frac{(\beta - y)^2}{2l}\right]\right\}, \tag{2.20}$$

which is known as the *spatial impulse response*, where the time-dependent exponent has been dropped for convenience. Thus, we see that the complex light field produced at the (α, β) coordinate system by the monochromatic wavefield $f(x, y)$ can be written as

$$g(\alpha, \beta) = \iint_{(x,y)} f(x, y) h_l(\sigma - \rho; k)\, dx\, dy, \tag{2.21}$$

which is the well-known *Kirchhoff's integral.* In view of the preceding equation, we see that the Kirchhoff's integral is, in fact, representing a convolution integral which can be written as

$$g(\alpha, \beta) = f(x, y) * h_1(x, y), \tag{2.22}$$

where the asterisk denotes the convolution operation,

$$h_l(x, y) = C \exp\left[i \frac{k}{2l}(x^2 + y^2)\right], \tag{2.23}$$

and C is a complex constant. Consequently, Eq. (2.22) can be represented by a block box diagram, as shown in Fig. 2.7. In other words, the complex wave field distributed over the (α, β) coordinate plane can be evaluated by the *convolution integral* of Eq. (2.21).

2.3.2. FOURIER TRANSFORMATION BY LENSES

It is well known that a two-dimensional Fourier transformation can be obtained with a positive lens. Fourier transform operations usually require

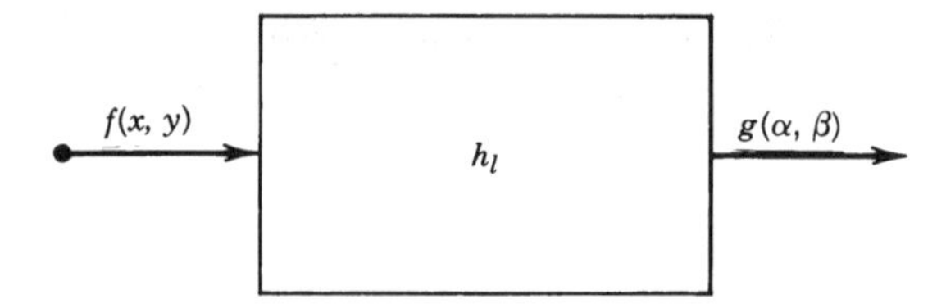

Fig. 2.7. Linear system representation.

complicated electronic spectrum analyzers or digital computers. However, this complicated transform can be performed extremely simply with a coherent optical system.

To perform Fourier transformation in optics, a positive lens must be inserted in a monochromatic wave field (Fig. 2.8). The action of the lens can convert a spherical wave front into a plane wave. Therefore, the lens must induce a phase transformation, such as

$$T(x, y) = C \exp\left[-i\frac{\pi}{\lambda f}(x^2 + y^2)\right], \tag{2.24}$$

where C is an arbitrary complex constant, and f is the focal length of the lens.

Let us now show the Fourier transformation by a lens, as illustrated in Fig. 2.8, in which a monochromatic wave field at input plane (ξ, η) is denoted by

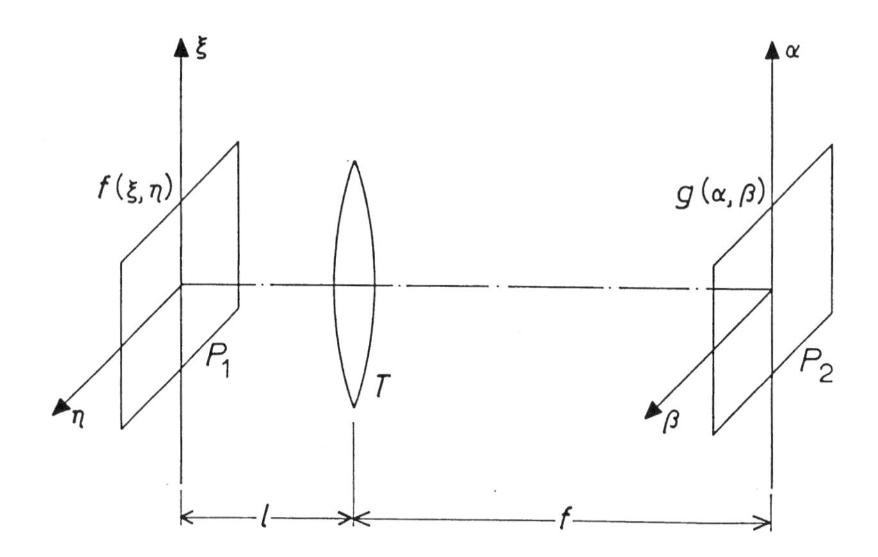

Fig. 2.8. Fourier transformation by a lens.

$f(\xi, \eta)$. By applying the Kirchhoff integral, the complex light distribution at (α, β) can be written as

$$g(\alpha, \beta) = C\{[f(\xi, \eta) * h_l(\xi, \eta)]T(x, y)\} * h_f(x, y),$$

where C is a proportionality complex constant, $h_l(\xi, \eta)$ and $h_f(x, y)$ are the corresponding spatial impulse responses, and $T(x, y)$ is the phase transform of the lens.

By a straightforward but tedious evaluation, we can show that

$$g(\alpha, \beta) = C_1 \exp\left[-i\frac{k}{2f}\frac{1-\mu}{\mu}(\alpha^2, \beta^2)\right] \cdot \iint f(\xi, \eta) \exp\left[-i\frac{k}{f}(\alpha\xi + \beta\eta)\right] d\xi\, d\eta, \tag{2.25}$$

which is essentially the Fourier transform of $f(\xi, \eta)$ associated with a quadratic phase factor, where $\mu = f/l$. If the signal plane is placed at the front focal plane of the lens; that is, $l = f$, the quadratic phase factor vanishes, which leaves an exact Fourier transformation,

$$G(p, q) = C_1 \iint f(\xi, \eta) \exp[-i(p\xi + q\eta)]\, d\xi\, d\eta, \tag{2.26}$$

where $p = k\alpha/f$ and $q = k\beta/f$ are the angular spatial frequency coordinates.

2.4. FOURIER TRANSFORM PROCESSING

There are two types of Fourier transform processors that are frequently used in practice: the Fourier domain (filter) processor (FDP), and the joint transform (spatial domain filter) processor (JTP), as shown in Figs. 2.9 and 2.10, respectively. The major distinction between them is that FDP uses a Fourier domain filter while JTP uses a spatial domain filter.

2.4.1. FOURIER DOMAIN FILTER

A Fourier domain spatial filter can be described by a complex amplitude transmittance function, as given by

$$H(p, q) = |H(p, q)| \exp[i\phi(p, q)]. \tag{2.27}$$

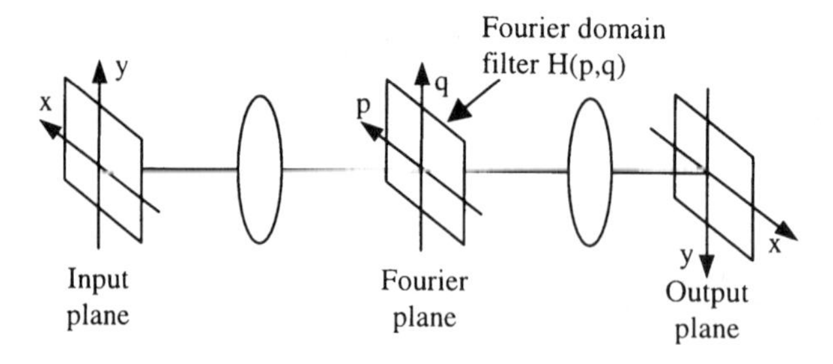

Fig. 2.9. Fourier domain processor (FDP). $H(p, q)$, Fourier domain filter.

In practice, optical spatial filters are generally *passive* types, for which the physically realizable conditions are imposed by

$$|H(p, q)| \leqslant 1 \tag{2.28}$$

and

$$0 \leqslant \phi(p, q) < 2\pi. \tag{2.29}$$

We note that such a transmittance function can be represented by a set of points within or on a unit circle in the complex plane, as shown in Fig. 2.11. The amplitude transmission of the filter changes with the transmission density,

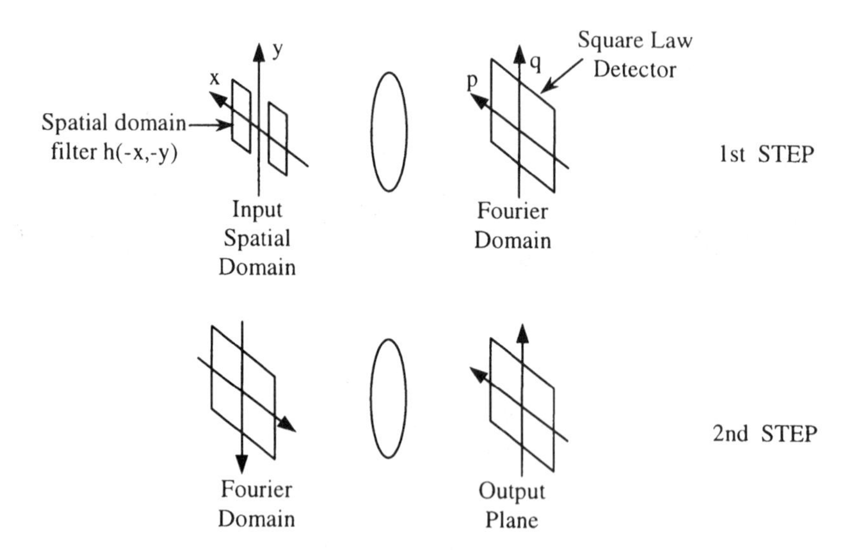

Fig. 2.10. Joint transform processor (JTP). $h(x, y)$, Spatial domain filter.

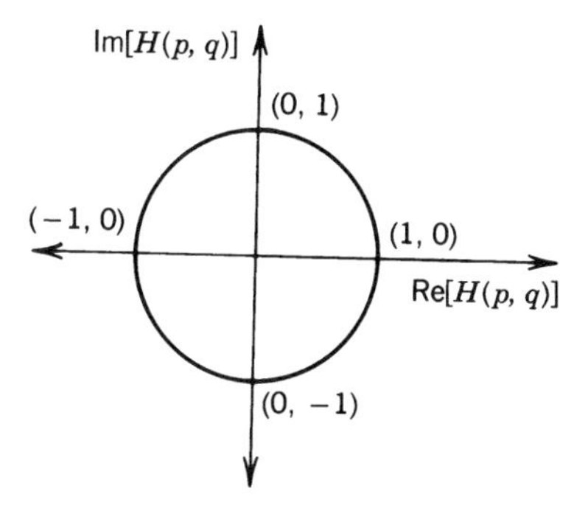

Fig. 2.11. Complex amplitude transmittance.

and the phase delay varies with the thickness of the spatial filter. Thus, a complex spatial filter in principle can be constructed by combining an amplitude filter and a phase-delay filter. However, in most cases, this would be very difficult to realize in practice.

Let us now discuss the technique developed by Vander Lugt for constructing a complex spatial filter using a holographic technique, as shown in Fig. 2.12. The complex light field over the spatial-frequency plane is

$$E(p, q) = F(p, q) + \exp(-i\alpha_0 p),$$

where $\alpha_0 = f \sin\theta$, f, the focal length of the transform lens, and

$$F(p, q) = |F(p, q)| \exp[i\phi(p, q)].$$

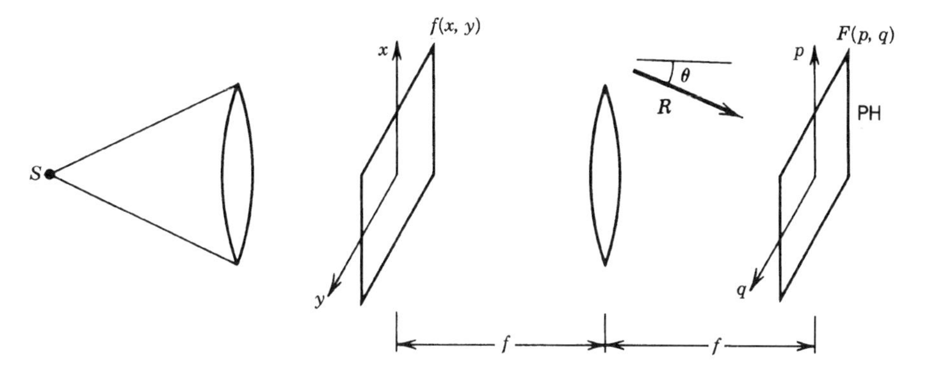

Fig. 2.12. Holographic construction of a Fourier domain matched filter. $f(x, y)$, input object transparency; R, reference plane wave; PH, photographic recording medium.

The corresponding intensity distribution over the recording medium is

$$I(p, q) = 1 + |F(p, q)|^2 + 2|F(p, q)| \cos[\alpha_0 p + \phi(p, q)]. \tag{2.30}$$

We assume that if the amplitude transmittance of the recording is linear, the corresponding amplitude transmittance function of the spatial filter is given by

$$H(p, q) = K\{1 + |F(p, q)|^2 + 2|F(p, q)| \cos[\alpha_0 p + \phi(p, q)]\}, \tag{2.31}$$

which is, in fact, a *real positive function.*

Remember also that, in principle, a complex Fourier domain filter can be synthesized with a *spatial light modulator* (SLM) using computer-generating techniques [1]. (This will be discussed in more detail when we reach the discussion of hybrid optical processing further on.)

2.4.1. SPATIAL DOMAIN FILTER

In a JTP we see that the (input) spatial function and the spatial domain filter are physically separated, as can be seen in Fig. 2.10. In other words a spatial domain filter can also be synthesized using the impulse response of the Fourier domain filter; that is

$$h(x, y) = \mathscr{F}^{-1}[H(p, q)].$$

Note that $h(x, y)$ can be a complex function and is also limited by the similar physical realizable conditions of a Fourier domain filter, such as

$$|h(x, y)| \leqslant 1 \tag{2.32}$$

$$0 \leqslant \phi(x, y) \leqslant 2\pi. \tag{2.33}$$

Needless to say, such a filter can be synthesized by the combination of an amplitude and a phase filter. In fact, such a filter can also be synthesized by computer-generation technique and then displayed on a spatial light modulator. A matched Fourier domain filter is given by

$$H(p, q) = KS^*(p, q), \tag{2.34}$$

where $S(p, q)$ is the signal (or target) spectrum. The corresponding impulse response is given by

$$h(x, y) = s(-x, -y), \tag{2.35}$$

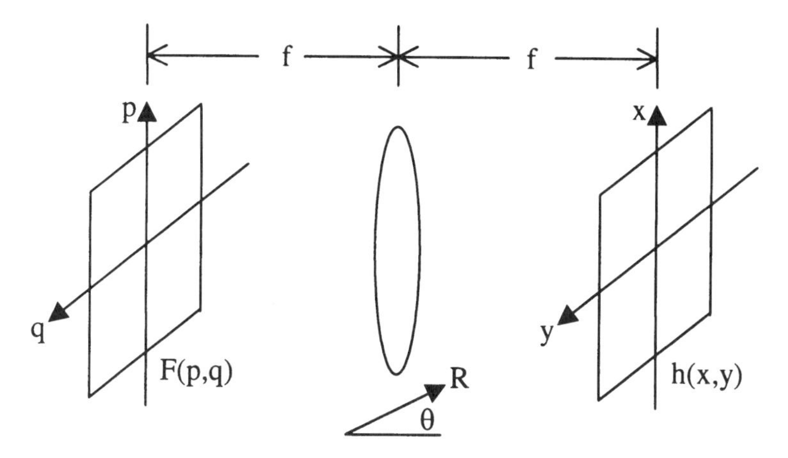

Fig. 2.13. Holographic construction of a spatial domain filter.

which is identical to the input signal function. Thus, we see that the spatial domain filter is, in fact, the signal (target) function. We further note that, in principle, a spatial domain filter can also be synthesized using the holographic technique, as illustrated in Fig. 2.13, provided the complex Fourier domain filter function is given. Similarly, the corresponding spatial domain filter can be shown as

$$h(x, y) = K\{1 + |h(x, y)|^2 + 2|h(x, y)| \cos[2\pi x_0 x + \phi(x, y)]\}, \qquad (2.36)$$

where

$$x_0 = \frac{\sin \theta}{\lambda}.$$

2.4.3. PROCESSING WITH FOURIER DOMAIN FILTERS

If the Fourier domain filter is inserted in the Fourier plane of a coherent optical processor (shown in Fig. 2.9), the output complex light distribution can be shown as

$$\begin{aligned} g(\alpha, \beta) = K[&f(x, y) + f(x, y) * f(x, y) * f^*(-x, -y) \\ &+ f(x, y) * f(x + \alpha_0, y) + f(x, y) * f^*(-x + \alpha_0, -y)] \qquad (2.37) \end{aligned}$$

where the asterisk and the superasterisk represent the convolution operation and the complex conjugation, respectively. In view of the preceding result, we

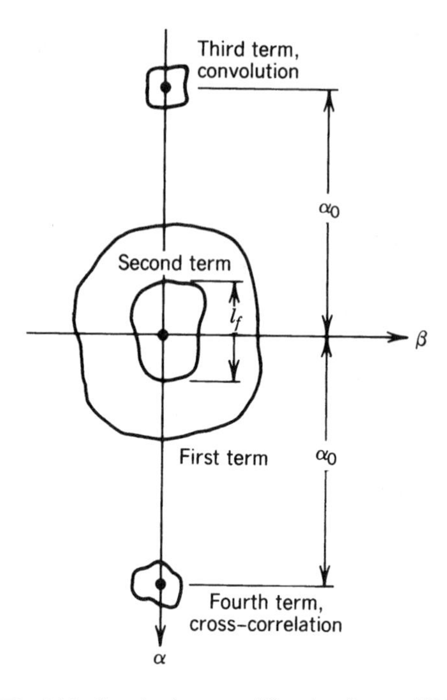

Fig. 2.14. Sketch of output diffraction from a FDP.

see that the first and second terms represent the zero-order diffraction, which appears at the origin of the output plane, and the third and fourth terms are the convolution and cross-correlation terms, which are diffracted in the neighborhood of $\alpha = -\alpha_0$, and $\alpha = \alpha_0$, respectively, as sketched in Fig. 2.14. To further show that the processing operation is *shift invariant*, we let the input object function translate to a new location; that is, $f(x - x_0, y - y_0)$. The correlation term of the preceding equation can then be written as

$$\iint f(x - x_0, y - y_0) f^*(x - \alpha - \alpha_0, y - \beta)\, dx\, dy = R_{11}(\alpha - \alpha_0 - x_0, \beta - y_0), \tag{2.38}$$

where R_{11} represents an autocorrelation function. In view of this result, we see that the target correlation peak intensity has been translated by the same amount, for which the operation is indeed shift invariant. A sketch of the output correlation peak as referenced to the object translation is illustrated in Fig. 2.15.

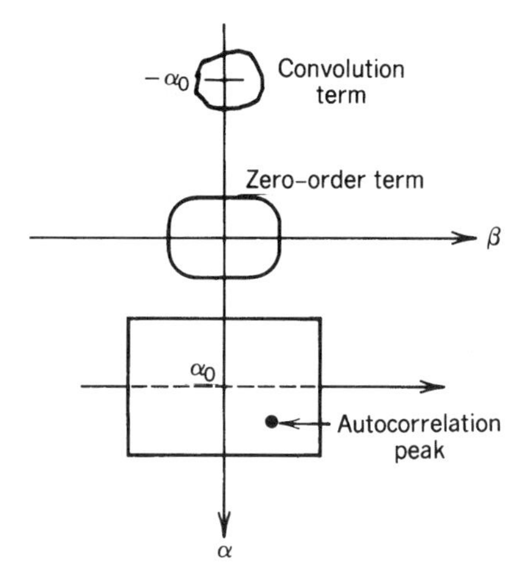

Fig. 2.15. Sketch of output diffraction from JTP.

2.4.4. PROCESSING WITH JOINT TRANSFORMATION

Let us now consider a joint transform processor (JTP), in which we assume that both object function $f(x, y)$ and (inverted) spatial impulse response $h(x, y)$ are inserted in the input spatial domain at $(\alpha_0, 0)$ and $(-\alpha_0, 0)$, respectively. By coherent illumination, the complex light distribution at the Fourier plane is given by

$$U(p, q) * = F(p, q) \exp(-i\alpha_0 p) + H^*(p, q) \exp(i\alpha_0 p), \tag{2.39}$$

where (p, q) represents the angular spatial frequency coordinate system and the asterisk represents the complex conjugate, $F(p, q) = \mathscr{F}[f(x, y)]$, and $H(p, q) = \mathscr{F}[h(x, y)]$.

The output intensity distribution by the square-law detection can be written as

$$\begin{aligned} I(p, q) &= |U(p, q)|^2 \\ &= |F(p, q)|^2 + |H(p, q)|^2 + F(p, q)H(p, q) \exp(-i2\alpha_0 p) \\ &\quad + F^*(p, q)H^*(p, q) \exp(i2\alpha_0 p), \end{aligned} \tag{2.40}$$

which we shall call the joint transform power spectrum (JTPS). By coherent readout of the JTPS, the complex light distribution at the output plane can be shown as

$$\begin{aligned} g(x, y) &= f(x, y) \otimes f(x, y) + h(-x, -y) \otimes h(-x, -y) \\ &\quad + f(x, y) * h(x, 2\alpha_0, y) + f(-x, -y) * h(-x, -2\alpha_0, -y), \end{aligned} \tag{2.41}$$

in which we see that the object function $f(x, y)$ convolves with the spatial impulse response $h(x, y)$ and are diffracted around $(-2\alpha_0, 0)$ and $(2\alpha_0, 0)$, respectively, where $\otimes$ and $*$ represent the correlation and convolution operations, respectively.

It is therefore apparent that the JTP can, in principle, perform all the processing operations that a conventional FDP can. The inherent advantages of using JTP are: (1) avoidance of complex spatial filter synthesis, (2) higher input space-bandwidth product, (3) lower spatial carrier frequency requirement, and (4) higher output diffraction efficiency, particularly by using the hybrid JTP, as will be seen later.

Matched filtering can also be performed with a JTP, which we shall call a joint transform correlator (JTC), as opposed to the Vander-Lugt correlator (VLC) described earlier. We assume that the two identical object functions $f(x - \alpha_0, y)$ and $f(x + \alpha_0, y)$ are inserted in the input plane of a JTP. The complex light distribution arriving at the square-law detector in Fourier plane P_2 will then be

$$E(p, q) = F(p, q)\exp(-i\alpha_0 p) + F(p, q)\exp(i\alpha_0 p), \tag{2.42}$$

where $F(p, q)$ is the Fourier spectrum of the input object. The corresponding irradiance at the input end of the square-law detector is given by

$$I(p, q) = 2|F(p, q)|^2[1 + \cos 2\alpha_0 p].$$

By coherent readout, the complex light distribution at the output plane can be shown as

$$\begin{aligned} g(\alpha, \beta) = 2f(x, y) \otimes f^*(x, y) + f(x, y) \otimes f^*(x - 2\alpha_0, y) \\ + f^*(x, y) \otimes f(x + 2\alpha_0, y), \end{aligned} \tag{2.43}$$

in which we see that two major autocorrelation terms are diffracted at $\alpha = 2\alpha_0$ and $\alpha = -2\alpha_0$, respectively. Notice that the aforementioned square-law converter or devices such as photographic plates, liquid crystal light valves, or charge-coupled device cameras, can be used.

To illustrate the *shift invariant* property of a JTC, we assume a target "B" is located at (α_0, y_0) and a reference image is located at $(-\alpha_0, 0)$, as shown in Fig. 2.16a. The set of input functions can be written as $f(x + \alpha_0, y - y_0)$ and $f(x + \alpha_0, y)$, respectively. Thus the complex light distribution at the input end of the square-law detector can be written as

$$E(p, q) = F(p, q)\{\exp[-i(\alpha_0 p + y_0 q)] + \exp[(i\alpha_0 p)\}.$$

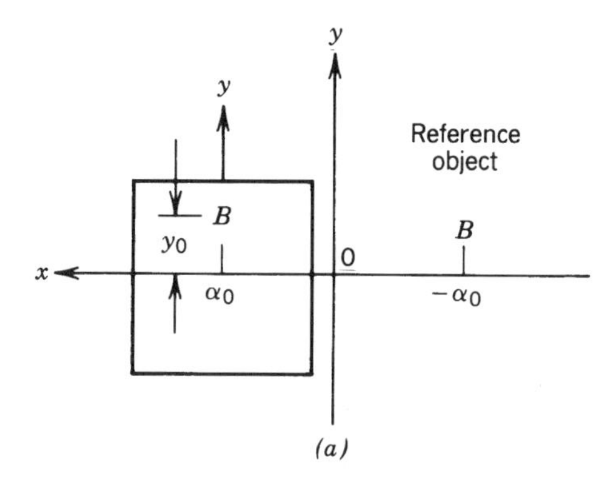

(a)

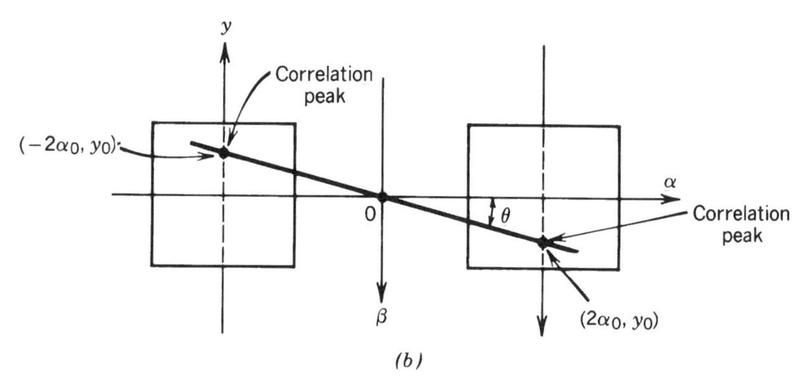

(b)

Fig. 2.16. Spatial invariant property of a JTC.

The corresponding JTPS can be shown as

$$I(p, q) = |F(p, q)|^2[1 + \cos(2\alpha_0 p + y_0 q)]. \tag{2.44}$$

By taking the inverse Fourier transform of $I(p, q)$, we have the following complex light distribution at the output plane:

$$\begin{aligned} g(\alpha, \beta) = 2f(x, y) \otimes f^*(x, y) + f(x, y) \otimes f^*(x - 2\alpha_0, y - y_0) \\ + f^*(x, y) \otimes f(x + 2\alpha_0, y + y_0), \end{aligned} \tag{2.45}$$

which is sketched in Fig. 2.16b. Thus, we see that the JTC preserves the *shift invariant property*.

2.4.5. HYBRID OPTICAL PROCESSING

It is trivial that a pure optical processor has certain severe constraints, which make some processings difficult or even impossible to carry out. Although an optical processor can be designed for specific purposes, it is difficult to make it *programmable* as a digital computer. Another major problem for optical processors is that they cannot make *decisions* as can their electronic counterparts. In other words, some of the deficiencies of the optical processor are the strong points of its electronic counterparts. For instance, accuracy, controllability, and programmability are the trademarks of digital computers. Thus, the concept of combining the optical system with its electronic counterparts is rather natural as a means of applying the rapid processing and parallelism of optics to a wider range of applications. We shall now use this concept to illustrate a couple of microcomputer-based hybrid optical processors that can achieve these objectives. The first approach to the hybrid optical processor is using the Fourier domain (or 4-f) system configuration, as shown in Fig. 2.17. We see that a programmable Fourier domain filter can be generated onto SLM2. The cross-convolution between the input scene and the impulse response of the Fourier filter can be captured by a charge-coupled device (CCD) camera. The detected signal can be either displayed on the monitor or a decision made by the computer. Thus, we see that a programmable real-time optical processor in principle can be realized.

The second approach to hybrid optical processing is using a joint transform configuration, as shown in Fig. 2.18, in which both the input object and the spatial impulse response of the filter function can be simultaneously displayed at the input SLM1. For instance, programmable spatial reference function can be generated side by side with the input object, such that the joint transform power spectrum (JTPS) can be detected by CCD1. By displaying the JTPS on the SLM2, via the microcomputer, cross-convolution between the input object

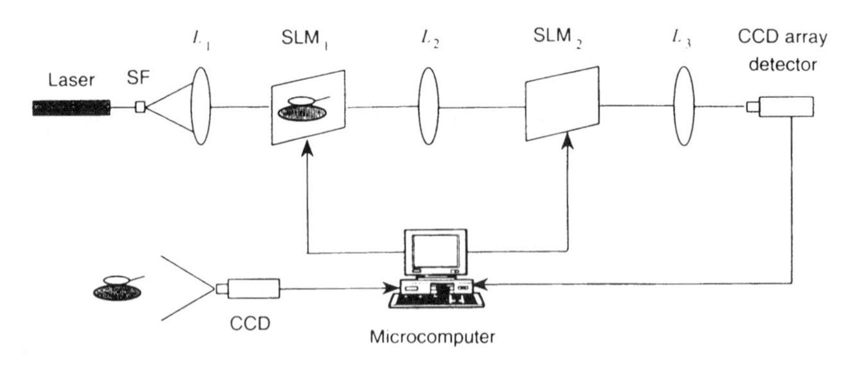

Fig. 2.17. A microcomputer-based FDP.

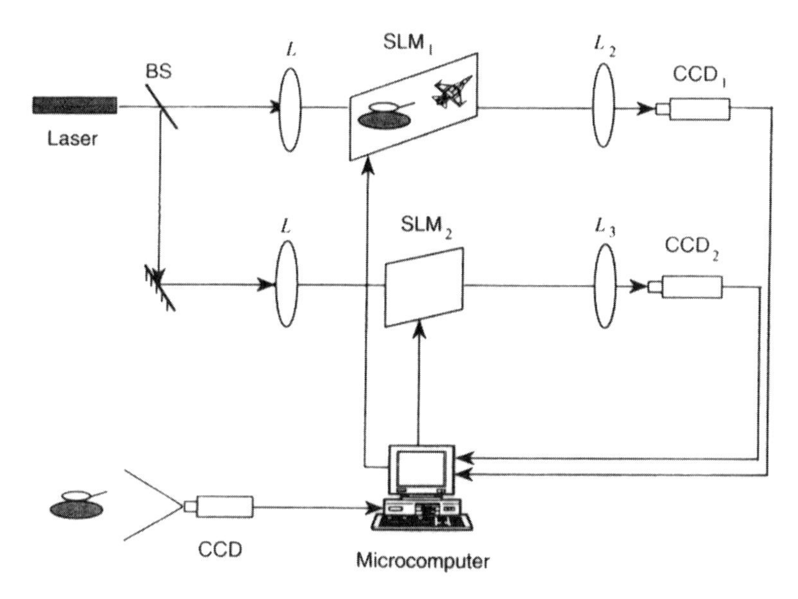

Fig. 2.18. A microcomputer-based JTP.

and the reference function can be obtained at the back focal plane of the Fourier transform lens L_3. Once again we see that a real-time hybrid optical processor can be constructed using the joint transform architecture.

2.5. IMAGE PROCESSING WITH OPTICS

The discovery of laser has prompted us to build more efficient optical systems for communication and signal processing. Most of the optical processing to date has been confined to either complete coherence or complete incoherence. However, a continuous transition between these two extremes is possible, as will be discussed later. In this section we examine a few applications under the coherent regime.

2.5.1. CORRELATION DETECTION

Let us consider the Fourier domain processor (FDP) of Fig. 2.17, in which an input object $f(x, y)$ is displayed at the input plane, the output complex light can be shown as

$$g(\alpha, \beta) = Kf(x, y) * h(x, y), \tag{2.46}$$

where $*$ denotes the convolution operation, K is a proportionality constant, and $h(x, y)$ is the spatial impulse response of the Fourier domain filter, which can be generated on SLM2. We note that a Fourier domain filter can be described by a complex amplitude transmittance such as

$$H(p, q) = |H(p, q)| \exp[i\phi(p, q)].$$

Let us further assume that a *holographic type* matched filter (as described in Sec. 2.4.1) is generated at SLM2, as given by

$$H(p, q) = K\{1 + |F(p, q)|^2 + 2|F(p, q)| \cos[\alpha_0 p + \phi(p, q)]\}, \quad (2.47)$$

where α_0/λ is the spatial carrier frequency. It is straightforward to show that the output complex light distribution can be written as

$$g(\alpha, \beta) = K[f(x, y) + f(x, y) * f(x, y) * f^*(-x, -y) + f(x, y) * f(x + \alpha_0, y) + f(x, y) * f^*(-x + \alpha_0, -y)]. \quad (2.48)$$

We see that third and fourth terms are the convolution and cross-correlation terms, which are diffracted in the neighborhood of $\alpha = \alpha_0$ and $\alpha = \alpha_0$, respectively.

If we assume the input object is embedded in an additive white Gaussian noise n; that is,

$$f'(x, y) = f(x, y) + n(x, y), \quad (2.49)$$

then the correlation term would be

$$R(\alpha, \beta) = K[f(x, y) + n(x, y)] * f^*(-x + \alpha_0, -y).$$

Since the cross-correlation between $n(x, y)$ and $f^*(-x + \alpha_0, -y)$ can be shown to be approximately equal to zero, the preceding equation reduces to

$$R(\alpha, \beta) = f(x, y) * f^*(-x + \alpha_0, -y), \quad (2.50)$$

which, in fact, is the autocorrelation detection of $f(x, y)$.

Notice that to ensure that the zero-order and the first-order diffraction terms will not overlap, α_0 is required that

$$\alpha_0 > l_f + \tfrac{3}{2} l_s, \quad (2.51)$$

where l_f and l_s are the spatial lengths in the x direction of the input scene (or frame) and the detecting signal $f(x, y)$, respectively.

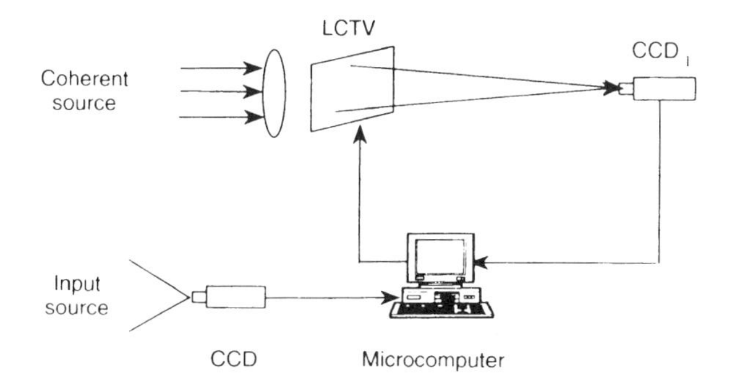

Fig. 2.19. A single SLM JTC.

Complex detection can be easily carried out by the joint transform correlator (JTC) shown in Figs. 2.18 and 2.19, in which a spatial domain filter is used, as given by

$$h(x, y) = Kf(x, y), \tag{2.52}$$

which is proportional to the detecting target (or object).

By referring to the preceding example, input target embedded in additive white noise, the JTC input displayed can be written as

$$f(x - \alpha_0, y) + f'(x + \alpha_0, y),$$

when $f' = f + n$ the input scene.

The corresponding joint transform power spectrum (JTPS), as detected by a charge-coupled detector, can be shown as

$$I(p, q) = |F|^2 + |F + N|^2 + F(F + N)^* \exp(-i2\alpha_0 p) + F^*(F + N) \exp(i2\alpha_0 p),$$

where F and N denote the target and the noise spectral distributions. If the JTPS is sent back for the correlation operations, the output light field can be shown as

$$g(\alpha, \beta) = (f + n)(f + n) + f \otimes f + (f + n) \otimes f(x - 2\alpha_0, y)$$
$$+ (f + n) \otimes f(x + 2\alpha_0, y).$$

Since n is assumed an additive white Gaussian noise with zero mean, the preceding equation reduces to

$$g(\alpha, \beta) = (\text{zero order term}) + f \otimes f(x - 2*, y) + f \otimes f(x + 2\alpha_0, y), \tag{2.53}$$

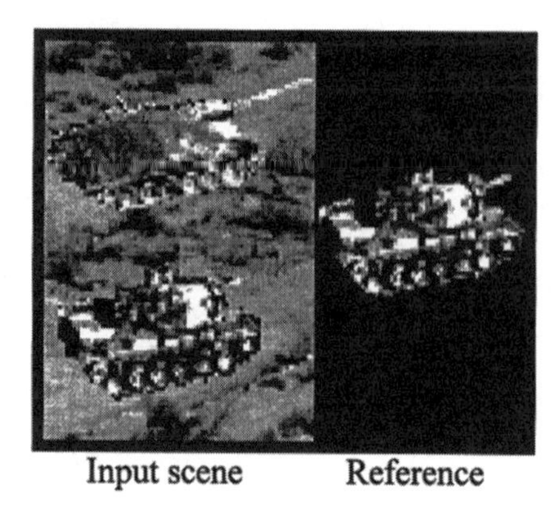

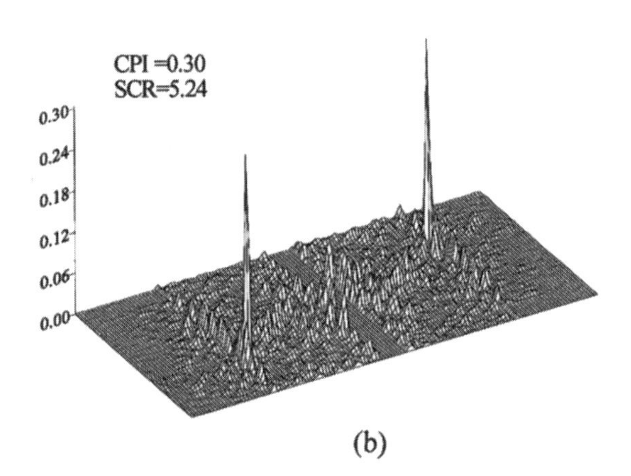

Fig. 2.20. (a) Input scene with a reference object, (b) Output correlation distribution.

in which we see that two cross-correlation terms are diffracted at $\alpha = 2\alpha_0$ and $\alpha = -2\alpha_0$, respectively.

Notice that to avoid overlapping with the zero diffraction, the separation between the input displayed objects functions must be adequately apart; that is,

$$2\alpha_0 > 2l_s, \tag{2.54}$$

where l_s is the width of the input scene.

An example obtained from the JTC detection is provided in Fig. 2.20, in which we see that distinctive correlation peaks are extracted from the random background noise.

2.5.2. IMAGE RESTORATION

One of the most interesting applications in optical processing is image restoration. As contrasted with the optimum correlation detection that uses the *maximum* signal-to-noise ratio criterion, for optimum image restoration one uses the *minimum* mean-square error criterion. Since this restoration spatial filter is restricted by the same physical constraints as Eqs. (2.28) and (2.29), we will discuss the synthesis of a practical filter that avoids the optimization. In other words, the example provided in the following is by no means *optimum*.

We know that a distorted image can be, in principle, restored by an *inverse filter*, as given by

$$H(p) = \frac{1}{D(p)}, \tag{2.55}$$

where $D(p)$ represents the distorting spectral distribution. In other words, a distorted image, as described in the Fourier domain, can be written as

$$G(p) = S(p)D(p),$$

where $G(p)$ and $S(p)$ are the distorted and undistorted image spectral distributions, respectively. Thus, we see that an image can be restored by inverse filtering, such as

$$S(p) = G(p)H(p). \tag{2.56}$$

Let us now assume that the transmission function of a linear smeared-point (blurred) image can be written as

$$f(\xi) = \begin{cases} 1, & -1/2\xi \leqslant \xi \leqslant 1/2\Delta\xi, \\ 0, & \text{otherwise}, \end{cases}$$

where $\Delta\xi$ is the smeared length. To restore the image, we seek for an inverse filter as given by

$$H(p) = \frac{1}{F(p)} = \frac{p\Delta\xi/2}{\sin(p\Delta\xi/2)}. \tag{2.57}$$

In view of the preceding equation, we quickly note that the filter is *not physically realizable*, since it has an infinite number of poles. This precisely corresponds to some information loss due to smearing. In other words, to retrieve the information loss due to blurring there is a price to pay in terms of *entropy*, which is usually very costly. Aside from the physical realizability, if we force ourselves to synthesize an approximate inverse filter which satisfies the physical realizable condition, then a practical, although not optimum, inverse filter can be synthesized, as given by

$$H(p) \approx A(p) \exp[i\phi(p)], \tag{2.58}$$

where $A(p)$ and $\phi(p)$ are the corresponding *amplitude* and *phase* filters.

In view of the Fourier spectral distribution of the distorted point image shown in Fig. 2.21, the restored Fourier spectra that we would like to achieve is the rectangular spectral distribution bounded by T_m and Δp. If we allow the distribution of the amplitude and the phase filters to vary within the physical realizable constraints as shown in Figs. 2.22a and 2.22b, the restored spectral distribution is the shaded areas. It is evident that the blurred image can be restored for some degrees of restoration error. By defining the *degree of restoration* as given by

$$\mathscr{D}(T_m) \text{ (percent)} = \frac{1}{T_m \Delta p} \int_{\Delta p} \frac{F(p)H(p)}{\Delta \xi} \, dp \times 100, \tag{2.59}$$

where Δp is the spatial bandwidth of interest, a plot can be drawn, as shown

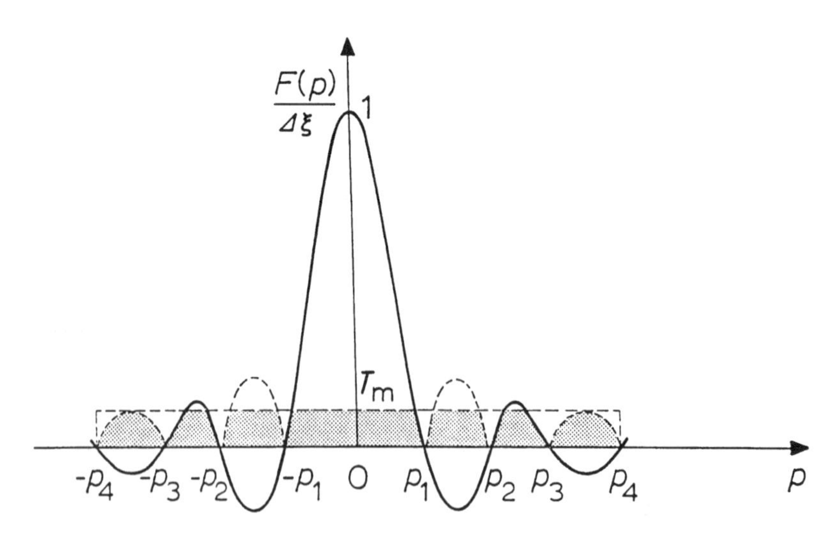

Fig. 2.21. The solid curve represents the Fourier spectrum of a linear smeared point image. The shaded area represents the corresponding restored Fourier spectrum.

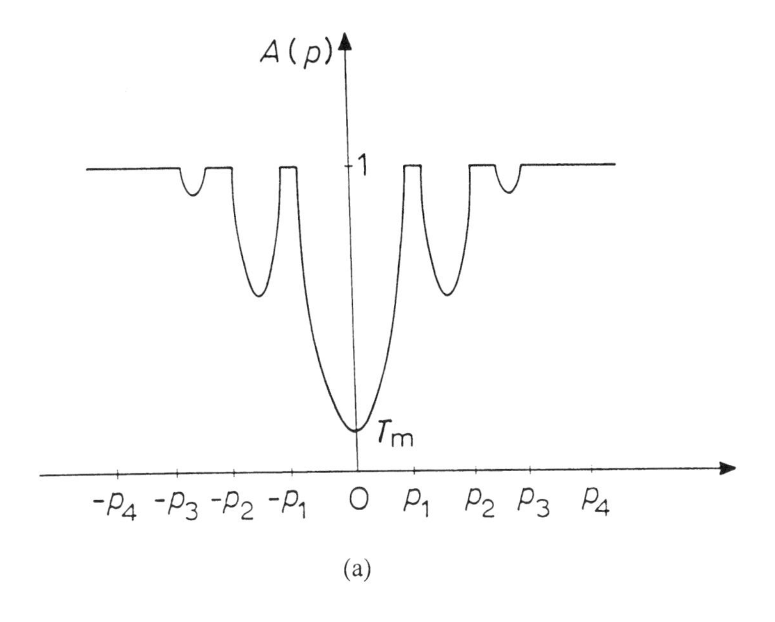

(a)

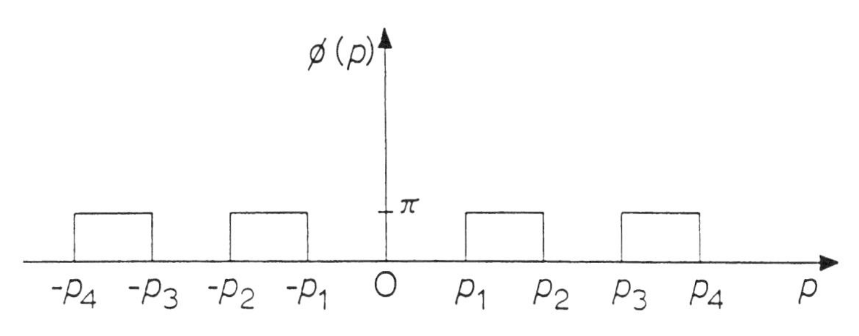

Fig. 2.22. (a) Amplitude filter, (b) Phase filter.

in Fig. 2.23. We see that the degree of restoration increases rapidly as T_m decreases. However, as T_m approaches zero, the transmittance of the inverse filter also approaches zero, which leaves no transmittance of the inverse filter. Thus, a perfect degree of restoration, even within the bandwidth, Δp, cannot actually be obtained in practice. Aside from this consequence, the weak diffracted light field from the inverse filter would also cause poor noise performance. The effects on image restoration due to the amplitude, the phase, and the combination of amplitude and phase filters are shown in Fig. 2.24. In view of these results, we see that using the phase filter alone would give rise to a reasonably good restoration result as compared with the complex filtering. This is the consequence of the image formation (either in the spatial or in the

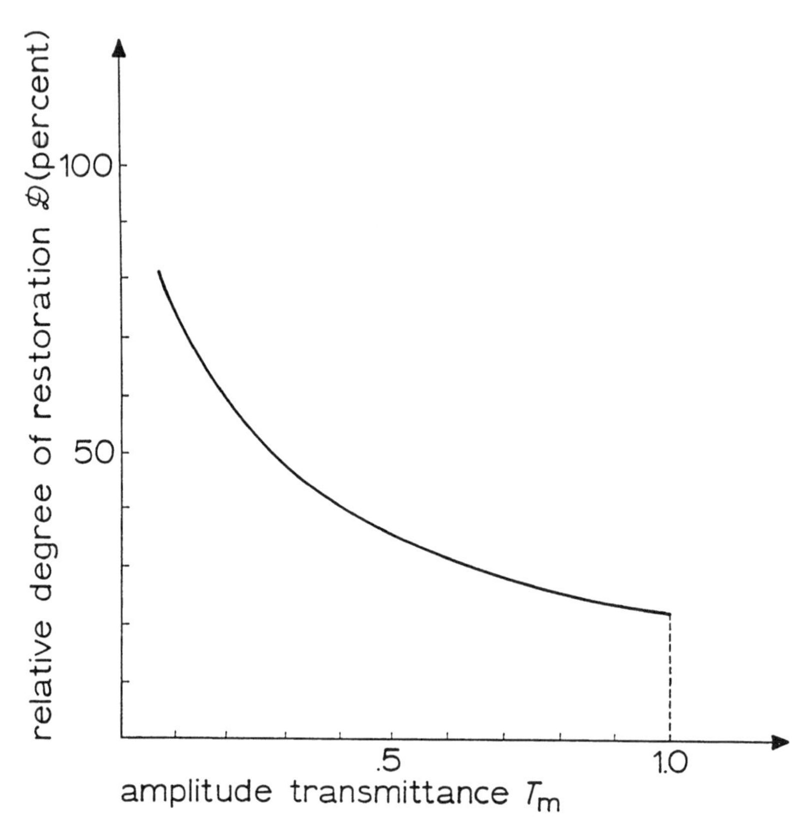

Fig. 2.23. Degree of restoration as a function of T_m.

Fourier domain); the phase distribution turns out to be a major quantity in the effect of image processing as compared with the effect due to amplitude filtering alone. In other words, in image processing, is as well as image formation, the amplitude variation, in some cases, can actually be ignored. A couple of such examples are optimum linearization in holography and phase-preserving matched filters.

Let us now provide an image restoration result we obtained from such an inverse filter, shown in Fig. 2.25, in which we see that a linear blurred image can indeed be restored. In addition, we have also seen that the restored image is embraced with *speckle noise*, also known as *coherent noise*. This is one of the major concerns of using coherent light for processing. Nevertheless, coherent noise can be actually suppressed by using an incoherent light source, as will be discussed in Sec. 2.8.

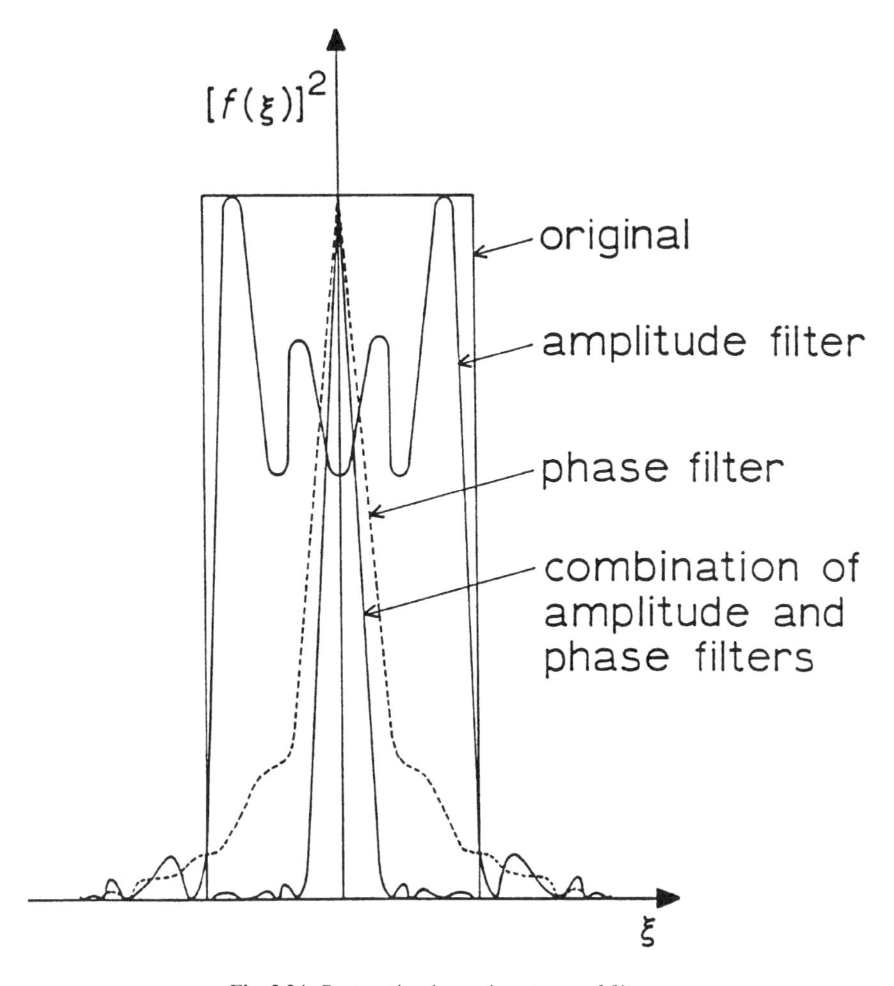

Fig. 2.24. Restoration by various types of filters.

Fig. 2.25. (a) Blurred image, (b) Deblurred image.

2.5.3. IMAGE SUBTRACTION

Another interesting application of optical processing is image subtraction. Image subtraction may be of value in many applications, such as urban development, highway planning, earth resources studies, remote sensing, meteorology, automatic surveillance, and inspection. Image subtraction can also apply to communication as a means of bandwidth compression; for example, when it is necessary to transmit only the differences among images in successive cycles, rather than the entire image in each cycle.

Let us assume two images, $f_1(x - a, y)$, $f_2(x + a, y)$, are generated at the input spatial domain SLM1 of Fig. 2.17. The corresponding joint transform spectra can be shown as

$$F(p, q) = F_1(p, q)e^{-iap} + F_2(p, q)e^{iap},$$

where $F_1(p, q)$ and $F_2(p, q)$ are the Fourier spectra of $f_1(x, y)$ and $f_2(x, y)$, respectively. If a bipolar Fourier domain filter,

$$H(p) = \sin(ap), \tag{2.60}$$

is generated in SLM2, the output complex light field can be shown as

$$g(\alpha, \beta) = C_1[f_1(x, y) - f_2(x, y)] + C_2[f_1(x - 2a, y) + f_2(x + 2a, y)], \tag{2.61}$$

in which we see that a subtracted image can be observed around the origin of the output plane. We note that the preceding image subtraction processing is, in fact, a combination of the joint transformation and the Fourier domain filtering. Whether by combining the joint transformation and Fourier domain filtering it is possible to develop a more efficient optical processor remains to be seen. Now consider an image subtraction result as obtained by the preceding processing strategy, as shown in Fig. 2.26. Once again, we see that the subtracted image is severely corrupted by coherent artifact noise, which is primarily due to the sensitivity of coherent illumination.

2.5.4. BROADBAND SIGNAL PROCESSING

An important application of optical processing is the analysis of a broadband signal. Because of the high space bandwidth product of optics, a one-dimensional large time–bandwidth signal can be analyzed by a two-dimensional optical processor.

In order to do so, a broadband signal is first raster-scanned onto the input SLM, as shown in Fig. 2.27. This raster-scanned process is, in fact, an excellent

PP E

I

APPLIED
OPTICS

(a)

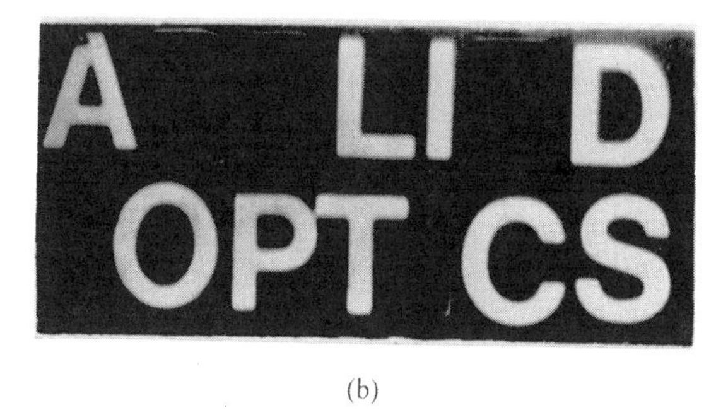

(b)

Fig. 2.26. (a) Images to be subtracted, (b) Subtracted image.

example of showing a time-to-spatial-signal conversion. In other words, a one-dimensional time signal can be converted into a two-dimensional spatial format for optical processing. If we assume the return sweep is adequately higher as compared with the maximum frequency content of the time signal, a two-dimensional raster-scanned format, which represents a long string of time signals, can be written as

$$f(x, y) = \sum_{n=1}^{N} f(x) f(y), \tag{2.62}$$

where $N = h/b$ is the number of scanned lines within the two-dimensional input

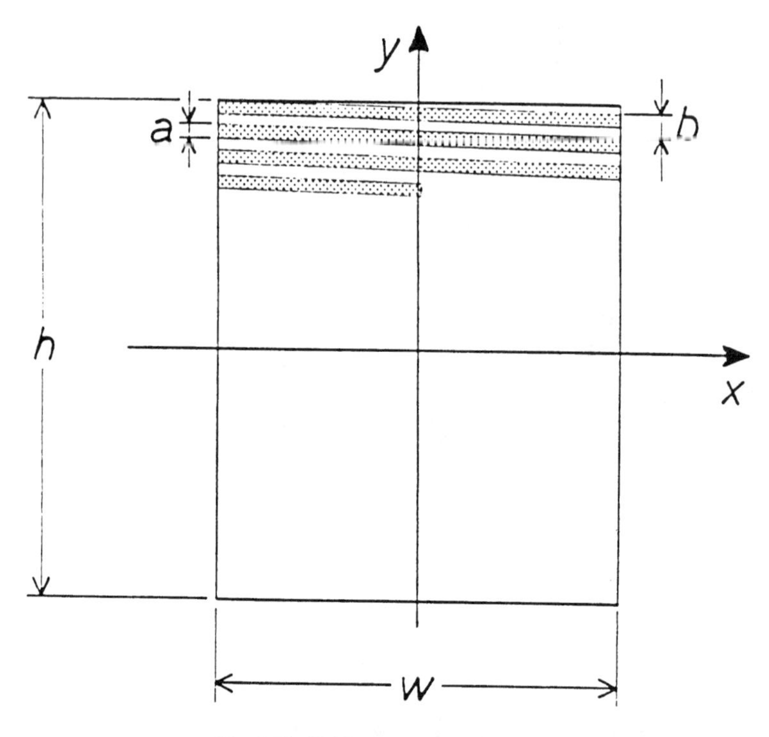

Fig. 2.27. SLM raster-scanned format.

format, $f(x)$ is the transmittance function proportioned to the time signal, as written by

$$f(x) = f\left[x + (2n - 1)\frac{w}{2}\right], \qquad -\frac{w}{2} \leqslant x \leqslant \frac{w}{2}, n = 1, 2, \ldots, N,$$

and

$$f(y) = \text{rect}\left[\frac{y - nb}{a}\right],$$

where rect [ ] represents a rectangular function.

By taking the Fourier transform of the input raster-scanned format the complex light distribution at the Fourier domain is given by

$$F(p, q) = C \sum_{n=1}^{N} \text{sinc}\left(\frac{qa}{2}\right)\left\{\text{sinc}\left(\frac{pw}{2}\right) * F(p) \exp\left[i\frac{pw}{2}(2n - 1)\right]\right\} \cdot \exp\left\{-i\frac{q}{2}[h - 2(n - 1)b - a]\right\}, \tag{2.63}$$

where C is a complex constant, and

$$F(p) = \int f(x')e^{-ipx'}\, dx',$$

$$x' = x + (2n - 1)w/2 \qquad \text{and} \qquad \operatorname{sinc} X \triangleq \frac{\sin X}{X}.$$

For simplicity of illustration, we assume $f(x') = \exp(ip_0 x)$ a complex sinusoidal signal, $w \approx h, b \approx a$, and $N \gg 1$; then the corresponding intensity distribution can be shown as

$$I(p, q) = K^2 \operatorname{sinc}\left[\frac{w}{2}(p - p_0)\right] \operatorname{sinc}^2\left(\frac{qa}{2}\right)\left(\frac{\sin N\theta}{\sin \theta}\right)^2, \tag{2.64}$$

in which the first sinc factor represents a *narrow* spectral line located at $p = p_0$. The second sinc factor represents a relatively *broad* spectral band in the q direction, which is due to the narrow channel width a. This last factor deserves special mention; for large values of N, it approaches a sequence of narrow pulses,

$$q = \frac{1}{b}(2\pi n - wp_0), \qquad n = 1, 2, \ldots \tag{2.65}$$

which yields a fine spectral resolution in the q direction.

In view of the spectral intensity distribution Eq. (2.64), we see that the first sinc factor is confined within a very narrow region along the p direction, due to large w. The half-width spread can be written as

$$\Delta p = \frac{2\pi}{w}, \tag{2.66}$$

which is equal to the resolution limit of the transform lens.

However, along the q direction, the intensity is first confined within a relatively broad spectral band, due to narrow channel width a, and then modulated by a sequence of narrow periodic pulses, as can be seen in Fig. 2.28. Thus, along the q axis modulation we can envision a series of spectral points which are located at $p = p_0$ and $q = 1/b(2\pi n - wp_0)$. As the input signal frequency changes, the position of the spectral points also changes by the amounts

$$dq = \frac{w}{b}\, dp_0. \tag{2.67}$$

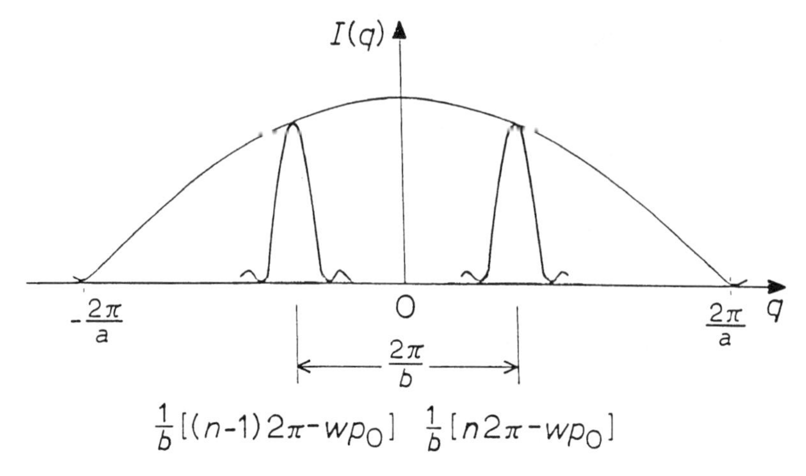

Fig. 2.28. A broad spectral band modulated by a sequence of narrow pulses.

In other words, the displacement in the q direction is proportional to the displacement in the p direction. Since the pulse width along the q direction decreases as the number of scan lines N increases, the output spectrum yields a frequency resolution equivalent to a one-dimensional processor for a continuous signal; that is, Nw long.

To avoid ambiguity in reading the output plane, all but one of the periodic pulses along the q axis should be ignored. This can be accomplished by masking out all the output plane except the region

$$-\frac{\pi}{b} \leqslant q \leqslant \frac{\pi}{b}, \qquad 0 \leqslant p \leqslant \infty, \tag{2.68}$$

as shown in Fig. 2.29. Since the periodic pulses are $2\pi/b$ apart, as the input signal frequency advances, one pulse leaves the open region at $q = -\pi b$, while another pulse is starting to enter the region at $q = \pi b$. Thus, we see that a bright spectral point will diagonally scan from top to bottom in the open region between $q = \pm\pi/b$, as the input frequency advances. In other words, a frequency locus as related to the input signal can be traced out as shown in the figure. We note that to remove the nonuniformity of the second sinc factor, a graded transparency can be placed at the output plane. Since the system can be performed on a continuous running basis, we see that as the SLM format moves up, a real-time spectrum analyzer can be realized. Notice that a space–bandwidth product greater than 10^7 is achievable within the current state of the art, if one uses high-resolution film instead of SLM.

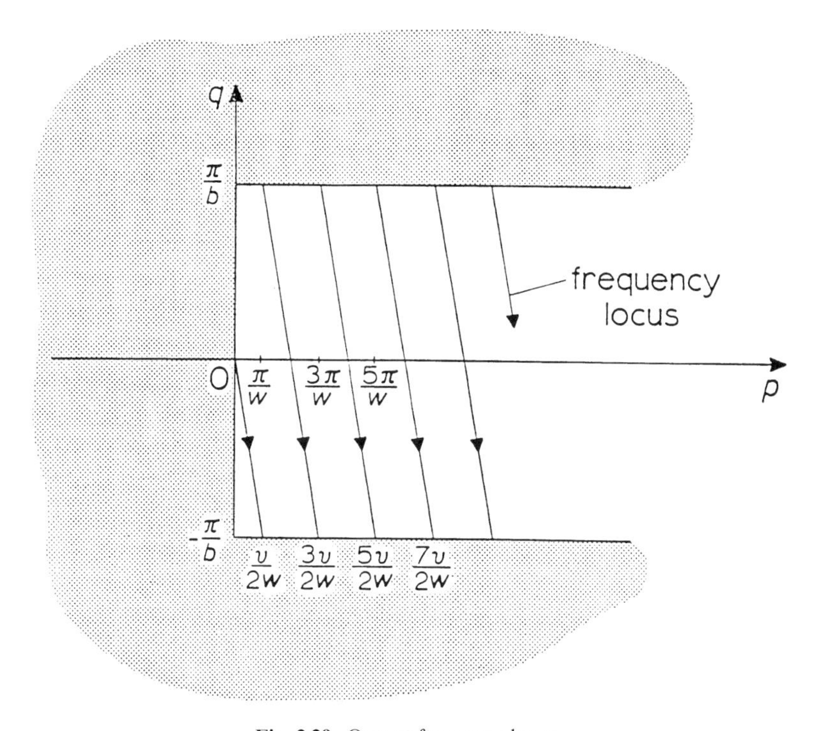

Fig. 2.29. Output frequency locus.

To conclude this section, we note that one interesting application of optical processing of broadband signals is its application to synthetic aperture radar. A broadband microwave signal is first converted in two-dimensional raster-scanned format, similar to the preceding example. If the raster-scanned format is presented at the input plane of a specially designed optical processor, an optical radar image can be observed at the output plane.

2.6. ALGORITHMS FOR PROCESSING

In the preceding we have addressed techniques that rely on linear spatial invariant operation, for which the output response of the processor can be derived from the input object function as it convolves with the spatial impulse response of the filter. There are several inherent constraints of the preceding processors, such as *scale* and *rotational variants*, *nonlinearity operation*, and others. We now discuss a few algorithms that can be used to alleviate some of those drawbacks.

2.6.1. MELLIN-TRANSFORM PROCESSING

Although complex signal detection is *shift invariant*, it is sensitive to rotation and scale variants. In other words, the scale of the input object for which the complex spatial filter is synthesized must be precisely matched. The miscaling problem can be alleviated by using a technique called the *Mellin-transform* technique.

The Mellin transformation has been successfully applied in time-varying circuits and in space-variant image restoration, as given by

$$M(ip, iq) = \iint_0^\infty f(\xi, \eta)\xi^{-(ip+1)}\eta^{-(iq+1)}\, d\xi\, d\eta. \tag{2.69}$$

The major obstacle to the optical implementation of Mellin transform is the nonlinear coordinator transformation of the input object. If we replace the space variables $\xi = e^x$ and $\eta = e^y$, the Fourier transform of $f(e^x, e^y)$ yields the *Mellin transform* of $f(\xi, \eta)$, as given by

$$M(p, q) = \iint_{-\infty}^\infty f(e^x, e^y) \exp[-i(px + qy)]\, dx\, dy, \tag{2.70}$$

where we let $M(ip, iq) = M(p, q)$ to simplify the notation. An *inverse Mellin transform* can also be written by

$$M^{-1}[M(p, q)] = f(\xi, \eta) = \frac{1}{2\pi}\iint_{-\infty}^\infty M(p, q)\xi^{ip}\eta^{iq}\, dp\, dq. \tag{2.71}$$

The preceding equation can be made equivalent to the inverse Fourier transform by replacing the variables $\xi = \exp(x)$ and $\eta = \exp(y)$.

The basic advantage of applying the Mellin transform to optical processing is its *scale-invariant property*. In other words, the Mellin transforms of two different-scale, but otherwise identical, spatial functions are *scale invariant*; that is,

$$M_2(p, q) = a^{i(p+q)}M_1(p, q),$$

where a is an arbitrary factor, $M_1(p, q)$ and $M_2(p, q)$ are the Mellin transforms of $f_1(x, y)$ and $f_2(x, y)$, respectively, and $f_1(x, y)$ and $f_2(x, y)$ are the identical but different-scale functions. From the preceding equation we see that the magnitudes of the Mellin transforms are

$$|M_2(p, q)| = |M_1(p, q)|, \tag{2.72}$$

which are of the same scaling. Thus, we see that the Mellin transform of an object function is *scale invariant*; that is,

$$|M[f(\xi, \eta)]| = |M[f(a\xi, a\eta)]|. \tag{2.73}$$

However, unlike the Fourier transform, the magnitude of the Mellin transform is not *shift invariant*; that is,

$$|M[f(x, y)]| \neq |M[f(x - x_0, y - y_0)]|. \tag{2.74}$$

In correlation detection, we assume that a Fourier domain filter of $M[f(x, y)] = M(p, q)$ has been constructed; for example, by using holographic technique as given by

$$H(p, q) = K_1 + K_2|M(p, q)|^2 + 2K_1K_2|M(p, q)| \cos[\alpha_0 p - \phi(p, q)],$$

where $M(p, q) = |M(p, q)| \exp[i\phi(p, q)]$. If the filter is inserted in the Fourier plane of which the input SLM function is $f(e^x, e^y)$, the output light distribution can be shown to be

$$\begin{aligned} g(\alpha, \beta) = K_1 f(e^x, e^y) &+ K_2 f(e^x, e^y) * f(e^x, e^y) * f^*(e^{-x}, e^{-y}) \\ &+ K_1K_2 f(e^x, e^y) * f(e^{x-\alpha_0}, e^y) \\ &+ K_1K_2 f(e^x, e^y) * f^*(e^{-x-\alpha_0}, e^{-y}), \end{aligned}$$

in which the last term represents the correlation detection, and it is diffracted at $\alpha = \alpha_0$. In other words, the implementation of a Mellin transform in a conventional linear processor requires a nonlinear coordinate transformation of the input object, as illustrated in the block diagram of Fig. 2.30.

2.6.2. CIRCULAR HARMONIC PROCESSING

Besides the scale-variant problem, the optical processor is also sensitive to object rotation. To mitigate this difficulty, we will discuss *circular harmonic* processing, for which the rotation variant can be alleviated. It is well known that if a two-dimensional function $f(r, \theta)$ is continuous and integrable over the region $(0, 2\pi)$, it can be expanded into a Fourier series, such as

$$f(r, \theta) = \sum_{m=-\infty}^{+\infty} F_m(r) e^{im\theta}, \tag{2.75}$$

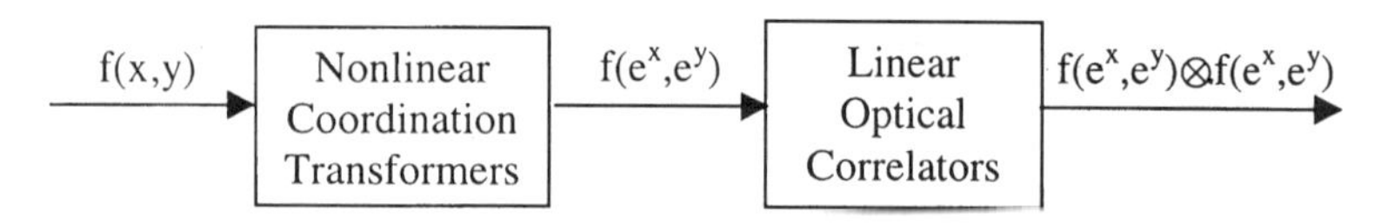

Fig. 2.30. A block diagram representation of Mellin transform.

where

$$F_m(r) = \frac{1}{2\pi} \int_0^{2\pi} f(r, \theta) e^{-im\theta} \, d\theta. \tag{2.76}$$

$F_m(r, \theta) = F_m(r)e^{im\theta}$ is called the mth-order *circular harmonic*.

If the object is rotated by an angle α, it can be written by

$$f(r, \theta + \alpha) = \sum_{m=-\infty}^{\infty} F_m(r) e^{im\alpha} e^{im\theta}. \tag{2.77}$$

Let us denote $f(x, y)$ and $f_\alpha(x, y)$ as object functions of $f(r, \theta)$ and $f(r, \theta + \alpha)$, respectively. By referring to matched filtering, when a rotated object $f_\alpha(r, \theta + \alpha)$ is applied to the input end of a Fourier domain process, the output light field can be evaluated, as given by

$$g_\alpha(x, y) = \int_{-\infty}^{+\infty} \int_{-\infty}^{+\infty} f_\alpha(\xi, \eta) f^*(\xi - x, \eta - y) \, d\xi \, d\eta. \tag{2.78}$$

It is apparent that if $\alpha = 0$, the autocorrelation peak appears at $x = y = 0$. By transforming the preceding convolution integral into a polar coordinate system, the center of correlation can be shown as

$$C(\alpha) = \int_0^{\infty} r \, dr \int_0^{2\pi} f(r, \theta + \alpha) f^*(r, \theta) \, d\theta. \tag{2.79}$$

In view of the definition of circular harmonic expansion of Eq. (2.75), we see that

$$C(\alpha) = \sum_{m=-\infty}^{+\infty} A_m e^{im\alpha}, \tag{2.80}$$

where

$$A_m = 2\pi \int_0^{\infty} |F_m(r)|^2 r \, dr.$$

Since the circular harmonic function is determined by different angles $m\alpha$ but not by a simple α, it is evident that object rotation poses severe problems for conventional matched filtering. Nevertheless, by using one of the circular harmonic functions as the reference functions, such as

$$f_{\text{ref}}(r, \theta) = F_m(r)e^{im\theta}, \tag{2.81}$$

the central correlation value between the target $f(r, \theta + \alpha)$ and the reference $f(r, \theta)$ can be written as

$$C(\alpha) = A_m e^{im\alpha}.$$

The corresponding intensity is independent from α, such as

$$|C(\alpha)|^2 = A_m^2, \tag{2.82}$$

for which we see that it is *independent* of the object orientation.

Needless to say, the implementation of circular harmonic processing can be by using either a Fourier domain filter or a spatial domain filter in a JTC. One of the major advantages of using a JTC is robust environmental factors. However, JTPS is input scene *dependent*, so the detection or correlation efficiency is affected by the input scene; for example, strong background noise, multi-input objects, and other problems, which cause poor correlation performance. Nevertheless, these disadvantages can be alleviated by using non-zero-order JTC, as will be described in Sec. 7.2.

2.6.3. HOMOMORPHIC PROCESSING

The optical processing discussed so far relies on linear spatial invariant operation. There are, however, some nonlinear processing operations that can be carried out by optics. One such nonlinear processing operation worth mentioning is *homomorphic processing*, described in Fig. 2.31. Note that homomorphic processing has been successfully applied in digital signal processing. In this section, we illustrate logarithmic processing for target detection in a multiplicative noise regime. Let us assume an input object is contaminated by a multiplicative noise, as written by

$$n(x, y)f(x, y). \tag{2.83}$$

Since matched filtering is *optimum* under an additive white Gaussian noise assumption, our first task is to convert the multiplicative noise to an additive noise by means of logarithmic transformation, such as

$$\log[n(x, y)f(x, y)] = \log n(x, y) + \log f(x, y). \tag{2.84}$$

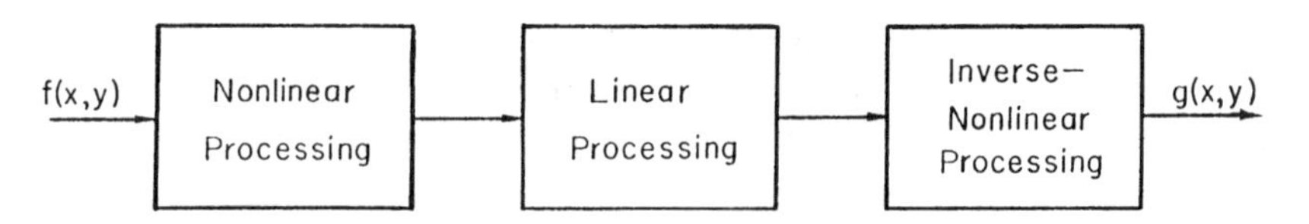

Fig. 2.31. A homomorphic processing system.

In order for a linear optical processor to perform the correlation operation, a Fourier domain filter or a spatial domain filter for the JTC should be made available, such as

$$H(p, q) = \{\mathscr{F}[\log f(x, y)]\}^*, \qquad \text{for FDP,}$$

and

$$h(x, y) = \log f(x, y), \qquad \text{for JTC.}$$

If the logarithmic transform input signal is displayed at the input plane of the optical correlator (either FDP or JTC), output correlation terms can be written as

$$\begin{aligned} g(x, y) = {} & [\log n(x, y)] * [\log f(-x + b, -y)]^* \\ & + [\log f(x, y)] * [\log f(-x + b, -y)]^*, \end{aligned}$$

in which the first term represents the cross-correlation between the $\log n$ and signal $\log f$, and the second term denotes the autocorrelation of $\log f$. Thus we see that high correlation peak detection can be obtained at the output plane. In order to make the processing more optimum, an additional step of converting the logarithmic noise into *white noise* is necessary. This step is generally known as the *prewhitening process*. Figure 2.32 shows a block box representation for the signal detection under multiplication regime, in which a linear optical processor (either FDP or JTC) is used within the system for the homomorphic processing.

2.6.4. SYNTHETIC DISCRIMINANT ALGORITHM

There are, however, techniques available to alleviate the rotational and scale-variant constraints in an optical processor. Aside from the distortion variant constraint, the spatial filters (Fourier and spatial domain) we have described are basically two-dimensional filters of the storage capacities of

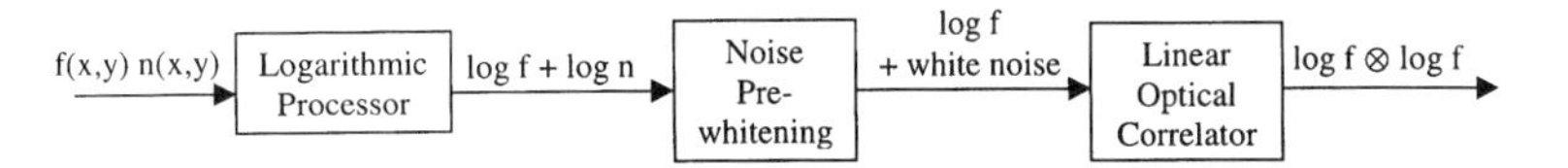

Fig. 2.32. A logarithmic processing system for multiplicative noise.

which were not exploited. In this section we discuss a type of *composite filter* in which a number of filters can be encoded in a single spatial filter. Nevertheless, composite filters can also be constructed by means of thick photorefractive crystals (a three-dimensional volume filter discussed later). In general, a composite filter can be constructed by means of a set of filter function $q = \{q_n\}$, which are used to correlate with a set of multivariant (or multiclass) objects $f = \{f_n\}$, $n = 1, 2, \ldots, N$, where N is the number of object orientations. We note that the sets of f and g could be entirely different objects. For simplicity, we assume the set $\{g_n\}$ represents different orientations of g and that $\{f_n\}$ belongs to the set $\{g_n\}$. We begin by expanding each orientation of the input object and filter function f and g in sets of *orthonormal* sets, as given by

$$f(x) = \sum_j a_j \phi_j(x), \tag{2.85}$$

$$g(x) = \sum_j b_j \phi_j(x), \tag{2.86}$$

where a and b are coefficients, and

$$\int \phi_j(x) \phi_i(x)\, dx = \delta_{ji}.$$

It is trivial that f and g can be represented by vectors in a vector space, such as

$$\hat{f} = (a_1, a_2, \ldots, a_n) \tag{2.87}$$

and

$$\hat{g} = (b_1, b_2, \ldots, b_n). \tag{2.88}$$

In terms of these expansions, the correlation of f and g can be shown as

$$R(\tau) = f(x) \otimes g(x) = \sum_i \sum_j a_j b_i \int \phi_j(x + \tau) \phi_i(x)\, dx. \tag{2.89}$$

Thus, at $t = 0$, we have

$$R(0) = \sum_j a_j b_j = \hat{f} \cdot \hat{g}, \tag{2.90}$$

which is the dot product of vectors $\hat{f}$ and $\hat{g}$.

The set of vectors $\hat{f}$, as well as $\hat{g}$, forms a hypervector space that yields *discriminant surfaces* which enable us to group the multiclass objects (i.e., the oriented vectors) apart from the false input objects. In other words, the shape of the discriminant surface decides the kind of average *matched filter* to be designed. For instance, the autocorrelation of an oriented input object f with respect to a specific oriented filter function g should lie within the discriminant surface.

For simple illustration, we assume only two object vectors are to be detected and all the other object vectors are to be rejected. By representing these input vectors as

$$\hat{f}_1 = (a_{11}, a_{12}) \qquad \text{and} \qquad \hat{f}_2 = (a_{21}, a_{22}),$$

a hyperplane resulting from $\hat{f}_1$ and $\hat{f}_2$ can be described by

$$\hat{f} \cdot \hat{h} = K,$$

where $\hat{h}$ represent the filter vector and K is a constant. In this illustration, $\hat{h}$ represents a vector perpendicular to the hyperplane form by $\hat{f}_1$ and $\hat{f}_2$, by which $K = (\hat{g} \cdot \hat{g})^{1/2}$. Thus, we see that the average filter function is a specific *linear combination* of $c_1 \hat{f}_1$ and $c_2 \hat{f}_2$, as written by

$$h = c_1 \hat{f}_1 + c_2 \hat{f}_2 = (c_1 a_{11} + c_1 a_{12})\phi_1 + (c_2 a_{21} + c_2 a_{22})\phi_2.$$

A *synthetic discriminant function (SDF) filter* can be described as a *linear combination* of a set of *reference functions*; that is,

$$h(x) = \sum_j c_j \phi_j(x), \tag{2.91}$$

for which the correlation output must be

$$R(\tau)|_{\tau=0} = \hat{g} \cdot \hat{h} = \sum_j b_{nj} c_j. \tag{2.92}$$

The remaining task is to find $\{\phi_j\}, n_{nj}$, then c_j, and finally $h(x)$, under the assumed acceptable correlation peaks $R(0)$. However, in the construction of the SDF filter, the *shift invariance* property is imposed; that is, all the autocorrelation peaks must occur at $\tau = 0$. These requirements ensure that we shift each g_n or ϕ_j to the correct input object location when we synthesized the filter. Such technique is acceptable, since the synthesis is an *off-line algorithm*. This off-line synthesis also provides us the flexibility of weighting each g_n when we are forming the SDF filter. The procedure is that we first form a correlation matrix for all the reference functions $\{g_n\}$, such as

$$R_{ij} = g_i \otimes g_j. \tag{2.93}$$

By using the *Gram-Schmidt expansion* for the orthonormal set $\{\phi_j\}$, we then have

$$\begin{aligned} \phi_1(x) &= g_1(x)/k_1 \\ \phi_2(x) &= [g_2(x) - C_{12_1}\phi_1(x)]/k_2 \\ &\vdots \\ \phi_n(x) &= \left[g_n(x) - \sum_{j=1}^{n-1} C_{n_j}\phi_j(x) \right] \Big/ k_n, \end{aligned} \tag{2.94}$$

where the k_n are normalization constants that are functions of the R_{ij} and where the c_{nj} are linear combinations of the R_{ij} with known weighting coefficients. This tells us that when the orthonormal set is determined, the coefficient b_{nj} can be calculated.

If we assume that all the autocorrelation peaks $R(0)$ are equal, then the weighting factors c_j can be evaluated and the desired SDF is therefore obtained. A block box diagram representation of the off-line SDF filter synthesis is shown in Fig. 2.33, in which a set of training images are available for the synthesis. Note that the implementation of the SDF filter can be in the Fourier domain for FDP or in the input domain for JTP. Needless to say, for the Fourier domain implementation, one uses the $H(p, q)$, instead of using the spatial domain filter $h(x, y)$ for the JTC.

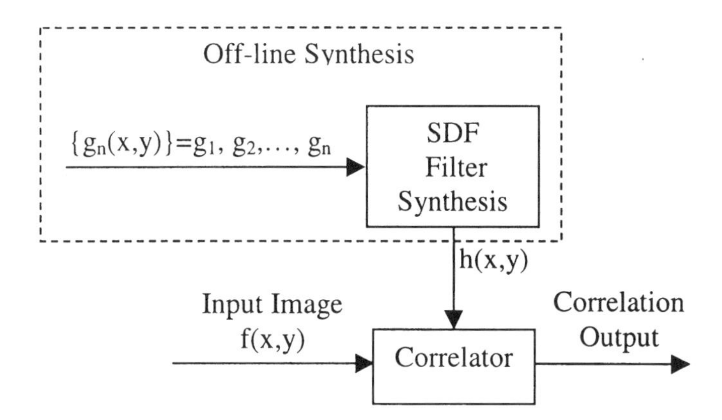

Fig. 2.33. A block diagram representation of the off-line SDF filter synthesis.

2.6.5. SIMULATED ANNEALING ALGORITHM

One of the major drawbacks of SDF filter synthesis is the high dynamic range requirement that deters its practical implementation on a commercially available SLM. There is, however, an alternative approach toward the synthesis of a composite filter; namely, the *simulated annealing* (SA) filter.

The SA algorithm is, in fact, a computational optimization process in which an energy function should be established based on certain optimization criteria. In other words, the state variables of a physical system can be adjusted until a global minimum energy state is established. For example, when a system variable u_1 is randomly perturbed by Δu, the change of the system energy can be calculated

$$\Delta E = E^{\text{new}} - E^{\text{old}}, \tag{2.95}$$

where $E^{\text{new}} = E(u_i + \Delta u_i)$ and $E^{\text{old}} = E(u_i)$. Notice that, if $\Delta E < 0$, the perturbation Δu_i is unconditionally accepted. Otherwise, the acceptance of Δu_i is based on the Boltzmann probability distribution $p(\Delta E)$; that is,

$$p(\Delta E) = \frac{1}{1 + \exp(\Delta E/kT)}, \tag{2.96}$$

where T is the temperature of the system used in the simulated annealing algorithm and k is the Boltzmann constant. The process is then repeated by randomly perturbing each of the state variables and slowly decreasing the system temperature T, which can avoid the system being trapped in a local minimum energy state. In other words, by continually adjusting the state variables of the system and slowly decreasing the system temperature T, a global minimum energy state of the system can be found. This is known as the *simulated annealing process*. We now illustrate synthesis of a *bipolar composite filter* (*BCF*) using the SA algorithm.

Let us consider two training images, called the *target* and *antitarget* set, as given by

$$\begin{cases} \text{target set} = \{t_m(x, y)\} \\ \text{antitarget set} = \{a_m(x, y)\}, \end{cases}$$

where $m = 1, 2, \ldots, M$, M is the number of training images. The purpose of designing a composite filter is to detect the *target* objects and to reject the antitarget objects, which are assumed similar to the target objects. Let us denote $W^t(x, y)$ and $W^a(x, y)$ as the desired correlation distributions for the target and antitarget objects, respectively. The mean-square error between the

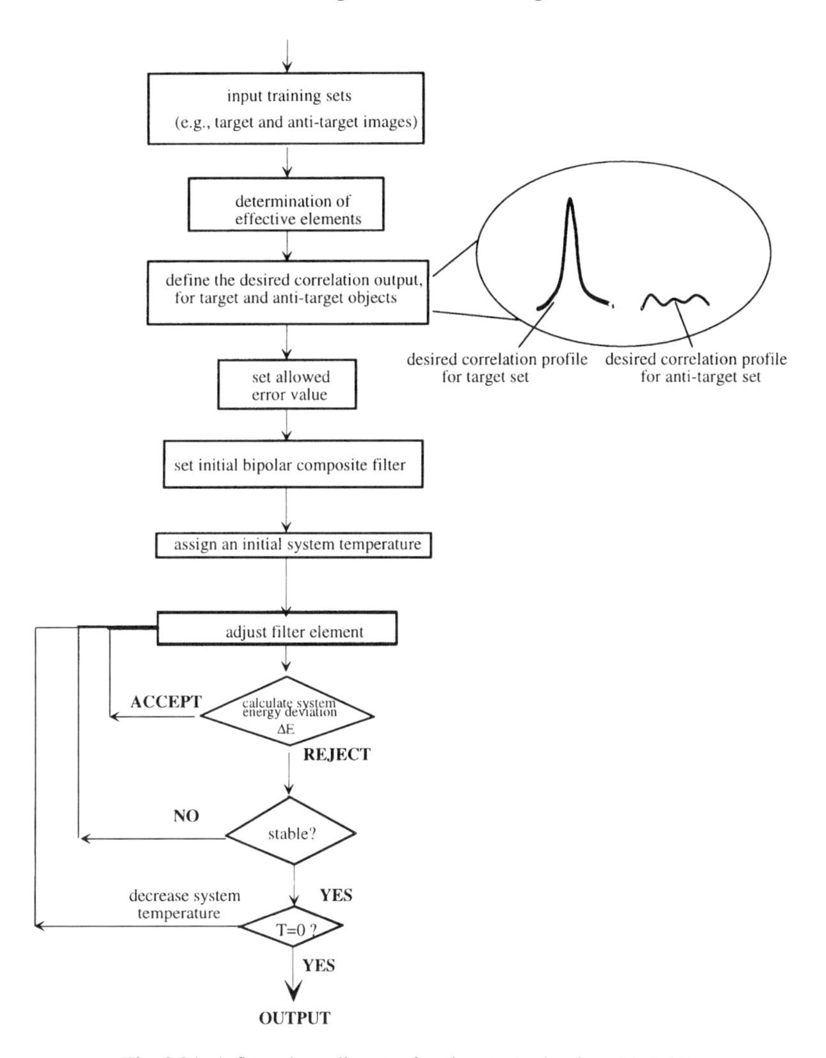

Fig. 2.34. A flow chart diagram for the synthesis of an SA BCF.

desired and the actual correlation distributions is defined as the energy function E of the system. The configuration of BCF can be viewed as the state variable of the system. The remaining task is to optimize the state variable of the system; that is, the configuration of the BCF, so that a global minimum energy state can be achieved.

A flow chart diagram showing the simulated annealing process for synthesizing a BCF as shown in Fig. 2.34 is described in the following:

1. Read the input data for the target and the antitarget train sets.
2. Determine the effective pixels of the BCF in the spatial domain. For example, in an effective pixel there is at least one training target or antitarget image the value of which is nonzero of this pixel. Notice that the noneffective pixels are equiprobable to either -1 or 1.
3. Assign different desired correlation distributions for the target training set and the antitarget train set, respectively. In fact, a sharp correlation profile can be assigned for the target training images and a null profile for the antitarget training set, as illustrated in the chart.
4. Set the allowable error value. For example, a 1% error rate is frequently used.
5. Assign an initial bipolar composite filter function (IBCF), which is assumed to have a random distribution of -1 and $+1$ values.
6. Set the initial system temperature T. We select the initial kT in the order of the desired correlation energy, as defined in step 3.
7. Calculate the energy function E for the initial system, which is the mean-square error between the desired correlation function and the correlation distribution obtained with the IBCF. We have

$$E = \sum_{m=1}^{M} \left(\iint \{[0^t_m(x, y) - W^t(x, y)]^2 + [0^a_m(x, y) - W^a(x, y)]^2\} \, dx \, dy \right), \tag{2.95}$$

where $0^t_m(x, y)$ and $0^a_m(x, y)$ are the target and antitarget correlation distributions for the mth training image obtained with IBCF, respectively. If we let $h(x, y)$ be the filter function of BCF, the correlation distribution with respect to target function t_m and antitarget a_m can be expressed as

$$\begin{cases} 0^t_m(x, y) = \iint h(x + x', y + y') t^*_m(x', y') \, dx' \, dy' \\ 0^a_m(x, y) = \iint h(x + x', y + y') a^*_m(x', y') \, dx' \, dy'. \end{cases} \tag{2.97}$$

8. By reversing the sign of a pixel (e.g., from -1 to $+1$) within the IBCF, a new system energy E' is established. Calculate the change of system energy ΔE of the system. If $\Delta E < 0$, the change sign of the pixel is unconditionally accepted. Otherwise, it will be accepted based on the Boltzmann probability distribution. Since $h(x, y)$ is a bipolar function, the iterative procedure for reversing the pixel processes is given by

$$\begin{cases} h^{(n+1)}(x, y) = -h^{(n)}(x, y), \ \Delta E < 0, \\ h^{(n+1)}(x, y) = h^{(n)}(x, y) \cdot \operatorname{sgn}\{\operatorname{ran}[p(\Delta E)] - p(\Delta E)\}, \ \Delta E \geqslant 0, \end{cases} \tag{2.98}$$

where n represents the nth iteration, ran $[\cdot]$ represents a uniformly distributed random function within the range $[0, 1]$, sign represents a sign function, and $p(\Delta E)$ is the Boltzmann distribution function.

9. Repeat step 8 for each pixel until the system energy E is stable. Then assign a reduced temperature T to the system and repeat step 8 again.
10. Repeat steps 8 to 9 back and forth until a global minimum E is found; the final $h(x, y)$ will be the desired BCF. In practice, the system will reach a stable state if the number of iterations is larger than $4N$, where N is the total number of pixel elements within the BCF.

2.7. PROCESSING WITH PHOTOREFRACTIVE OPTICS

Besides electronically addressable SLMs, photorefractive (PR) optics also plays an important role in real-time optical processing. Because of their volume storage capability, PR materials have been used to synthesize *large capacity composite filters*. This section briefly discusses the PR effect and some of its processing capabilities.

2.7.1. PHOTOREFRACTIVE EFFECT AND MATERIALS

Photorefractive effect is a phenomenon in which the local refractive index of a medium is changed by a spatial variation of light intensity. In other words, PR materials generally contain donors and acceptors that arise from certain types of impurities of imperfections. The acceptors usually do not directly participate in the photorefractive effect, whereas the donor impurities can be ionized by absorbing photons. Upon light illumination, donors are ionized, by which electrons are generated in the conduction band. By drift and diffusion transport mechanisms, these charge carriers are swept into the dark regions in the medium to be trapped. As a consequence, a space charge field is built up which induces a change in the refractive index.

We now consider how two coherent plane waves with equal amplitude interfere within the photorefractive medium, as illustrated in Fig. 2.35. The intensity of the interference fringes is given by

$$I(x) = I_0\left[1 + \cos\left(\frac{2\pi x}{\Lambda}\right)\right], \tag{2.99}$$

where I_0 is a constant intensity, $\Lambda = \lambda/\sin\theta$ is the period of the fringes, and λ is the wavelength of the light beams. In the bright regions close to the

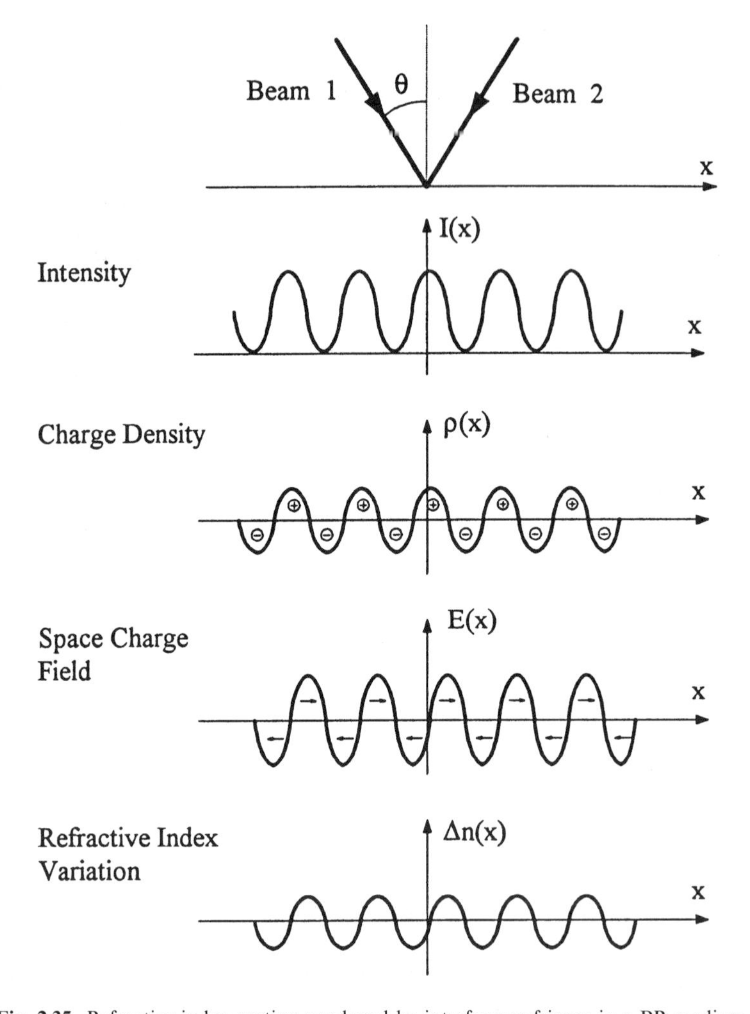

Fig. 2.35. Refractive index grating produced by interference fringes in a PR medium.

maximum intensity, photoionized charges are generated by the absorption of photons. These charge carriers diffuse away from the bright region and leave behind positively charged donor impurities. If these charge carriers are trapped in the dark region, they will remain there because there is almost no light to reexcite them. This leads to the charge separation as shown in Fig. 2.35. Due to the periodic intensity distribution within the PR medium, a space charge is created with the dark regions negatively charged and the bright regions positively charged. The space charge continues building up until the diffusion current is counterbalanced by the drift current. The space charges produce a space charge field which is shifted by a quarter of a period (or $\pi/2$ in phase) relative to the intensity pattern. Owing to *Pockel's effect*, the space charge field induces a change in the refractive index, as given by

$$\Delta n(t) = \Delta n_{\text{sat}} \int_0^t \frac{A_1 A_2^* e^{(t'-t)/\tau} \, dt'}{I_0}, \tag{2.100}$$

where A_1 and A_2 are the amplitude of the two incident waves, Δn_{sat} is the saturation index amplitude, and τ is the time constant. We note that Δn_{sat} and τ are material-dependent parameters.

The two beams can, in fact, carry spatial information (e.g., images) with them. The interaction of the two beams then generates a volume index hologram in the PR medium. While an optical beam propagates through the medium, it undergoes Bragg scattering by the volume hologram. If the Bragg scatterings are perfectly *phase matched*, a strong diffraction beam will reconstruct the spatial information. The formation and diffraction of the dynamic holograms within the PR medium can be explained by nonlinear optical wave mixing (as briefly discussed in the next section).

PR effect has been found in a large variety of materials. The most commonly used photorefractive materials in optical processing applications fall into three categories: electro-optic crystals, semi-insulating compound semiconductors, and photopolymers.

Lithium niobate ($LiNbO_3$), barium titanate ($BaTiO_3$), and strontium barium niobate (SBN $Sr_{(1-x)}Ba_xNb_2O_6$) are by far the three most efficient electro-optic crystals exhibiting PR effects at low intensity levels. Iron-doped $LiNbO_3$ has a large index modulation due to the *photovoltaic effect*. It is also available in relatively large dimensions. For instance, a 3 cm^3 sample used to record 5000 holograms has been reported. Because of its strong mechanical qualities, $LiNbO_3$ has been extended to near infrared (670 nm) wavelengths by adding Ce dopants.

High diffraction efficiency can be achieved with $LiNbO_3$ crystals without the involvement of PR effect. This eliminates the phase distortion of the image beam and is preferred in some applications. However, $BaTiO_3$ crystals are not available in large dimensions. A $5 \times 5 \times 5$ mm^3 sample is considered to be big. Moreover, in order to reach the maximum index modulation, the sample must be cut at a certain angle of the crystallographic *c*-axis. This cut is difficult, and it reduces the size of the sample. A phase transition exists around 13°C; therefore, the crystal must always be kept above this temperature.

SBN has a large electro-optic coefficient, which can be reached without a special crystal cut. The phase transition (which can be tuned by doping) is far from room temperature. It can also be subject to electric or temperature fixing. The optical quality of SBN obtained up to now is still poorer as compared with $LiNbO_3$ and $BaTiO_3$ crystals, however.

Several semi-insulating semiconductors, such as gallium arsenide (GaAs), gallium phosphite (GaP), and idium phosphite (InP), have demonstrated PR effect and have been used in optical processing systems. A prominent feature of these semiconductor crystals is their fast response to optical fields; i.e., the small value of time constant τ at low light intensity. Typically, a submillisecond response time can be achieved with GaAs at a modest laser power intensity of 100 mW/cm^2, which is one or two orders of magnitude faster than the more conventional materials such as $BaTiO_3$ under the same conditions. The spectral response range of these materials is in the near infrared wavelength. This can be an advantage or a disadvantage, depending upon the applications. One of the problems of these semiconductor PR materials is the disparity among between different samples. The PR effect of the same material varies considerably among suppliers as well as from one ingot to another.

Photopolymer is a new type of PR material. Its PR effect was first discovered in the early 1990s. Recently a photopolymer based on photoconductor poly (N-vinylcarbazalo) doped with a nonlinear optical chromophore has been developed; it exhibits better performance than most inorganic PR crystals. Due to its relatively small thickness (typically tens of micrometers), only a limited number of images can be multiplexed. However, the photopolymer is more suitable for the implementation of 3-D optical disks, which can store a huge amount of data.

2.7.2. WAVE MIXING AND MULTIPLEXING

There are two generic configurations for optical pattern recognition systems using photorefractive materials; namely, *two-wave mixing* and *four-wave mixing*. Depending upon the coherence among the read and write beams, four-wave mixing configurations can be further classified into degenerated four-wave mixing and nondegenerated four-wave mixing.

In a two-wave mixing configuration, shown in Fig. 2.36, two coherent beams intersect in a PR medium and create an index grating. The Bragg scattering involved in two-wave mixing is very similar to the readout process in holography. If one lets beam A_2 be the *reference beam*, the variation of the PR index can be written as

$$\Delta n(x) \propto A_1^* A_2 e^{-i\mathbf{K}\cdot\mathbf{r}} + A_1 A_2^* e^{i\mathbf{K}\cdot\mathbf{r}}, \tag{2.101}$$

where

$$\mathbf{K} = \mathbf{k}_2 - \mathbf{k}_1,$$

and $\mathbf{k}_1$ and $\mathbf{k}_2$ are wave vectors of the two beams, respectively. When such a

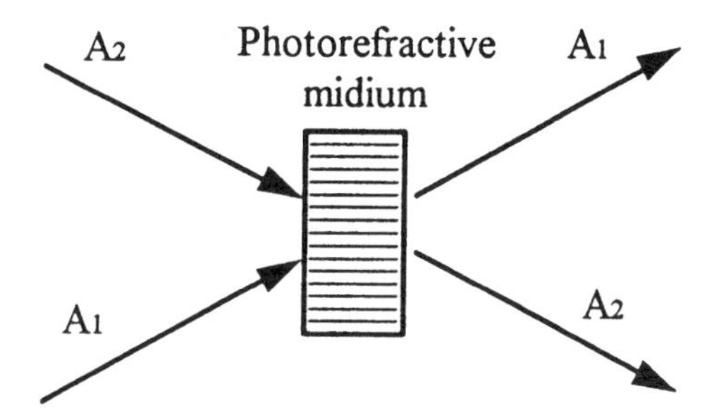

Fig. 2.36. A two-wave mixing configuration.

hologram is illuminated by the reference beam $A_2e^{-i\mathbf{k}_2\cdot\mathbf{r}}$, the diffracted beam is given by

$$0(x) = \eta A_1 A_2^* A_2 e^{-i\mathbf{k}_1\cdot\mathbf{r}}, \tag{2.102}$$

where η is the diffraction efficiency. Notice that the phase of A_2 cancels out and the diffracted beam is the reconstruction of the object beam $A_1^{-i\mathbf{k}_1\cdot\mathbf{r}}$. Similarly, the reference beam A_2 can be reconstructed by illuminating the hologram with object beam A_1.

In addition to holographic analogy, two-wave mixing in most PR crystals exhibits amplification, which is a unique feature not available in conventional holography. This occurs most efficiently in crystals where the dynamic PR index grating is 90° out of phase with respect to the intensity interference grating, as can be seen in Fig. 2.35. The energy exchange is unidirectional, with the direction of the energy flow determined by the crystal parameters, such as the crystal orientation and the sign of the photoionized charge carriers. Customarily, the beam that loses energy is labeled as the *pump beam*, and the beam that becomes amplified is called the *probe beam*. Because the light energy is coupled from one beam to another, two-wave mixing is also known as *two-beam coupling*.

In the four-wave mixing configuration, two coherent beams write an index hologram and a third beam reads the hologram, creating the fourth (i.e., output) beam by diffraction, as illustrated in Fig. 2.37. To satisfy the Bragg condition, the third (read) beam must be counterpropagating relative to one of the two writing beams. If the read beam has the same wavelength as the writing beams, the configuration is called *degenerate four-wave mixing*; if the wavelengths of the read beam and write beams are different, it is called *nondegenerate four-wave mixing*. Although degenerate four-wave mixing has been used in most of the applications demonstrated so far, nondegenerate four-wave mixing may be used in some cases where the nondestructive reading of the hologram is required. This can be achieved by choosing a reading wavelength beyond the spectral response range of the photorefractive medium.

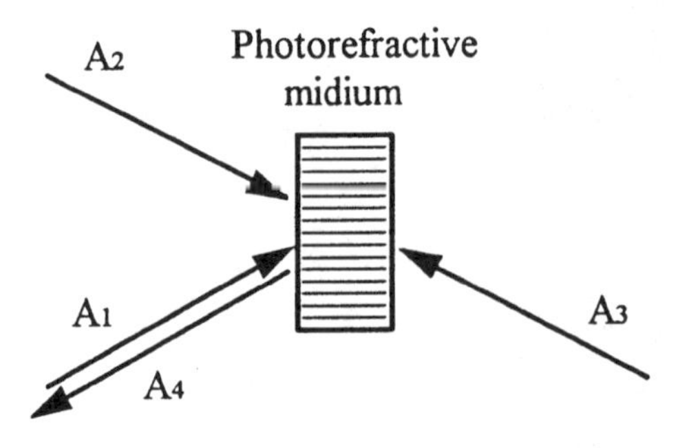

Fig. 2.37. A four-wave mixing configuration.

For example, if the index hologram is generated by the two writing beams A_1 and A_2 and this recorded hologram is illuminated by a counterpropagating reading beam $A_3 e^{i\mathbf{k}_2 \cdot \mathbf{r}}$, the diffraction can be written by

$$0(x) = \eta A_1^* A_2 A_3 e^{i\mathbf{k}_1 \cdot \mathbf{r}}, \tag{2.103}$$

in which we see that the diffracted beam counterpropagates with respect to the writing beam A_1. If both the writing beam A_2 and reading beam A_3 are *plane waves*, the diffraction is a *time-reserved replica of* A_1. Therefore, four-wave mixing provides a convenient method for the generation of *phase conjugate waves*. In general, however, all three waves can carry spatially modulated signals and the amplitude of the diffracted beam represents the multiplication of these three images. By suitably manipulating the diffracted beam, many image processing operations can be implemented.

Beams A_1 and A_2 write a *transmission-type hologram* in the PR medium. If the beams A_1 and A_2 are used in the writing process, a *reflection-type hologram* is formed, which can be read with beam A_2 and generates the same diffraction beam as given by the preceding equation. Note that the transmission and reflection modes have different response times selectivities and diffraction efficiencies. They are suitable for different applications.

Multiple images can be stored in a single piece of PR medium via wave multiplexing. Once these images are stored, they can be retrieved and can serve as a library of reference images for signal detection and pattern recognition. Due to the parallel readout and fast access capabilities of wave mixing storage, an input image can be compared with all the stored reference images at very high speed. The three most commonly used multiplexing schemes in volume holographic storage are angular multiplexing, wavelength multiplexing, and phase-code multiplexing. All three multiplexing options are based on *Bragg-selective readout* of thick holograms (discussed later).

In an angular multiplexing scheme, the address of each image is represented by the incident angle of the reference beam. To change the angle of the reference beam, a mirror mounted on a rotating step motor can be utilized. For rapid access to all of the stored images, acousto-optic cells can be used to deflect the reference beam. Note that two acousto-optic cells must be used to compensate for the *Doppler shift* in frequency.

For wavelength multiplexing, both the object and reference beams are fixed and only their wavelengths are changed. The first demonstration of wavelength multiplexing with PR materials was made to record three primary-color holograms from a color object. The simultaneous replay of the three holograms reconstructs the colored image. The application of wavelength multiplexing has stimulated the development of solid-state tunable laser diodes and specially doped PR crystals that are sensitive to laser diode wavelength range.

In phase-code multiplexing, the reference beam consists of multiple plane wavefronts. The relative phases among all these wavefronts are adjustable and represent the addresses of the stored images. Each image can be retrieved by illuminating the holograms with the exact same phase code used for recording the image. The merits of phase-code multiplexing include fast access, high light efficiency, and the elimination of beam steering.

2.7.3. BRAGG DIFFRACTION LIMITATION

Bragg diffraction limitation in a thick PR crystal can be explained with the **k** vector diagram, as depicted in Fig. 2.38. The recorded spatial grating vector **k** is

$$\mathbf{k} = \mathbf{k}_0 - \mathbf{k}_1, \tag{2.104}$$

where $\mathbf{k}_0$ and $\mathbf{k}_1$ are the writing wave vectors. If the recorded hologram is read out by a wave vector $\mathbf{k}_2$ (where the scattered wave vector is denoted by $\mathbf{k}_3$), then the optical path difference (OPD) of the scattered light from two points within the crystal can be written as

$$\text{OPD} = \mathbf{k}\cdot\mathbf{r} - (\mathbf{k}_3 - \mathbf{k}_2)\cdot\mathbf{r} = \Delta\mathbf{k}\cdot\mathbf{r}, \tag{2.105}$$

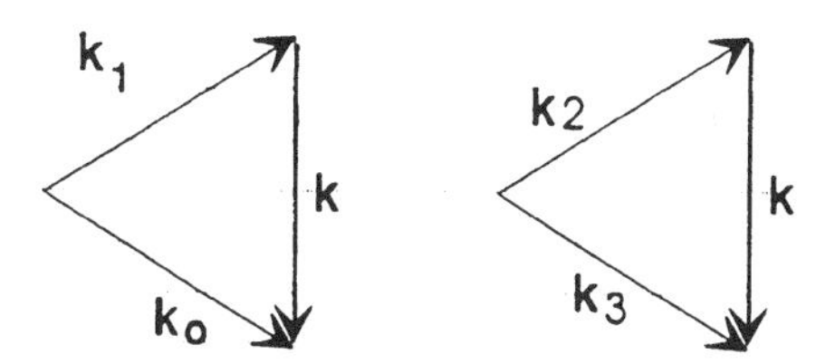

Fig. 2.38. Bragg diffraction vectors.

where **r** represents the displacement vector between two points and

$$\Delta\mathbf{k} = \mathbf{k}_0 - \mathbf{k}_1 + \mathbf{k}_2 - \mathbf{k}_3 \tag{2.106}$$

is known as the *dephasing wave vector* and corresponds to the *Bragg diffraction (dephasing) limitation.* In other words, if the reading beam is not matched with the writing beam, the OPD will be nonzero. The field of scattered light from the crystal, as a function of the (scattered) wave vector $\mathbf{k}_3$, can be expressed as

$$A(\mathbf{k}_3) = \int_v \exp(i\Delta\mathbf{k}\cdot\mathbf{r})\,dV, \tag{2.107}$$

where the integration is over the entire volume of the crystal. We note that the preceding equation is valid under the weak diffraction condition, such that multiple diffraction within the crystal can be neglected.

In view of the preceding equation, we see that the performance of the PR filter will be severely limited by the Bragg diffraction. Since the angular and the wavelength selectivities of a multiplexed matched filter are limited by the Bragg diffraction condition, the shift tolerance of the matched filter (related to the Bragg mismatch) is affected by these selectivities, as will be discussed in the following subsection.

2.7.4. ANGULAR AND WAVELENGTH SELECTIVITIES

Let us now consider the unslanted spatial filter synthesis for a transmission-type and a reflection-type spatial filter, as depicted in Fig. 2.39. By using Kogelnik's *coupled-wave theory*, the normalized diffraction efficiency can be shown to be

$$\eta_t = \frac{\sin^2(v_t^2 + \xi_t^2)^{1/2}}{1 + \xi_t^2/v_t^2}, \qquad \eta_r = \frac{1}{1 + \dfrac{1 - \xi_r^2/v_r^2}{\sinh^2(v_r^2 - \xi_r^2)^{1/2}}} \tag{2.108}$$

where the subscripts t and r denote the *transmission-type* and the *reflection-type* holographic filters,

$$v_t = \frac{\pi\Delta nd}{\lambda\cos\theta}, \qquad \xi_r = \frac{2\pi nd\sin\theta}{\lambda}\Delta\theta,$$

$$v_r = \frac{\pi\Delta nd}{\lambda\sin\theta}, \qquad \xi_r = \frac{2\pi nd\cos\theta}{\lambda}\Delta\theta,$$

n and Δn are the refractive index and the amplitude of its variation, d is the

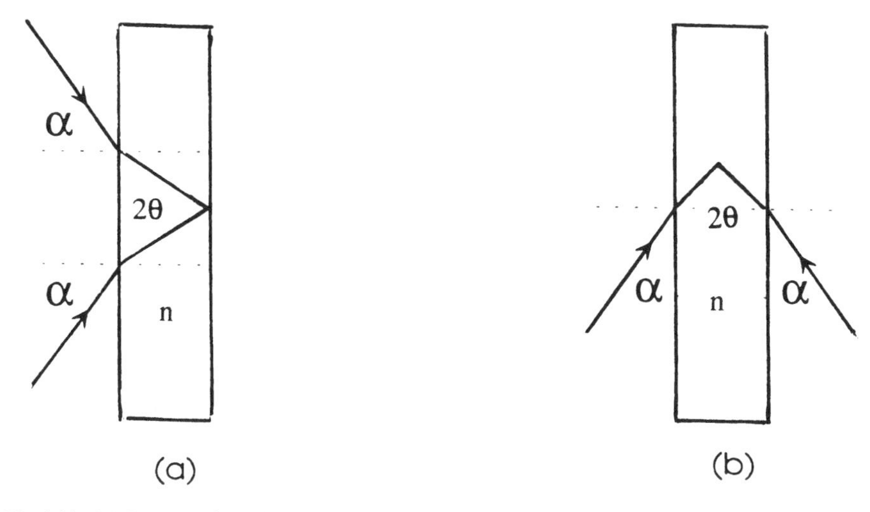

Fig. 2.39. (a) Construction of a transmission-type filter, (b) Construction of a reflection-type filter.

thickness of the crystal, and θ and $\Delta\theta$ are the internal construction angle and its variation. Under the assumption of *weak coupling*; i.e., $|v| \ll |\xi|$, the normalized diffraction efficiency η can be approximated as

$$\eta_t \approx v_t^2 \operatorname{sinc}^2 \xi_t, \qquad \eta_r \approx v_r^2 \operatorname{sinc}^2 \xi_r. \tag{2.109}$$

Based on preceding equations, the angular selectivity for the transmission-type and the reflection-type holographic filters can be shown to be

$$\left\{\frac{1}{\Delta\alpha}\right\}_t = \frac{\sin\alpha\cos\alpha}{\sqrt{n^2 - \sin^2\alpha}}\frac{d}{\lambda}, \qquad \left\{\frac{1}{\Delta\alpha}\right\}_r = \frac{\sin\alpha\cos\alpha}{\sqrt{n^2 - \cos^2\alpha}}\frac{d}{\lambda}, \tag{2.110}$$

where 2α is the external construction angle, as shown in Fig. 2.39. The variation of the angular selectivities is plotted in Fig. 2.40a.

Similarly, the wavelength selectivity for the transmission-type and the reflection-type holographic filters can be shown as

$$\left\{\left|\frac{\lambda}{\Delta\lambda}\right|\right\}_t = 1 + \frac{2}{\dfrac{\sin\alpha/n}{\sin\left[\sin^{-1}\left(\dfrac{\sin\alpha}{n}\right) - \dfrac{\lambda}{d\sin\alpha}\right]^{-1}}} \tag{2.111}$$

$$\left\{\left|\frac{\lambda}{\Delta\lambda}\right|\right\}_r = \frac{(n^2 - \cos^2\alpha)^{1/2}}{\lambda}d, \tag{2.112}$$

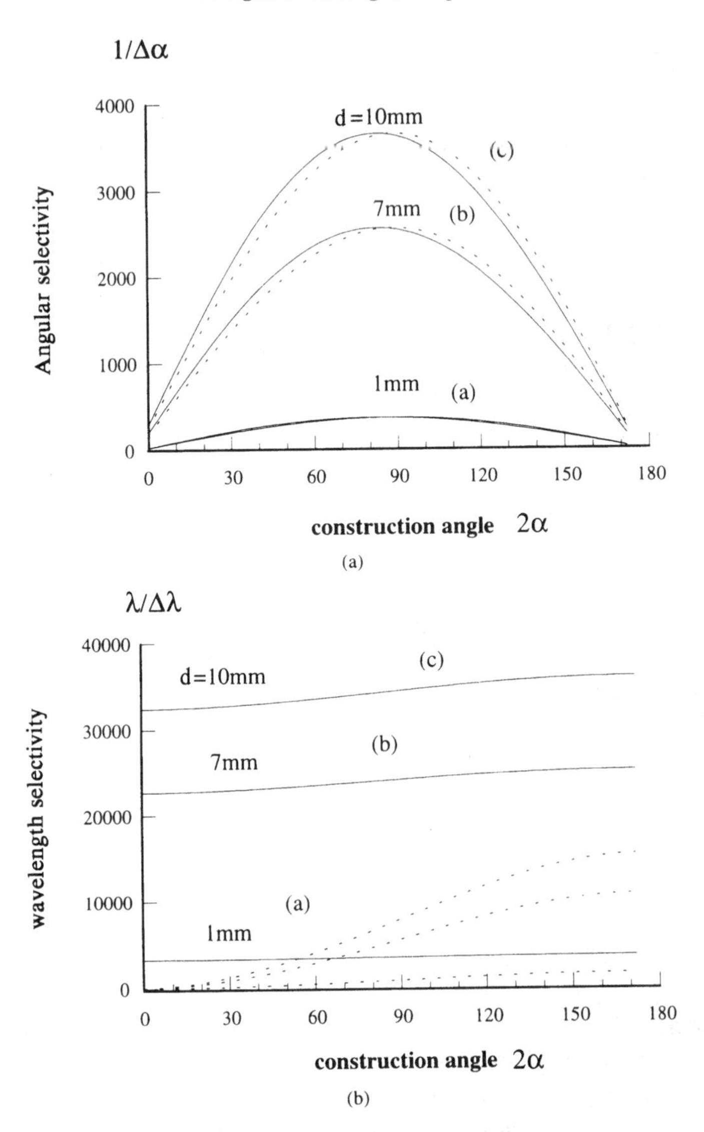

Fig. 2.40. (a) Angular selectivity, (b) Wavelength selectivity. Dash line for transmission type and solid line for reflection type.

selectivities of which are plotted in Fig. 2.40b. We see that the wavelength selectivity for the reflection-type hologram is quite uniform and higher than the transmission type. Although the wavelength selectivity for both the reflection type and the transmission type increases as the thickness of the crystal filter increases, the increase is far more rapid for the reflection type than for the transmission type. It is therefore apparent that the reflection-type filter is a better choice for the application of the wavelength-multiplexed filter.

2.7.5. SHIFT-INVARIANT LIMITED CORRELATORS

In a Fourier domain–matched filtering system, a shift of the input target at the input plane will cause a change in the readout angle at the Fourier plane where the PR filter is located. When the change of the readout angle is large, the readout beam intensity decreases rapidly due to high angular selectivity (i.e., the Bragg diffraction limitation). In other words, the higher the angular selectivity of the PR filter, the lower its shift tolerance will be. Thus, to optimize the shift invariance in a thick PR filter, a minimum angular selectivity is needed. However, from Fig. 2.40 we see that the wavelength selectivity for the reflection-type filter has its highest value at $2\alpha = 180^\circ$ where the angular selectivity is minimum. Therefore, a reflection-type wavelength-multiplexed PR filter will be the best choice for two major reasons, large storage capacity and optimum shift tolerance.

We now investigate the shift invariance of three commonly used PR–based correlators, the Vandelugt correlator (VLC), the joint transform correlator (JTC), and the reflection-type correlator (RC). First, let us consider the VLC (shown in Fig. 2.41) in which a point light source located at position x_0 produces a plane reference beam. Within the PR crystal, this plane wave can be described by vector $\mathbf{k}_0 = |\mathbf{k}|\cos\alpha\hat{\mathbf{z}} - |\mathbf{k}|\sin\alpha\hat{\mathbf{u}}$, where α is the intersection angle between wave vector $\mathbf{k}_0$ and the optical axis inside the crystal, and $\hat{\mathbf{u}}$ and $\hat{\mathbf{z}}$ are the transversal and the longitudinal unit wave vectors, respectively. By referring to the well-known *Snell's law*, we have $\sin\alpha = (\sin\theta)/n$, and

$$\cos\alpha = (1 - \sin^2\alpha)^{1/2} = \left(1 - \frac{\sin^2\theta}{n^2}\right)^{1/2},$$

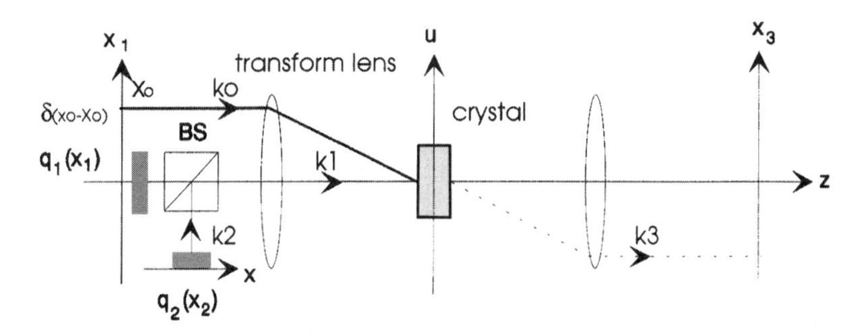

Fig. 2.41. A PR–based VLC.

where θ is the intersection angle outside the crystal, and n is the refractive index of the crystal. Under the paraxial approximation, and $\alpha \ll 1$, wave vector $\mathbf{k}_0$ can be shown as

$$\mathbf{k}_0 = \frac{x_0}{\lambda f}\hat{\mathbf{u}} + \frac{n}{\lambda}\left(1 - \frac{x_0^2}{2n^2 f^2}\right)\hat{\mathbf{z}},$$

where λ is the wavelength of the light source, and f is the focal length of the lens. Similarly, the recording object wave vector $\mathbf{k}_1$, reading wave vector $\mathbf{k}_2$, and diffracted wave vector $\mathbf{k}_3$ can be shown as

$$\mathbf{k}_1 = -\frac{x_1}{\lambda f}\hat{\mathbf{u}} + \frac{n}{\lambda}\left(1 - \frac{x_1^2}{2n^2 f^2}\right)\hat{\mathbf{z}},$$

$$\mathbf{k}_2 = -\frac{x_2}{\lambda f}\hat{\mathbf{u}} + \frac{n}{\lambda}\left(1 - \frac{x_2^2}{2n^2 f^2}\right)\hat{\mathbf{z}},$$

$$\mathbf{k}_3 = -\frac{x_3}{\lambda f}\hat{\mathbf{u}} + \frac{n}{\lambda}\left(1 - \frac{x_3^2}{2n^2 f^2}\right)\hat{\mathbf{z}}.$$

With reference to momentum conservation and the infinite extension in the u direction, we have $\Delta k_u = 0$. By substituting preceding wave vectors into Eq. (2.106) and using the condition $\Delta k_u = 0$, the *dephasing wave vector* for the *transversal* and the *longitudinal directions* can be shown to be

$$\Delta k_u = x_1 - x_0 + x_3 - x_2 = 0, \tag{2.113}$$

$$\Delta k_z = \frac{1}{2n\lambda f^2}(x_1^2 - x_0^2 + x_3^2 - x_2^2) = \frac{1}{n\lambda f^2}(x_1 - x_0)(x_0 - x_3). \tag{2.114}$$

If readout beam $\mathbf{k}_2$ is shifted; that is, $x_2 = x_1 - S$, then we have

$$\Delta k_z = \frac{1}{n\lambda f^2}(x_1 - x_0)S. \tag{2.115}$$

At the input plane x_1 the reference and object beams are represented by $q_0(x_0) = \delta(x_0 - X_0)$ and $q_1(x_1)$, respectively. If the crystal filter is read out by a shifted object, represented by $q_2(x_2) = q_1(x_1 - S)$, the output correlation peak intensity can be calculated by Eq. (2.107) over all $\mathbf{k}_0$ and $\mathbf{k}_1$ with weighting factors $q_0(x_0)$ and $q_1(x_1)$, as given by

$$I(x_3) = \left| \int_{-\infty}^{\infty} dx_0 \int_{-\infty}^{\infty} dx_1 \int_{-\infty}^{\infty} du \int_{-d/2}^{d/2} dz \cdot \delta(x_0 - X_0) q_1^*(x_1 - S) q_1(x_1) \exp(i\Delta\mathbf{k}\cdot\mathbf{r}) \right|^2. \tag{2.116}$$

By substituting Eq. (2.115) into Eq. (2.116), the output correlation peak intensity as a function of the shift distance can be shown to be

$$R(S) = \left| \int |q_1(x_1)|^2 \operatorname{sinc}\left(\frac{\pi d}{n\lambda} \frac{S(x_1 - X_0)}{f^2} \right) dx_1 \right|^2. \tag{2.117}$$

Thus, we see that the intensity is modulated by a broad sinc factor with a width equal to

$$W = \frac{2n\lambda f^2}{Sd}.$$

In order to keep the target object within this width, the sinc factor has to be sufficiently broad compared with the width of the target and the location of the input reference point source; that is,

$$\frac{W}{2} \geqslant X_0 + \frac{X}{2},$$

where X is the width of the input target. In order to avoid an overlapping situation, the location of this point source should be $X_0 \geqslant X5/2$, by which we have $W \geqslant 6X$. It follows that the object shift constraint is

$$S \leqslant \frac{n\lambda f^2}{3Xd}, \tag{2.118}$$

which is inversely proportional to the thickness of the PR filter. In other words, the thinner the PR filter, the higher the shift tolerance will be. However, the thinner the PR filter, the lower will be the storage capacity. If the product of the target width and the maximum permissible shift is defined as the figure of merit (FOM) for the shift tolerance, the FOM for the VLC can be shown to be

$$\mathrm{FOM}_{\mathrm{VLC}} = XS_{\max} = \frac{n\lambda}{3d} f^2, \tag{2.119}$$

which is inversely proportional to the thickness of the PR filter, and proportional to λ and the square of the focal length.

Let us now consider a PR–crystal–based JTC as shown in Fig. 2.42, where q_0 and q_1 are the reference and input targets. The corresponding Bragg

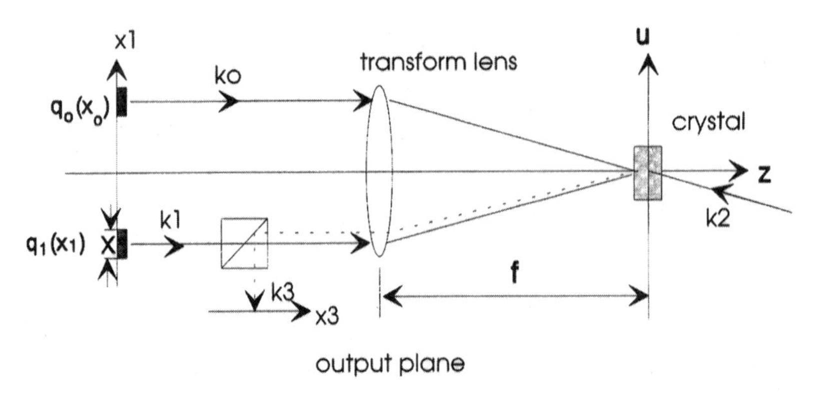

Fig. 2.42. A PR–based JTC.

diffraction wave vectors can be written as

$$\mathbf{k}_0 = -\frac{x_0}{\lambda f}\hat{\mathbf{u}} + \frac{n}{\lambda}\left(1 - \frac{x_0^2}{2n^2 f^2}\right)\hat{\mathbf{z}},$$

$$\mathbf{k}_1 = -\frac{x_1}{\lambda f}\hat{\mathbf{u}} + \frac{n}{\lambda}\left(1 - \frac{x_1^2}{2n^2 f^2}\right)\hat{\mathbf{z}},$$

$$\mathbf{k}_2 = -\frac{x_2}{\lambda f}\hat{\mathbf{u}} + \frac{n}{\lambda}\left(1 - \frac{x_2^2}{2n^2 f^2}\right)\hat{\mathbf{z}},$$

$$\mathbf{k}_3 = -\frac{x_3}{\lambda f}\hat{\mathbf{u}} + \frac{n}{\lambda}\left(1 - \frac{x_3^2}{2n^2 f^2}\right)\hat{\mathbf{z}}.$$

Similarly, one can show that the output correlation peak intensity is given by

$$R(S) = \left|\int |q_1(x_0 - h)|^2 \operatorname{sinc}\left[\frac{\pi d}{n\lambda}\frac{(x_0 - h)(2h - S)}{f^2}\right] dx_0\right|^2. \qquad (2.120)$$

Again, we see that the peak intensity is modulated by a broad sinc factor due to the Bragg diffraction limitation. To keep the target within the width of the sinc factor, W must be broader than the width of the input targets, i.e.,

$$W = \frac{2n\lambda f^2}{(2h - S)d} \geqslant X.$$

For JTC the maximum distance $2h$ between the two input objects can be determined by setting $W = X$ and $S = 0$, for which we have

$$2h_{\max} = \frac{2n\lambda f^2}{Xd}.$$

When the readout beam is shifted closer to the input object by a distance S, the distance between the zeroth-order and the first-order diffraction will be $2h - S$. To avoid overlapping between the two diffraction orders, the separation between them should be greater than twice the target X; that is,

$$2h - S \geqslant 2X, \tag{2.121}$$

for which the maximum permissible target translation is

$$S_{\max} = \frac{2n\lambda f^2}{Xd} - 2X. \tag{2.122}$$

The shift-tolerance FOM for the JTC can be written as

$$\mathrm{FOM}_{\mathrm{JTC}} = XS_{\max} = \frac{2n\lambda f^2}{d} - 2X^2, \tag{2.123}$$

which is inversely proportional to the thickness of the PR crystal and decreases rapidly as the width of the target X increases.

Similarly, for the RC correlation shown in Fig. 2.43, the Bragg diffraction wave vectors can be written as

$$\mathbf{k}_0 = -\frac{n}{\lambda}\hat{\mathbf{z}},$$

$$\mathbf{k}_1 = -\frac{x_1}{\lambda f}\hat{\mathbf{u}} + \frac{n}{\lambda}\left(1 - \frac{x_1^2}{2n^2f^2}\right)\hat{\mathbf{z}},$$

$$\mathbf{k}_2 = -\frac{x_2}{\lambda f}\hat{\mathbf{u}} + \frac{n}{\lambda}\left(1 - \frac{x_2^2}{2n^2f^2}\right)\hat{\mathbf{z}},$$

$$\mathbf{k}_3 = \frac{x_3}{\lambda f}\hat{\mathbf{u}} - \frac{n}{\lambda}\left(1 - \frac{x_3^2}{2n^2f^2}\right)\hat{\mathbf{z}},$$

where $\mathbf{k}_0$ and $\mathbf{k}_1$ are the writing vectors, and $\mathbf{k}_2$ and $\mathbf{k}_3$ are the reading and the diffracted wave vectors, respectively. With reference to momentum conserva-

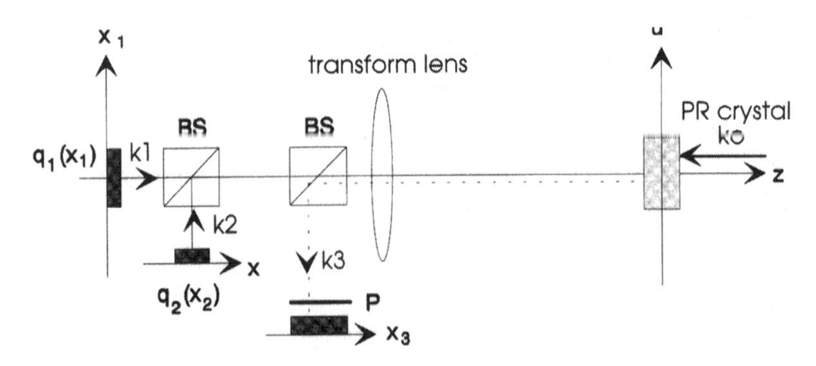

Fig. 2.43. A PR–based reflection-type wavelength-multiplexed correlator (RC).

tion, the transversal and longitudinal dephasing wave vectors can be written as

$$\Delta k_u = \frac{x_1}{\lambda} - \frac{x_2 + x_3}{\lambda} = 0, \qquad \Delta k_z = \frac{1}{n\lambda} \frac{x_3(x_1 - x_3)}{f^2}.$$

Similarly, the correlation peak intensity as a function of the shift distance S can be shown to be

$$R(S) = \left| \int |q_1(x_1)|^2 \operatorname{sinc}\left[\frac{\pi d}{n\lambda} - \frac{S(x_1 - S)}{f^2} \right] dx_1 \right|^2. \tag{2.124}$$

Again, to maintain the target within the sinc factor, we have

$$\frac{W}{2} \geqslant S + \frac{X}{2},$$

for which we have

$$SX \leqslant \frac{2n\lambda}{d} f^2 - 2S^2. \tag{2.125}$$

Since the shift variable S can be positive or negative, S should be maintained within half the maximum permissible translation $S_{\max}$. Thus, by substituting $S = S_{\max}/2$, we obtain

$$\mathrm{FOM}_{\mathrm{RC}} = XS_{\max} = \frac{4n\lambda}{d} f^2 - S_{\max}^2, \tag{2.126}$$

which is inversely proportional to the thickness of the PR filter and decreases as the maximum permissible shift distance increases.

In view of the preceding results, for moderate width of the input target, we see that RC performs better in terms of shift tolerance than both VLC and JTC.

2.8. PROCESSING WITH INCOHERENT LIGHT

The use of a coherent light enables the use of optical processing to carry out complex amplitude processing, which offers a myriad of applications. However, coherent processing also suffers from coherent artifact noise, which limits its processing capabilities. To alleviate these limitations, we discuss methods to exploit the coherence contents from an incoherent source for complex amplitude processing. Since all physical sources are neither strictly coherent nor strictly incoherent, it is possible to extract their inherent coherence contents for coherent processing.

2.8.1. EXPLOITATION OF COHERENCE

Let us begin with the exploitation of spatial coherence from an extended incoherent source. By referring to the conventional optical processor shown in Fig. 2.44, the *spatial coherence* function at the input plane can be written as

$$\Gamma(x_2 - x_2') = \iint \gamma(x_1) \exp\left[i2\pi \frac{x_1}{\lambda f}(x_2 - x_2')\right] dx_1, \tag{2.127}$$

which is the well-known *Van Citter-Zernike theorem*, where $\gamma(x_1)$ is the extended source, f is the focal length of the collimated lens, and λ is the wave-

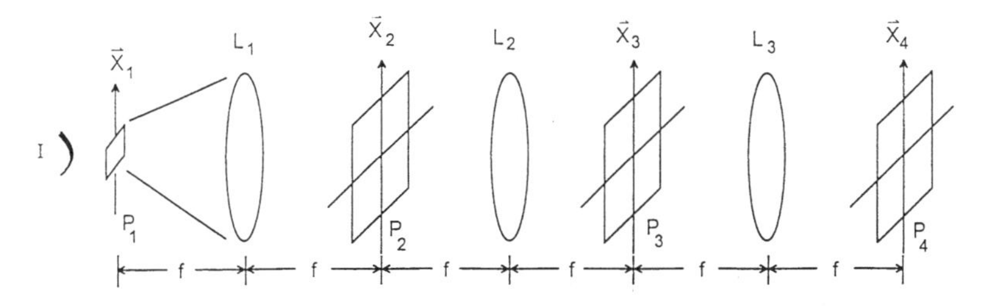

Fig. 2.44. Incoherent source optical processor: I_1, Incoherent source; L_1, collimating lens; L_2 and L_3, achromatic transformation lenses; P_1, source encoding mask; P_2, input plane; P_3, Fourier plane; and P_4, output plane.

length of the extended source. Thus, we see that the spatial coherence at the input plane and the source-encoding intensity transmittance form a *Fourier transform pair*, as given by

$$\gamma(x_1) = \mathscr{F}[\Gamma(x_2 - x_2')], \qquad \text{and} \qquad \Gamma(x_2 - x_2') = \mathscr{F}^{-1}[\gamma(x_1)], \quad (2.128)$$

where $\mathscr{F}$ denotes the Fourier transform operation. In other words, if a specific spatial coherence requirement is needed for certain information processing, a *source encoding* can be performed. The source-encoding $\gamma(x_1)$ can consist of apertures of different shapes or slits, but it should be a positive real function that satisfies the following *physically realizable constraint*:

$$0 \leqslant \gamma(x_1) \leqslant 1. \quad (2.129)$$

For the exploitation of *temporal coherence*, note that the Fourier spectrum is linearly proportional to the wavelength of the light source. It is apparently not capable of (or is inefficient at) using a broadband source for complex amplitude processing. To do so, a narrow-spectral–band (i.e., temporally coherent) source is needed. In other words, the spectral spread of the input object should be confined within a small fraction fringe spacing of the spatial filter, which is given by

$$\frac{p_m f \Delta\lambda}{2\pi} \ll d, \quad (2.130)$$

where d is the fringe spacing of the spatial filter, p_m is the upper angular spatial frequency content of the input object, f is the focal length of the transform lens, and $\Delta\lambda$ is the spectral bandwidth of the light source. In order to have a higher temporal coherence requirement, the spectral width of the light source should satisfy the following constraint:

$$\frac{\Delta\lambda}{\lambda} \ll \frac{\pi}{hp_m}, \quad (2.131)$$

where λ is the center wavelength of the light source, $2h$ is the size of the input object transparency, and $2h = (\lambda f)/d$.

There are ways to exploit the temporal coherence content from a broadband source. One of the simplest methods is by dispersing the Fourier spectrum, which can be obtained by placing a sampling grating at the input domain. For example, if the input object is sampled by a phase grating as given by

$$f(x_2)T(x_2) = f(x_2)\exp(ip_0 x_2),$$

then the corresponding Fourier transform would be

$$F(p, q) = F\left(x_3 - \frac{\lambda f}{2\pi} p_0\right),$$

in which we see that $F(p, q)$ is smeared into rainbow colors at the Fourier domain. Thus, a high temporal coherence Fourier spectrum within a narrow-spectral–band filter can be obtained, as given by

$$\frac{\Delta\lambda}{\lambda} = \frac{4p_m}{p_0} \ll 1. \tag{2.132}$$

Since the spectral content of the input object is dispersed in rainbow colors, it is possible to synthesize a set of narrow-spectral–band filters to accommodate the dispersion, as illustrated in Fig. 2.45a.

On the other hand, if the spatial filtering is a 1-D operation, it is possible to construct a fan-shaped spatial filter to cover the entire smeared Fourier spectrum, as illustrated in Fig. 2.45b. Thus, we see that a high degree of temporally coherent filtering can be carried out by a simple white light source. Needless to say, the (broadband) spatial filters can be synthesized by computer-generated techniques.

In the preceding we have shown that spatial and temporal coherence can be exploited by spatial encoding and spectral dispersion of an incoherent source. We have shown that complex amplitude processing can be carried out with either a set of 2-D narrow-spectral–band filters or with a 1-D fan-shaped broadband filter.

Let us first consider that a set of narrow-spectral–band filters is being used, as given by

$$H_n(p_n, q_n; \lambda_n), \qquad \text{for } n = 1\,2, \ldots, N,$$

where (p_n, q_n) represents the angular frequency coordinates and λ_n is the center wavelength of the narrow-width filter. It can then be shown that the output light intensity would be the incoherent superposition of the filtered signals, as given by

$$I(x, y) \cong \sum_{n=1}^{N} \Delta\lambda_n |f(x, y; \lambda_n) * h(x, y; \lambda_n)|^2, \tag{2.133}$$

where * denotes the convolution operation, $f(x, y; \lambda_n)$ represents the input signal illuminated by λ_n, $\Delta\lambda_n$ is the narrow spectral width of the narrow-

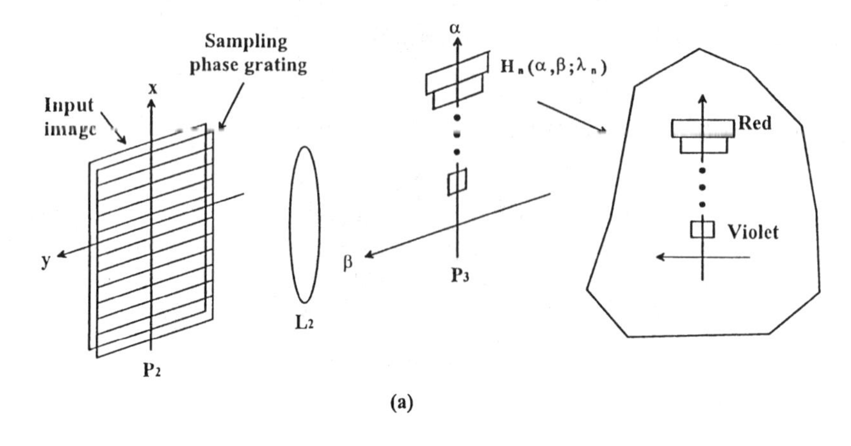

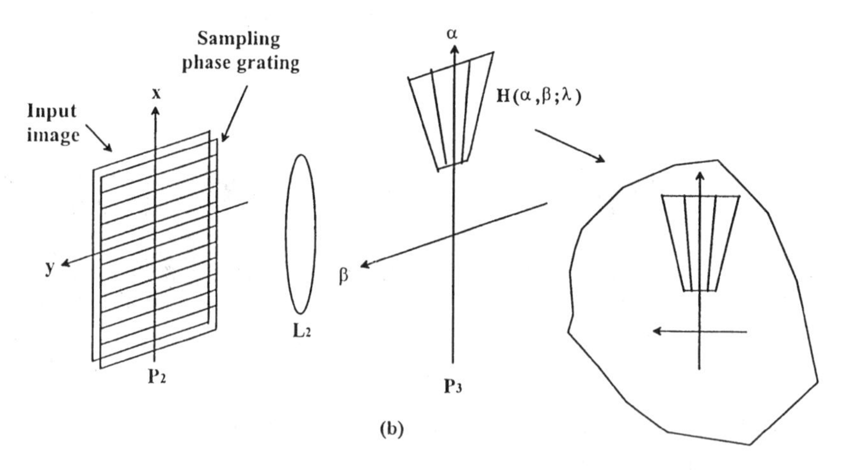

Fig. 2.45. Broad spectral filtering. (a) Using a set of narrow–spectral-band filters, (b) Using a 1-D fan-shaped broadband filter.

spectral–band filter, and $h(x, y; \lambda_n)$ is the spatial impulse response of $H_n(p_n, q_n; \lambda_n)$; that is,

$$h_n(p_n, q_n; \lambda_n) = \mathscr{F}^{-1}[H(p_n q_n; \lambda_n)].$$

Thus, we see that by exploiting the spatial and the temporal coherence content, an incoherent source processor can be made to process the information in complex amplitude as a coherent processor. Since the output intensity distribution is the sum of mutually incoherent image irradiances, the annoying coherent artifact noise can be avoided.

On the other hand, if the signal processing is a 1-D operation, the information processing can be carried out with a 1-D fan-shaped broadband filter. The output intensity distribution can be shown as

$$I(x, y) = \int_{\Delta\lambda} |f(x, y; \lambda) * h(x, y; \lambda_n)|^2 \, d\lambda, \tag{2.134}$$

where the integral is over the entire spectral band of the light source. Again, we see that the output irradiance is essentially obtained by incoherent superposition of the entire spectral-band–image irradiances, by which the coherent artifact noise can be avoided. Since one can utilize a conventional white light source, the processor can indeed be used to process polychromatic images. The advantages of exploiting the incoherent source for coherent processing are that it enables the information to be processed in complex amplitude as a coherent processor, and it is capable of suppressing the coherent artifact noise as an incoherent processor.

2.8.2. SIGNAL PROCESSING WITH WHITE LIGHT

One interesting application of coherent optical processing is the restoration of blurred images, as described in Sec. 2.5.2, by means of inverse filtering. Deblurring can also be obtained by a *white light processor*, as described in the preceding section.

Since smeared image deblurring is a 1-D processing operation, inverse filtering takes place with respect to the smeared length of the blurred object. Thus, the required spatial coherence depends on the smeared length instead of the entire input object. If we assume that a spatial coherence function is given by

$$\Gamma(x_2 - x_2') = \operatorname{sinc}\left\{\frac{\pi}{\Delta x_2}(x_2 - x_2')\right\},$$

as shown in Fig. 2.46a, the source-encoding function can be shown as

$$\gamma(x_1) = \operatorname{rect}\left(\frac{x_1}{w}\right),$$

where Δx_2 is the smeared length, $w = (f\lambda)/(\Delta x_2)$ is the slit width of the encoding aperture (as shown in Fig. 2.46b, and

$$\operatorname{rect}\left(\frac{x_1}{w}\right) = \begin{cases} 1, & -\frac{w}{3} \leqslant x_1 \frac{w}{2} \\ 0, & \text{otherwise} \end{cases}$$

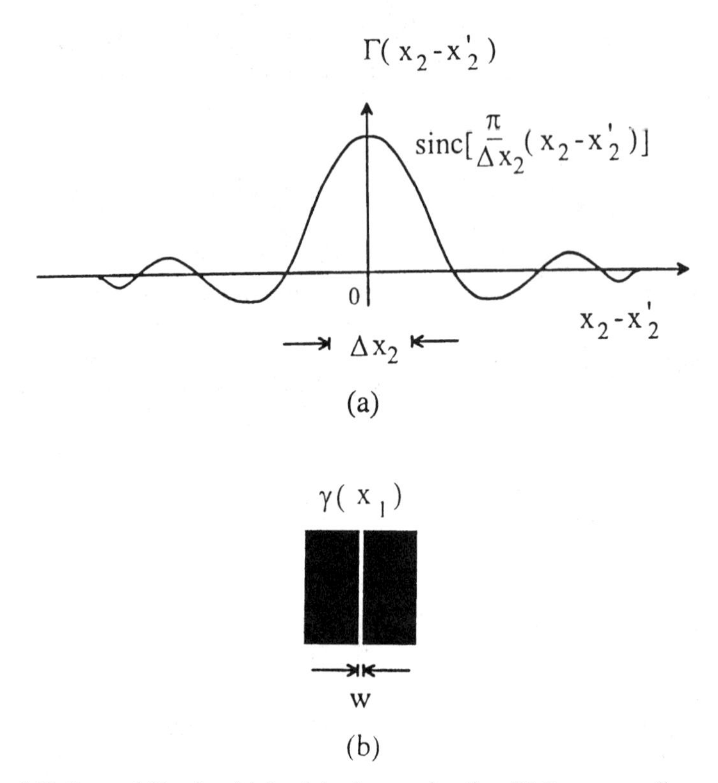

Fig. 2.46. Image deblurring. (a) Spatial coherence function, (b) Source encoding mask.

As for the temporal coherence requirement, a sampling phase grating is used to disperse the Fourier spectrum in the Fourier plane. Let us consider the temporal coherence requirement for a 2-D image in the Fourier domain. A high degree of temporal coherence can be achieved by using a higher sampling frequency. We assume that the Fourier spectrum dispersion is along the y axis. Since the smeared image deblurring is a 1-D processing, a *fan-shaped* broadband spatial filter to accommodate the smeared Fourier spectrum can be utilized. Therefore, the sampling frequency of the input phase grating can be determined by

$$p_0 \geqslant \frac{4\lambda p_m}{\Delta\lambda},$$

in which λ and $\Delta\lambda$ are the center wavelength and the spectral bandwidth of the light source, respectively. p_m is the y-axis spatial frequency limit of the blurred image.

(a)

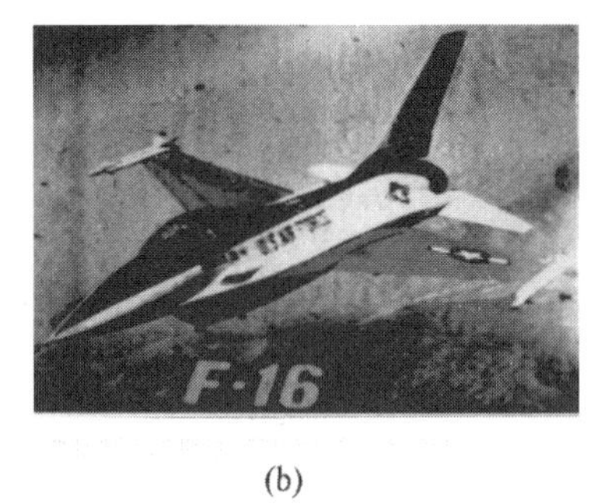

(b)

Fig. 2.47. Restoration of blurred image. (a) A black-and-white blurred (color) image, (b) Deblurred (color) image.

Figure 2.47a shows a blurred color image due to linear motion. By inserting this blurred transparency in the processor of Fig. 2.44 a deblurred color image is obtained, as shown in Fig. 2.47b. Thus, we see that by properly exploiting the coherence contents, complex amplitude processing can be obtained from an incoherent source. Since the deblurred image is obtained by incoherent integration (or superposition) of the broadband source, the coherent artifact can be suppressed. In addition, by using white light illumination, the polychromatic content of the image can also be exploited, as shown in this example.

Let us provide another example: an image subtraction with white light processing. Since the spatial coherence depends on the corresponding point pair of the images to be subtracted, a strictly broad spatial coherence function is not required. Instead, a point-pair spatial coherence function is actually needed. To ensure the physical reliability of the source-encoding function, we let the point-pair spatial coherence function be

$$\Gamma(x_2 - x_2') = \frac{\sin[N(\pi/h)(x_2 - x_2')]}{N\sin[(\pi/h)(x_2 - x_2')]}\operatorname{sinc}\left[\frac{\pi w}{hd}(x_2 - x_2')\right],$$

where $2h$ is the main separation of the two input image transparencies. $N \gg 1$ and $w \ll d$, Γ converge to a sequence of narrow pulses located at $(x_2 - x_2') = nh$, as shown in Fig. 2.48a. Thus, a high degree of coherence among the corresponding point pair can be obtained. By Fourier transforming the preceding equation the source-encoding function can be shown as

$$\lambda(x_1) = \sum_{n=1}^{N} \operatorname{rect}\left(\frac{x_1 - nd}{w}\right),$$

where w is the slit width, and $d = (\lambda f)/h$ is the separation between the slits. The

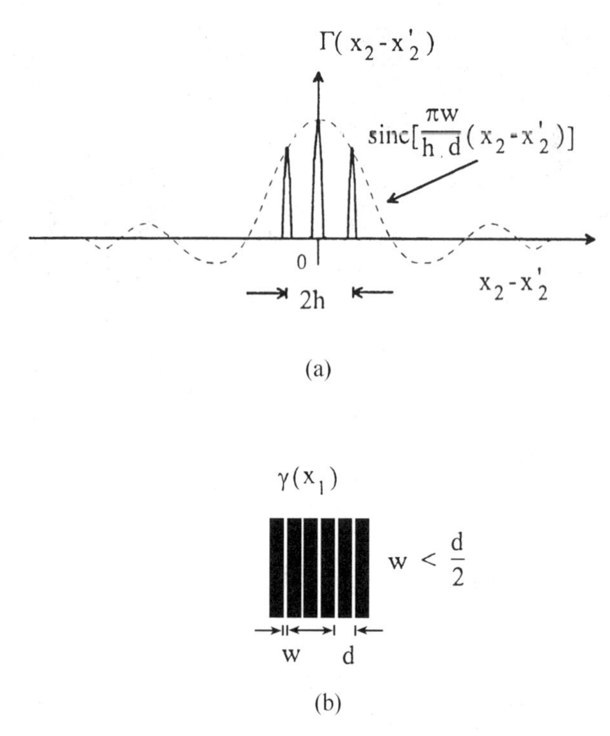

Fig. 2.48. Image substration. (a) Point-pair coherence function, (b) Source encoding.

source-encoding mask is, in fact, represented by N equally spaced narrow slits, as shown in Fig. 2.48b.

Since the image subtraction is a 1-D processing operation, the Fourier domain filter should be a *fan-shaped* broadband sinusoidal grating, as given by

$$G = \frac{1}{2}\left[1 + \sin\left(\frac{2\pi xh}{\lambda f}\right)\right].$$

Figure 2.49a shows two input color images, and the output subtracted image as obtained by the preceding method is shown in Fig. 2.49b. Again, we see that the coherent artifact noise has been suppressed and the polychromatic content of the subtracted image is exploited.

2.8.3. COLOR IMAGE PRESERVATION AND PSEUDOCOLORING

One interesting application of white light processing is color image preservation. By spatial sampling, a color image can be encoded in a black-and-white

(a)

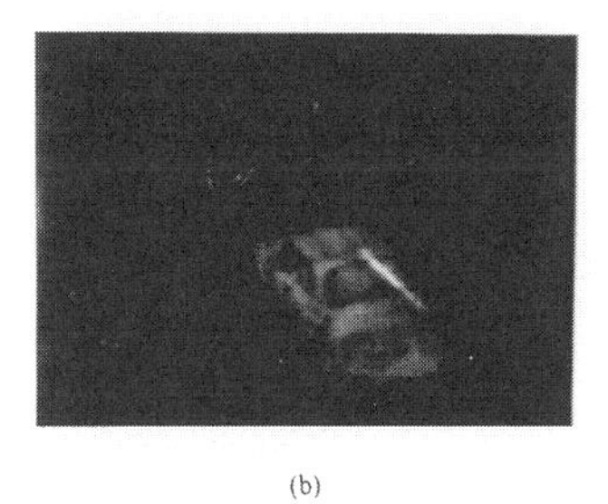

(b)

Fig. 2.49. Color image subtration. (a) Input (color) images, (b) Subtrated (color) image.

transparency. To avoid the output *moire* fringe pattern, we sample the primary color images in orthogonal directions as given by

$$T(x, y) = K\{T_r(x, y)[1 + \operatorname{sgn}(\cos \omega_r y)] + T_b(x, y)[1 + \operatorname{sgn}(\cos \omega_b x)] \\ + T_g(x, y)[1 + \operatorname{sgn}(\cos \omega_g x)]\}^{-\gamma}$$

where K is a proportionality constant; T_r, T_b, and T_g are the red, blue, and green color image exposures; ω_r, ω_b, and ω_g are the respective carrier spatial frequencies; γ is the film gamma; and

$$\operatorname{sgn}(\cos x) \triangleq \begin{cases} 1, & \cos x > 0, \\ 0, & \cos x = 0, \\ -1, & \cos x < 0. \end{cases}$$

Notice that the diffraction efficiency can be improved by bleaching the encoded transparency. Nevertheless, if we insert the encoded transparency at the input domain of a white light processor, the complex light distribution at the Fourier domain can be shown as

$$
\begin{aligned}
S(\alpha, \beta; \lambda) \approx\ & \hat{T}_r\left(\alpha, \beta \pm \frac{\lambda f}{2\pi}\omega_r\right) + \hat{T}_b\left(\alpha \pm \frac{\lambda f}{2\pi}\omega_b, \beta\right) + \hat{T}_g\left(\alpha \pm \frac{\lambda f}{2\pi}\omega_g, \beta\right) \\
& + \hat{T}_r\left(\alpha, \beta \pm \frac{\lambda f}{2\pi}\omega_r\right) * \hat{T}_b\left(\alpha \pm \frac{\lambda f}{2\pi}\omega_b, \beta\right) \\
& + \hat{T}_r\left(\alpha, \beta \pm \frac{\lambda f}{2\pi}\omega_r\right) * \hat{T}_g\left(\alpha \pm \frac{\lambda f}{2\pi}\omega_g, \beta\right) \\
& + \hat{T}_b\left(\alpha \pm \frac{\lambda f}{2\pi}\omega_b, \beta\right) * \hat{T}_g\left(\alpha \pm \frac{\lambda f}{2\pi}\omega_g, \beta\right)
\end{aligned}
$$

where $\hat{T}_r$, $\hat{T}_b$, and $\hat{T}_g$ are the Fourier transforms of T_r, T_b, and T_g, respectively. By proper color filtering of the smeared Fourier spectra, a true color image can be retrieved at the output image plane, as given by

$$
I(x, y) = T_r^2(x, y) + T_b^2(x, y) + T_g^2(x, y), \tag{2.135}
$$

which is a superposition of three primary encoded color images.

Many of the images obtained in various scientific applications are gray-level density images; for example, scanning electron micrographs, multispectral-band aerial photographic images, X-ray transparencies, infrared scanning images, and others. However, humans can perceive details in color better than in gray levels; in other words, a color-coded image can provide better visual discrimination.

We now describe a density pseudocolor encoding technique for monochrome images. We start by assuming that a gray-level transparency (called T_1) is available for pseudocoloring. Using the contact printing process, a negative, and a product (called T_2 and T_3, respectively) image transparency can be made. It is now clear that spatial encoding onto monochrome film can be accomplished by the same method used for color image preservation, for which the encoded transparency is given by

$$
\begin{aligned}
T(x, y) = K\{ & T_1(x, y)[1 + \operatorname{sgn}(\cos \omega_1 y)] + T_2(x, y)[1 + \operatorname{sgn}(\cos \omega_2 x)] \\
& + T_3(x, y)[1 + \operatorname{sgn}(\cos \omega_3 x)]\}^{-\gamma}.
\end{aligned}
$$

Similarly, if the encoded transparency is inserted at the input white light

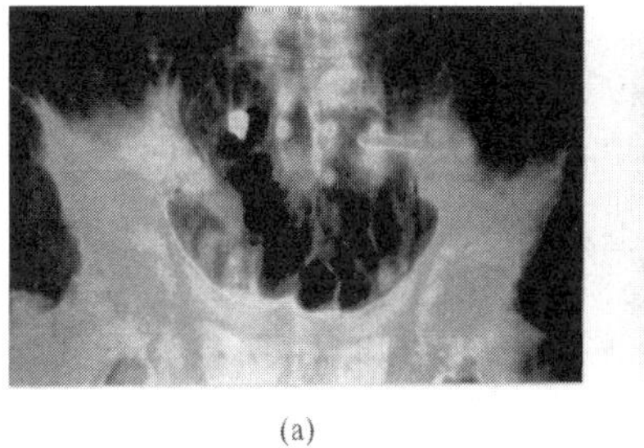
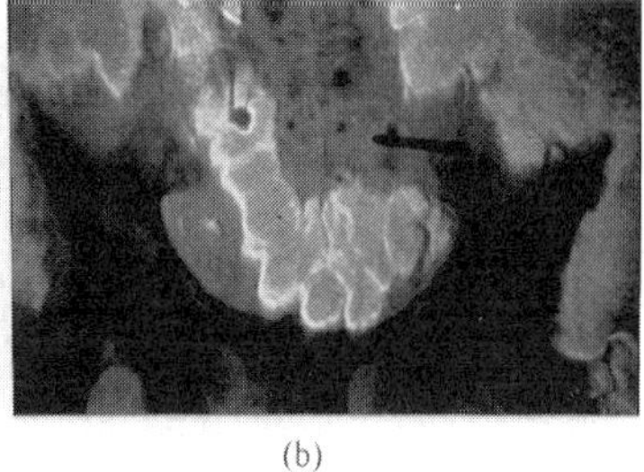

(a) (b)

Fig. 2.50. Pseudocoloring. (a) Black-and-white X-ray image, (b) Density (color)-coded image.

processor, the complex light field at the Fourier domain is

$$\begin{aligned}
S(\alpha, \beta; \lambda) \approx \hat{T}_1\left(\alpha, \beta \pm \frac{\lambda f}{2\pi}\omega_1\right) &+ \hat{T}_2\left(\alpha \pm \frac{\lambda f}{2\pi}\omega_2, \beta\right) + \hat{T}_3\left(\alpha \pm \frac{\lambda f}{2\pi}\omega_3, \beta\right) \\
&+ \hat{T}_1\left(\alpha, \beta \pm \frac{\lambda f}{2\pi}\omega_1\right) * \hat{T}_2\left(\alpha \pm \frac{\lambda f}{2\pi}\omega_2, \beta\right) \\
&+ \hat{T}_1\left(\alpha, \beta \pm \frac{\lambda f}{2\pi}\omega_1\right) * \hat{T}_3\left(\alpha \pm \frac{\lambda f}{2\pi}\omega_3, \beta\right) \\
&+ \hat{T}_2\left(\alpha \pm \frac{\lambda f}{2\pi}\omega_2, \beta\right) * \hat{T}_g\left(\alpha \pm \frac{\lambda f}{2\pi}\omega_3, \beta\right).
\end{aligned}$$

By proper color filtering in the Fourier spectral, a pseudocolor image can be observed at the output plane.

An example of a pseudocolor-coded image is depicted in Fig. 2.50. It is indeed easier and more pleasant to visualize a color-coded image than a black-and-white image.

2.9. PROCESSING WITH NEURAL NETWORKS

Electronic computers can solve classes of computational problems faster and more accurately than the human brain. However, for cognitive tasks, such as pattern recognition, understanding and speaking a language, and so on, the human brain is much more efficient. In fact, these tasks are still beyond the reach of modern electronic computers. A neural network consists of a collection of processing elements called *neurons*. Each neuron has many input signals but only one output signal, which is fanned out to many pathways connected to other neurons. These pathways interconnect with other neurons to form a network called a *neural network*. The operation of a neuron is determined by a transfer function that defines the neuron's output as a function of the input signals. Every connection entering a neuron has an adaptive coefficient called a *weight* assigned to it. The weight determines the interconnection strength among neurons, and they can be changed through a learning rule that modifies the weights in response to input signals. The learning rule allows the response of the neuron to change, depending on the nature of the input signals. This means that the neural network *adapts* itself and organizes the information within itself, which is what we term *learning*.

2.9.1. OPTICAL NEURAL NETWORKS

Roughly speaking, a one-layer neural network of N neurons should have N^2 interconnections. The transfer function of a neuron can be described by a nonlinear relationship such as a step function, making the output of a neuron either zero or one (binary), or a sigmoid function, which gives rise to analog values. The state of the ith neuron in the network can be represented by a *retrieval equation*, as given by

$$u_i = f\left\{\sum_{j=1}^{N} T_{ij}u_j - \theta_i\right\},$$

where u_i is the activation potential of the ith neuron, T_{ij} is the *interconnection weight matrix* (IWM) or *associative memory* between the jth neuron and the ith neuron, θ_i is a phase bias, and f is a nonlinear processing operator. In view of the summation within the retrieval equation, it is essentially a *matrix-vector outer-product* operation, which can be optically implemented.

Light beams propagating in space will not interfere with each other, and optical systems have large space–bandwidth products. These are the traits of optics that prompted the optical implementation of neural networks (NNs). An optical NN using a liquid-crystal TV (LCTV) SLM is shown in Fig. 2.51, in which a lenslet array is used for the interconnection between the IWM and the input pattern. The transmitted light field after LCTV2 is collected by an imaging lens, focusing at the lenslet array and imaging onto a CCD array detector. The array of detected signals is sent to a thresholding circuit and the final pattern can be viewed at the monitor, and it can be sent back for the next iteration. The data flow is primarily controlled by the microcomputer, such that this hybrid optical neural network is indeed an *adaptive* processor.

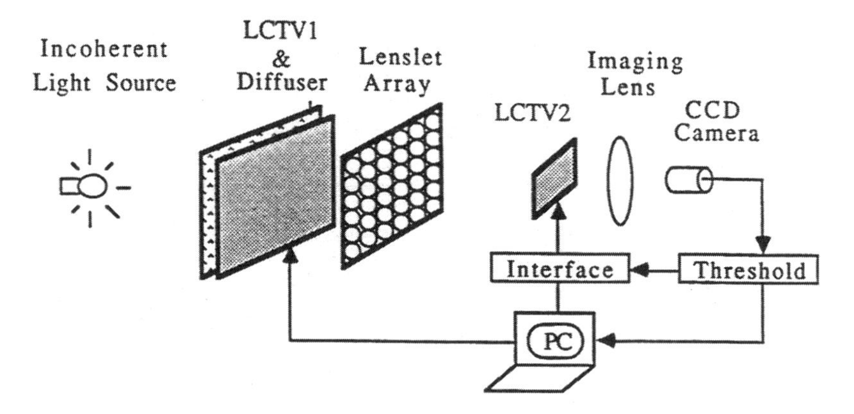

Fig. 2.51. A hybrid optical neural network.

2.9.2. HOLPFIELD MODEL

One of the most frequently used neural network models is the *Hopfield* model, which allows the desired output pattern to be retrieved from a distorted or partial input pattern. The model utilizes an associative memory retrieval process equivalent to an iterative thresholded matrix-vector outer-product expression, as given by

$$V_i = \begin{cases} 1, & V_i \to 1, \quad \sum_{j=1}^{N} T_{ij} V_j \geqslant 0 \\ 0, & V_i \to 0, \quad < 0 \end{cases} \tag{2.136}$$

where V_i and V_j are binary output and binary input patterns, respectively, and the *associative memory matrix* is written as

$$T_{ij} = \begin{cases} \sum_{m=1}^{N} (2V_i^m - 1)(2V_j^m - 1), & i \neq j \\ 0, & i = j \end{cases} \tag{2.137}$$

where V_i^m and V_j^m are ith and jth elements of the mth binary vectory.

The Hopfield model depends on the outer-product operation for constructing the associated memory, which severely limits the storage capacity and often causes failure in retrieving similar patterns. To overcome these shortcomings, neural network models, such as back propagation, orthogonal projection, and others have been used. One of the important aspects of neural computing is

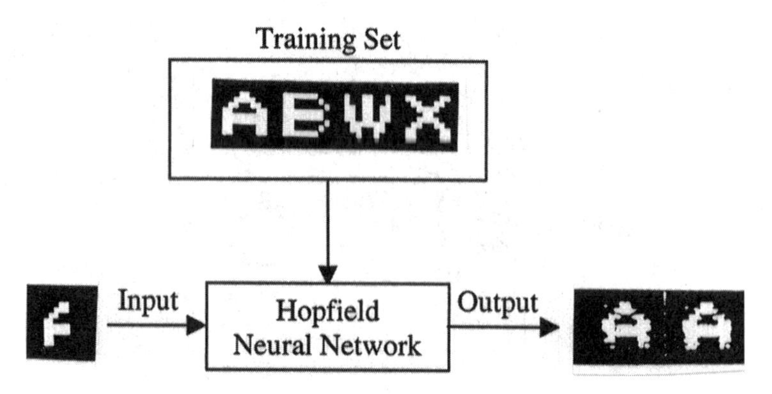

Fig. 2.52. A block box representation of a Hopfield NN.

the ability to retrieve distorted and partial inputs. For example, a partial image of A is presented to the Hopfield NN, as shown in Fig. 2.52. By repeated iteration, we see that a recovered A is converged at the output end.

2.9.3. INTERPATTERN ASSOCIATION MODEL

Although the Hopfield neural network can retrieve erroneous or partial patterns, the construction of the Hopfield neural network is through *intrapattern association*, which ignores the association among the stored exemplars. In other words, Hopfield would have a limited storage capacity and it is not effective or even capable of retrieving similar patterns. One of the alternative approaches is called *interpattern association* (IPA) neural network. An example illustrating the IPA relationship is shown in Fig. 2.53, in which we assume that Tony and George are identical twins. Nevertheless, we can identify them quite easily by their special features: hair and mustache. Therefore, it is trivial by using a simple logic operation that an IPA neural network can be constructed.

For simplicity, let us consider a three-overlapping pattern as sketched in Fig. 2.54, where the common and the special subspaces can be defined. If one uses the following logic operations, an IPA neural network can be constructed:

$$I = A \wedge \overline{(B \vee C)}, \qquad II = B \wedge \overline{(A \vee C)}, \qquad III = C \wedge \overline{(A \vee B)}$$

$$IV = (A \wedge B) \wedge \overline{C}, \qquad V = (B \wedge C) \wedge \overline{A}, \tag{2.138}$$

$$VI = (C \wedge A) \wedge \overline{B}, \qquad VII = (A \wedge B \wedge C) \wedge \varnothing$$

where $\wedge$, $\vee$, and $\overline{}$ stand for AND, OR, and NOT logic operation, and $\varnothing$ denotes the empty set.

Fig. 2.53. Concept of interpattern association (IPA).

If the interconnection weights are assigned equal to 1, -1, and zero, for excitory, inhibitory, and null interconnections, then a tristate IPA neural net can be constructed. For instance, in Fig. 2.55a, pixel one is the common pixel among patterns A, B, and C, pixel two is the common pixel between A and B, pixel three is the common pixel between A and C, and pixel four is a special pixel, which is also an exclusive pixel with respect to pixel two. Thus, by applying the preceding logic operations, a tristate *neural network* can be constructed as shown in Fig. 2.55b, and the corresponding IPA–IWM is shown in Fig. 2.55c.

For comparison of the IPA and Hopfield models, we have used an 8×8 neuron NN for the tests. The training set is the 26 English letters lined up in sequence based on their similarities. Figure 2.56 shows the error rate as a function of stored letters. We see that the Hopfield model becomes unstable to

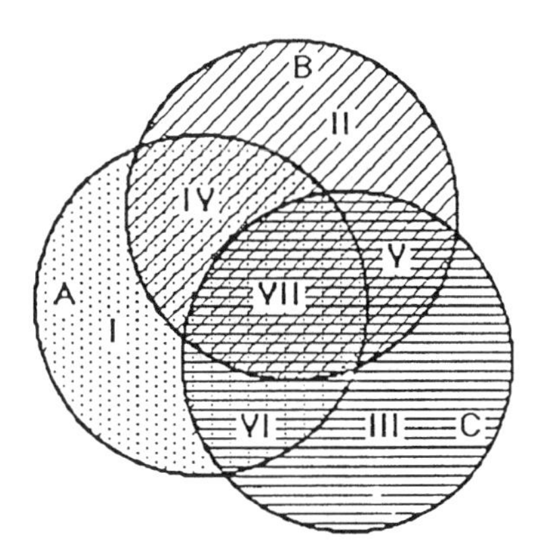

Fig. 2.54. Common and special subspaces.

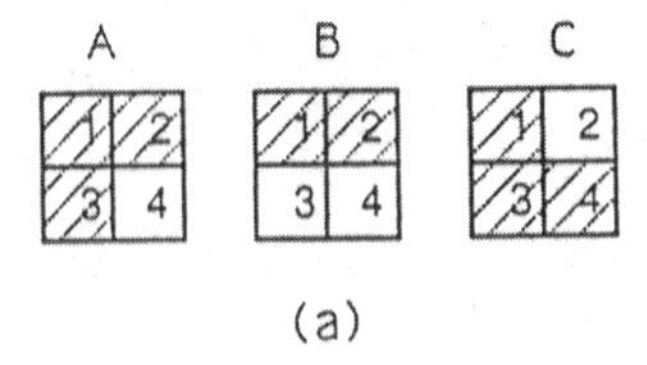

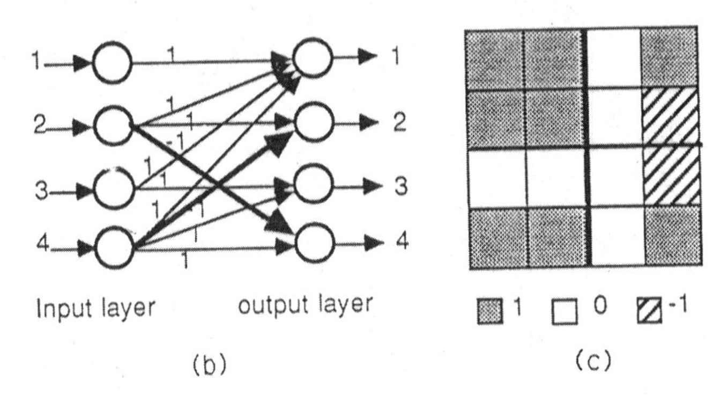

Fig. 2.55. Construction of IPA NN. (a) Three reference patterns, (b) One-layer NN, (c) Associative memory.

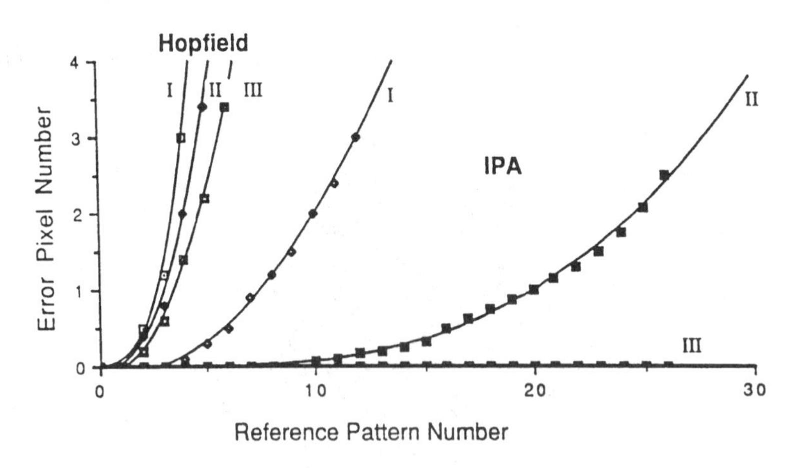

Fig. 2.56. Comparison of the IPA and the Hopfield models. Training sets; 26 capital English alphabets; 8×8 NN.

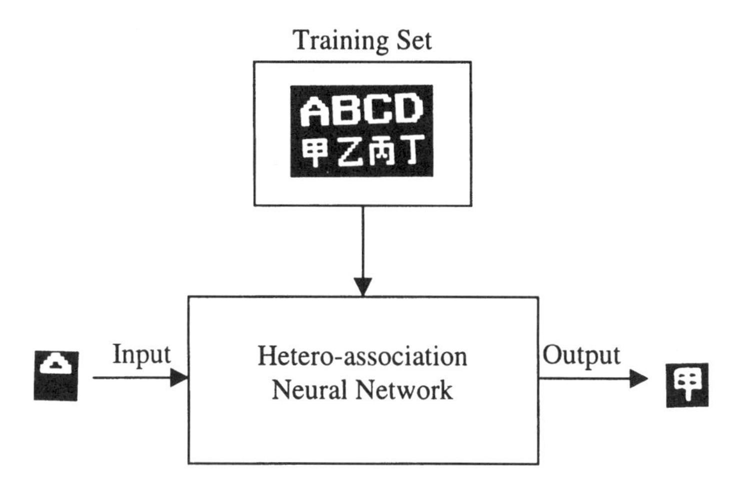

Fig. 2.57. A block box representation of a heteroassociation NN.

about 4 stored letters, whereas the IPA model is quite stable for all 26 letters. Even for a 10% input noise level, it can retrieve effectively up to 12 stored letters. As for noiseless input, the IPA model can, in fact, produce correct results for all 26 stored letters.

Furthermore, pattern translation can also be accomplished using a hetero-association IPA neural network. Notice that by using similar logic operations among input–output (translation) patterns, a *heteroassociative* IWM can be constructed. An example of the heteroassociation NN is shown in Fig. 2.57. The objective is to translate a set of English alphabets to a set of Chinese characters. We see that if A is presented to the heteroassociation NN, a corresponding Chinese character is converged at the output end.

REFERENCES

1 F. T. S. Yu and S. Jutamulia, *Optical Signal Processing, Computing and Neural Networks*, Wiley-Interscience, New York, (1992).

2. A. VanderLugt, *Signal Detection by Complex Spatial Filtering*, IEEE Trans. Inform. Theory, IT-10, 139 (1965).

3. L. J. Cutrona et al., *On the Application of Coherent Optical Processing Techniques to Synthetic Aperture Radar*, Proc. IEEE, 54, 1026 (1966).

4. D. Casacent and D. Psaltis, *Scale Invariant Optical Correlation Using Mullin Transforms*, Opt. Commun., 17, 59 (1976).

5. C. F. Hester and D. Casacent, *Multivariant Technique for Multiclass Pattern Recognition*, Appl. Opt., 19, 1758 (1980).

6. W. H. Lee, *Sampled Fourier Transform Hologram Generated by Computer*, Appl. Opt. 9, 639 (1970).
7. S. Yin et al., *Design of a Bipolar Composite Filter Using Simulated Annealing Algorithm*, Opt. Lett., 20, 1409 (1996).
8. F. T. S. Yu, *White-Light Optical Processing*, Wiley-Interscience, New York, (1985).
9. P. Yeh, *Introduction to Photorefractive Nonlinear Optics*, Wiley, New York, (1993).
10. A. Yariv and P. Yeh, *Optical Waves in Crystal*, Wiley, New York, (1984).
11. J. J. Hopfield, *Neural Network and Physical System with Emergent Collective Computational Abilities*, in Proc. Natl. Acad. Sci., 79, 2554 (1982).
12. D. Psaltis and N. Farhat, *Optical Information Processing Based on an Associative Memory Model of Neural Nets with Thresholding and Feedback*, Opt. Lett., 10, 98 (1985).
13. F. T. S. Yu and S. Jutamulia, ed., *Optical Pattern Recognition*, Cambridge University Press, Cambridge, UK, (1998).
14. F. T. S. Yu and S. Yin, *Photorefractive Optics*, Academic Press, Boston, (2000).

EXERCISES

2.1 Two mutually coherent beams are impinging on an observation screen. Determine the intensity ratio by which maximum visibility can be observed.

2.2 By referring to Young's experiment, we assume that a 45° angle monochromatic line source of infinite extend is illuminating on a diffraction screen, as shown in Fig. 2.58.

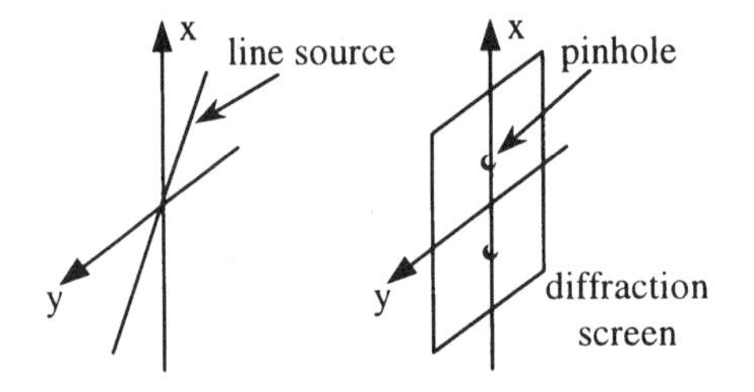

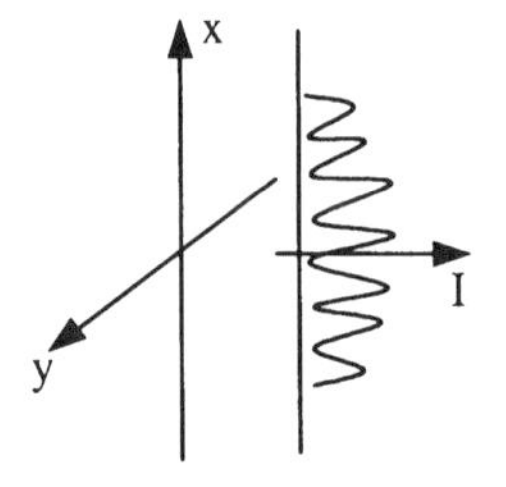

Fig. 2.58.

(a) What would be the degree of spatial coherence on the diffraction screen?

(b) Sketch a 3-D plot of $|\gamma|$ over the $(|x - x'|, |y - y'|)$ plane.

2.3 Consider an electromagnetic wave propagated in space, as shown in Fig. 2.59. Assume that the complex fields of point 1 and point 2 are given by

$$u_1(t) = 3\exp[i\omega t] \qquad \text{and} \qquad u_2(t) = 2\exp[i(\omega t + \varphi)],$$

where ω and φ are the time frequency and the phase factors, respectively.

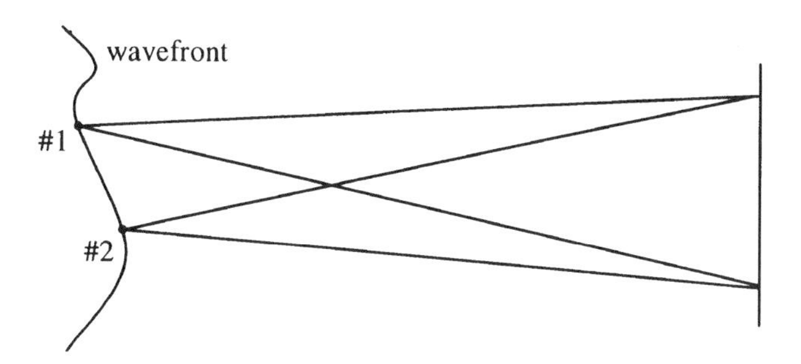

Fig. 2.59.

(a) Calculate the mutual coherence function.
(b) What is the degree of coherence between u_1 and u_2?

2.4 Consider the spatial impulse response of an input–output optical system, as given by

$$h(x, y) = \operatorname{rect}\left(\frac{x}{\Delta x}\right) - \operatorname{rect}\left(\frac{x}{\Delta x}\right).$$

(a) If the optical system is illuminated by a spatially limited incoherent wavefront given by

$$f(x, y) = \operatorname{rect}\left(\frac{x}{\Delta x}\right),$$

calculate the corresponding irradiance at the output plane.
(b) If the spatially limited illumination of the optical system is a coherent wavefront, compare the corresponding complex light field at the output plane.
(c) Sketch the irradiance of parts (a) and (b), respectively, and comment on their major differences.

2.5 A Fourier domain filter is given by

$$H(p, q) = \exp[i(\alpha_0 p + \beta_0 q)],$$

where α_0 and β_0 are arbitrary positive constants. Assume that the complex amplitude transmittance at the input plane of the FDP is

$$f(x, y) = e^{i\phi(x, y)}.$$

Calculate the output responses under incoherent and coherent illuminations, respectively.

2.6 Given an input–output linear spatially invariant optical system under strictly incoherent illumination,
   (a) Derive the incoherent transfer function in terms of the spatial impulse response of the system.
   (b) State some basic properties of the incoherent transfer function.
   (c) What are the cutoff frequencies of the system under coherent and incoherent illuminations?

2.7 Consider an optical imaging system that is capable of combining two Fraunhofer diffractions resulting from input objects $f_1(x, y)$ and $f_2(x, y)$, as shown in Fig. 2.60.
   (a) Compute the output irradiances under the mutual coherent and the mutual incoherent illuminations, respectively.
   (b) Comment on the results of part (a).

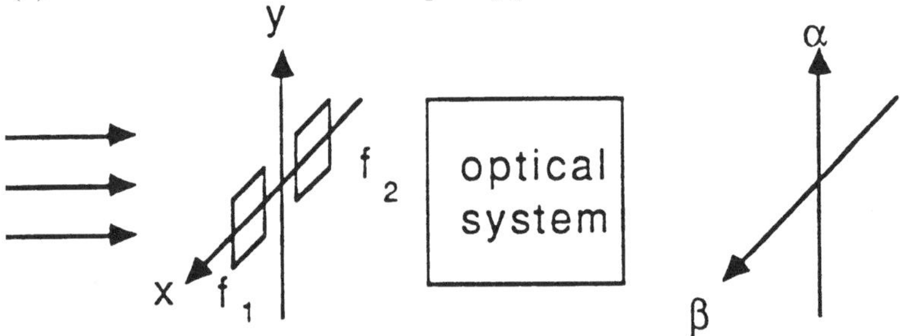

Fig. 2.60.

2.8. Assume that a coherent light distribution on the (α, β) spatial coordinate system is shown in Fig. 2.61. Use it to evaluate the complex light field at the (x, y) plane.

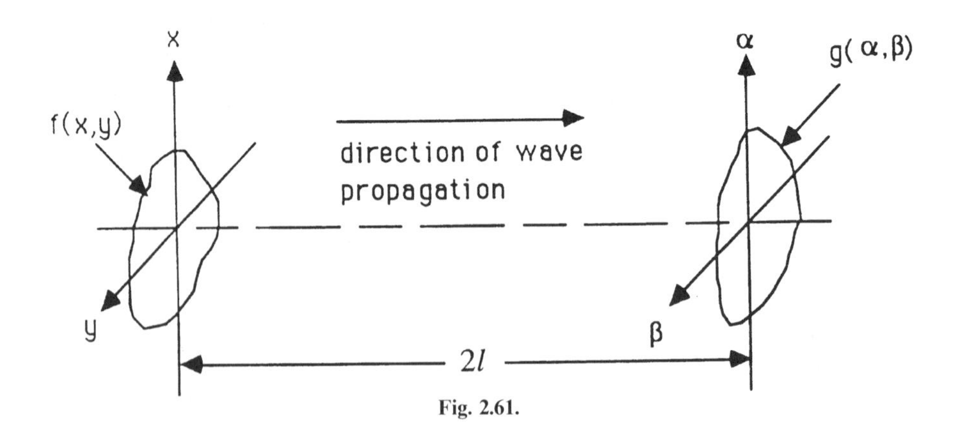

Fig. 2.61.

(a) Draw an analog system diagram to represent this problem.
(b) If $g(\alpha, \beta)$ is given by

$$g(\alpha, \beta) = \exp\left[i\frac{\pi}{\lambda l}(\alpha^2 + \beta^2)\right],$$

evaluate $f(x, y)$.
(c) We assume a transparency is inserted at (x, y) domain, for which the amplitude transmittance is given by

$$T(x, y) = \exp\left[-i\frac{2\pi}{\lambda l}(x^2 + y^2)\right].$$

Calculate the complex wave field impinging on the transparency.

2.9 A diffraction screen is normally illuminated by a monochromatic plane wave of $\lambda = 500$ nm. It is observed at a distance of 30 cm from the diffraction screen, as shown in Fig. 2.62.
(a) Draw an analog system diagram to evaluate the complex light field at the observation screen.
(b) Calculate the separations between the slits, by which the irradiance at the origin of the observation screen would be the minimum.
(c) Compute the irradiance of part (b).

2.10 Consider a double-convex cylindrical lens, as shown in Fig. 2.63.
(a) Evaluate the phase transform of the lens.
(b) If the radii of the first and the second surfaces of the lens are given by $R_1 = 10$ cm and $R_2 = 15$ cm, respectively, and the refractive index of the lens is assumed to be $n = 1.5$, calculate its focal length.
(c) If the lens is submerged in a tank of fresh water in which the refractive index of the water $n = 1.33$, determine the effective focal length in the water.

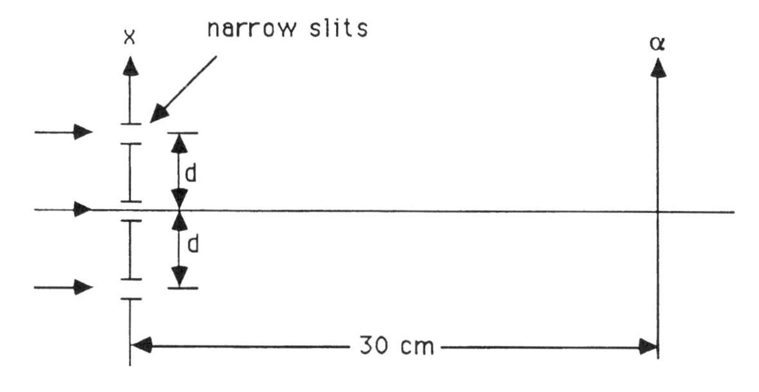

Fig. 2.62.

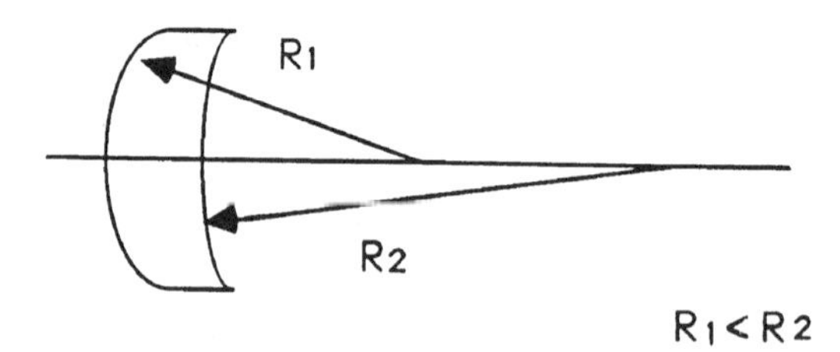

Fig. 2.63.

2.11 Assume that an optical processor is illuminated by a monochromatic plane wave, as shown in Fig. 2.64, where the amplitude transmittance of the transparency is denoted by $f(\xi, \eta)$, and $l < f$.
(a) Calculate the output complex light field at the back focal length of the transform lens.
(b) Determine the output irradiance.
(c) What is the *effective* focal length of this Fourier transform system?
(d) Show that the scale of the Fourier spectrum is proportional to the effective focal length.

2.12. Assume that the amplitude transmittance of an object function is given by

$$f(x, y)[1 + \cos(p_0 x)],$$

where p_0 is an arbitrary angular carrier spatial frequency, and $f(x, y)$ is

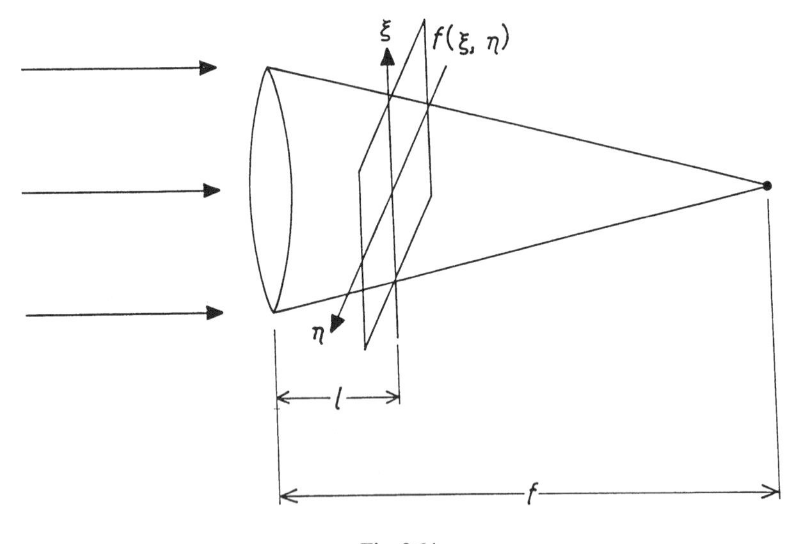

Fig. 2.64.

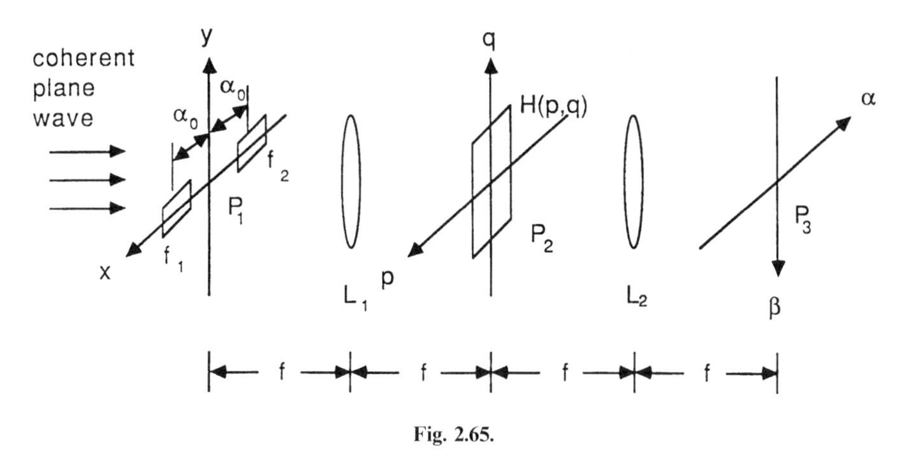

Fig. 2.65.

assumed to be spatial-frequency limited. If this object transparency is inserted at the input plane of the FDP,

(a) Determine the spectral distribution at the Fourier plane.

(b) Design a stop-band filter for which the light distribution at the output plane will be $f(x, y)$.

2.13 Consider the coherent optical processor shown in Fig. 2.65. The spatial filter is a one-dimensional sinusoidal grating.

$$H(p) = \tfrac{1}{2}[1 + \sin(\alpha_0 p)],$$

where α_0 is an arbitrary constant that is equal to the separation of the input object functions $f_1(x, y)$ and $f_2(x, y)$. Compute the complex light field at the output plane P_3.

2.14 A method for synthesizing a complex spatial filter is shown in Fig. 2.66a. If a signal transparency $s(x, y)$ is inserted at a distance d behind the transform lens, and the amplitude transmittance of the recorded filter is linearly proportional to the signal transparency,

(a) What is the spatial carrier frequency of the spatial filter?

(b) If the spatial filter is used for signal detection, determine the appropriate location of the input plane, as shown in Fig. 2.66b.

2.15 Suppose that the input object functions on an FDP is a rectangular grating of spatial frequency p_0, as shown in Fig. 2.67.

(a) Evaluate and sketch the spectral content of the input object at the Fourier domain.

(b) If we insert a small half-wave plate at the origin of the Fourier domain, sketch the light distribution at the output plane and comment on your observation.

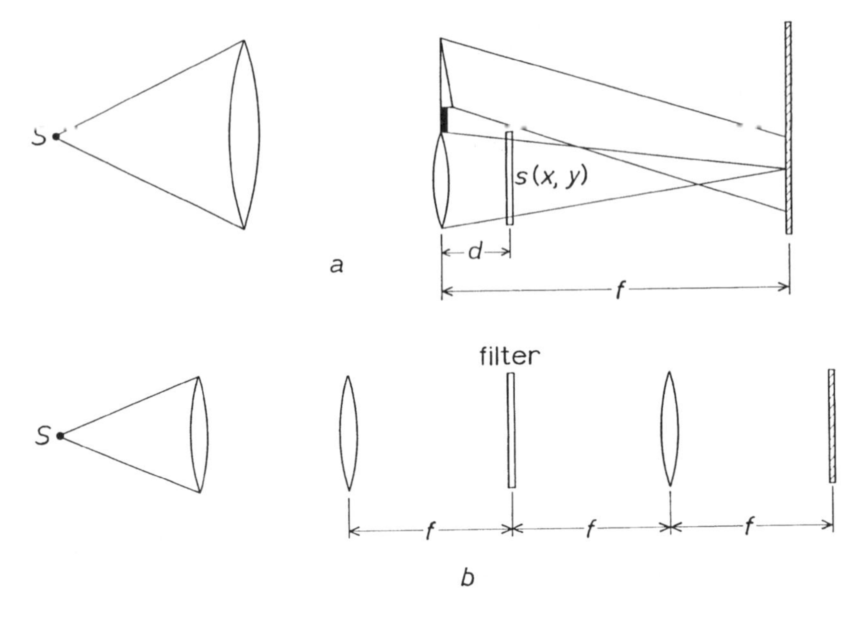

Fig. 2.66.

2.16 The transaxial location of an input transparency introduces a quadratic phase factor in a spatial filter synthesis.

(a) Show that the quandratic phase factors produced in the optical setup of Fig. 2.68 would be mutually compensated.

(b) If the recorded $H(p, q)$ (joint transform filter) is inserted back at the input plane of Fig. 2.68, calculate the output complex light field.

(c) If $H(p, q)$ is inserted at the front of the transform lens, what would be the scale of the output light field?

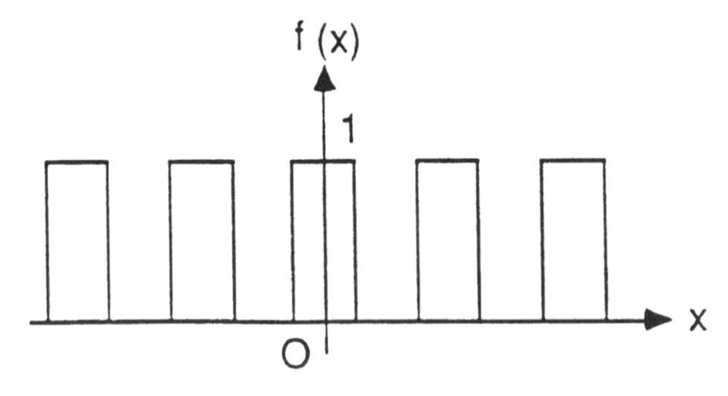

Fig. 2.67.

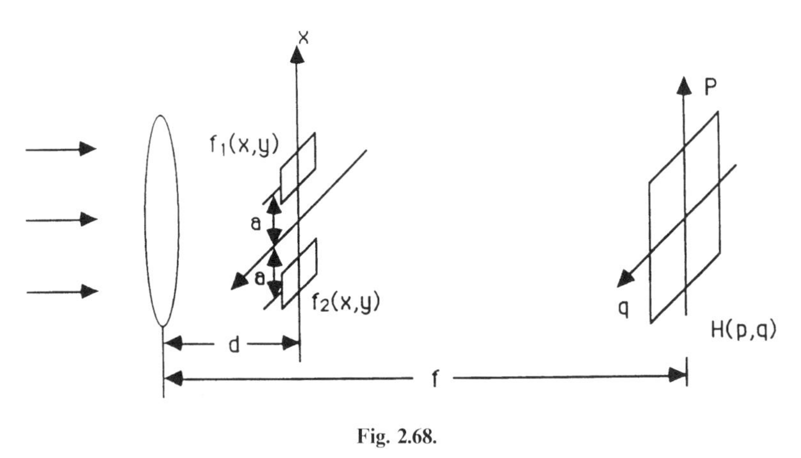

Fig. 2.68.

2.17 With reference to the JTP of Fig. 2.18, if the input object is imbedded in an additive white Gaussian noise with zero means; that is $f(x, y) + n(x, y)$, show that the JTC is an optimum correlator.

2.18 Consider the microcomputer-based FDP of Fig. 2.17. Assume that the operation speed is mostly limited by the addressing time of the SLMs and the response time of the CCD array detector. If the addressing time of the SLMs is 60 frames/sec and the response time of the CCD detector array is 50 frames/sec, calculate the operation speed of the processor.

2.19 Refer to the hybrid FDP and the JTP of Figs. 2.17 and 2.18, respectively.
(a) State the basic advantages and disadvantages of these two processors.
(b) What would be their required spatial carrier frequencies?
(c) What would be their input spatial coherence requirements?

2.20 Suppose that the addressing time of the SLMs and the response time of the CCDs are given by 1/60 sec and 1/50 sec, respectively.
(a) Calculate the operating speeds for the JTC and the single SLM JTC, respectively.
(b) Comment on your results with respect to the one you obtained for the FDP in Exercise 2.18.

2.21 Suppose the size SLM is given as $4 \times 4\,\text{cm}^2$. If the spatial-frequency bandwidth of the signal is 10 lines/mm, the focal length of the transform lens is 500 mm, and the wavelength of the light source is 600 nm,
(a) Compute the optimum number of joint transform power spectra that can be replicated within the device's panel.
(b) Assume that the replicated JTPs are illuminated by temporal coherent but spatially partial-coherent light. What would be the spatial coherence requirement of the source?
(c) Show that the signal-to-noise ratio of the correlation peak improved by using spatially partial-coherent illumination.

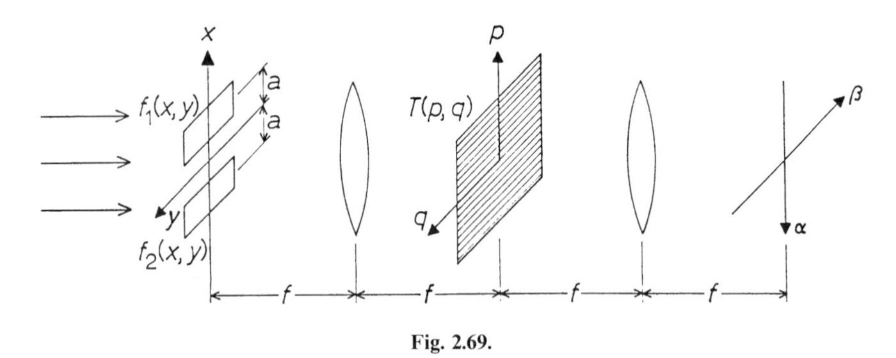

Fig. 2.69.

(d) Calculate the accuracy of the detection under strictly coherent and spatially partial-coherent illuminations, respectively.

(e) Compute the correlation peak intensities of part (d).

2.22 Consider the JTP in Sec. 2.4.4.

(a) Design a phase-reversal function to sharpen a linear smeared image.

(b) Calculate the output complex light distribution.

(c) Show that a reasonable degree of restoration may be achieved by using the phase-reversal technique.

2.23 Consider the complex image subtraction of Fig. 2.69. If we let one of the input transparencies be an open aperture; say $f_2(x, y) = 1$,

(a) Calculate the output irradiance distribution around the optical axis.

(b) On the other hand, if $f_1(x, y) = \exp[i\phi(x, y)]$ is a pure phase object, show that the phase variation of $f_1(x, y)$ can be observed at the output plane.

2.24 Consider a multisource coherent processor for image subtraction, as shown in Fig. 2.70. We assume that the two input object transparencies are separated by a distance of $2h_0$.

(a) Determine the grating spatial frequency with respect to the spacing of these coherent point sources.

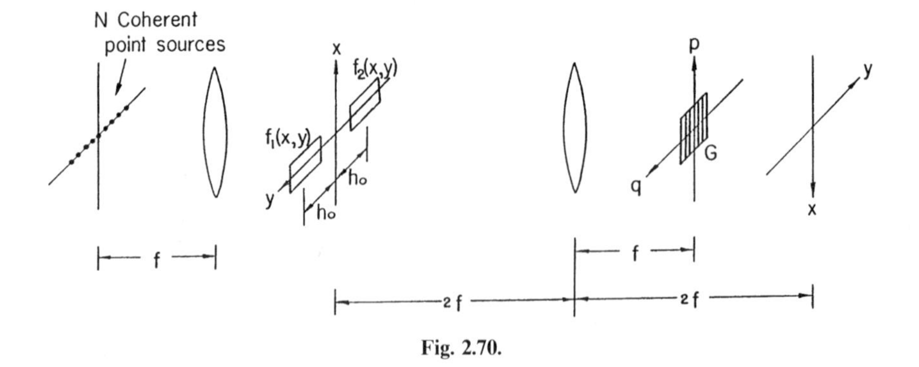

Fig. 2.70.

(b) Evaluate the intensity distribution of the subtracted image at the output plane.

2.25 Refer to the broadband spectrum analyzer of Sec. 2.5.4.
(a) Determine the frequency resolution of the analyzer.
(b) For a large number of scan lines, compute the number of resolution elements at the output plane.
(c) Compute the space–bandwidth product of the wide-band spectrum analyzer.

2.26 The spatial resolution of a CRT scanner is 2 lines/mm and the raster-scan aperture is assumed 10×13 cm. If the CRT is used for wide-band intensity modulation,
(a) Compute the space–bandwidth product of the CRT.
(b) If the time duration of the processing signal is 2 sec and we assume that the separation among the scanned lines is equal to the resolution of the CRT, determine the highest temporal frequency limit of the CRT.

2.27 Consider a two-dimensional rectangular impulse function as given by

$$f(x, y) = \text{rect}\left(\frac{x}{a}\right)\text{rect}\left(\frac{x}{b}\right).$$

(a) Evaluate the Mellin transform of $f(x, y)$.
(b) Sketch the result of part (a).
(c) Show that the Mellin transform is indeed a scale invariant transformation.
(d) By preprocessing $f(x, y)$, draw a coherent optical system that is capable of performing the Mellin transformation.

2.28 Given a circularly symmetric function, such as

$$f(r) = \begin{cases} 1, & r \leqslant R_0 \\ 0, & \text{otherwise.} \end{cases}$$

where $r = \sqrt{x^2 + y^2}$,
(a) Expand the circular harmonic series at the origin of the (x, y) coordinate system.
(b) Assuming that the center of the circular disk of part (a) is located at $x = 0$, $y = R_0$, expand the circular harmonic series of this disk at $x = 0$, $y = 0$.

2.29 Given the spokelike function shown in Fig. 2.71; that is,

$$f(r, \theta) = \text{rect}\left[\frac{\theta}{\theta_0} + \frac{1}{2}\right] * \sum_{n=1}^{N} \delta(\theta - n\theta_s),$$

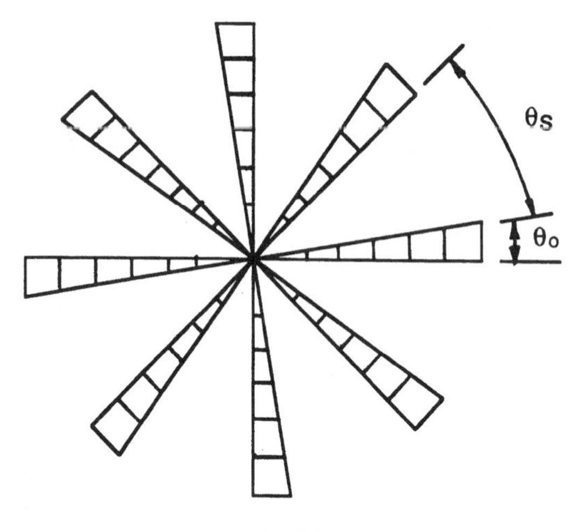

Fig. 2.71.

where $\theta_0 \leqslant \theta_s$ and $\theta_s = 2\pi/N$, evaluate the circular harmonic expansion of $f(r, \theta)$ around the origin. *Hint:* Assume that $\theta_0 = \theta_s$. Since θ_s is very small, N can be assumed infinitely large.

2.30 (a) Discuss the synthesis of an optical circular harmonic filter.
(b) Show the optical implementation in a JTC and in a VLC, respectively.

2.31 With reference to the nonlinear processing system of Fig. 2.31, show that a higher correlation peak intensity may be obtained using a circular homomorphic filtering system.

2.32 We have shown that phase distortion introduced by an SLM in a JTC can be compensated for by using a nonlinear photorefractive crystal. Discuss in detail whether phase distortion can indeed be compensated.

2.33 By using the four-wave mixing technique,
(a) Draw an optical architecture to show that a contrast reversal image can be performed.
(b) Repeat part (a) for image edge enhancement processing.

2.34 Sketch a two-wave mixing architecture, and show that wavefront-distortion compensation can be accomplished with the optical setup.

2.35 A 1-mm-thick $LiNbO_3$ photorefractive crystal is used for a reflection-type matched filter synthesis. Assume that the writing wavelength is $\lambda = 500$ nm and that the refractive index of the crystal is $n = 2.28$.
(a) Calculate the minimum writing-wavelength separation for a wavelength-multiplexed filter.
(b) If the object size has a width of 2 mm, and the focal length of the

transform lens is 100 mm, what would be the allowable transversal shift of the object?

2.36 An input transparency contains several 5 mm targets. If targets are to be detected with the white light processor of Fig. 2.44,
(a) Calculate the required source size. Assume that the focal length of the transform lenses is $f = 500$ mm.
(b) If the spatial frequency of the input object is assumed to be 10 lines per millimeter, calculate the width of the spatial filter H_n and the required spatial frequency of the sampling phase grating.

2.37 Assume that the light source of Fig. 2.44 is a point source. If we ignore the input signal transparency,
(a) Determine the smeared length of the Fourier spectra as a function of the focal length f of the achromatic transform lens and the spatial frequency p_0 of the sinusoidal phase grating $T(x)$.
(b) If the spatial frequency of the diffraction grating is $p_0 = 80\pi$ rad/mm and $f = 30$ cm, compute the smeared length of the rainbow color spectrum.

2.38 Referring to the preceding exercise, if the white light source is a uniformly circular extended source of diameter D,
(a) Determine the size of the smeared Fourier spectra as a function of f and p_0.
(b) If $D = 2$ mm, $p_0 = 80\pi$ rad/mm, and $f = 30$ cm, determine the precise size of the smeared Fourier spectra.

2.39 Consider a spatially incoherent processor, as depicted in Fig. 2.72. We assume that the spatial coherence function at the input plane is given by

$$\Gamma(|x - x'|) = \text{rect}\left[\frac{|x - x'|}{\Delta x}\right],$$

which is independent of y.
(a) Evaluate the transmittance function of the source encoding mask.

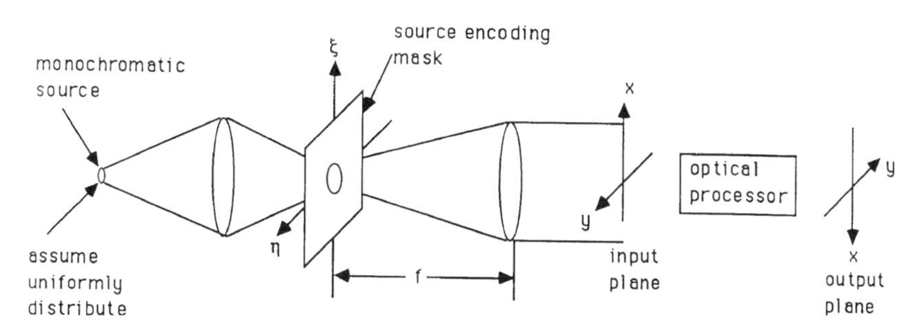

Fig. 2.72.

(b) Is the source encoding mask of part (a) physically realizable? Give the reason for your answer.

(c) Design a physically realizable source encoding mask that has a coherence distance equal to $2\Delta x$.

2.40 Refer to the white light photographic image deblurring of Sec. 2.8. If the linear smeared length is about 1 mm,

(a) What are the spatial and temporal coherence requirements for the deblurring process with an incoherent source?

(b) In order to achieve the spatial coherence requirement, what is the requirement of the light source?

(c) In order to obtain the required temporal coherence, what would be the minimum sample frequency of the diffraction grating $T(x)$?

2.41 Consider the image subtraction with the encoded source discussed in Sec. 2.8. Assume that the spatial frequencies of the input object transparencies are about 10 lines/mm.

(a) Calculate the temporal coherence requirement.

(b) If the main separation of the two input object transparencies is 20 mm and the focal length of the achromatic transform lenses is 300 mm, design a spatial filter and a source encoding mask to produce point-pair spatial coherence for the processing operation.

(c) Calculate the visibility of the subtracted image.

2.42 Assume that an integrated-circuit mask is a two-dimensional cross-grating–type pattern. The corresponding Fourier spectrum may be represented in a finite region of a spatial frequency plane, as depicted in Fig. 2.73.

(a) Design a white light optical processing technique for the integrated-circuit mask inspection.

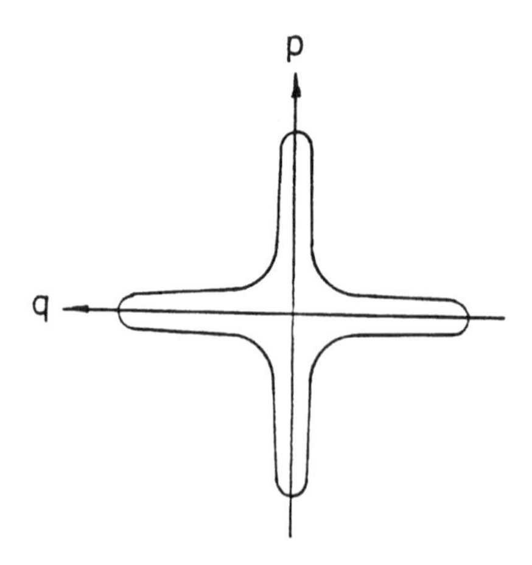

Fig. 2.73.

(b) Carry out a detailed analysis to show that defects can be easily detected through color coding.

2.43 The amplitude transmittance function of a multiplexed transparency, with a positive and a negative image, is given by

$$t(x, y) = t_1(x, y)(1 + \cos p_0 x) + t_2(x, y)(1 + \cos q_0 y).$$

(a) Evaluate the smeared spectra at the Fourier plane.

(b) If the focal length of the transform lens is $f = 300$ mm, and the sampling frequencies are $p_0 = 80\pi$ and $q_0 = 60\pi$ rad/mm, compute the smearing length of the Fourier spectra. Assume that the spectral bandwidth of the white light source is limited by 350 to 750 nm.

(c) Design a set of transparent color filters by which the density (i.e., gray levels of the image) can be encoded in pseudocolors at the output plane.

(d) Compute the irradiance of the pseudocolor image.

2.44 To illustrate the noise immunity of the partially coherent system of Fig. 2.44, we assume that the narrow spectral band filters are contaminated with additive white Gaussian noise; that is, $H(\alpha, \beta) + n(\alpha, \beta)$. Show that the output signal-to-noise ratio improves when the white light source is used.

2.45 Refer to the LCTV optical neural network (ONN) of Fig. 2.51.

(a) Calculate the space–bandwidth product (SBP) and the resolution requirement for a 16×16-neuron network.

(b) Repeat part (a) for an $N \times N$–neuron net.

(c) Calculate the operation speed of the ONN.

(d) To improve the operation speed, show that one can either increase the number of neurons or increase the frame rate of the SLM.

2.46 Four binary patterns are represented by a Venn diagram in Fig. 2.74. Since excitory, inhibitory, and null interconnections can be determined by logical operation, describe the relationships among neurons in the

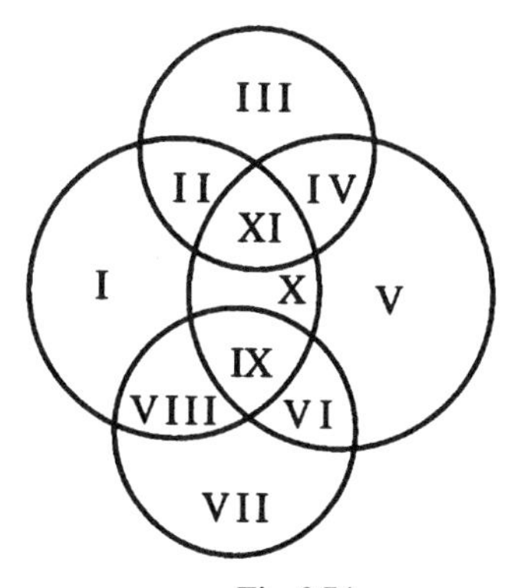

Fig. 2.74.

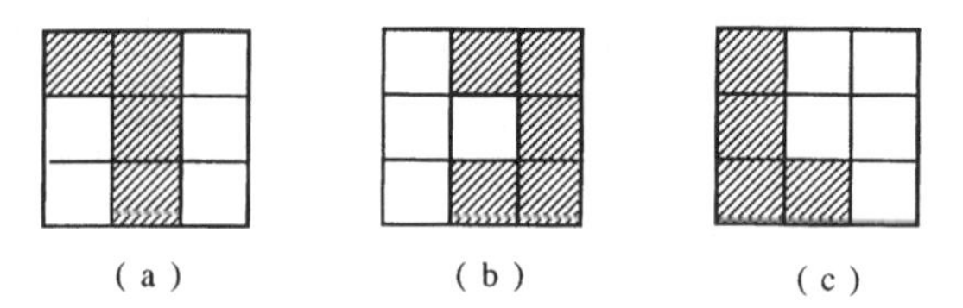

Fig. 2.75.

following subsets:
(a) I and II.
(b) I and V.
(c) VI and X.
(d) IX and XI.
(e) III and IX.

2.47 A set of the training patterns shown in Fig. 2.75 are to be stored in a neural network.
(a) Use a Hopfield model to construct a one-level neuron network.
(b) With reference to part (a), construct the memory matrix.
(c) Determine the interconnections by logical operation.
(d) Construct a one-level IPA neural net.
(e) Compare the IPA model with the Hopfield model.

2.48 Refer to the set of training images shown in Fig. 2.76. Synthesize a synthetic discriminant function filter (SDF).

2.49 Consider the target set of Fig. 2.76. Synthesize a simulated annealing bipolar composite filter (SABCF).

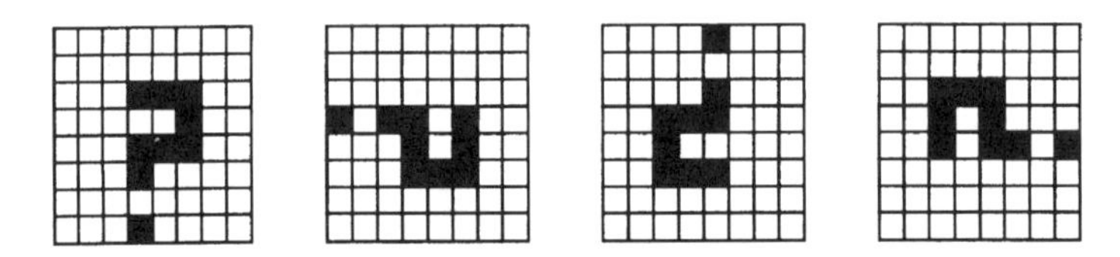

Fig. 2.76.

2.50 Use Fig. 2.76 as the target set and Fig. 2.77 as the antitarget set, and synthesize a SABCF. In view of the results obtained in Exercises 2.48 and 2.49, comment on the SDF and SABCF results.

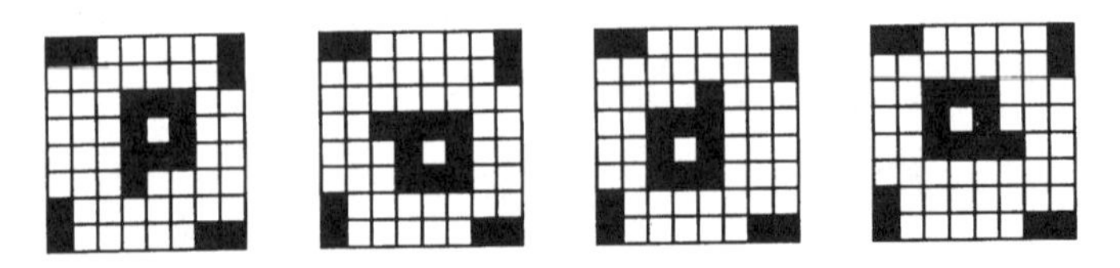

Fig. 2.77.

Chapter 3 | Communication With Optics

Shizhuo Yin
THE PENNSYLVANIA STATE UNIVERSITY

Light has been used for communication for more than five thousand years. In ancient times, information was transmitted by blocking or transmitting light. In recent years, with the advent of optic fibers and corresponding components such as semiconductor lasers and receivers, fiber-optic communication has become the core technology for long haul broadband communications. In this chapter, we briefly introduce the basic concepts of fiber-optic communications.

3.1. MOTIVATION OF FIBER-OPTIC COMMUNICATION

With the exponential growth in transmission bandwidth usage, in particular (due to the rapid growth of Internet data traffic), there is a huge demand on transmission bandwidth. Fiber-optic communication is the best candidate to fulfill this broad bandwidth due to its following unique features:

1. *Huge bandwidth due to extremely high carrier frequency*

The light signal has a very high carrier frequency. For example, at the infrared communication window, assuming that $\lambda = 1500$ nm, the corresponding carrier frequency, f, is as high as

$$f = c/\lambda = \frac{3 \times 10^8 \text{ m/s}}{1500 \times 10^{-9} \text{ m}} = 2 \times 10^{14} \text{ (Hz)}. \tag{3.1}$$

This extremely high carrier frequency makes fiber-optic communication suitable for broadband communications.

2. *Very low loss*

Fiber-optic communication can have a very low loss. The loss can be as low as 0.2 dB/Km, which makes long-distance communications possible. For example, the total loss for 100-km fiber is as low as 20 dB, so there is no need for amplification or regeneration for 100-km fiber. Note that at high frequency, traditional copper wire has a very large loss.

3. *Light weight and compact size*

For high-speed communication, coaxial cable is very heavy and cumbersome. However, fiber cable has a very compact size and light weight. The diameter of optical fiber (core plus cladding) is as small as 125 μm, which is about twice the diameter of human hair.

4. *Highly secure and immune to the external electromagnetic interference*

The dielectric nature of optical fiber makes it immune to external electromagnetic interference and highly secure.

Example 3.1. Assume that an optical fiber has a length $L = 1$ km and attenuation constant $\alpha = -0.2$ dB/km. What is the output power of the optical fiber if the input power is 1 W?

Solve: Based on the definition of attenuation, we have

$$P_{\text{out}} = P_{\text{in}} \cdot 10^{\alpha \cdot L/10} = 1 \cdot 10^{-0.2 \cdot 1/10} = 10^{-0.02} \approx 0.96\,(\text{W})$$

From Example 3.1, we can see that more than 95% of power energy can be reserved in the optical fiber after propagating 1 km. Again, this shows the low-attenuation feature of optical fiber.

3.2. LIGHT PROPAGATION IN OPTICAL FIBERS

3.2.1. GEOMETRIC OPTICS APPROACH

The basic principle of light propagating in optic fiber can be illustrated by the simple geometric optics principle, i.e., total internal reflection.

The basic structure of optical fiber can be described by Fig. 3.1, which includes three layers: (1) core layer; (2) cladding layer; and (3) protection jacket layer. The core and cladding have refractive index n_1 and n_2, respectively. To

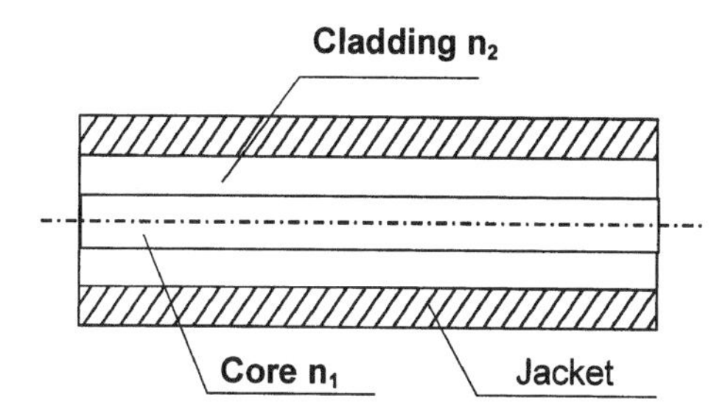

Fig. 3.1. Basic structure of optical fiber.

ensure light propagation in the fiber, the core must have a refractive index larger than that of the cladding so that the total internal reflection can happen. Let us look at a simple step index fiber, as shown in Fig. 3.2, where the core has a refractive index n_1, the cladding has a refractive index n_2, the outside has a refractive index n_0, the incident angle from the outside to the fiber is θ_i, the refractive angle of θ_i is θ_r, and the internal incident angle from the core to the cladding is ϕ. Assume that the light is coupled into the fiber at point A and reaches to the intersection point between the core and the cladding at point B. The minimum angle that can result in total internal reflection at point B, ϕ_c, is

$$\phi_c = \sin^{-1}\left(\frac{n_2}{n_1}\right). \tag{3.2}$$

In this critical angle case, the incident angle θ_i, denoted as θ_{ic}, can be described in terms of n_1 and n_2 from the following geometric relationship (as shown in Fig. 3.2),

$$\phi_c + \theta_r = 90^\circ, \tag{3.3}$$

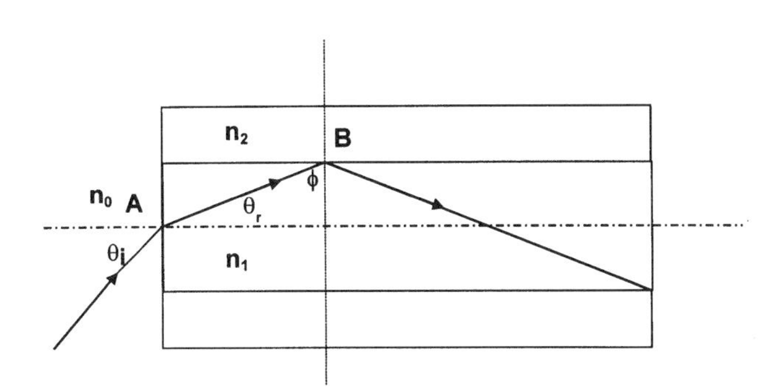

Fig. 3. 2. Basic principle of light propagation in optical fiber: total internal reflection.

and Snell's law,

$$n_0 \sin \theta_{ic} = n_1 \sin \theta_r. \tag{3.4}$$

Substituting Eqs. (3.2) and (3.3) into Eq. (3.4), we have

$$n_0 \sin \theta_{ic} = \sqrt{n_1^2 - n_2^2} \tag{3.5}$$

Note that $n_0 \sin \theta_{ic}$ is called the numerical aperture (NA) of the fiber and θ_{ic} is called the acceptance angle of the fiber. When the incident angle is $\theta_i \leqslant \theta_{ic}$, the light can propagate in the optical fiber without severe attenuation since the total internal reflection happens at the intersection surface between the core and the cladding. However, when $\theta_i > \theta_{ic}$, the leakage happens at the intersection surface between the core and the cladding, which results in severe attenuation. This is the basic principle of how light can propagate in optical fiber from a geometric optics point of view. Since the fiber length is very long, there are many incidences of reflection. To ensure low attenuation, we need 100% perfect reflection. A little bit of loss in each reflection can result in huge attenuation after many reflections, illustrated in the following example.

Example 3.2. Consider a step index fiber with parameters $n_1 = 1.475$, $n_2 = 1.460$, $n_0 = 1.000$, and core radius $a = 25\ \mu\text{m}$.

(a) Calculate the maximum incident angle and numerical aperture (NA) of the fiber.
(b) Under the maximum incident angle, how many total reflections happen for a 1-km–long fiber?
(c) If the power loss is 0.01% for each reflection, what is the total loss of this 1-km–long fiber (in dB)?

Solve:

(a) $\text{NA} = n_0 \sin \theta_i = \sqrt{n_1^2 - n_2^2} = \sqrt{1.475^2 - 1.460^2} = 0.21.$
(b) From the relationship $n_0 \sin \theta_i = n_1 \sin \theta_r$, as shown in Fig. 3.2, the internal refractive angle, θ_r, can be calculated as

$$\theta_r = \sin^{-1}\left(\frac{n_0 \sin \theta_i}{n_1}\right) = \sin^{-1}\left(\frac{0.21}{1.475}\right) = 8.19^\circ.$$

Then, the propagation length for each reflection is $2a/\tan \theta_r$. Thus, the total number of reflections is

$$N = \frac{L}{2a/\tan \theta_r} = \frac{10^3\ \text{m}}{2 \cdot 25 \cdot 10^{-6}\ \text{m}} \tan 8.19^\circ = 2.88 \cdot 10^6 \text{ times},$$

where L is the length of the fiber.

(c) The total loss (in dB) is

$$\text{Loss} = 10 \cdot \log_{10} \frac{P_{\text{out}}}{P_{\text{in}}} = 10 \cdot \log_{10}(1 - 10^{-4})^{2.88 \cdot 10^6} = -1238\,\text{dB}.$$

Again, this example tells us that we need 100% reflectivity for each reflection. A reflectivity of 99.99% is absolutely not acceptable.

So far, we have explained how light can propagate in optical fiber from a geometric optics point of view. The next question we need to answer is how high the bandwidth can be, which can be estimated as follows: Assume that the fiber length is L. The minimum time, t_{min}, that light can pass through this fiber corresponds to the case of incident angle $\theta_i = 0°$. Mathematically, t_{min} is given by

$$t_{\text{min}} = \frac{L}{v} = \frac{L}{c/n_1} = \frac{n_1 L}{c}. \tag{3.6}$$

The maximum time, t_{max}, corresponds to the critical incident angle case. In this case, the traveling distance becomes

$$L_{\text{max}} = \frac{L}{\sin \phi_c} = \frac{L}{n_2/n_1} = \frac{n_1 L}{n_2}. \tag{3.7}$$

Thus, the maximum traveling time is

$$t_{\text{max}} = \frac{L_{\text{max}}}{v} = \frac{n_1 L/n_2}{c/n_1} = \frac{n_1^2 L}{n_2 c}. \tag{3.8}$$

The traveling time difference, Δt, is

$$\Delta t = t_{\text{max}} - t_{\text{min}} = \frac{n_1 L}{c}\left(\frac{n_1}{n_2} - 1\right). \tag{3.9}$$

Note that this traveling time difference results in a basic limitation on the maximum bandwidth that can be used for transmission. To avoid confusion among differing time information, the maximum bandwidth, B, is

$$B = \frac{1}{\Delta t} = \frac{1}{n_1 L/c(n_1/n_2 - 1)}. \tag{3.10}$$

For a quantitative feeling about Eq. (3.10), let us look at the following example.

Example 3.3. A step index optical fiber has core refractive index $n_1 = 1.5$, cladding refractive index $n_2 = 1.485$, length $L = 1$ km. Calculate the maximum bit rate for this fiber.

Solve: The maximum bit rate, B, for this fiber is

$$B = \frac{1}{\Delta t} = \frac{1}{n_1 L/c(n_1/n_2 - 1)} = \frac{1}{1.5 \cdot 10^3\,\mathrm{m}/3 \cdot 10\,\mathrm{m/s}(1.5/1.485 - 1)} \approx 20\,\mathrm{Mbps}$$

From this example, one can see that B is much less than the optical carrier frequency; that is, in the order of 10^{14} Hz. To alleviate this problem, we need to reduce the time difference among different propagating routes. Fortunately, there is a case in which only one route is allowed to propagate in the fiber (i.e., single mode fiber) so that much higher bandwidth can be achieved. However, the simple geometric optics theory cannot fully explain this phenomenon, which must be clarified by the more precise wave-optics theory, described in the next section.

3.2.2. WAVE-OPTICS APPROACH

Since the transversal dimension of optical fiber is comparable with its wavelength, the more precise wave-optics theory is needed to explain all the phenomena happening in the fiber; in particular, the mode theory.

The wave-optics approach starts with the well-known Maxwell equation. Optical fiber is a dielectric material so the free charge density is $\rho = 0$ and free current density $\bar{j} = 0$. In addition, assume the light wave is a harmonic wave. Note that the general case can be treated as the weighted summation based on the Fourier transform for the linear system. Under these assumptions, the electric field of the light field, $\bar{E}$, satisfies the following wave equation:

$$\nabla^2 \bar{E} + n^2 k_0^2 \bar{E} = 0, \tag{3.11}$$

where $k_0^2 = \omega^2/C^2$ is the wave number, ω is the angular frequency, C is the light speed in vacuum, $n = \sqrt{\mu_r \varepsilon_r}$ is the refractive index of the fiber material—and it may be a function of angular frequency (i.e., $n = n[\omega]$), and ∇^2 is the Laplacian operator. Due to the cylindrical symmetry of the optical fiber, as shown in Fig. 3.3, it is convenient to solve Eq. (3.11) under cylindrical coordinates. Note that Eq. (3.11) is a vector differential equation. For simplicity, let us deal with the z component of the electric field, E_z, first. In this case, Eq. (3.11) becomes the following simplified scalar differential equation:

$$\nabla^2 E_z + n^2 k_0^2 E_z = 0. \tag{3.12}$$

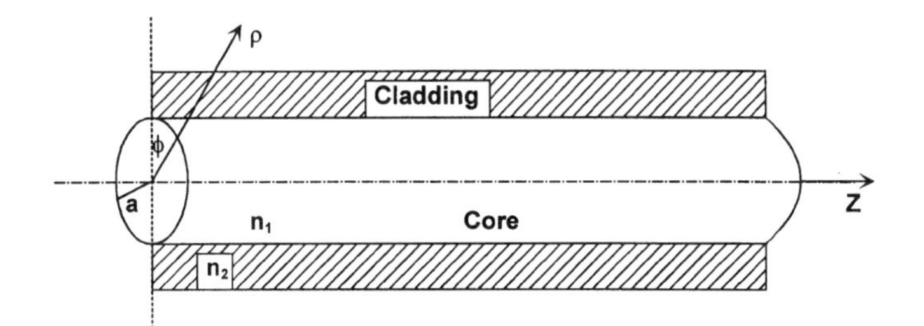

Fig. 3.3. Description of optical fiber under cylindrical coordinate.

Under the cylindrical coordinate as illustrated by Fig. 3.3, Eq. (3.12) can be written as

$$\left(\frac{1}{\rho}\frac{\partial}{\partial\rho}\left(\rho\frac{\partial}{\partial\rho}\right)+\frac{1}{\rho^2}\frac{\partial^2}{\partial\phi^2}+\frac{\partial^2}{\partial z^2}\right)E_z(\rho,\phi,z)+n^2k_0^2E_z(\rho,\phi,z)$$
$$=\left(\frac{1}{\rho}\frac{\partial}{\partial\rho}+\frac{\partial^2}{\partial\rho^2}+\frac{1}{\rho^2}\frac{\partial^2}{\partial\phi^2}+\frac{\partial^2}{\partial z^2}\right)E_z(\rho,\phi,z)+n^2k_0^2E_z(\rho,\varphi,z)$$
$$=\frac{\partial^2E_z(\rho,\phi,z)}{\partial\rho^2}+\frac{1}{\rho}\frac{\partial E_z(\rho,\phi,z)}{\partial\rho}+\frac{1}{\rho^2}\frac{\partial^2E_z(\rho,\phi,z)}{\partial\phi^2}+\frac{\partial^2E_z(\rho,\phi,z)}{\partial z^2}$$
$$+\,n^2k_0^2E_z(\rho,\phi,z)=0. \tag{3.13}$$

Equation (3.13) is a partial differential equation, which contains three variables (ρ, ϕ, z). Since this is a linear differential equation, it can be solved by the method of variable separation; that is,

$$E_z(\rho,\phi,z)=F(\rho)\cdot\Phi(\phi)\cdot Z(z). \tag{3.14}$$

Each function of $F(\rho)$, $\Phi(\phi)$, and $Z(z)$ satisfies one ordinary differential equation. Substituting Eq. (3.14) into Eq. (3.13), the following three equations can be derived:

$$\frac{d^2Z(z)}{dz^2}+\beta^2Z(z)=0, \tag{3.15a}$$

$$\frac{d^2\Phi(\phi)}{d\phi^2}+m^2\Phi(\phi)=0, \tag{3.15b}$$

$$\frac{d^2F(\rho)}{d\rho^2}+\frac{1}{\rho}\frac{dF(\rho)}{d\rho}+\left(n^2k_0^2-\beta^2-\frac{m^2}{\rho^2}\right)F(\rho)=0, \tag{3.15c}$$

where m is an integer and β is a constant. The solution for Eq. (3.15*a*) is

$$Z(z) = e^{i\beta z}, \tag{3.16}$$

which represents how light propagates in the z direction. Thus, the constant β is often called the propagation constant. The solution for Eq. (3.15*b*) is

$$\Phi(\phi) = e^{im\phi}, \tag{3.17}$$

which represents how the light field changes along the angular direction. Due to the periodic nature of the angular function, i.e., $\Phi(\phi) = \Phi(\phi + 2\pi)$, m must be an integer. Equation (3.15*c*) is a little bit complicated. For the step index fiber, an analytical solution can be obtained. As shown in Fig. 3.3, the refractive index distribution for a step index fiber can be expressed as

$$n(\rho) = \begin{cases} n_1, & \rho \leqslant a \\ n_2 & \rho > a, \end{cases} \tag{3.18}$$

where a is the radius of the fiber core. Substituting Eq. (3.18) into Eq. (3.15*c*), the following equations are obtained:

$$\frac{d^2F(\rho)}{d\rho^2} + \frac{1}{\rho}\frac{dF(\rho)}{d\rho} + \left(n_1^2k_0^2 - \beta^2 - \frac{m^2}{\rho^2}\right)F(\rho) = 0, \quad \rho \leqslant a \tag{3.19a}$$

$$\frac{d^2F(\rho)}{d\rho^2} + \frac{1}{\rho}\frac{dF(\rho)}{d\rho} + \left(n_2^2k_0^2 - \beta^2 - \frac{m^2}{\rho^2}\right)F(\rho) = 0, \quad \rho > a. \tag{3.19b}$$

Equations (3.19*a*) and (3.19*b*) can be further simplified by defining two new constants:

$$\kappa^2 = n_1^2k_0^2 - \beta^2, \tag{3.20a}$$

$$\gamma^2 = \beta^2 - n_2^2k_0^2, \tag{3.20b}$$

Substituting Eqs. (3.20*a*) and (3.20*b*) into Eqs. (3.19*a*) and (3.19*b*), we get

$$\begin{cases} \dfrac{d^2F(\rho)}{d\rho^2} + \dfrac{1}{\rho}\dfrac{dF(\rho)}{d\rho} + \left(\kappa^2 - \dfrac{m^2}{\rho^2}\right)F(\rho) = 0, & \rho \leqslant a \quad (3.21a) \\ \dfrac{d^2F(\rho)}{d\rho^2} + \dfrac{1}{\rho}\dfrac{dF(\rho)}{d\rho} - \left(\gamma^2 + \dfrac{m^2}{\rho^2}\right)F(\rho) = 0, & \rho > a. \quad (3.21b) \end{cases}$$

Equation (3.21*a*) is the well-known Bessel equation and Eq. (3.21*b*) is the modified Bessel equation. The solutions of these two equations are the Bessel functions. Thus, $F(\rho)$ can be expressed as

$$F(\rho) = \begin{cases} A \cdot J_m(\kappa\rho) + B \cdot Y_m(\kappa\rho), & \rho \leqslant a \\ C \cdot K_m(\gamma\rho) + D \cdot I_m(\gamma\rho), & \rho > a, \end{cases} \tag{3.22}$$

where J_m is the mth order first kind Bessel function, Y_m is the mth order second kind Bessel function, K_m is the mth order modified second Bessel function, I_m is the mth order modified first kind Bessel function, and A, B, C, and D are constants.

When $\rho \to 0$, $Y_m(\kappa\rho) \to \infty$. Since light energy cannot be infinite in the real world, B must be zero (i.e., $B = 0$). Similarly, when $\rho \to \infty$, $I_m(\gamma\rho) \to \infty$. Again, since light energy cannot be infinitely large, D must be zero (i.e., $D = 0$). Thus, Eq. (3.22) can be simplified into

$$F(\rho) = \begin{cases} A \cdot J_m(\kappa\rho), & \rho \leqslant a \\ C \cdot K_m(\gamma\rho), & \rho > a. \end{cases} \tag{3.23}$$

The Bessel functions $J_m(\kappa\rho)$ and $K_m(\gamma\rho)$ can be found by looking at Bessel function tables or calculated by computers from series expressions, as given by

$$J_m(x) = \sum_{n=0}^{\infty} \frac{(-1)^n}{n!(n+m)!} \left(\frac{x}{2}\right)^{2n+m}, \tag{3.24a}$$

$$Y_m(x) = \frac{2}{\pi}\left[\left(\gamma + \ln\frac{\pi}{2}\right) J_m(x) - \frac{1}{2}\sum_{n=0}^{m-1} \frac{(m-n-1)!}{n!}\left(\frac{x}{2}\right)^{2n-m}\right] + \frac{2}{\pi}\left[\frac{1}{2}\sum_{n=0}^{\infty} (-1)^{n+1} \frac{\phi(n) + \phi(m+n)}{n!(m+n)!}\left(\frac{x}{2}\right)^{2n+m}\right], \tag{3.24b}$$

$$\phi(k) = \sum_{j=1}^{k} \frac{1}{j}, \tag{3.24c}$$

$$K_m(x) = \frac{\pi}{2} i^{m+1} [J_m(ix) + iY_m(ix)]. \tag{3.24d}$$

Substituting Eqs. (3.16), (3.17), and (3.23) into Eq. (3.14), we can get the final solution of light field E_z,

$$E_z(\rho, \phi, z, t) = \begin{cases} AJ_m(\kappa\rho)e^{im\phi}e^{i\beta z}e^{i\omega t}, & \rho \leqslant a \\ CK_m(\gamma\rho)e^{im\phi}e^{i\beta z}e^{i\omega t}, & \rho > a. \end{cases} \tag{3.25}$$

Then, by using the Maxwell equation, we can get H_z, E_ρ, E_ϕ, H_ρ, and H_ϕ.

The constants κ and γ can be found by using the following boundary condition between the core and the cladding surfaces. Mathematically, it can be expressed as

$$\begin{aligned} E_z(\kappa\rho)|_{\rho=a} &= E_z(\gamma\rho)|_{\rho=a} \\ \left.\frac{\partial E_z(\kappa\rho)}{\partial\rho}\right|_{\rho=a} &= \left.\frac{\partial E_z(\gamma\rho)}{\partial\rho}\right|_{\rho=a}. \end{aligned} \tag{3.26}$$

Substituting Eq. (3.25) into Eq. (3.26), we obtain

$$A\cdot J_m(\kappa a) = C\cdot K_m(\gamma a), \tag{3.27a}$$

$$A\cdot\kappa\cdot J'_m(\kappa a) = C\cdot\gamma\cdot K'_m(\gamma a), \tag{3.27b}$$

where symbol prime indicates differentiation with respect to the argument. Dividing Eq. (3.27a) by Eq. (3.27b), we get following formular (so-called dispersion relationship):

$$\frac{J_m(\kappa a)}{\kappa\cdot J'_m(\kappa a)} = \frac{K_m(\gamma a)}{\gamma\cdot K'_m(\gamma a)}. \tag{3.28}$$

To understand Eq. (3.28), let us substitute Eq. (3.20) into Eq. (3.28). Then, Eq. (3.28) becomes

$$\frac{J_m(\sqrt{n_1^2k_0^2-\beta^2}\,a)}{\sqrt{n_1^2k_0^2-\beta^2}\cdot J'_m(\sqrt{n_1^2k_0^2-\beta^2}\,a)} = \frac{K_m(\sqrt{\beta^2-n_2^2k_0^2}\,a)}{\sqrt{\beta^2-n_2^2k_0^2}\cdot K'_m(\sqrt{\beta^2-n_2^2k_0^2}\,a)}. \tag{3.29}$$

Equation (3.29) determines the possible values of propagation constant β, as discussed in the following:

1. Since the term $e^{i\beta z}$ represents the light propagation in z direction, the β is called the propagation constant. For nonattenuation propagation, β must be a real number. For simplicity, assume that the light only propagates in one direction and β must be larger than zero for the selected propagation directions (i.e., $\beta > 0$).
2. The conditions $\kappa^2 = n_1^2k_0^2 - \beta^2 > 0$ and $\beta > 0$ result in $\beta < n_1k_0$. The conditions of $\gamma^2 = \beta^2 - n_2^2k_0^2 > 0$ and $\beta > 0$ result in $\beta > n_2k_0$. Thus, the overall constraint on β is

$$n_2k_0 < \beta < n_1k_0. \tag{3.30}$$

3. Multiplying $1/k_0$ on both sides of Eq. (3.30), we get

$$n_2 < \frac{\beta}{k_0} < n_1. \tag{3.31}$$

$\bar{n} = \beta/k_0$ is defined as the effective refractive index of the fiber for a light field with propagation constant β. From Eq. (3.31), it can be seen that this refractive index is larger than the cladding refractive index n_2 and smaller than the core refractive index n_1.

4. For a given m, we may get a set of solutions for β, denoted by n ($n = 1, 2, 3, \ldots$). Thus, we may get many possible propagation constants β_{mn} corresponding to different m and n. Since m and n are integers, β_{mn} are discrete numbers. Each β_{mn} corresponds to one possible propagation mode. For example, β_{01} represents one mode and β_{11} represents another mode.
5. To find out the number of modes that propagate in the fiber, first let us define an important parameter—normalized frequency, V,

$$V = \sqrt{(\kappa^2 + \gamma^2)k_0^2 a^2} = \sqrt{k_0^2 a^2 (n_1^2 - n_2^2)} = k_0 a \sqrt{n_1^2 - n_2^2} = k_0 a \cdot \mathrm{NA}. \tag{3.32}$$

There is only one solution of β in Eq. (3.29) when $V < 2.405$. In other words, there is only one possible propagation mode in the fiber in this case. This is the so-called single mode fiber case. Since there is only one mode propagating in the fiber, there is no intermodal dispersion in this type of fiber, so that much higher bandwidth can be achieved in a single mode fiber for long-haul communications. When V is larger, it can also be shown that the number of modes existing in the fiber is about

$$N = \frac{V^2}{2}. \tag{3.33}$$

This corresponds to the multimode fiber case.

Example 3.4. Compute the number of modes for a fiber the core diameter of which is 50 μm. Assume that $n_1 = 1.48$, $n_2 = 1.46$, and operating wavelength $\lambda = 0.82\ \mu$m.

Solve:

$$V = \frac{2\pi a}{\lambda}\sqrt{n_1^2 - n_2^2} = \frac{2\pi \cdot 50\ \mu m/2}{0.82\ \mu m}\sqrt{1.48^2 - 1.46^2} = 46.45.$$

Since $V \gg 1$, we can use the approximation formular $N = V^2/2$ to calculate the number of modes in the fiber.

$$N = \frac{V^2}{2} = \frac{46.65^2}{2} = 1089.$$

Thus, there are over a thousand modes propagating in the common multimode fiber.

Example 3.5. What is the maximum core radius allowed for a glass fiber having $n_1 = 1.465$ and $n_2 = 1.46$ if the fiber is to support only one mode at a wavelength of 1250 nm?

Solve: The single mode operating condition is $V = 2\pi a/\lambda(\sqrt{n_1^2 - n_2^2} \leqslant 2.405)$. Thus, the maximum radius, $a_{\max}$, is

$$a_{\max} = \frac{2.405 \cdot \lambda}{2\pi\sqrt{n_1^2 - n_2^2}} = \frac{2.405 \times 1.25\,\mu\text{m}}{2\pi\sqrt{1.465^2 - 1.46^2}} = 3.96\,\mu\text{m}.$$

This result tells us that the radius of a single mode fiber is very small.

Example 3.6. An optical fiber has a radius $a = 2\,\mu$m, $n_2 = 1.45$, relative refractive index difference $\Delta = 0.01$, and operating wavelength $\lambda = 1.288\,\mu$m. Calculate the propagating constant, β, and effective refractive index of the fiber, $\bar{n}$.

Solve: Based on the definition of relative refractive index difference $\Delta = (n_1 - n_2)/n_1$, we get

$$n_1 = \frac{n_2}{1 - \Delta} = \frac{1.45}{1 - 0.01} = 1.4646.$$

The wave number

$$k_0 = \frac{2\pi}{\lambda} = \frac{2\pi}{1.288\,\mu\text{m}} = 4.878\,\mu\text{m}^{-1}.$$

The normalized frequency

$$V = \frac{2\pi a}{\lambda}\sqrt{n_1^2 - n_2^2} = \frac{2\pi \cdot 2\,\mu\text{m}}{1.288\,\mu\text{m}}\sqrt{1.4646^2 - 1.45^2} = 2.016 < 2.405.$$

Thus, there is only one mode propagating in the fiber. This is the single mode fiber case. The propagating constant β can be found by solving Eq. (3.29); i.e.,

$$\frac{J_m(\sqrt{n_1^2k_0^2 - \beta^2}\,a)}{\sqrt{n_1^2k_0^2 - \beta^2} \cdot J_m'(\sqrt{n_1^2k_0^2 - \beta^2}\,a)} = \frac{K_m(\sqrt{\beta^2 - n_2^2k_0^2}\,a)}{\sqrt{\beta^2 - n_2^2k_0^2} \cdot K_m'(\sqrt{\beta^2 - n_2^2k_0^2}\,a)}.$$

Our fiber is a single mode fiber, which has only a fundamental mode propagating along the fiber that corresponds to $m = 0$. Substituting $m = 0$ into the above equation, we get

$$\frac{J_0(\sqrt{n_1^2 k_0^2 - \beta^2}\, a)}{\sqrt{n_1^2 k_0^2 - \beta^2} \cdot J_0'(\sqrt{n_1^2 k_0^2 - \beta^2}\, a)} = \frac{K_0(\sqrt{\beta^2 - n_2^2 k_0^2}\, a)}{\sqrt{\beta^2 - n_2^2 k_0^2} \cdot K_0'(\sqrt{\beta^2 - n_2^2 k_0^2}\, a)}.$$

This equation can be further simplified by using the identity relationship of Bessel functions; i.e., $J_0'(x) = J_1(x)$ and $K_0'(x) = K_1(x)$. Then, we get

$$\frac{J_0(\sqrt{n_1^2 k_0^2 - \beta^2}\, a)}{\sqrt{n_1^2 k_0^2 - \beta^2} \cdot J_1(\sqrt{n_1^2 k_0^2 - \beta^2}\, a)} = \frac{K_0(\sqrt{\beta^2 - n_2^2 k_0^2}\, a)}{\sqrt{\beta^2 - n_2^2 k_0^2} \cdot K_1(\sqrt{\beta^2 - n_2^2 k_0^2}\, a)}. \quad (3.34)$$

Equation (3.34) is a transcendental equation, which does not have an analytical solution. To find propagation constant β, the graph method is employed. The MathCAD program is used to draw both the left and right parts of Eq. (3.34) as a function of β and the intersection point of these two curves gives the propagation constant β, as shown in Fig. 3.4. The propagation constant $\beta = 7.103$ and the effective refractive index is $\bar{n} = \beta/k_0 = 1.456$, which is smaller than n_1 and larger than n_2; that is consistent with the theoretical analysis.

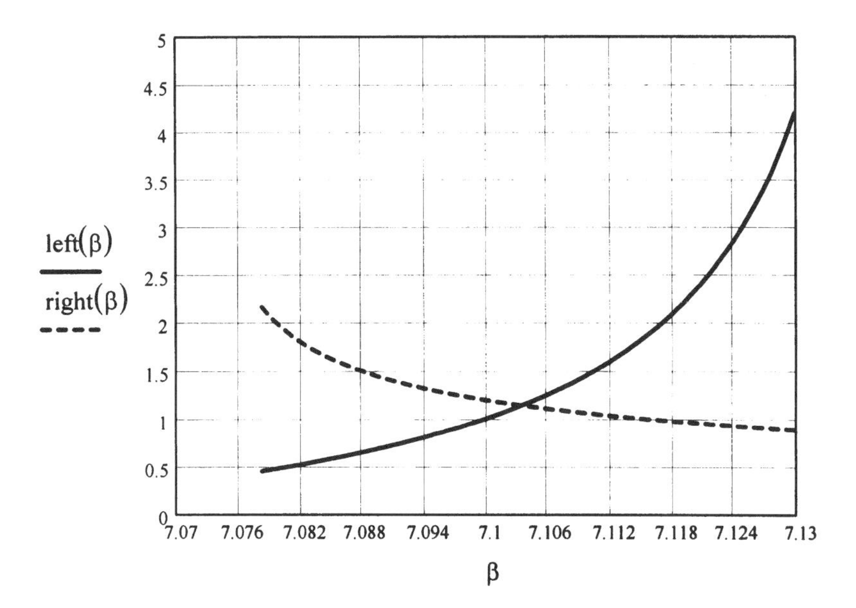

Fig. 3.4. Curves of left and right part of Eq. (3.34) as a function of β.

3.2.3. OTHER ISSUES RELATED TO LIGHT PROPAGATING IN OPTICAL FIBER

3.2.3.1. *Attenuations in Optical Fiber*

As discussed in Secs. 3.2.1 and 3.2.2, we know that light can be kept in optical fiber based on the principle of total internal reflections. However, some mechanisms can cause losses in optical fiber. Figure 3.5 shows attenuation as a function of wavelength. When the operating wavelength $\lambda < 1.3\,\mu m$, the loss mainly comes from the Rayleigh scattering; that is, $\propto 1/\lambda^4$. However, when $\lambda > 1.6\,\mu m$, the loss will become bigger and bigger due to infrared absorption. There is also a loss peak around $\lambda = 1.4\,\mu m$ that is mainly due to the absorption of $-OH$. Thus, to minimize the loss, current communication systems are operated in the low-loss windows centered at $1.3\,\mu m$ and/or $1.55\,\mu m$.

Note that recently it has been reported that this water absorption feature has been almost completely removed for Lucent All Wave fiber, so a much wider communication window (from 1.3 to 1.5 mm) can be used.

3.2.3.2. *Fabricating Optical Fibers*

Optical fibers are made by drawing a fiber mother rod that is called a *preform*, as shown in Fig. 3.6, which has a typical diameter of a few centimeters. The central part of the rod has a higher refractive index, obtained by doping GeO_2. This part corresponds to the fiber core. The rod is heated to about

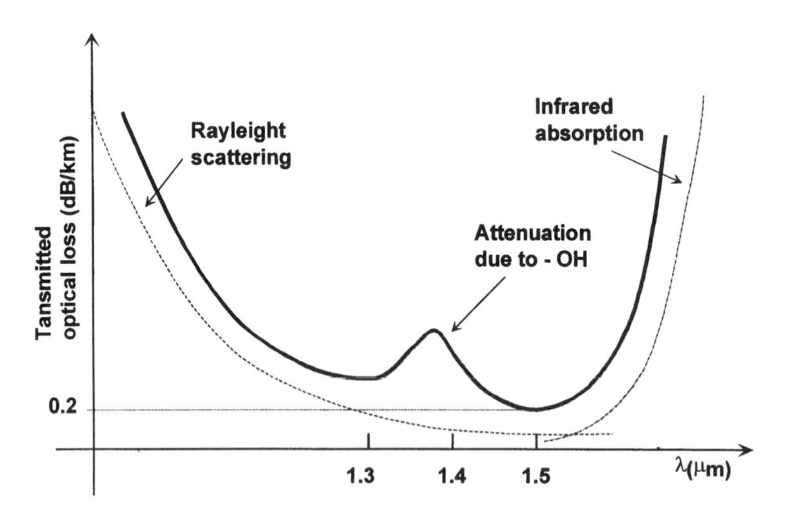

Fig. 3.5. Optical loss (or attenuation) in optical fiber.

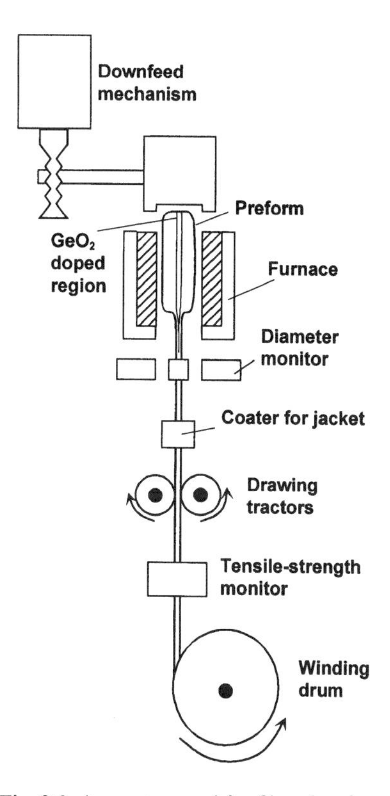

Fig. 3.6. Apparatus used for fiber drawing.

2000°C by a furnace for the glass fiber. The dimensional and mechanical properties of the optical fibers are all determined by the drawing process. The typical core diameters are 4–9 μm for single mode fiber and 50 μm for multimode fiber. The typical cladding outside diameter is about 125 μm for both single mode and multimode fibers. In order to get good quality, during manufacturing the fiber diameter is feedback controlled by varying the drawing speed using fiber diameter monitoring signals. The drawing speed is typically 0.2–0.5 m/s. The monitoring signals are usually obtained by a contactless laser measurement method. A laser light focuses on a fiber and the fiber diameter is measured by processing the scattering laser light. Using the drawing control technique and the precise preform fabrication technique, precision of fiber dimensions to about 0.1 μm has been achieved. This is very important for passive optical components, because fiber dimensions determine some of their performance for many components. For example, the accuracy of fiber core and outer diameter is critical for determining a splice loss. For the purpose of protection, the fiber drawing is accompanied by fiber coating (plastic coating).

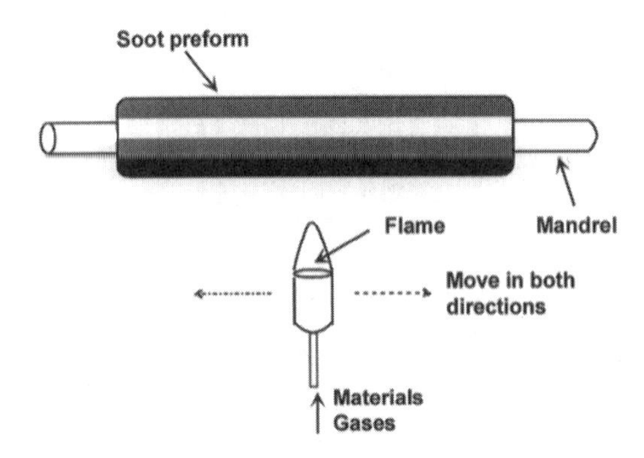

Fig. 3.7. Fiber preform OVD fabrication method.

The typical diameter of the coated fiber is about 250 μm, although it can be as large as 900 μm when multiple coatings are used.

The preform is made using several vapor deposition methods. Typical methods are the outer vapor deposition (OVD) method, as shown in Fig. 3.7, the modified chemical vapor deposition (MCVD) method, and the vapor-phase axial deposition (VAD) method. In the OVD method, submicron-sized particles generated by a flame are deposited on a rotating rod (mandrel). The mandrel is removed from a soot preform before a sintering process and the soot preform is dehydrated. The removal of OH (the ion with a combination of oxygen and hydrogen) is important to lower transmission loss. In the sintering process, a soot preform changes to a glass preform.

3.2.3.3. *Transversal Patterns for Single Mode Fiber*

As discussed in the previous sections, the electric field, E_z, has a distribution described by Eq. (3.25). Since the single mode fiber is the most widely used case, we focus our discussion on the single mode fiber field. This corresponds to the fundamental mode; i.e., $m = 0$. Substituting $m = 0$ into Eq. (3.25), the normalized transversal distribution of E_z is

$$E_z(\rho) = \begin{cases} \dfrac{J_0(\kappa\rho)}{J_0(\kappa a)}, & \rho \leqslant a \\ \dfrac{K_0(\gamma\rho)}{K_0(\gamma a)}, & \rho > a. \end{cases} \tag{3.35}$$

To make the calculation easier, a simplified empirical formular was developed for the case of $1.2 < V < 2.4$. In this case, the normalized electric field $E_z(\rho)$

can be expressed as

$$E_z(\rho) = e^{-\rho^2/W^2}, \qquad W = a\left(0.65 + \frac{1.619}{V^{3/2}} + \frac{2.879}{V^6}\right), \tag{3.36}$$

which is basically Guassian function. To understand Eqs. (3.35) and (3.36), let us look at the following example.

Example 3.7. A single mode silica fiber has a radius $a = 2.6\ \mu m$, core refractive index $n_1 = 1.465$, the cladding refractive index $n_2 = 1.45$, and operating wavelength $\lambda = 1.55\ \mu m$.

(a) Draw the transversal electric field distribution $E_z(\rho)$ for both exact formular (i.e., Eq. [3.35]) and empirical Gaussian approximation formular (i.e., Eq. [3.36]).
(b) Redo part (a) if the fiber radius is changed to $a = 1.2\ \mu m$.

Solve: (a) First, the parameters in Eqs. (3.35) and (3.36) are calculated. In case (a), the normalized frequency is

$$k_0 = \frac{2\pi}{\lambda} = \frac{2\pi}{1.55\ \mu m} = 4.054\ \mu m^{-1},$$

$$V = \frac{2\pi a\sqrt{n_1^2 - n_2^2}}{\lambda} = \frac{2\pi \cdot 2.6\ \mu m}{1.55\ \mu m}\sqrt{1.465^2 - 1.45^2} = 2.204$$

The propagation constant, β, is calculated using the graphic approach as described in Example (3.6). It is found that $\beta = 5.907$. Then, the parameters κ and γ are calculated:

$$\kappa = \sqrt{\mathrm{n}_1^2 k_o^2 - \beta^2} = \sqrt{1.465^2 \cdot 4.054^2 - 5.907^2} = 0.825.$$

Similarly, we get

$$\gamma = \sqrt{\beta^2 - n_2^2 k_0^2} = \sqrt{5.907^2 - 1.45^2 \cdot 4.054} = 0.586.$$

Based on these parameters, $E_z(\rho)$ was drawn using the MathCAD program, as shown in Fig. 3.8*a*. From Fig. 3.8, it can be seen that the difference between precise formular and empirical formular is very small. This confirms that the Gaussian approximation is very good when $1.2 < V < 2.4$. In our case $V = 2.204$ is indeed within this range.

(b) Based on the new radius, $a = 1.2\ \mu m$, we have recalculated the electric field distribution, as shown in Fig. 3.8*b*. There is a substantial difference

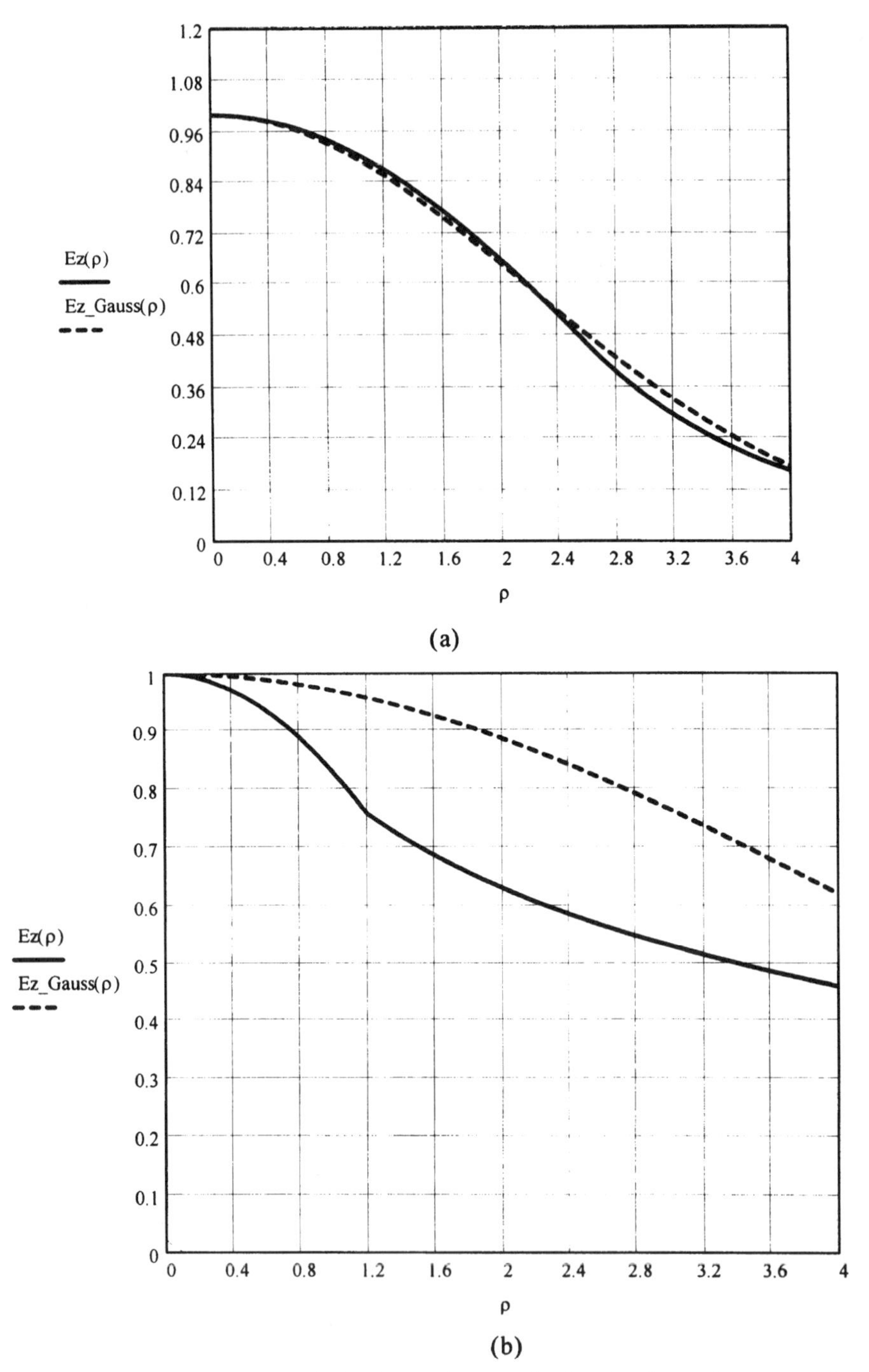

Fig. 3.8. (a) Transversal electric field distribution in the radial direction for good approximation case. Solid line: exact form; dash line: Gaussian approximation. (b) Transversal electric field distribution in the radial direction for bad approximation case. Solid line: exact form; dash line: Gaussian approximation.

between the two curves. Note that in this case, $V = 1.02$ is not within the range [1.2, 2.4]. Again, this confirms the requirement of $1.2 < V < 2.4$ when Eq. (3.36) is used.

Before the end of this section, we would like to provide percentage power of energy inside the core, η, under Gaussian approximation.

$$\eta = \frac{P_{\text{core}}}{P_{\text{cladding}}} = \frac{\int_0^{2\pi} d\phi \int_0^a |E_z(\rho)|^2 \rho \, d\rho}{\int_0^{2\pi} d\phi \int_0^\infty |E_z(\rho)|^2 \rho \, d\rho} = 1 - e^{-2a^2/w^2}, \tag{3.37}$$

where w is determined by Eq. (3.36). Thus, w is in fact a function of normalized frequency V. It can be calculated that when $V = 2$, about 75% light energy is within the core. However, when V becomes smaller, the percentage of light energy within the core also becomes smaller.

3.2.3.4. *Dispersions for Single Mode Fiber*

Dispersion in fiber optics is related to the bit rate or bandwidth of fiber-optic communication systems. Due to dispersion, the narrow input pulse will broaden after propagating in an optical fiber. As discussed in the previous sections, different modes may have different propagating constants, β_{mn}. Thus, for a multimode fiber, a narrow input pulse can generate different modes, which propagate at different speeds. Thus, the output pulse broadens, as illustrated in Fig. 3.9. This type of dispersion is called intermodal dispersion, which is large. For example, for a step index fiber with $n_1 = 1.5$, $(n_1 - n_2)/n_1 = 0.01$, and length $L = 1$ km, the width of the output pulse can be as wide as

$$\tau = \frac{n_1 L}{c}\left(\frac{n_1}{n_2} - 1\right) = 50 \text{ ns}$$

(as discussed in the geometric optics approach section). The corresponding

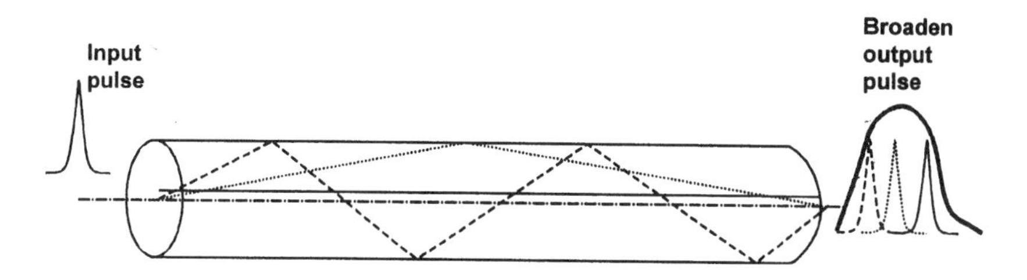

Fig. 3.9. Illustration of intermodal dispersion for multimode fiber.

maximum bit rate is $B \approx 1/\tau = 20$ Mbps. This is not that high; thus, multimode fiber is not suitable for high-speed long-haul communication.

Fortunately, when the normalized frequency is $V < 2.4$, we can have single mode fiber operation. In this case, there is only one mode; there will not be intermodal dispersion. Thus, the expected dispersion will be significantly smaller. However, the dispersion is still not zero. In this case, we have intramodal dispersion due to material dispersion and waveguide dispersion.

3.2.3.4.1. *Material Dispersion for Single Mode Fiber*

Since refractive index is a function of wavelength (i.e., $n = n[\lambda]$), different wavelength light will propagate at different speeds (i.e., $v(\lambda) = c/n[\lambda]$). Since any real light source always has a finite size spectral bandwidth $\Delta\lambda$, we will have dispersion. This type of dispersion is called material dispersion.

For long-haul fiber-optic communication, the single longitudinal mode diode laser is used as the light source, which has a very narrow spectral line width ($\Delta\lambda \leqslant 0.1$ nm). Thus, a very low material dispersion can be achieved.

The mathematical description of material dispersion is

$$D_m = \Delta\tau_m = -\frac{L}{c}\lambda\frac{d^2n(\lambda)}{d\lambda^2}\Delta\lambda. \tag{3.38}$$

Example 3.8. A pure silica fiber has a length $L = 1$ km. Use Eq. (3.38) to calculate the pulse broadening, $\Delta\tau$, due to the material dispersion and maximum bit rate of a laser light source that has spectral line width $\Delta\lambda = 2$ nm, operating wavelength $\lambda = 1.55$ μm, and

$$\frac{d^2n(\lambda)}{d\lambda^2} = -0.00416491\ (\mu\text{m}^{-2}).$$

Solve:

$$\begin{aligned}\Delta\tau &= -\frac{L}{c}\lambda\frac{d^2n(\lambda)}{d\lambda^2}\\ \Delta\lambda &= -\frac{10^3\ \text{m}}{3\times10^8\ \text{m/s}}\times 1.55\ \mu\text{m}\times(-0.00416491\ \mu\text{m}^{-2})\times 2\times10^{-3}\ \mu\text{m}\\ &\approx 40\ \text{ps}.\end{aligned}$$

The corresponding maximum bit rate is $B = 1/\Delta\tau \approx 25$ Gbps.

3.2.3.4.2. *Waveguide Dispersion for Single Mode Fiber*

Besides material dispersion, we also have waveguide dispersion. Waveguide dispersion comes from the fact that the propagation constant β depends on normalized frequency V that depends on λ even when $n(\lambda)$ is a constant. The waveguide dispersion, D_w, can be written as

$$D_w = -\frac{L}{c} \cdot n_2 \cdot \Delta \cdot \left(\frac{\Delta\lambda}{\lambda}\right) \cdot \left(V \frac{d^2(bV)}{dV^2}\right), \tag{3.39}$$

where

$$\Delta = (n_1 - n_2)/n_2,$$

$$b = \frac{\beta/(k_0) - n_2}{n_1 - n_2} \quad \text{(normalized propagation constant)},$$

$$V \frac{d^2(bV)}{dV^2} \approx 0.08 + 0.549 \cdot (2.834 - V)^2.$$

Example 3.9. For a single mode step index fiber, we have $n_1 = 1.4508$, $n_2 = 1.446918$, $a = 4.1\ \mu\text{m}$, $\lambda = 1560$ nm, $\Delta\lambda = 1$ nm, and $L = 1$ km.

Solve:

Step 1: Calculate the normalized frequency

$$V = \frac{2\pi a}{\lambda} \sqrt{n_1^2 - n_2^2} = 1.751.$$

Step 2: Calculate

$$V \frac{d^2(bV)}{dV^2} \approx 0.08 + 0.549 \cdot (2.834 - V)^2 = 0.724.$$

Step 3: Calculate

$$D_w = -\frac{L}{c} \cdot n_2 \cdot \Delta \cdot \left(\frac{\Delta\lambda}{\lambda}\right) \cdot \left(V \frac{d^2(bV)}{dV^2}\right) = -6\ \text{ps}.$$

The total dispersion for the single mode fiber is the summation of material and waveguide dispersions. Figure 3.10 shows the total dispersion as a function of operating wavelength for the standard silica fiber under the conditions of $\Delta\lambda = 1$ nm and $L = 1$ km. For the pure silica single mode fiber, the dispersion is on the order of tens of ps/nm-km, which is much smaller than that of multimode fiber (tens of ns/nm-km). When $\lambda \approx 1.31\ \mu\text{m}$, we have zero disper-

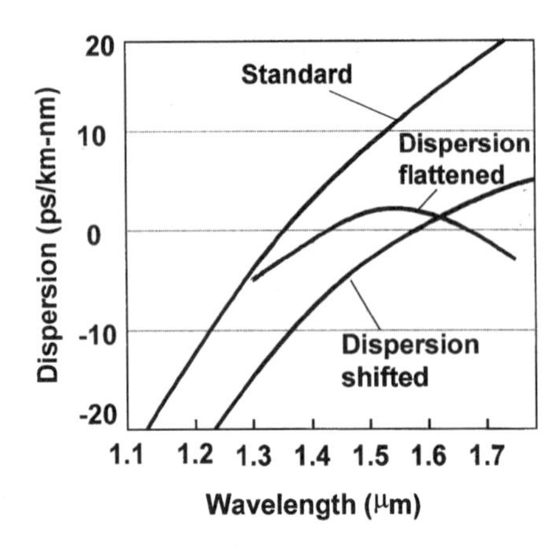

Fig. 3.10. Dispersion of silica fiber as a function of operating wavelength for different types of single mode fibers.

sion. Since the waveguide dispersion contribution D_w depends on fiber parameters, such as the core radius and the index difference Δ, it is possible to design the fiber such that the zero dispersion point is shifted to the vicinity of 1.55 μm, as shown in Fig. 3.10. Such fibers are called dispersion-shifted fibers. It is also possible to tailor the waveguide contribution such that the total dispersion is relatively small over a wide wavelength range (e.g., 1.3 to 1.6 μm). Such fibers are called dispersion flattened fibers, as shown in Fig. 3.10.

3.3. CRITICAL COMPONENTS

3.3.1. OPTICAL TRANSMITTERS FOR FIBER-OPTIC COMMUNICATIONS—SEMICONDUCTOR LASERS

We would like to find out what kind of light sources are required for fiber-optic communications. The light sources to be used for fiber-optic communications must have the following unique properties:

- Proper output wavelengths—centered at $\lambda = 1.3$ μm range or $\lambda = 1.5$ μm so that lower attenuations can be achieved.
- Narrow bandwidth $\Delta\lambda$ so that low intramodal dispersion can be achieved.
- High energy efficiency and compact size.

Based on the above requirements, semiconductor lasers or laser diodes are the best candidates because they have output spectral range from visible to infrared, narrow bandwidth (<1 nm), compact size, and high energy efficiency ($>50\%$).

A semiconductor laser consists of a forward-biased p–n junction. When the forward current through the diode exceeds a critical value known as the threshold current, optical gain in the resonator due to stimulated emissions overcomes the losses in the resonator, leading to net amplification and eventually to steady-state laser oscillation, as described in detail in the following.

Under the forward bias, free electrons and holes will move back to the depleted region. Thus, there is a chance for "recombination." This recombination process releases the energy in the "light" form → generate light. In terms of band gap theory, there are two allowed energy levels existing in the material, which are called the conduction band and the valence band. In the conduction band, electrons are not bound to individual atoms so that they are free to move. In the valence band, unbound holes are free to move. The generated photon energy is determined by the band gap between the conduction and valence bands. Mathematically, this can be written as

$$h\nu = E_c - E_v, \tag{3.40}$$

where h is the Planck's constant $h = 6.63 \times 10^{-34}\,\mathrm{J \cdot s}$, ν is the output light frequency, and E_c and E_v are the energy of conduction and valence bands, respectively. Figure 3.11 shows the basic structure of a semiconductor laser. The polished side surfaces form the resonant cavity so that a particular wavelength can be amplified.

Table 3.1 lists several types of materials used to fabricate semiconductor lasers and their corresponding operating wavelength range.

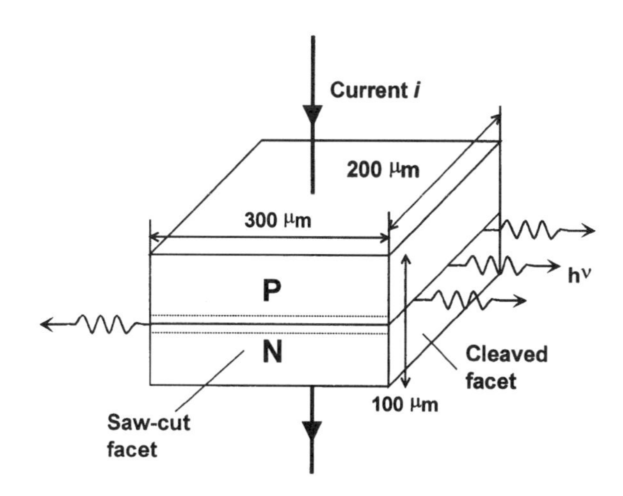

Fig. 3.11. Basic structure of the semiconductor laser.

Table 3.1

Material	Wavelength range (μm)	Band gap energy (eV)
GaInP	0.64–0.68	1.82–1.94
GaAs	0.9	1.4
AlGaAs	0.8–0.9	1.4–1.55
InGaAs	1.0–1.3	0.95–1.24
InGaAsP	0.9–1.7	0.73–1.35

To enhance the performance of the semiconductor laser, a heterostructure is generally used, as shown in Fig. 3.12. For example, for the AlGaAs–based laser, a heterostructure consists of a thin layer of GaAs sandwiched between two layers of *p*- and *n*-doped AlGaAs. With this structure, under the forward-bias case, a large concentration of accumulated carriers in the thin GaAs layer can be formed that leads to a large number of carrier recombinations and photon emissions so that medium gain can be achieved. In addition, the refractive index of GaAs is larger than that of AlGaAs, so this thin layer may also serve as an optical waveguide. The single transversal mode operation can be achieved when the thickness of the waveguide is thin enough; i.e., the normalized frequency V satisfies the following condition:

$$V = \frac{2\pi d}{\lambda}\sqrt{n_1^2 - n_2^2} < \pi, \tag{3.41}$$

where n_1 and n_2 are the refractive index of the GaAs layer and the AlGaAs layers, respectively. Thus, by introducing a thin GaAs layer, a good-quality optical beam can be achieved.

To further reduce the spectral linewidth of the output laser beam, $\Delta\lambda$, so that lower intramodal dispersion can be achieved, besides employing the single transversal mode structure, the single longitudinal mode operation is also

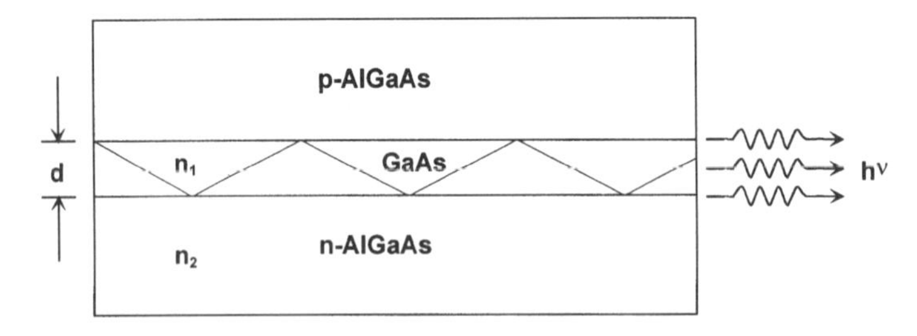

Fig. 3.12. Basic structure of the heterostructure semiconductor laser.

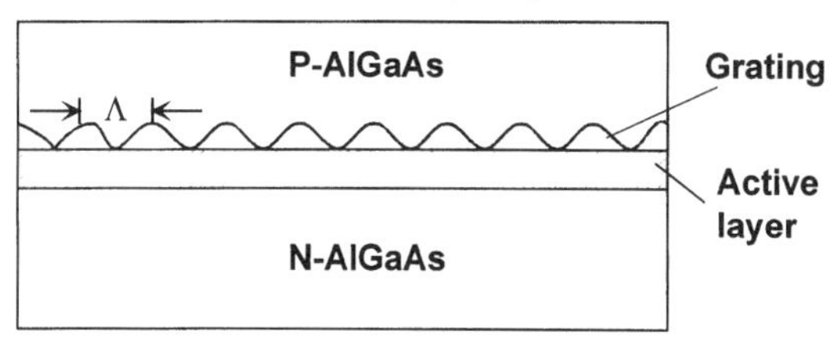

Fig. 3.13. Basic structure of the distributed feedback laser.

involved by using a wavelength-selective reflection. In general, a grating (a period structure) is fabricated into the laser cavity, as shown in Fig. 3.13. The only resonant wavelength is the wavelength that satisfies the Bragg condition; i.e.,

$$2n\Lambda = m\lambda_B, \tag{3.42}$$

where m represents the order of diffraction, λ_B represents the wavelength that satisfies the Bragg condition (Eq. [3.42]), Λ represents the period of the grating, and n is the refractive index of the cavity. A spectral linewidth less than <0.1 nm can be achieved by using this distributed feedback structure.

Example 3.10. The band gap of the $Ga_{0.8}Al_{0.2}As$ semiconductor laser = 1.6734 eV. Calculate the output wavelength of this laser.

Solve:

$$E_g = h\nu = h\frac{c}{\lambda} \Rightarrow \lambda = \frac{hc}{E_g} = 0.74\ \mu\text{m}.$$

Example 3.11. The forward current through a GaAsP red LED emitting at 670 nm wavelength is 30 mA. If the internal quantum efficiency of GaAsP is 0.1, what is the optical power generated by LED?

Solve:

$$\frac{i}{e} \rightarrow \text{number of electrons/second}$$

$$\eta\frac{i}{e} \rightarrow \text{number of photons/second}$$

$$h\frac{c}{\lambda} \rightarrow \text{energy of each photon.}$$

Thus,

$$P_{\text{out}} = \eta \left(\frac{i}{e} \right) \frac{hc}{\lambda} = 5.5\,\text{mW represents energy of photons/second} \rightarrow \text{power.}$$

Example 3.12. A semiconductor laser has a heterostructure. If the cavity length $L = 0.2$ mm and the refractive index $n_1 = 3.5$, the spectral gain width of the material is $\Delta\lambda_g = 5$ nm. In addition, $n_2 = 3.4$ and $\lambda = 1.5\,\mu$m. Calculate

(a) The number of the longitudinal mode for this laser.
(b) The maximum thickness, d, to maintain the single transversal mode.
(c) If the first-order Bragg reflection happens, what is the required period of the distributed feedback Bragg grating?

Solve:

(a) The spectral space between the adjacent longitudinal modes is

$$\Delta\lambda_s = \frac{\lambda^2}{2n_1 L} = 1.6\,\text{nm}.$$

Thus, the number of longitudinal mode N_L is

$$N_L = \frac{\Delta\lambda_s}{\Delta\lambda_g} = \frac{5\,\text{nm}}{1.6\,\text{nm}} = 3\ \text{modes}.$$

$$V\ (b) \frac{2\pi d_{\max}}{\lambda}\sqrt{n_1^2 - n_2^2} = \pi \rightarrow d_{\max} = \frac{\lambda}{2\sqrt{n_1^2 - n_2^2}} = \frac{1.5\,\mu\text{m}}{2\sqrt{3.5^2 - 3.4^2}} = 0.9\,\mu\text{m}$$

(c) Since $m = 1$ we have $2n_1\Lambda = \lambda_B$. Thus,

$$\Lambda = \frac{\lambda_B}{2n_1} = \frac{1.5\,\mu\text{m}}{2 \cdot 3.5} = 0.21\,\mu\text{m}.$$

3.3.2. OPTICAL RECEIVERS FOR FIBER-OPTIC COMMUNICATIONS

In fiber-optic communication applications, an optical receiver is a device that converts input light signals into electronic signals. There are many types of optical receivers (also called *photodetectors*). However, the most widely used ones for fiber-optic communication are semiconductor optical receivers, including the PIN photodetector and the Avalanche photodetector.

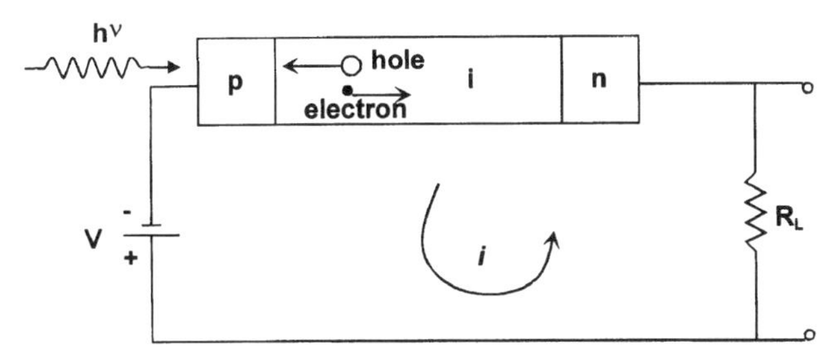

Fig. 3.14. Structure of the PIN photodetector.

3.3.2.1. *Principle of PIN Photodetector*

Figure 3.14 shows the basic structure of a PIN photodetector, which consists of an intrinsic semiconductor layer sandwiched between *p*-doped and *n*-doped layers. This is why it is called a PIN photodetector. In contrast to the optical transmitter, the photodetector is reversibly biased. This reverse bias can increase the thickness of the depleted region, which in turn results in a large internal electric field.

The basic process of light detection can be described as follows:

1. Light is incident on the PIN photodetector.
2. If the photon energy, $h\nu$ is greater than the band gap of the semiconductor, it can be absorbed, generating electron–hole pairs.
3. Under the reverse bias, the electron–hole pairs generated by light absorption are separated by the high electric field in the depletion layer; such a drift of carriers induces a current in the outer circuit.

3.3.2.2. *Principle of Avalanche Photodetector (APD)*

Figure 3.15 shows the basic structure of the APD. The difference between PIN and APD photodetectors is that the APD is a photodiode with an internal current gain that is achieved by having a large reverse bias.

In an APD the absorption of an incident photon first produces electron–hole pairs just like in a PIN. The large electric field in the depletion region causes the charges to accelerate rapidly. Such charges propagating at high velocities can give a part of their energy to an electron in the valence band and excite it to the conducting band. This results in an additional electron–hole pair. This process leads to avalanche multiplication of the carriers.

For avalanche multiplication to take place, the diode must be subjected to large electric fields. Thus, in APDs, one uses several tens of volts to several hundreds of volts of reverse bias.

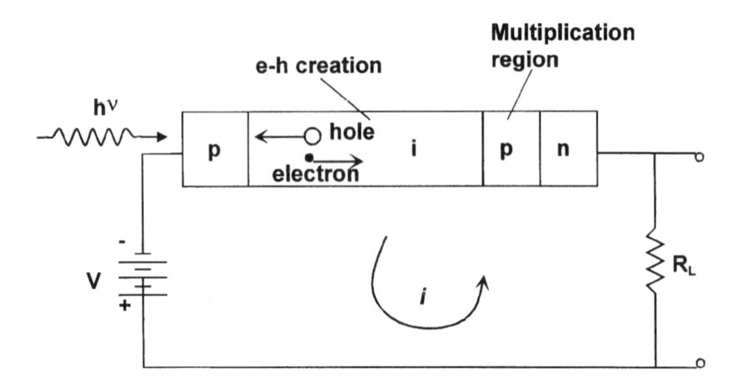

Fig. 3.15. Structure of APD.

3.3.2.3. *Important Parameters of Photodetectors*

The important parameters of photodetectors include:

1. Quantum efficiency

$$\eta = \frac{h\nu}{q} \frac{I_p}{P_{\text{in}}},$$

   where I_p is photogenerated current and P_{in} is the input light power.
2. Responsivity

$$R = \frac{\eta q}{h\nu}.$$

3. Response time.
4. Wavelength response of the photodetector. Since photo energy must be larger than the band gap, the detectors have wavelength response. Photodetectors fabricated for different materials can only measure different wavelength ranges.
5. Noise response of the photodetector. The conversion from light to electric current is accompanied by the addition of noise. There are two main types of noises, shot noise and thermal noise. Shot noise arises from the fact that an electric current is made up of a stream of discrete charges; namely, electrons, which are randomly generated. Thus, even when a photodetector is illuminated by constant optical power, due to the random generation of electron–hole (e–h) pairs, the current will fluctuate randomly around an average value determined by the average optical

power. Mathematically, the mean square of shot noise–determined current is

$$\langle i_{\mathrm{NS}}^2 \rangle = 2 \cdot e \cdot (I + I_d)\Delta f, \tag{3.43}$$

where I is the average current generated by the detector, e is the electron charge unit, I_d is the dark current (i.e., the current arisen from thermally generated carriers without light illumination), and Δf is the bandwidth over which the noise is being considered. Thermal noise (also referred to as Jonshon noise or Nyquist noise) arises in the load resistor of the photodiode circuit due to random thermal motion of electrons. Mathematically, it can be written as

$$\langle i_{\mathrm{NT}}^2 \rangle = \frac{4k_B T \cdot \Delta f}{R_L}, \tag{3.44}$$

where $k_B = 1.381 \times 10^{-23}$ J/K is the Boltzmann constant, T is the absolute temperature, and R_L is the resistance of the loading resistor. The thermal noise is proportional to the absolute temperature. This is why, in some cases, to reduce noise for very low light signal detection, we put the photodetector in liquid nitrogen (i.e., $T = 77$ K).

Table 3.2 summaries typical performance characteristics of detectors.

Example 3.13. The band gap energy of a PIN photodetector is 0.73 eV. Obtain the corresponding cutoff wavelength.

Solve:

$$E_g = h\nu = h\frac{c}{\lambda_c} \rightarrow \lambda_c = \frac{hc}{E_g} = \frac{6.63 \times 10^{-34}\,\mathrm{J \cdot s} \cdot 3 \times 10^8\,\mathrm{m/s}}{0.73 \cdot 1.6 \times 10^{-19}\,\mathrm{J}} = 1.7\,\mu\mathrm{m}.$$

Table 3.2

Material	Silicon		Germanium		InGaAs	
Types	PIN	APD	PIN	APD	PIN	APD
Wavelength range (nm)	400–1100		800–1800		900–1700	
Peak (nm)	900	830	1550	1300	1300 1550	1300 1550
Responsivity (A/W)	0.6	77–139	0.65–0.7	3–28	0.6–0.8	
Quantum efficiency (%)	65–90	77	50–55	55–75	60–70	60–70
Gain (M)	1	150–250	1	5–40	1	10–30
Bias voltage (−V)	45–100	220	6–10	20–35	5	<30
Dark current (nA)	1–10	0.1–1.0	50–500	10–500	1–20	1–5
Capacitance (pF)	1.2–3	1.3–2	2–5	2–5	0.5–2	0.5
Rise time (ns)	0.5–1	0.1–2	0.1–0.5	0.5–0.8	0.06–0.5	0.1–0.5

Example 3.14. A silicon PIN photodiode operates at 850 nm. Assume that the input optical power is 1 mW with a responsivity of 0.65 A/W and the detector bandwidth is 100 MHz.

(a) Calculate the photocurrent.
(b) Calculate the shot noise current.
(c) Calculate the thermal noise current if $T = 300$ K and $R_L = 500\,\Omega$.
(d) Calculate the signal-to-noise ratio of this detector (in dB).

Solve:

(a) $I_p = R \cdot P_{\text{in}} = 0.65\ \text{A/W} \cdot 10^{-6}\ \text{W} = 0.65\ \mu\text{A}.$

(b) $\sqrt{\langle i_{\text{NS}}^2 \rangle} = \sqrt{2 \cdot e \cdot I_p \cdot \Delta f} = \sqrt{2 \cdot 1.6 \times 10^{-19}\ \text{C} \cdot 0.65 \times 10^{-6}\ \text{A} \cdot 10^8\ \text{Hz}} = 4.5\ \text{nA}.$

(c)
$$\sqrt{\langle i_{\text{NT}}^2 \rangle} = \sqrt{\frac{4kB \cdot T \cdot \Delta f}{R_L}}$$
$$= \sqrt{\frac{4 \times 1.381 \times 10^{-23}\ \text{J/K} \times 300\ \text{K} \times 10^8\ \text{Hz}}{500\ \Omega}} = 57.5\ \text{nA}.$$

(d)
$$\text{SNR} = \frac{I_p^2}{\langle i_{\text{NS}}^2 \rangle + \langle i_{\text{NT}}^2 \rangle} = \frac{650^2}{4.5^2 + 57.5^2} = 128 = 21\ \text{dB}.$$

3.3.3. OTHER COMPONENTS USED IN FIBER-OPTIC COMMUNICATIONS

Besides optical fibers, transmitters, and receivers, there are also other components needed in fiber-optic networks, including optical couplers, switchers, amplifiers, isolators, dispersion compensators, and more. Due to space limitations, we will not describe each component in detail this chapter. Interested readers can learn more in special books that deal with fiber-optic components.

3.4. FIBER-OPTIC NETWORKS

3.4.1. TYPES OF FIBER-OPTIC NETWORKS CLASSIFIED BY PHYSICAL SIZE

In terms of physical size, networks can be classified into three categories:

1. Local Area Networks (LANs). LAN is up to approximately 2 kilometers total span, such as Ethernets, token rings, and token buses.

2. Metropolitan Area Networks (MANs). MAN is up to approximately 100 kilometers, such as telephone local exchange environments or cable television distribution systems.
3. Wide Area Networks (WANs). WAN can be thousands of kilometers.

3.4.2. PHYSICAL TOPOLOGIES AND ROUTING TOPOLOGIES RELEVANT TO FIBER-OPTIC NETWORKS

We need to know the requirements of physical topologies, which include:

1. Scalability. The term *scalability* means the ability to expand the network to accommodate many more nodes than the number in the initial installation.
2. Modularity. The term *modularity* means the ability to add just one more node.
3. Irregularity. The term *irregularity* means that the topology should not be forced artificially into some unusual, highly stylized pattern that may not meet the user's requirement.

Physical topologies take many forms. The most widely used topologies include start, ring, and bus as shown in Fig. 3.16. Optical or electronic switches can be added to optical networks to realize the routing operation.

3.4.3. WAVELENGTH DIVISION MULTIPLEXED OPTICS NETWORKS

To fully employ extremely high carrier frequency (i.e., 10^{14} Hz) of light, multiplexing techniques are used. The most widely used multiplexing technique is called wavelength division multiplexing (WDM), which is based on the unique property of light, as described in the following.

Optical beams with different wavelengths propagate without interfering with one another, so several channels of information (each having a different carrier wavelength) can be transmitted simultaneously over a single fiber. This scheme, called wavelength division multiplexing (WDM), increases the information-carrying capacity by a number of w, where w is the number of wavelengths used in the optics networks. The critical components needed for WDM optics networks include the optical multiplexer and the optical demultiplexer. An optic multiplexer couples wavelengths with different light widths from individual sources to the transmitting fiber, as illustrated in Fig. 3.17. An optical demultiplexer separates the different carriers before photodetection of the individual signals, as shown in Fig. 3.18. The important parameters for optic multiplexers and demultiplexers are *insertion loss* and *cross talk*. The insertion

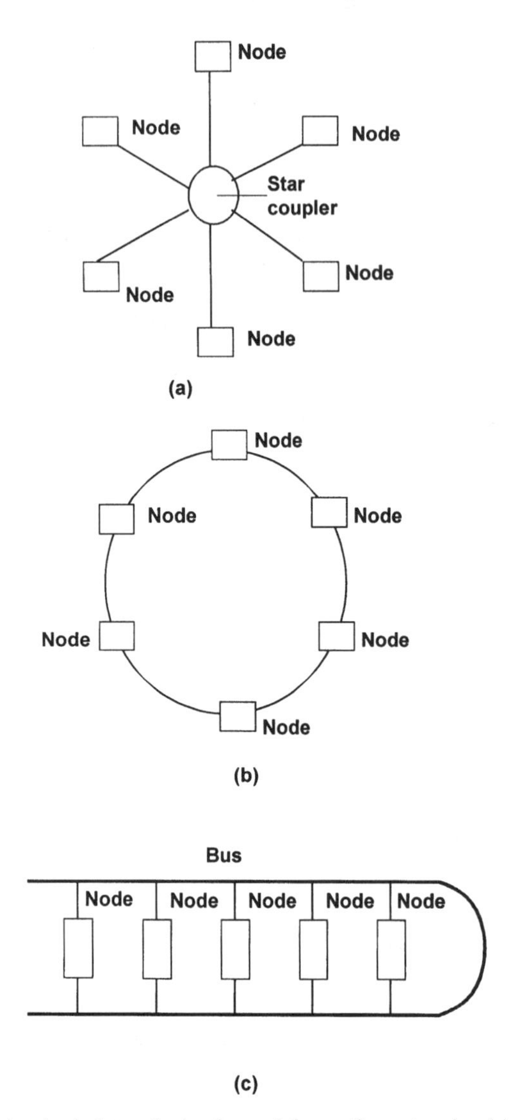

Fig. 3.16. Some basic physical topologies favored for optics networks. (a) Star, (b) Ring, (c) Reentrant bus.

loss has to be as small and as uniform as possible over different wavelength channels. Cross talk is light attenuation measured at an unintended port. The attenuation is as large as possible in this case. The best case is zero output at the unintended port due to the large attenuation. Multiplexers have been designed to accommodate numerous channels (more than 100 will be used

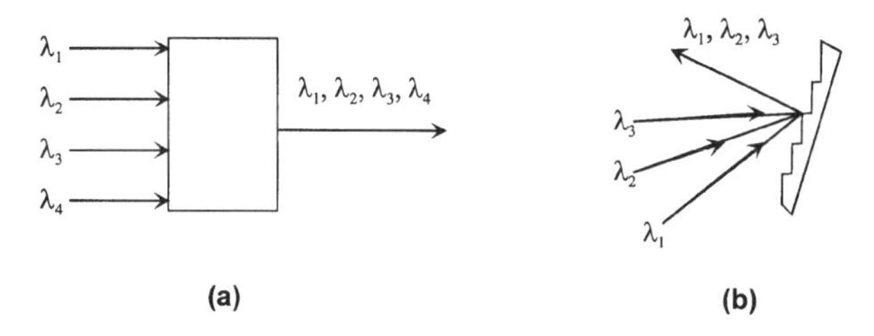

Fig. 3.17. Optical multiplexer for WDM optics networks. (a) Block diagram for optical multiplexer. (b) An example of physical implementation of optical multiplexer with blazed grating.

soon), with bandwidths and spacings under 1 nm. When there are more than just a few (e.g., >40) WDM channels, the system is referred to as dense wavelength division multiplexing (DWDM).

Figure 3.19 illustrates a 1 Tb/s WDM optical network used in the real system, which includes different wavelength sources (within c band and L band), polarization control (pc), waveguide grating routers (WGRs), a polarization beam splitter, an ultrawideband amplifier (UWBA), dispersion compensation fiber, a bandpass filter, a photo receiver, and a bit error rate (BER) monitor. Figure 3.20 illustrates the corresponding output signal spectrum for this 1 Tb/s experiment. Figure 3.21 shows the basic process of a digital fiber-optic communication link. To achieve a low bit error rate, amplification and dispersion compensation are necessary components.

3.4.4. TESTING FIBER-OPTIC NETWORKS

To ensure good performance, it is very important to test the functioning of optics networks. The most important parameter of a digital system is the rate at which errors occur in the system. A common evaluation is the bit error ratio

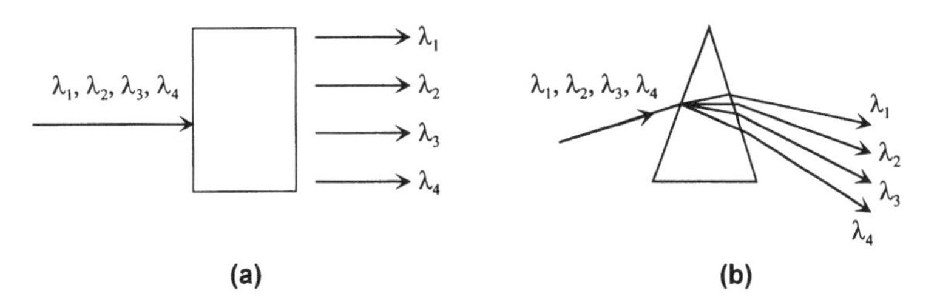

Fig. 3.18. Optical demultiplexer for WDM optics networks. (a) Block diagram for optical demultiplexer. (b) An example of physical implementation of optical demultiplexer with prism.

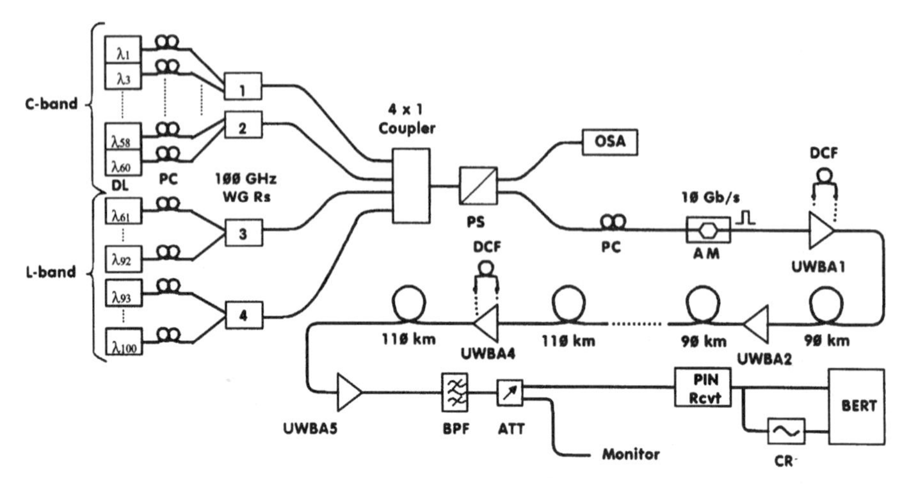

Fig. 3.19. Illustration of 1 Tbps WDM optical networks.

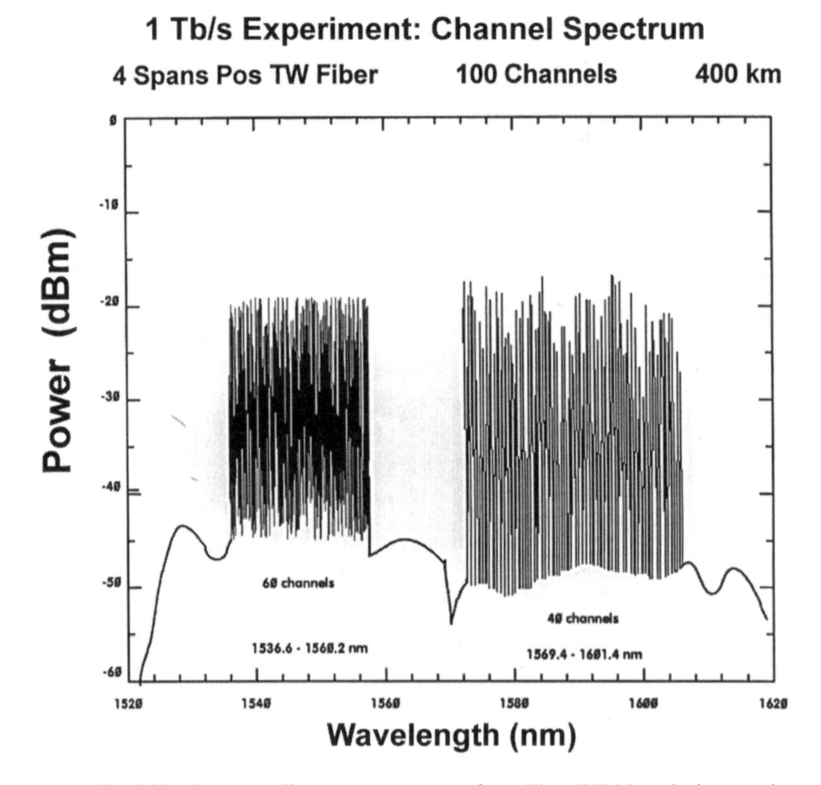

Fig. 3.20. Corresponding output spectrum for 1 Tbps WDM optical networks.

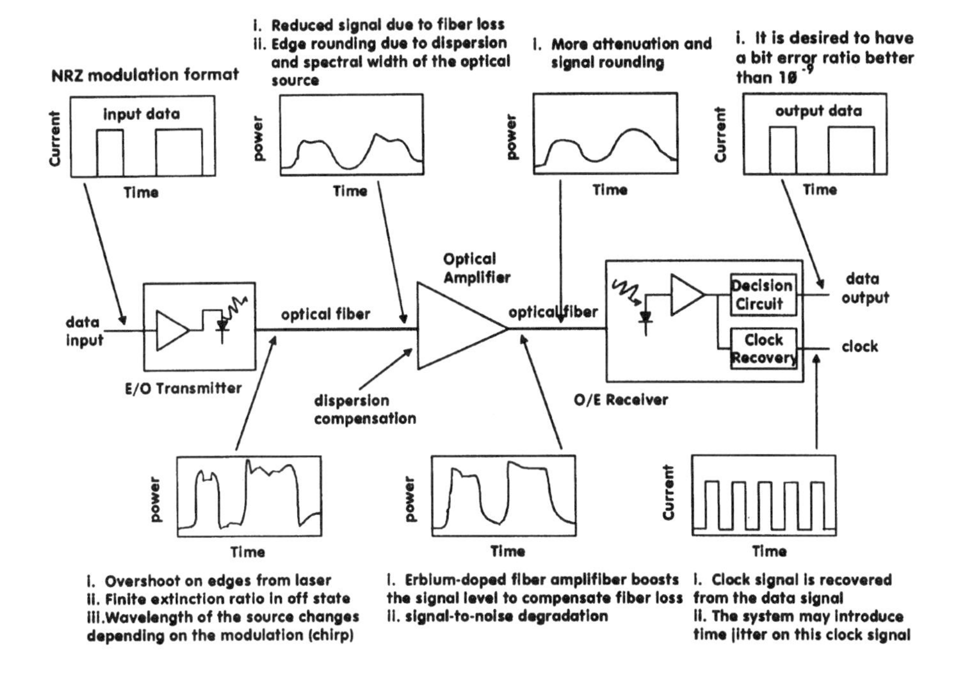

Fig. 3.21. Basic process for a digital fiber-optic communication link.

(BER) test, as shown in Fig. 3.22. A custom digital pattern is injected into the system. It is important to use a data pattern that simulates data sequences most likely to cause system errors. A pseudorandom binary sequence (PRBS) is often used to simulate a wide range of bit patterns. The PRBS sequence is a random sequence of bits that repeats itself after a set number of bits. A common pattern is $2^{23} - 1$ bits in length. The output of the link under test is compared to the known input with an error detector. The error detector records the number of errors and then ratios this to the number of bits transmitted. A BER of 10^{-9} is often considered the minimum acceptable bit error ratio for telecommunication applications. A BER of 10^{-13} is often considered the minimum acceptable bit ratio for data communications.

Bit error ratio measurements provide a pass/fail criteria for the system and can often identify particular bits that are in error. It is then necessary to troubleshoot a digital link to find the cause of the error or onto find the margin of performance that the system provides. Digital waveforms at the input and output of the system can be viewed with a high-speed oscilloscope to identify and troubleshoot problem bit patterns. In general, an eye diagram is used.

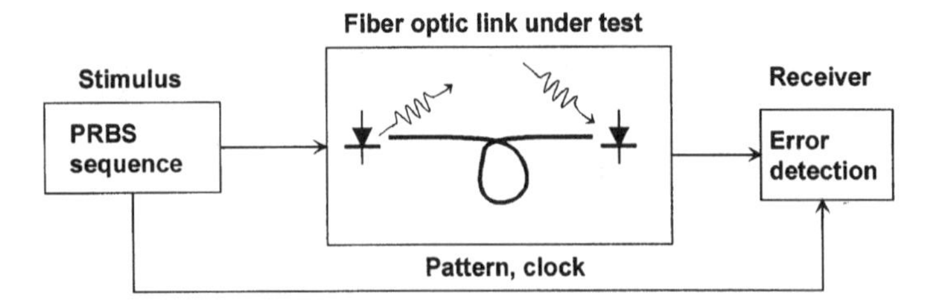

Fig. 3.22. Bit error ratio measurements and functional test.

Another link evaluation is clock jitter measurement. A perfect clock waveform would have a uniform bit period (unit interval) over all time. The fiber-optic system can add variability to the unit interval period, referred to as jitter. Jitter causes bit errors by preventing the clock recovery circuit in the receiver from sampling the digital signal at the optimum instant in time. Jitter originates primarily from noise generated by the regenerator electronic components.

REFERENCES

3.1 Govind P. Agrawal, *Fiber-Optic Communication Systems*, John Wiley & Sons, Inc., New York, 1997.
3.2 Stamatios V. Kartalopoulos, *Introduction to DWDM Technology*, SPIE Optical Engineering Press, New York, 2000.
3.3 Norio Kashima, *Passive Optical Components for Optical Fiber Transmission*, Artech House, Boston, 1995.
3.4 Joseph C. Palais, *Fiber-Optic Communications*, Prentice Hall, New Jersey, 1998.
3.5 Paul E. Green, Jr., *Fiber-Optic Networks*, Prentice Hall, New Jersey, 1993.
3.6 Ajoy Ghatak and K. Thyagar Ajan, *Introduction to Fiber Optics*, Cambridge Press, New York, 1998.

EXERCISES

3.1 Calculate the carrier frequency of optical communication systems operating at 0.88, 1.3, and 1.55 μm. What is the photon energy (in eV) in each case? (1 eV $= 1.6 \times 10^{-19}$ J).

3.2 A 1.55 μm fiber-optic communication system is transmitting digital signals over 100 km at 2 Gb/s. The transmitter launches 2 mW of average power into the fiber cable, having a net loss of 0.3 dB/km. How many photons are incident on the receiver during a single 1 bit ? (assume equal one and zeros).

3.3 A bare silica fiber has $n_1 = 1.46$, $n_2 = 1.0$, and radius $a = 25$ μm. Under the maximum angle of incidence,

(a) Calculate the number of reflections that would take place in traversing a kilometer length of the fiber.

(b) Assuming a loss of only 0.01% of power at each reflection at the core–cladding interface, calculate the corresponding loss in dB/km.

3.4 Consider a single mode optical fiber with $n_1 = 15$ and $n_2 = 1.485$ at wavelength $\lambda = 0.82$ μm. If the core radius is 50 μm, how many modes can propagate? Repeat if the wavelength is changed to 1.2 μm (Hint: Students can use the approximation formula for large V number.)

3.5 Consider a step index fiber with $n_1 = 1.45$, $\Delta = 0.003$, and $a = 3$ μm.

(a) For $\lambda = 0.9$ μm, calculate the value of V. Is this single mode fiber? Explain why or why not.

(b) For $\lambda = 1.55$ μm, recalculate the effective refractive index, $\bar{n}$, for this fiber. (Hint: A simple computer program made from MathCAD, MatLAB, C++ may be used.)

3.6 A single mode fiber has $n_1 = 1.465$, $n_2 = 1.45$, operating at $\lambda = 1.55$ μm. Draw the transversal mode field dispersion for (a) radius $a = 2.3$ μm and (b) $a = 1.3$ μm.

3.7 In a fiber-optic communication system, the light source is an LED operating at $\lambda = 0.85$ μm with a spectral width of 30 nm. Calculate the material dispersion $\Delta\tau$ in 1 km due to material dispersion and the maximum bit rate for this fiber.

3.8 Repeat Prob. 3.7 if the light source becomes a laser with a spectral width of 1 nm.

3.9 Consider a step index silica fiber operating at 1550 nm. If we know that $n_2 = 1.45$, $\Delta = 0.00905$, $a = 2.15$ μm, and the light source spectral bandwidth $\Delta\lambda = 1$ nm, calculate the waveguide dispersion for this single mode fiber 1 km long. (Hint: The approximation formula

$$V \frac{d^2(bV)}{dV^2} \approx -0.08 + 0.549 \times (2.834 - V)^2$$

can be used.)

3.10 Describe the principles of the light-emitting diode and the diode laser.

3.11 Sketch the output light spectrum of the LED, conventional diode laser, and DFB diode laser.

3.12 A GaAs laser diode has 1.5 nm gain linewidth and a cavity length of 0.5 mm. Sketch the output spectrum, including the emitted wavelengths and the number of longitudinal modes. Assume that $n = 3.35$, $\lambda = 0.9$ μm.

3.13 Calculate the maximum allowed thickness of the active layer of single transversal mode operation of a 1.3 μm semiconductor laser. Assume that $n_1 = 3.5$ and $n_2 = 3.2$.

3.14 Compute the grating space for an InGaAsP DFB laser diode operating at 1.3 μm. Give results for both the first- and second-order diffraction.

3.15 Plot the cutoff wavelength (in μm) as a function of the bandgap energy (in eV) for a PIN photodiode for wavelengths from 0.4 to 1.6 μm.

3.16 Compute the responsivity of a Si photodiode operating at 1.55 μm and having a quantum efficiency of 0.7.

3.17 In Prob. 3.16, how much optical power is needed by this detector to produce 20 nA photo current?

3.18 Consider a silicon PIN photodiode with responsivity $R = 0.65$ A/W, $R_L = 1000$ W, operating at 850 nm and at room temperature (i.e., $T = 300$ K). If the incident optical power is 500 nW and the bandwidth is 100 MHz,
(a) Compute the signal current.
(b) Compute the shot noise current.
(c) Compute the thermal noise current.
(d) Compute the signal-to-noise ratio (in dB).

3.19 Sketch a four-channel, full-duplex WDM network. Indicate the location of the transmitters, receivers, multiplexers/demultiplexers, and so on.

3.20 Describe the DWDM optics networks.

3.21 Describe how to test the bit error ratio of optics networks.

Chapter 4 Switching with Optics

Yang Zhao

Department of Electrical and Computer Engineering
Wayne State University
Detroit, MI 48202

Switches are one of the most important devices employed for manipulating optical signals, and are used in optical communication networks, optical displays, and light modulations. In the past several decades, many types of optical switches have been proposed and developed. A few books and monographs in this field have been published in the last several years [1, 2, 3]. This chapter does not attempt to review the field of optical switches. Instead, several switching devices are discussed as examples to introduce the concept and to illustrate the principles and applications of the field. Since the main advantages of using optical switches are ultrafast switching speed of more than 50 GHz and massive parallelism, we will examine three kinds of optical switches with these two advantages: ultrafast all-optical switches using nonlinear optics, fast electro-optic modulators to convert electric data to optical ones, and massive parallel switches using microelectromechanical systems.

Optical switches can be classified into two configurations, as shown in Fig. 4.1. One is an on–off switch in which the input is connected to one output port. The other is a routing switch in which the input is connected to two or more output ports. On–off switches are mainly used in modulation, light valves, and displays, while routing switches are used in connecting many nodes in networks. There are several ways to control an optical switch, and the performance of the switch largely depends on the control mechanism. Traditional control mechanisms include electro-optical effect, acousto-optic effect, magneto-optic effect, thermo-optic effect, piezoelectric effect, and electro-mechanical actuation. Switches based on these mechanisms have a speed well

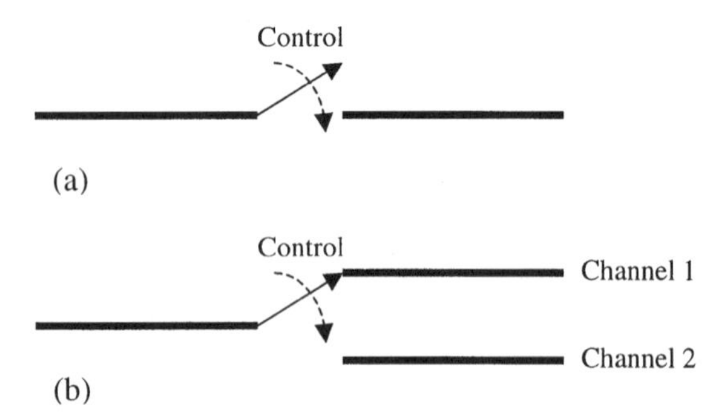

Fig. 4.1. Two configurations of optical switches. (a) On-off switch, (b) Routing switch.

below what is required in optical communication systems (50 Gbits/s). For application in free-space switching fabrics, these switches are bulky and difficult to integrate into a compact unit.

Recent technological developments have led to three very important types of optical switching devices. The first one, ultrafast all-optical switches, is based on nonlinear optical effects. It has many applications in digital optical signal processing, and can have a switching speed over 100 Gbits/s. The second one is fast electro-optic modulators using $LiNbO_3$ or semiconductor quantum wells. The third is based on a microelectromechanical system (MEMS). In contrast to conventional optomechanical systems that use bulky parts, these switching elements are batch fabricated by micromachining techniques. They are therefore smaller, lighter, faster, and cheaper than their bulk counterparts, and can be monolithically integrated with other optical components. While the speed of these switches is still slow, they can form large arrays in a compact unit and be used in networks with massive parallel connections.

4.1. FIGURES OF MERITS FOR AN OPTICAL SWITCH

There are several basic parameters used for evaluating the performance of an optical switch. These include on–off ratio, bandwidth or switching time, insertion loss, power consumption, and cross talk between channels.

The on–off ratio (also called contrast ratio) is the ratio of the maximum transmitted light intensity I_{max} to the minimum transmitted light intensity I_{min}, and is often described in decibels:

$$R_{on-off} = 10\log(I_{max}/I_{min}). \tag{4.1}$$

This ratio measures the quality of generated data by the switch and is related to the eventual error rate of the transmission system. An ideal switch would have an infinite on–off ratio ($I_{min} = 0$).

Switching time (τ) measures how fast the switch can perform, and is defined as the time required for switching the output intensity from 10% to 90% of I_{max}. It is related to the -3 dB bandwidth ($\Delta\nu$)

$$\Delta\nu = 0.35/\tau \text{ Hz}. \tag{4.2}$$

Insertion loss (L) describes the fraction of power lost when the switch is placed in the system. The insertion loss does not include the additional loss during switching, and is defined as

$$L(\text{dB}) = 10 \log P_{out}/P_{in}. \tag{4.3}$$

where P_{out} is the transmission power when the switch is not in the system, and P_{in} is the transmitted power when the switch is in the system and adjusted to provide the maximum transmission.

Power consumption is defined as the power consumed by the switch during operation. The consumed power will eventually turn into heat, and limit the number of switches or other devices can be put on a system unit. It will also set a demand on the power supply.

While the above parameters measure the performance of both on–off and routing switches, the final parameter, cross talk, only applies to the routing switches. It describes how effective a signal is isolated between two unconnected channels. Consider that a routing switch has one input and two outputs. When the input is connected to output channel 1 of the output, cross talk in this case describes how much of the input signal appears on channel 2. It is defined as

$$\text{Cross talk (dB)} = 10 \log(I_2/I_1), \tag{4.4}$$

where I_1 is the output intensity in the connected channel, and I_2 is the intensity in the unselected channel. Ideally, I_2 should be zero.

The above parameters depend on the material and configurations used in switching devices. In some switching configurations, wavelengths and polarization states of the signal and control beams may also affect these parameters. In these cases, the switches are called wavelength and polarization dependent.

4.2. ALL-OPTICAL SWITCHES

All-optical switches are nonlinear optical devices the output characteristics of which are controlled by the intensity of the input signal or by a separate optical signal for self-switching and controlled switching, respectively (Fig. 4.2).

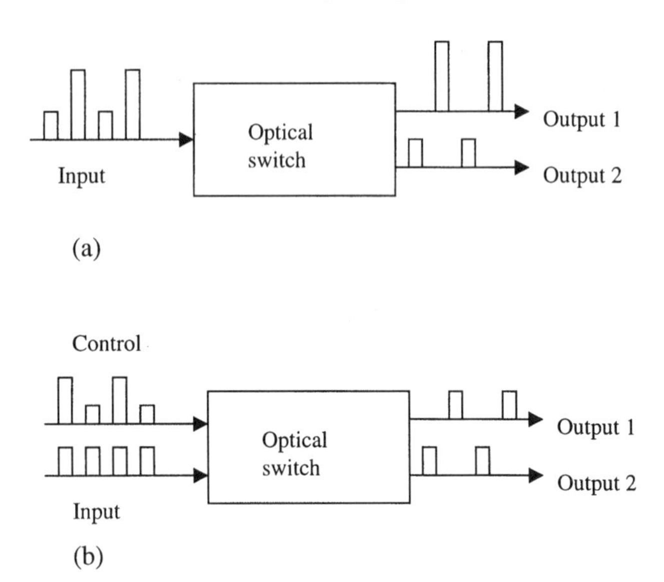

Fig. 4.2. All-optical switches. (a) Self-switching, (b) Controlled switching.

In the last two decades, nonlinear optical effects have been used to switch and route optical signals or perform logic operations on them in a number of ways. Switches with various materials and configurations have been proposed and demonstrated. This section discusses the principles of several all-optical switching devices, their material requirements, and their applications.

4.2.1. OPTICAL NONLINEARITY

The key element in all-optical switches is a medium with significant nonlinear optical effect [4]. The main materials property is an intensity-dependent refractive index $n(I)$, where I is the total intensity of the optical field in the medium. The nonlinear optical effect used in all-optical switching is the third-order effect, or the Kerr effect, where

$$n = n_0 + n_2 I. \tag{4.5}$$

Here n_0 is the linear refractive index and n_2 is the nonlinear refractive coefficient. For many optical materials, n_2 is very small. For example, silica glass has $n_2 = 3 \times 10^{-20}\,\mathrm{m^2/W}$. Therefore, we cannot directly observe any change in refractive index at low light intensity.

Nonlinear optical effect can be observed much more easily when we study the phase shift induced by the nonlinear refractive index. The phase shift of a medium is given by

$$\Delta\phi = \frac{2\pi}{\lambda} nl = \frac{2\pi}{\lambda} n_0 L + \frac{2\pi}{\lambda} n_2 I L, \tag{4.6}$$

where λ is the wavelength and L is the length of the medium over which the phase shift is accumulated. The phase shift due to nonlinear effect can be significant when L is large, even though n_2 is small. Therefore, it is a logical choice that all-optical switches are based on nonlinear phase shift.

For high-speed switching applications, nonlinear media are required to have a high nonlinear coefficient, high transparency at the operation wavelength, and fast response time. A general rule for selecting nonlinear media is that a change of π in phase can be achieved at a practical optical power level [5]. Since materials with a high nonlinear coefficient normally have high loss, it appears that most passive nonlinear materials are not suitable for high-speed switching applications. The best nonlinear medium is active semiconductor amplifiers, to be discussed in Sec. 4.2.4.3.

Refractive index can also be controlled by electro-optical effects, where the refractive index of the medium depends on an externally applied electric field. Electro-optical switches based on this effect will be discussed in Sec. 4.3.

4.2.2. ETALON SWITCHING DEVICES

A nonlinear optical medium in a Fabry–Perot etalon perhaps is the first optical switching device [6]. The scheme creates a system that can reduce the length required to produce a nonlinear phase shift, since the intracavity intensity can be much higher than the input intensity. When used for self-switching, the system has feedback, and exhibits optical bistability which has two possible stable output states for given input power. Optical bistability has many applications in optical signal routing, image processing, and set–reset operations.

This switching scheme is shown in Fig. 4.3. It consists of two mirrors with reflectivities R_1 and R_2, respectively, separated by a distance D. The medium in the etalon is assumed to be lossless, and have a refractive index n.

Let us consider the case of linear medium ($n_2 = 0$) and normal light incidence from the left-hand side. The transmittance of the etalon is given by [7]

$$T = I_{out}/I_{in} = A/(1 + F\sin^2\delta), \tag{4.7}$$

where $A = (1 - R_1)(1 - R_2)/(1 - \bar{R})^2$, $F = 4R/(1 - \bar{R})^2$, $\bar{R} = (R_1 R_2)^{1/2}$, $\delta = (2\pi/\lambda)(n_0 + n_2 I)D$ is the phase delay between the mirrors.

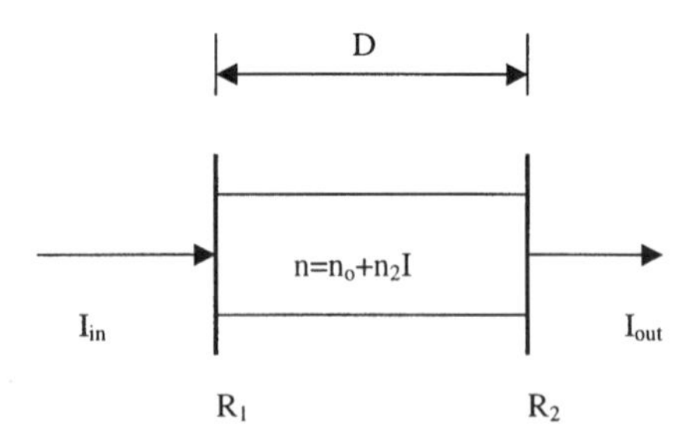

Fig. 4.3. A nonlinear Fabry–Perot etalon.

Obviously T is a function of δ. Figure 4.4 plots T versus δ for $R_1 = R_2 = R$. The maximum transmittance is unity and the minimum transmittance approaches zero as R approaches unity. Therefore, the switching operation can be realized by changing n. If the etalon is initially set at the maximum T, change in δ will switch T to a low value.

4.2.2.1. *Self-switching*

When a nonlinear medium is placed in the etalon, the phase delay becomes

$$\delta = (2\pi/\lambda)(n_0 + n_2 I)D, \tag{4.8}$$

where I is the total intensity in the etalon. In this case, the transmittance in Eq. (4.7) is a function of total intensity. On the other hand, the total intensity is related to the input intensity

$$I = I_{in}T/B, \tag{4.9}$$

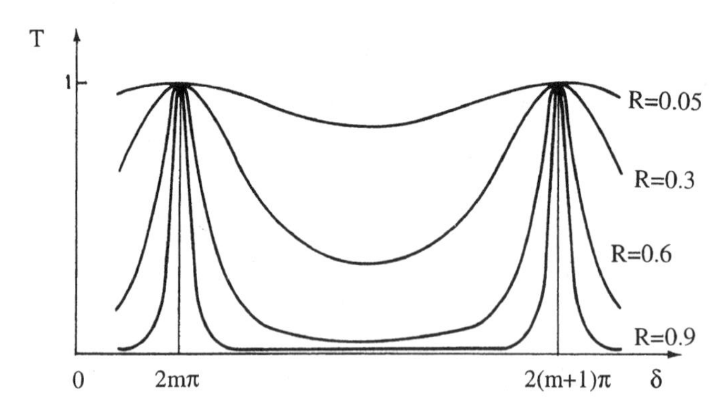

Fig. 4.4. Transmittance of a Fabry–Perot etalon as a function of δ for $R_1 = R_2 = R$.

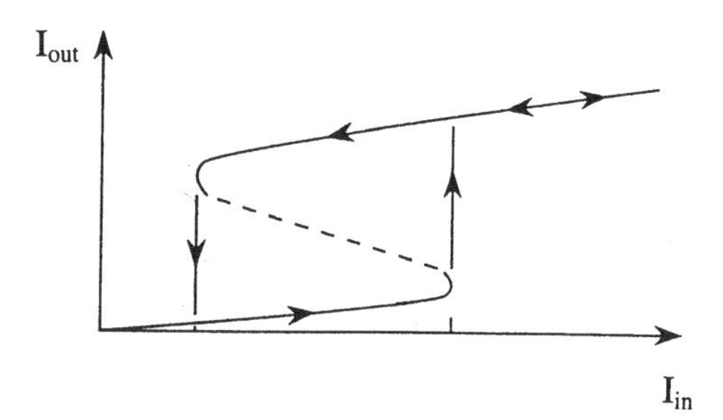

Fig. 4.5. Bistable characteristics of a nonlinear etalon.

where $B = D(1 - R_2)/(1 + R_2)$. Equations (4.7) and (4.9) have to be solved simultaneously in order to find T as a function of incident intensity I_{in}. With proper choices of parameters, the output intensity has bistable solutions as a function of I_{in} (Fig. 4.5). As we can see, in optical bistable devices, the output intensity can be controlled by the input intensity.

Optical bistability has been observed in a number of schemes using various materials. For switching applications of optical bistability, a medium with high n_2 must be used. This usually means materials with resonant nonlinearity. Unfortunately, resonant nonlinear optical media are always associated with high loss. For compact devices and compatibility with existing integrated optics, semiconductors are the preferred nonlinear materials. The nonlinearity in semiconductors is high and the loss can be compensated for if semiconductor amplifiers are used.

A typical application of optical bistable devices is all-optical set–reset operation. In set–reset operation, output intensity is controlled by a narrow additive input pulse. The principle of the all-optical set–reset operation is shown in Fig. 4.6. The input intensity has an initial bias (level B in Fig. 4.6). A "set" pulse (S) added to the input intensity will switch the output intensity to a "high" state (I_S), while a "reset" pulse (R) switches the output to a "low" state (I_R). Experimental implementations of all-optical set–reset operation have been reported in nonlinear etalons and distributed feedback waveguides.

4.2.2.2. *Controlled Switching*

Controlled switching devices can also be implemented using a separated control signal to change the intensity. In this case, δ in Eq. 4.8 becomes

$$\delta = \frac{2\pi}{\lambda}(n_0 + n_2 I_c)D \tag{4.10}$$

where I_c is the intensity of the control beam. The transmittance T can be

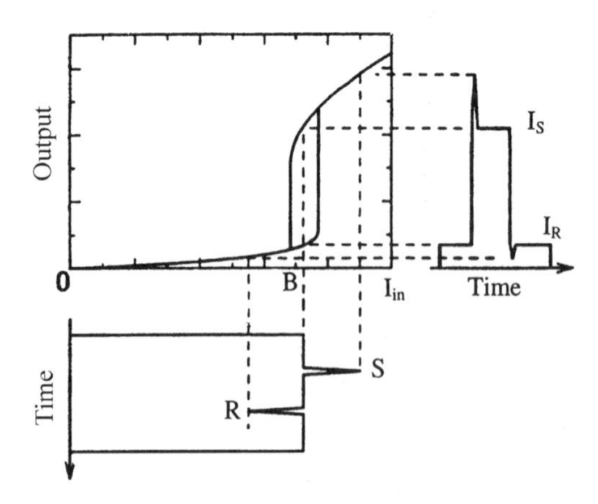

Fig. 4.6. Set-rest operation in an optically bistable system.

controlled by the control intensity I_c. The required amount of change in δ for switching depends on the linewidth of the transmission peak.

The control signal should be at a different wavelength or polarization, or different incident angle from the data signal, such that it can be separated from the data signal at the output.

4.2.3. NONLINEAR DIRECTIONAL COUPLER

A nonlinear directional coupler is an intensity-dependent routing switch. Figure 4.7 shows the structure of a directional coupler. It consists of two waveguides placed closely to one another for a finite distance. These waveguides are coupled together through their overlapped evanescent fields. They

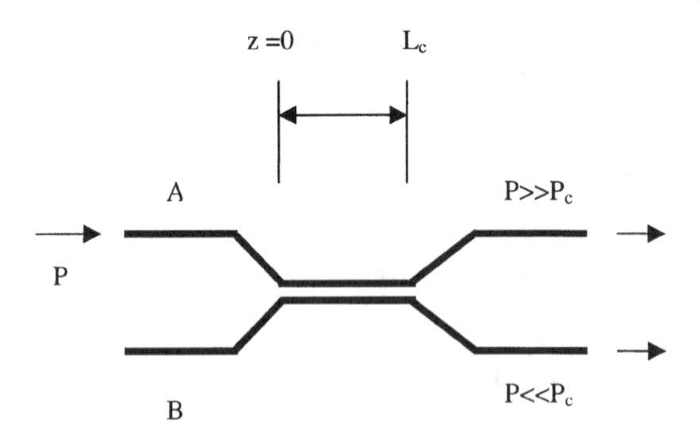

Fig. 4.7. Nonlinear directional coupler.

have a coherent interaction and periodically exchange power. Assume these waveguides are single mode. The fields in the core of the waveguide can be written as

$$\begin{aligned} E_a(x, y, z) &= A(z)\varepsilon_a(x, y)e^{-j\beta_a z} \\ E_b(x, y, z) &= B(z)\varepsilon_b(x, y)e^{-j\beta_b z}, \end{aligned} \tag{4.11}$$

where β_a and β_b are the propagation constants in the two waveguides. The coupling origins from the polarization perturbation form the presence of the evanescent field of the adjacent waveguide. For different propagation constants in A and B, we have coupled mode equations for field amplitudes A and B [7]:

$$\begin{aligned} \frac{dA}{dz} &= -j\kappa B - i\beta_a A \\ \frac{dB}{dz} &= -j\kappa A - i\beta_b B, \end{aligned} \tag{4.12}$$

where κ is the coupling coefficient which depends on the refractive indices n_1, n_2, geometry of the waveguide, and separation s. With input to waveguide A only; i.e., $A(0) = 1$ and $B(0) = 0$, the solution to Eq. (4.12) is

$$\begin{aligned} A(z) &= [\cos(gz) - j\Delta/(2g)\sin(gz)]e^{-j(\beta_a - \Delta)z} \\ B(z) &= [-j(\kappa/g)\sin(gz)]e^{-j(\beta_b - \Delta)z}. \end{aligned} \tag{4.13}$$

Here $g^2 = \kappa^2 + \Delta^2$ and the phase mismatch $\Delta = (\beta_a - \beta_b)/2$.

In terms of normalized power P_a and P_b in the two waveguides,

$$\begin{aligned} P_b &= (\kappa^2/g^2)\sin^2(gz) \\ P_a &= 1 - P_b. \end{aligned} \tag{4.14}$$

4.2.3.1. *Self-switching*

Self-switching occurs when the waveguides are made of nonlinear materials with refractive index $n = n_0 + n_2 I$. The coupled equations, however, become more complicated, and are given as

$$\begin{aligned} \frac{dA}{dz} &= -j\kappa B - i\beta_a A + \gamma|A|^2 A \\ \frac{dB}{dz} &= -j\kappa A - i\beta_b B + \gamma|B|^2 B. \end{aligned} \tag{4.15}$$

Here γ is the coefficient, depending on the nonlinearity of the material and the

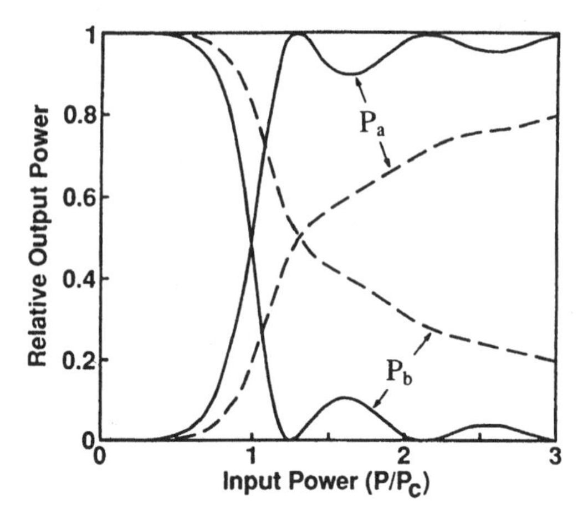

Fig. 4.8. Output power from waveguides A (P_a) and B (P_b) as a function of the input power P/P_c in self-switching. Solid curves: continuous-wave input signal. Dashed curves: soliton signal output. [9].

modal power in the waveguide. If the input signal with power P is sent into waveguide A, the output power in waveguide A can be described as [8]

$$\frac{P_a}{P} = \frac{1}{2}\left\{1 + cn\left[\frac{\pi z}{L_c}\middle|\left(\frac{P}{P_c}\right)^2\right]\right\}, \tag{4.16}$$

where P is the total input power, $cn(x|m)$ is the Jacobi elliptic function, L_c is the coupling length which is a function of the geometry and separation between the waveguides, $P_c = \lambda A_e/(n_2 L_c)$ is the critical power corresponding to the power needed for a 2π nonlinear phase shift in a coupling length, and A_e is the effective area of the two waveguides. For lossless waveguides, the output power P_b is $P - P_a$. Figure 4.8 shows the output power P_a and P_b as a function of the input power P for constant intensity and solitons when $z = L_c$ [9]. When P increases, more power is switched from waveguide A to waveguide B.

When $P \ll P_c$, the result in Eq. (4.16) is reduced to the case of a linear directional coupler (with $\beta_1 = \beta_2$); i.e.,

$$P_a/P = \tfrac{1}{2}\{1 + \cos[\pi z/L_c]\}. \tag{4.17}$$

Compared with Eq. (4.14), we see that $g = \pi/(2L_c)$.

The above results are for continuous-wave signals. For pulses, the situation is more complicated, and Eq. (4.15) becomes a partial differential equation involving both time and space. The solution to the partial differential equation can only be obtained through numerical analysis. It should be noted that a pulse contains a distribution of power, ranging from zero at the edge to peak power at the center. Therefore, different parts of the pulse are switched by different amounts, which may lead to pulse breakup. Several schemes have been developed to minimize this problem. One approach is to use temporal solitons, where the nonlinear effect for pulse breakup is balanced by dispersion. However, solitons require special conditions for their existence. More details on this topic as well as fundamentals of solitons can be found in [1].

NLDC has been realized using nonlinear waveguides using GaAs and other semiconductors as well as dual-core optical fibers.

4.2.3.2. *Controlled Switching*

Controlled switching can also be implemented. In this case, a separate control beam can be coupled into one of the nonlinear waveguides, and the refractive index of this waveguide depends on the intensity of the control signal. Since the input is an optical beam independent from the signal, the result is the same as the linear case, except that output in waveguides A and B is controlled by the nonlinear phase shift which affects Δ in Eq. (4.13). Assume that the two nonlinear waveguides are identical, and the control beam was sent into one of the waveguides. In this case, the phase mismatch (Δ) is proportional to $(2\pi/\lambda)n_2 I_c$, where I_c is the control intensity. Figure 4.9 shows the output power in waveguides A and B as a function of Δ (normalized to κ) for $z = L_c$.

Note that this scheme for controlled switching has also been realized using electro-optical media, such as $LiNbO_3$ and GaAs. In these cases the phase shift is controlled by an external electric field [7], as discussed in Sec. 3.

4.2.4. NONLINEAR INTERFEROMETRIC SWITCHES

The most successful form of all-optical switches is the nonlinear interferometer. The Sagnac, Mach–Zehnder, and Michaelson configurations have all been used for implementing optical switches. The basic operation is the same for all interferometric configurations where the nonlinear effect creates an additional intensity-dependent phase shift between the two arms. The signal to be switched is split between the arms of the interferometer. The output intensity of the interferometer depends on the total phase and, therefore, can be controlled using an external control signal. If we have two beams (1 and 2) with different optical paths, $n_1 r_1$ and $n_2 r_2$

$$
\begin{aligned}
E_1 &= A_1 \cos[(2\pi/\lambda)n_1 r_1 + \phi_1] \\
E_2 &= A_2 \cos[(2\pi/\lambda)n_2 r_2 + \phi_2],
\end{aligned}
\tag{4.18}
$$

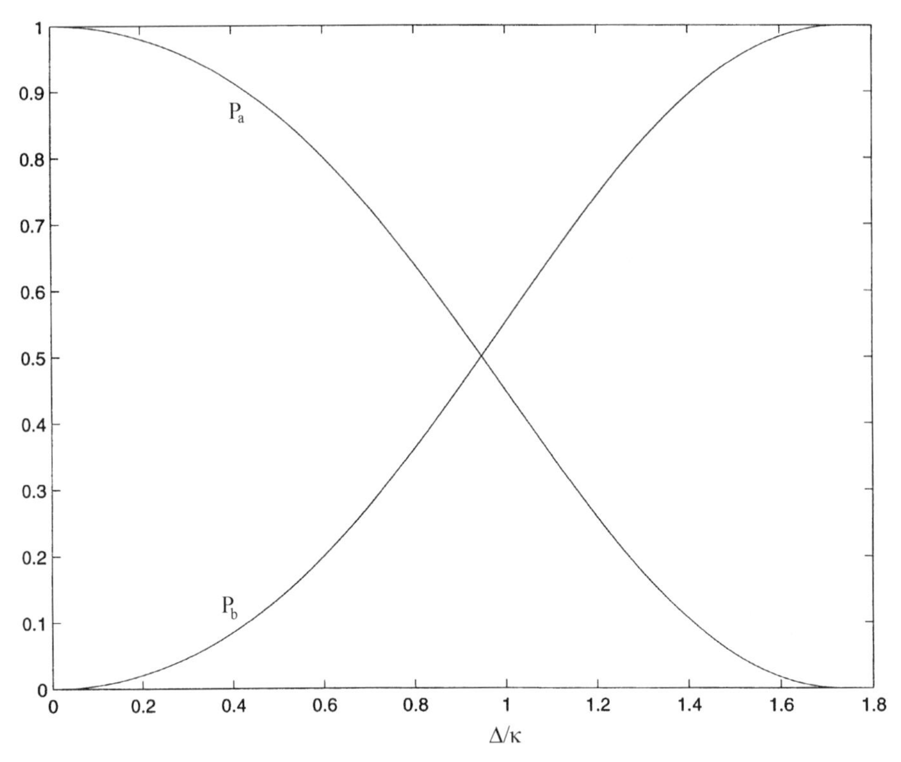

Fig. 4.9. Dependence of output power from waveguides A and B as a function of nonlinear phase mismatch Δ.

the output intensity from an interferometer is

$$I = I_1 + I_2 + 2(I_1 I_2)^{1/2} \cos[(2\pi/\lambda)(n_1 r_1 - n_2 r_2) + \phi_1 - \phi_2], \qquad (4.19)$$

where I_1 and I_2 are the intensity for beams 1 and 2 respectively. The required phase shift in δ in this case is only π to switch the total intensity from constructive to destructive interference, whereas it is 2π as in a NLDC. However, the response of an interferometer follows a squared sinusoid function of the phase shift and is less sharp than an NLDC.

4.2.4.1. *Self-switching Nonlinear Optical Loop Mirror*

Phase shift can be induced by changing n_1, r_1, n_2, or r_2. This can be achieved using nonlinear optical media. However, if the nonlinear effect is small, long nonlinear interaction length is needed for the arm of an interferometer, in order to reduce switching energy. Since the interferometer has to be stable to within a fraction of a wavelength, it is difficult to use two physically separate fibers to implement a long interferometer. The best choice, therefore, is a nonlinear Sagnac interferometer or a nonlinear optical loop mirror (NOLM) where the two counterpropagating directions serve as the arm of the interferometer (Fig. 4.10). The configuration is very stable since both arms have exactly the same optical path length. The phase difference in the two counterpropagating beams comes from the nonlinear effect of the medium. For clockwise field 1, the phase shift after the loop is $|E_1|^2 n_2 L$, and for the counterclockwise field it is $|E_2|^2 n_2 L$. Therefore, if E_1 is different from E_2; i.e., the coupling ratio of the 2×2 input/output coupler is not $\frac{1}{2}$, there will be phase difference in these two counterpropagating beams.

The operation of NOLM can be described as follows. Assume that the 2×2 coupler has a power coupling ratio q:$(1 - q)$. Then the fields E_1, E_2, E_3, and E_4 in Fig. 4.10 are coupled through the equations [10]:

$$\begin{aligned} E_3 &= \sqrt{q}\,E_1 + j\sqrt{1-q}\,E_2 \\ E_4 &= j\sqrt{1-q}\,E_1 + \sqrt{q}\,E_2. \end{aligned} \tag{4.20}$$

Consider the case of a single input at port 1, $E_1 = E_{in}$ and $E_2 = 0$. We have

$$\begin{aligned} E_3 &= \sqrt{q}\,E_{in} \\ E_4 &= j\sqrt{1-q}\,E_{in}. \end{aligned} \tag{4.21}$$

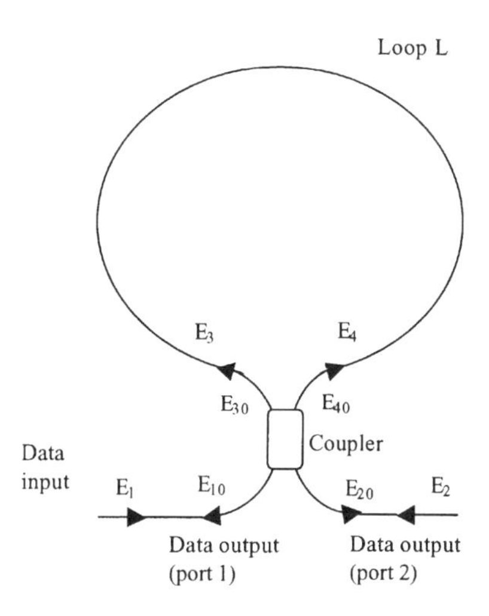

Fig. 4.10. A nonlinear optical loop mirror.

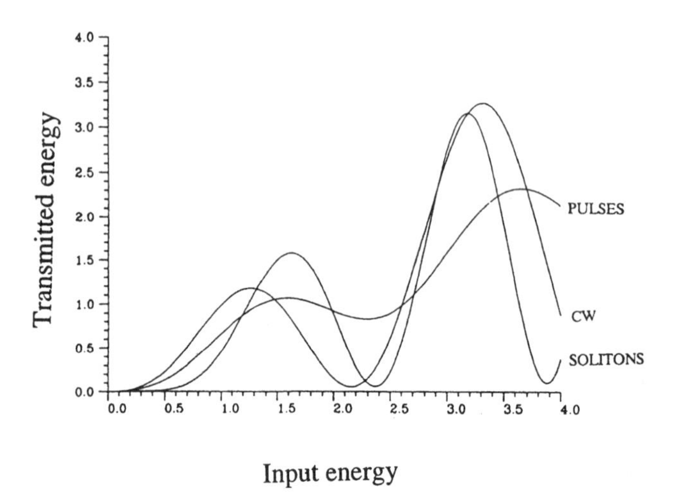

Fig. 4.11. The transmission $|E_{20}|^2/|E_{in}|^2$ of a NOLM as a function in input energy for a continuous-wave signal, nonsoliton pulse signal, and soliton signal. [10].

After they travel around the loop of the length L, the fields E_{30} and E_{40} are given by

$$\begin{aligned} E_{30} &= \sqrt{q}\, E_{in} \exp(jq|E_{in}|^2 2\pi n_2 L/\lambda) \\ E_{40} &= j\sqrt{1-q}\, E_{in} \exp[j(1-q)|E_{in}|^2 2\pi n_2 L/\lambda]. \end{aligned} \tag{4.22}$$

Using a similar relationship for the output coupling, we obtain

$$|E_{20}|^2 = |E_{in}|^2\{1 - 2q(1-q)(1 + \cos[(1-2q)|E_{in}|^2 2\pi n_2 L/\lambda])\}. \tag{4.23}$$

The extremes of the output occur when q is not equal to $\frac{1}{2}$, and

$$(1-2q)|E_{in}|^2 2n_2 L/\lambda = m. \tag{4.24}$$

We can find the required switching energy from this equation. When m is odd, all the power emerges from port 2. When m is even, E_{20} has a minimum transmitted power, which is equal to the linear output power ($n_2 = 0$), given by

$$|E_{20}|^2 = |E_{in}|^2[1 - 4q(1-q)]. \tag{4.25}$$

This amount determines the switching contrast ratio. The transmission functions $|E_{20}|^2/|E_{in}|^2$ for both a continuous-wave signal and a soliton signal are shown in Fig. 4.11 [10].

4.2.4.2. *NOLM with External Control*

In the NOLM discussed above, the output is controlled by the intensity of the input. In many applications, the input is a message signal the power of which is weak and cannot be changed. In this case an external control signal has to be used to switch the NOLM, as shown in Fig. 4.12. The nonlinear refractive index change induced by the powerful control signal is experienced fully by the copropagating signal pulse but to a much lesser degree by a counterpropagating signal pulse. This difference in phases of the copropagating and counterpropagating signals results in a relative phase shift between these counterpropagating signal pulses. The induced phase shift should be π in order to effect complete switching.

Assuming that there is no nonlinear effect induced by the input signal E_{in}, the total intensity output is found from Eq. (4.23) as

$$|E_{20}|^2 = |E_{in}|^2\{1 - 2q(1 - q)(1 + \cos[(1 - 2q)I_{ext}2\pi n_2 D/\lambda])\}. \quad (4.26)$$

Here I_{ext} is the intensity of the control beam, and D is the length of the nonlinear medium. In fiber NOLM, $D = L$, and when a semiconductor optical amplifier (SOA) is used as the nonlinear medium, D is the length of the SOA. The control signal should be at a different wavelength or polarization from the data signal, and can be coupled into and separated from the data signal through a wavelength or polarization coupler.

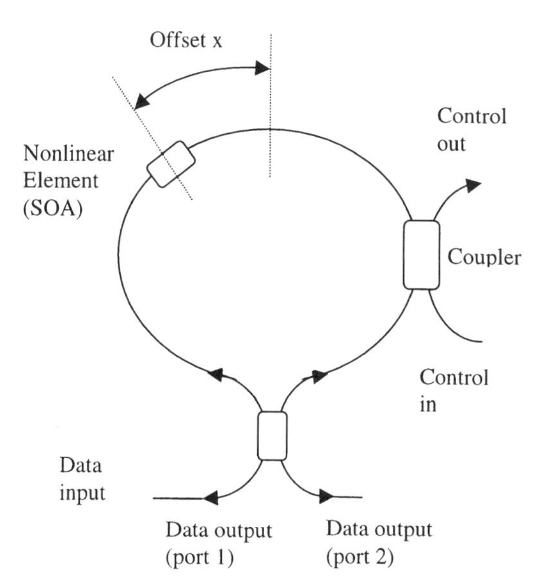

Fig. 4.12. Asymmetrical NOLM with a nonlinear element and an external control.

Controlled switching can also be realized via soliton collisions (or overlap). When copolarized solitons with different group velocities collide (or pass through each other), an offset in arrival time and phase occurs for both solitons relative to the case when there is no soliton collision. For small-frequency difference, phase shift greater than π can be induced by the collision [1]. This phase shift can be used in the interference of two solitons, and the output can be changed from zero to unity, depending on whether or not a collision occurs.

4.2.4.3. *Semiconductor Amplifier as a Nonlinear Medium*

NOLM was first realized using optical fiber as a nonlinear medium. Very fast switching processing is possible in a glass fiber interferometer because the response and relaxation times of the intrinsic nonlinear optical refraction in glass are a few femtoseconds. The main drawbacks stem from the small optical nonlinearity. For example, to produce a π nonlinear phase shift at a control power level of 100 mW, the required length of the fiber at 1.55 μm wavelength is around 200 km.

For an interferometer with long optical path, the stability is significantly degraded. It is also difficult to integrate such an interferometer into other optical devices.

A good nonlinear material is needed in order to implement optical switches. The essential material requirements are sufficient large nonlinear coefficients and low optical loss coefficients such that switching operation can achieve at a practical optical power level. The best nonlinear medium to date for switching applications appears to be semiconductor optical amplifiers (SOA) [11].

A SOA is a laser diode with the reflectivity of the end faces minimized. Strong nonlinearity of SOAs was discovered when associated effects, such as frequency chirping, were easily detected. In fact, much work has been done to minimize the nonlinear effects in SOAs. The nonlinearity in SOAs is a resonant effect. If an optical beam with photon energy slightly larger than the bandgap is incident on an SOA, it is amplified through stimulated emission. The amplification saturates as the conduction band is depopulated. The change in gain coefficient due to saturation causes the variations in refractive index, as described by the Kramer–Kronig relations. The change in refractive index is proportional to the optical intensity, provided the gain is not fully depleted. The refractive nonlinearity of SOAs is 10^8 times larger than that of silica fiber, and optical loss is not an issue since the power of the control and data signals is amplified.

An interesting configuration using a SOA in a NOLM is an asymmetrical arrangement as shown in Fig. 4.12 [12]. A data pulse is split into clockwise and counterclockwise components by the input coupler. Since the SOA is placed asymmetrically with an *offset* x in the loop, the clockwise pulse will arrive at the SOA before the counterclockwise component. The difference in the arrival time (τ) depends on the amount of *offset* x. The role of the control signal is to create a refractive index change by rapidly sweeping out some of the gain of the SOA. Therefore, a pulse passing through the SOA before the control pulse will experience different phase shift from the pulse passing through the SOA after the control pulse. The dynamics of the phase change in the SOA are shown in Fig. 4.13. With an offset, a control pulse, for example C_1, can be timed to arrive at the SOA after the clockwise pulse (A_1) has passed the SOA but before the counterclockwise pulse (B_1) has arrived. If the interferometer is constructed such that, in the absence of the control signal, no input signal appears at the output port, and the control pulse induces a phase difference of π, the recombined clockwise and counterclockwise pulses will have constructive interference and emerge from the output. Subsequent data pulse pairs (A_2, B_2; and A_3, B_3) experience a similar gain and refractive index in the SOA that is slowly recovering. Therefore, the pulses in each of these pairs have a nearly identical phase shift passing through the SOA and no signal appears at the output port.

A unique feature of this asymmetric scheme is that the switching window is determined by the offset x, instead of the relaxation time of the nonlinear effect.

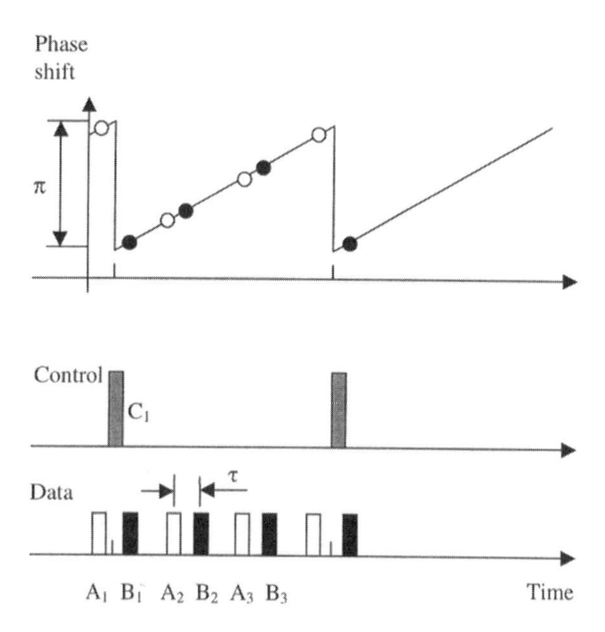

Fig. 4.13. The dynamics of the phase change in the SOA in asymmetric NOLM. $\tau = x/v$ is the time delay between the two counterpropagating pulses, where x is the offset in Fig. 4.12 and v is the velocity of the pulse in the loop. Black symbols represent counterclockwise pulses and white symbols represent clockwise pulses.

As a result, a very high speed pulse train can be demultiplexed down to a lower base rate. In addition, the base rate in this scheme can be as high as 100 GHz, much higher than the value (a few GHz) limited by the carrier lifetime in SOA.

The reasons for this are as follows [13].

First, it takes only a small amount of carrier density change to create an additional phase shift of *p*. For example, for a 500-μm–long InGaAsP SOA and a wavelength of 1.5 μm, a π phase shift corresponds to a change in refractive index of 10^{-3}. The rate of change of refractive index with electron–hole pair density is

$$\frac{dn}{dN}(\mathrm{cm}^3) = 2 \times 10^{-20}. \tag{4.27}$$

Therefore, a carrier population change (ΔN) of 10^{17} cm^{-3} is needed to have π phase shift, compared with a full population inversion of 10^{18} cm^{-3}. In other words, it is not necessary to have a full decay in the SOA to have a π phase shift. There would be a phase shift of several π during the complete decaying of the gain and refractive index in SOA.

Second, the time needed for carrier injection via electric current can be accomplished at a typical rate of 4×10^{-3} cm^{-3} per picosecond. Therefore, the time needed for replenishment of 10^{17} cm^{-3} is only 25 ps. With a higher injection current and hence a shorter carrier lifetime, it was demonstrated that a phase shift of π can be achieved in 10 ps, implying a base rate as high as 100 GHz.

This scheme has found important applications in high-speed demultiplexing and digital optical 3R regeneration (reamplification, reshaping, and retiming). Digital demultiplexers (Fig. 4.14) with 100 Gbit/s clock rate and optical digital regenerator (Fig. 4.15) operating at 160 Gbits/s have been demonstrated [11].

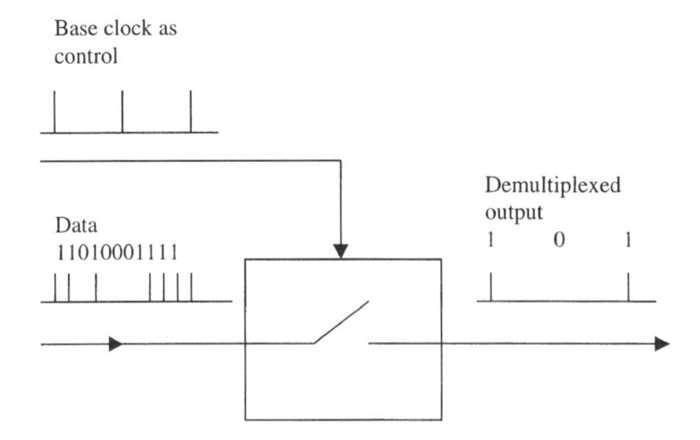

Fig. 4.14. Digital optical demultiplexers.

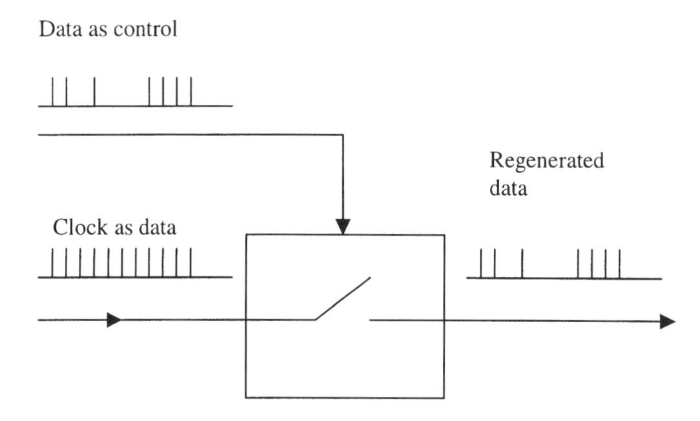

Fig. 4.15. Optical digital regenerator.

4.3. FAST ELECTRO-OPTIC SWITCHES: MODULATORS

Modulators are devices that alter one of the detectable characteristics (amplitude, phase, or frequency) of a coherent optical wave according to an external electrical signal. They are essential components for converting electrical data into optical ones in optical communication systems. Generally speaking, we can make a fast optical modulator by using an electrical pulse to induce a dielectric change in the medium through which the (carrier) optical signal is to pass. In electro-optic (EO) modulators, an external electrical signal modulates the gain, absorption, or refractive index of the medium. The output optical signal, therefore, is altered according to the external electrical signal.

Research on electro-optic modulators started with bulk devices based on the electro-optic effect that changes the optical properties of the medium. Modulation bandwidths on the order of gigahertz were achieved. However, there are limitations related to diffraction effects and drive power in bulk devices with small transverse dimensions. Thin film and waveguide modulators were investigated both for their compatibility with fibers and for their potentials for optical integration. Optical waveguide modulators normally need much smaller drive power than bulk ones. In this type of modulator, the optical beam is coupled into a waveguide the refractive index of which is higher than that of the substrate and of any covering layer. Materials suitable for waveguide modulators are those that possess desired optical properties (for example, a good electro-optic merit figure), and are also capable of being configured in a waveguide. There are two main types of waveguide structures, planer waveguides and stripe waveguides. Waveguides have been produced by sputtering, epitaxial layers, ion implantation, ion exchange, and diffusion; the main substrate materials include glass, semiconductors (both GaAs and InP), and $LiNbO_3$.

In this section we consider thin film and waveguide modulators based on the electro-optic effect. The main characteristics and the performance of $LiNbO_3$ and semiconductor devices will be discussed. The materials presented here are based on several publications [8, 14, 15, 16].

4.3.1. DIRECT MODULATION OF SEMICONDUCTOR LASERS

Direct modulation of diode lasers offers a simple approach to the generation of coded optical pulses [17]. In this case, an external electrical signal modulates the gain in the medium. Since the minimum loss in conventional silica fibers occurs at around 1.55 μm wavelength and the zero-dispersion wavelength is near 1.3 μm, semiconductor laser sources, which have small size and matched wavelengths, have attracted attention as the most promising candidates for application in this field. Another advantage of diode lasers over other types of lasers is that the optical signal can be directly modulated by the drive current. The principle of the modulation can be illustrated using Fig. 4.16. The output laser power has a linear relationship with the drive current if the drive current ($I = I_0$) is above the threshold and below the saturation point. In this case, when the drive current changes, the laser output changes according to the current signal. The dynamic response and modulation behavior of diode lasers have been extensively studied by many researchers. The results of experimental

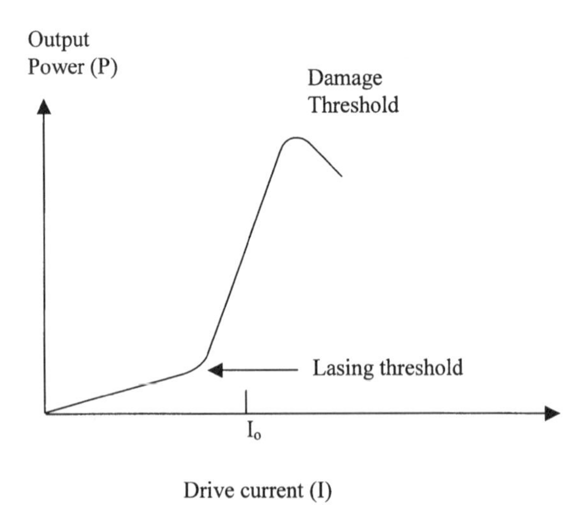

Fig. 4.16. Output power of a laser diode as a function of drive current.

and theoretical investigations have been presented in numerous publications. Considerable attention has been paid to increasing the modulation bandwidth of diode lasers. In this section, we discuss the factors that limit the modulation bandwidth of diode lasers.

4.3.1.1. *Small-Signal Modulation Response*

When the modulation depths of the drive current and optical output are substantially less than 1, we call this small-signal modulation. A simplified two-port model of a high-speed diode laser [18] can be used to study small-signal modulation response. The laser model is shown in Fig. 4.17. The laser is divided into three subsections: (1) the package (or mount) parasitics; (2) the parasitics associated with the laser chip; and (3) the intrinsic laser. The modulating drive current is time dependent, and is in the form of $I(t) = I_0 + I_m f_p(t)$, where I_0 is the bias current, I_m is the magnitude of the modulating current, and $f_p(t)$ is the shape of the current pule. This current will affect the output optical power P and the frequency chirp $\Delta\nu(t)$ which describe the variation of the laser frequency due to modulation. Parasitics associated with the package include bond-wire inductance and capacitance between the input terminals. These parasitics can be substantially decreased by the monolithic integration of the laser with its drive circuitry. Chip parasitics include resistance associated with the semiconductor material surrounding the active region and stray capacitance. The package parasitics are considered to be linear circuit elements, while the chip parasitics are nonlinear with values depending on the input current. In the frequency domain, parasitics cause a high-frequency roll-off in the small-signal response. In the time domain, parasitics result in a slowing-down of fast transients of the drive current waveform. The dynamic response of the overall laser is a combination of the responses of the parasitics and the intrinsic laser.

(a) Bandwidth limit due to intrinsic laser: Let us first consider the dynamic response of the intrinsic laser. A considerable amount of information can be

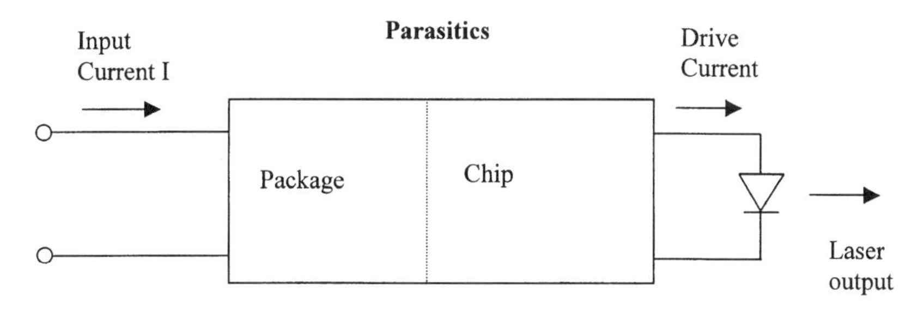

Fig. 4.17. A two-port model of a laser diode.

obtained by a small-signal analysis of the rate equations. The small-signal response of the intrinsic laser is obtained by linearizing laser rate equations and can be found [19] as

$$R_i = \frac{\omega_0^4}{(\omega^2 - \omega_0^2)^2 + \omega^2\gamma^2}, \tag{4.28}$$

where γ is the damping rate, and ω_0 is the resonance frequency is given by

$$\omega_0 = \frac{g_0 S}{\tau_{\text{ph}}}, \tag{4.29}$$

where g_0 is the small gain coefficient, S is the steady-state photon density, and τ_{ph} is the photon lifetime. Figure 4.18 illustrates the general form of $R_i(\omega)$ for two values of S [17]. There is a universal relationship between the resonant frequency and the damping rate. It has been demonstrated both experimentally and theoretically [20] that

$$\gamma = K\omega_0^2 + C_0. \tag{4.30}$$

The K factor is often taken as a figure of merit for high-speed diode lasers since the maximum possible intrinsic modulation bandwidth Δf is determined solely by this factor [21]:

$$\Delta f = 0.23/\text{K}. \tag{4.31}$$

In fact, the K factor depends on several parameters of the laser and the measured values are found to be as low as 0.005 to 0.01 ns. This implies that

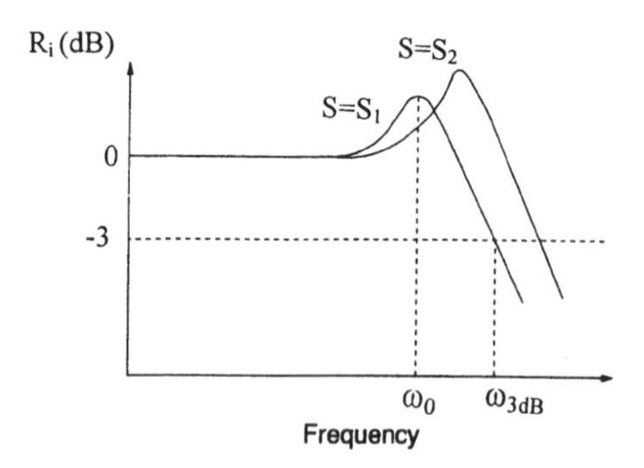

Fig. 4.18. Small-signal response of the intrinsic diode laser for $S = S_1$ and $S = S_2$ ($S_2 > S_1$).

the maximum possible modulation bandwidth in lasers is in excess of 23 to 46 GHz [20, 21]. In practice, this bandwidth is usually limited by RC parasitics, device heating, and maximum power-handling capability of the laser facets.

(b) Bandwidth limits due to parasitics: The dynamic response of the intrinsic laser is reduced by parasitics. The parasitic element limitation has been comprehensively discussed previously [18, 20, 21].

To discuss the effect of parasitics, the laser is treated as an electrical element and an equivalent circuit is established for the intrinsic laser in conjunction with the bond wire, the mount elements, and the transmission line used for impedance matching. Then, practical equivalent circuits can be developed for diode lasers with different designs. A number of low-parasitic devices for high-speed operation have been demonstrated [22]. One scheme uses a semi-insulating substrate for decreasing shunt capacitance. Another type of low-parasitic device is called the constricted mesa or mushroom stripe laser. Several features in these devices help minimize shunt capacitance. The 3-dB bandwidth due to parasitics in properly designed laser diodes can be as high as 20 to 25 GHz [20, 21, 22].

In general, the effect of parasitics can be represented by using the electrical response R_e of the diode laser in the following simple form [21]:

$$R_e = m^2 \frac{1}{1 + (\omega/\omega_{RC})^2} \frac{1}{1 + (\omega/\omega_{ph})^2}, \tag{4.32}$$

where m represents low-frequency modulation efficiency, and ω_{RC} and ω_{pn} describe the high-frequency roll-off caused by the series resistance and the diffusion capacitance of the p–n junction, respectively. The overall laser frequency response, therefore, should be $R = R_i R_e$. However, detailed analyses of any particular laser structure will produce more complicated equations [18, 21, 23].

The 3-dB direct modulation bandwidth is mainly limited by the parasitics in practical devices. For instance, a 25-GHz modulation bandwidth has been measured in long-wavelength MQW lasers, whereas the maximum possible intrinsic bandwidth determined by the K factor of the laser has been obtained over 40 GHz [24].

4.3.1.2. *Large-Signal Effect*

The results presented in the previous sections are concerned with the small-signal modulation response of diode lasers. However, in high–bit-rate digital optical fiber communication systems, the on–off ratio in the laser output often should be as large as possible to reduce the bit-error rate and avoid excessive power penalty [22]. This leads to the requirement of large-

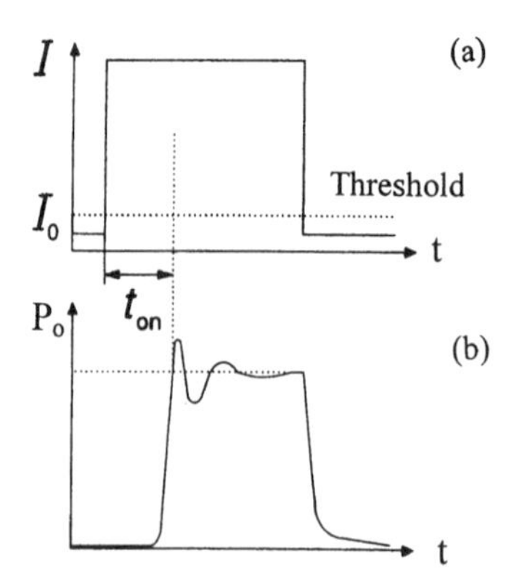

Fig. 4.19. Large-signal response of a laser diode. (a) Input current waveform, (b) Laser output power waveform.

signal modulation. Large-signal modulation can be realized by biasing the laser close to the threshold and modulation with current pulses of large amplitude. Large-signal modulation response is generally worse compared to small-signal response [22, 25]. Large-signal modulation can also cause severe frequency chirping of the laser.

Due to the highly nonlinear optical properties of diode lasers, large-signal dynamic response can be quite complex. The output optical waveform depends strongly on the frequency and amplitude of the input current, and harmonic and intermodulation distortion can be significant [18, 22, 25]. The large-signal behavior of high-speed lasers has been investigated both theoretically and experimentally for a variety of modulation schemes, including gain switching and conventional pulse code modulation [17, 25]. Figure 4.19 shows an idealized rectangular current drive waveform and the corresponding large-signal response of a laser [17]. The dc current I_0 can either be slightly below or above the threshold. The turn-on time of the laser t_{on} is an important parameter that affects the maximum achievable bit rate in digital systems. The turn-on time is a function of I_0 that increases while the current increases. It is clear that turn-on behavior is improved provided the laser is bias above the threshold. Rate-equations analysis shows that t_{on} can drop by an order of magnitude from several hundreds of picoseconds to 30 to 60 ps when I_0 is varied from below the threshold to 30 to 40% above the threshold [18]. Even though the output optical pulse is not an exact replica of the applied electrical pulse, deviations are small enough that semiconductor lasers can be used for data transmission up to bit rates of about 10 Gb/s.

Note that direct current modulation results in the simultaneous amplitude modulation (AM) and frequency modulation (FM) of the laser emission. This originates from the refractive index variations in the laser at the same time that the optical gain changes as a result of carrier density variations. The interdependence between AM and FM is governed by the linewidth enhancement factor. As a result, spectral broadening and chirping exist, caused by the current modulation. This usually affects the spectral stability of the output optical signal.

4.3.2. EXTERNAL ELECTRO-OPTIC MODULATORS

4.3.2.1. Electro-optic Devices

The principle of electro-optic modulators is based on the linear electro-optic effect (Pockels effect), which induces a change in the refractive index proportional to an externally applied electrical field E. The magnitude of the electro-optic effect depends on the electro-optic coefficients of the medium and the applied electrical field. Using common notations [7], the refractive index of the crystal can be described by the index ellipsoid. The linear changes in the coefficients of the index ellipsoid due to an applied field of components E_j ($j = 1, 2, 3$ corresponding to x, y, z components, respectively) can be expressed by [7]

$$\Delta\left(\frac{1}{n^2}\right)_i = \sum_{j=1}^{3} T_{ij} E_j \qquad i = 1, 2, \ldots, 6, \quad j = 1, 2, 3, \tag{4.33}$$

where T_{ij} is the electro-optic tensor. In the case of $LiNbO_3$, if a field E is applied along the z axis, the induced index change seen by an optical field polarized along the z direction is given by

$$\Delta n = \frac{n_e^3}{2} T_{33} E_z, \tag{4.34}$$

where n_e is the corresponding extraordinary index. It should be noted that the EO effect depends on the orientation of the crystal and the direction of the applied field. The coefficient T_{33}, which corresponds to the stronger electro-optic effect for $LiNbO_3$ with applied field in z direction, is about $30 \times 10^{-6}\ \mu m/V$. In GaAs, the higher index change is seen by an optical field polarized along the x direction and the corresponding parameter T_{13} is about $1.4 \times 10^{-6}\ \mu m/V$. InP has about the same value.

The electro-optic effect is a fast process, since it is mainly related to electronic lattice transitions. The response time of the index change approaches electronic lattice relaxation times, which range from 10^{-13} to 10^{-14} s. Therefore, potential modulators with a few hundreds GHz bandwidth can be built.

The change in refractive index induced by the EO effect will produce a phase change to the optical beam passing through the crystal. The total phase change over an interaction length L (equal to the electrode length) is a function of the refractive index change Δn. It is expressed by

$$\Delta\phi = \frac{2\pi \Delta n L}{\lambda_0}, \tag{4.35}$$

where λ_0 is the free-space wavelength.

Electrically induced index change can be directly used for phase modulation. Amplitude modulation can be achieved via phase modulation, either by using interferometric techniques (Mach–Zehnder modulator, balanced bridge switch) or by phase-matched control in directional couplers.

4.3.2.1.1. Waveguide Phase Modulator

A typical waveguide phase modulator with two electrode configurations is shown in Fig. 4.20. If the electrodes are placed on either side of the waveguide (Fig. 4.20a), the horizontal component of the electric field is used. If one electrode is placed directly over the waveguide (Fig. 4.20b), the vertical component of the field is used. The crystal orientation must be chosen to use the largest electro-optic coefficient. For a $LiNbO_3$ modulator, the largest electro-optic coefficient, T_{33}, should be used, and the orientation of the crystal is shown in Fig. 4.20a for the TE wave polarized in the plane of the substrate.

$LiNbO_3$ waveguides are commonly fabricated using the Ti in-diffusion process. In the process, Ti stripes with a thickness of 50–60 nm and a width of 5 μm are evaporated onto a $LiNbO_3$ crystal and in-diffused at a proper temperature for a few hours. This creates waveguides with Gaussian index distribution in depth. The maximum index increase at the surface of the $LiNbO_3$ waveguide is typically a few hundredths.

Using the scheme shown in Fig. 4.20a, the relationship between the effective electro-optically induced index change and the applied voltage can be expressed as

$$\Delta n = \frac{n_e^3}{2} T_{33} \frac{V}{d} \Gamma, \tag{4.36}$$

where d is the interelectrode gap and Γ is the overlap integral between the applied electric field and the optical mode. The phase change over an interaction length L is thus expressed as

$$\Delta\phi = \frac{\pi n_e^3}{\lambda_0} T_{33} \frac{VL}{d} \Gamma. \tag{4.37}$$

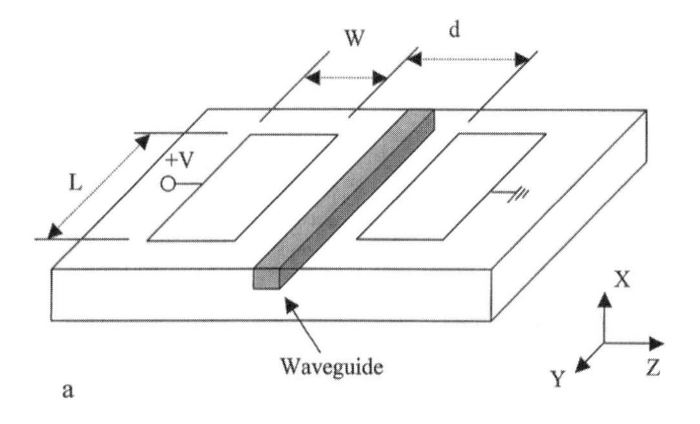

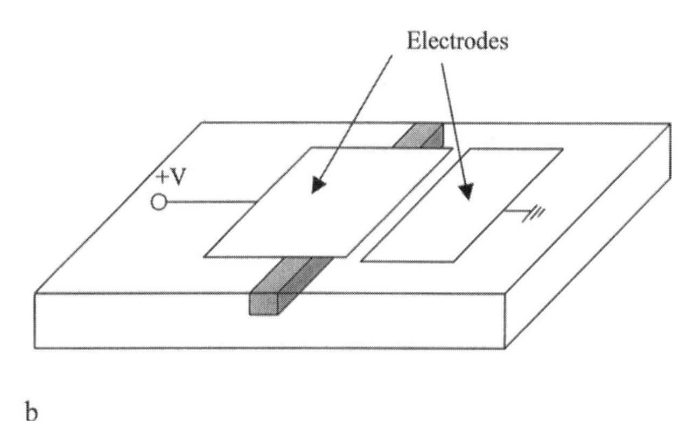

Fig. 4.20. Waveguide phase modulator with two different electrode arrangements.

Setting $|\Delta\phi| = \pi$, Eq. (4.37) can be inverted to provide the half-wave voltage-length product,

$$V_\pi L = \frac{\lambda_0 d}{n_e^3 \Gamma T_{33}}. \tag{4.38}$$

4.3.2.1.2. Bandwidth Considerations

In many practical applications, we need modulators with high bandwidth for high-speed communications. There are several factors limiting the highest modulation bandwidth for different configurations of modulators. The main parameters characterizing a waveguide modulator are the modulation bandwidth and the drive voltage required to obtain a given modulation depth.

Typically, the required drive voltage is traded off by the modulation bandwidth. Therefore, the ratio of drive power to modulation bandwidth is usually adopted as a figure of merit of the device. The drive voltage depends upon the type of modulator, the electrode geometry, the optical wavelength, and the overlap geometry. In the case of lumped electrode modulators, where the electrode length is much smaller than the modulating radio frequency wavelength, modulation speed is constrained by the larger of the electrical or optical transit time and the RC charging time, where C is the electrode capacitance, including parasitics, and R is the resistance value providing matching to driver impedance. For the electrode configuration shown in Fig. 4.20a, the capacitance, which is proportional to the electrode length L, is usually the main limiting factor. It is given by [14, 15]

$$C = \frac{\varepsilon_0 L}{\pi}(1 + \varepsilon/\varepsilon_0) \ln\left(\frac{4W}{d}\right). \tag{4.39}$$

In this case, the bandwidth of the modulator is

$$\Delta f = 1/(2\pi RC). \tag{4.40}$$

Obviously, one way to increase Δf is to use a small L. However, L cannot be arbitrarily reduced because the required drive voltage scales inversely with L, given by

$$L = \frac{\lambda_0 d}{n_e^3 \Gamma T_{33} V_\pi}. \tag{4.41}$$

The goal in modulator design is a high bandwidth and a low driving voltage at the same time. Modulator performance can be characterized appropriately by introducing the voltage–bandwidth figure of merit: $F = V_\pi/B$. This parameter is proportional to the voltage–length product $V_\pi L$ and inversely proportional to the bandwidth–length product (BL). Here BL depends on the modulator geometrical structure. For $LiNbO_3$ modulators with lumped electrodes, a realistic value of the bandwidth–length product is 2.2 GHz-cm, for a load of 50 Ω.

The bandwidth limitations in lumped electrode modulators can be overcome by using modulator structures based on traveling wave (TW) electrodes, the basic scheme of which is shown in Fig. 4.21. The electrode is designed as a matched transmission line, and the radio-frequency drive is fed collinearly with the propagating optical signal. In this case the bandwidth is limited by the transit, time difference between the radio-frequency drive and the optical signal [8, 14], resulting in

$$B = \frac{1.4c}{\pi L n_{RF}(1 - n/n_{RF})}, \tag{4.42}$$

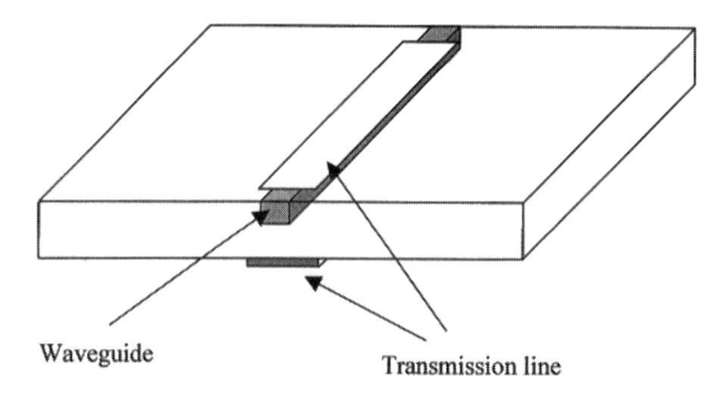

Fig. 4.21. Waveguide phase modulator using traveling wave electrodes.

where c is the light velocity in free space, and n and n_{RF} are the optical and radio-frequency effective indexes, respectively. Therefore, if $n = n_{RF}$, the electrode length can be made arbitrarily long, which allows low drive voltage without affecting modulation bandwidth. In $LiNbO_3$, however, the optical and radio-frequency indexes are quite different ($n = 2.2$ and $n_{RF} = 4.3$), the TW electrode offers an improvement of a factor of about 3 in the length–bandwidth product and in the power–bandwidth ratio over a lumped electrode of identical dimension (length, width, and gap) [26]. Several other techniques have been proposed for achieving effective velocity matching, in spite of the different values of the indexes, such as dielectric loading, radio-frequency phase reversal, or periodical withdrawal of the modulating signal along the electrode length. Recently, a new scheme for achieving 100 GHz modulation bandwidth has been proposed [27] using a phase–velocity matching approach.

4.3.2.1.3. *Waveguide Intensity Modulator*

A number of applications require optical intensity modulation, and there are several techniques to convert phase modulation into intensity modulation. Waveguide intensity modulators can be produced by means of directional couplers and interferometric configurations (Y-branch interferometer, Mach–Zehnder interferometer). These schemes are similar to the ones that discussed in Sec. 4.2. The basic structure of the Mach–Zehnder interferometric modulator is shown in Fig. 4.22. An input wave, split into equal components, propagates over the two arms of the interferometer, which are sufficiently separated to avoid coupling. If no phase shift is introduced between the interferometer arms, the two components combine in phase at the output 3-dB coupler and then propagate in the output waveguide. For a relative π-phase shift, the two components combine out of phase at the output and then cancel each other. This type of $LiNbO_3$ intensity modulator typically exhibits an insertion loss of -4 to 6 dB and is suitable for high-speed modulation. EO

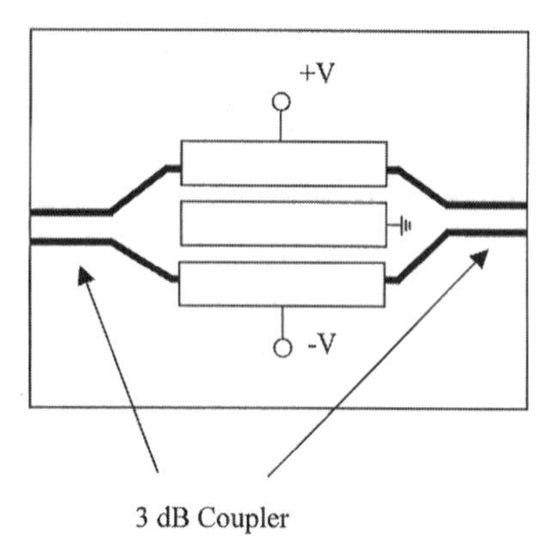

Fig. 4.22. Top view of a Mach–Zehnder interferometric intensity modulator.

modulators operating at various wavelengths with a bandwidth of up to several tens of GHz and an operating voltage of a few volts have been demonstrated [28]. High-speed EO modulators using superconducting electrodes have also been also studied [29].

Although $LiNbO_3$ modulators have been successful in some practical applications, they have several limitations. The main disadvantage is the devices' strong polarization dependence, since light with different polarizations will "see" different elements of the EO coefficients. This means that a simple $LiNbO_3$ cannot simultaneously switch inputs with different polarizations. Other disadvantages include limited optical bandwidth, difficulty in integration with semiconductor lasers and amplifiers, and periodic light output versus control voltage.

4.3.2.2. *Electroabsorptive Modulator*

Another way to modulate an optical field with an electric field is through electroabsorption in semiconductors. With proper design, this type of modulator can be polarization insensitive. These modulators are based on effects of absorption change induced by an external field. Most commonly used effects are the Franz–Keldysh effect and the quantum-confined Stark effect (QCSE). The FK effect can occur in all semiconductors; the QCSE occurs only in quantum-well (QW) semiconductors.

In the FK effect, the absorption edge of a semiconductor shifts towards the long-wavelength direction in the presence of an electric field. Figure 4.23 shows the absorption spectrum for a semiconductor sample with and without an applied field. Without the field, the absorption coefficient shows the typical increase for optical energies that equal or exceed the bandgap of the material.

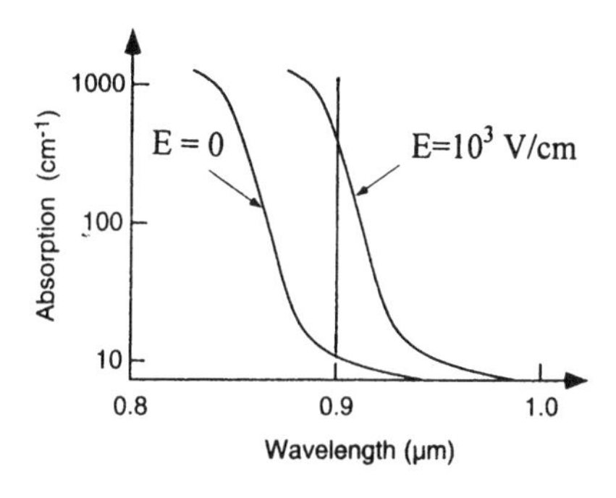

Fig. 4.23. Absorption spectra for a GaAs sample with and without an external field.

Over a range of wavelengths, the absorption coefficient increases from 10 cm^{-1} to over 10^3 cm^{-1}. An optical signal at wavelength λ_0 ($\lambda_0 = 0.9$ μm in Fig. 4.23) will experience a significant change in absorption when an external field is applied. The total change in the transmitted signal depends on the change in the absorption coefficient and the path length through the modulator. Note that the electroabsorptive effect associated with the FK mechanism in bulk materials is usually quite limited because the large electric field leads to the smearing out of the absorption edge as electron and hole wave functions have diminished spatial overlap with increasing field [30].

4.3.2.2.1. Multiple–Quantum-Well Modulator

The multiple–quantum-well (MQW) modulator is based on the QCSE in quantum wells (QW) [31, 32, 33]. The EA modulator consists of MQW layers and, when unperturbed, it is transparent to the light signal. However, under proper electrical bias, the quantum-well and barrier energies are distorted so that absorption rises sharply at energies just above the MQW band gap. Optical absorption change in the quantum well via QCSE is relatively large. In practice this means we can make small and efficient optical modulators using quantum wells. The QCSE, like other electroabsorption mechanisms in semiconductors, is very fast. In fact, there are no intrinsic speed limitations on the mechanism itself until time scales well below a picosecond. In practice, speed is limited only by the time taken to apply voltage to the quantum wells, which is typically limited by resistor-capacitance limits of the external circuit.

There are two commonly used schemes for MQW modulators. One is the transverse (also called surface-normal) modulator, where light comes in and out of the surface of a semiconductor chip. The other is the MQW waveguide modulator. Using a surface-normal scheme, it is possible to make two-dimensional arrays of quantum-well optical modulators.

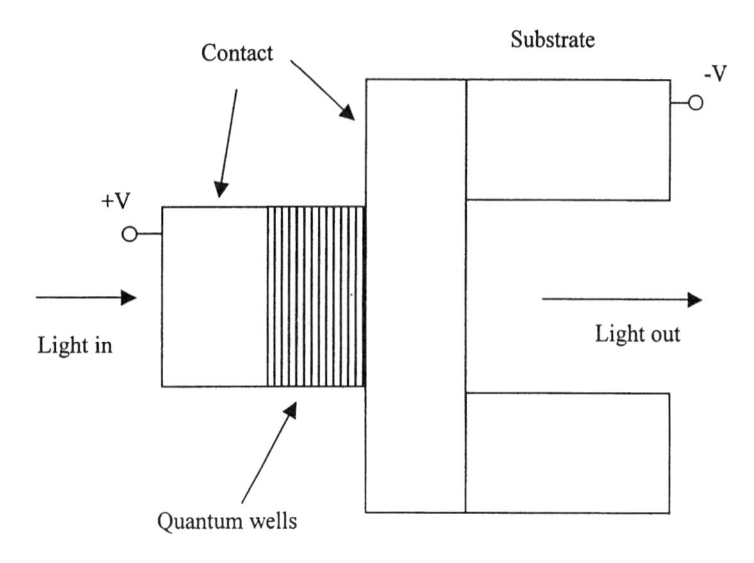

Fig. 4.24. Transverse multiple–quantum-well modulator.

Figure 4.24 shows an example of a transverse modulator structure. In a simplified scheme, a quantum-well structure can be viewed as a thin heterostructure of low-bandgap material (GaAs), the thickness of which is roughly 10 nm (about 30 atomic layers), sandwiched between two layers of higher-bandgap material (AlGaAs). The band diagram of the structure is shown in Fig. 4.25a. Electrons and holes tend to be localized in the region of lower-

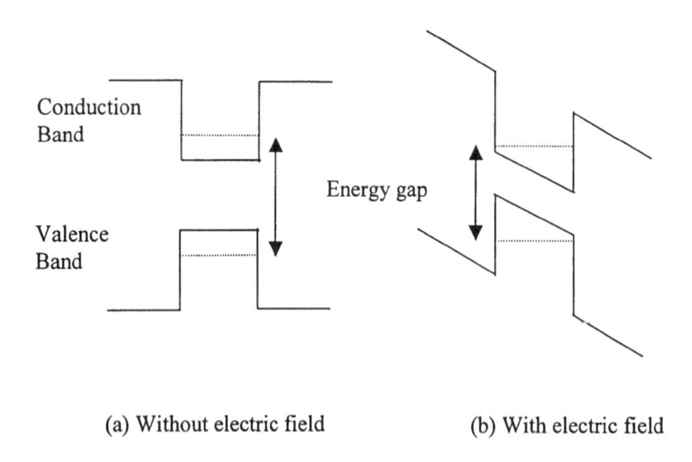

Fig. 4.25. Band diagram of a quantum well (a) with and (b) without external electrical field.

bandgap material. Because of thin wells, the behavior of electrons and holes in these two-dimensional (2D) potentials alters their density-of-states functions in such a way that the absorption edge is sharpened. In addition, the motion of electrons and holes is constrained by their confinement, leading to discrete energy levels in the wells. Because of the 2D nature of the electron–hole gas in quantum wells, exciton binding energy is increased with respect to the bulk semiconductor so as to make excitons observable at room temperature in the absorption spectrum of MQW.

The principle of MQW modulators can be illustrated using Fig. 4.25. If an electric field is applied perpendicularly to the quantum-well layers, the energy levels in the wells change and modify the zero-point energies of the particles. This effect, called QCSE, arises because of the difference in the potential wells seen by the particles. The electron and hole wave functions are modified to reduce the zero-point energies so as to decrease the effective bandgap of the quantum wells. As a consequence, the increase in the electric field applied to the wells reduces the energy required to generate electron–hole pairs, so that the exciton absorption peaks move toward lower energy. The optical transmission spectra for varying applied voltage for a GaAs/AlGaAs single quantum well is shown in Fig. 4.26.

One of the most effective ways to apply the necessary voltage to the quantum wells is to make a diode with quantum wells in the middle. As the diode is reverse biased, the electric field is applied perpendicular to the quantum-well layers. In a reverse-biased diode, the necessary field can be applied without having any current flowing, which makes this a particularly efficient device. The device shown in Fig. 4.26 is made using GaAs and AlGaAs semiconductor materials. The modulator works best typically at wavelengths

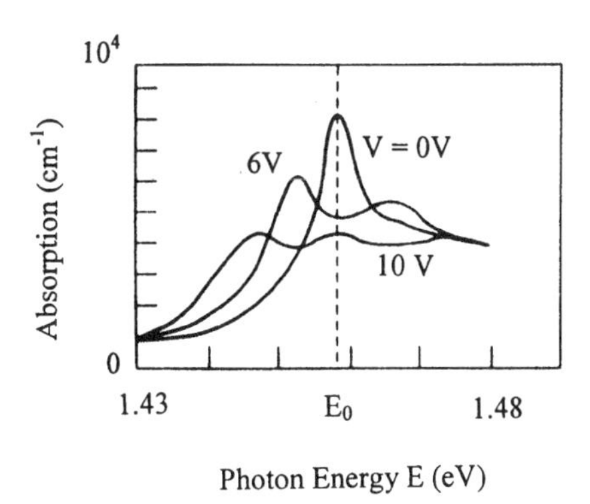

Fig. 4.26. QCSE shift of absorption spectrum for a GaAs quantum well sample with different external fields [31]. Photon energy E is related to optical wavelength through $\lambda(\mu m) = 1.242/E$ (eV).

of about 850 nm, in the near-infrared region of the spectrum. At this wavelength, the AlGaAs material is transparent, so there is no optical loss in the AlGaAs contact regions. GaAs itself is opaque at these wavelengths, so the substrate has to be removed to make a transmission modulator. Devices can be made with various other semiconductor materials for various operating wavelengths in the infrared region, and some devices have been demonstrated in the visible region.

In a typical transverse modulator, there are about 50 to 100 quantum wells. This gives a total thickness of roughly 1 micron for the quantum-well region. The entire diode structure will be a few microns thick altogether. A typical device might have an operating voltage of 2–10 V. An important parameter characterizing the behavior of this device as an intensity modulator is the on–off ratio $R_{\text{on-off}}$, expressed as

$$R_{\text{on-off}} = \exp(\Delta\alpha L), \tag{4.43}$$

where $\Delta\alpha$ is the maximum achievable change in the absorption coefficient and L is the light path length through the electroabsorptive material. The values of $\Delta\alpha$ for MQW modulators are on the order of 10^5/cm. Within these parameters, the optical transmission of the modulator might change from 50 to 20% as the voltage is applied.

Sometimes, it is more convenient to make a modulator that works in reflection rather than transmission. This can be done by incorporating a mirror into the diode structure, so that the light reflects back through the quantum wells. The advantage of this is that the light makes two passes through the quantum-well material, which increases the contrast of the optical modulation. Such reflective modulation is also convenient when mounting the devices on electronic chips, since it allows conventional chip mounting without having to make transparent electronic chips and mounts.

Despite the large value of $\Delta\alpha$ available in MQWs, only limited on–off ratio values are achieved in transverse modulator geometry, because of the limited interaction length. The best way to increase the effective interaction length is to use a waveguide modulator, shown in Fig. 4.27. The propagation direction of the light lies in the plane of the MQW layers, which is part of the optical waveguide. The interaction length can be as long as necessary to obtain the desired on–off ratio. Since the quantum wells usually represent only a fraction of the overall waveguide, and the rest of the waveguide is composed of material without electroabsorption effect, the on–off ratio must be modified as follows:

$$R_{\text{on-off}} = \exp(\Gamma\Delta\alpha L), \tag{4.44}$$

where Γ is the overlap between the quantum wells and the optical mode of the waveguide, the value of which is in the range 0.01–0.3 for typical waveguide modulators.

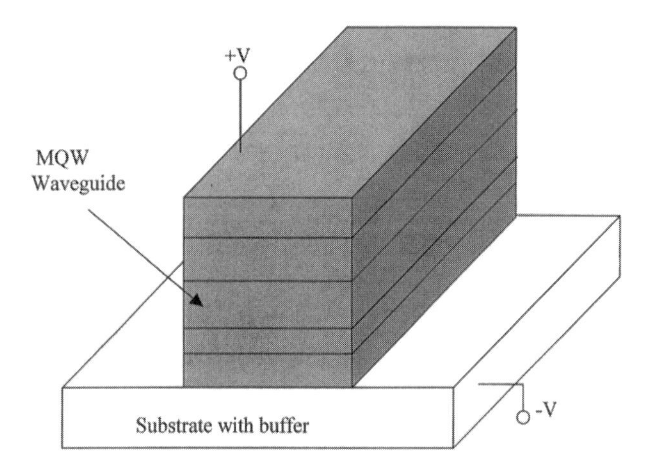

Fig. 4.27. Waveguide multiple–quantum-well modulator.

The speed of MQW modulators is limited by the resistor-capacitance limits of the external circuits. Both time- and frequency-domain measures have been performed to characterize the response of MQW modulators. With a simplified model, the 3-dB bandwidth bias is expressed as

$$\Delta f = 1/(2\pi RC), \tag{4.45}$$

where R is the source resistance and C is the modulator capacitance. The frequency response can be increased by decreasing the device capacitance. One way is to reduce the device area to the smallest practical value. The capacitance can also be decreased at the expense of the drive voltage by increasing the thickness of the MQW layer in the p–i–n junction. However, as this thickness increases, the drive voltage required to obtain a given electric field increases linearly. Therefore, a trade-off must be made between device bandwidth and drive voltage. Speeds of 40 GHz have been demonstrated with drive voltage around 2 V [34].

As in the case for $LiNbO_3$ modulators, bandwidth of EA MQW modulators can be increased by using traveling wave (TW) electrodes (Fig. 4.21). MQW modulators with 50 GHz bandwidth, 15 dB on–off ratio, and <2 V drive voltage have been demonstrated using such a configuration [35].

Several effects contribute to device insertion loss. The main effect is residual absorption by the semiconductor material in the maximum transmission state due to loss of band tails of the quantum wells or free carrier absorption in the doped layers. Residual absorption depends on the materials of the QWs and the operating wavelength. Another loss effect is the one associated with reflection off its facets. For semiconductors of interest, a typical value of the refractive index is around 3.5. Therefore, the reflection off a single air-to-semiconductor or semiconductor-to-air interface at normal incidence is about 30%. This loss can be eliminated by integrating the modulator with the laser source. Other approaches, including antireflection coatings, can be used to reduce this loss to values to around 0.1 dB.

Main advantages of MQW devices in comparison with $LiNbO_3$ modulators include (1) the feasibility of monolithically integrated structures combining several functions onto the same device, (2) polarization-insensitive operation, and (3) low drive voltage. Integrated semiconductor laser–modulator structures have been reported with high on–off ratios, low drive voltage, and high bandwidth [34, 36].

4.4. OPTICAL SWITCHING BASED ON MEMS

In addition to ultrafast serial processing speed, optical switches can also be used for massive parallel connections. If we build a 1000×1000 array with each switch operating on a moderate 1 μs cycle time, terahertz processing speed can be realized. Switches using waveguides (including optical fibers) utilize long interaction lengths and confined high-optical intensity to build up phase changes for switching. However, a high degree of parallelism is unlikely using waveguides, since large arrays of switching fabric are difficult to implement. Therefore, these switches are mainly used for ultra–high-speed serial processing and small arrays of crossbar routing networks. Optical bistable etalons have the potential for massive parallel switching networks. In fact, switching arrays have been fabricated and tested using multiple quantum wells and self-electro-optic devices (SEED) for signal processing applications. However, current bistable devices require a holding power of at least 10 mW. For a 1000×1000 array, the required power would be 10 kW, and most of the power would be absorbed by the device and converted into heat. This would prevent the use of current bistable devices for practical applications.

An attractive scheme is switching devices using microelectromechanical systems (MEMS). MEMS are integrated microdevices or systems combining electrical and mechanical components fabricated using integrated circuit (IC)–compatible batch-processing techniques. MEMS are fabricated using microengineering and have a size ranging from micrometers to millimeters. These systems can sense, control, and actuate on the micro scale and function individually or in arrays to generate effects on the macro scale. MEMS can be used to provide robust and inexpensive miniaturization and integration of simple elements into more complex systems. Current MEMS applications include accelerometers; pressure, chemical, and flow sensors; micro-optics; optical scanners; and fluid pumps [37, 38].

4.4.1. MEMS FABRICATIONS

MEMS can be fabricated using micromachining techniques. There are three main micromachining techniques currently in use for making MEMS devices: bulk micromachining, surface micromachining and LIGA [37, 38, 39].

Bulk silicon micromachining is one of the best-developed micromachining techniques. Silicon is the primary substrate material used in production microelectronic circuitry, and so it is the most suitable candidate for the eventual production of MEMS. Traditional silicon processing techniques, including photolithography, etching, and packaging, are used in this method. Bulk micromachining mainly uses isotropic and/or anisotropic etching to make devices with deep cavities, such as pressure sensors and accelerometers. A drawback of traditional bulk micromachining is that the geometries of the structures are restricted by the aspect ratios inherent in the fabrication methods, as the devices are large.

To get greater complexity out of bulk micromachining, fusion bonding techniques were developed for virtually seamless integration of multiple wafers. The process starts with separate fabrication of various elements of a complex system, and subsequently assembles them. It relies on the creation of atomic bonds between two wafers to produce a single layer with the same mechanical and electrical properties.

The second one is surface micromachining. This technique is very simple, powerful, and versatile. Many optical elements, including moving micromirrors, microgratings, and other components used in optical switching, have been made using this technique. It has a thin-film sacrificial layer selectively added to and removed from the Si substrate. The mechanical parts of the devices are made using the thin film, and the Si is often used for interface circuitry. The process starts with thin films, usually polysilicon, silicon dioxide, and nitride, as mask and sacrificial layers. Sacrificial etching is the basis of surface micromachining. A soluble layer is grown or deposited for later removal from beneath other patterned deposited materials. Since the patterned materials left behind are separated from the substrate by the thickness of the removed sacrificial layer, they form freestanding thin-film mechanical structures. Springs and suspended structures can be constructed with lengths in the plane of wafer much greater than their widths.

The third technique is called LIGA. The acronym LIGA comes from the German name for the process (lithography, electroplating, and molding). Tall structures with submicrometer resolution can be formed with this technique. Microcomponents can be made from a variety of materials using this method. The LIGA process begins by generating a photoresist pattern using X-ray lithography on a conductive substrate. The generated space can be preferentially electroplated right to the brim, yielding a very accurate negative replica of the original resist pattern. This replica can further be used as a mold for low viscosity polymers. After curing, the mold is removed, leaving behind a microreplica of the original patterns. The main drawback of this technique is the need for a synchrotron for collimated X-ray.

4.4.2. ELECTROSTATIC ACTUATORS

Microactuators are essential parts of MEMS optical switches. A wide variety of actuation mechanisms have been researched in the MEMS field. These include electrostatic, electromagnetic, piezoelectric, and thermomechanical. While there is continuing research work going on in this field, it appears that for optical switching, electrostatic actuation is an effective technique. At microsize, it is easy to produce a high electric field for actuation. For example, an electric field of 3×10^6 V/m can be generated by applying $3V$ voltage across a 1-micrometer gap. Using surface micromachining, the interface circuitry in the silicon wafer can used to produce various patterns of control signals for desired actuation of many elements in arrays. Other advantages of electrostatic actuation include simplicity in design, fast response, and low power consumption.

A simple electrostatic actuator is a parallel-plate capacitor (Fig. 4.28). Assuming the area of the plates is much greater than the separation x, its capacitance is given by

$$C = \frac{\varepsilon A}{x}, \tag{4.46}$$

where ε is the dielectric permitivity of the medium between the plates, and A

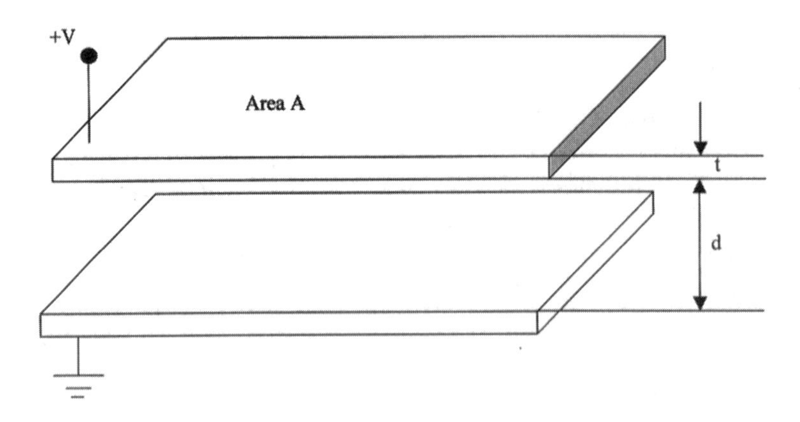

Fig. 4.28. A parallel-plate capacitor.

is the area of the plates. Corresponding to a voltage V applied to the capacitor, an electrostatic potential energy exists, given by

$$U = \tfrac{1}{2}CV^2. \tag{4.47}$$

This potential energy represents the energy required to prevent the opposite charged parallel plates from collapsing into each other. The Coulumb force of attraction is expressed as the negative gradient of the potential energy

$$F = -\nabla U. \tag{4.48}$$

Since

$$U = \frac{eAV^2}{2x}, \tag{4.49}$$

Then

$$F = \frac{\varepsilon AV^2}{2x^2}. \tag{4.50}$$

The Force drops as $1/x^2$. If the top plate is free to move, then the Coulumb force will make it approach to the bottom plate, and the gap, x, will decrease; that is, the attraction force will drive the gap closure, until it is balanced by a force produced by a spring or other supports. This leads to moving parallel plate actuators (Fig. 4.29), deformable membrane actuators (Fig. 4.30), and torsion mirrors (Fig. 4.31).

To generate large forces, which will do more work for many devices, a large change of capacitance with distance is required. This has led to the development of electrostatic comb drives. Comb drives are particularly popular with surface-micromachined devices. They consist of many interdigitated fingers

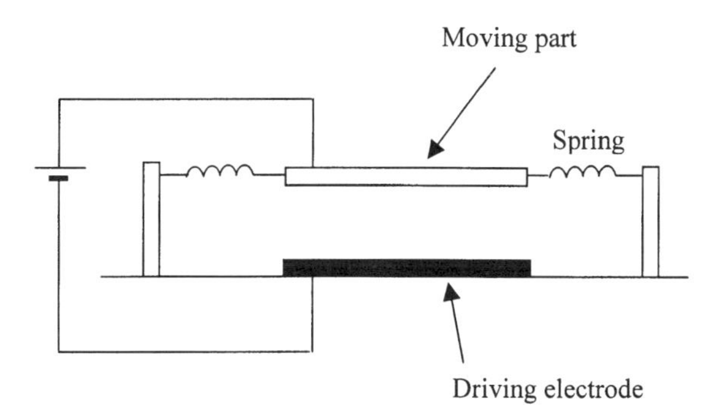

Fig. 4.29. A moving parallel-plate actuator.

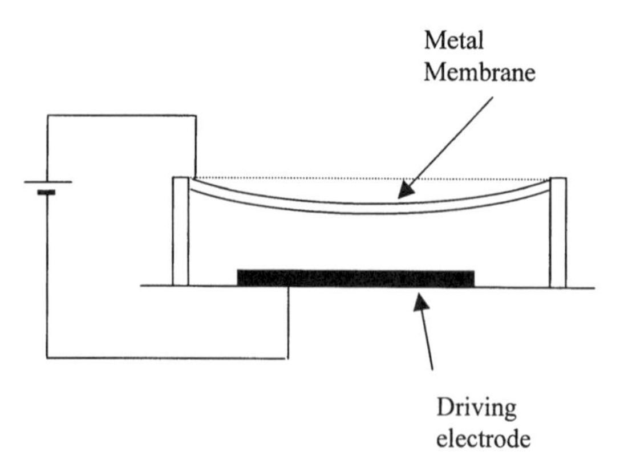

Fig. 4.30. A deformable membrane actuator.

(Fig. 4.32). When voltage is applied, an attractive force develops between the fingers, which move together. The increase in capacitance is proportional to the number of fingers. So, to generate large forces, large numbers of fingers can be used.

To obtain the voltage-displacement relation for the electrostatic comb drive, we assume that both the left and right elements are constrained from moving and that the left element is biased at a fixed voltage V. The capacitance for a single finger face across the gap is given by

$$C_{\text{single}} = \frac{\varepsilon A}{g}, \tag{4.51}$$

where the area is given by

$$A = t(L - x). \tag{4.52}$$

Since each finger has two sides, it follows that each finger has two capacitors. For an n-finger upper actuator, we have $2n$ capacitors. The total capacitance is

$$C_{\text{comb}} = 2n\frac{\varepsilon t(L - x)}{g}. \tag{4.53}$$

Using Eqs. (4.47), (4.48), and (4.53) we obtain the driving force

$$F = n\varepsilon\frac{t}{g}V^2. \tag{4.54}$$

If we set the left element free to move, this force will control the movement

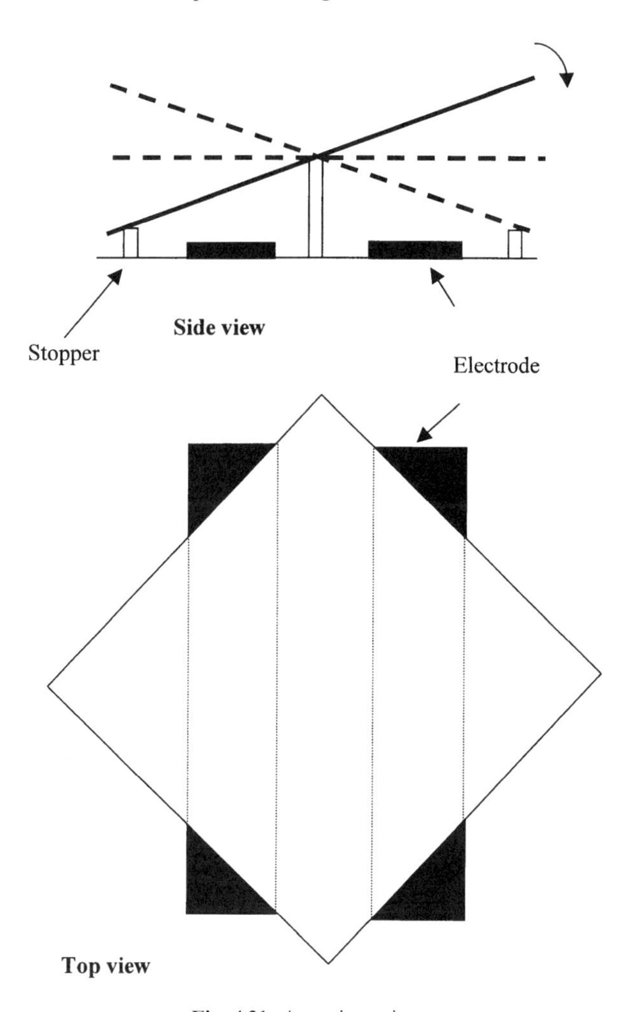

Fig. 4.31. A torsion mirror.

according to the applied voltage *V*. By comparing force with that in the parallel capacitor, we found that the force in the comb-drive device is a constant independent of the displacement x, and we can increasing the force by increase the number of fingers in the comb, while maintaining a reasonable driving voltage *V*.

Comb drives have become one of the most commonly used drivers in making moving microelements, including moving micromirrors driven by the comb drive through hinges and push-rods.

One drawback of comb drives is that fringing fields also give rise to forces out of the plane, which can lead to levitation of the actuator away from the

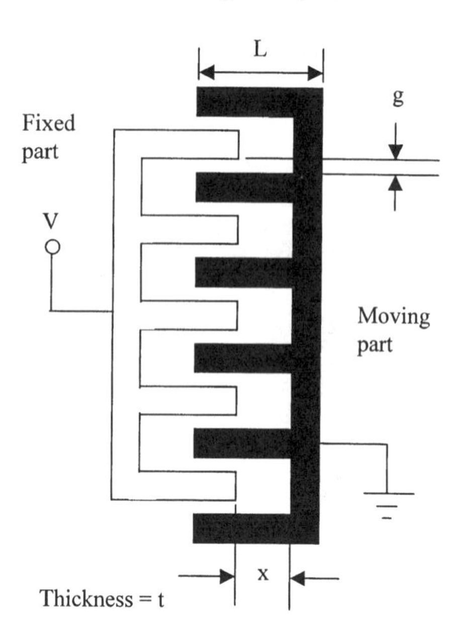

Fig. 4.32. A comb drive with interdigitated fingers.

substrate. In addition, there may be lateral instability depending on how the actuator is supported. If lateral stiffness is insufficient, the upper actuator will be attracted sideways and fingers may stick to the fixed ones.

4.4.3. MEMS OPTICAL SWITCHES

Optical switches based on MEMS have several advantages. They can form very large arrays in a very compact volume with low signal attenuation. They also have high contrast and low cross talk, and are wavelength and polarization independent. Quite a few MEMS switching schemes have been proposed and demonstrated. We will discuss two examples of these switches.

4.4.3.1. *Deformable Diffraction Gratings*

Deformable diffraction gratings are an array of microelectromechanical switching devices that manipulate light by diffraction [40]. Figure 4.33 shows a cross section of one pixel for a switched and nonswitched state. Electrostatically deflectable microbridges are made from silicon nitride that is deposited in tension over a silicon dioxide sacrificial spacer. The bridges are overcoated with aluminum for high reflection. The air gaps are formed by using an isotropic wet etch to selectively remove the sacrificial spacer.

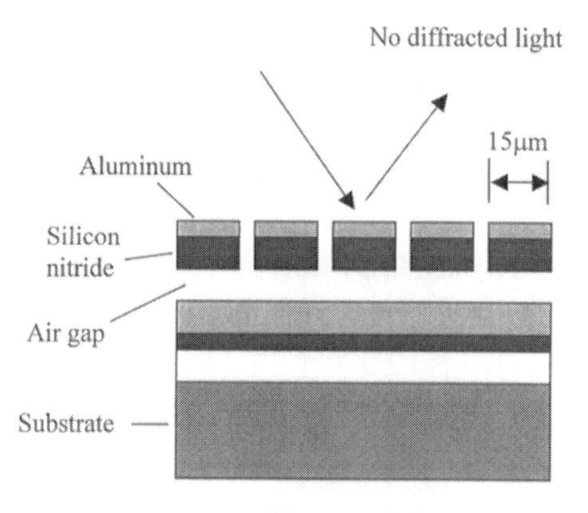

Non-switched state

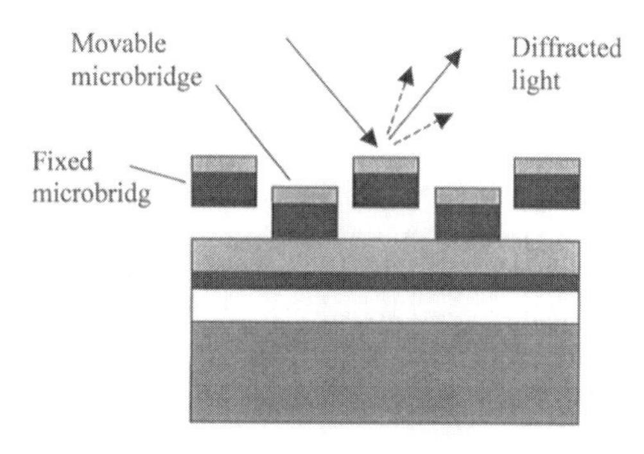

Switched state

Fig. 4.33. Deformable diffraction gratings.

The array is addressed by a set of row and column electrodes. Every other microbridge in the pixel is addressable. The others are held at a fixed-bias voltage so that they cannot be switched. When a pixel is selected by the combined effect of the row and column address voltages, the air gap voltage of the selected microbridges exceeds a threshold level. The movable bridges deflected through one-quarter of the wavelength of the incident light and touch down onto the substrate. They remain there, electromechanically latched, as long as a minimum holding voltage is maintained by the electrode.

Light incident onto a switched pixel is strongly diffracted because the optical path difference upon reflection between the pairs of microbridges is one-half the wavelength. For the nonswitched state, the microbridges are coplanar and the light is specularly reflected.

The size of the bridges is on the order of 80 μm $\times$ 15 μm. Because the inertia of the microbridges is small and they only need to move a small distance, switching time from the nondiffraction state to the diffraction state is on the order of 20 nanoseconds, and required driving voltage is 3.5 V. With this fast switching time and the latching property of the microbridges, this type of switching device has many potential applications. Optical modulators, light valves, and projection displays have been demonstrated using this technology.

4.4.3.2. *Micromirror Arrays*

Micromirror arrays have been used for image processing and display applications. Several examples of micromirror-based crossbar switches have already been demonstrated. The main benefits of these switching arrays include large numbers of optical channels, low signal attenuation, and a very compact volume integrated on a chip. With proper design, the integrated switch system may also be capable of detecting and identifying the data content of incoming optical channels and reconfiguring the switching pattern accordingly.

The backbone of crossbar switches is a two-dimensional $N \times N$ array of micromirrors. Micromirrors are ultrasmall movable structures fabricated by using micromachining. They are normally electrostatically actuated. The most commonly used ones are torsion mirrors (Fig. 4.31). Each micromirror is attached on a torsion hinge which allows rotational movement of the mirror. The rotation is controlled individually to assume either a reflective or non-reflective position. Therefore, the switch array can perform any arbitrary switching between the N incoming channels and N output channels.

An interesting and important device in micromirror arrays is the digital mirror device (DMD) which has been developed by Texas Instruments [41]. The DMD is a micromechanical spatial light modulator array which consists of a matrix of tiny torsion mirrors (16-μm base) supported above silicon addressing circuitry by small hinges attached to support posts. Each mirror can be made to rotate 10° about its axis by applying a potential difference between the mirror and the addressing electrode. The DMD is now commercialized for high-luminance TV projection. Arrays of up to 2048 $\times$ 1152 with full addressing circuitry are available. The response time of each mirror is 10 μs, and the addressing voltage is 5 V. In the wake of developments by Texas Instruments, several research groups are now working on DMD–like devices for switching purposes.

For switches in cross-connection of optical fiber arrays and optical routing, micromirrors with large rotation angles (45–90°) are required. With larger rotation angle, the switching time and driving voltages become larger. One example is the 45° vertical torsion mirror for 2 $\times$ 2 switching array (Fig. 4.34) [42]. Currently, the required driving voltage is 80 V and the switching time is 80 μm. Better versions of optical switches arrays with higher density, lower

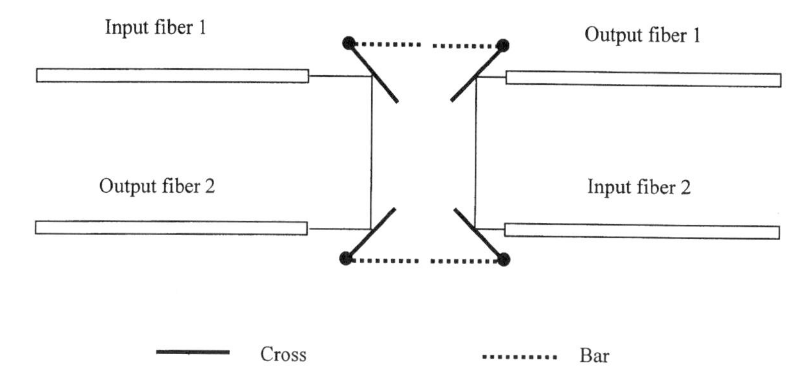

Fig. 4.34. A 2×2 switching array using micromirrors.

driving voltage, and lower switching time need to be developed in order for MEMS optical switches to have practical applications.

Another issue regarding MEMS optical switches is the need for proper control circuitry to sample the incoming optical signals and to perform the switching according to the data content of each channel. In this case, the incoming optical signals should be redirected on an array of optical sensors, which are built into the substrate, by using an optical beam splitter. Each optical sensor would continuously detect the incoming data stream and send this information to a built-in decision and control unit. Based on the extracted information in data headers or specific patterns in the data stream, the control unit will identify the data content of each channel and reconfigure the switch matrix accordingly. This capability offers truly adaptive switching among a number of incoming channels, where a change in data content of an incoming channel during operation will automatically result in redirection of the outgoing channels.

4.4.3.3. *MEMS Switches without Moving Parts*

Most photonic switches based on MEMS technology have movable micromirrors. Recently, an unusual optical switch without any moving parts was proposed and fabricated [43]. The switch uses total internal reflection to route the optical beam. It combines inkjet technology and planar waveguides consisting of silica and silicon sections. The silica section includes optical waveguides intersected, at the cross points, by trenches filled with index-matching fluid. The waveguides and trenches form a proper angle so that total internal reflection will occur when the refractive index of the waveguide is larger than that of the trench. The silicon section includes small heaters as thermal actuators. The heaters are located near the cross points of the

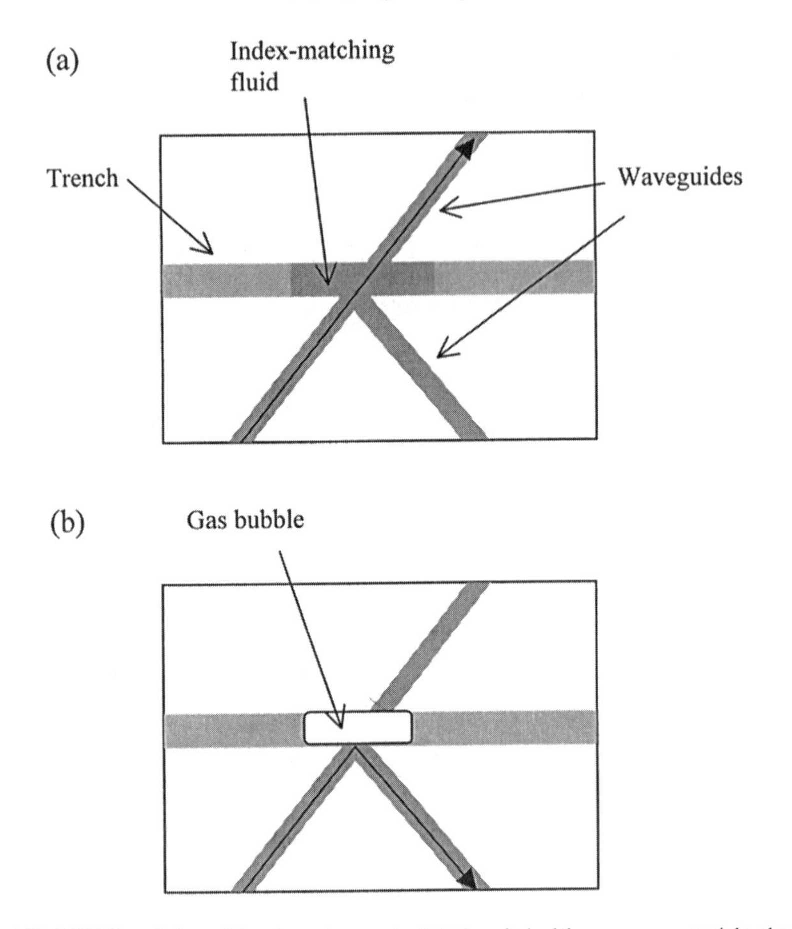

Fig. 4.35. MEMS switches without moving parts. (a) Unswitched beam passes straight through the trench. (b) Switched beam bounced into the new waveguide by total internal reflection.

waveguides. Inkjet technology uses small heaters near the index-matching fluid to create a bubble at that spot. When the heaters are not turned on, the trenches are filled with the index-matching fluid. The beam passes unimpeded straight across the trench and continues into the solid waveguide (Fig. 4.35a). When the optical signal needs to be rerouted, a bubble is created by the heater. The gas bubble occupies the trench near the waveguide. In this case, total internal reflection occurs since light comes at a sufficiently oblique angle from a higher index medium (waveguide) to a lower index medium (gas bubble in the trench). The beam reflects off the surface of the bubble, and moves in a new direction into a different waveguide (Fig. 4.35b).

This type of switche has several interesting features. Cross talk in this tech nology is very low compared with traditional planar waveguide technology.

The commercial 32 × 32 port switch by Agilent Technologies has a specification of −50 dB cross talk. The switching operation is digital, and switching time for the device is ~10 ms. Typical operating voltages are 15 and 5 V DC. Optical alignment can be easily performed, and, once manufactured, no alignment change is necessary. The switch has no moving mechanical parts, which may translate to longer lifetimes and better reliability. The switch architecture is flexible and allows signal ports to be added and dropped for free. By contrast, in a beam-steering MEMS switch additional mirror pairs must be added to accomplish these functions, resulting in a larger switch.

4.5. SUMMARY

This chapter discussed three very important types of optical switching devices: ultrafast all-optical switches based on nonlinear optical effects, fast electro-optic modulators, and massive parallel switching devices using micro-electromechanical systems.

In ultrafast optical switches, three different schemes were discussed. They are nonlinear Fabry–Perot etalons, nonlinear directional couplers, and nonlinear optical loop mirrors. Nonlinear Fabry–Perot etalons are the first all-optical switches experimentally demonstrated. They consist of a Fabry–Perot cavity within a nonlinear optical medium. Switches arrays can be formed for parallel signal processing. However, current etalon switching devices require a high holding power, which prevents them from practical applications. NLDC has high contrast and can be implemented using integrated waveguides as well as optical fibers. However, it requires a 2π nonlinear phase shift to perform the switching, and it is difficult to form large arrays. NOLM is an interferometric switching device. It requires a π nonlinear phase shift to perform switching. The contrast is not as high as that in NLDC. It can be implemented using optical fiber or a semiconductor optical amplifier as a nonlinear element. Asymmetric arrangement of SOA in a NOLM has proven to be a very effective scheme to perform ultrafast switching for digital optical signal processing.

Fast electro-optic modulators are essential devices to convert electrical data into optical ones in optical communication systems. These devices are made by using an electrical pulse to induce a dielectric change in the medium through which the (carrier) optical signal is to pass. Three schemes were discussed in this chapter. Direct modulation of diode lasers is a simple and direct approach to the generation of coded optical pulses. External electrical signal modulates the gain in a laser diode. The main advantage of this scheme is that it can generate modulated optical signals without the use of a separate modulator; therefore the unit is very compact. Modulation bandwidths up to several tens of GHz have been achieved. However, direct current modulation produces simultaneous amplitude modulation and frequency modulation of the laser emission. As a result, spectral broadening and chirping exist, which usually affect the spectral stability of the output optical signal. Electro-optic modulators utilize the linear electro-optic effect, where a change in the refractive index is induced by an externally applied electrical field. Modulation bandwidths up to 100 GHz are potentially possible with a drive voltage of several volts. The main disadvantages include the strong polarization dependence of the devices, limited optical bandwidth, and difficulty in integration with semiconductor lasers and amplifiers. Electroabsorptive modulators using MQWs can be easily integrated with semiconductor lasers and amplifiers. Bandwidths up to 50 GHz have been achieved and the drive voltage is less than 2 V. Main concerns include strong wavelength dependence of the devices and insertion loss.

MEMS optical switches are optomechanical switching devices using micromechanical and optical elements. These elements are fabricated using micromachining techniques, and the switches are usually actuated electrostatically. Two schemes of MEMS switches were discussed in this chapter. Deformable diffraction gratings are arrays of microgratings with two states, diffraction and nondiffraction, controlled by an external voltage. The switching device has a response time of 20 ns, and can form large arrays for massive parallel signal processing and optical displays. Micromirror switching arrays consist of two-dimensional arrays of torsion mirrors. Each mirror has a size of about 16 μm. The mirrors can rotate and redirect incoming optical beams to the desired direction. Currently, to rotate a mirror by 45°, requires a driving voltage of 80 V, and a switching time of 80 μm.

REFERENCES

4.1 M. N. Islam, *Ultrafast Fiber Switching Devices and Systems*, Cambridge Press, Oxford, 1992.
4.2 J. Midwinter, Ed., *Photonics in Switching*, Academic Press, London, 1993.
4.3 H. T. Mouftah and J. M. H. Elmirghani, Ed., *Photonic Switching Technology: Systems and Networks*, New York, IEEE Press, 1999.
4.4 N. Bloembergen, *Nonlinear Optics*, World Scientific, Singapore, 1996; R. Boyd, Nonlinear Optics, Academic Press, Boston, 1992.
4.5 G. I. Stegeman and E. M. Wright, 1990, "All-optical Waveguide Switching," *Opt. Quant. Electron.*, 22, 95.
4.6 H. M. Gibbs, *Optical Bitability*, Academic Press, Orlando, 1985.
4.7 A. Yariv, *Optical Electronics*, Oxford Press, New York, 1997.
4.8 S. M. Jensen, 1982, "The Nonlinear Coherent Coupler, *IEEE J. Quant. Electron.*, QE-18, 1580.
4.9 S. R. Friberg et al., "Femtosecond Switching in a Dual-Core-Fiber Nonlinear Coupler, *Opt. Lett.*, 13, 904.
4.10 N. J. Doran, and D. Wood, 1988, "Nonlinear-Optical Loop Mirror, *Opt. Lett.*, 13, 56; K. J. Blow, N. J. Doran, and B. K. Nayar, 1989, "Experimental Demonstration of Optical Soliton Switching in an All-Fiber Nonlinear Sagnac Interferometer," *Opt. Lett.*, 14, 754.

4.11 D. Cotter et al., 1999, "Nonlinear Optics for High-Speed Digital Information Processing," *Science*, 286, 1523.

4.12 A. W. O'Neill and R. P. Webb, 1990, "All-Optical Loop Mirror Switch Employing an Asymmetric Amplifier/Attenuator Combination," *Electron Lett.*, 26, 2008; M. Eiselt, 1992, "Optical Loop Mirror with Semiconductor Laser Amplifier,"*Electron. Lett.*, 28, 1505; J. P. Sokoloff et al., 1993, "A Terahertz Optical Asymmetric Demultiplexer," *IEEE Photon. Technol., Lett.*, 5, 787.

4.13 R. J. Manning et al., 1997, "Semiconductor Laser Amplifiers for Ultrafast All-Optical Signal Processing," *J. Opt. Soc. Am.*, B-14, 3204.

4.14 R. C. Alferness, 1982, "Waveguide Electro-Optic Modulators," *IEEE Trans. Microwave Theory Tech.*, MTT-30, 1121.

4.15 R. E. Tench et al., 1987, "Performance Evaluation of Waveguide Phase Modulators for Coherent Systems at 1.3 and 1.5 μm," *J. Lightwave Tech.*, 5, 492.

4.16 L. Thylen, 1988, "Integrated Optics in $LiNbO_3$: Recent Developments in Devices for Telecommunications," *J. Lightwave Tech.*, 6, 847.

4.17 P. Vasil'ev, *Ultrafast Diode Lasers: Fundamentals and Applications*, Boston, Artech House, Boston, 1995.

4.18 R. S. Tucker, 1985, "High-Speed Modulation of Semiconductor Lasers," *J. Lightwave Tech.*, 3, 1180.

4.19 K. Petermann, *Laser Diode Modulation and Noise*, Kluwer, Dordrecht, 1988.

4.20 R. Nagarajan et al., 1992, "High Speed Quantum-Well Lasers and Carrier Transport Effects,"*IEEE J. Quant. Elect.*, QE-28; M. C. Tatham et al., 1992, "Resonance Frequency, Damping, and Differential Gain in 1.5 μm Multiquantum-Well Lasers," *IEEE J. Quant. Elect.*, QE-28, 408.

4.21 R. Olshansky et al., 1987, "Frequency Response of 1.3 μm InGaP High Speed Semiconductor Lasers," *IEEE J. Quant. Elect.*, QE-23, 1410.

4.22 K. Y. Lau and A. Yariv, 1985, "Ultra-Fast Speed Semiconductor Lasers," *IEEE J. Quant. Elect.*, QE-21, 121; R. T. Huang, 1992, "High Speed Low-Threshold InGaAsP Semi-Insulating Buried Crescent Lasers with 22 GHz Bandwidth," *IEEE Photon. Tech. Lett.*, 4, 293.

4.23 R. S. Tucker and D. J. Pope, 1983, "Microwave Circuit Models of Semiconductor Injection Lasers," *IEEE Trans. Microwave Theory Tech.*, MTT-31, 289.

4.24 P. A. Morton, 1992, "25 GHz Bandwidth 1.55 μm GaInAsP *p*-Doped Strained Multiquantum-Well Lasers," *Electron Lett.*, 28, 2156.

4.25 G. P. Agrawal, *Fiber-Optic Communication Systems*, Wiley, New York, 1992.

4.26 W. K. Burns et al., 1998, "Broad-Band Reflection Traveling-Wave $LiNbO_3$ Modulator," *IEEE Photon. Tech. Lett.*, 10, 805; R. C. Alferness, S. K. Korotky, and E. A. J. Marcatili, "Velocity-Matching Techniques for Integrated Optic Traveling Wave Switching/Modulators," *IEEE J. Quant. Electron.*, QE-20.

4.27 J. B. Khurgin, J. U. Kang, and Y. J. Ding, 2000, "Ultrabroad-Bandwidth Electro-Optic Modulator Based on a Cascaded Bragg Grating," *Opt. Lett.*, 25, 70.

4.28 G. K. Gopalakrishnan et al., 1992, "40 GHz Low Half-Wave Voltage Ti:$LiNbO_3$ Intensity Modulator," *Electron. Lett.*, 28, 826; K. Noguchi, H. Miyazawa, and O. Mitomi, 1994, "75 GHz Broadband Ti:$LiNbO_3$ Optical Modulator with Ridge Structure," *Electron. Lett.*, 12, 949.

4.29 K. Yoshida, Y. Kanda, and S. Kohjiro, 1999, "A Traveling-Wave-Type $LiNbO_3$ Optical Modulator with Superconducting Electrodes," *IEEE Trans. Microwave Theory Tech.*, MTT-47, 1201.

4.30 P. N. Butcher and D. Cotter, *The Elements of Nonlinear Optics*, Cambridge University Press, Cambridge, 1990.

4.31 D. A. B. Miller, 1990, "Quantum Well Optoelectronic Switching Devices," *Int. J. High Speed Electron.*, 1, 19.

4.32 D. A. B. Miller, 1990, "Quantum-Well Self-Electro-Optic Effect Devices,"*Opt. Quant. Electron.*, 22, S61.
4.33 A. L. Lentine and D. A. B. Miller, 1993, "Evolution of the SEED Technology: Bistable Logic Gates to Optoelectronic Smart Pixels," *IEEE J. Quant. Electron.*, QE-29, 655.
4.34 D. A. Ackerman, P. R. Prucnal, and S. L. Cooper, 2000, "Physics in the Whirlwind of Optical Communications," *Physics Today*, September, 30.
4.35 K. Kawano et al., 1997, "Polarization-Insensitive Traveling-Wave Electrode Electroabsorption (TW-EA) modulator with bandwidth over 50 GHz and driving voltage Less Than 2 V *Electron Lett.*, 33, 1580; V. Kaman, 1999, "High-Speed Operation of Traveling-Wave Electroabsorption Modulator," *Electron. Lett.*, 35, 993.
4.36 H. Takeuchi et al., 1997, "NRZ Operation at 40 Gb/s of a Compact Module Containing an MQW Electroabsorption Modulator Integrated with a DFB Laser," *IEEE Photon. Tech. Lett.*, 9, 572.
4.37 K. D. Wise, Ed., 1998, (MEMS), *IEEE Proceedings*, 86, 1687.
4.38 M. C. Wu, N. F. De Rooij, and H. Fujita, Ed., 1999, *IEEE J. Selected Topics Quant. Electron.*, 5, 2.
4.39 N. Maluf, *An Introduction to Microelectromechanical Systems Engineering*, Artech House, UK, 1999.
4.40 O. Solgaard, F. S. A. Sandejas, and D. M. Bloom, 1992, "Deformable Grating Optical Modulator," *Opt. Lett.*, 17, 688.
4.41 L. J. Hornbeck, 1998, "From Cathode Rays to Digital Micromirrors: A History of Electronic Projection Display Technology," *TI Technical J.*, July–September issue, 7.
4.42 S.-S. Lee, L.-S. Huang, C.-J. Kim, and M. C. Wu, 1999, "Free-Space Fiber-Optic Switches Based on MEMS Vertical Torsion Mirrors," *J. Lightw. Technol.*, 17, 7.
4.43 J. Fouquet et al., 1997, "Total Internal Reflection Optical Switches Employing Thermal Activation," US Patent 5,699,462; J. Fouquet et al., 2000, "Fabrication of a Total Internal Reflection Optical Switch with Vertical Fluid Fill-Holes," US Patent 6,055,344.

EXERCISES

4.1 Write a computer program to solve Eqs. (4.3) and (4.5), and generate curves similar to Fig. 4.4. Assume the nonlinear medium in the etalon is GaAs with $n_2 = 2 \times 10^{-6}\,\text{cm}^2/\text{W}$.
4.2 Study the effect of loss on the performance of a nonlinear etalon. Assume the nonlinear medium has a loss coefficient of α (cm^{-1}). Derive the results similar to Eqs. (4.3), (4.4), and (4.5).
4.3 Derive Eq. (4.8) from Eq. (4.7) using the initial conditions $A(0) = 1$, and $B(0) = 0$.
4.4 Solve Eq. (4.7) using initial condition $A(0) = 1/2$ and $B(0) = 1/2$.
4.5 Show that Eq. (4.10) reduces to Eq. (4.8) when $P \ll P_c$. (Hint: Find the property of the elliptic function $cn(x|m)$ for special m values.)
4.6 Study the effect of loss on the performance of an NLDC. Assume the nonlinear medium has a loss coefficient of $\alpha = 1\ \text{cm}^{-1}$, and coupling length $L_c = 1$ cm. Obtain the results similar to Fig. 4.9.
4.7 Derive Eq. (4.12) from Eq. (4.11) assuming $E_1 = E_{in}$ and $E_2 = 0$.

4.8 Find the required length of optical fiber (made of silica glass) in order to have π nonlinear phase shift at a control power level of 100 mW at 1.55 μm wavelength.

4.9 The $LiNbO_3$ crystal has a EO tensor given by

$$\begin{bmatrix} 0 & -T_{22} & T_{13} \\ 0 & T_{22} & T_{13} \\ 0 & 0 & T_{33} \\ 0 & T_{51} & 0 \\ T_{51} & 0 & 0 \\ -T_{22} & 0 & 0 \end{bmatrix},$$

where for $\lambda = 633$ nm, $T_{13} = 9.6 \times 10^{-12}$, $T_{22} = 6.8 \times 10^{-12}$, $T_{33} = 30.9 \times 10^{-12}$, and $T_{51} = 32.6 \times 10^{-12}$ m/V. What is the half-wave electric field for a $LiNb0_3$ modulator where light polarizes along the y axis, and the modulating electric field is applied along the z axis? Assume $\lambda =$ 633 nm and the length of the modulator $L =$ 4 mm.

4.10 Consider the case described in Problem 9. How large an electric field E_Z is required to change the index of refraction by 0.0001?

4.11 The sensitivity of a device is defined as the incremental change of the output per unit incremental change of the input. Find the sensitivity of an interferometric EO intensity modulator using the Mach–Zehnder configuration (Fig. 4.22), if the half-wave voltage is $V_\pi = 10$ V.

4.12 An optical intensity modulator uses two integrated electro-optic phase modulators and a 3-dB directional coupler, as shown in Fig. 4.P1. The input wave is split into two waves of equal amplitudes, each of which is phase modulated, reflected from a mirror, and phase modulated once

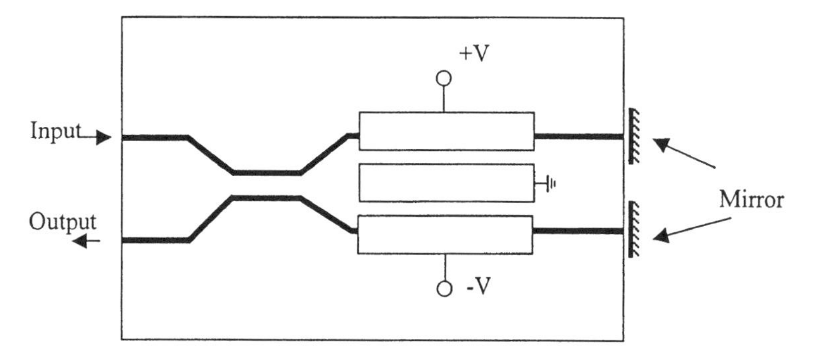

Fig. 4.P1.

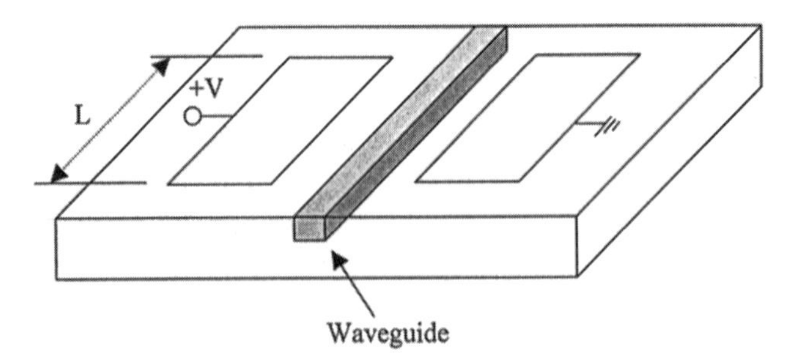

Fig. 4.P2.

more; then the two returning waves are added by the directional coupler to form the output wave. Derive an expression for the intensity transmittance of the device in terms of the applied voltage, the wavelength, the dimensions, and the physical parameters of the phase modulator.

4.13 Consider the waveguide phase modulator shown in Fig. 4.P2.
(a) Assume that only the vertically polarized wave is to be modulated at $\lambda = 1.15\ \mu m$. What is the correct orientation for the GaAs crystal if the applied field is as shown?
(b) Assume that the electric field strength in the waveguide is equal to 100 (V/mm). What length should the electrodes be to produce a $\pi/2$ phase shift?

The EO tensor of GaAs at $\lambda = 1.15\ \mu m$ is given below.

$$\begin{bmatrix} 0 & 0 & 0 \\ 0 & 0 & 0 \\ 0 & 0 & 0 \\ T_{41} & 0 & 0 \\ 0 & T_{41} & 0 \\ 0 & 0 & T_{41} \end{bmatrix},$$

where $T_{41} = 1.43 \times 10^{-12}$ m/V.

4.14 If you only had the device described in Problem 4.13 in your lab, and you wanted to make a polarization rotator using it, how would you do it?

4.15 An intensity modulator is built based on the concept of turning a single-mode waveguide on and off via the electro-optic effect. Consider the semiconductor structure shown in Fig. 4.P3. The top layer of GaAs is lightly doped, and is $5\ \mu m$ thick. The substrate is heavily doped. A Schottky barrier is placed on the surface for a distance L. (Due to the light doping of the top layer, a reverse-biased field will develop most of

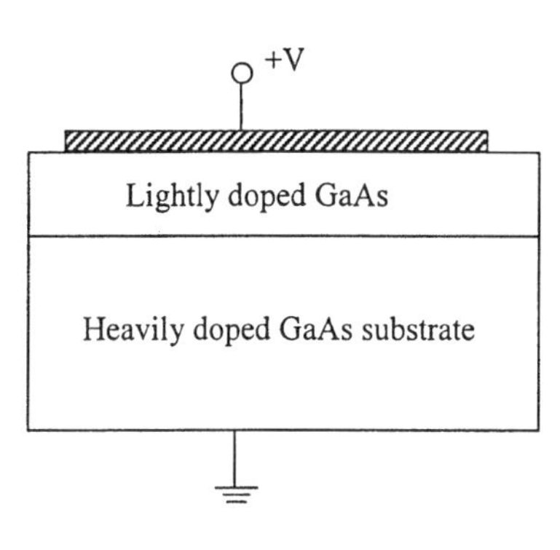

Fig. 4.P3.

the field in the thin layer, as in the case of reverse-biased transverse MQW modulators.) Consider a wavelength of 1.15 μm.

(a) What orientation should the crystal be if a positive voltage is to increase the index of the top layer?

(b) How much voltage is required to increase the index sufficiently to cause the top layer to become a waveguide for the lowest-order mode of an asymmetric waveguide? (Hint: Find the condition for a single-mode waveguide first.)

4.16 Design a $LiNb0_3$ integrated optic intensity modulator using the Mach–Zehnder interferometer shown in Fig. 4.22. Select the orientation of the crystal and the polarization of the guided wave for the smallest half-wave voltage V_π. Assume that the active region has length $L = 1$ mm and width $d = 5\,\mu$m, the wavelength is $\lambda = 633$ nm, the refractive indices are $n_0 = 2.29$, $n_e = 2.17$, and the electro-optic coefficients are as given in Problem 1.

4.17 A Franz–Keldysh modulator is built using GaAs. If the contrast ratio between "on" and "off" is to be 10 dB and the operating wavelength is 900 nm, how thick should the GaAs device be made? (Use the data in Fig. 4.23.)

4.18 Use the data in Fig. 4.23, and ignore the loss of surface reflection. What is the minimum insertion loss for the device in Problem 4.17?

4.19 Design a $N \times N$ ($N > 4$) switching array by using micromirrors. Specify the configurations of the micromirrors.

Chapter 5 Transformation with Optics

Yunlong Sheng[1] and Yueh OuYang[2]
[1]Department of Physics, Laval University,
Québec, Canada
[2]Department of Physics, Chinese Military Academy,
Kaohsiung 830 Taiwan

Lightwaves are one of the most important sources of information for human beings, who acquire 85% of information by vision and 15% by audition. An optical system, whether for imaging or nonimaging, performs mapping from the input plane to the output plane with an extremely high information throughput at the speed of light. In most cases, this mapping is a two-dimensional (2D) transform.

Mathematical transforms are closely related to the history of the development of optics. In the 1940s Duffieux introduced the Fourier transform to optics [1]. This work and earlier works by Abbe and Rayleigh in the beginning of the 20th century [2, 3], and many other later works by Marechal and O'Neill in the 1950s [4, 5], and Leith and Van De Lugt in the 1960s constituted the foundation of a new branch of optics science: Fourier optics [6, 7]. The invention of the laser in 1960 created huge interest in coherent and incoherent optical systems. Fourier optics with the concept of Fourier transform and spatial-frequency spectrum analysis, is now a fundamental basis of optical system analysis and design. Apart from imaging systems, many new optical systems have been proposed and developed that perform a variety of transformations for optical information processing, communication, and storage.

In this chapter, we discuss the relation between mathematical transformations and optics. All optical systems can be considered as systems which perform mapping, or transformation, from the input plane to the output plane. After a brief review of the Huygens–Fresnel diffraction, Fresnel transform, and

Fourier transform, which describe optical propagation and diffraction in free-space optical systems, we introduce three transforms: the physical wavelet transform, the Wigner distribution function, and the fractional Fourier transform. These recently developed transforms also describe optical propagation and diffraction but from different perspectives. The Wigner distribution function was discussed in Sec. 1.5.5 as a signal analysis tool. In Sec. 5.6 we show the Wigner distribution function with a geometrical interpretation; this leads to Wigner distribution optics in Sec. 5.8. Also, we discuss the Hankel transform in terms of the Fourier transform in the polar coordinate system.

There are a large number of optical processors and systems developed for implementation of a variety of transforms for optical signal processing and pattern recognition. These optical transforms are discussed in different chapters of the book, such as the Mellin transform in Sec. 2.6.1, the circular harmonic transform in Sec. 2.6.2, the homomorphic transform in Sec. 2.6.3 and many other optical processing algorithms and neural networks implemented with optical correlators and other optical processors.

In this chapter we include the radon transform, which is widely used in medical image processing; and the geometric transform, which has a variety of applications. The Hough transform is a type of geometric transform, also discussed.

The collection of optical transforms in this chapter is far from complete. Readers interested in transformation optics can find many books [8, 9] and technical journals on the subject.

5.1. HUYGENS–FRESNEL DIFFRACTION

As early as 1678, Huygens suggested for interpreting optical diffraction that each element on a wavefront could be the center of a secondary disturbance, which gives rise to a spherical wavelet [10], and that the wave front at any later time is the envelope of all such wavelets. Later, in 1818, Fresnel extended Huygens's hypothesis by suggesting that the wavelets can interfere with one another, resulting in the Huygens–Fresnel principle, which is formulated as

$$E(r) = \frac{1}{j\lambda}\int_{\Sigma} ds E(\boldsymbol{r}')\frac{\exp(ik(\boldsymbol{r}-\boldsymbol{r}'))}{|\boldsymbol{r}-\boldsymbol{r}'|}\cos(\boldsymbol{n},(\boldsymbol{r}-\boldsymbol{r}')), \tag{5.1}$$

where $E(\boldsymbol{r})$ is the complex amplitude of the optical field in the three-dimensional (3D) space, $\boldsymbol{r}$ is the position vector in the 3D space, and $\boldsymbol{r}'$ is the position vector at the aperture Σ, where the integral is taken over, $k = 2\pi/\lambda$ with the wavelength λ, and $\cos(\boldsymbol{n},(\boldsymbol{r}-\boldsymbol{r}'))$ is a directional factor, which is the cosine of the angle between the normal of the wavefront at the aperture and the direction of the radiation.

The Huygens–Fresnel formula was found to be the solution of the Helmhotz equation with the Rayleigh–Sommerfeld's Green function [11]. The Helmhotz equation results from the Maxwell equations for a monochromatic component of the light field.

In Eq. (5.1) one considers only one scalar component $E(\boldsymbol{r})$ of the complex amplitude of the vector electric field. When the aperture dimension is much larger than the wavelength and the observation is far from the aperture, one can consider the components E_x, E_y, and E_z of the vector electric field as independent; they can then be computed independently by the scalar Helmhotz equation. The Huygens–Fresnel formula is still the foundation of optical diffraction theory.

5.2. FRESNEL TRANSFORM

5.2.1. DEFINITION

Mathematically the Fresnel transform is defined as

$$F(u) = \int f(x) \exp\lfloor(i\pi(u - x)^2\rfloor dx, \tag{5.2}$$

where the notation $\int()$ represents the integrals with the limits from $-\infty$ to $+\infty$. The inverse Fresnel transform is given by

$$f(x) = \int F(u) \exp\lfloor -i\pi(u - x)^2\rfloor du. \tag{5.3}$$

5.2.2. OPTICAL FRESNEL TRANSFORM

In relation to optics, the Fresnel transform describes paraxial light propagation and diffraction under the Fresnel approximation. We now describe the optical field in an input and an output plane, which are normal to the optical axis. Let x denote the position in the aperture plane, which is considered as the input plane, and x' denote the position in the output plane, where the diffracted pattern is observed. Let z denote the distance from the aperture to the output plane. (Although both the input and output planes of an optical system are 2D, we use 1D notations throughout this chapter for the sake of simplicity of the formula. Generalization of the results to 2D is straightforward.)

Under the paraxial condition, the distance z is much larger than the size of the aperture Σ, $z \gg \max(x)$, so that the direction factor in the Huygens–Fresnel formula in Eq. (5.1) may be neglected and the denominator $|\boldsymbol{r} - \boldsymbol{r}'|$ in the right-hand side of Eq. (5.1) is approximated by the distance z. The phase factor in the integrand on the right-hand side of Eq. (5.1) is approximated, in the 1D notation for the input plane x and output plane x', by $(\boldsymbol{r} - \boldsymbol{r}') = z(1 + (x - x')^2/2z^2)$; this results in

$$E(u) = \frac{e^{ikz}}{j\lambda z} \int E(x) \exp(\pi i(x - x')^2/\lambda z)\, dx, \tag{5.4}$$

where the phase shift, $\exp(ikz)$, and the amplitude factor $1/z$ in the front of the integral in Eq. (5.4) are associated with wave propagation over the distance z, and are constant over the output plane at a given propagation distance z. The paraxial approximation that leads to Eqs. (5.1) to (5.4) is referred to as the Fresnel approximation.

Hence, the diffracted field $E(u)$ is the Fresnel transform of the complex amplitude $E(x)$ at the aperture, where $u = x'/\lambda z$ is the spatial frequency. The optical diffraction under the Fresnel paraxial approximation is described by the Fresnel transform. On the other side, we can consider light propagation through an aperture as an implementation of the mathematical Fresnel transform.

One can expand the quadratic phase on the right-hand side of Eq. (5.4) and rewrite Eq. (5.4) as

$$E(u) = \frac{e^{ikz}}{j\lambda z} \exp(i\pi x'^2/\lambda z) \int E(x) \exp(-i2\pi x x'/\lambda z) \exp(i\pi x^2/\lambda z)\, dx. \tag{5.5}$$

In Eq. (5.5) the quadratic phase terms, $\exp(i\pi x'^2/\lambda z)$ and $\exp(i\pi x^2/\lambda z)$, describe the spherical wavefronts of radius z in the input and output planes, respectively, as shown in Fig. 5.1. A spherical wavefront passing through an aperture

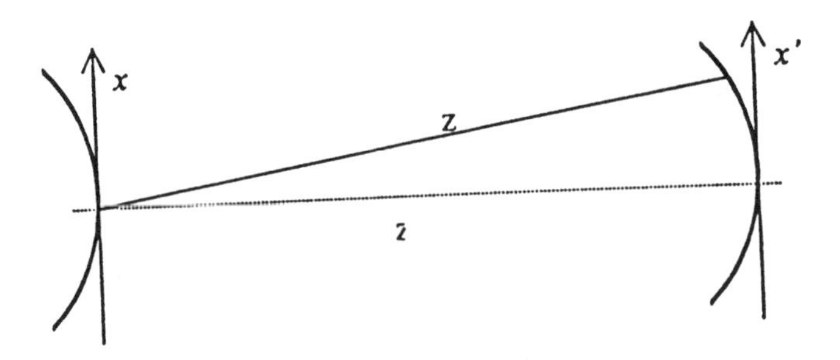

Fig. 5.1. Optical Fresnel transform.

would have the complex amplitude $E(x)\exp(i\pi x^2/\lambda z)$ in the aperture plane. According to Eq. (5.5), this complex amplitude field is exactly Fourier transformed, as described by the integral on the right-hand side of Eq. (5.5), and then multiplied by a quadratic phase factor, $\exp(i\pi x'^2/\lambda z)$, resulting in $E(u)$ in the output plane x'. The $E(u)\exp(i\pi x'^2/\lambda z)$ is the representation in the output plane of a spherical wavefront with the radius of z. In this sense one can say that the exact Fourier transform exists between the complex amplitudes on two spherical surfaces in the input and output planes in the Fresnel transform scheme.

5.3. FOURIER TRANSFORM

When the diffraction distance from the aperture to the diffracted pattern, z, tends toward infinity, the Fresnel diffraction becomes the Fraunhofer diffraction. When the condition

$$z \gg x^2/\lambda \tag{5.6}$$

is satisfied, the quadratic phase factor $\exp(i\pi x^2/\lambda z)$ in the integral on the right-hand side of Eq. (5.5) can be removed. Then, the Fresnel diffraction formula becomes the Fourier transform

$$E(u) = \int E(x)\exp(-2\pi iux/\lambda z)\,dx. \tag{5.7}$$

For instance, when $\lambda = 0.6\ \mu\text{m}$ and the aperture is of radius of 25 mm, to satisfy the condition in Eq. (5.6), z should be much longer than 1000 m, which is not practical.

The Fraunhofer diffraction at an infinite distance is implemented usually by means of a converging lens, which brings the Fraunhofer diffraction from infinity to its focal plane. There are basically three schemes for the implementation of the optical Fourier transform:

1. The input is placed in the front of a converging lens. The quadratic phase introduced by the lens cancels the quadratic phase factor $\exp(i\pi x^2/\lambda f)$ in Eq. (5.5) and we obtain the Fourier transform in the back focal plane of the lens at a distance $z = f$, where f is the focal length of the lens, with a quadratic phase factor $\exp(i\pi x'^2/\lambda f)$. This phase factor has no effect, if we are only interested in the intensity of the Fourier transform.
2. The input is placed in the front focal plane of a converging lens, and the exact Fourier transform is obtained in the back focal plane of the lens. (For more details see [9]).

3. The input is placed in a converging beam behind a converging lens of a focal length f at any distance $d < f$ from the focal plane of the lens, to where the beam converges. The Fourier transform is obtained in the focal plane of the lens [9] with the Fourier spectrum scaled as $u = x'/\lambda d$ and multiplied by the quadratic phase factor $\exp(i\pi x'^2/\lambda d)$, where x' is the coordinate in the Fourier plane.

The physics behind the optical Fourier transform is the Fraunhofer diffraction. The lens is used to remove the quadratic phase factor in the Fresnel diffraction integral. However, some people used to call the lens as the Fourier transform lens. Note that it is the Fraunhofer diffraction, not the lens, that implements the optical Fourier transform.

5.4. WAVELET TRANSFORM

The wavelet transform has been introduced in the last fifteen years for multiresolution and local signal analysis, and is widely used for nonstationary signal processing, image compression, denoising, and processing [12, 13].

5.4.1. WAVELETS

The wavelet transform is an expansion of a signal into a basis function set referred to as wavelets. The wavelets $h_{s,t}(t)$ are generated by dilations and translations from a reference wavelet, also called the mother wavelet, $h(t)$:

$$h_{s,\tau}(t) = \frac{1}{\sqrt{s}} h\left(\frac{t-\tau}{s}\right), \tag{5.8}$$

where $s > 0$ is dilation and τ is the translation factors. The wavelet transform of a signal $f(t)$ is defined as the inner product in the Hilber space of L^2 norm:

$$W_f(s, \tau) = \langle h_{s,\tau}(t), f(t)\rangle = \frac{1}{\sqrt{s}} \int h^*\left(\frac{t-\tau}{s}\right) f(t)\, dt, \tag{5.9}$$

which can be considered a correlation between the signal and the dilated wavelet, $h(t/s)$. The normalization factor $1/\sqrt{s}$ in the definition of the wavelet in Eq. (5.8) is such that the energy of the wavelet, which is the integral of the squared amplitude of the wavelet, does not change with the choice of the dilation factor s.

The Fourier transform of the wavelet is

$$H_{s,\tau}(\omega) = \int \frac{1}{\sqrt{s}} h\left(\frac{t-\tau}{s}\right) \exp(-j\omega t)\, dt = \sqrt{s}\, H(s\omega)\,(\exp - j\omega\tau), \quad (5.10)$$

where $H(\omega)$ is the Fourier transform of the basic wavelet $h(t)$. In the frequency domain the wavelet is scaled by $1/s$, multiplied by a phase factor $\exp(-j\omega\tau)$ and by a normalization factor $s^{1/2}$.

5.4.2. TIME–FREQUENCY JOINT REPRESENTATION

According to Eq. (5.9), the wavelet transform of a one-dimensional (1D) signal is a two-dimensional (2D) function of the scale s and the time shift τ. The wavelet transform is a mapping of the 1D time signal to a 2D time–scale joint representation of the signal. The time–scale joint wavelet representation is equivalent to the time–frequency joint representation, which is familiar in the analysis of nonstationary and fast transient signals.

A signal is stationary if its properties do not change during the course of the signal. Most signals in nature are nonstationary. Examples of nonstationary signals are speech, radar, sonar, seismic, electrocardiographic signals, music, and two-dimensional images. The properties, such as the frequency spectrum, of a nonstationary signal change during the course of the signal.

In the case of the music signal, for instance, the music signal can be represented by a 1D time function of air pressure, or equivalently by 1D Fourier transform of the air pressure function. We know that a music signal consists of very rich frequency components.

The Fourier spectrum of a time signal is computed by the Fourier transform, which should integrate the signal from minus infinity to plus infinity in the time axis. However, the Fourier spectrum of the music signal must change with time. There is a contradiction between the infinity integral limits in the mathematical definition of the Fourier transform frequency and the nonstationary nature of the music signal. The solution is to introduce the local Fourier transform and the local frequency concept. Indeed, nonstationary signals are in general characterized by their local features rather than by their global features.

A musician who plays a piece of music uses neither the representation of the music as a 1D time function of the air pressure, nor its 1D Fourier transform. Instead, he prefers to use the music note, which tells him at a given moment which key of the piano he should play. The music note is, in fact, the time–frequency joint representation of the signal, which better represents a nonstationary signal by representing the local properties of the signal.

The time–frequency joint representation has an intrinsic limitation; the product of the resolutions in time and frequency is limited by the uncertainty principle:

$$\Delta t \Delta \omega \geqslant 1/2.$$

This is also referred to as the Heisenberg inequality. A signal cannot be represented as a point of infinitely small size in time–frequency space. The position of a signal in time–frequency space can be determined only within a rectangle of $\Delta t \Delta \omega$.

5.4.3. PROPERTIES OF WAVELETS

The wavelet transform is of particular interest for analysis of nonstationary and fast-transient signals, because of its property of localization in both time and frequency domains. In the definition of the wavelet transform in Eq. (5.9), the kernel wavelet functions are not specified. This is the difference between the wavelet transform and many other mathematical transforms, such as the Fourier transform. Therefore, when talking about the wavelet transform one must specify what wavelet is used in the transform.

In fact, any square integrable function can be a wavelet, if it satisfies the admissibility and regularity conditions. The admissible condition is obtained as

$$c_h = \int \frac{|H(\omega)|^2}{|\omega|} d\omega < +\infty, \tag{5.11}$$

where $H(\omega)$ is the Fourier transform of the mother wavelet $h(t)$. If the condition in Eq. (5.11) is satisfied, the original signal can be completely recovered by the inverse wavelet transform. No information is lost during the wavelet transform. The admissible condition implies that the Fourier transform of a wavelet must be zero at the zero frequency

$$|H(\omega)|\,|_{\omega=0} = 0$$

and, equivalently, in the time domain the wavelet must be oscillatory, like a wave, to have a zero mean:

$$\int h(t)\,dt = 0. \tag{5.12}$$

Regularity of the wavelet is not an obligated condition but is usually required, because regularity leads to the localization of the wavelet in both time and frequency domains. One measure of wavelet regularity requires the wavelets to have the first $n + 1$ moments up to the order n, equal to zero as [14]:

$$M_p = \int t^p h(t)\, dt = 0 \qquad \text{for } p = 0, 1, 2, \ldots, n.$$

According to the admissibility condition given in the preceding equation the wavelet must oscillate to have a zero mean. If at the same time the wavelet is regularly satisfying the zero high-order moment condition, then the wavelet also should have fast decay in the time domain. As a result, the wavelet must be a "small" wave with a fast-decreasing amplitude, as described by the name *wavelet*. The wavelet is localized in the time domain.

By using the Taylor expansion of the signal $f(t)$ the wavelet transform defined in Eq. (5.9) can be written as

$$W_f(s, 0) = \frac{1}{\sqrt{s}}\left(\sum_p f^{(p)}(0) \int \frac{t^p}{p!} h(t/s) + \cdots + dt\right),$$

where $f^{(p)}(0)$ denotes the pth derivative of $f(t)$ at $t = 0$, and the integrals on the right-hand side of the above equation are the moments of pth orders of the wavelet. Hence, we have

$$W_f(s, 0) = \frac{1}{\sqrt{s}}\left[f(0)M_0 s + \frac{f'(0)}{1!} M_1 s^2 + \frac{f''(0)}{2!} M_2 s^3 + \cdots + \frac{f^{(n)}(0)}{n!} M_n s^{n+1} + \cdots +\right].$$

According to the regularity condition, the high-order moments of the wavelet up to M_n are all equal to zero. Therefore, the wavelet transform coefficient decays with decreasing of the scale s, or increasing of $1/s$, as fast as $s^{n+(3/2)}$ for a smooth signal $f(t)$, whose derivatives of orders higher than the $(n + 1)$th order have finite values.

The regularity of the wavelet leads to localization of the wavelet transform in frequency. The wavelet transform is therefore a local operator in the frequency domain. According to the admissible condition, the wavelet already must be zero at the zero frequency. According to the regularity condition, the wavelet must have a fast decay in frequency. Hence, the wavelet transform is a bandpass filter in the Fourier plane, as described in Eq. (5.10).

5.5. PHYSICAL WAVELET TRANSFORM

The wavelet transform is a multiresolution local signal analysis and was developed mainly in the 1980s. Historically, the term *wavelet* was first introduced by Huygens in 1678 to describe the secondary waves emitted from the wavefront in his interpretation of diffraction. It is of interest to seek a possible fundamental link [15] between Huygens wavelets and the wavelet transform developed some 300 years later. At first glance, the Huygens wavelets and the mathematical wavelets are different. The latter are designed to be local in both the space and frequency domains and to permit perfect reconstruction of the original signal. The wavelets must satisfy the admissibility condition and have regularity.

However, as we shall show in the following, the Huygens wavelets are, in fact, electromagnetic wavelets, proposed recently by Kaiser [16], and the Huygens–Fresnel diffraction is, in fact, a wavelet transform with the electromagnetic wavelets [17]. The electromagnetic wavelets are solutions of the Maxwell equations and satisfy the conditions of the wavelets. The optical diffraction described by the fundamental Huygens–Fresnel principle is a wavelet transform with the electromagnetic wavelet.

In the following subsection we introduce the electromagnetic wavelet and show that the Huygens–Fresnel diffraction is the wavelet transform with the electromagnetic wavelet. (Understanding this subsection requires a basic knowledge of electrodynamics [18].)

5.5.1. ELECTROMAGNETIC WAVELET

An electromagnetic field is described in the space–time of 4D coordinates $r = (\boldsymbol{r}, r_0) \in \boldsymbol{R}^4$, where $\boldsymbol{r}$ is the 3D space vector and $r_0 = ct$, with c the speed of light and t the time. In the Minkowsky space–time the Lorentz transform invariant inner product is defined as

$$r^2 = r_0^2 - |\boldsymbol{r}|^2.$$

The corresponding 4D frequencies are the wave–number vector $p = (c\boldsymbol{p}, p_0) \in \boldsymbol{R}^4$, where $\boldsymbol{p}$ the 3D spatial frequency. For free-space propagation in a uniform, isotropic, and homogeneous dielectric medium the Maxwell equations are reduced to the wave equations, and it is easy to show that the solutions of the wave equation are constrained on the light cone C in the 4D frequency space: $p^2 = p_0^2 - c^2|\boldsymbol{p}|^2 = 0$, so that $p_0 = \pm\omega$ with $\omega/c = |\boldsymbol{p}|$, where ω is the temporal frequency.

The electromagnetic wavelet is defined in the frequency domain as [16]

$$H_{\xi,s}(p) = 2(\omega/c)^2\theta(s\omega/c)e^{-s\omega/c}e^{-ip\xi}, \tag{5.13}$$

where $\xi = (\bar{\xi}, t')$ is a 4D translation of the wavelet in the 4D space–time, s is the scalar scaling factor with a dimension of $[M]$, $\theta(s\omega/c)$ is the step function defined as

$$\theta(a) = \begin{cases} 1 & \text{if } a \geqslant 0 \\ 0 & \text{if } a < 0, \end{cases}$$

and $\exp(-ip\xi)$ is the linear phase related to the translation ξ of the wavelet. The electromagnetic wavelet defined in Eq. (5.13) is localized in the temporal frequency axis by an one-sided window

$$2(\omega/c)^2\theta(s\omega/c)e^{-s\omega/c}.$$

There is no window in the spatial-frequencies domain in the definition of the electromagnetic wavelet. However, the spatial and temporal frequencies are related by the light cone constraint $\omega/c = |\boldsymbol{p}|$.

Expression of the electromagnetic wavelet in the space–time domain is obtained with the inverse Fourier transform of Eq. (5.13).

$$h_{\xi,s}(r) = \int_{C^+} \frac{d^3\boldsymbol{p}}{2(2\pi)^3(\omega/c)} 2(\omega/c)^2 e^{-s\omega/c} \exp[ip\cdot(r - \xi)], \tag{5.14}$$

where C^+ is the light cone with the positive frequencies with $\omega > 0$ and $s\omega/c > 0$, and $d^3\boldsymbol{p}/(16\pi^3\omega/c)$ is the Lorentz-invariant measure. To compute the integral in Eq. (5.14), one needs to first put the translation $\xi = 0$, that results in the reference wavelet, or mother wavelet, $h_{0,s}(r)$, which is not translated but only scaled by the scale factor s. Then, put $r = 0$ to consider the wavelet $h_{0,s}(0)$ and compute the integral on the right-hand side of Eq. (5.14), which is an integral over a light cone in the 3D spatial-frequency space. For computing this integral an invariance under the Lorentz transform should be used. Finally, $h_{0,s}(r)$ can be obtained from $h_{0,s}(0)$ as [16]

$$h_{0,s}(r) = \frac{3c^2(s - it)^2 - |\mathbf{r}|^2}{\pi^2(c^2(s - it)^2 + |\boldsymbol{r}|^2)^3} \tag{5.15}$$

and the $h_{\xi,s}(r)$ is obtained from the reference wavelet $h_{0,s}(r)$ by introducing translations in space–time as

$$h_{\xi,s}(r) = h_{0,s}(r - \xi).$$

According to Eq. (5.15) the electromagnetic wavelet is localized in space–time. At the initial time $t = 0$ the reference wavelet with no translation $\xi = 0$

is located at the origin $|\boldsymbol{r}| = 0$, because it decays in the 3D space approximately as $|\boldsymbol{r}|^{-4}$, so that is well localized in the 3D space. When time progresses, the wavelet spreads in the space from the origin as a spherical wave in an isotropic and homogeneous medium.

The electromagnetic wavelet is a solution of the Maxwell equations, because the light cone constraint and the invariance under the conformal group are respected in the mathematical treatment of the wavelet. In addition, a direct substitution of the expression in Eq. (5.15) into the scalar wave equation has shown that this wavelet is a solution of the wave equation [19].

5.5.2. ELECTROMAGNETIC WAVELET TRANSFORM

The wavelet transform of a scalar electromagnetic field $f(x)$ is an inner product among the field and the wavelets, both in the frequency domain

$$W_f(\xi, s) = \frac{1}{(2\pi)^3} \int_C \frac{d^3\boldsymbol{p}}{2(\omega/c)} \frac{1}{(\omega/c)^2} H^*_{\xi,s}(p) \hat{f}(p), \tag{5.16}$$

where $d^3\boldsymbol{p}/(16\pi^3\omega/c)$ is the Lorentz invariant measure and the Fourier transform of $f(r)$ is expressed as $F(p) = 2\pi\delta(p^2)\hat{f}(p)$ in order to show that $F(p)$, the Fourier transform of the electromagnetic field, must be constrained by the light cone constraint in the frequency space, $p^2 = 0$, and $\hat{f}(p)$ is the unconstrained Fourier transform.

The inner product defined in Eq. (5.16) is different from conventional inner product. Beside the the Lorentz invariant measure in the integral, the integrant on the right-hand side of Eq. (5.16) contains an additional multiplication term: $1/(\omega/c)^2$. This is because Eq. (5.16) is an inner product of two electromagnetic fields: the signal field and the electromagnetic wavelet. Like all physical rules and laws, the electromagnetic wavelet transform, should describe permanencies of nature, and should be independent of any coordinate frame; i.e. invariant-to-coordinate-system transforms, such as the Lorentz transform. In fact, in 4D space–time the inner product must be defined as that in Eq. (5.16). To understand the definition of the inner product of two electromagnetic fields in the frequency domain as that in Eq. (5.16), we recall that according to electrodynamics [18] electrical and magnetic fields are neither independent from each other, nor invariant under the Lorentz transformation. Instead, the wave equation for the vector potential and the scalar potential takes covariant forms with the Lorentz condition. The square of the 4D potential $(|a_0(p)|^2 - |\boldsymbol{a}(p)|^2)$, where $a_0(p)$ is the scalar potential and $\boldsymbol{a}(p)$ is the 3D vector potential in the frequency domain, is invariant under the Lorentz transformation. It can be shown [16] that the unconstrained Fourier transform of the electromagnetic field is related to the 3D potential vector by a relation as $\hat{f}(p) = 2p_0\boldsymbol{a}(p)$, so that the integral

$$\frac{1}{(2\pi)^3}\int_C \frac{d^3\boldsymbol{p}}{2(\omega/c)}\frac{1}{(\omega/c)^2}|\hat{f}(p)|^2$$

defines a norm of solutions of the Maxwell equations that are invariant under the Lorentz transformation, because $\hat{f}(p)/p_0$ with $p_0 = \pm\omega$ has a dimension of the vector potential $\boldsymbol{a}(p)$. The square module of the latter is invariant under the Lorentz transformation, and the Lorentz-invariant inner products must be defined according to the norm of field. That is the reason for the factor $1/(\omega/c)^2$ on the right-hand side of Eq. (5.16), which is the definition of the electromagnetic wavelet transform. On the other hand, in the definition of the electromagnetic wavelet, we introduced a factor $(\omega/c)^2$ in Eq. (5.13). As a result, the wavelet transform defined in Eq. (5.16) is invariant under the Lorentz transformation. The inner products defined in Eq. (5.16) form a Helbert space where the wavelet transform is defined.

The scalar electromagnetic field $f(r)$ may be reconstructed from the inverse wavelet transform by [16]

$$f(r) = \int_E d^3\bar{\xi}\, ds h_{\xi,s}(r) W_f(\xi, s), \tag{5.17}$$

where the Euclidean region E is the group of spatial translations and scaling, which act on the real-valued space–time with $\bar{\xi} \in \boldsymbol{R}^3$ and $s \in \boldsymbol{R} \neq 0$. The scalar wavelet transform is defined on E, which is a Euclidean region in the Helbert space.

According to Eq. (5.17), an electromagnetic field is decomposed as a linear combination of the scaled and shifted electromagnetic wavelets. In this wavelet decomposition in the space–time domain, the integral on the right-hand side of Eq. (5.17) is a function of the time translation t' with a component of $\xi = (\bar{\xi}, t')$, while the signal $f(r)$, on the left-hand side of Eq. (5.17), is not a function of the time translation t'. We shall show below that, in fact, $t' = 0$ because the translations in the time and space domains are not independent from each other, according to the electromagnetic theory [18].

The wavelet transform with the electromagnetic wavelet is a decomposition of an electromagnetic field into a linear combination in the scaled and shifted electromagnetic wavelet family, $h_{\xi,s}(r)$. However, we must note that the electromagnetic wavelet is a solution of the wave equation; its space and time translations cannot be arbitrary, but must be constrained by the light cone in the space–time $c^2t'^2 = |\bar{\xi}|^2$.

Let us consider, among the whole set of the electromagnetic wavelets, which are translated in 4D space–time, two wavelets that are translated by $(t'_1, \overline{\xi}_1)$ and $(t'_2, \overline{\xi}_2)$, respectively, as two events. Their squared interval in 4D space time is

$$d_{1,2}^2 = c^2[(t - t'_1) - (t - t'_2)]^2 - |(\boldsymbol{r} - \overline{\xi}_1) - (\boldsymbol{r} - \overline{\xi}_2)|^2,$$

which is Lorentz invariant. When $c^2(t'_2 - t'_1)^2 = |\overline{\xi}_2 - \overline{\xi}_1|^2$ and $d_{1,2} = 0$, one wavelet propagating at the speed c results in the other wavelet. The two wavelets are the same event. This lightlike separation is not relevant. When $d_{1,2}^2 > 0$, one can Lorentz-transform the two events into a rest reference frame where $\overline{\xi}_1 = \overline{\xi}_2$, so that the two wavelets are at the same location, but are observed at different instances t'_1 and t'_2. This is the timelike separation. When $d_{1,2}^2 < 0$, one can Lorentz-transform the two events into a rest reference frame where $t'_1 = t'_2$. One observes the two wavelets at the same instant. This is the spacelike separation. Since the Euclidean region E supports full spatial translations, the wavelets are spacelike separated. They are all observed at the same time. There should be no time translation of the wavelets, $t' = 0$, in the wavelet transform of Eq. (5.17).

5.5.3. ELECTROMAGNETIC WAVELET TRANSFORM AND HUYGENS DIFFRACTION

We now apply the wavelet transform to a monochromatic optical field.

$$f(\boldsymbol{r}, t) = E(\boldsymbol{r})e^{j\omega_0 t}$$

with the positive temporal frequency $\omega_0 > 0$. The unconstrained Fourier transform of the monochromatic field is

$$\hat{f}(\boldsymbol{r}, \omega_0) = 2(\omega/c) \int_{\omega/c = |\boldsymbol{p}| = \omega_0/c} E(\boldsymbol{r}) \exp(-j\boldsymbol{p} \cdot \boldsymbol{x}) d^3\boldsymbol{r}.$$

The wavelet transform of the monochromatic field is then, according to Eqs. (5.16) and (5.13),

$$W_f(\overline{\xi}, s) = E(\overline{\xi})e^{j\omega_0 t'}\theta(\omega_0 s)e^{-s\omega_0}, \tag{5.18}$$

where $t' = 0$, if we consider the wavelet family is spacelikely separated, as discussed in the preceding. We take the temporal Fourier transform in both sides of the inverse wavelet transform shown in Eq. (5.17), and let the time-frequency $\omega = \omega_0$

$$\int_{-\infty}^{\infty} f(\bar{\xi}, t)e^{-j\omega_0 t}\, dt = \int_E d\bar{\xi}\, ds \int_{-\infty}^{\infty} dt W_f(\bar{\xi}, s) h_{\bar{\xi},s}(\boldsymbol{r}, t) e^{-j\omega_0 t}. \tag{5.19}$$

On the right-hand side of Eq. (5.19) only the wavelet $h_{\bar{\xi},s}(\boldsymbol{r}, t)$ is a function of time t. The temporal Fourier transform of the electromagnetic wavelet described in Eq. (5.15) can be computed by the contour integral or by the Laplace transform that yields

$$\int_{-\infty}^{\infty} h_{\bar{\xi}s}(\boldsymbol{r}, t)e^{-j\omega t}\, dt = \frac{1}{4\pi^2 c^3 j} \frac{e^{j\omega|\boldsymbol{r}-\bar{\xi}|/c} - e^{-j\omega|\boldsymbol{r}-\bar{\xi}|/c}}{|\boldsymbol{r}-\bar{\xi}|} \theta(s\omega)\omega^2 e^{-s\omega}. \tag{5.20}$$

Combining Eqs. (5.18), (5.19), and (5.20) we obtain

$$E(\boldsymbol{r}) = \frac{1}{4\pi^2 c^3 j} \int_E d^3\bar{\xi}\, ds E(\bar{\xi})\theta(s\omega_0)\omega_0^2 e^{-2s\omega_0} \frac{e^{j\omega_0|\boldsymbol{r}-\bar{\xi}|/c} - e^{-j\omega_0|\boldsymbol{r}-\bar{\xi}|/c}}{|\boldsymbol{r}-\bar{\xi}|}.$$

The integration with respect to s may be computed as

$$\int_{-\infty}^{\infty} ds\theta(s\omega_0)e^{-2s\omega_0} = \frac{1}{2\omega_0}$$

because $\omega_0 > 0$. Hence, we have finally

$$E(\boldsymbol{r}) = \frac{1}{4\pi^2 c^3 j} \int_{R^3} d^3\bar{\xi} E(\bar{\xi}) \frac{e^{j2\pi|\tau-\bar{\xi}|/\lambda_0} - e^{-2\pi|\boldsymbol{r}-\bar{\xi}|/\lambda_0}}{|\boldsymbol{r}-\bar{\xi}|}, \tag{5.21}$$

where the wavelength $\lambda_0 = 2\pi c/\omega_0$. According to Eq. (5.21) the complex amplitude $E(\boldsymbol{r})$ of a monochromatic optical field is reconstructed from a superposition of the monochromatic spherical wavelets, the centers of which are at the points $\boldsymbol{r} = \bar{\xi}$ and the amplitudes of which $E(\bar{\xi})$ vary as a function of 3D space translation $\bar{\xi}$. The coherent addition of all those spherical wavelets forms the monochromatic electromagnetic field $E(\boldsymbol{r})$. Equation (5.21) is indeed the expression of the Huygens principle. Only the directional factor in the Huygens–Fresnel formula is absent in Eq. (5.21). Another difference from the Huygens–Fresnel formula is that Eq. (5.21) describes the convergent and divergent spherical wavelets, which are incoming to and outgoing from the localization point $\bar{\xi}$. The incoming wavelets were not considered in the Huygens–Fresnel formula.

Thus, we have shown that the optical diffraction is a wavelet transform of a monochromatic optical field with electromagnetic wavelets. We have decomposed a monochromatic optical field into electromagnetic wavelet basis and shown that the complex amplitude of the field is reconstructed by the inverse wavelet transform, which is equivalent to the Huygens–Fresnel formula. Hence, in the case of monochromatic field, electromagnetic wavelets are the monochromatic spherical wavelets proposed by Huygens, and wavelet decomposition gives the expression of the Huygens principle.

5.6. WIGNER DISTRIBUTION FUNCTION

5.6.1. DEFINITION

The Wigner distribution function is a mapping of a signal from the space-coordinate system to the space-frequency joint coordinates system. The space-frequency joint representation is useful for analysis of nonstationary and transient signals, as discussed in Sec. 5.4.2. The Wigner distribution of a function $f(x)$ is defined in the space domain as

$$W_f(x, \omega) = \int f\left(x + \frac{x'}{2}\right) f^*\left(x - \frac{x'}{2}\right) \exp(-j\omega x')\, dx', \tag{5.22}$$

which is the Fourier transform of the product of a signal $f(x/2)$ dilated by a factor of 2 and the inverted signal $f^*(-x/2)$ also dilated by a factor of 2. Both the dilated signal and its inverse are shifted to the left and right, respectively, by x along the x-axis. By the definition in Eq. (5.22) the Wigner distribution function is a nonlinear transform and is a second-order or bilinear transformation. The Wigner distribution of a 1D signal is a 2D function of the spatial-frequency ω and the spatial shift x in the space domain.

The Wigner distribution function can be also defined in the frequency domain and is expressed as

$$W_f(\omega, x) = \frac{1}{2\pi} \int F\left(\omega + \frac{\omega'}{2}\right) F^*\left(\omega - \frac{\omega'}{2}\right) \exp(jx\omega')\, dv, \tag{5.23}$$

where $F(\omega)$ is the Fourier transform of $f(x)$. From Eqs. (5.22) and (5.23) one can see that the definitions of the Wigner distribution functions in the space and frequency domains are symmetrical.

5.6.2. INVERSE TRANSFORM

The projection of the Wigner distribution function $W_f(x, \omega)$ in the space–frequency joint space along the frequency ω-axis gives the square modulus of the signal $f(t)$, because according to Eq. (5.22) the projection of $W_f(x, \omega)$ along the ω-axis is

$$\int W_f(x, \omega)\, d\omega = \iint f\left(x + \frac{x'}{2}\right) f^*\left(x - \frac{x'}{2}\right) \exp(-j\omega x')\, dx'\, d\omega = 2\pi |f(x)|^2.$$

The projection of $W_f(x, \omega)$ in the space–frequency joint space along the space axis x gives the square modulus of the Fourier transform $F(\omega)$ of the signal, because according to Eq. (5.23) the projection along the x-axis is

$$\int W_f(x, \omega)\, dx = \frac{1}{2\pi} \iint F\left(\omega + \frac{\omega'}{2}\right) F^*\left(\omega - \frac{\omega'}{2}\right) \exp(j\omega' x)\, d\omega'\, dx = |F(\omega)|^2.$$

In addition, we have the energy conservation of the Wigner distribution function in the space–frequency joint representation:

$$\frac{1}{2\pi} \int W_f(x, \omega)\, dx\, d\omega = \frac{1}{2\pi} \int |F(\omega)|^2\, d\omega = \int |f(x)|^2\, dx.$$

5.6.3. GEOMETRICAL OPTICS INTERPRETATION

Let $f(x)$ denote the complex amplitude of the optical field. Its Wigner distribution function $W_f(x, \omega)$ describes the propagation of the field in the space–frequency joint representation, where the frequency ω is interpreted as the direction of a ray at point x.

The concept of spatial frequency is introduced in the Fourier transform. In the optical Fourier transform described in Eq. (5.7), the spatial frequency is defined as $\omega - 2\pi u/\lambda z$, where u is the spatial coordinate in the Fourier plane. Therefore, u/z is the angle of ray propagation. In this case, the spatial frequency ω can be considered the local frequency. When an optical field is represented in the space–frequency joint space, its location x and local frequency must obey the uncertainty principle. One can never represent a signal with an infinite resolution in the space–frequency joint space, but can only determine its location and frequency within a rectangle of size

$$\Delta x \Delta \omega \geqslant 1/2.$$

This Heisenberg inequality familiar in quantum mechanics can be easily interpreted when one considers the local frequency as the ray propagation direction. According to the physical diffraction when the light is localized by a pinhole of a size Δx in the space, it would be diffracted within a range of diffraction angles, $\Delta\omega$, which is inversely proportional to the pinhole size Δx.

5.6.4. WIGNER DISTRIBUTION OPTICS

The Wigner distribution function can be used to describe light propagation in space in a similar way as geometrical ray optics does.

In Wigner distribution function interpretation, a parallel beam having only a single spatial frequency $f(x) = \exp(i\omega_0 x)$ is a horizontal line normal to the ω-axis in the space–frequency joint space, because

$$W_f(x, \omega) = \int \exp(i\omega_0(x+x'/2)) \exp(-i\omega_0(x-x'/2)) \exp(i\omega x')\, dx' = \delta(\omega-\omega_0).$$

A spatial pulse passing through a location x_0 is described as $f(x)=\delta(x-x_0)$, its Wigner distribution function is a vertical line normal to the x-axis in the space–frequency joint space, because

$$W_f(x, \omega) = \int \delta(x - x_0 + x'/2)\delta(x - x_0 - x'/2) \exp(i\omega x')\, dx' = \delta(x - x_0).$$

In both cases we keep the localization of the signal in frequency and space, respectively.

The Wigner distribution function definition in space, Eq. (5.22), and in frequency, Eq. (5.23), which are completely symmetrical. The former Wigner distribution function of $f(x)$ is computed in space, resulting in $W_f(x, \omega)$, and the latter Wigner distribution function of the Fourier transform $F(\omega)$ is computed in frequency, resulting in $W_f(\omega, x)$. In the two equations, the roles of x and ω are interchanged. The Wigner distribution function is then rotated by 90° in the space–frequency joint representation by the Fourier transform of the signal.

When the optical field $f(x)$ passes through a thin lens, it is multiplied by a quadratic phase factor, $\exp(-i\pi x^2/\lambda f)$, where f is the focal length of the lens. The Wigner distribution function of the optical field behind the lens becomes

$$\begin{aligned} W_f(x, \omega) &= \int f(x+x'/2) \exp(-i\pi(x+x'/2)^2)\, f^*(x-x'/2) \\ &\quad \times \exp(-i\pi(x-x'/2)^2) \exp(-i\omega x')\, dx' \\ &= W_f(x, \omega + (2\pi/\lambda f)x). \end{aligned}$$

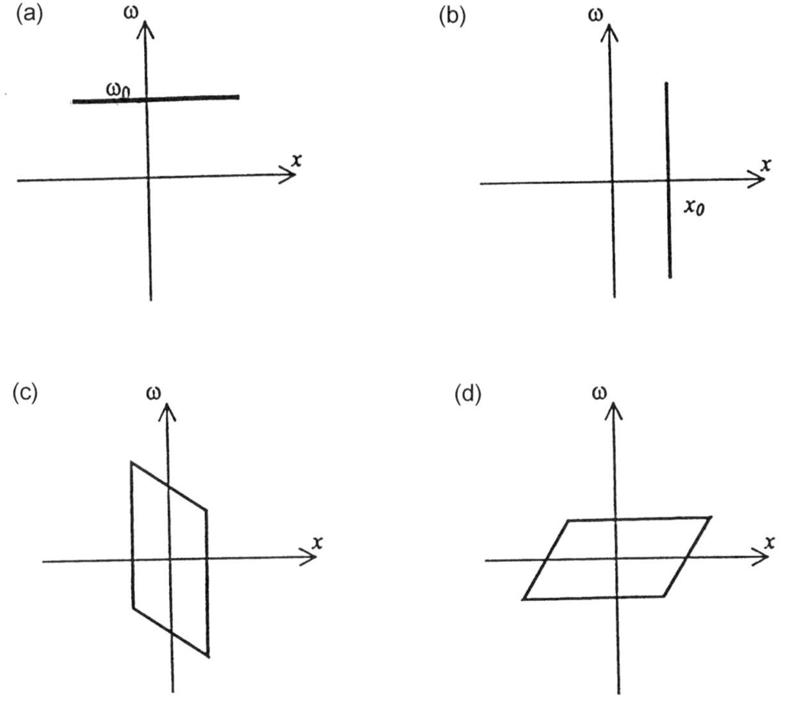

Fig. 5.2. Space–frequency joint representation of an optical field by the Wigner Distribution Function (a) A plane with single spatial frequency ω_0, (b) A beam passing through a pinhole at x_0, (c) A beam passing through a lens, (d) A beam propagating in free space over a distance z.

This is the original Wigner distribution function sheared in the ω direction.

When the optical field $f(x)$ simply propagates in free space over a distance z, its Fourier transform will be multiplied by the transfer function of free-space propagation, which is a phase factor [11], and becomes

$$F(\omega)\exp(i2\pi z\sqrt{1-(\omega\lambda/2\pi)^2}/\lambda) \approx F(\omega)\exp(i2\pi z/\lambda)\exp(i\lambda z\omega^2/4\pi),$$

where the term $\exp(i2\pi z/\lambda)$ in the right-hand side of the equation, corresponding to a phase shift related to the propagation over distance z, can be disregarded. Then the Wigner distribution function becomes

$$\begin{aligned}W_f(\omega, x) &= \int F(\omega + \omega'/2)\exp(-i\lambda z(\omega + \omega'/2)^2/4\pi)F(\omega - \omega'/2)\\ &\quad \times \exp(-i\lambda z(\omega - \omega'/2)^2/4\pi)\exp(-ix\omega')\,d\omega'\\ &= W_f(\omega, x - \lambda z\omega/2\pi).\end{aligned}$$

This is the original Wigner distribution function sheared in the x direction.

When an image is magnified by the optical system with a factor of s, $f(x)$ becomes $f(x/s)$, and the Wigner distribution function becomes

$$W_f(x, \omega) = \int f((x + x'/2)/s) f((x - x'/2)/s) \exp(-i\omega x')\, dx'$$
$$= sW_f(x/s, s\omega).$$

The original Wigner distribution function is then dilated along the x-axis in the space and is shrunk along the ω-axis in the frequency.

In geometrical optics, the relation between the input and the output of an optical system is commonly described by

$$\begin{bmatrix} \xi \\ v \end{bmatrix} = M \begin{bmatrix} x \\ \omega \end{bmatrix},$$

where (ξ, v) and (x, ω) are the position and the propagation direction of the rays in the output and input planes, respectively, and M is the ray propagation matrix. According to geometrical optics, optical systems, such as those involving the Fourier transform, lens, free space, and magnifier, have ray propagation matrices M as

$$M_F = \begin{bmatrix} 0 & -1 \\ 1 & 0 \end{bmatrix} \qquad M_f = \begin{bmatrix} 1 & 1 \\ 2\pi/\lambda f & 1 \end{bmatrix}$$
$$M_z = \begin{bmatrix} 1 & -z\lambda/2\pi \\ 0 & 1 \end{bmatrix} \qquad M_s = \begin{bmatrix} s & 0 \\ 0 & 1/s \end{bmatrix},$$

respectively. These optical systems implement affine transforms in the (x, ω) space.

According to the preceding calculation for the Wigner distribution functions, when an optical field passes through an optical system, its Wigner distribution function does not change the values, but they are modified by an affine coordinate transform, which corresponds exactly to that given by the ray propagation matrices in geometrical optics. One of the links between physical optics and geometrical optics is then established. Both show that an optical system performs affine transforms in the space frequency joint space, in which each location corresponds to the position and orientation of rays.

According to ray optics, when optical systems are cascaded, the ray propagation matrices are simply multiplied. The cascaded optical system is described by an overall ray propagation matrix. Similarly, in terms of the Wigner distribution function, subsystems perform the affine transforms of the

Wigner distribution functions. The overall system performs a cascade of the affine transformations of the Wigner distribution functions.

5.7. FRACTIONAL FOURIER TRANSFORM

The Fourier transform is among the most widely used transforms in applications for science and engineering. The fractional Fourier transform has been introduced in an attempt to get more power and a wider application circle from the Fourier transform.

5.7.1. DEFINITION

In the framework of the fractional Fourier transform, the ordinary Fourier transform is considered a special case of unity order of the generalized fractional order Fourier transform. When defining such a transform, one wants this fractional Fourier transform to be a linear transform and the regular Fourier transform to a fractional Fourier transform of unity order. More important, a succession of applications of the fractional Fourier transforms should result in a fractional Fourier transform with a fractional order, which is the addition of the fractional orders of all the applied fractional Fourier transforms; i.e. in the case of two successive fractional Fourier transforms

$$FT^{\alpha}FT^{\beta} = FT^{(\alpha+\beta)}, \tag{5.24}$$

where FT^{α} denotes the fractional Fourier transform of a real valued order α.

The additive fractional order property can be obtained when one looks for the eigenfunction of the transform. For this purpose, one considers an ordinary differential equation

$$f''(x) + 4\pi^2[(2n+1)/2\pi - x^2]f(x) = 0. \tag{5.25}$$

Taking the Fourier transform of Eq. (5.25) and using the properties of the Fourier transforms of derivatives and moments one can show that

$$F''(u) + 4\pi^2[(2n+1)/2\pi - u^2])F(u) = 0,$$

where $F(u) = FT[f(x)]$ is the Fourier transform of $f(x)$. Hence, the Fourier transform of the solution also is the solution of the same differential as Eq. (5.25). Indeed, the solution of Eq. (5.25) is the Hermite–Gaussian function [20], expressed as

$$\psi_n(x) = \frac{2^{1/4}}{\sqrt{2^n n!}} H_n(\sqrt{2\pi}\, x)\exp(-\pi x^2),$$

where H_n is the Hermite polynomial of order n. Hence, the Hermite–Gaussian function is an eigenfunction of the Fourier transform [21] satisfying the eigenvalue equation such that

$$FT(\psi_n(x)) = \exp(-in\pi/2)(\psi_n(u)), \tag{5.26}$$

where u is the frequency. The $\exp(-in\pi/2) = i^{-n}$ is the eigenvalue of the Fourier transform operator.

In order to define the fractional Fourier transform one expands the validity of Eq. (5.26) to meet a real value α, which can be inserted in the power of the eigenvalue, such that the fractional Fourier transform of the eigenfunction $\psi_n(x)$ can be written as

$$FT^{\alpha}(\psi_n(x)) = \exp(-in\alpha\pi/2)\psi_n(u).$$

In this condition, the additive fractional order property, Eq. (5.24), can be observed.

One can define the fractional Fourier transform as an operator that has the Hermite–Gaussian function as an eigenfunction, with the eigenvalue of $\exp(-in\alpha\pi/2)$. The Hermite–Gaussian functions form a complete set of orthogonal polynomials [20]. Therefore, an arbitrary function can be expressed in terms of these eigenfunctions,

$$f(x) = \sum_{n=0}^{\infty} A_n \psi_n(x),$$

where the expansion coefficients of $f(x)$ on the Hermite–Gaussian function basis are

$$A_n = \int \psi_n(x) f(x)\, dx.$$

The definition of the fractional Fourier transform can then be cast in the form of a general linear transform

$$FT^{\alpha}[f(x)] = \sum_{n=0}^{\infty} \exp(-in\alpha\pi/2) A_n \psi_n(u).$$

The fractional Fourier transform applied to an arbitrary function $f(x)$ is expressed as a sum of the fractional Fourier transforms applied to the eigenfunctions $\psi_n(x)$ multiplied with a coefficient A_n, which is the expansion coefficient of $f(x)$ on a Hermite–Gaussian function basis. The eigenfunctions

$\psi_n(x)$ obey the eigenvalue equations of the Fourier transform. Using the expression for A_n, we have

$$FT^{\alpha}[f(x)] = \int B_{\alpha}(x, u) f(x)\, dx, \tag{5.27}$$

with the αth order fractional Fourier transform kernel functions

$$\begin{aligned} B_{\alpha}(x, u) &= \sum_{n=0}^{\infty} \exp(-in\alpha\pi/2)\psi_n(x)\psi_n(u) \\ &= 2^{1/2} \exp[-\pi(x^2 + u^2)] \sum_{n=0}^{\infty} \frac{\exp(-in\alpha\pi/2)}{2^n n!} H_n(\sqrt{2\pi}\, x) H_n(\sqrt{2\pi}\, u). \end{aligned}$$

After a number of algebraic calculations, one obtains

$$\begin{aligned} B_{\alpha}(x, u) \equiv {} & \frac{\exp[-i\pi(\hat{\phi} - \alpha)/4]}{\sqrt{\sin(\alpha\pi/2)}} \\ & \times \exp[i\pi(x^2 \cot(\alpha\pi/2) - 2xu/\sin(\alpha\pi/2)) + u^2 \cot(\alpha\pi/2)], \end{aligned} \tag{5.28}$$

where the fractional order $0 < |\alpha| < 2$ and $\hat{\phi} = \mathrm{sgn}(\sin(\alpha\pi/2))$. When $\alpha = 0$ and $\alpha = \pm 2$, the kernel function is defined separately from Eq. (5.28) as

$$B_0(x, u) \equiv \delta(x - u) \quad \text{and} \quad B_{\pm 2}(x, u) \equiv \delta(x + u),$$

respectively.

5.7.2. FRACTIONAL FOURIER TRANSFORM AND FRESNEL DIFFRACTION

Let us consider now the Fresnel integral, Eq. (5.5), which describes propagation of the complex amplitude $f(x)$ under the Fresnel paraxial approximation

$$f(x') = \frac{e^{ikz}}{j\lambda z} \int f(x) \exp[-i\pi(x^2 - 2\pi x x' + x'^2)/\lambda z]\, dx, \tag{5.29}$$

where $f(x)$ and $f(x')$ are the complex amplitude distribution in the input and output plane, respectively, which are separated by a distance z.

One now can compare the Fresnel diffraction, Eq. (5.29), and the fractional Fourier transform, Eq. (5.27), and find out that the Fresnel diffraction can be formulated as the fractional Fourier transform. Let $p_1(x)$ and $p_2(x')$ denote the complex amplitude distributions on the two spherical surfaces of the radii, R_1 and R_2, respectively. The complex field amplitude in the planes x and x' are expressed with

$$f(x) = p_1(x)\exp(i\pi x^2/\lambda R_1)$$
$$f(x') = p_2(x')\exp(i\pi x'^2/\lambda R_2),$$

respectively, where λ is the wavelength. One introduces the dimensionless variables, $v \equiv x/s_1$ and $v' \equiv x'/s_2$, where s_1 and s_2 are scale factors. In order to describe the Fresnel diffraction by the relationship between the complex field amplitudes in the two spherical surfaces, we rewrite Eq. (5.29) as

$$p_2(v') = \frac{\exp(i2\pi z/\lambda)}{j\lambda z} s_1 \int p_1(v)\exp\left[\frac{i\pi}{\lambda z}(g_1 s_1^2 v^2 - 2s_1 s_2 v v' + g_2 s_2^2 v'^2)\right] dv,$$

where

$$g_1 \equiv 1 + z/R_1$$
$$g_2 \equiv 1 - z/R_2.$$

Comparing this Fresnel integral formula with the definition of the fractional Fourier transform, Eqs. (5.27) and (5.28), we conclude that the complex amplitude distribution on a spherical surface of radius R_2 in the output, $p_2(v')$, is the fractional Fourier transform of that on a spherical surface of radius R_1 in the input, $p_1(v)$, i.e.,

$$p_2(v') = \frac{\exp(i2\pi z/\lambda)\exp\lfloor i\pi(\hat{\phi} - \alpha)/4\rfloor}{j\lambda z}(\sin(\alpha\pi/2))^{1/2} s_1 FT^{\alpha}[p_1(v)],$$

provided that

$$g_2 s_2^2/\lambda z = g_1 s_1^2/\lambda z = \cot(\alpha\pi/2) \tag{5.30}$$

and

$$s_1 s_2/\lambda z = 1/\sin(\alpha\pi/2). \tag{5.31}$$

Given the reference spherical surfaces R_1, R_2, and z we can compute g_1 and g_2 and the scales s_1 and s_2 and the fractional order α, according to the preceding relations.

A special case concerns the Fresnel diffraction of an aperture with a unit amplitude illumination and the aperture complex amplitude transmittance $t(x)$.

In this case, $R_1 \to \infty$, $g_1 = 1$; we let the scale parameter s_1 in the input be chosen freely. The fractional order is determined by Eq. (5.30) as a function of the propagation distance z:

$$\alpha\pi/2 = \arctan(\lambda z/s_1^2) \tag{5.32}$$

and s_2, R_2 also can be computed from Eq. (5.30). Hence, from the aperture $z = 0$ to $z \to \infty$ at every distance z, the complex amplitude distribution on the reference surface of radius R_2 is the fractional Fourier transform of the transmittance $t(x)$, computed according to Eqs. (5.30) and (5.31) where the fractional order α increases with the distance z, from 0 to 1, according to (5.32). At the infinite distance, $z \to \infty$, we have $\alpha \to 1$ according to Eq. (5.32), the Fresnel diffraction becomes the Fraunhorff diffraction, and the fractional Fourier transform becomes the Fourier transform. Notice that if we measure the optical intensity at distance z, the phase factor associated to the reference spherical surface with radius R_2 has no effect.

5.8. HANKEL TRANSFORM

Most optical systems, such as imaging systems and laser resonators, are circularly symmetrical, or axially symmetrical. In these cases, the use of the polar coordinate system is preferable, and the 2D Fourier transform expressed in the polar coordinate system will lead to the Hankel transform, which is widely used in the analysis and design of optical systems.

5.8.1. FOURIER TRANSFORM IN POLAR COORDINATE SYSTEM

The Hankel transform is also referred to as the Fourier–Bessel transform. Let $f(x, y)$ and $F(u, v)$ denote a 2D function and its Fourier transform, respectively, such that

$$F(u, v) = \iint f(x, y) \exp(-i2\pi(ux + vy))\, dx\, dy.$$

The Fourier transform is then expressed in the polar coordinates, denoted by (r, θ) in the image plane, and by (ρ, ϕ) in the Fourier plane, respectively, as

$$F(\rho, \phi) = \int_0^{2\pi} \int_0^{\infty} f(r, \theta) \exp(-i2\pi r\rho \cos(\theta - \phi)) r\, dr\, d\theta. \tag{5.33}$$

If the input image is circularly symmetric, $f(r, \theta)$ is independent of θ; we have

$$F(\rho, \phi) = \int_0^\infty f(r)r\,dr \int_0^{2\pi} \exp(-i2\pi r\rho\cos(\theta - \phi))\,d\theta = \int_0^\infty f(r)J_0(2\pi r\rho)r\,dr,$$

where $J_0()$ is the Bessel function of the first kind and zero order. Hence, the Fourier transform $F(\rho, \phi)$ is the Hankel transform of the zero order of the input image. According to the zero order Hankel transform, the Fourier transform $F(\rho, \phi)$ of a circularly symmetric input image is independent of ϕ and is also circularly symmetric. An example is when an optical beam is passed through a disk aperture; the Fraunhofer diffraction pattern of the aperture is the Airy pattern, which is computed as the Hankel transform of the zero order of the disk aperture.

In general $f(r, \theta)$ is dependent of θ. In this case, we can compute the circular harmonic expansions of $f(r, \theta)$ and $F(\rho, \phi)$, which are the one-dimensional Fourier transforms with respect to the angular coordinates, as

$$f_m(r) = \frac{1}{2\pi}\int_0^{2\pi} f(r, \theta)\exp(-im\theta)\,d\theta$$

and

$$F_m(\rho) = \frac{1}{2\pi}\int_0^{2\pi} F(\rho, \phi)\exp(-im\phi)\,d\phi,$$

where m is an integer. $f_m(r)$ and $F_m(\rho)$ are referred to as the circular harmonic functions of the input image and its Fourier transform. Because both $f(r, \theta)$ and $F(\rho, \phi)$ are periodic functions of period 2π, $f_m(r)$ and $F_m(\rho)$ are, in fact, the coefficients in the Fourier series expansions as

$$f(r, \theta) = \sum_{-\infty}^{\infty} f_m(r)\exp(im\,\theta)$$

and

$$F(\rho, \phi) = \sum_{-\infty}^{\infty} F_m(\rho)\exp(im\phi).$$

Hence, we can find the relationship between the circular harmonic functions of the input image and that of its Fourier transform, which is obtained by computing the angular Fourier transforms with respect to ϕ in both sides of Eq. (5.33) and replacing the input image $f(r, \theta)$ with its circular harmonic expansion; that results in

$$F_m(\rho) = \frac{1}{(2\pi)^2} \int_0^{2\pi} \exp(-im\phi)\, d\phi \int_0^{2\pi} \int_0^{\infty} \sum_{n=-\infty}^{\infty} f_n(r) \exp(in\theta)$$

$$\times \exp(i2\pi r\rho \cos(\theta - \phi)) r\, dr\, d\theta \tag{5.34}$$

$$= \int_0^{\infty} f_m(r) J_m(2\pi r\rho) r\, dr,$$

where the Bessel function of the first kind and of order m is represented by

$$J_m(x) = \frac{i}{2\pi} \int_0^{2\pi} \exp(im\alpha) \exp(ix \cos(\alpha))\, d\alpha$$

with $\alpha = \theta - \phi$.

The second equality in Eq. (5.34) is the definition of the nth order Hankel transform. Hence, a 2D input function and its Fourier transform are related in such a way that the Hankel transform of the circular harmonic function of the input function is the circular harmonic function of its Fourier transform.

5.8.2. HANKEL TRANSFORM

Mathematically, the Hankel transform of a function $f(r)$ is expressed as

$$H_m(\rho) = \int_0^{\infty} f(r) J_m(2\pi r\rho) r\, dr. \tag{5.35}$$

If $m > -1/2$, then it can be proved that the input function can be recovered by the inverse Hankel transform as

$$f(r) = \int_0^{\infty} H_m(\rho) J_m(2\pi r\rho) \rho\, d\rho. \tag{5.36}$$

All the optical systems that perform the optical Fourier transform, as described in Sec. 5.3, perform the Hankel transform as described in Eq. (5.35), where $f(r)$ is in fact the mth order circular harmonic function of the input image, and $H_m(\rho)$ and is the mth order circular harmonic function of the Fourier transform of the input image.

5.9. RADON TRANSFORM

The Radon transform describes projection operation in computed tomography (CT), radio astronomy, and nuclear medicine [22]. In tomography the object of interest is 3D, which is sliced by a set of planes of projection.

5.9.1. DEFINITION

Let us consider one particular projection plane, where the object slice is described as $f(r, \theta)$ in a polar coordinate system. Radiation passes through the object under an angle of incidence, and is then recorded by a 1D photodetector array. The detected signal is a collection of the line integrals taken along the optical paths, which is denoted by $L(R, \phi)$, where R is the distance from the origin of the coordinate system to the path and ϕ is the angle of the normal path relative to the $\theta = 0$ axis, as shown in Fig. 5.3. The radiation is rotated for a large number of incident angles ϕ. The phtodetector array follows the rotation of the radiation and collects a set of projections, which may be regarded as a 2D function, and is called the shadow, represented by the Radon transform as

$$g(R, \phi) = \int_0^{2\pi} \int_0^{\infty} f(r, \theta)\delta[r \cos(\phi - \theta) - R] r\, dr\, d\theta, \tag{5.37}$$

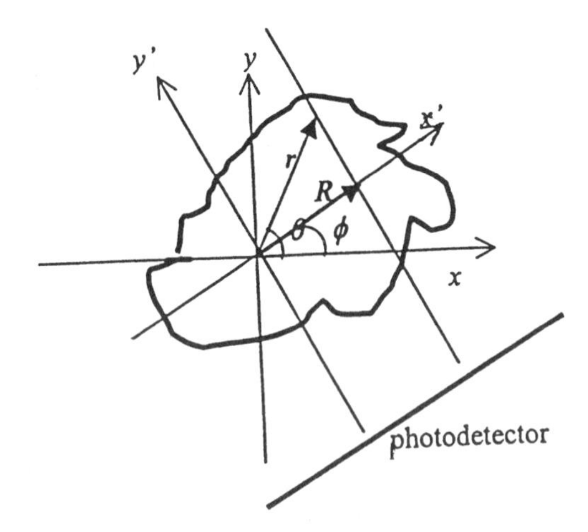

Fig. 5.3. Optical Radon transform.

where δ is the Dirac delta function. This delta function incorporates the parallel projection constraint such that the position $\boldsymbol{r}$ would kept on the path $\boldsymbol{r}\cdot\boldsymbol{R}/R = R$, where the direction of $\boldsymbol{R}$ is normal to the path, as shown in Fig. 5.3.

One property of the Radon transform is symmetry as,

$$g(-R, \phi) = g(R, \phi + \pi).$$

5.9.2. IMAGE RECONSTRUCTION

It is important now to examine the reconstruction of an object from its Radon transform. This is the inverse Radon transform. One method of reconstruction is based on the central slice theorem. The 1D Fourier transform of the projection $g(R, \phi)$ with respect to R produces the Fourier transform of the object $f(r, \theta)$, because using Eq. (5.37) this 1D Fourier transform becomes

$$\begin{aligned} \int g(R, \phi) \exp(-i2\pi\rho R)\, dR &= \int_0^{2\pi}\int_0^{\infty} f(r, \theta) \exp(-i2\pi u\rho r \cos(\phi - \theta)) r\, dr\, d\theta \\ &= \iint f(x, y) \exp(-i2\pi(ux + vy))\, dx\, dy = F(u, v), \end{aligned} \tag{5.38}$$

where $x = r\cos\theta$, $y = \sin\theta$ and $u = \rho\cos\phi$, $v = \rho\sin\phi$. The object can then be recovered from $F(x, y)$ by the inverse Fourier transform.

In the projection plane in image space, if for a given radiation projection direction we rotated the Cartesian coordinate system (x, y) by the angle ϕ to (x', y'), the projection $R = x'$, and integral (5.37) become 1D, as

$$g(R, 0) = \int f(x', y')\, dy'. \tag{5.39}$$

Equivalently, if we rotate the Fourier transform in the frequency space by an angle ϕ, the frequency components (u, v) become (u', v'), and the integral in Eq. (5.38) becomes

$$\int g(R, 0) \exp(-i2\pi\rho R)\, dR = \int f(x', y') \exp(-i\pi u'x')\, dx'\, dy' = F(u', 0). \tag{5.40}$$

This result shows that at each given projection angle, the 1D Fourier transform of the projection of a 2D object slice is directly one line through the 2D Fourier transform of the object slice function itself. Equation (5.40) is referred to as the central-slice theorem. When rotating the angle ϕ the 1D Fourier transform of the projection, $g(R, \phi)$ gives the entire 2D Fourier transform of the object slice. At each projection angle, the Fourier transform of the object slice is sampled by a line passing through the origin (hence, the term *central slice*).

As a result, the inverse Radon transform with Eq. (5.38) gives the entire 2D Fourier transform of the object, and the object itself. However, this reconstruction requires a coordinate transform and an interpolation of the Fourier spectrum from the polar coordinates to the Cartesian coordinates.

The above-mentioned direct Fourier transform method is the simplest in concept. Various methods exist in practice for reconstructing an image from the shadow or Radon transform, such as the filtered back-projection algorithm and the circular harmonic reconstruction algorithms using the Hankel transform [23, 24].

5.10. GEOMETRIC TRANSFORM

The geometric transformation, or coordinate transformation, is primarily used for simplifying forms of mathematical operations, such as equations and integrals. For example, it is easier to handle the volume integral of a sphere body in the spherical coordinate system than in the rectangular coordinate system. Geometric transformations can be implemented by an optical system using a phase hologram [25, 26, 27]. The optical geometrical transform is useful for redistributing light illumination [28], correcting aberrations and compensate for distortions in imaging systems, beam shaping [29], and for invariant pattern recognition [30]. Recently, the coordinate transformations also have been applied to analyze surface-relief and multiplayer gratings by reforming Maxwell electromagnetic equations [31].

5.10.1. BASIC GEOMETRIC TRANSFORMATIONS

The geometric transformation is a change of the coordinate system and is defined as

$$F(u) = \int f(x)\delta[x - \varphi^{-1}(u)]\, dx, \tag{5.41}$$

where $f(x)$ is the input function and $\varphi^{-1}(u)$ is defined by a desired mapping from a coordinate system x to another coordinate system u with

$$u = \varphi(x).$$

Hence, the inverse coordinate transform is

$$x = \varphi^{-1}(u).$$

The original input can be recovered from its geometrical transform as

$$F(u) = f[\varphi^{-1}(u)] = f(x). \tag{5.42}$$

(a) *Shift transformation.* This transformation is the simplest one of the geometric transformations. The mapping is $u = \varphi(x) = x + a$ and $x = \varphi^{-1}(u) = u - a$. Hence, the geometrical transformed function becomes $F(u) = f(u - a)$. The transformation shifts the input function to a distance, a, along the positive x direction.

(b) *Scaling transformation.* When the mapping is in the form $u = \varphi(x) = ax$, the geometric transform becomes the scaling transform. The input function is magnified by a factor a, and the scaling transformed function becomes $F(u) = f(u/a)$.

(c) *Logarithmic transformation.* This transformation is used in scale-invariant pattern recognition (see Exercise 5.12). The coordinate transform is $u = \varphi(x) = \ln x$ and the inverse transform is $x = \varphi^{-1}(u) = e^u$. Hence, the geometric transformed function becomes $F(u) = f(e^u)$.

(d) *Rotation transformation.* This process is better described in 2D formula, where we define two mappings as

$$u = \varphi_1(x, y) = x \cos\theta + y \sin\theta$$

$$v = \varphi_2(x, y) = -x \sin\theta + y \cos\theta.$$

The geometrical transform of the input pattern $f(x, y)$ is

$$F(u, v) = f[(u \cos\theta - v \sin\theta), (u \sin\theta + v \cos\theta)],$$

which is found to be the original pattern rotated by an angle $-\theta$.

(e) *Polar coordinate transformation.* This transformation is defined by the mappings from the Cartesian coordinate system to the polar coordinate system as

$$r = \varphi_1(x, y) = \sqrt{x^2 + y^2}$$

$$\theta = \varphi_2(x, y) = \tan^{-1}\frac{y}{x}.$$

After the polar coordinate transform, a pattern $f(x, y)$ turns out to be

$$F(r, \theta) = f(r \cos \theta, r \sin \theta).$$

This transformation has been used for rotational-invariant pattern recognition.

Example 5.1

Find the transformation between two functions:

$$f(x, y) = x^2 + 4y^2 \quad \text{and} \quad g(x, y) = 5x^2 + 5y^2 - 6xy.$$

Solution

From the contours determined by $f(x, y) = 4$ and $g(x, y) = 4$ in Fig. 5.4, we can see that the transformation involves rotation and scaling. Assuming $f(x, y)$ is the rotation and scaling transformation of $g(x, y)$, we have

$$\begin{aligned} f(u, v) &= g\left[\left(\frac{u \cos \theta - v \sin \theta}{a}\right), \left(\frac{u \sin \theta + v \cos \theta}{a}\right)\right] \\ &= 5\left(\frac{u \cos \theta - v \sin \theta}{a}\right)^2 + 5\left(\frac{u \sin \theta + v \cos \theta}{a}\right)^2 \\ &\quad -6\left(\frac{u \cos \theta - v \sin \theta}{a}\right)\left(\frac{u \sin \theta + v \cos \theta}{a}\right) \\ &= u^2 + 4v^2. \end{aligned}$$

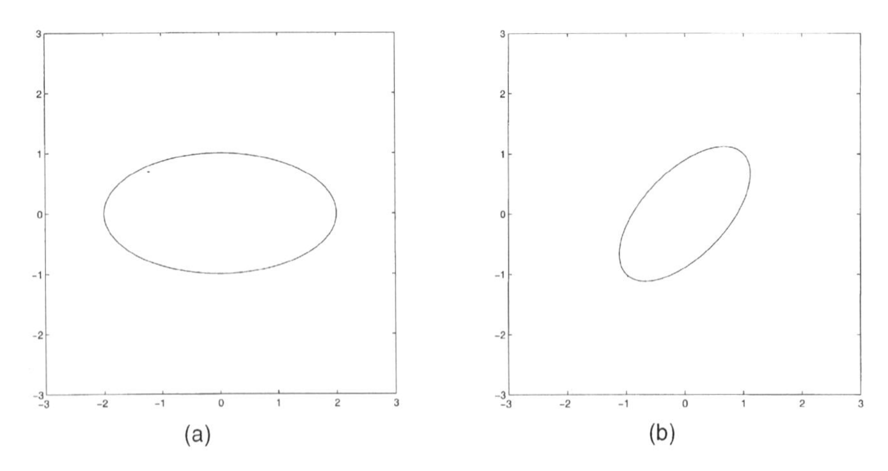

Fig. 5.4. (a) The contour of the functions $f(x, y) = x^2 + 4y^2 = 4$, (b) the contour of the functions $g(x, y) = 5x^2 + 5y^2 - 6xy = 4$.

The solution for this equation is $\theta = \pi/4$ and $a = \sqrt{2}/2$. Therefore, the mappings in the geometric transformation are

$$u = \varphi_1(x, y) = \frac{\sqrt{2}}{2}(x \cos\theta + y \sin\theta)$$

$$v = \varphi_2(x, y) = \frac{\sqrt{2}}{2}(-x \sin\theta + y \cos\theta).$$

Example 5.2

The log–polar transformation is a combination of the logarithmic transformation and the polar coordinate transformation with the mapping as

$$u = \varphi_1(x, y) = \ln\sqrt{x^2 + y^2} = \ln r$$

$$v = \varphi_2(x, y) = \tan^{-1}\frac{y}{x} = \theta.$$

It is a transformation from the Cartesian coordinate system to the polar coordinate system; then the logarithmic transformation is applied to the radial coordinate in the polar coordinate.

Derive the log–polar transformation of the two functions in Example 5.1.

Solution

We first process the polar transform by using inverse mapping, $x = r\cos\theta$ and $y = r\sin\theta$; that leads to

$$F_p(r, \theta) = f(r\cos\theta, r\sin\theta) = (r\cos\theta)^2 + 4(r\sin\theta)^2 = r^2 + 3r^2\sin^2\theta$$

$$G_p(r, \theta) = g(r\cos\theta, r\sin\theta) = 5(r\cos\theta)^2 + 5(r\sin\theta)^2 - 6r^2\cos\theta\sin\theta$$

$$= 2r^2 + 6r^2\sin^2\left(\theta - \frac{\pi}{4}\right).$$

We then execute the transformation by using $r = \exp u$ and $v = \theta$

$$F_{lp}(u, v) = F_p(\exp u, v) = (\exp u)^2 + 3(\exp u)^2\sin^2 v = \exp(2u) + 3\exp(2u)\sin^2 v$$

$$G_{lp}(u, v) = G_p(\exp u, v) = 2(\exp u)^2 + 6(\exp u)^2\sin^2\left(v - \frac{\pi}{4}\right)$$

$$= \exp[2(u - \ln\sqrt{2})] + 3\exp[2(u - \ln\sqrt{2})]\sin^2\left(v - \frac{\pi}{4}\right).$$

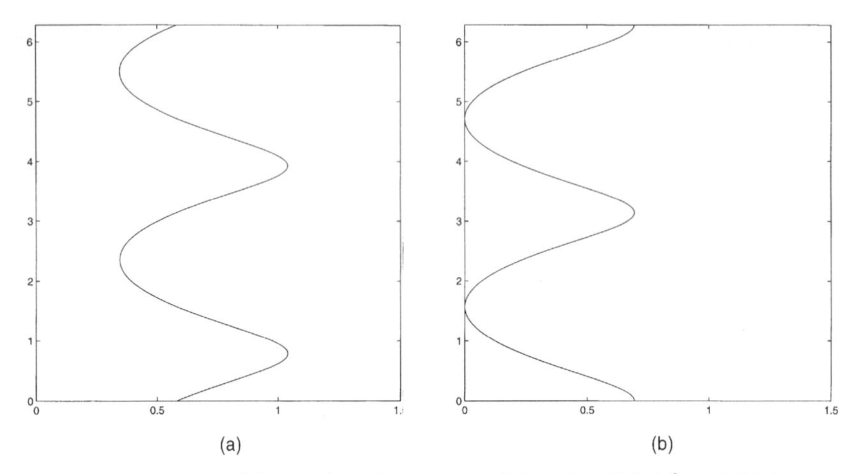

Fig. 5.5. (a) The contour of the functions $F_{lp}(u, v) = \exp(2u) + 3\exp(2u)\sin^2 v = 4$, (b) the contour of the functions $G_{lp} = \exp[2(u - \ln\sqrt{2})] + 3\exp[2(u - \ln\sqrt{2})]\sin^2(v - \pi/4) = 4$.

The two functions are shifted with respect to each other in the log–polar coordinate system. The contours of $F_{pl}(u, v) = 4$ and $G_{pl}(u, v) = 4$ are shown as Fig. 5.5. Casasent *et al.* use the log–polar transform for rotation and scale-invariant optical pattern recognition [30].

5.10.2. GENERALIZED GEOMETRIC TRANSFORMATION

Some geometric transformations are not simply reversible, because the mapping or the inverse mapping is not unique. For example, in the coordinate transformation $u = \varphi(x) = x^2$, the mapping is not invertible, because the inverse mapping may be not unique as $x = \varphi^{-1}(u) = \sqrt{u}$ or $x = \varphi^{-1}(u) = -\sqrt{u}$, and is a one-to-two mapping. The geometric transformation must be a two-to-one mapping. Two points in the x-plane could be mapped into a single point in the u-plane.

In order to handle this two-to-one mapping, we extend the geometric transformation definition to

$$F(u) = f(\sqrt{u}) + f(-\sqrt{u}).$$

In general, when the mapping is many points to one point, such as

$$\varphi(x) = u, \qquad \text{as} \quad x \in A, \tag{5.43}$$

generalized geometric transformations should be defined as

$$F(u) = \int_A f(x)\,dx. \tag{5.44}$$

Example 5.3

The ring-to-point geometric transformation is useful to transform concentrically distributed signals into a linear detector array, by mapping rings in the x–y plane to the points along the v axis in the u–v plane. We execute the polar transformations first. For an input $f(u, v)$ we have

$$F_p(r, \theta) = f(r\cos\theta, r\sin\theta).$$

According to the definition of the generalized geometric transformation, the ring-to-point transform can be written as

$$F_{rp}(r) = \oint F_p(r, \theta) r\,d\theta,$$

where $r = \sqrt{x^2 + y^2}$ and the integral is computed along a circle of radius r.

5.10.3. OPTICAL IMPLEMENTATION

Many techniques exist to implement geometric transformations with optics. A typical optic system for geometric transformation is shown in Fig. 5.6. The input image is represented by a transparency and is illuminated by a collimated coherent beam. A phase mask is placed behind the input plane, which implements the geometric transformation.

The complex amplitude transmittance of the phase mask

$$t = \exp[j\phi(x, y)]$$

is computed in such a way that the required coordinate system mapping is proportional to its derivatives as:

$$\begin{aligned} u = \varphi_1(x, y) &= \frac{\lambda f}{2\pi}\frac{\partial\phi(x, y)}{\partial x} \\ v = \varphi_2(x, y) &= \frac{\lambda f}{2\pi}\frac{\partial\phi(x, y)}{\partial y}. \end{aligned} \tag{5.45}$$

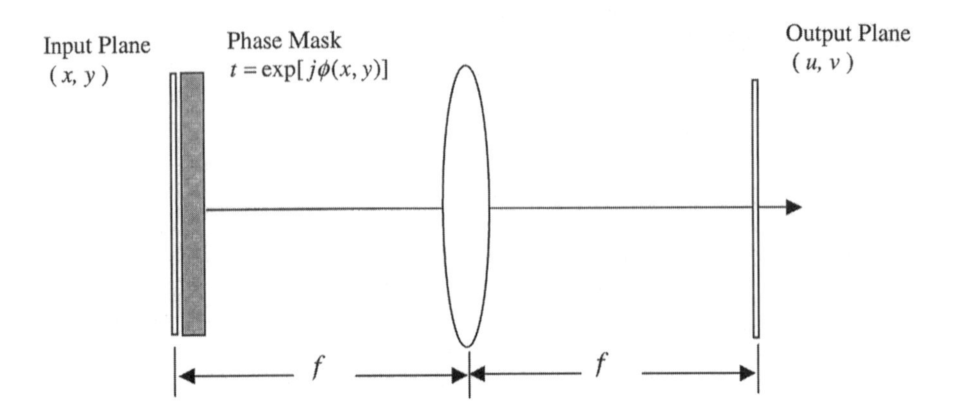

Fig. 5.6. The optical system for implementing geometric transformation.

Bryngdahl was the first to analyze this system [25]. In the optical system shown in Fig. 5.6, the Fourier transform of the product of the input and the phase mask is exactly produced in the output plane as

$$G(u, v) = \iint f(x, y) \exp(j\phi(x, y)) \exp\left[-j2\pi\left(\frac{u}{\lambda f}x + \frac{v}{\lambda f}y\right)\right] dx\,dy. \quad (5.46)$$

The concept of the stationary phase [32] is useful when computing this Fourier transform. We rewrite Eq. (5.46) as

$$G(u, v) = \iint f(x, y) \exp\left[j\frac{2\pi}{\lambda} h(x, y, u, v)\right] dx\,dy,$$

with

$$h(x, y\, u, v) = \frac{\lambda\phi(x, y)}{2\pi} - \frac{u}{f}x - \frac{v}{f}y. \quad (5.47)$$

In this type of the integral, as the wavelength λ is several orders of magnitude smaller than the coordinates u, v and x, y, the phase factor h can vary very rapidly over x and y; as a result, the integral can vanish. The integral can be fairly well approximated by contributions from some subareas around the saddle points (x_0, y_0) where the derivatives of h are equal to zero and the phase factor is stationary:

$$\left.\frac{\partial h}{\partial x}\right|_{(x_0, y_0)} = \left.\frac{\partial h}{\partial y}\right|_{(x_0, y_0)} = 0 \quad (5.48)$$

because only in the regions of vicinity of the saddle points can significant contributions to the value of the integral be established. From Eqs. (5.41) and (5.48) we see that the required mapping from Eq. (5.45) is satisfied at the saddle points.

In general, there is more than one saddle point. Therefore, the x–y plane is subdivided into subareas, each containing only one saddle point. The output intensity distribution only is taken into consideration; the geometric transformed image is

$$|G(u, v)| \propto |f[\varphi_1^{-1}(u), \varphi_2^{-1}(v)]|.$$

This technique can be also understood as the phase mask,

$$\exp(j\pi h(x, y; u, v)/\lambda),$$

acts as a set of local gratings or local prisms which diffract $f(x, y)$ to $G(u, v)$ at a set of subareas in the x–y plane where the phase is stationary.

Example 5.4

Find the phase mask amplitude transmittance for logarithmic transformation.

Solution

The mappings of logarithmic transformation are

$$u = \varphi_1(x) = \ln x$$

$$v = \varphi_2(y) = \ln y.$$

As mentioned in Eq. (5.45), the phase of the mask $\phi(x, y)$ bears the relation

$$\frac{\partial \phi(x, y)}{\partial x} = \frac{2\pi}{\lambda f} \ln x$$

$$\frac{\partial \phi(x, y)}{\partial y} = \frac{2\pi}{\lambda f} \ln y.$$

Comparing these two integrals of the above equations, we get

$$\phi(x, y) = \frac{2\pi}{\lambda f}(x \ln x - x + y \ln y - y).$$

The amplitude transmittance for logarithmic transformation is

$$t(x, y) = \exp\left[j\frac{2\pi}{\lambda f}(x \ln x - x + y \ln y - y) \right].$$

5.11. HOUGH TRANSFORM

The Hough transform is a specific geometrical transform (see Sec. 5.10) and is a special case of the Radon transform (see Sec. 5.9). Although the definition of the Hough transform can be derived from the Radon transform, the basic distinction between them is that the Radon transform deals with the projection of an image into various multidimensional subspaces, while the Hough transform deals with the coordinate transform of an image. The Hough transform is useful for detecting curves in an image and tracking moving targets, and for many image-processing and pattern-recognition applications [33].

5.11.1. DEFINITION

A straight line in the Cartesian coordinate system can be Hough-transformed as given by [34]

$$\begin{aligned} H(\theta, r) &= \iint f(x, y)\delta(r - x\cos\theta - y\sin\theta)\, dx\, dy \\ &= \begin{cases} 1, & \text{for } (\theta_1, r_1) \\ 0, & \text{otherwise,} \end{cases} \end{aligned} \tag{5.49}$$

where

$$f(x, y) = \begin{cases} 1, & (x, y) \in r_1 = x\cos\theta_1 + y\sin\theta \\ 0, & \text{otherwise} \end{cases}$$

is a straight line, as shown in Fig. 5.7. Thus, we see that a straight line is transformed into a point (θ_1, r_1) in the parameter space (θ, r).

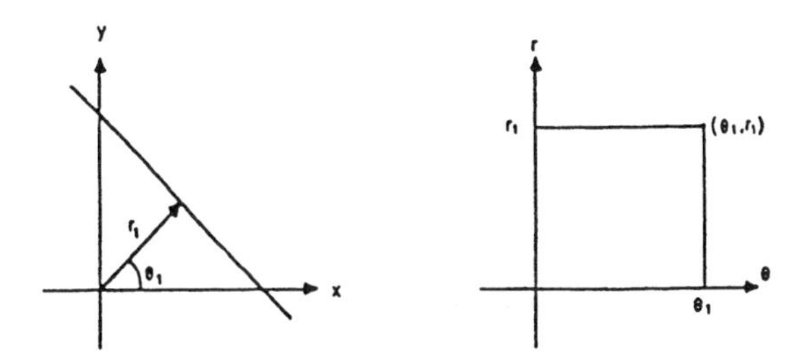

Fig. 5.7. A straight line in (x, y) plane is Hough-transformed into a point in the parameter space (θ, r),

On the other hand, an arbitrary point $\delta(x - x_0, y - y_0)$ in the (x, y) coordinate space can be Hough-transformed as given by

$$\begin{aligned} H(\theta, r) &= \iint \delta(x - x_0, y - y_0)\delta(r - x\cos\theta - y\sin\theta)\,dx\,dy \\ &= \delta(r - x_0\cos\theta - y_0\sin\theta), \end{aligned}$$

which describes a pseudo-sinusoidal function in the (θ, r) plane, as shown in Fig. 5.8. In fact, a point in the (x, y) plane can be represented as an intersection of many straight lines. Each line is mapped into a point in the (θ, r) plane according to its orientation and its distance to the origin of the (x, y) coordinate system. All those points constitute a pseudo-sinusoidal curve.

5.11.2. OPTICAL HOUGH TRANSFORM

The optical system performing the Hough transform as suggested by Gindi and Gmitro is shown in Fig. 5.9. The optical system images an input object

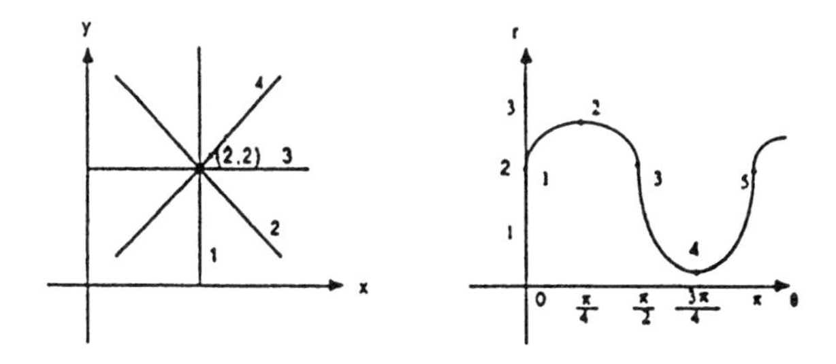

Fig. 5.8. A point in the (x, y) plane is Hough transformed into a pseudo-sinusoidal curve in the (θ, r) parameter space, where the points numbered by 1, 2, 3, 4, 5 correspond to the straight lines in the (x, y) plane.

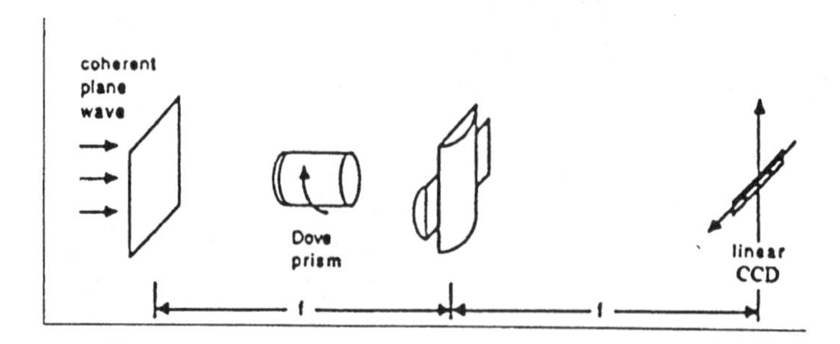

Fig. 5.9. A coherent optical system to perform the Hough transform.

and performs a one-dimensional Fourier transform in the vertical and the horizontal directions by means of two cylindrical lenses, respectively. Thus, by picking up the zero-order Fourier spectrum with a CCD camera, a line of Hough transform can be obtained for a given value of θ. By rotating the Dove prism, θ can be scanned angularly, so that the Hough transformation can be picked up by the linear CCD detector array at the scanning rate [35].

Although optical implementation for straight-line mappings with Hough transforms works well to some extent, the space variance of the Hough transform may create serious problems for generalized transformation. Moreover, the usefulness of the Hough transform depends on the accuracy of mapping, and the performance of the optical system depends on the input signal-to-noise ratio and clutter; there is also the multiobject problem. Most Hough transforms have been performed in digital image processing.

REFERENCES

5.1 P. M. Duffieux, *The Fourier Transform and Its Applications to Optics*, John Wiley & Sons, New York, second edition, 1983.

5.2 Ernst Abbe, 1873, "Archiv Mikroskopische," *Anat.*, vol. 9, 413.

5.3 Load Rayleigh, 1896, "On the Theory of Optical image, with Special References to the Microscope," *Phil. Mag.* 5, 43, 167.

5.4 André Maréchal, and Maurice Françon, *Diffraction Structure des Images*, Masson & Cie Éeiteurs, Paris, 1970.

5.5 E. L. O'Neill, 1956, "Spatial Filtering in Optics," *IRE Trans. Inform. Theory*, IT-2, 56–65.

5.6 Emmett N. Leith, James A. Roth, 1977, "White light optical processing and holography," *Applied Optics*, Vol. 16, no. 9, 2565.

5.7 A. B. VanderLugt, *Optical Signal Processing*, John Wiley & Sons, New York, 1992.

5.8 B. E. A. Saleh and M. O. Freeman, "Optical Transformations," in *Optical Signal Processing*, J. Horner, ed., Academic, New York, 1987.

5.9 F. T. S. Yu, *Optical Information Processing*, Wiley-Interscience, New York, 1983.

5.10 M. Born and E. Wolf, *Principle of Optics*, 6th edition, Pergamon, Oxford, 1980.

5.11 J. W. Goodman, *Introduction to Fourier Optics*, McGraw-Hill, New York, 1968.
5.12 I. Daubechie, 1988, "Orthogonal Bases of Compactly Supported Wavelets," *Comm. Pure Appl. Math.*, vol. 41, 909–996.
5.13 C. K. Chui, *An Introduction to Wavelets*, Academic, San Diego, 1992.
5.14 Y. Sheng, "Wavelet Transform," in *The Transforms and Applications Handbook*, A. D. Poulariskas, ed. 2nd edition, CRC and IEEE Press, Boca Raton, 2000.
5.15 Y. Li, H. Szu, Y. Sheng, and H. J. Caulfield, 1996, "Wavelet Processing and Optics," *Proceedings of the IEEE*, vol. 84 no. 5, 720–732.
5.16 G. Kaiser, *A Friendly Guide to Wavelet*, Birkhauser, Boston, 1994.
5.17 Y. Sheng, S. Deschens, and J. Caulfield, 1998, "Monochromatic Electromagnetic Wavelet and Huygens Principle," *Applied Optics*, Vol. 27, no. 5, 828–833.
5.18 J. D. Jackson, *Classical Electrodynamics*, 3rd edn., John Wiley & Sons, New York, 1999.
5.19 J. Y. Lu and F. Greenleaf, 1992, "Nondiffraction X-Wave-Exact Solution to Free-Space Scalar Wave Equation and Their Finite Aperture Realization," *IEEE Trans. Ultra. Ferro. Freq. Control*, vol. 39, no. 1, 19–31.
5.20 M. Abramowitz and I. A. Stegun, *Handbook of Mathematical Functions*," Dover, New York, 1972.
5.21 H. M. Ozaktas, B. Barshan, D. Mendlovic, and L. Onural, 1994, "Convolution, Filtering, and Multiplexing in Fractional Fourier Domains and Their Relation to Chirp and Wavelet Transforms," *J. Opt. Soc. Am. A*, vol.11, no. 2, 547–559.
5.22 H. H. Barrett, "The Radon Transform and Its Applications," in *Progress in Optics XXI*, E. Wolf, ed., Elsevier, New York, 1984.
5.23 H. H. Barrett and W. Swindell, *Radiology Imaging*, Academic, New York, 1981.
5.24 E. W. Hasen, "Theory of Circular Harmonic Image Reconstruction," *J. Opt. Soc. Am.* vol. 71, no. 4, 304–308.
5.25 O. Bryngdahl, 1974, "Geometrical Transformations in Optics," *J. Opt. Soc. Am.* 64, 1092.
5.26 J. N. Cederquist and A. M. Tai, 1984, "Computer-Generated Holograms for Geometric Transformations," *Appl. Opt.* 23, 3099.
5.27 D. Wang, A. Peer, A. A. Friesem, and A. W. Lohmann, 2000, "General Linear Optical Coordinate Transformations," *J. Opt. Soc. Am. A*, 17, 1864.
5.28 J. R. Leger and W. C. Goltsos, 1992, "Geometric Transformation of Linear Diode-Laser Arrays for Longitudinal Pumping of Solid-State Lasers," *IEEE J. Quantum Electron.* 28 1088.
5.29 Y. Arieli, N. Eisenberg, A. Lewis, and I. Glaser, 1997, "Geometrical-Transformation Approach to Optical Two-Dimensional Beam Shaping," *Applied Optics*, 36, 9129.
5.30 D. Casasent and D. Psaltis, 1976, "Position, rotation, and scale invariant optical correlation," *Appl. Opt.* Vol. 15, no. 7, 1795.
5.31 J. P. Plumey and G. Granet, 1999, "Generalization of the Coordinate Transformation Method with Application to Surface-Relief Gratings," *J. Opt. Soc. Am. A*, 16, 508.
5.32 M. Born and E. Wolf, *Principles of Optics*, Pergamon, New York, 1965.
5.33 T. Szoplik, "Line Detection and Directional Analysis of Images," in *Optical Processing and Computing*, H. H. Arsenault, T. Szoplic, and B. Macukow, eds., Academic, New York, 1989.
5.34 P. V. C. Hough, "Methods and Means for Recognizing Complex Patterns," U.S. Patent 3-069-654, 1962.
5.35 F.T.S. Yu and S. Jutamulia, *Optical Signal Processing, Computing, and Neural Networks*, John Wiley, New York, 1992, pp. 102–106.

EXERCISES

5.1 Demonstrate the inverse Fresnel transform in Eq. (5.3) using the definition of the Fresnel transform in Eq. (5.2).

5.2 Prove a spherical wavefront of a radius R, described as

$$\exp(i\pi(x^2 + y^2)/\lambda R)$$

in the x–y plane in the paraxial condition.

5.3 When using the local frequency concept to analyze the nonstationary signal, the windowed Fourier transform, or Gabor transform, is widely used.
  (a) Find the mathematical definition of the Gabor transform.
  (b) Prove the inverse Gabor transform for recovering the original signal.
  (c) Compare the Gaussian window used in the Gabor transform with the wavelet window.

5.4 The wavelet transform is mainly used for detecting irregularities in the signal. The wavelet transform of a regular signal can be all zero. When a signal is regular, its $(n + 1)$ order derivative and the derivatives of orders higher than $(n + 1)$ are equal to zero. What property should the wavelet have in order to ensure the wavelet transform coefficients of this signal are all equal to zero ?

5.5 Prove that any electromagnetic waves, which are solutions of the Maxwell equations, should be limited in the light-cone

$$p^2 = p_0^2 - c^2|\boldsymbol{p}|^2 = 0.$$

5.6 According to the expression in Eq. (5.15) of the electromagnetic wavelets in the 4D space–time, write down expressions for:
  (a) The reference electromagnetic wavelet (mother wavelet).
  (b) The family of wavelets with dilations and scaling in 4D space–time.

5.7 Show that the electromagnetic wavelets given by Eq. (5.15) can be decomposed into

$$h(x) = h^-(x) + h^+(x),$$

where $x = x(\mathbf{r}, t)$ and

$$h^-(x) = \frac{1}{2\pi^2 c^4 |\mathbf{r}/c|} \frac{1}{[|\mathbf{r}/c| + t + i]^3}$$

$$h^+(x) = \frac{1}{2\pi^2 c^4 |\mathbf{r}/c|} \frac{1}{[|\mathbf{r}/c| - t - i]^3}.$$

$h^-(x)$ converges to the origin and $h^+(x)$ diverges from the origin.

5.8 Show the two wavelet components $h^-(x)$ and $h^+(x)$ are solutions of the two equations

$$-\partial_t^2 h^-(x) + \nabla^2 h^-(x) = -\frac{2\delta(\mathbf{r})}{i\pi(1-it)^2}$$

$$-\partial_t^2 h^+(x) + \nabla^2 h^+(x) = \frac{2\delta(\mathbf{r})}{i\pi(1-it)^2}.$$

The right-hand side of the two equations represents the elementary courant at the origin, which is generated by the converging wavelet $h^-(x)$ and then generates the diverging wavelet $h^+(x)$, which is, in fact, the Huygens wavelet.

5.9 Write the Wigner distribution function of a summation of two signals, $f(x) + g(x)$, and show the cross-talk term, which indicates the bilinear property of the transform.

5.10 Show that the Fourier transform of a circular harmonic function of an image is equal to the circular harmonic function of the Fourier transform of the image.

5.11 Show that the fractional Fourier transform can be optically implemented by using a segment of certain length of the gradient index fiber lens.

5.12 Prove that Mellin transforms can be implemented by using Fourier transforms after some geometric transformation.

5.13 Show that the phase mask must satisfy Eq. (5.47) with the saddle-point method in the implementation of the mapping of the coordinate system in Eq. (5.45).

5.14 Find the phase-mask amplitude transmittance for logarithmic polar transformation.

Chapter 6 Interconnection with Optics

Ray T. Chen and Bipin Bihari
MICROELECTRONICS RESEARCH CENTER,
THE UNIVERSITY OF TEXAS, AUSTIN

6.1. INTRODUCTION

Extraordinary developments during the last decade in microelectronics fabrication technologies have provided the means to achieve submicron feature size and accommodate millions of logic gates in a small volume. This shrinking feature size in Si VLSI circuits manifests into short transit time and switching speeds within the processor chip, allowing extremely fast intrinsic signal speeds. However, preserving the integrity of the signals generated by these novel devices while communicating among other chips on a system or PC board level is becoming extremely difficult by electrical means (i.e., using conventional metallic transmission lines and interconnects), and requires employment of bulky, expensive terminated coaxial interconnections. High packaging density and long on-board communication distances (>10 cm) make even state-of-the-art electrical interconnects unrealistic in some cases [1]. Several critical problems exist for electrical interconnects in high-speed, large-area, massive signal transmission. These include:

1. Impedance mismatch caused by multistage electrical fanouts.
2. Transmission reflection (noise) at each fanout junction as well as at the end of each transmission line.
3. Electromagnetic interference with other interconnection lines and other interconnection layers.
4. High transmission loss resulting in a large driving power [2, 3].

These fundamental problems translate into wide interconnection time bandwidths, large clock and signal skews, and large RC time constants. Even the distributed line RLC time constant is often too large for chip-to-chip interconnects and higher-level hierarchies. These factors are the cause of serious bottlenecks in the most advanced electronic backplane interconnect prototypes, such as IBM's backplane, in which the bottleneck occurs at 150 Mbit/sec. For existing high-speed buses, electronic limitations are even more pronounced. For example, the VMEbus serial bus (VSB) is designed to transfer data at 3.2 Mbit/sec. However, its speed degrades to 363 Kbits/sec when the two communication points are separated by 25-m [4].

Sematech and the SIA (Semiconductor Industry Association) have published road maps for the growth of electronics industries in the future. Based on these road maps, even with the advancement of copper wiring and low-K (dielectric constant) materials, electrical interconnection still is likely to experience a major bottleneck in the very near future [5]. Finding an optical means to overcome this problem is a necessity. Implementation of optical components to provide high-speed, long-distance (>10 cm) interconnects has already been a major thrust for many high-performance systems where electrical interconnects failed to provide the bandwidth requirement. GHz personal computers have already hit consumer markets. However, the slowness of transmitting signals off the processor chip; for example, processor to memory, makes the system bus speed (~ 133 MHz) significantly slower than the clock speed. As a result, the bottleneck is the off-processor interconnection speed rather than the on-chip clock rate that provides the references for arbitrating the data and signal processing. Further upgrading the bus speed by electrical means is difficult.

Continuous increase of bus speed is a challenging task for the microelectronics industry due to the required distance and packing density. The speed limit becomes more stringent as the interconnection distance increases. For example, the dispersion-limited 1 GHz speed limit for an electrical interconnect is confined to lengths not longer than a few millimeters, and the 100 MHz speed limit holds for an interconnect length of only a few centimeters. Employing optical interconnects for upgrading system bus speed has been widely discussed in the computer industry. However, the major concern regarding incorporating optical bus into high-performance microelectronic devices and systems such as the board-level system bus is packaging incompatibility. Transmission of optical signals can provide tens of Gbit/sec with an interconnection distance well above 10 cm, which is at least an order of magnitude higher than electrical interconnects. However, electrical-to-optical and optical-to-electrical signal conversions impose serious problems in packaging and in decreasing the latency of data processing. For example, conventional microelectronic device interfaces may not be easily and inexpensively altered to incorporate optical interconnects. While the speed advantage promised by optical interconnects will certainly provide handsome payoffs to the computer industry, how to accommodate this in the existing systems architecture cost effectively remains a major issue.

Machine-to-machine interconnection has already been significantly replaced by optical means such as optical fibers, and interconnection distances from 1 m to 10^6 m have been realized. The present challenge is for interprocessor–memory interconnections, where the transmission bus line such as backplane–bus represents the most serious problem for upgrading system performance. Therefore, major research thrusts in optical interconnection are in the backplane and board level where the interconnection distance, the associated parasitic RLC effects, and the large fanout-induced impedance mismatch impose the most serious problems in fulfilling bandwidth requirements. Optical interconnection, in general, has been widely thought to be a better alternative for upgrading system performance. Optical interconnect approaches can be divided into two main categories, free-space and guided-wave optical interconnects. A number of free-space approaches have been proposed and demonstrated. However, reliability and packaging compatibility of these demonstrated optical interconnect systems [6, 7, 8] have impeded the integration of optical interconnects into a real system. For example, the board-level optical interconnections reported in [6] all use hybrid approach where both electronic and optoelectronic components are located at the surface of the board. Such an approach makes packaging difficult and costly. Furthermore, the employment of free space instead of guided-wave optical interconnections reported in [6, 7, 8] makes the system vulnerable in harsh environments. On the other hand, the guided-wave approach offers several inherent advantages over the free-space approach. For example, in guided-wave interconnects all optical components, including source and detector, can be fully imbedded in the multilayer board, making them less sensitive to environmental variables, and also leaving the top surface of the board available for other necessary electronic components. This chapter describes the research being carried out at Microelectronics Research Center, University of Texas at Austin, and its collaborators on similar guided-wave interconnects based on polymer waveguides functioning as optical transmission bus lines.

To make optical interconnects acceptable to computer companies, it is important to make the insertion of optics compatible with the standard PC board fabrication process so that technology improvement can be achieved in a cost-effective manner. The polymer-based optical bus can be employed to upgrade interconnection speed and distance. High parallelism can be realized by implementing a linear optical waveguide array. A board-level polymeric channel waveguide array–based optical bus incorporating both transmitting (electrical-to-optical) and receiving (optical-to-electrical) functions within the embedded interconnection layers of the three-dimensionally (3-D) integrated multilayer PC board involving both electrical and optical interconnections is

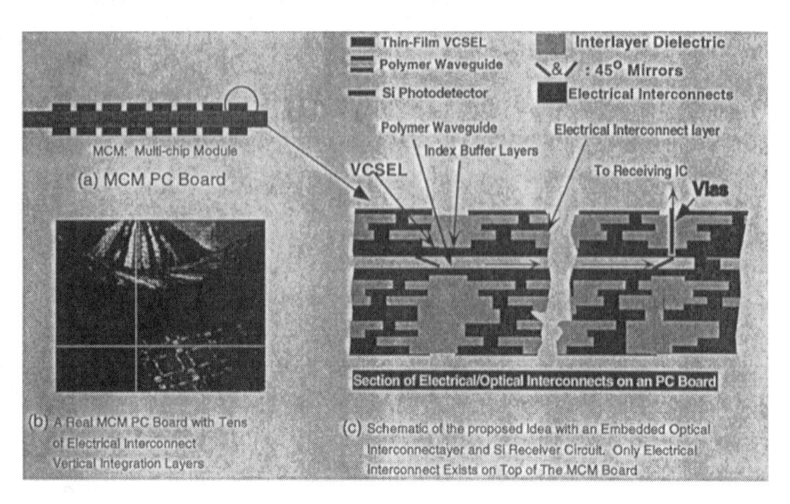

Fig. 6.1. Fully embedded optical interconnection using thin-film transmitter and receiver within a PC board.

illustrated in Fig. 6.1. All the elements involved in providing high-speed optical communications within one board are shown. These include a vertical cavity surface-emitting laser (VCSEL), surface-normal waveguide couplers, a polyimide-based channel waveguide functioning as the physical layer of optical bus, and a photoreceiver. The driving electrical signal to modulate the VCSEL and the demodulated signal received at the photoreceiver are all through electrical vias connecting to the surface of the PC board. In this approach, all the areas of the PC board surface are occupied by electronics and therefore one only observes performance enhancement due to the employment of optical interconnections but does not worry about the interface problem between electronic and optoelectronic components, unlike conventional approaches. The approach described here presents two major opportunities. The first is to accomplish the requirement of planarization necessary to provide three-dimensional (3D) on-board interconnect integration through *vias* to fulfill the required interconnection density. The board-level optical interconnection layers can be sandwiched between electrical interconnection layers. Assurance of the flatness and compatibility of the insertion of optical layers is crucial to ensure a technology transferable to the computer industry. The second opportunity is to provide compatible optical-to-electrical and electrical-to-optical conversions, which are the major concerns for system packaging. This approach will use fully embedded transmitters (electrical-to-optical) and receivers (optical-to-receiver) within the 3-D integrated board [9, 10, 11]. For a multilayer interconnection the optical components, including lasers, waveguides, couplers, and detectors, can be fully embedded. The input electrical signal driving the lasers and the output electrical signal from the photodetector can be realized using conventional *vias* common to all electrical interconnects at the board level. It is clear from the schematic shown in Fig. 6.1 that the achievement of planarization helps the stacking of other electrical interconnection layers on top of the optical layer while the full embedding of the transmitters (lasers) and receivers (detectors) makes the packaging 100% compatible with microelectronic packaging on the surface of the PC board. In this case, only electrical signals exist on the surface of the PC board and therefore significantly ease optoelectronic packaging problems. The fully embedded structure makes the insertion of optoelectronic components into microelectronic systems much more realistic, considering the fact that the major stumbling block for implementing optical interconnections onto high-performance microelectronics is packaging incompatibility.

The research findings of the necessary building blocks for the architecture shown in Fig. 6.1, polyimide-based waveguides, waveguide couplers, high-speed thin-film transmitters using VCSELs, and thin-film receivers operating at 850 nm are described sequentially in this chapter. A board-level 1-48 clock signal distribution for the Cray-T-90 supercomputer board is demonstrated with a speed of 6 Gbit/sec. Finally, a polymer-based optical bus structure is illustrated that has a full compatibility with existing board-level IEEE–standardized buses such as VME bus and FutureBus.

6.2. POLYMER WAVEGUIDES

6.2.1. POLYMER MATERIALS FOR WAVEGUIDE FABRICATION

To provide system integration using guided-wave optical interconnections, polymer-based material has many exclusive advantages. It can be spin-coated on a myriad of substrates with a relatively great interconnection distance. Many organic polymers are attractive microelectronic and optoelectronic materials with potential applications as interlayer dielectric, protective overcoat, α-ray shielding, optical interconnects, or even conductive electrical interconnects. In addition to their ease of processing, they possess favorable electrical and mechanical properties such as high resistivity, low dielectric constant, light weight, and flexibility. Table 6.1 compares the advantages of polymer-based devices with that of inorganic materials such as GaAs and $LiNbO_3$.

The required interconnection distance makes polymeric material, especially polyimides, the best candidate for guided-wave applications. Inorganic materials such as $LiNbO_3$, GaAs, and InP are good for small-size optical circuits ($\sim$cm). However, board-level optical interconnects require interconnection

Table 6.1

Advantages of Polymers as Compared to Common Inorganic Photonic Materials

Features	Polymer based	GaAs	$LiNbO_3$
Channel waveguide	Yes	Yes	Yes
Waveguide propagation loss	<0.1 dB/cm	0.2–0.5 dB/cm	<0.1 dB/cm
OEIC size	Unlimited	Limited	Limited
Channel waveguide packaging density (channels/cm)	Up to 1250	500	333
Implementation on other substrates	Easy	Difficult	Difficult
Fabrication cost	Low	High	High

distances of tens of centimeters. Transparency at the operating wavelength, thermal and mechanical stability, and compatibility with Si–CMOS processes are the main requirements for a given polymeric material to be successfully used in optoelectronic devices. Table 6.2 lists the optical properties of some of the materials investigated in our laboratories, including photolime gel, commercially available polyimides, and BCB (Cyclotenes). These polymers have great potential for optical waveguide applications in optoelectronic devices due to their favorable properties such as Si–CMOS process compatibility, planarization, and patternability. Although photolime gel has a relatively low glass transition temperature (~160°C), it is a very good test polymer due to its low cost and ease of handling. Moreover, it forms a graded index waveguide structure, and therefore waveguides can be formed on any suitable substrate.

Table 6.2

Optical Properties of Photolime Gel, Ultradel Polyimides (Amoco Chemicals), and Cyclotenes (DOW Chemicals)

	Refractive index (n) @			Optical loss dB/cm @		
Material	633 nm	850 nm	1300 nm	633 nm	850 nm	1300 nm
Photolime gel	1.54	—	—	0.3	0.3	—
Ultradel 9020	1.55	—	1.523	1.29	0.34	0.43
Ultradel 9021	—	—	1.536	1.04	0.13	0.34
Ultradel 7501	1.58	—	1.554	1.57	—	0.38
Ultradel 4212	1.61	—	—	21.1	3.24	0.38
Ultradel 1608	1.71	1.68	1.667	9.57	2.16	0.37
Cyclotene 5021	1.56	—	—	—	—	—
Cyclotene 3022	1.558	1.55	1.54	—	—	—

On the other hand, Ultradel polyimides and Cyclotenes (BCB) are well known for applications in electronic circuit boards as protective dielectric layers. Further, they form good-quality optically transparent thin films. Compatibility with the microelectronic fabrication process, however, is the thermal requirement for the polymer material to survive wire-bonding and metal deposition processes. Ultradel polyimides from Amoco Chemicals stand out in this aspect, as they exhibit thermal stability to above 400°C.

6.2.2. FABRICATION OF LOW-LOSS POLYMERIC WAVEGUIDES

In order to realize high-performance waveguide-based optical interconnects, the optical polymeric waveguide must be designed and fabricated for low scattering and bending losses. To fabricate such polymeric waveguide circuits, we investigated three waveguide fabrication technologies [12–15]:

- *Compression-molding technique,*
- *VLSI lithography technique, and*
- *Laser-writing technique.*

Figure 6.2 shows the schematic diagrams of these three waveguide fabrica-

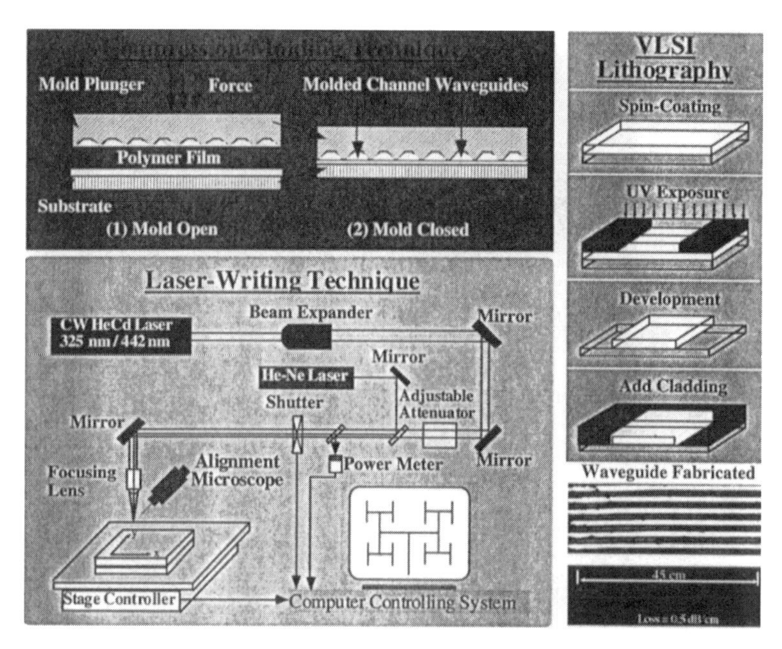

Fig. 6.2. Schematic diagrams of three polymeric waveguide fabrication technologies developed at Radiant Research, Inc.

tion technologies. The experimental results indicate that high-performance polymeric waveguide circuits with waveguide propagation loss less than 0.02 dB/cm can be produced by using laser-writing technique. Compression-molding technique has demonstrated its uniqueness in producing three-dimensional (3D) tapered waveguide circuits, crucial for obtaining efficient optical coupling between the input laser diode and the waveguide circuit. Mass-producible waveguides with excellent repetitiveness have been obtained by using VLSI lithography technique, originally developed for fabricating very large scale integrated circuits on silicon wafers.

6.2.2.1. *Compression-Molding Technique*

High-performance polymeric waveguides can be obtained at low cost with compression-molding technique. The process of compression molding is described by reference to Fig. 6.2. A two-piece mold provides a cavity having the shape of the target polymer-based channel waveguide array. The mold is heated to a desired temperature that is often above the glass transition temperature. An appropriate amount of molding material, polymer waveguide film in this case, is loaded into the substrate. The molding process is conducted by bringing two parts of the mold together under pressure. The polymer film, softened by heat, is thereby welded into the shape of the stamp. The molding process is performed during the phase-transition period within which the polymer film is deformable. The compression-molding technique has produced polymeric waveguides with shape and sizes typically unachievable by other methods.

A 45-cm–long polymer-based compression-molded channel waveguide on a glass substrate is shown in Fig. 6.3. The light propagation is shown by employing a microprism to couple a HeNe laser beam (0.6328 mm) into the waveguide. Waveguide propagation losses as low as 0.5 dB/cm at 632.8 nm have been obtained in these waveguides. The channel waveguide shown in Fig. 6.3 had a rib width (W) of 110 μm, a groove depth (T_2) of 8 μm, and a cladding

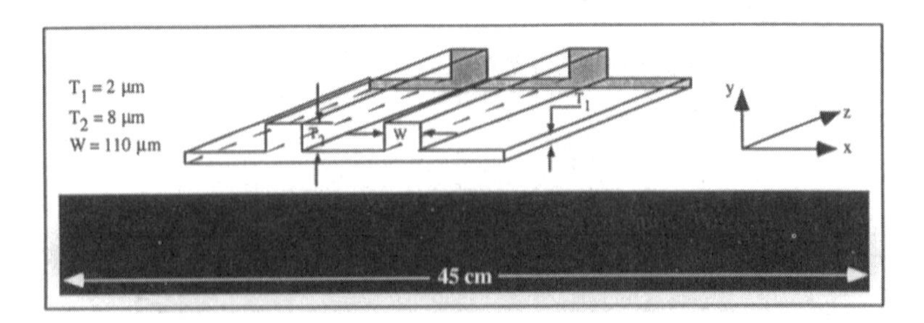

Fig. 6.3. A 45 cm long compression-molded polymeric waveguide working at 632.8 nm.

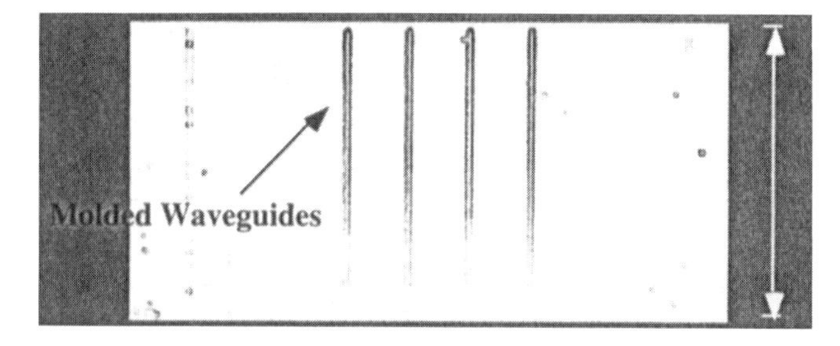

Fig. 6.4. Photograph of compression-molded 3D tapered polymeric waveguides.

layer thickness (T_1) of 2 μm. Another excellent example of waveguides fabricated by compression-molding technique is an array of 3-D tapered waveguides with large-end cross section of 100 μm $\times$ 100 μm, and small-end cross section of 5 μm $\times$ 5 μm, as shown in Fig. 6.4.

A small section of the molded polymer waveguide is shown in Fig. 6.5, where the 3D tapering is clearly indicated. The waveguide thus fabricated demonstrated multiple modes without a cover cladding. However, it exhibits single mode operation at the small end if a polymeric cladding layer is further spin-coated on it. Such tapered waveguides have been proposed to bridge the mode mismatch between two optoelectronic devices having different shapes and sizes. The one drawback of the compression molding technique is the initial cost of fabrication of the mold plunger. However, for large production quantities it may turn out to be most cost effective as the same plunger can be used again and again.

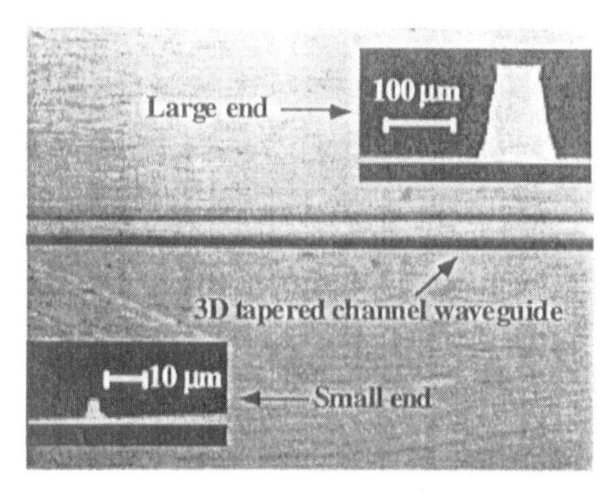

Fig. 6.5. Photo of a small section of the molded 3D linear tapered waveguide. The cross sections of the two ends are also shown in the figure.

6.2.2.2. *Laser-Writing Technique*

The laser-beam waveguide writing system has also been investigated for fabricating high performance polymer-based channel waveguides. The laser-beam writing system, shown in Fig. 6.2, consists of a dual-wavelength HeCd laser ($\lambda_1 = 325$ nm and $\lambda_2 = 442$ nm), beam-shaping optics, an electronic shuttle, and a computer-controlled X-Y-Z translation stage with a stroke of $30 \times 30 \times 2.5$ cm^3. The stage translation speed is continuously adjustable below 1.0 cm/s. The positioning resolution is 0.5 μm and 0.01 μm for the X-Y axes and Z-axis, respectively. The Z-stage is employed to precisely control the focused laser beam sizes. Figure 6.6 shows a polymeric waveguide H-tree structure fabricated on a silicon wafer.

The laser-beam writing technique is a straightforward process with minimal equipment and steps; uses dry, precoated and quality-controlled materials, and is amenable to large area exposures, as opposed to etching, molding, or embossing procedures. The ability to progressively laminate and bond successive layers to build up polymeric waveguide structures provides a significant and essential attribute for performance, applicability, and manufacturability. Using pressure- and temperature-controlled laminators with precoated materials adjusted for adhesion properties, bubble-free high-quality multilayered bonded structures can be created.

6.2.2.3. *VLSI Lithography Technique*

The most commonly used technique for research and development of polymer waveguide circuits is based on conventional photolithography orig-

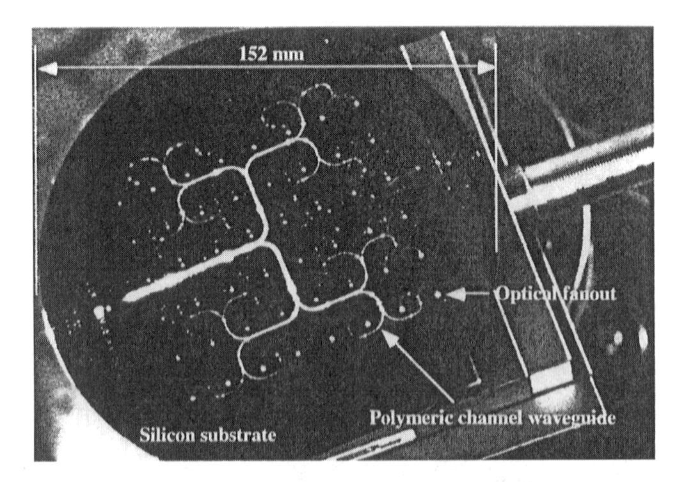

Fig. 6.6. Photograph of a polymeric waveguide H-tree fabricated on a silicon wafer.

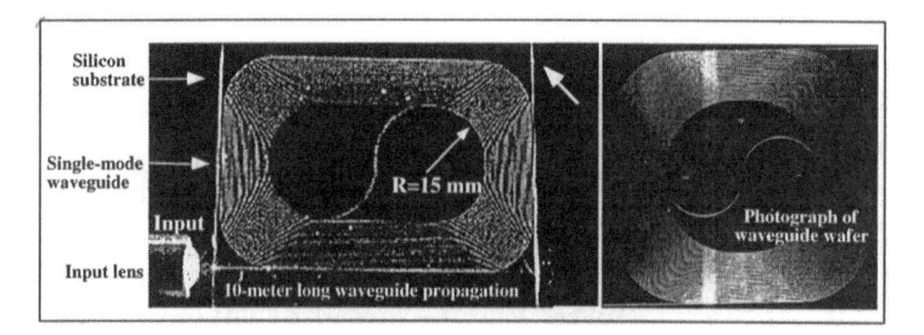

Fig. 6.7. Photograph of a ten-meter–long polymeric waveguide circuit fabricated using photolithography.

inally developed for the microelectronics industry. The standard VLSI lithography techniques provide easy and reproducible results. Since the length of waveguides is defined by photolithography, the waveguide length can be precisely controlled with accuracy in the submicron range. Figure 6.7 shows a 10-meter–long polymeric waveguide circuit fabricated by this technique. The waveguide propagation loss is about 0.2 dB/cm measured at $\lambda = 1064$ nm.

A number of waveguide structures using polyimide planarization and the above-described photolithography and laser-writing procedures have been constructed. Figure 6.8 shows the cross-section scan of a typical waveguide. Figure 6.9 shows microscope pictures of the various waveguide components. Figure 6.9(a) shows a 3dB 1-to-2 waveguide splitting structure. Figure 6.9(b) shows the end portion of a waveguide. Figure 6.9(c) and (d) show the tapered and curved waveguides.

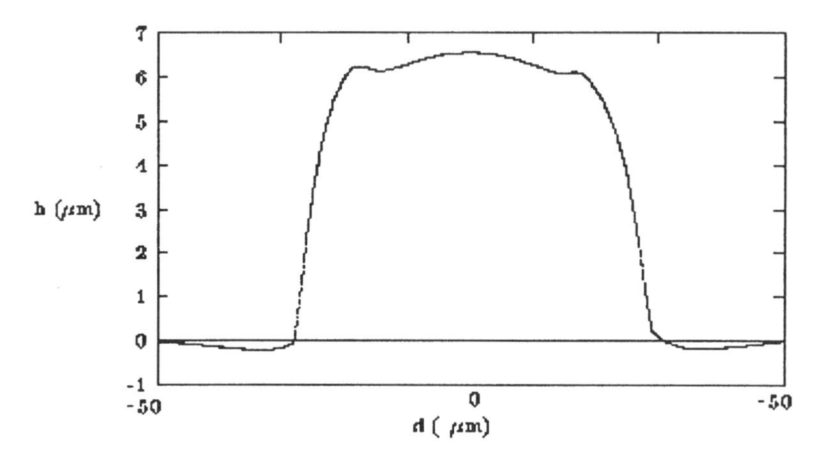

Fig. 6.8. Cross-section scan of polyimide waveguide fabricated.

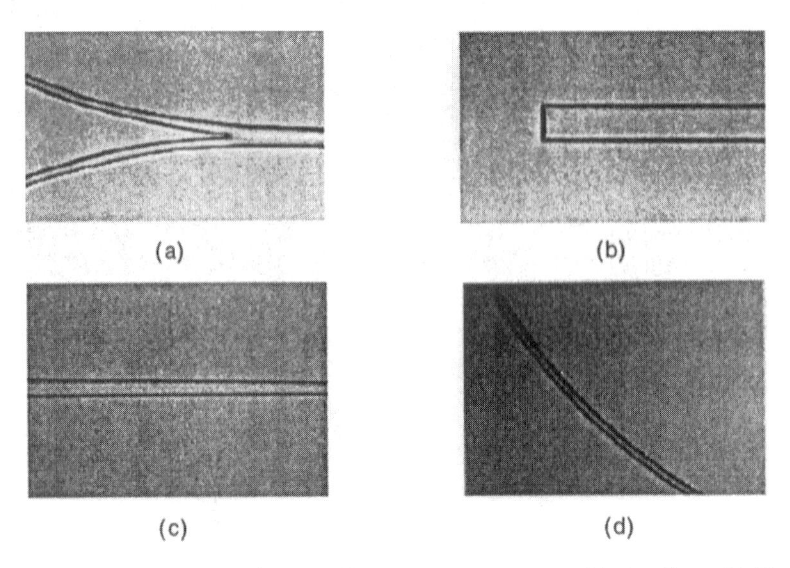

Fig. 6.9. Microscope pictures of waveguide structure components. (a) A splitter, (b) The end portion of a waveguide, (c), Tapered waveguide, (d) Curved waveguide.

6.2.3. WAVEGUIDE LOSS MEASUREMENT

Waveguide losses are important parameters in determining total insertion loss for transmitters and receivers. Low-loss waveguides significantly ease optoelectronic packaging. However, there is no simple technique available to measure waveguide propagation loss with reasonable accuracy for integrated optical waveguides that are fabricated on substrates. So far the most widely used method is the sliding-prism measurement. In this technique, the optical coupling prism is slid along the streak in the waveguide and the ratio of light coupled in and out of the waveguide is measured as a function of the propagation length. A second technique employs a moving fiber probe, in which the optical fiber is traced along the streak and the light scattered out of the waveguide is coupled into the fiber probe. These methods, however, suffer from lack of accuracy and reproducibility because of the mechanical nature of the measurement technique. The sliding-prism technique will also damage the waveguide.

This section describes a semiautomatic method for quickly measuring optical loss using a video camera combined with a laser beam analyzer. This method does not require any mechanical alignment, leading to accurate and reproducible measurements, and can be used with all kinds of waveguides employed in this research. This technique is routinely employed in our laboratories to characterize the propagation properties of polymer-based waveguides fabricated. This method exploits the light streak scattered out of the optical waveguide for the propagation-loss measurement. A prism–film coupler is used to excite the desired guided mode. The waveguide under test is mounted on a precise six-axis prism coupling stage. A optical stop is used to prevent light scattered from the edge of the input prism from overlapping the light streak. A laser-beam profile analyzer (such as the Spiricon LBA-100A) is set to observe the light streak of the excited mode in the waveguide from the front. The output video signal from the camera is analyzed by the system to provide waveguide propagation loss.

The peak intensity variations along the streak can be determined by scanning along the propagation direction, and the loss value can be directly acquired from the longitudinal change. However, this 1-D scanning applies only to the straight guides; also, setting the sampling line along the streak requires precise adjustment. Therefore, transverse scan is performed repeating the same procedure along the streak. This procedure provides a 2-D intensity distribution.

Figure 6.10 shows an image of the waveguide coupling using a prism. Both the waveguide and prism were mounted on a prism-coupling stage. A Ti-sapphire laser with an operating wavelength of 850 nm was employed. Figure 6.11 shows the 2-D light intensity profile along the streak. Repeating the integration of the data along each sampling line, the longitudinal variation of the mode power in the waveguide was obtained, as shown by the dots in Fig. 6.12. The solid line is the least-mean-squares fit to a decreasing exponential; the slope of this line yields the power loss coefficient. In this case the propagation loss of the TE_0 mode is 0.21 dB/cm at 850 nm and 0.58 dB/cm at 632.8 nm.

In summary, this technique provides a nondestructive waveguide loss measurement technique with excellent accuracy and sensitivity. This technique has been used to measure the waveguide loss over a wide range (as low as <0.2 dB/cm to as high as >10 dB/cm).

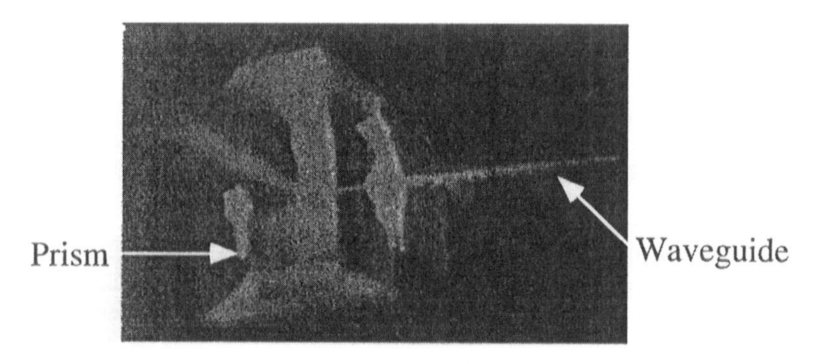

Fig. 6.10. Photograph of polymer-based waveguide coupling using a prism.

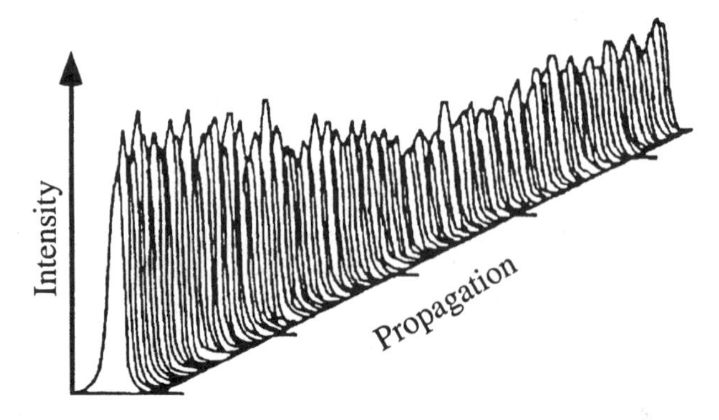

Fig. 6.11. Two-dimensional light intensity profiles for a polymer-based waveguide with TE_0 mode excitation.

6.3. THIN-FILM WAVEGUIDE COUPLERS

Efficient input and output coupling is an important factor to be addressed for reasonable performance from optoelectronic interconnects. Employment of conventional coupling techniques utilizing prisms and lenses tends to be costly, bulky, and very inconvenient from the packaging point of view and often puts severe restrictions on planarization. To efficiently couple optical signals from vertical cavity surface-emitting lasers (VCSELs) to polymer waveguides and then from waveguides to photodetectors, two types of waveguide couplers, grating couplers and 45° waveguide-coupling mirrors, can be employed. There are tilted-profile grating couplers and 45° waveguide micromirrors. This section gives a brief description of research results for both approaches.

6.3.1. SURFACE-NORMAL GRATING COUPLER DESIGN AND FABRICATION

For the targeted applications, the planarized surface-normal coupler is the most crucial element that needs thorough investigation. There are a large number of publications about grating design [16–20]. However, the surface-normal coupling scenario in optical waveguides has not been carefully investigated thus far. We are interested in a tilted grating profile in a planar structure within a thin waveguide layer. The tilted grating profile greatly enhances coupling efficiency in a desired direction. The gratings under investigation are fabricated by well-established planar microfabrication processes, such as photo and holographic lithographs. It is very important to conduct a theoretical investigation first to determine the required grating profile par-

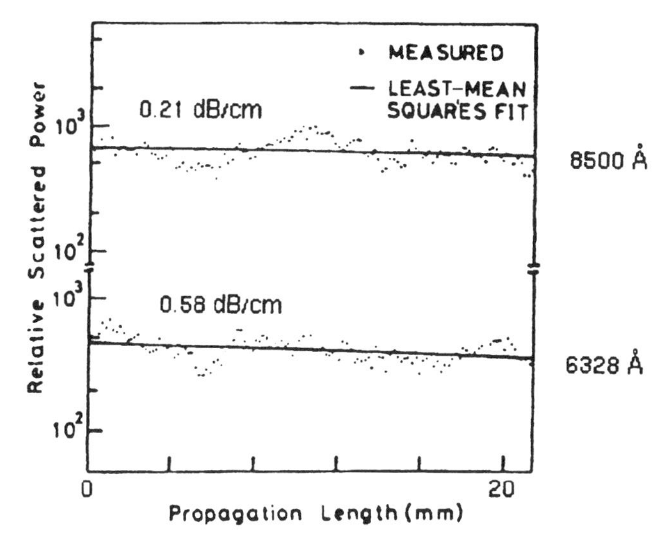

Fig. 6.12. Scattered power versus the propagation length for a measured waveguide with TE_0 mode launching at the wavelength of (a) 850 nm, (b) 632.8 nm.

ameters, such as tooth height, tooth width, and tilted angle, before starting any manufacturing work. We have developed a new analytical method to simulate the planarized tilted grating for waveguide coupling.

The phenomenon of grating-coupled radiation is widely used in guided-wave optical interconnects. Very often coupling in a specific direction is required. To achieve this unidirectional coupling, the tilted grating profile can be used. Coupling efficiency can then be evaluated. A very important aspect of manufacturing such couplers is the tolerance interval of the profile parameters, such as tooth height, width, tilted angle, and so on. Gratings with very tight intervals would be very difficult to manufacture and are practically useless. Problem of gratings design have been described in many publications [16]. However, described numerical methods work well only when the grating profile is relatively shallow, and fail when grating becomes deep. Thus, new analytical methods which provide high numerical accuracy were developed. Grating couplers with tilted parallelogram profiles demonstrate, as will be shown below, very high coupling efficiency in certain directions.

6.3.1.1. *Theoretical Formulation*

Consider the grating structure shown in Fig. 6.13. For simplicity, we derive the equation for the TE mode; however, in the case of TM mode, the resulting equation is similar. A transverse electric (TE) mode guided E field is along the

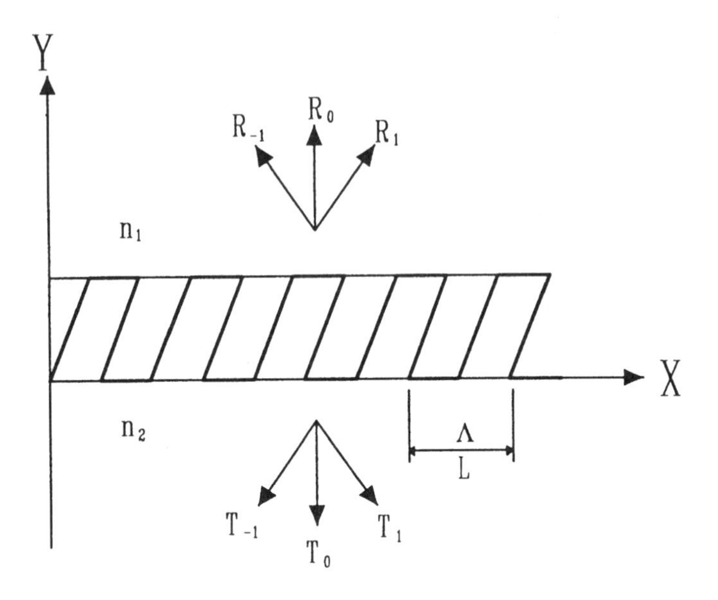

Fig. 6.13. The grating structure under consideration.

z-direction and satisfies:

$$\frac{\partial^2 E_z}{\partial x^2} + \frac{\partial^2 E_z}{\partial y^2} + k^2(x, y)E_z = 0, \tag{6.1}$$

where the time dependence $\exp(-iwt)$ has been omitted, $k(x, y)$ is the wave-vector with $k(x, y) = k_0 n(x, y)$, $k_0 = 2\pi/\lambda$, λ is the free-space wavelength, $n(x, y)$ is the index of refraction.

We can express field E_z as a Floquet's infinite summation of partial waves:

$$E_z(x, y) = \sum_{m=-\infty}^{\infty} E_m(y) \exp(i\alpha_m x), \tag{6.2}$$

where $\alpha_m = \alpha_0 + m(2\pi/\Lambda)$ and Λ is the grating period, α_0 is a phase constant. Outside the groove region, in the area $y > h$, using Rayleigh expansion we can write

$$E_{m1}(y) = \delta_m \exp(i\beta_1 y) + R_m \exp(i\beta_{m1} y), \tag{6.3}$$

where $\beta_{m1} = \sqrt{k_1^2 - \alpha_{m1}^2}$, $\delta_m = 1$ when $m = 0$ and 0, otherwise.

Below the groove region, in the area $y > 0$, the electric field can be written for each corresponding m as:

$$E_{m3}(y) = T_m \exp(-\beta_{m2} y) \tag{6.4}$$

where

$$\beta_{m2} = \sqrt{k_2^2 - \alpha_{m2}^2}$$

Inside groove region $0 < y < h$, $k^2(x, y)$ is periodic in the x direction and can be represented by a Fourier series:

$$k^2(x, y) = \sum_{m=-\infty}^{\infty} C_m(y) \exp(2\pi imx/\Lambda). \tag{6.5}$$

Upon substituting Eqs. (6.2) and (6.5) into Eq. (6.1) and collecting all terms with the same x-dependence, we can obtain an equation in the region $0 < y < h$:

$$\frac{\partial^2 E_{m2}(y)}{\partial y^2} - \alpha_m^2 E_{m2}(y) + \sum_q C_{m-q}(y) E_{q2}(y) = 0. \tag{6.6}$$

Equation (6.6) can be written in a matrix form of

$$E_2'' = V(y) E_2, \tag{6.7}$$

where E_2 is a column whose elements are E_{m2} and $V(y)$ is a known square matrix whose elements are defined by:

$$\text{Art in box} = \frac{\text{mentioned in manuscript}}{??????????} \tag{6.8}$$

The solution to Eq. (6.7) is subject to the boundary condition of E_z and $\partial E_z/\partial y$ being continuous at $y = h$ and $y = 0$. Taking into account the boundary conditions we get:

$$E_m'(0) + i\beta_{m2} E_m(0) = 0$$

$$E_m'(h) - i\beta_{m1} E_m(h) = 2i\beta\delta_m \exp(-i\beta h).$$

This can be written in matrix form as follows:

$$\begin{aligned} E'(h) + U_h E(h) &= S \\ E'(0) + U_0 E(0) &= 0, \end{aligned} \tag{6.9}$$

where U_0, U_h are diagonal known matrices whose diagonal elements are:

$$[U_0]_m = i\beta_{m2}, \qquad [U_h]_m = -i\beta_{m1}$$

and S is a column matrix which in fact reduces to a number $[S]_m = -2i\beta \exp(-i\beta h)$, $m = 0$.

Solution

In the case when rectangular profile Fourier coefficients in Eq. (6.5) are constants:

$$C_{Rm}(y) = \frac{i}{2\pi n}(k_2 - k_1)\left[\exp\left(-2\pi i m \frac{t}{\Lambda}\right) - 1\right]. \tag{6.10}$$

For the known height y, the tilted groove refractive index distribution can be considered as the rectangular groove distribution by making a shift along direction x as shown in Fig. 6.14: $d(y) = y \tan \varphi$, where φ is the tilt angle.

According to the rules of Fourier transformation, a coordinate shift in the original function creates a phase shift in the transformation result. Thus, the tilted-profile coefficients can be written as:

$$C_m(y) = \exp(-2i\pi m y \tan(\varphi)/\Lambda) C_{Rm} = \exp(\gamma m y) C_{Rm}. \tag{6.11}$$

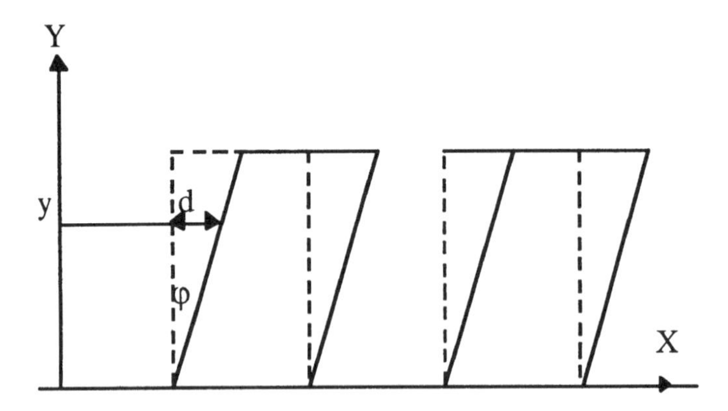

Fig. 6.14. Consideration of a tilted grating as a shifted rectangular grating.

Matrix $V(y)$, in the case of a tilted grating, can be written as follows:

$$V(y) = Q(y)V_R Q^*(y), \tag{6.12}$$

where Q is diagonal matrix the elements of which are $[Q]_n = \exp(\gamma yn)$, $n = 1, 2, \ldots,$ and the sign * means complex conjugation. Now, Eq. (6.7) can be rewritten as

$$E''(y) = Q(y)V_R Q(y)E(y). \tag{6.13}$$

We can introduce a new matrix variable:

$$D(y) = Q(y)E(y). \tag{6.14}$$

Derivatives of the Q matrix can be found:

$$Q'(y) = PQ(y), \qquad Q''(y) = P^2 Q(y), \tag{6.15}$$

where P is a constant diagonal matrix,

$$[P]_n = ni\gamma.$$

Upon substituting Eqs. (6.13), (6.14) into (6.12) we have an equation

$$D'' + 2PD' + (P^2 - V_R)D = 0. \tag{6.16}$$

This equation is a second-order differential equation with constant coefficients, and its general solution is

$$D = \exp(W_1 y)A + \exp(W_2 y)B, \tag{6.17}$$

where exp is the exponential function from matrix (computation of this function will be discussed below), A, B are constant matrices that are determined from boundary condition. In Eq. (6.17) matrix multiplication is noncommutative; therefore, order of its fractions is important W_1, W_2 matrices are roots of a second-order matrix equation:

$$W^2 + 2PW + (P^2 - V_R) = 0 \tag{6.18}$$

The boundary condition in Eq. (6.9) can be written as

$$\begin{aligned} D'(h) + (U_h - P)D(h) &= Q(h)S \\ D'(0) + (U_0 - P)D(0) &= 0. \end{aligned} \tag{6.19}$$

Using Eqs. (6.17) and (6.19) we can find constant matrices A, B:

$$\begin{aligned} A &= (W_1 + U_h - P)^{-1}(W_2 + U_h - P)B \\ B &= [(W_2 + U_h - P)\exp(W_2 h) \\ &\quad - (W_1 + U_h - P)\exp(W_1 h)(W_1 + U_h - P)^{-1}(W_2 + U_h - P)]^{-1} Q_h S \end{aligned} \tag{6.20}$$

The above equation gives an exact analytical solution in the case of a tilted grating. In the case of a rectangular grating, W_1 and W_2 matrices are

$$W_1 = W_2 = W^{1/2}.$$

To calculate exponential function in Eq. (6.20) we can factorize W_1 and W_2 matrices:

$$W_m = Z_m^{-1}\Omega_m Z_m \quad (m = 1, 2),$$

where Ω_m is a diagonal matrix the elements of which are the eigen-values of W_m, and Z_m is an orthogonal matrix whose columns are eigen-vectors of W_m. Using exponential function series and matrix factorization we can write:

$$\begin{aligned} \exp(W_m h) &= \sum_{k=0}^{\infty} \frac{(W_m h)^k}{k!} = \sum_{k=0}^{\infty} \frac{(Z_m^{-1} W_m Z_m)^k h^k}{k!} = Z_m^{-1}\left[\sum_{k=0}^{\infty} \frac{(W_m h)^k}{k!}\right] Z_m \\ &= Z_m^{-1}\exp(W_m h) Z_m. \end{aligned}$$

Thus, the exponential function in Eq. (6.20) is: $\exp(W_m h) = Z_m^{-1}\hat{\Omega}_m Z_m$, where $\hat{\Omega}_m$ is a diagonal matrix the elements of which are $[\hat{\Omega}]_n = \exp(\Omega_n h)$. Coefficients R_n and T_n in Eqs. (6.3), (6.4) can be found from condition of E_z being continued at $y = h$ and $y = 0$. Energy in the nth diffracted, reflected, and transmitted orders can be found

$$\begin{aligned} E_R &= R_n^* R_n \cos\alpha_{n1} \\ E_T &= T_n^* T_n \cos\alpha_{n2} \end{aligned} \tag{6.21}$$

where * means complex conjugate.

To check the precision of numerical calculations the energy balance can be used. This criteria takes the form

$$\sum_{n1} E_R + \sum_{n2} E_T = 1.$$

6.3.1.1. *Numerical Results*

The numerical results presented below are obtained utilizing Eqs. (6.20) and (6.21). The diffraction grating period is selected to make the angle of the first diffraction order α_1 equal to a defined value (for example, the bouncing angle of a waveguide) and cut all other higher-diffraction orders. In reflected light, only zero-order diffraction exists. All linear sizes are measured in wavelength value. The diffraction grating has a period 0.9 of the wavelength value. Substrate refractive index is assumed to be 1.5. Tooth width is selected to be one-half of the grating period. Figure 6.15 is a plot of coupling efficiency in different diffraction orders versus tooth height (in μm) in the case of a rectangular grating profile. It can be seen that coupling efficiency in this case is low. Figure 6.16 is a plot of coupling efficiency in different diffraction orders versus tooth height (in μm) in the case of a tilted grating profile. The tilt-angle of 32° was defined to obtain maximum coupling efficiency in the first diffraction order for the predefined first-order diffraction angle. This plot shows high coupling efficiency in the first diffraction order. In Fig. 6.17 coupling efficiency is shown as a function of tooth tilt-angle for the optimal tooth height of 1.1 μm. Figure 6.18 is a plot of coupling efficiency versus tooth width. The tooth width is changing from 0 to 0.9 μm, which is the grating period value. Tooth tile-angle

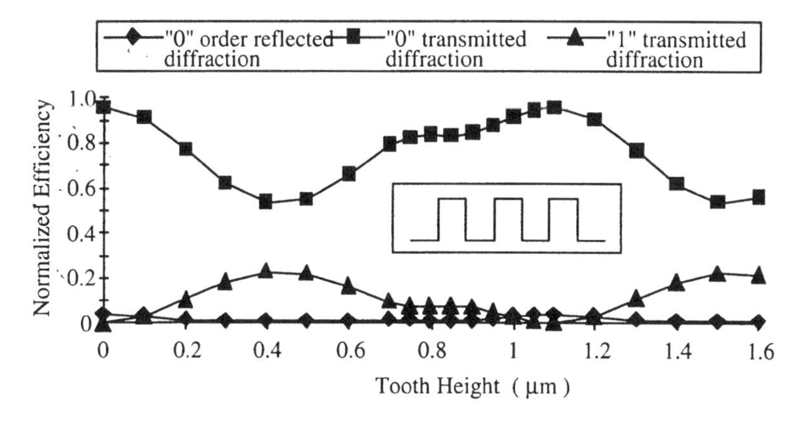

Fig. 6.15. Plot of coupling efficiency versus tooth height for a rectangular grating profile.

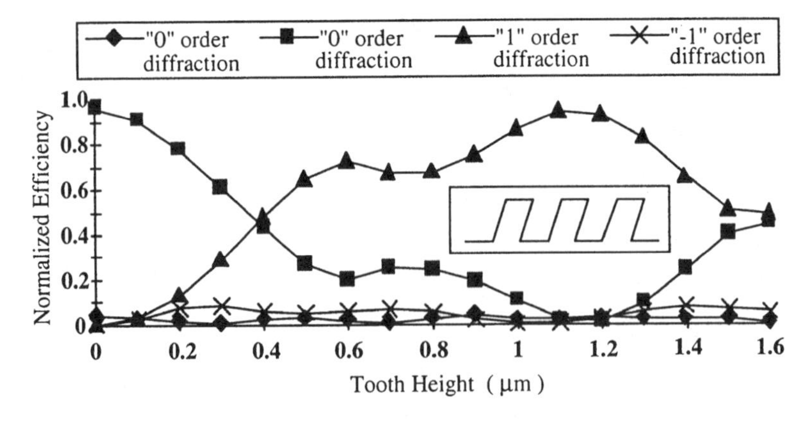

Fig. 6.16. Plot of coupling efficiency versus tooth height for a tilted grating.

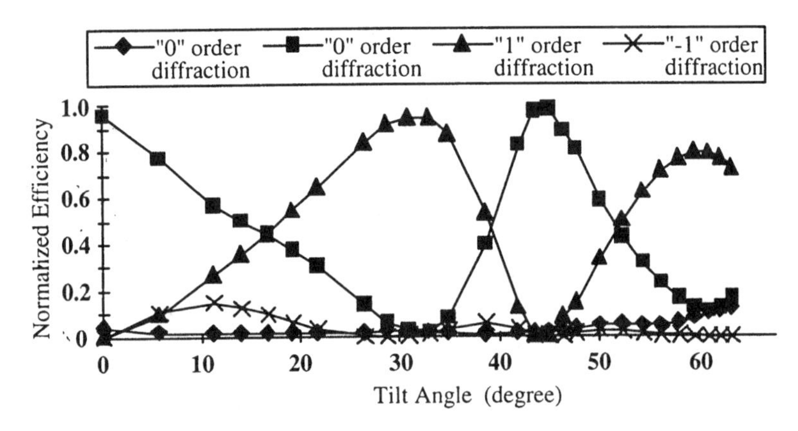

Fig. 6.17. Plot of coupling efficiency versus grating tilt angle for the optimal tooth height of 1.1 μm.

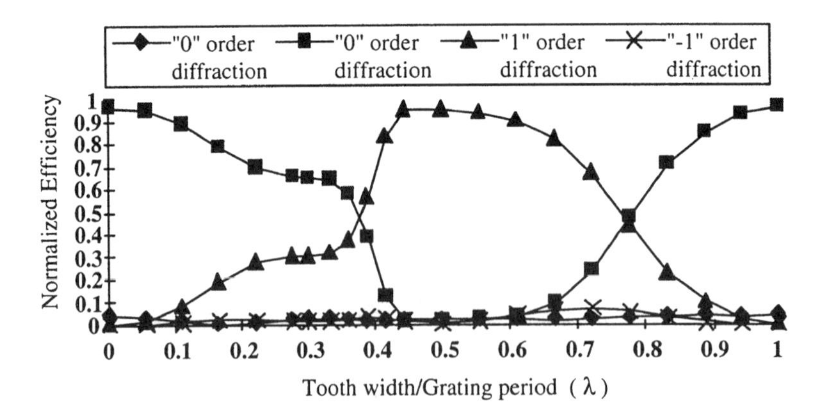

Fig. 6.18. Plot of coupling efficiency versus tooth width.

equals the optimal one and tooth height is also optimal. When tooth width equals zero or period value, only zero-order reflection and refraction radiation exists. When tooth width is between 0.4 μm and 0.6 μm, coupling efficiency into the first diffraction order is very high. This value diapason allows the gratings to be made with a relatively large tooth width tolerance interval. In the above case, allowable error equals $0.1/0.5 = 20\%$, which is very good. The results demonstrate optimum values of gratings parameters to maximize coupling in one diffraction order.

6.3.1.2. *Experimental Results*

6.3.1.2.1. Tilted Grating Profile Fabrication

It is clear that to provide effective free-space–to-waveguide and waveguide-to–free-space conversions, the microstructure of the surface-normal grating coupler shall be tilted to provide the needed phase-matching condition at one waveguide propagating direction. The tilted angle of the grating corrugation determines the vertical component of the grating K vector to be built to provide the required phase-matching condition. In the following, fabrication of the tilted waveguide grating and some experimental results are described. The schematic diagram for the fabrication process is shown in Fig. 6.19.

The polyimide waveguide can be fabricated on different substrates, such as PC board, Si, glass, and others, by spin coating. To fabricate polyimide waveguides, an A600 primer layer was spin-coated first on the substrate with a spin speed of 5000 rpm, and prebaked at 90°C for 60 seconds. The Amoco polyimide 9120D was then spin-coated with a speed of 2000 rpm. A final curing at 260°C in nitrogen atmosphere was carried out for more than three hours. Typical thickness of the waveguide was 7 μm. The planar waveguide has also

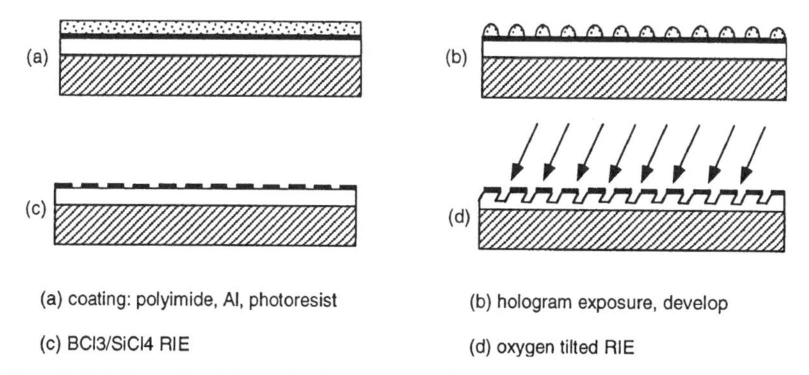

Fig. 6.19. Schematic of the fabrication process of tilted grating in polyimide waveguide.

been successfully fabricated on Si substrate by inserting a high index polyimide layer ($n = 1.56$–1.76) between the 9020D cladding layers.

To form the tilted grating pattern on the polyimide waveguide, we used a reactive ion etching (RIE) process with a low oxygen pressure of 2KPa to transfer the grating pattern on the aluminum layer to the polyimide layer. In order to get the tilted profile, a Faraday cage was used [21]. Microstructures of the tilted grating having a periodicity varying from 0.5 mm to 3 mm have been fabricated.

In order to fabricate the grating coupler by reactive ion etching (RIE), a thin aluminum metal mask must be fabricated on top of the polyimide-based planar guide. For this purpose, a 500 Angstroms aluminum layer was coated on top of the waveguide by electron beam evaporation, followed by a thin layer of 5206E photoresist. The grating patterns on photoresist were recorded by interfering two beams of the $\lambda = 442$ nm He-Cd laser line. The recording geometry is shown in Fig. 6.20. In order to record a grating with a period of Λ, the cross-angle θ of the two interfering beams was determined through the formula of $\sin(\theta/2) = (\lambda/\Lambda)$. The laser intensity was about 2 mW/cm^2 and the recording time varied from 1 minute to 2 minutes. After the sample development, a postbake at 120°C for 30 minutes was carried out. Two typical SEM pictures of the cross section of the photoresist patterns under different exposure and development times are shown in Fig. 6.21. It was also found that increasing exposure or development time any further might not enhance the contrast of the grating pattern on the photoresist.

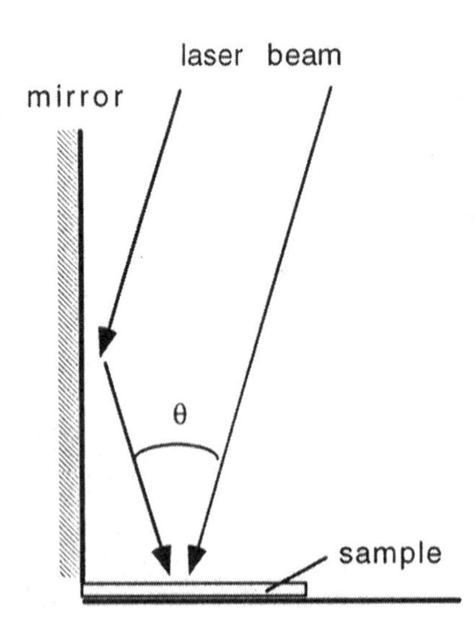

Fig. 6.20. Schematic for recording a hologram on photoresist.

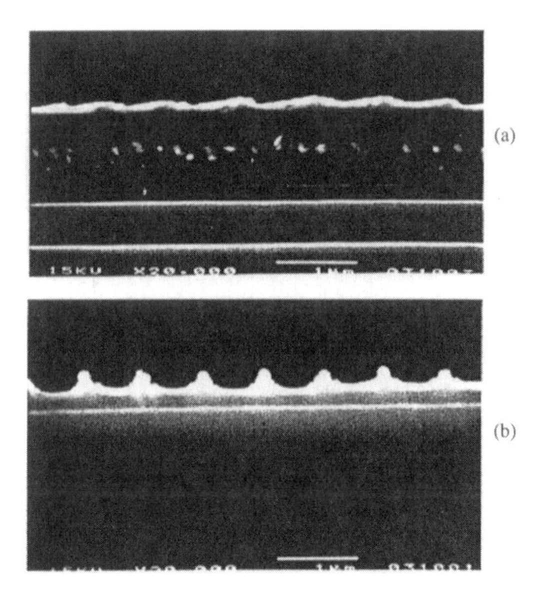

Fig. 6.21. SEM pictures of the cross section of photoresist grating patterns. (a) 1 minute exposure, 30 seconds development. (b) 1 minute exposure, 60 seconds development.

To transfer the photoresist grating patterns to aluminum, we used RIE to etch the aluminum in the opening window of the photoresist pattern. The gases used were $BCl_3/SiCl_4$ with a pressure of 20 millitorr. However, there were still some photoresist residuals in the grating grooves, which could block the aluminum RIE process. In order to clean these residuals, an additional step of RIE etching using oxygen was applied before removing the Al layer. The conditions used in the experiment were RIE power of 150 W, oxygen pressure of 10 millitorr, and oxygen flow rate of 15 sccm. The resulting characteristic etch rate of photoresist under this condition is 2000 Å/min.

To form the tilted grating pattern on the polyimide waveguide, we used a RIE process with a low oxygen pressure of 10 millitorr to transfer the grating pattern on the aluminum layer to the polyimide layer. The characteristic etching rate of 9120D under the power of 100 W is shown in Fig. 6.22, which indicates an etching rate of 0.147 μm/min. In order to get the tilted profile, a Faraday cage [21] was used. The sample inside the cage was placed at an angle of 32° with respect to the incoming oxygen ions. The final step was to remove the aluminum mask by another step of RIE process. The microstructure of the tilted grating is shown in Fig. 6.23 from a scanning electron microscope (SEM) picture. The grating period can be turned from 0.6 μm to 4 μm by changing the recording angles of the two-beam interference. From the theoretical result in Fig. 6.18, one can see that the coupling efficiency can be increased if the refractive index difference between the guiding layer and the cladding layer is

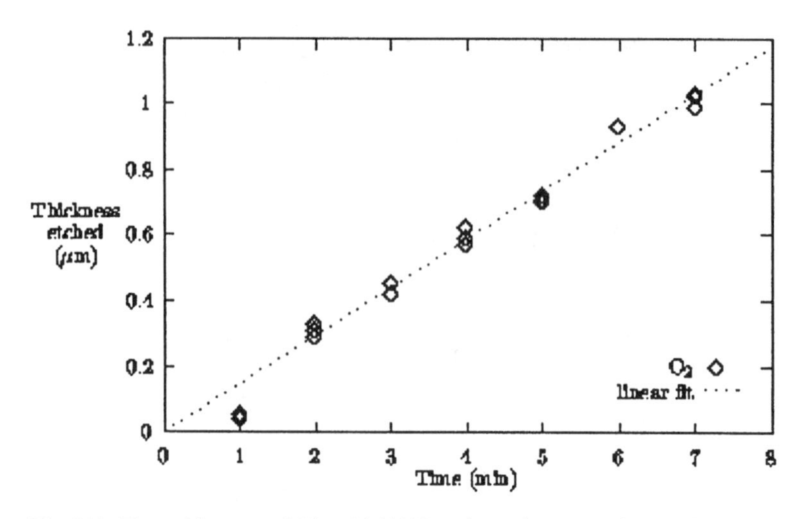

Fig. 6.22. The etching rate of Ultradel 9120D polymer by oxygen in a RIE process.

increased. Following the same fabrication processes as above, tilted gratings have been successfully made on other high-index polymers such as 1608D polyimide. Figure 6.24 is a SEM picture of the profile of the tilted grating on 1608D polyimide.

6.3.1.2.2. Input and Output Coupling Results

The schematic of coupling a surface-normal input light into a waveguide using the device fabricated is shown in Fig. 6.25(a), together with an experi-

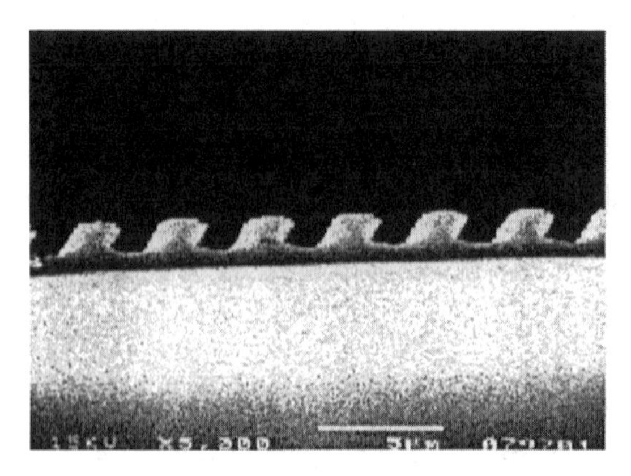

Fig. 6.23. SEM picture of the 4-micron period tilted grating.

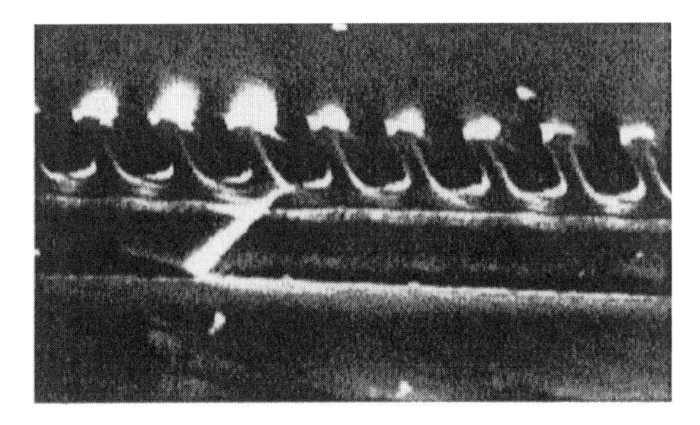

Fig. 6.24. SEM picture of the grating on 1608D polyimide.

mental photograph in Fig. 6.25(b). The grating is designed to surface-normally couple the laser beam into the 9120D polyimide waveguide with an operating wavelength at 1.3 μm. The coupling into the planar waveguide with unidirectional propagation can be clearly observed with a measured efficiency of 5%. The coupling efficiency is relatively low because the difference between the refractive index of the guiding and cladding layers of the waveguide is very small ($\sim$0.015) in the case of 9120D polyimide. This experimental result is consistent with theoretical calculation based on the coupled-mode theory, as shown in the curve in Fig. 6.18. Furthermore, the calculated results indicate

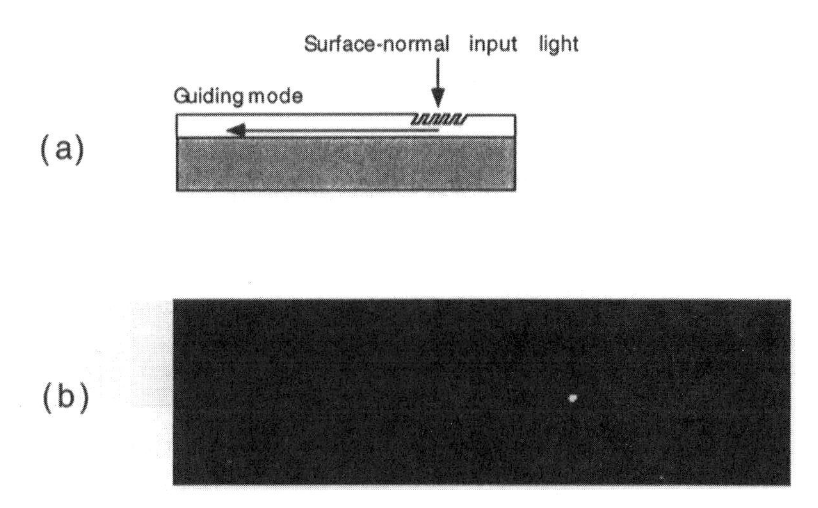

Fig. 6.25. (a) The schematic of coupling of a surface-normal input light into a waveguide using the tilted grating. (b) The experimental photograph of coupling a surface-normal input laser into the 9120D polyimide waveguide.

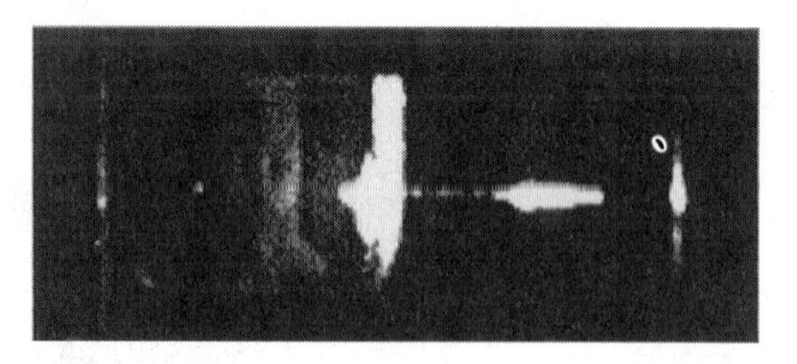

Fig. 6.26. Photograph of surface-normal output coupling.

that increasing the refractive index difference and optimizing the microstructure of the grating could significantly improve coupling efficiency.

The demonstration of surface-normal output coupling is shown in Fig. 6.26. The 1.3 μm laser light was coupled into the guided mode by prism coupling and coupled out of the waveguide by grating. The length of output coupling grating is $\sim$3 mm. From the surface scattered light intensity of the guiding path, the output coupling efficiency is estimated close to 100%. This is simply due to the longer interaction length associated with the longer output grating length. If multiple output grating couplers are employed along the waveguide and the grating parameters (i.e., depth, length, etc.) are adjusted such that the coupling efficiencies of the output gratings increase along the waveguide propagation direction, the uniform multistage optoelectronic interconnect can be realized.

6.3.2. 45° SURFACE-NORMAL MICROMIRROR COUPLERS

Another method of waveguide coupling utilizes total internal reflection mirrors, which are relatively wavelength insensitive and can be easily manufactured using reactive ion etching and photolithography. The input coupling efficiency of a micromirror coupler can be higher than 90% when a profile-matched input source is employed.

Figure 6.27 shows the main processing steps for fabricating the polymer channel waveguide array and 45° TIR micromirror couplers. Two-layer structured channel waveguides are first formed using standard processing steps described in the previous section. The 45° TIR micromirror couplers are formed within the channel waveguide by the reactive ion etching (RIE) technique. The detailed procedure is described below

6.3.2.1. *Fabrication Procedure*

1. Formation of the Mask to Perform RIE Process

RIE (reactive ion etching) is used to form the 45° slanted surface acting as the micromirror coupler at the end of the waveguide channel. To make the RIE

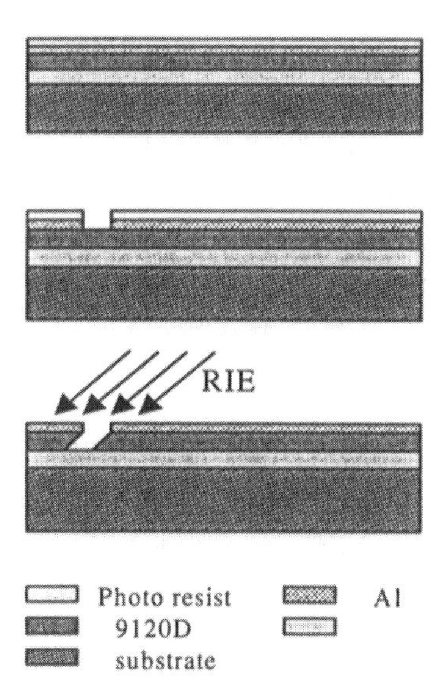

Fig. 6.27. Procedure to fabricate a channel waveguide with 45° TIR micromirror coupler.

etching selective, a aluminum hard mask is formed covering the whole sample before RIE etching. The procedure of forming the RIE mask is as follows:

1. The sample is coated with a 300-nm–thick layer of aluminum film using a CHA e-beam metal deposition system.
2. The aluminum layer is patterned by photolithography. A thick photoresist is required to cover the height difference between the top of the channel and the substrate. AZ9260 is a good photoresist for this purpose. It is important to ensure that the photoresist covers the entire sample surface without bubbles or other defects. A mask having an array of $50\,\mu m \times 50\,\mu m$ square windows with a spacing of $250\,\mu m$ is carefully aligned with the waveguide channel array at the position where the 45° micromirror couplers are to be formed. UV-exposure is performed and pattern is developed using appropriate developers.
3. The window pattern in the photoresist layer must be transferred to the aluminum layer. To do this, the sample is soaked in aluminum etchant, washed with DI water, and blow-dried. The aluminum within the windows must be completely etched away.
4. Finally, the photoresist is removed by soaking the sample in photoresist stripper until the residual AZ9260 is dissolved; washing with DI water and blow-drying leaves a layer of aluminum with square windows at the waveguide ends. This patterned aluminum layer acts as a hard mask in the RIE process.

II. Reactive Ion Etching

RIE processes are combinations of both physical (sputtering) and chemical etching actions. Basically, the physical component of the etching process provides better anisotropy on the side walls, but results in inferior surface quality; whereas, chemical etching tends to produce smooth etched surfaces but creates curved side walls. The extent of each etching action can be adjusted through optimizing the RIE conditions to obtain the best result; i.e., to have both straight side walls and smooth surface quality. To make physical etching dominant, the conditions can be selected with low pressure, high RF power, and use of a less reactive ion. Similarly, chemical etching can be enhanced through increasing the pressure, lowering the RF power, or using more reactive ions. In our process, the optimized RIE conditions are chamber pressure 15 mTorr, RF power 150 W, and O_2 (etchant gas) flow rate 10-SCCM (cubic centimeters per minute at standard temperature and pressure). A special holder is used to hold the sample with a 45° slope angle with respect to the electrode in the RIE chamber. The ion stream coming down vertically to the sample bombards the polymer through the windows in the aluminum film at a 45° angle with respect to the substrate. The aluminum mask does not wear down and protects the other parts against RIE. Two parallel 45° slanted side walls are formed in the desired position and direction under each opening window after RIE etching. A Faraday cage is used to cover the sample and the holder during the etching. A modified Faraday cage having a 45° tilted grid surface, which can change the direction of the ion stream by 45°, can be used to put the sample horizontally on the electrode and cover the area to be etched with the cage, with the tilt surface perpendicular to the direction along which the micromirror is to be etched. With this method, a larger sample can be processed and etched more uniformly.

III. Aluminum Removal

In the final step, the aluminum mask is removed through wet etching.

The micromirror couplers fabricated by the above procedures were examined using α-step and SEM. Usually with the conditions given above, 120 minutes of etching results in a 7-micron etching depth, which is sufficient to terminate the waveguide layer with a 45° TIR micromirror coupler. The SEM (scanning electronic microscope) is used to inspect the etching profile and the surface quality. The sample to be inspected is cut into small pieces to meet the size requirement of the SEM specimen holder. The sample is tilted to a proper angle to view the cross section and obtain the information of the etched profile. The quality of the 45° sloped surface can also be checked through SEM under a proper viewing angle and magnification. The SEM micrograph of the cross section of a 45° microcoupler is shown in Fig. 6.28 and a photograph showing a waveguide mirror coupler acting as a input coupler is shown in Fig. 6.29.

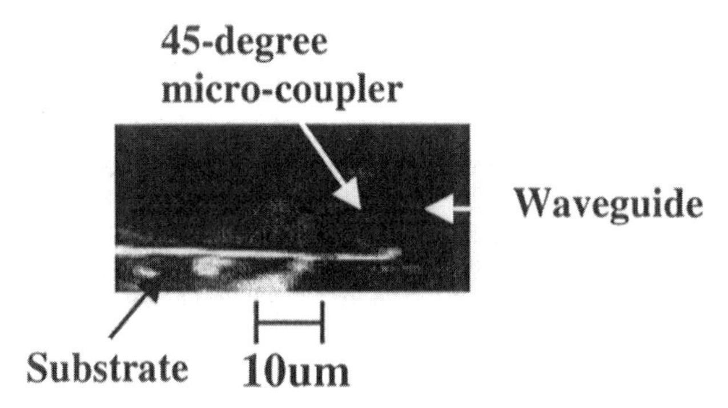

Fig. 6.28. THE SEM micrograph of the cross section of the 45° microcoupler.

6.3.2.2. *Output Profile of Micromirror Coupler*

The output profiles from one of the 45° microcouplers are shown in Fig. 6.30(a), (b), and (c). These figures correspond to $z = 100\,\mu\text{m}$, $z = 1$ mm and $z = 5$ mm, respectively, where z is the distance from upper surface of the coupler to the point of observation. The output coupling efficiency of this 45° microcoupler is nearly 100%. The half width at half maximum (HWHM) of the output profile at the microcoupler is about $60\,\mu\text{m}$, which is comparable with the active region of a silicon-based photodetector having a bandwidth of 6 GHz [22]. If the photodetectors are mounted close to the microcouplers, most of the light can reach the photodetectors and thus the coupler-to-detector coupling efficiency can be very high.

The output profile from the 45° microcoupler can be determined using diffraction theory. According to Fresnel approximation [23, 24], the near field distribution $U(x, y)$ is given by:

$$U(x, y) = \frac{\exp(jkz)}{jz\lambda} \int_{-\infty}^{\infty} U(\xi, \eta) \exp\left\{\left(j \frac{k}{2z} [(x - \xi)^2 + (y - \eta)^2]\right\} d\xi\, d\eta,$$

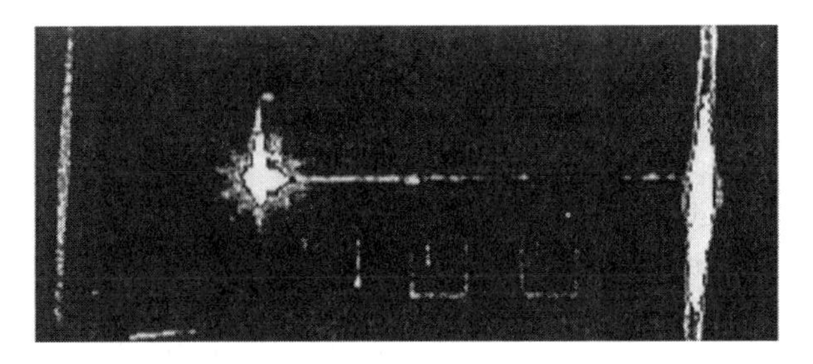

Fig. 6.29. A planar polyimide waveguide with waveguide mirror.

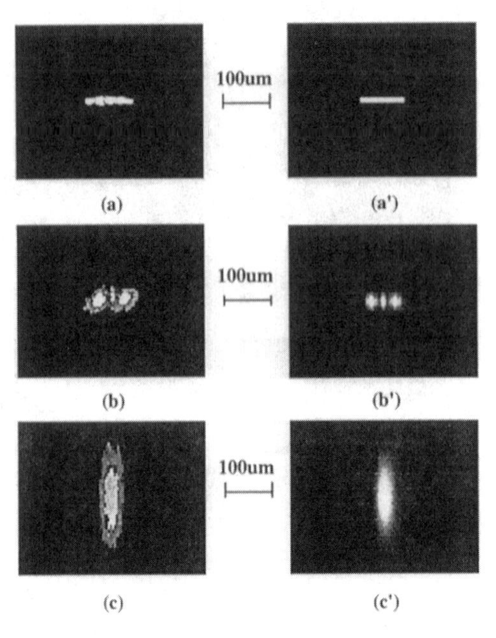

Fig. 6.30. Output profiles from the 45° surface-normal microcoupler. (a) $z = 100\,\mu m$ (experiment), (b) $z = 1$ mm (experiment), (c) $z = 5$ mm (experiment), (a′) $z = 100\,\mu m$ (theory), (b′) $z = 1$ mm (theory), (c′) $z = 5$ mm (theory).

where $U(\xi, \eta)$ is the complex amplitude of the excitation at point (ξ, η) on the 45° microcoupler, $U(x, y)$ the complex amplitude of the observed field at point (x, y), k the magnitude of the wave vector, and z the distance from the upper surface of the 45° microcoupler to the point of observation. However, direct integration based on Fresnel approximation fails for a small z due to the fast oscillations of the Fresnel factor. A modified convolution approach was used to calculate the $U(x, y)$ and output profiles. Theoretical output profiles at $z = 100\,\mu m$, $z = 1$ mm, and $z = 5$ mm from the 45° microcoupler are shown in Fig. 6.30(a′), (b′), and (c′), respectively. In these calculations, the input to the microcoupler was assumed to be the fundamental mode of the waveguide. This is a reasonable assumption, since the most of the energy in a multimode waveguide remains confined to the fundamental propagating mode. Note that the input is in the TEM_{00} mode. Compared to Fig. 6.30(c′), the side lobes in Fig. 6.30(c) are not clear due to the poor contrast of the image. As z becomes larger, the output light diverges faster in the direction that corresponds to the smaller dimension of the 45° microcoupler. There is a good agreement between the theoretically simulated and experimentally observed output profiles.

Figure 6.31 shows a SEM micrograph of the mirror coupler integrated to an array of channel waveguides. The measured output coupling efficiency is

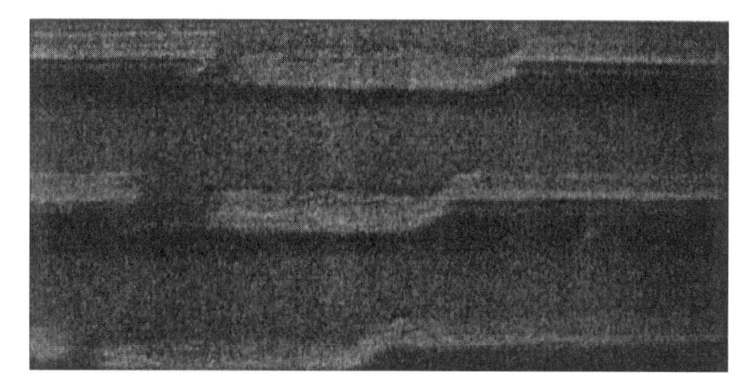

Fig. 6.31. A SEM picture of a mirror coupler integrated with a polymer channel waveguide and input light coupling in a planar polyimide waveguide with a waveguide mirror.

nearly 100% for a waveguide mirror coupler due to total internal reflection. Considering the implementation of thin-film vertical cavity surface-emitting lasers and Si-photodetectors onto a board involving 3-D interconnection layers, we need to use a very short working distance ($\sim$ few μm to tens of μm) in the surface-normal direction. This restriction makes the 45° waveguide mirrors the best approach for this purpose.

6.4. INTEGRATION OF THIN-FILM PHOTODETECTORS

In the context of board-level optical interconnect applications, we determine the thin-film MSM photodetector to be the most appropriate because it can provide a very high demodulation speed, due to the fast transit time of electron–hole pairs, and it can be directly integrated onto target systems. Moreover, compared to the vertical structure of a PIN diode, the structure of the MSM photodetector makes it much easier to integrate it with microwave planar waveguide circuits. Recent research on distributed traveling-wave MSM photodetectors reveals the possibility of making optoelectronic hybrid integrated circuits, where optical channel waveguide and microwave coplanar waveguide (CPW) coexist on the same substrate [25]. This is an important feature for the architectural design of optical interconnects, because it can provide better integration level and simplify the configuration.

Further, MSM structure can function as a switching device by biasing it appropriately [26]. This specific attribute can be extremely helpful to reconfigure the interconnect topology. The switching capability is a consequence of the energy-band structure of the MSM junction, in which a potential well

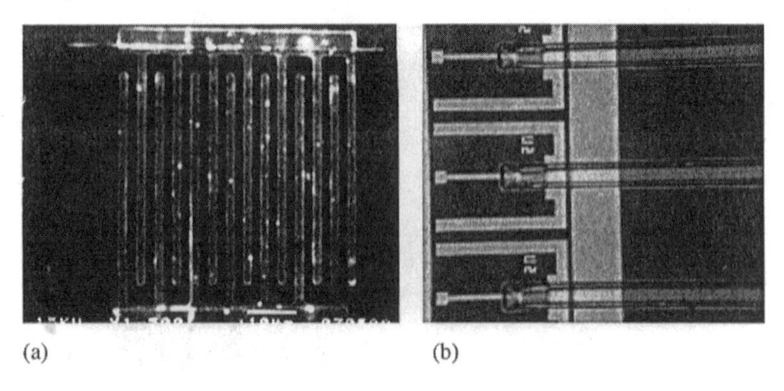

(a) (b)

Fig. 6.32. (a) SEM photograph of the MSM photodetector with 2 μm finger width and spacing. (b) Polymer channel waveguide array aligned with MSM photodetector array.

exists. To overcome this potential well to produce photoresponse, the bias voltage applied to the detector has to be larger than a threshold value that is determined by

$$V_{\text{bias}} = \frac{q_e \cdot N_d \cdot d^2}{2\varepsilon},$$

where N_d is the donor impurity density in n-type semiconductor region, d is the width of the depletion region, q_e is the electron charge.

MSM photodetectors have been fabricated on thin Si wafers and GaAs substrates with rough back surfaces. The thin film ensures that the photogenerated electron–hole pairs are only created in the high field region, making high-speed operation feasible. Also, the rough back surface scatters the light and traps it inside the thin film to compensate for otherwise low quantum efficiency. Figure 4.32(a) shows 2-μm fine lines of the electrode pattern employed for MSM detector fabrication. Since we intend to develop an optoelectronic interconnection employing this structure, we integrated the polyimide channel waveguide array with a 1 × 12 GaAs MSM photo detector array through 45° TIR micromirror couplers. First, we fabricated the high-speed MSM photodetector arrays on the GaAs substrate. The fabrication procedures for 1 × 12 MSM photodetector arrays are as follows. First, the 100-nm–thick SiO_2 was deposited on the surface of semi-insulating LEC–grown GaAs (100) wafer by plasma-enhanced chemical deposition (PECVD). Then the interdigitized electrode pattern was formed by conventional photolithography technique and part of the SiO_2 was etched away using 1:6 oxide etchant. The interdigitized gold electrodes were formed by first depositing 100 nm gold directly on the surface of the GaAs wafer using electron beam evaporation and then lifting it off to form the Schottky contacts. The interdigitized contact fingers have the same 2-μm width and spacing, resulting in a relatively photosensitive area of 50%. Finally, the metal pad for probing the device was fabricated by evaporating 100 nm Cr/Au. Each photodetector fabricated has an active area of 50 μm × 50 μm. Following the detector array fabrication the polymer channel waveguide array is fabricated on the substrate-containing photodetector array such that the active area of detectors overlaps with the waveguide output couplers. A photograph of a portion of the integrated system is shown in Fig. 6.32(b). Then 45° TIR micromirror couplers are formed on each channel at the position directly above the active area of the photodetector. Such integration facilitates precise alignment of the light coupler with the receiver, and helps avoid difficult optical alignment and packaging processes.

The DC I–V characteristic of the MSM photodetector was measured to obtain the dark current before integrating with the polyamide waveguide array. We used a HP4145B semiconductor parameter analyzer to measure the dark current of the device. The device fabricated has a low dark current of 4.6 pA under 5 V biasing voltage. The DC I–V characteristic of one of the photodetectors is shown in Fig. 6.33. To characterize the integrated system, we polished the input ends of the channel waveguides and used a single mode fiber to butt-couple the light into the waveguide. When the light propagating in the channel waveguide arrives at the 45° TIR micromirror couplers, it is reflected onto the MSM photodetector beneath the coupler. The frequency dependence of the photoresponse of the MSM photodetector and polyimide waveguide is measured by using an HP 8702 spectrum analyzer operating over a frequency

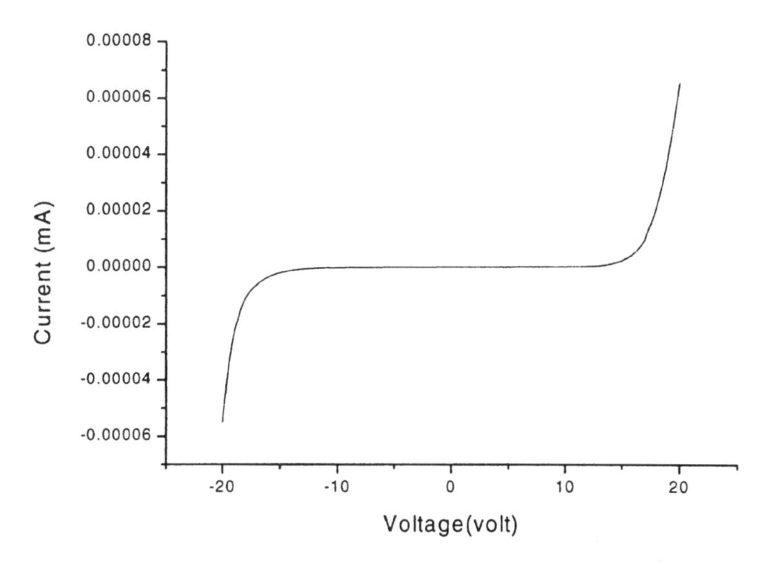

Fig. 6.33. I–V characteristic of one of the photodetectors prior to integration.

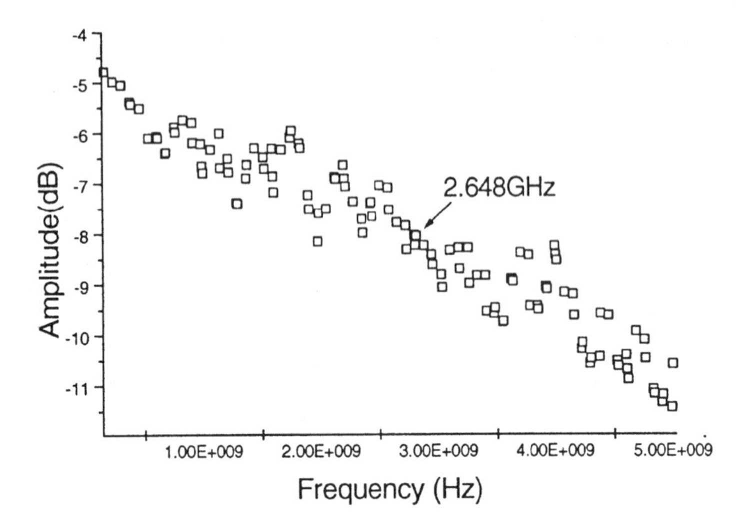

Fig. 6.34. The frequency response of integrated MSM photodetector and polyimide waveguide with 45° TIR mirror. The 3 dB bandwidth is 2.648 GHz.

range from 9 kHz to 26 GHz. The light source for frequency response measurement is a Ti:sapphire mode-locked laser operating at 800 nm with a pulse width of about 150 fs and a repetition rate of 76 MHz. The laser power is attenuated from its original 200 mW to 10 mW before it is coupled into the waveguide array. An ampere meter is serially connected between the contact probe and the biasing DC power supply. A 40 GHz ground–signal–ground (GSG) microwave probe is used for the device contact. The biasing voltage is provided by a DC power supply (Model E3611A, HP, Inc.) through a 100 k–18 GHz biasing tee (Model 5550B, Picosecond, Inc.). The pulsed laser light is butt-coupled into the input end of the waveguide array through a bare, cleaved-end, single mode fiber. At the output end the guided light is reflected from the 45° microcoupler onto the active area of the photodetector. The optical input pulses and the complete measurement system are calibrated using a 30 GHz bandwidth photodetector module (Model D-15, Newport, Inc.). The 3dB bandwidth of a single MSM detector is 2.648 GHz, as shown in Fig. 6.34.

6.5. INTEGRATION OF VERTICAL CAVITY SURFACE-EMITTING LASERS (VCSELs)

We chose 850 nm to be the operating wavelength because of the availability of high-performance vertical cavity surface-emitting lasers (VCSELs) [27]. These laser devices, compared to conventional edge-emitting lasers, offer a very low threshold current with much less temperature sensitivity, moderate optical power (few mW), very high direct modulation bandwidth (>14 GHz), wide operating temperature range (-55 to $+125^{\circ}$C), and ease of packaging in an array configuration due to their unique surface-normal output nature. To provide a high coupling efficiency using the proposed waveguide couplers, the control of the emitting aperture and of the wavelength are pivotal and make VCSELs the best choice due to the large separation of the adjacent longitudinal modes implied by the short cavity length. To provide the required fanouts with a bit-error rate of 10^{-9} at the required speed, a VCSEL with enough modulated power is needed to accommodate all losses including the -3 dB power margin. Planar configuration of VCSELs allows these devices to be fabricated and wafer-scale tested with conventional microelectronics manufacturing processes. The unique surface-normal emitting nature of the device allows us to use exactly the same packaging scheme for coupling light from a VCSEL into a waveguide as that used for coupling light from a waveguide into a photodetector.

The important aspects of the VCSEL for fully embedded board-level optical interconnects applications are its very thin laser cavity, typically less than $10\,\mu$m, and its surface-normal emitting characteristic. Table 6.3 gives a Epi-layer structure of a 3 QW (quantum-well) VCSEL. If the substrate of the VCSEL is removed, a very thin VCSEL enables the formation of a fully embedded optical interconnection, as depicted in Fig. 6.1. However, VCSELs in this embedded scheme are surrounded with a thermal insulator such as polymer; therefore, generated heat builds up within the embedded layer. This necessitates consideration of efficient heat-removal methods. Intrinsically, heat generated in a thin VCSEL will be more rapidly transported to the surfaces of the VCSEL than in the case of a thick VCSEL. A forced heat-removal method using via and thermoelectric cooler (TEC) can be the best choice for effective cooling. Via, which feed electrical current to the VCSEL, can be used simultaneously as a heat conduction path.

For a parallel channel interconnection, an 850-nm linear VCSEL array with 12 independently addressable lasers with 250-μm separation will be employed. To incorporate the VCSEL array in the fully embedded architecture, the VCSEL array has to be thin enough to build such a 3-D structure. There are two methods to make thin-film VCSELs; chemical mechanical polishing and epitaxial liftoff [28, 29, 30]. A mechanically polished thin VCSEL array is shown in Fig. 6.35(a) and (b). Figure 6.35(a) shows a section of the 12-channel VCSELs array, and Fig. 6.35(b) shows the SEM picture of the mechanically polished thin VCSEL array where the thickness of the GaAs is measured to be $42\,\mu$m. An n-ohmic contact was fabricated on the polished side using Cr(40 nm)/Au(200 nm) and annealed at 350°C for 20 sec.

Substrate removal using epitaxial liftoff is based on extremely selective etching ($>10^{9}$) of AlAs in dilute hydrofluoric acid. The epitaxial liftoff process

Table 6.3

Layer Structure of a VCSEL

Layer	Al(x)GaAs fraction [x]	Optical thickness	Physical thickness [μm]	Dopant	Doping level [cm^{-3}]	Type	Comments
GaAs	0.00		625.0000			n^+	Substrate
AlGaAs	0.90		0.0400	Si	1-3E18	n	Sacrificial layer
GaAs	0.00		0.0500	Si	1-3E18	n	Buffer
AlGaAs	Grade		0.0100	Si	1-3E18	n	1/2 n-DBR
AlGaAs	0.90	$\lambda/4$		Si	1-3E18	n	
AlGaAs	Grade		0.0100	Si	1-3E18	n	
AlGaAs	0.15	$\lambda/4$		Si	1-3E18	n	n-DBR
AlGaAs	Grade		0.0100	Si	1-3E18	n	32 ×
AlGaAs	0.90	$\lambda/4$		Si	1-3E18	n	
AlGaAs	0.60		0.0920	—	—	nid	Spacer
AlGaAs	0.30		0.0100	—	—	nid	Barrier
GaAs	0.00		0.0070	—	—	nid	Quantum well
AlGaAs	0.30		0.0100	—	—	nid	Barrier
GaAs	0.00	$4\lambda/4$	0.0070	—	—	nid	Quantum well
AlGaAs	0.30		0.0100	—	—	nid	Barrier
GaAs	0.00		0.0070	—	—	nid	Quantum well
AlGaAs	0.30		0.0100	—	—	nid	Barrier
AlGaAs	0.60		0.0920	—	—	nid	Spacer
AlGaAs	Grade		0.0100	C	3-5E18	p	
AlGaAs	0.90	$\lambda/4$		C	3-5E18	p	
AlGaAs	0.98		0.0200	C	3-5E18	p	p-DBR
AlGaAs	0.98		0.0200	C	3-5E18	p	
AlGaAs	Grade		0.0180	C	3-5E18	p	
AlGaAs	0.15	$\lambda/4$		C	3-5E18	p	
AlGaAs	Grade		0.0090	C	3-5E18	p	
AlGaAs	Grade		0.0090	C	3-5E18	p	
AlGaAs	0.90	$\lambda/4$		C	3-5E18	p	
AlGaAs	Grade		0.0090	C	3-5E18	p	p-DBR
AlGaAs	Grade		0.0090	C	3-5E18	p	21 ×
AlGaAs	0.15	$\lambda/4$		C	3-5E18	p	
AlGaAs	Grade		0.0090	C	3-5E18	p	
AlGaAs	0.90	$\lambda/4$		C	3-5E18	p	
AlGaAs	Grade		0.0090	C	3-5E18	p	p-DBR
AlGaAs	Grade		0.0090	C	3-5E18	p	
AlGaAs	Grade		0.0090	C	3-5E18	p	
AlGaAs	0.15	$\lambda/4$		C	3-5E18	p	
GaAs	0.00		0.0200	Zn	> 5.0E19	p^+	Cap layer

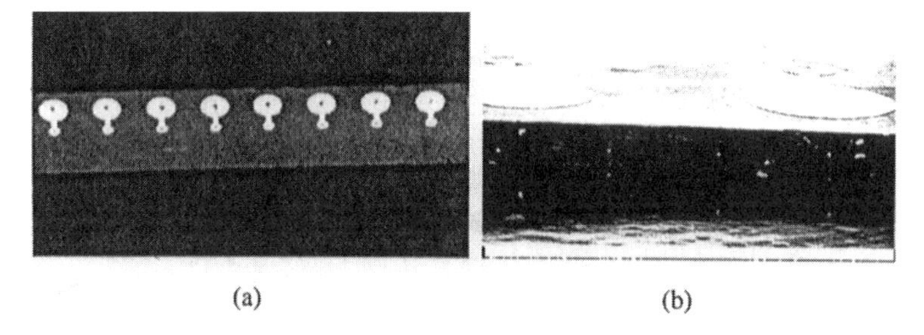

Fig. 6.35. (a) SEM picture of a section of a linear thin-film VCSEL array with 12 elements. (b) Mechanically polished VCSELs (substrate thickness $\sim 42\,\mu$m, and emitting aperture $15\,\mu$m).

is pictorially shown in Fig. 6.36. The epitaxial liftoff method has the advantages of reproducibility and mass productivity. To adapt the epitaxial liftoff method, VCSEL devices should be fabricated on the top of the AlAs sacrificial layer, which is located between the bottom DBR and the GaAs substrate.

Epitaxial liftoff process for VCSELs (with reference to Fig. 6.36)

1. Etch a donut-shaped 2-μm mesa using wet etchant. It defines oxidizing regions to confine current. In the next step, lateral oxidization in a furnace (465°C, water vapor introduced) is performed until the desired laser aperture is reached ($15\,\mu$m diameter).

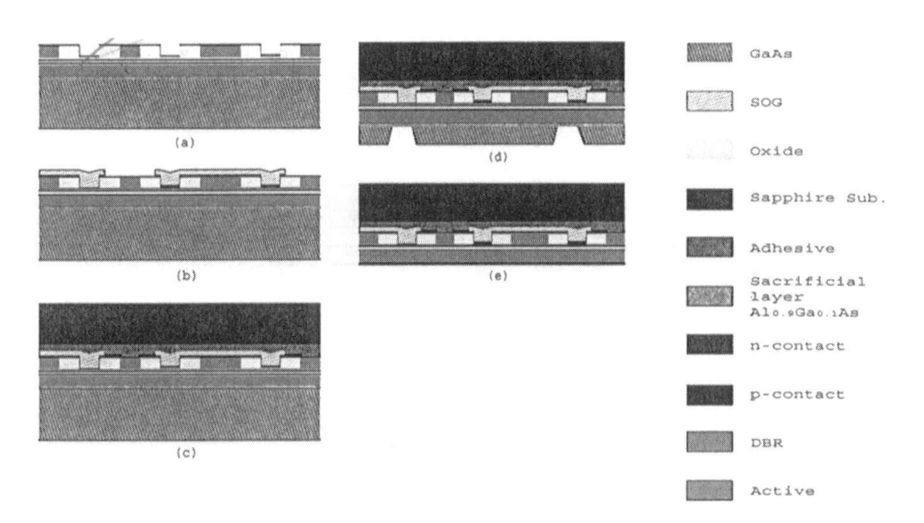

Fig. 6.36. Epitaxial lift-off process for VCSEL integration.

2. The donut-shaped mesa is filled with SOG (Spin On Glass) to provide low contact pad capacitance (it ensures high-frequency operation), better planarization for metalization, and enhanced mechanical strength (helpful in thin VCSEL applications).
3. This step involves metalization and binding to a temporary superstrate (sapphire). Next, the p-contact is formed by depositing a layer of (Ti/Pt/Au), and the VCSEL is attached on a temporary sapphire superstrate using Crybond^{tm}.
4. In this step the GaAs wafer is lapped till 250 μm (625 μm → 250 μm) and back side etch holes are etched.
5. Finally, substrate is removed by introducing HF through etched hole from backside and etching the sacrificial layer away.

Output laser power and current versus voltage characteristics of the VCSEL before and after polishing down to 42-μm thickness were measured, and the result is shown in Figs. 6.37 and 6.38. The VCSEL array was mounted on gold-coated substrate using an InGa eutectic alloy. Threshold current and threshold bias voltage are measured before thinning to be 5.8 mA and 1.7 V, respectively, after thinning they should be 6.0 mA and 1.2 V, respectively. After polishing to its final thickness, there is no significant change in optoelectrical characteristics. The L–I curves showed that there was no kink. The current versus voltage (I–V) curve showed that the series resistance after thinning (200 Ω) was smaller than that of the original (300 Ω) at threshold voltages.

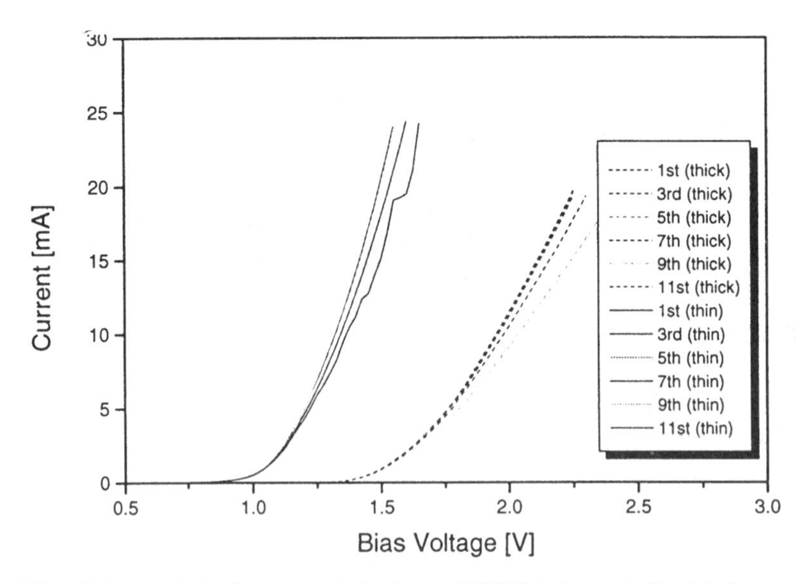

Fig. 6.37. I–V characteristic of a commercially obtained VCSEL before and after thinning process.

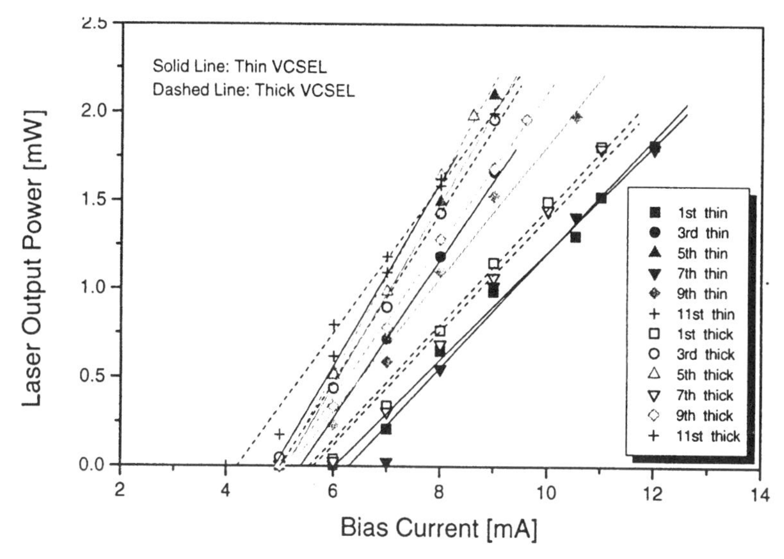

Fig. 6.38. Laser output as a function of bias current before and after thinning process.

Slight increase of threshold current after thinning is due to the increasing of dislocation density during mechanical polishing of VCSEL. However, the slope efficiency is almost the same.

If the epitaxial liftoff method is employed to remove substrate, substrate will be removed without any increase in the dislocation density. The series resistance of the VCSEL was reduced from 300 to 200 Ω because of the reduced substrate thickness. Series resistance of the VCSEL can cause serious problems in embedded-type packaging. Resistance generates heat, increasing the temperature of the VCSEL; this may cause a drop in emitting efficiency. Thermal problems in packaging are relieved by reducing series resistance and by attaching a heat sink. However, a large heat sink cannot be used in embedded-type packaging. Further, the VCSEL is surrounded by poor thermal conductors such as polyimides or other dielectric materials. Therefore, small series resistance is necessary for embedded-type packaging because it generates less heat. Another merit of this approach (using thin substrate) is that heat is removed faster through a thin substrate.

6.6. OPTICAL CLOCK SIGNAL DISTRIBUTION

For a multiprocessor computer system, such as a Cray T-90 supercomputer, it is difficult to obtain high-speed (>500 MHz) synchronous clock distribution using electrical interconnections due to large fanouts (48×2) and long interconnection lengths (>15 cm) [31–35]. A fanout chip is required to provide massive electrical fanout. The synchronous global clock signal distribution is highly desirable to simplify the architecture and enable higher-speed performance. High-speed, large-area massive fanout optoelectronic interconnects may overcome many of the problems associated with electrical interconnects in this interconnection scenario [31–38]. An array of novel optical interconnect architecture has been proposed and then demonstrated by earlier researchers [38, 39, 40], which may partially satisfy the above requirements for a massive clock signal distribution in intraboard and interboard hierarchies.

This section describes a guided-wave optoelectronic interconnect network for optical clock signal distribution for a board-level multiprocessor system. For comparison, the electrical interconnect network of the clock signal distribution for a Cray T-90 supercomputer at the board level currently employed and the corresponding implementation of optical interconnection are shown in Fig. 6.39(a) and 6.39(b), respectively. Figure 6.39(a) shows the existing 500 MHz 1-to-48 clock signal distribution (one side) realized in one of the 52 vertically integrated layers within the Cray T-90 supercomputer board. To further upgrade the clock speed, an appropriate optical interconnect scheme, shown in Fig. 6.39(b) has to be incorporated to minimize electrical interconnection–induced unwanted effects. An integrated board-level optoelec-

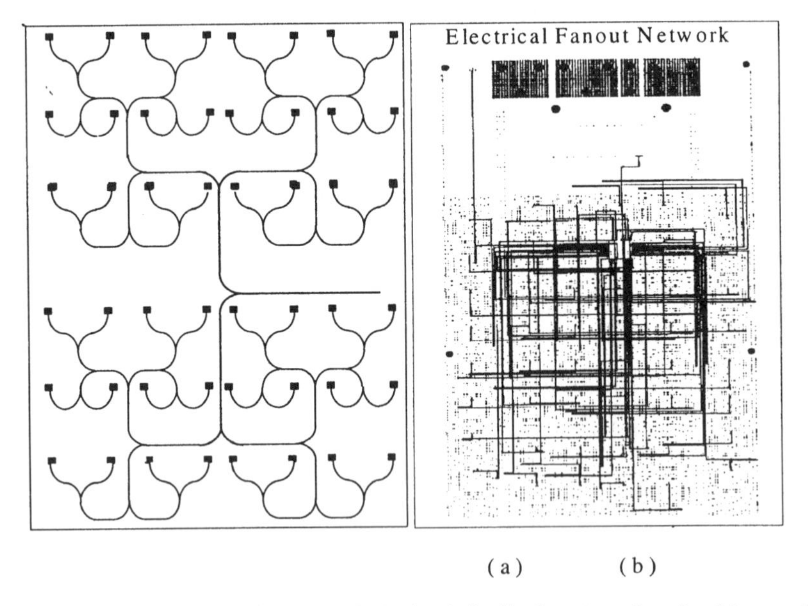

Fig. 6.39. Schematic diagrams of massive clock signal distribution networks using (a) an optical waveguide H-tree and (b) an electrical transmission line network.

tronic interconnection layer is constructed using the building blocks mentioned earlier. Such a guided-wave optoelectronic interconnect network is inserted into the Cray supercomputer boards to become an additional optical interconnection layer among many other electrical interconnection layers. As a result, the future supercomputer system may have a clock speed more than ten times higher than current systems.

Our approach is to construct an additional optoelectronic interconnection layer (OIL) for the high-speed optical clock signal distribution using polymer-based guided-wave devices. The selection of the guided-wave approach is mainly based on system alignment and reliability concerns. Si-CMOS process compatibility and planarization of the OIL are the two major technical concerns. As shown in Fig. 6.39(a), a polymer-based waveguide H-tree system is employed to replace the existing electrical fanout interconnect network shown in Fig. 6.39(b). The optical clock signal delivered by an optical fiber is coupled into the OIL using an input surface-normal waveguide coupler, and distributed throughout the board by the polymer-based channel waveguide network. The distributed optical clock signal at each fanout end will be coupled into a corresponding photodetector by an output surface-normal grating coupler.

The building blocks required to facilitate such an optical H-tree system (Fig. 6.39(a)) include high performance low-loss polymer-based channel waveguides, waveguide couplers, 1-to-2 3 dB waveguide splitters and 90° curved waveguide bends. These components allow the formation of a waveguide H-tree for the required optical clock signal distribution, where all the optical paths have the same length to minimize the clock skew problem. The employment of optical channel waveguides and surface-normal waveguide couplers provides a compact, mechanically reliable system. Due to the massive fanouts (48) over a large area, waveguide propagation loss must be minimized while waveguide grating coupling efficiency has to be maximized. These two factors are very important to ensure enough optical power at the end of the photodetectors for high-speed operation. While fiber-optics technology has been successfully implemented among cabinets as replacements for coaxial cable point-to-point link, its application inside a cabinet on the board-level system is severely limited due to the bulkiness of fibers and fiber connectors and significant labor and cost involved in parallelism of the interconnects. Recent achievements in plastic fiber–based optical interconnects have demonstrated an optical interconnect from a centralized light source to 1-to-many direct fanouts within one board. Several Gbit/sec optical signal transmission has been demonstrated experimentally [41, 42, 43]. While having the advantages of low loss and easy implementation of the optical layer, this type of interconnect has an intrinsic drawback in intraboard optical interconnections. The alignments of laser to fiber and fiber to detector are difficult to achieve. In the context of optical clock signal distribution, alignment of 48 plastic fiber to 48 independently addressed photodetectors cannot be easily achieved. Polymer optical waveguide technology, on the other hand, is particularly suitable for intraboard interconnects applications for its large area waveguide formation, and each lithography layer is precisely aligned. This can be viewed as an optical equivalent of electrical printed wiring board technology in which the fabrication cost is independent of the interconnect functionality and complexity. In order to take advantage of the polymer waveguides, we must be capable of implementing optoelectronic devices such as laser diodes and photodetectors in the same PC board package. The polymer waveguides can be fabricated, integrated, and packaged into the board with the fully embedded architecture shown in Fig. 6.1; the insertion of optical interconnects becomes an acceptable approach to upgrade microelectronics-based high-performance computers.

An optical H-tree fanout network based on polyimide channel waveguides was fabricated for optical clock signal distribution in a Cray multiprocessor supercomputer board. The original design of the Cray T-90 supercomputer was based on an external laser diode modulating at 500 MHz which went through a 1-to-32 waveguide star coupler to provide a system clock to 32 different boards of the supercomputer. At the edge of each board, the optical clock was converted to an electrical clock signal. The 1-to-48 fanouts of all 32 board were realized through delay-equalized electrical transmission lines. The communication distance of these clock lines within each board is as long as 32.6 cm, which makes the further upgrade of the clock speed to GHz level unrealistic. To release such a bottleneck, we demonstrate an optical clock signal distribution using the building blocks mentioned in the previous sections. Polyimide waveguides are employed as the physical layer to bridge the system clock optically to the chip level. As a result, the distance of electrical interconnection to realize the system clock signal distribution at the GHz level will be minimized at the chip level instead of board level. The H-tree structure is selected to equalize the propagation delays of all 48 fanouts. Due to the relatively short interconnection distance, the waveguide is multimode with a cross section of 50 μm wide and 10 μm deep. The horizontal dimension of the waveguide is to match the 50-μm multimode glass fiber, and the vertical dimension can be matched with a three-dimensional tapered waveguide [13]. Both tilted grating couplers and 45° TIR mirror couplers are fabricated to efficiently couple light in and out of the H-tree waveguide structure. A 45° TIR mirror provides better coupling efficiency and a shorter interaction length to fulfill such a coupling. Figure 6.40(a) shows the broadcasting of the optical signal through the H-tree structure at 632.8 nm. Forty-eight well defined light spots are coming out surface-normally. Figure 6.40(b) uses a fabricated 1-to-48 H-tree waveguide structure operating at 850 nm. All 48 surface-normal fanouts are provided through 45° TIR waveguide mirrors. Based on the loss measurement, a VCSEL with enough modulated power is needed to compensate -2 dB coupling loss, -6 dB waveguide propagation loss, -17 dB fanout loss, -3 dB bending loss, and -3 dB power margin. The H-tree waveguide structure is

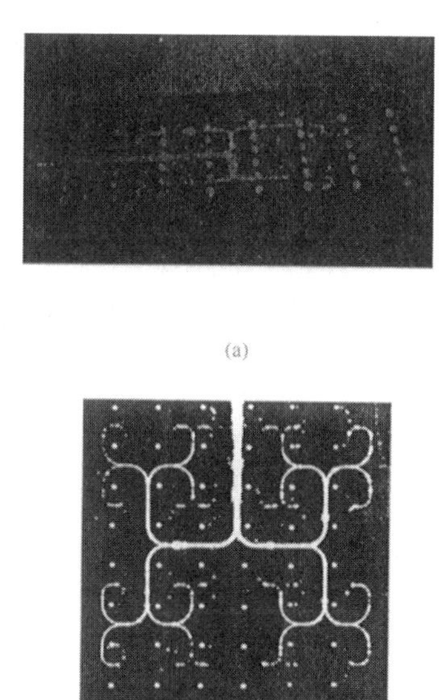

(a)

(b)

Fig. 6.40. Polymer waveguide H-tree showing optical broadcasting at (a) 632.8 nm and (b) 850 nm.

employed to equalize the propagation delays of different locations where Si-based clock detectors were located. The process of investigating the feasibility of further reducing the bending losses by using curved waveguides, which have lower bending losses than the present structure, is being studied.

To ensure the polyimide waveguide material can function as the physical layers of optical bus and also the interlayer dielectric, a multilayer test structure is made. Three electrical interconnection layers and three metal layers are formed. Each polyimide layer is 37 μm thick and metalized vias are formed to provide vertical interconnect involving different layers. A cutaway perspective of the MCM structure and a micrograph of the MCM cross section showing multilayer polyimide thin-film dielectrics with metalized electrical vias is shown in Fig. 6.41, where the planarized electrical interconnection is shown in white. The bottom layer is the PC board.

6.7. POLYMER WAVEGUIDE-BASED OPTICAL BUS STRUCTURE

Unlike the optical interconnect, the electrical interconnect on the board level has two serious problems that significantly limit the data transfer speed. They are signal propagation time delay and skew between parallel bus links.

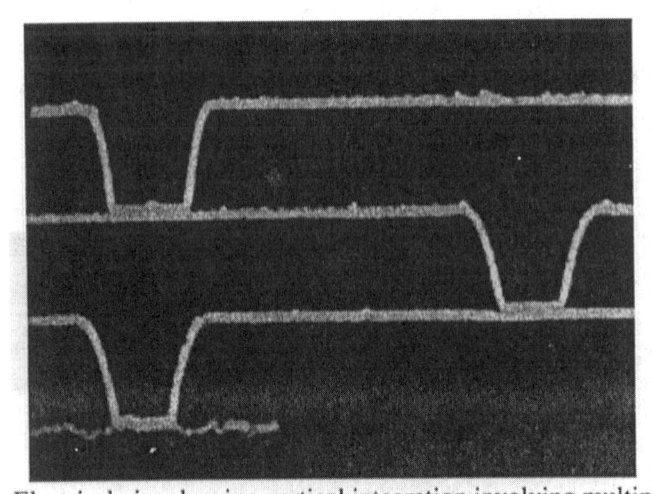

Fig. 6.41. Electrical vias showing vertical integration involving multiple polyimide layers.

These problems become more significant when the linear dimension of a board increases.

Propagation time does not affect the maximum data rate of an uncompelled asynchronous block transfer (such as the source-synchronous block transfer [SSBLT] proposed for addition to the VME bus standard). However, it does limit other types of bus transactions: address transfers, handshake single-word transfers, bus contention, and so on. Estimates have been made by Sweazey [44] of the sustained throughput; i.e., the data transfer rate averaged over a time that is long compared to the duration of a single transaction. Assuming that bus overhead is 200 ns per read and 100 ns per write operation, and assuming reads outnumber writes by 2 to 1, Sweazey calculated the sustained throughput as a function of block transfer speed (burst speed) and of the number of bytes per transfer. For 64-byte transfers, the calculated sustained throughput is 196 MB/sec for a burst rate of 400 MB/sec, and 384 MB/sec for infinitely rapid block transfers. The propagation speed for the electronic bus is at present greatest for backplane-transceiver logic (BTL) backplanes such as FutureBus: about 0.18 c (c is the speed of light in vacuum), giving a 15 ns round-trip time for a 40 cm backplane. This cannot decrease by much, since it is based on the extremely low driver capacitance of 5 pf/driver provided by BTL.

Uncompelled block transfers are limited by bus line skew. The principal cause of this is speed variations (time jitter) between transceiver chips. This jitter is at least 5 ns, even for a well-designed set of transceivers. This means that there will be a total skew between data lines and strobes of up to 20 ns from transmission to receiver latching. In addition, there may be skew in the transmission lines themselves, due to unequal capacitive loading, unequal distances to ac grounds, or for some other reason. (FutureBus transmission lines are purposely skewed to ensure that data arrive before strobe.) These skews limit the attainable transfer rate to 40 mega transfers/sec or 160 MB/sec for a 32-bit bus. Electronic bus lines are not typically terminated in matched impedance, since this would require the drive currents to be too high. Therefore, the bus line will not settle until all end reflections have subsided (several round-trip times). By contrast, polymer bus lines may be terminated in antireflection coatings, suppressing end reflections and reducing settling time to zero.

6.7.1. OPTICAL EQUIVALENT FOR ELECTRONIC BUS LOGIC DESIGN

Before discussing an optical backplane design in its entirety, we must present optical equivalents of necessary bus components, such as bidirectional transmission lines, stubs, transmitters, and receivers. Further, optical equivalents of line voltage, logic levels, and open-collector and tristate line driving and receiving must be derived. Once these issues are resolved, the way will be clear for defining an optical bus that is fully compatible with existing IEEE–standardized bus protocol.

The optical equivalent of a PC board trace is a polymer-based optical waveguide. A very important consequence is the ability to provide modulation using VCSELs and demodulation using photoreceivers for the same board where the polymer waveguide is located. An unloaded (no boards attached) PC board trace has a typical signal propagation speed on the order of 0.6 c. The speed drops to below 0.2 c for a fully loaded bus line. The polymer, which forms the optical waveguide, has an index of refraction $n \cong 1.5$. The optical signal propagation speed is $c/n \cong 0.67$ c, similar to that of the unloaded electronic bus line.

It is important to note, however, that there is no optical analogue to driver capacitance from attached boards, which causes loading of electronic bus lines. Therefore, the optical signal speed retains the same high value regardless of the presence or absence of line drivers in the system. This means that the optical bus round-trip delay time will be lower by a factor of 3 than that of the electronic bus. A connection to an electronic bus line takes the form of a stub or tee junction in the PC board trace; usually, such a stub connects to a line transceiver. The optical equivalent of a stub is high-efficiency waveguide couplers, such as tilted gratings or 45° TIR waveguide mirrors, which allow light from a second waveguide to be coupled into the optical bus line, and low-efficiency coupling, used to couple light out of the bus for detection. The

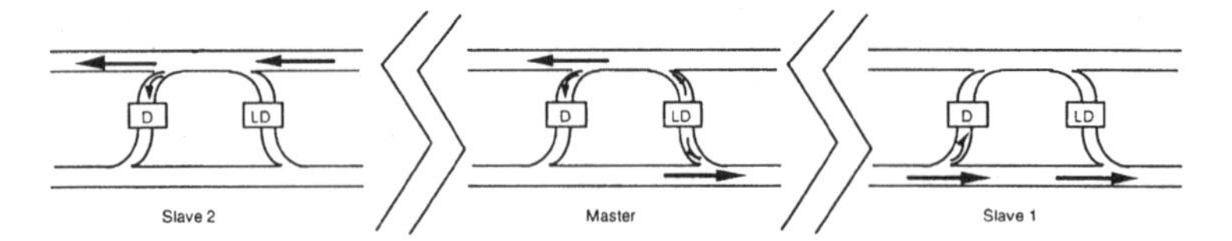

Fig. 6.42. The optical equivalent of a bidirectional electronic bus line driven by open-collector drivers (3-D coupling involving TIR holograms is not shown). Communication between one master (e.g., processor) and two slaves (e.g., memories) is clearly indicated (boards are not shown). Commands from the master to slaves 1 and 2 are carried out using the bottom and the top polymer bus lines, respectively. In this specific scenario, the master is broadcasting signals that are received by slaves 1 and 2, and the high power margin of the operation is preserved.

key feature of the couplers is that while light is injected from the stub with high efficiency, light propagating in the bus line and passing the coupling is almost unaffected by it ($<1\%$ fanout). This is not the case, however, for light propagating in the other direction, which suffers high losses at the coupler.

Because of this, an optical waveguide with stubs attached is necessarily unidirectional. The optical equivalent of an electronic bus line thus involves two parallel optical waveguides, each carrying light in the opposite direction as depicted in Fig. 6.42. Optical waveguide signals can be detected at much lower levels than the level of transmission. For instance, a VCSEL can easily provide 5 mW modulated power, while a photodetector (e.g., a p–i–n diode) can detect a 5 μW signal at 5 Gbit/sec. This implies high fanout capability; i.e., many receivers can be connected using low-efficiency couplings to a bus line driven by one transmitter. Figure 6.42 shows the optical equivalent of a single bidirectional electronic bus line. The drive current provided by each electronic transceiver powers the corresponding laser diode, the output of which is split and injected into both waveguides. Each photodiode detects light from either waveguide, since the low-efficiency couplings lead to waveguide segments which are merged with a unidirectional coupler. Each photodiode current powers the corresponding electronic receiver.

The scheme in Fig. 6.42 may be considered fully equivalent to an electronic bus line driven by open-collector drivers, terminated in pull-up resistors, if the following identification is made:

- The state in which no light is present on either waveguide (no laser diode is operating) corresponds to the unasserted electronic line which is pulled high by the pull-up resistors; and
- The state in which there is light in both waveguides (one or more laser diodes are operating) corresponds to the asserted (low-level) electronic line.

Note that there is no optical effect corresponding to the wire–OR glitch.

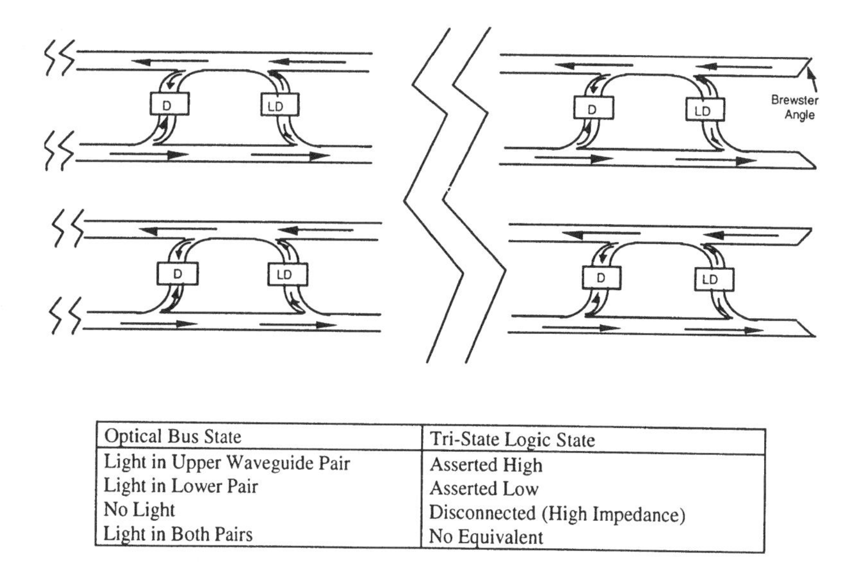

Optical Bus State	Tri-State Logic State
Light in Upper Waveguide Pair	Asserted High
Light in Lower Pair	Asserted Low
No Light	Disconnected (High Impedance)
Light in Both Pairs	No Equivalent

Fig. 6.43. The optical equivalent of tristate bidirectional electronic bus line.

When two diodes are on, and one turns off, every detector continues to receive light from the other as long as it remains on. This is because the optical powers from two diodes simply add in the waveguide; any detected nonzero power (above a characteristic noise threshold) corresponds to the asserted line state.

The analogue to resistive transmission line termination is treatment of the ends of the optical waveguides on a PC board so that the reflected power is zero. This is done by implementing an antireflection coating or by fabricating the waveguide ends at the Brewster angle, with an absorptive beam dump outside the waveguides. If this optical isolation is done, settling-time effects are removed. Note that the optical bus line just described has exactly two states: light present (low) and light absent (high). This suffices, as just described, to represent a two-state open-collector–driven line exactly. However, it is insufficient to represent a tristate–driven line: a tristate driver has an asserted-high state, which is distinguishable from the disconnected state. Where the tristate line must be exactly emulated, the corresponding optical bus line can consist of a pair of the lines just described, that is, four optical waveguides, with a separate laser diode and a separate photodiode for each pair (see Fig. 6.43). In this case, the state with no light in any waveguide represents the disconnected state (no device asserting) as before, while the state with light present in the top pair (for example) indicates the asserted-high state, and that with light in the lower pair, the asserted-low state. Note that this scheme overrepresents the tristate line, as there is a fourth state with light in both pairs of waveguides. The optoelectronic circuit can be integrated into a fully embedded optical interconnection layer, as shown in Fig. 6.1. Each bus line consists of two optical waveguides, one OIC per plug-in board, couplings and stub waveguides, and power and ground lines (metal traces) to power the OICs.

6.8. SUMMARY

This chapter described necessary building blocks of the board-level guided wave optical interconnection. All elements involved in providing high-speed optical communications within one board were demonstrated. These include a vertical cavity-surface emitting laser (VCSEL), surface-normal waveguide couplers, a polyimide-based channel waveguide functioning as the physical layer of optical bus, and a photoreceiver. The driving electrical signal to modulate the VCSEL and the demodulated signal received at the photoreceiver can be applied through electrical vias connecting to the surface of the PC board. In such an approach, all the areas of the PC board surface are occupied by electronics and therefore one only observes performance enhancement due to the employment of optical interconnections but does not worry about the interface problem between electronic and optoelectronic components, unlike conventional approaches.

A 1-to-48 optical clock signal distribution network for a supercomputer board was described. Further experimental results on a 12-channel linear array of thin-film polyimide waveguides, VCSELs (42 μm, can be made as thin as 8 μm), and silicon MSM photodetectors (10 μm) suitable for a fully embedded implementation are provided. Two types of waveguide couplers, tilted gratings and 45° total internal reflection (TIR) mirrors, are fabricated within the polyimide waveguides. A waveguide bus architecture was presented which provides bidirectional broadcasting transmission of optical signals. Such a structure is equivalent to such IEEE–standardized bus protocols as VME bus and FutureBus.

ACKNOWLEDGMENTS

Authors will like to acknowledge the research contributions from the Lei lin, Chulchae choi, Yujie Liu, Linghui Wu, M. Dubinovski, Feming Li, Dr. Bing Li, and Dr. Suning Tang. We also acknowledge the research funds provided by Radiant Photonics, ONR, BMDO, DARPA, AFRL, the ATP program of the State of Texas, 3M Foundation, Dell Computer, Cray Research, GE, Honeywell and MCC.

REFERENCES

6.1 See, for example, many papers in *The Critical Review in Optoelectronic Interconnects and Packaging*, Ray Chen and Pete Guilfoyle, Eds., 1996, Vol. CR-62.

6.2 D. P. Seraphim and D. E. Barr, 1990, "Interconnect and Packaging Technology in the 90's," *Proc. SPIE*, Vol. 2. 1390, pp. 39–54.

6.3 ■ Neugerbauer, R. O. Carlson, R. A. Fillion, and T. R. Haller, "Multichip Module Designs for High-Performance Applications," in *Multichip Modules, Compendium of 1989 Papers*, pp. 149–163, International Electronic Packaging Society, ■, 1989.

6.4 Steve Heath, Chapter 3, *VMEbus User's Handbook*, CRC Press, Inc., Florida, 1989.

6.5 Robert Haveman, "Scaling and Integration Challenges for Cu/low *k* Dielectrics," workshop on processing for ULSI: transistors to interconnects, April 22, 1999, Austin Texas.

6.6 M. R. Feldman, 1990, "Holographic Optical Interconnects for Multichip Modules," *Proceedings SPIE*, vol. 1390, pp. 427–433.

6.7 Michael Feldman, Iwona Turlik, and G. M. Adema, "Microelectronic Module Having Optical and Electrical Interconnects," US Patent, No. 5638469, 1997.

6.8 D. Plant, B. R. Robertson, H. S. Hinton, G. C. Boisset, N. H. Kim, Y. S. Liu, M. R. Otazo, D. R. Rolston, A. Z. Shang, and W. M. Robertson, 1995, "Microchannel–Based Optical Backplane Demonstrators Using FET-SEED Smart Pixel Array," *Proceedings of SPIE*, Vol. 2400, pp. 170–174, Canada, June 22–24, 1997.

6.9 C. Tocci and J. Caufield, *Optical Interconnects, Foundation and Applications*, Artech House, Boston, 1994.

6.10 Scott Hinton, 1993, "Progress in Systems Based on Free Space Optical Interconnection," *SPIE* Vol. 1849, pp. 2–3.

6.11 David Plant, B. Robertson, and Scott Hinton, "Optical, Optomechanical, and Optoelectronic Design and Operational Testing of a Multistage Optical Backplane Demonstration System," IEEE Third International Conference on Massively Parallel Processing Using Optical Interconnections, Maui, Hawaii, 1996.

6.12 Ray T. Chen, Suning Tang, T. Jannson, and J. Jannson, 1993, "A 45-cm–long Compression-Molded Polymer-Based Optical Bus," *Appl. Phys. Lett.* vol. 63, pp. 1032–1034.

6.13 Linghui Wu, Feiming Li, Suning Tang, Bipin Bihari, and Ray T. Chen, 1997, "Compression-Molded Three-Dimensional Tapered Polymeric Waveguides for Low-Loss Optoelectronic Packaging" *IEEE Photon. Technol. Lett.*, Vol. 9, No. 12, pp. 1601–1603.

6.14 Suning Tang, Ting Li, F. Li, Michael Dubinovsky, Randy Wickman, and Ray T. Chen, 1996, "Board-Level Optical Clock Signal Distribution Based on Guided-Wave Optical Interconnects in Conjunction with Waveguide Hologram," *Proc. SPIE*, vol. 2891, pp. 111–117.

6.15 Linghui Wu, Feiming Li, Suning Tang, Bipin Bihari, and Ray T. Chen, 1997, "Compression-Molded Three-Dimensional Tapered Polymeric Waveguides for Low-Loss Optoelectronic Packaging," *IEEE Photonics Technol. Lett.*, vol. 9, no. 12, pp. 1601–1603.

6.16 Toshiaki Suhara and Hiroshi Nishihara, 1996, "Integrated Optics Components and Devices Using Periodic Structures," *IEEE J. of Quantum Electronics*, vol. QE-22, no. 6, pp. 845–867.

6.17 D. Brundrett, E. Glystis, and T. Gaylord, 1994, "Homogeneous Layer Models for High-Spatial-Frequency Dielectric Surface-Relief Gratings," *Appl. Opt.*, vol. 33, no. 13, pp. 2695–2760.

6.18 D. Y. Kim, S. K. Tripathy, Lian Li, and J. Kumar, 1995, "Laser-Induced Holographic Surface-Relief Gratings on Nonlinear Optical Polymer Films," *Appl. Phys. Lett.*, vol. 66, no. 10, pp. 1166–1168.

6.19 Ray T. Chen, Feiming Li, Micheal Dubinovsky, and Oleg Ershov, 1996, "Si-Based Surface-Relief Polygonal Gratings for 1-to-many Wafer-Scale Optical Clock Signal Distribution," *IEEE Photonics Technol. Lett.*, vol. 8, no. 8.

6.20 R. K. Kostuk, J. W. Goodman, and L. Hesselink, 1987, "Design Considerations for Holographic Optical Interconnects," *Appl. Opt.*, vol. 26, pp. 3947–3953.

6.21 G. D. Boyd, L. A. Coldren, and F. G. Storz, 1980, "Directional Reactive ion Etching at Oblique Angles," *Appl. Phys. Lett.*, vol. 36, no. 7, pp. 583–585.

6.22 Jianhua Gan, Linghui Wu, Hongfa Luan, Bipin Bihari, and Ray T. Chen, 1999, "Two-Dimensional 45° Surface-Normal Microcoupler Array for Guided-Wave Optical Clock Distribution," *IEEE Phot. Tech. Lett*, vol. 11, pp. 1452–1454.

6.23 M. Sypek, 1995, "Light Propagation in the Fresnel Region: New Numerical Approach," *Optics Communications*, vol. 116, 43–48.

6.24 J. W. Goodman, *Introduction to Fourier Optics*, 2nd edition, pp. 63–89, McGraw-Hill Publishers, New York, 1996.

6.25 M. Saiful Islam, 1999, "Distributed Balanced Photodetectors for High-Performance RF Photonic Links," *IEEE Phot. Tech. Letts.*, vol. 11, no. 4, pp. 457–459.

6.26 Bing Li, Suning Tang, Ninhua Jiang, Zan Shi, and Ray T. Chen, 2000, "Switching Characteristic of Wide-Band MSM and PIN Photodetectors for Photonic-Phased Array Antenna," *Proc. SPIE*, 3952-A13.

6.27 O. Wada, *Optoelectronic Integration: Physics, Technology, and Applications*, Norwell, Massachusetts, Kluwer Academic, 1994.

6.28 Y. Sasaki, T. Katayama, T. Koishi, K. Shibahara, S. Yokoyama, S. Miyazaki, and M. Hirose, 1999, "High-Speed GaAs Epitaxial Liftoff and Bonding with High Alignment Accuracy Using a Sapphire Plate," *J. Electrochem. Soc.*, 146(2), pp. 710–712.

6.29 B. D. Dingle, M. B. Spitzer, R. W. McClelland, J. C. C. Fan, and P. M. Zavracky, 1993, "Monolithic Integration of a Light Emitting Diode Array and a Silicon Circuit Using Transfer Process," *Appl. Phys. Lett.* 62(22), pp. 2760–2762.

6.30 E. Yablonovitch, T. Sands, D. M. Hwang, I. Schnitzer, T. J. Bmitter, S. K. Shastry, D. S. Hill, and J. C. C. Fan, 1991, "Van der Walls Bonding of GaAs on Pd Leads to a Permanent, Solid-Phase-Topotaxial Metallurgical Bond," *Appl. Phys. Lett.* 59(24), pp. 3159–3161.

6.31 J. W. Goodman, F. I. Leonberger, S. Y. Kung, and R. A. Athale, 1984, "Optical Interconnections for VLSI systems," *Proc. IEEE* 72, pp. 850–866.

6.32 M. R. Feldman, S. C. Esener, C. C. Guest, and S. H. Lee, 1988, "Comparison between Optical and Electrical Interconnects Based on Power and Speed Considerations," *Appl. Opt.*, vol. 27, pp. 1742–1751.

6.33 Founad E. Kiamilev, Philippe Marchand, Ashok V. Krishnamoorthy, Sadik C. Esener, and Sing H. Lee, 1993, "Performance Comparison between Optoelectronic and VLSI Multistage Interconnects Networks," *IEEE J. of Light. Technol.*, vol. 9, pp. 1674–1692.

6.34 Paola Cinato and Kenneth C. Young, Jr., 1993, "Optical Interconnections within Multichip Modules," *Opt. Eng.*, vol. 32, pp. 852–860.

6.35 Bradley D. Clymer and Joseph W. Goodman, 1986, "Optical Clock Distribution to Silicon Chips," *Opt. Eng.*, vol. 25, pp. 1103–1108.

6.36 Suning Tang, Ray T. Chen, and Mark Peskin, 1994, "Packing Density and Interconnection Length of a Highly Parallel Optical Interconnect Using Polymer-Based Single Mode Bus Arrays," *Opt. Eng.*, vol. 33, pp. 1581–1586.

6.37 Ray T. Chen, 1993, "Polymer-Based Photonic Integrated Circuits," *Opt. Laser Tech.*, vol. 25, pp. 347–365.

6.38 Ray. T. Chen, H. Lu, D. Robinson, Michael Wang, Gajendra Savant, and Tomasz Jannson, 1992, "Guided-Wave Planar Optical Interconnects Using Highly Multiplexed Polymer Waveguide Holograms," *IEEE J. Light. Technol.*, vol. 10, pp. 888–897.

6.39 R. W. Wickman, 1996, "Implementation of Optical Interconnects in GigaRing Supercomputer Channel," *Proc. SPIE*, vol. CR62, pp. 343–356.

6.40 Suning Tang and Ray T. Chen, 1994, "1-to-42 Optoelectronic Interconnection for Intramultichip-Module Clock Signal Distribution," *Appl. Phys. Lett.*, vol. 64, pp. 2931–2933.

6.41 Y. Li, J. Popelek, L. J. Wang, Y. Takiguchi, T. Wang, and K. Shum "Clock Delivery Using Laminated Polymer Fiber Circuits," *J. Opt. A: Pure and Appl. Opt.*, vol. 1, pp. 239–243.
6.42 Y. Li, T. Wang, J.-K. Rhee, and L. J. Wang, 1998, "Multigigabits per Second Board-Level Clock Distribution Schemes Using Laminated End-Tapered Fiber Bundles," *IEEE Photonics Tech. Lett.*, vol. 10, pp. 884–886.
6.43 Y. Li and T. Wang, 1996, "Distribution of Light Power and Optical Signals Using Embedded Mirrors Inside Polymer Optical Fibers," *IEEE Photonics Tech. Lett.*, vol. 8, pp. 1352–1354.
6.44 P. Sweazey, "Limits of Performance of Backplane Buses," in *Digital Bus Handbook*, J. De Giacomo, Ed., McGraw-Hill, New York, 1990.
6.45 Ray T. Chen, 1994, "VME Optical Backplane Bus for High Performance Computer," *Jpn. J. Optoelec. Dev. Tech.*, vol.9, pp. 81–94.

EXERCISES

6.1 Explain why single mode waveguides can support high-speed optical signals over longer length as compared to multimode waveguides.

6.2 What are the factors limiting the transmission speed in an optoelectronic interconnect? In your opinion, which are most crucial and why?

6.3 What is the phase-matching condition in the context of waveguide grating couplers?
   (a) Show the phase matching in grating couplers with tilted grating profile.
   (b) What do you expect if the gratings are not tilted?

6.4 The figure below shows the cross section of a channel waveguide with a 45° micromirror coupler, which can surface-normally couple the light from the VCSEL underneath into the waveguide. The refractive indices of the bottom cladding layer, the waveguide layer, and the top cladding layer are n_1, n_2, and n_3, respectively. The thickness of each layer is shown in the figure. Calculate the reflectivity of the 45° micro-mirror. Assume the 45° tilted surface is perfectly smooth and uniform, the mirror area covers the aperture of the VCSEL, and the divergence is negligible.

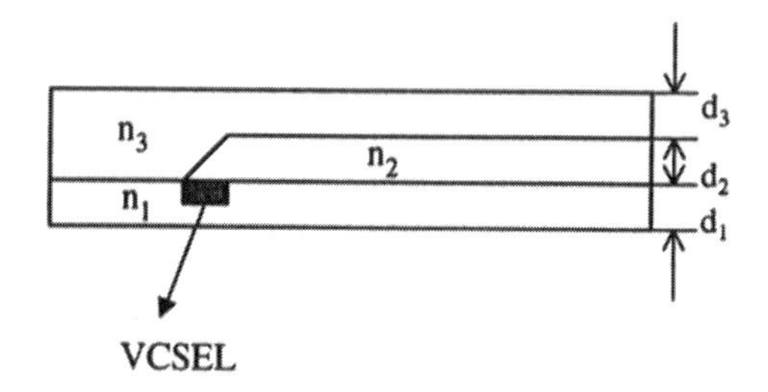

6.5 In the question above assume $n_1 = n_3 = 1.5$, and $n_2 = 1.58$. Estimate the optimum mirror angle for coupling.

6.6 The figure below shows a structure of a multilayer optical interconnect. The output power of the VCSEL is p_0, the power of the light coupled into the waveguide n_1 through the 45° mirror m_1 is p_1, and p_2 is the light power coupled into waveguide n_2 through mirror m_2. The refractive index and thickness of each layer are shown in the figure. Design a structure so that $p_1/p_0 = 20\%$ and $p_2/p_0 = 50\%$.

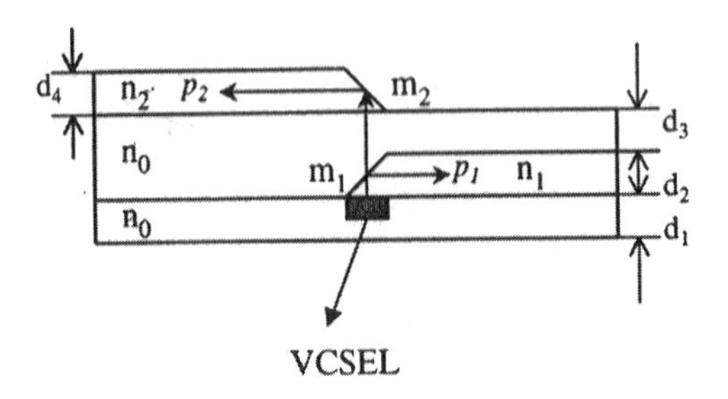

6.7 Find the atomic density for the zinc-blende silicon. The crystal lattice of silicon is the face-centered cubic (FCC) illustrated in the figure below. The lattice constant of silicon at room temperature is 5.43 Å. Though different sizes and colors are used for the sphere for illustration purposes, they actually represent the same kind of silicon atoms.

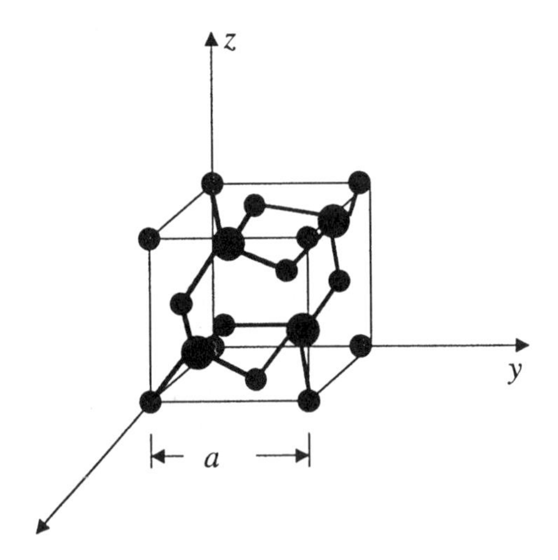

6.8 Find the surface atomic density for the (100) plane of zinc-blende silicon (illustrated below).

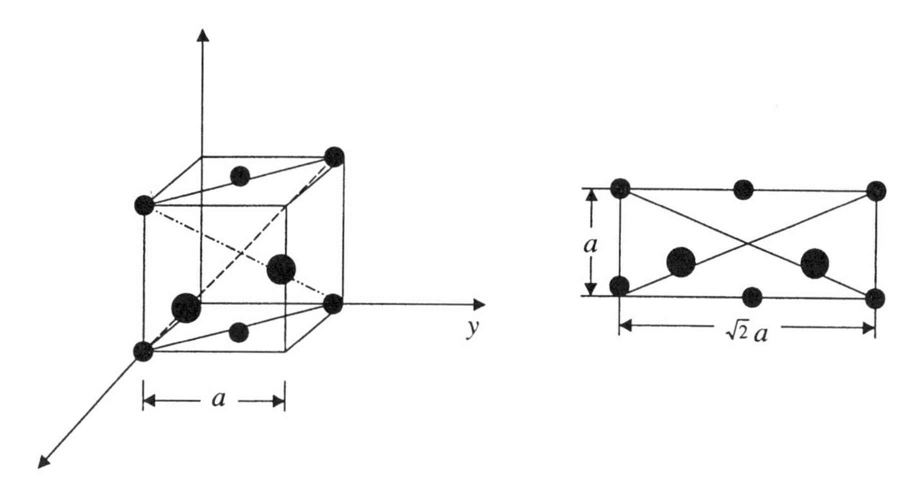

6.9 A Ge detector is to be used for an optical communication system using a GaAs laser with emission energy of 1.43 eV. Calculate the depth of the detector needed to be able to absorb 90% of the optical signal entering the detector. The absorption coefficient α is given to be $2.5 \times 10^4\ \text{cm}^{-1}$.

6.10 An optical intensity of $10\ \text{W/cm}^2$ at a wavelength of $0.75\ \mu\text{m}$ is incident on a GaAs detector. Calculate the rate at which electron–hole pairs will be produced at this intensity at 300 K. If the electron–hole combination time is 10^{-9} s, calculate the excess carrier density. The absorption coefficient of GaAs at $0.75\ \mu\text{m}$ is $7 \times 10^3\ \text{cm}^{-1}$. $0.75\ \mu\text{m}$ wavelength is equivalent to a photon of 1.65 eV.

6.11 Consider a long silicon p–n junction that is reverse biased with a reverse-bias voltage 2 V. The diode has the following parameters (all at 300 K):

Diode area,	$A = 10^4\ \mu\text{m}^2$
p-side doping	$N_a = 2 \times 10^{16}\ \text{cm}^{-3}$
n-side doping	$N_d = 10^{16}\ \text{cm}^{-3}$
Electron diffusion coefficient	$D_n = 20\ \text{cm}^2/\text{s}$
Hole diffusion coefficient	$D_p = 12\ \text{cm}^2/\text{s}$
Electron minority carrier lifetime	$\tau_n = 10^{-8}\ \text{s}$
Hole minority carrier lifetime	$\tau_p = 10^{-6}\ \text{s}$
Electron–hole pair generation rate	$G_L = 10^{22}\ \text{cm}^{-3}\ \text{s}^{-1}$

Calculate the photocurrent.

6.12 Consider a typical avalanche photodiode with the following parameters:

Incident optical power	$P^*_{op}A = 50\,\text{mW}$
Efficiency	$\eta_{det} = 90\%$
Optical frequency	$\nu = 4.5 \times 10^{14}$ Hz
Breakdown voltage	$V_B = 35$ V
Diode voltage	$V = 34$ V
Dark current	$I_0 = 10\,\text{nA}$
Parameter n' for the multiplication	$= 2$

Assume that the series resistance is negligible. Calculate:

(a) Multiplication factor.
(b) Photoflux.
(c) Photocurrent.

Chapter 7 | Pattern Recognition with Optics

FRANCIS T. S. YU
PENNSYLVANIA STATE UNIVERSITY

The roots of optical pattern recognition can be traced back to Abbe's work* in the 1870s, when he developed a method that led to the discovery of spatial filtering to improve the resolution of microscopes. However, optical pattern recognition was not actually appreciated until the complex spatial filtering work of Vander Lugt† in the 1960s. Since then, techniques, architectures, and algorithms have been developed to construct efficient optical systems for pattern recognition.

Basically, however, there are two approaches in the optical implementation of pattern recognition the correlation approach and the neural net approach. In the correlation approach, there are two frequently used architectures: the Vander Lugt correlator (VLC) and the joint-transform correlator (JTC), discussed briefly in Chapter 2. In this chapter, we address some of the basic architectures, techniques, and algorithms that have been applied to pattern recognition. The pros and cons of each approach will be discussed. Because of recent technical advances in interface devices (such as electronically addressable SLMs, nonlinear optical devices, etc.), new philosophies and new algorithms have been developed for the design of better pattern recognition systems.

* Reference 1.
** Reference 2.

7.1. BASIC ARCHITECTURES

7.1.1. CORRELATORS

In terms of correlation detection, optical implementation for pattern recognition can be accomplished either by using Fourier-domain complex matched filtering or spatial-domain filtering. Correlators that use Fourier-domain matched filtering are commonly known as VLCs contrast to spatial-domain filtering in the JTC. The basic distinction between them is that the VLC depends on Fourier-domain spatial filter synthesis (e.g., Fourier hologram), whereas the JTC depends on spatial-domain (impulse-response) filter synthesis. In other words, the complex spatial detection of the Vander Lugt arrangement is input scene *independent*, while the joint-transform method is input scene *dependent*. The basic optical setup of these two types of correlators is depicted in Figs. 7.1 and 7.2, as repeated here from Chapter 2 for convenience. A prefabricated Fourier-domain matched filter $H(p, q)$ is needed in the VLC, whereas a matched filter is not required in the JTC but a spatial-domain impulse response $h(x, y)$ is needed. Although the JTC avoids spatial filter synthesis problems, it generally suffers from lower detection efficiency, particularly when applied to multitarget recognition or targets imbedded in intense background noise. Nonetheless, the JTC has many merits, particularly when interfaced with electronically addressable SLMs.

The JTC has other advantages, such as higher space–bandwidth product, lower carrier frequency, higher index modulation, and suitability for real-time implementation. Additional disadvantages include inefficient use of illuminating light, larger transform lens, stringent spatial coherence requirements, and the small size of the joint transform spectrum. Nonetheless, these shortcomings can be overcome by using hybrid architectures, as will be discussed later.

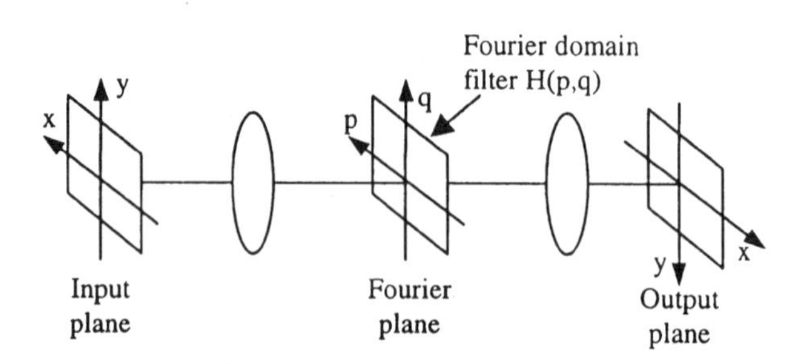

Fig. 7.1. A Vander Lugt correlator (VLC).

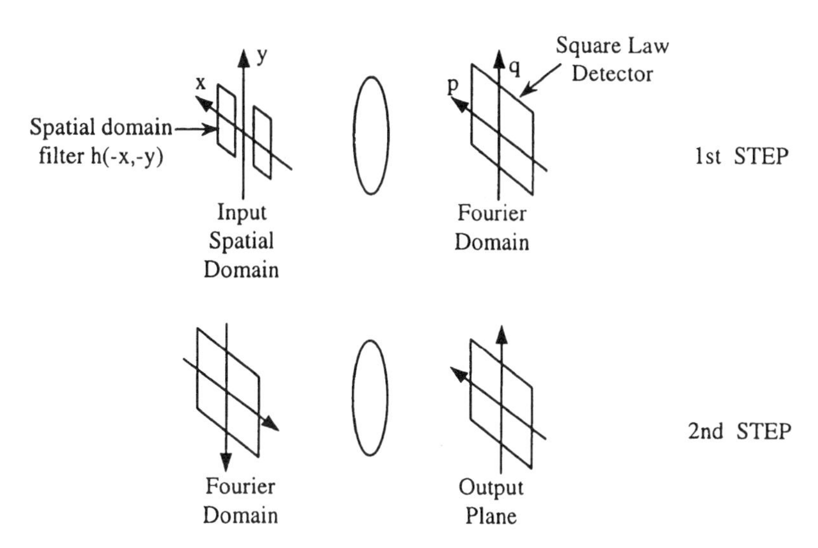

Fig. 7.2. A joint-transform correlator (JTC).

7.1.2. NEURAL NETWORKS

Digital computers can solve some classes of computational problems more efficiently than the human brain. However, for cognitive tasks, such as pattern recognition, a 3-year-old child can perform better than a computer. This task of pattern recognition is still beyond the reach of modern digital computers.

There has been much interest in the implementation of associative memory using optics, much of it centered around the implementation of neural networks (NNs). The associative memory is a process in which the presence of a complete or partial input pattern directly results in a predetermined output pattern. The neural network shown in Fig. 7.3 is one of the possible implementations for associative memory. A neural network attempts to mimic the real structure and function of neurons, as shown in Fig. 7.4. However, it is not necessary to exactly mimic a human brain in order to design a special-purpose cognitive machine, just as it is not necessary to exactly imitate a bird in order to make a flying machine. In view of the associative-based neuron model of Fig. 7.4, it is, in fact, a matrix–vector multiplier for which a single-layer NN can be implemented by optics, as will be shown later.

7.1.3. HYBRID OPTICAL ARCHITECTURES

It is apparent that a purely optical system has drawbacks which make certain tasks difficult or impossible to perform. The first problem is that optical

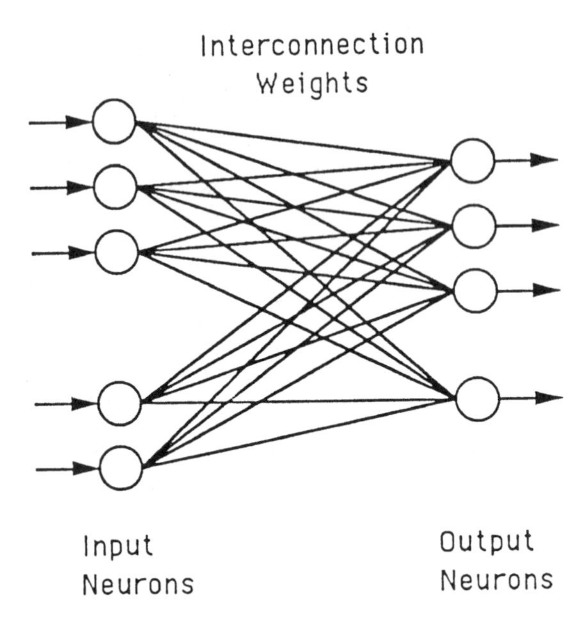

Fig. 7.3. A single-layer fully interconnected neural network.

systems are difficult to program, in the sense of programming general-purpose digital computers. A purely optical system can be designed to perform specific tasks (analogous to a hard-wired electronic computer), but cannot be used when more flexibility is required. A second problem is that a system based on Fourier optics is naturally analog, which often makes great accuracy difficult to achieve. A third problem is that optical systems by themselves cannot easily

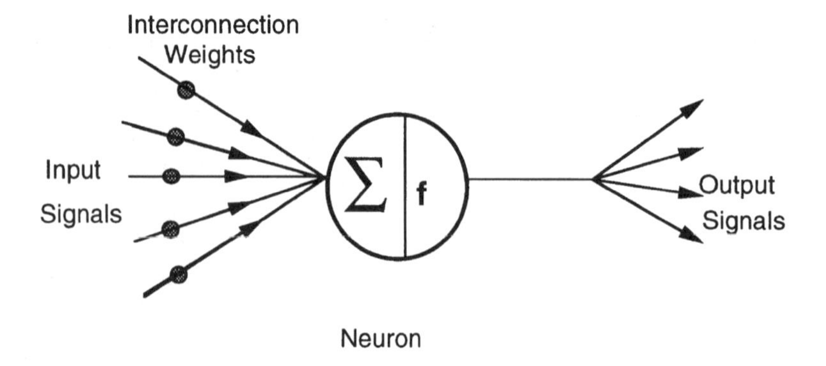

Fig. 7.4. A mathematical model of a neuron.

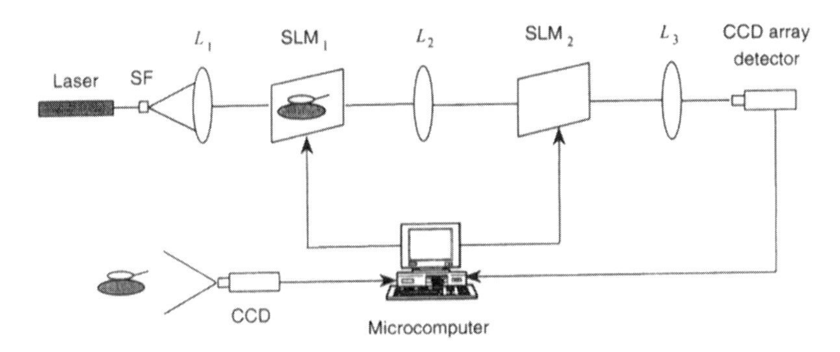

Fig. 7.5. A hybrid–optical VLC.

be used to make decisions. Even the simplest type of decision making is based on the comparison of an output with a stored value. Such an operation cannot be performed optically without the intervention of electronics.

Many deficiencies of optical systems happen to be strong points in their electronic counterparts. For instance, accuracy, controllability, and programmability are some obvious traits of digital computers. Thus, the idea of combining an optical system with its electronic counterpart is rather natural as a means of applying the rapid processing speed and parallelism of optics to a wider range of applications.

We will show a few commonly used hybrid–optical architectures, as applied to pattern recognition. For convenient discussion, these architectures are repeated in Figs. 7.5, 7.6, and 7.7, respectively. In view of these figures, we see that optical–electronic interfacing devices such as SLMs and CCD cameras are used. Remember, the operations of VLC and JTC are basically the same. There are, however, some major distinctions between them. For example, the spatial filter synthesis (Fourier hologram) in the VLC is *independent* of the input scene,

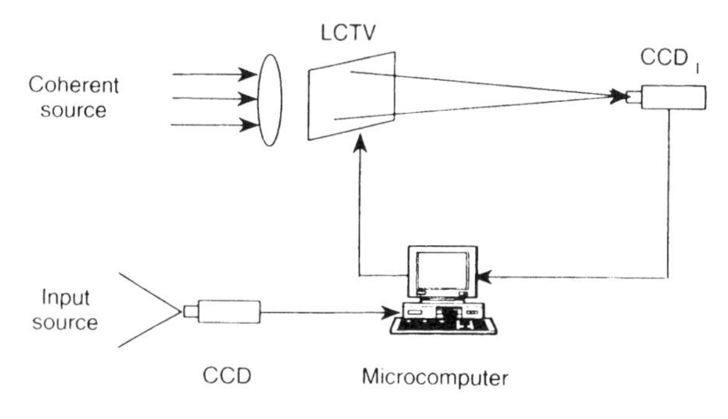

Fig. 7.6. A hybrid JTC.

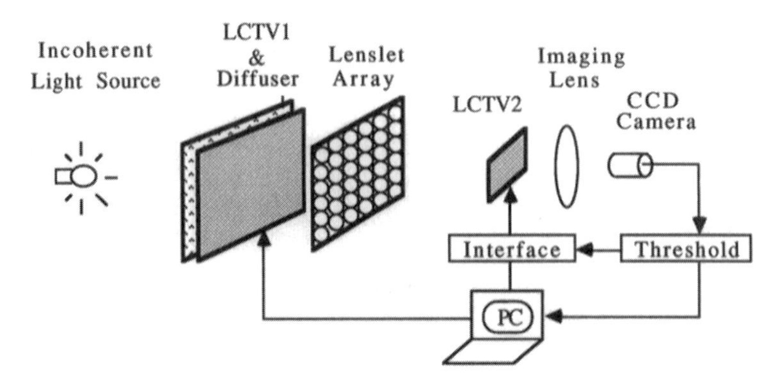

Fig. 7.7. An LCTV optical neural network.

whereas the joint-transform filter is *dependent*. Since the joint-transform hologram is dependent on the input scene, nonlinear filtering is difficult and may lead to false alarms and poor performance in the presence of noise.

The performance of the LCTV–based NN needs to be mentioned. The lenslet array provides the interconnection between the memory matrix or the interconnection weight matrix (IWM) and the input pattern. The transmitted light field after LCTV2 is collected by an imaging lens, focusing at the lenslet array and imaging onto a CCD array detector or camera. The array of detected signals is sent to a threshholding circuit and the final pattern can be viewed at the TV monitor or sent back for the next iteration. The data flow is primarily controlled by the microcomputer. Thus, the LCTV–based NN is indeed an *adaptive* optical NN.

Another hybrid architecture using an optical disk is worth mentioning, as shown in Fig. 7.8. As we know, a single input pattern might require correlation with a huge library of reference patterns before a match is found. Therefore, the lack of large-capacity storage devices that can provide high-speed readout is still a major obstacle to practical optical correlation systems. Optical disks (ODs), developed in recent years as mass-storage media for many consumer products, are excellent candidates for this task. We shall now illustrate an OD–based JTC for the continuing effort to develop a practical pattern recognizer. The system employs an electrically addressable SLM to display the input pattern, and an OD to provide a large volume of reference patterns in the JTC, shown in Fig. 7.8. A target captured by a video camera is first displayed on SLM1. A beam expansion/reduction system is then used to reduce the size of the input m times. One of the reference patterns on the OD is read out in parallel and magnified m times by another set of expansion lenses. The joint transform is done by transform lens FL1 and the joint-transform power spectrum (JTPS) recorded on the write side of SLM2. After FL2 Fourier

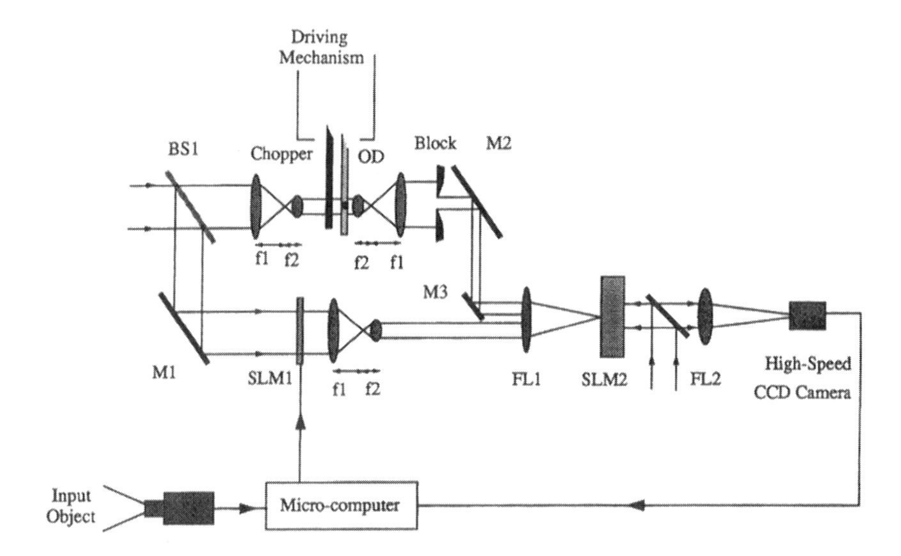

Fig. 7.8. An optical-disk–based JTC OD.

transforms the JTPS, the resulting correlation is recorded by a CCD camera. The operation can be seen as the OD advances to the next reference pattern, and so on, until a match is found.

If binary patterns of 200×200 pixels are to be recorded, each pattern will occupy an $0.2 \times 0.2\,\text{mm}^2$ on the OD. If a 0.01 mm spacing is assumed between adjacent blocks, then more than 27,000 reference patterns can be recorded on a single 120 mm diameter OD. To estimate the sequential access time for a block of $0.2 \times 0.2\,\text{mm}^2$ on a standard OD, assume that the average revolution speed is 1122 r/min and the average radius is 40 mm, and that all the images are scanned sequentially on each consecutive band of the OD. The minimum access time is approximately 40 μs. To further estimate the operating speed of this system, assume that the write side of an optically addressed ferroelectric liquid SLM is used as the square-law detector, the response time of which is between 10 μs and 155 μs. Thus, we see that to complete one correlation process should take 40–155 μs. This is equivalent to performing more than 6400 correlations per second, a number that can hardly be matched using current electronic counterparts.

Advances in photorefractive (PR) materials have stimulated interest in phase-conjugate correlators for pattern recognition. Although SLMs can be used to display complex spatial filters, current state-of-the-art SLMs are low-resolution and low-capacity devices. On the other hand, PR materials offer real-time recording, high resolution, and massive storage capacity; all desirable traits for multiplexed matched filter synthesis. Thick PR material has high

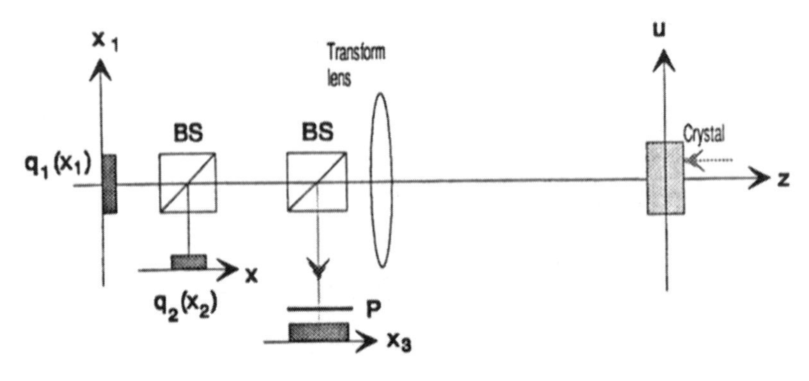

Fig. 7.9. A wavelength-multiplexed RC.

storage capacity, but the shift invariant property is limited by the Bragg diffraction condition. This can be minimized by using a thinner crystal; however, diffraction efficiency and storage capacity is substantially reduced. For high storage capacity, high diffraction efficiency, and a large shift invariance, a reflection-type wavelength-multiplexed PR matched filter can be used. A reflection-type matched filter correlator (RC) is depicted in Fig. 7.9, in which a z-cut PR crystal is used for matched filter synthesis. The matched filter is recorded by combining the Fourier spectrum of an object beam $q_1(x_1)$ with a reference plane wave from the opposite direction. Correlation can be done by inserting an input object $q_2(x_2)$ at plane x. The output correlation distribution can then be observed at plane x_3. In order to separate the reading beam $q_2(r_2)$ from the writing beam, the reading beam can be made *orthogonally polarized* to the writing beam by using a polarized beam splitter. The insertion of a polarizer P at the front of plane x_3 is to prevent the polarized writing beams from reaching the output plane.

7.1.4. ROBUSTNESS OF JTC

One of the major distinctions between VLC and JTC is that VLC uses a Fourier-domain filter and JTC uses a spatial-domain filter. We note that the VLC filter is *input signal independent*, while JTPS is *input signal dependent*, which produces poor diffraction efficiency under intense bright background or a multitarget input scene. Nevertheless, these shortcomings can be easily mitigated by the removal of the zero-order diffraction by means of computer intervention, as will be discussed later. The major advantages of the JTC are the flexibility to operate and robustness to environmental perturbation.

If we assume the Fourier-domain matched filter is transversely misaligned;

for example, by $\Delta\alpha$ and $\Delta\beta$, in the Fourier plane, as given by

$$H(\alpha - \Delta\alpha, \beta - \Delta\beta), \tag{7.1}$$

the output correlation distribution can be shown as

$$R(x, y) = \left\{ f(x, y) \exp\left[i \frac{2\pi}{\lambda f}(x\Delta\alpha + y\Delta\beta) \right] \right\} \otimes F(x, y), \tag{7.2}$$

where $\otimes$ denotes the correlation operation. Thus, we see that the larger the transversal misalignment of the Fourier-domain filter, the higher the degradation of signal detection will be.

On the other hand, JTC will not pose any major problem of the spatial-domain filter $h(x, y)$ transversal misalignment, since the fringe visibility of the JTPS is independent of the filter translation. As for the longitudinal misalignment of a VLC, the effects can be evaluated by referring to Fig. 7.10. We assume that the input object and the filter plane are longitudinally displaced by δ_1 and δ_2, respectively. The corresponding complex light distribution at the Fourier plane can be evaluated by the following integral equation:

$$\begin{aligned} g(\alpha, \beta) \approx C \exp\left[i \frac{\pi\delta_1}{\lambda(f^2 - \delta_1\delta_2)}(\alpha^2 + \beta^2) \right] \\ \cdot \iint f(x, y) \exp\left[i \frac{\pi\delta_2}{\lambda(f^2 - \delta_1\delta_2)}(x^2 + y^2) \right] \\ \cdot \exp\left[i \frac{2\pi}{\lambda f_{\text{eff}}}(\alpha x + \beta y) \right] dx\, dy, \end{aligned} \tag{7.3}$$

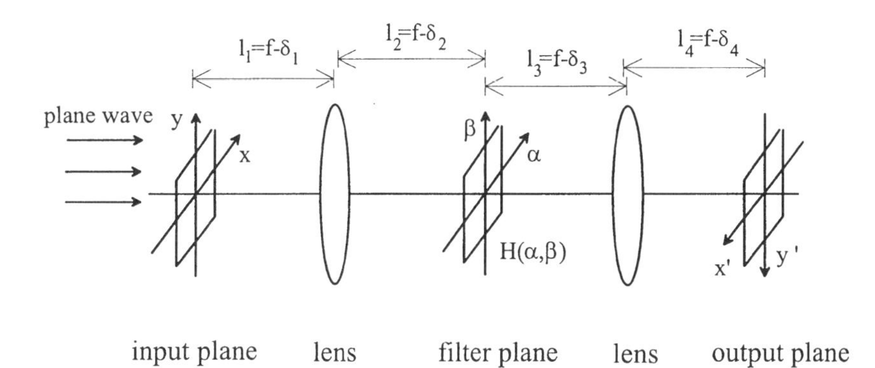

Fig. 7.10. Misalignment in the VLC setup.

in which we see that a quadratic phase factor is introduced, where

$$f_{\text{eff}} = f - \frac{\delta_1 \delta_2}{f} \approx f.$$

As long as the misalignment δ_s are sufficiently small, the longitudinal filter misalignment in a VLC is not as severe as the effect due to transversal misalignment. In other words, by using long focal length of the transform lenses, one would have a better longitudinal displacement tolerance. Since JTC is insensitive to longitudinal input plane alignment and transversal filter alignment, we see that JTC is more robust to the environmental factors, as compared with VLC. Additionally, JTC is more flexible to operate and the spatial-domain matched filter synthesis is easier to synthesize. Our discussion on pattern recognition by means of correlation detection will be mostly addressed to the JTC architectures.

7.2. RECOGNITION BY CORRELATION DETECTIONS

7.2.1. NONCONVENTIONAL JOINT-TRANSFORM DETECTION

The JTC has higher space–bandwidth product, lower carrier frequency, higher index modulation, and suitability for real-time implementation, but also displays inefficient use of illumination and a required larger transform lens, stringent spatial coherence, and small size of the joint-transform spectrum. A quasi-Fourier–transform JTC (QFJTC) that can alleviate some of these limitations is shown in Fig. 7.11, in which the *focal depth* or *focal tolerance* is given by

$$\delta \leqslant 2\lambda \left(\frac{f}{b} \right)^2. \tag{7.4}$$

To illustrate the shift-invariant property of the QFJTC, an input object such as that shown in Fig. 7.12a is used. The JTPS is recorded as a photographic transparency, which can be thought of as a joint-transform hologram (JTH). By performing a Fourier transform of the recorded JTH with coherent light, the cross-correlation distribution can be viewed as shown in Fig. 712b, where autocorrelation peaks indicating the location of the input character G are detected. Representative experimental results obtained using the QFJTC with the input object and reference functions are shown in Fig. 7.13a. Three JTHs for $\delta = 0$, $\delta = f/10 = 50$ mm, and $\delta = f/5 = 100$ mm are shown in Fig. 7.13b,

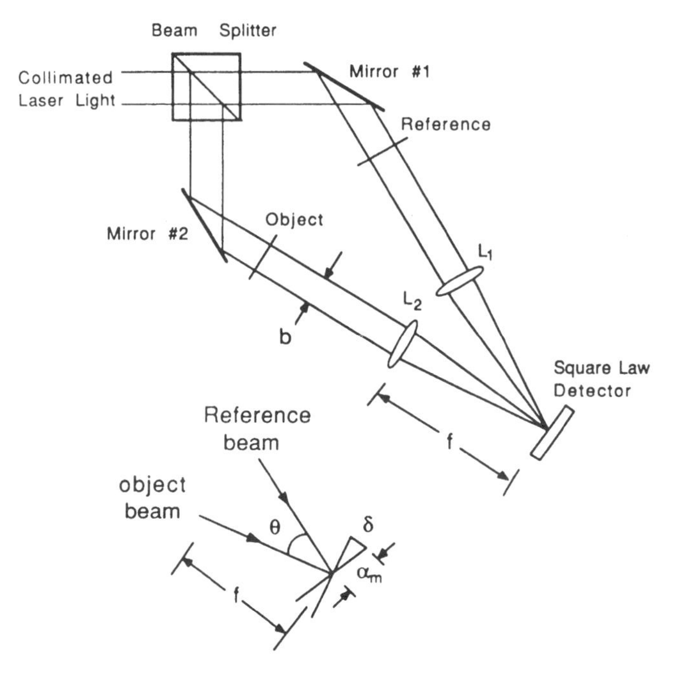

Fig. 7.11. A quasi-Fourier transform JTC (QFJTC).

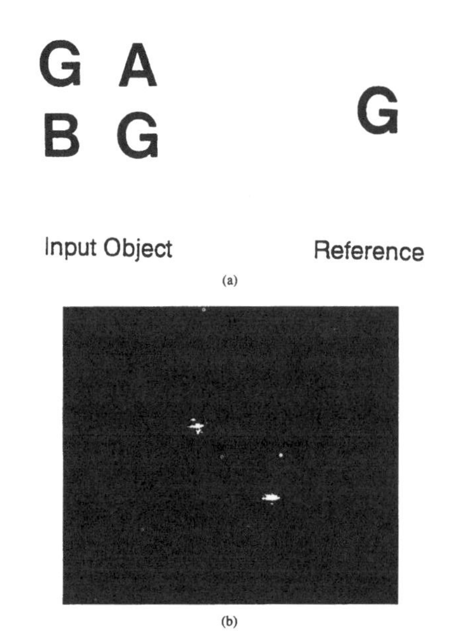

Fig. 7.12. (a) Input object and reference function, (b) Output correlator distribution.

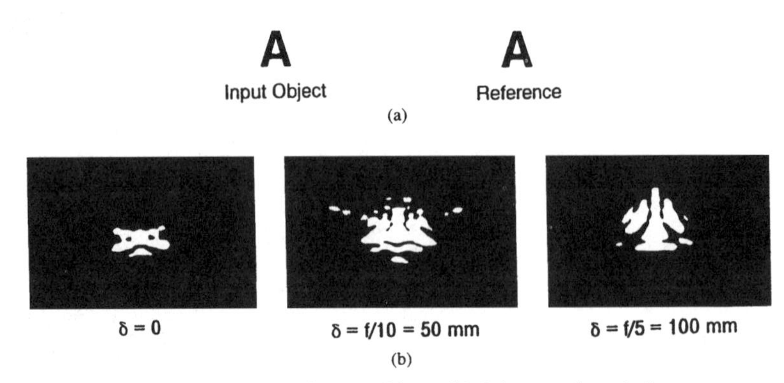

Fig. 7.13. (a) Input and reference objects, (b) Joint-transform holograms.

which illustrates that the size of the JTP enlarges as δ increases. Figure 7.14 illustrates that the correlation peak intensity increases as δ increases, while the size of the correlation spot decreases as δ increases. Thus, we see that the QFJTC can indeed improve the signal-to-noise ratio (SNR) and the *accuracy* of detection.

Moreover, to alleviate the lower diffraction efficient in a JTC, the readout correlation operation can be easily modified, as shown in Fig. 7.15, in which

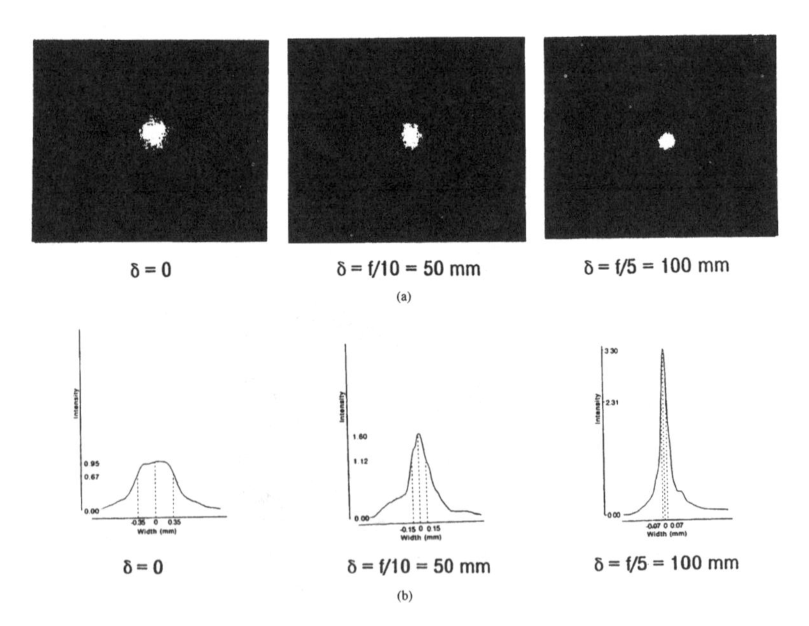

Fig. 7.14. (a) Output correlation spots, (b) Output correlation distributions.

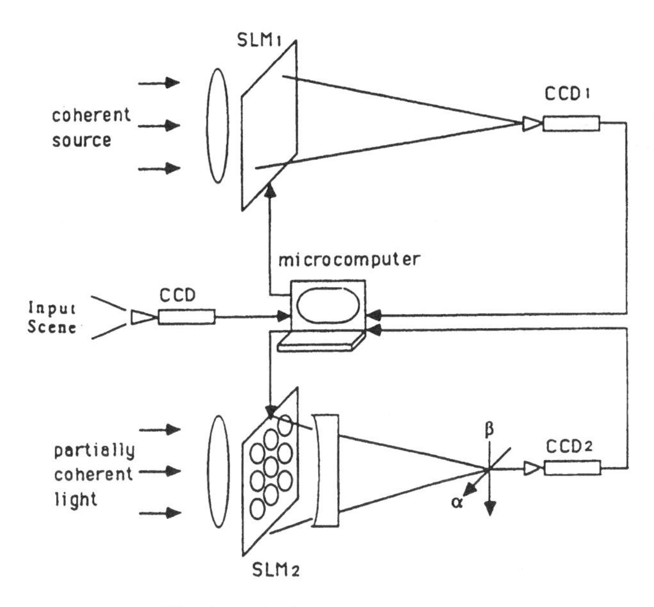

Fig. 7.15. A high-light–efficient JTC.

we see that the JTPS as detected by the joint-transform operation can be actually replicated into an $N \times N$ spectral array on SLM2. The corresponding amplitude transmittance distribution of the replicated JTPSs can be written as

$$T(p, q) = 2 \sum_{n=1}^{N} \sum_{m=1}^{N} |F(p - nd, q - md)|^2 \{1 + \cos[2x_0(p - nd)]\}. \quad (7.5)$$

We see that the overall correlation diffraction efficiency can be written as

$$\varepsilon = N^2 \varepsilon_m \frac{\iint [f(x, y) \otimes f(x, y)]^2 \, dx \, dy}{1/R \iint |u(x, y)|^2 \, dx \, dy}, \quad (7.6)$$

which increases about N^2 times.

Let us provide a result, which was obtained by replicating the JTPS into 3×3 arrays. The correlation spots under the coherent and partial coherent readouts are shown in Fig. 7.16. We see that the correlation intensity under the coherent readout is higher and sharper than the one obtained under the partial coherent readout. However, the output signal-to-noise ratio seems to be lower as compared with the partial coherent case. As discussed in Chapter 2, JTCs can perform all the processing operations that an FDP can offer. The high-efficiency JTC can also be used as a generalized image processor.

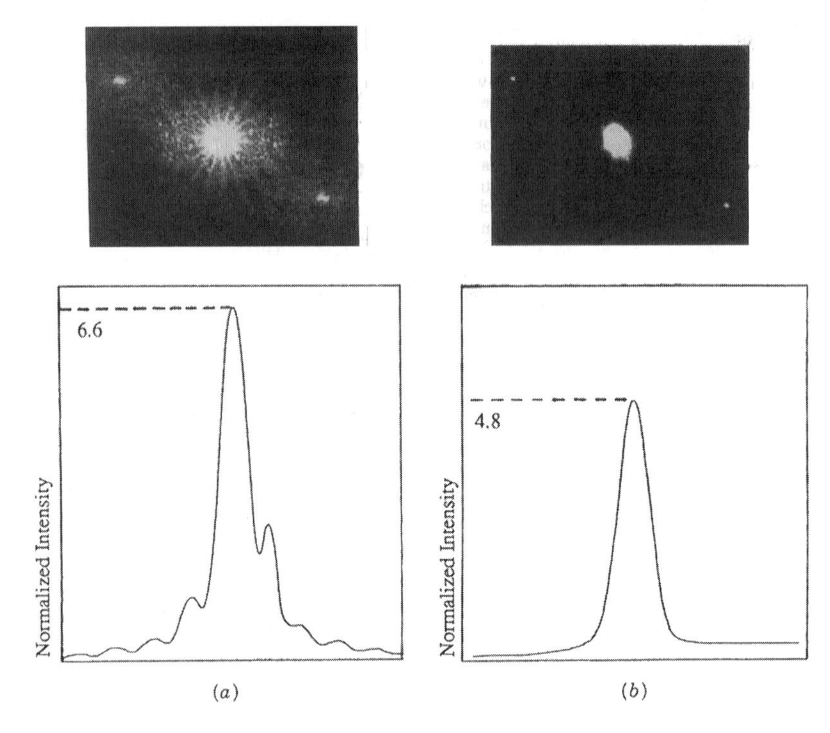

Fig. 7.16. Output irradiance obtained with a 3 × 3 replicated JTPS array. (a) Using coherent readout, (b) Using partial coherent readout.

7.2.2. NONZERO-ORDER JOINT-TRANSFORM DETECTION

The most important aspect of using hybrid optical architecture is the exploitation of the parallelism of optics and the flexibility of the electronic computer. JTC is known to be more robust with easy implementation, but the conventional JTC suffers from poor detection efficiency, particularly for multi-target and high–background noise environments. These drawbacks are primarily due to the existence of the zero-order spectra. The zero-order spectrum can be easily removed, however, by simple computer intervention.

Let us look at the JTPS, as obtained from the joint-transform duty cycle as given by

$$\begin{aligned} I(p, q) = {} & |F(p, q)|^2 + |R(p, q)|^2 + F(p, q)R^*(p, q) \\ & \cdot \exp(-2ix_0p) + F^*(p, q)R(p, q) \\ & \cdot \exp(+2ix_0p), \end{aligned} \tag{7.7}$$

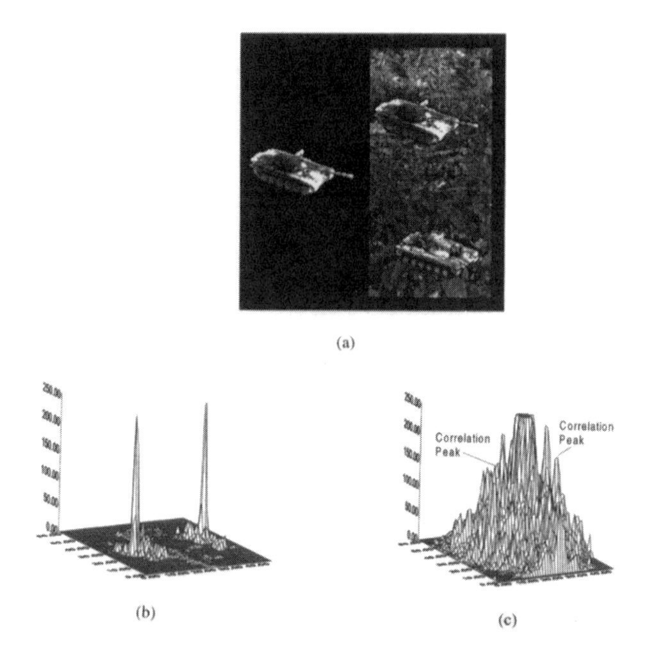

Fig. 7.17. (a) Objects input to the JTC, (b) Output correlation profile from the NOJTC, (c) Output correlation profile from the CJTC.

where $F(p, q)$ and $R(p, q)$ are the Fourier spectra of $f(x, y)$ and $r(x, y)$, respectively. It is apparent that if one removes the zero-order JTPS, the on-axis diffraction can be avoided. Since the zero-order correlations are derived from the power spectra of $f(x, y)$ and $r(x, y)$, these terms can be eliminated from the recorded JTPS. For example, if we record the power spectral distributions of $|F(p, q)|^2$ and $|R(p, q)|^2$ before the joint-transform operation, then these two zero-order power spectra can be removed by computer intervention before being sent to the correlation detection operation. Notice that the nonzero-order JTPS (NOJTPS) is a *bipolar* function that requires a write-in *phase-modulated* SLM for the correlation operation. Thus, the output complex light field can be written as

$$g(\alpha, \beta) = r^*(\alpha - 2x_0, \beta) \otimes f(\alpha, \beta) + r(-\alpha - 2x_0, -\beta) \otimes f^*(-\alpha, -\beta), \quad (7.8)$$

in which we see that the zero-order diffractions have been removed.

Let us show a result obtained from the NOJTC, as depicted in Fig. 7.17, in which the input scene represents a set of targets embedded in a noisy terrain. The input correlation peak intensities as obtained from the NOJTC are shown in Fig. 7.17b; a pair of distinctive correlation peaks can be easily observed. For comparison, the output light distribution as obtained from the conventional JTC (CJTC) is provided in Fig. 7.17c. We see that the correlation peaks have been overwhelmed by the zero-order diffraction.

In summary, besides the removal of the zero-order diffraction, there are other benefits of using NOJTC; for example, high diffraction efficiency and better utilization of SLM pixel elements. Despite these benefits, there is a small price to pay; NOJTC requires additional steps to capture and store the zero-order power spectra.

7.2.3. POSITION-ENCODING JOINT-TRANSFORM DETECTION

The JTC can perform the convolution of two functions without using a Fourier-domain filter, and it can also be used as a general optical processor. Even though the JTC usually has a lower detection efficiency, the architecture has certain merits. For instance, it does not have the stringent filter alignment problem as does VLC. It is suitable for real-time implementation and more robust to environment perturbation. To realize a spatial-domain filter in a JTC, complex function implementation is often needed. It is, however, possible to obtain complex-valued reference functions with an amplitude-modulated SLM.

It is well known that a real function can be decomposed into $c_1\phi_0 + c_2\phi_1$, and a complex function can be decomposed into $c_1\phi_0 + c_2\phi_{2/3} + c_3\phi_{4/3}$, where c_1, c_2, and c_3 are nonnegative coefficients, and $\phi_k = \exp(i\pi k)$ are elementary phase vectors. The decompositions can be optically realized with *position encoding*, as illustrated in Fig. 7.18.

A proof-of-concept experiment is shown in Fig. 7.19. Position-encoded letters "*E*" and "*F*" are shown in Fig. 7.19a and the corresponding output correlation distribution is shown in Fig. 7.19b. The autocorrelation peaks for detecting *F* can readily be seen; they were measured to be about twice as high

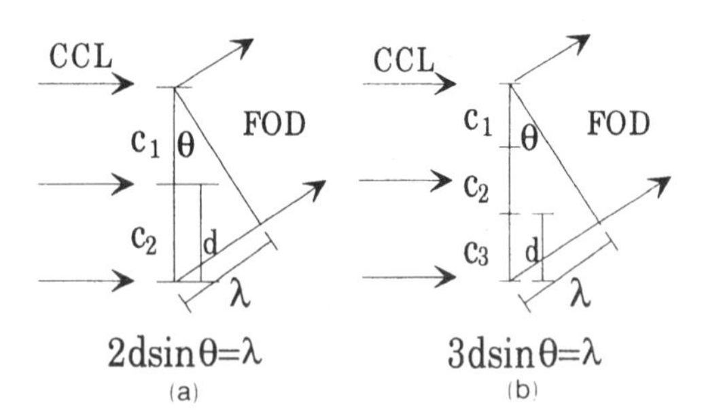

Fig. 7.18. Optical realization of (a) real value and (b) complex value representations with a position encoding method. CCL: coherent coliminated light; FOD: first-order diffraction.

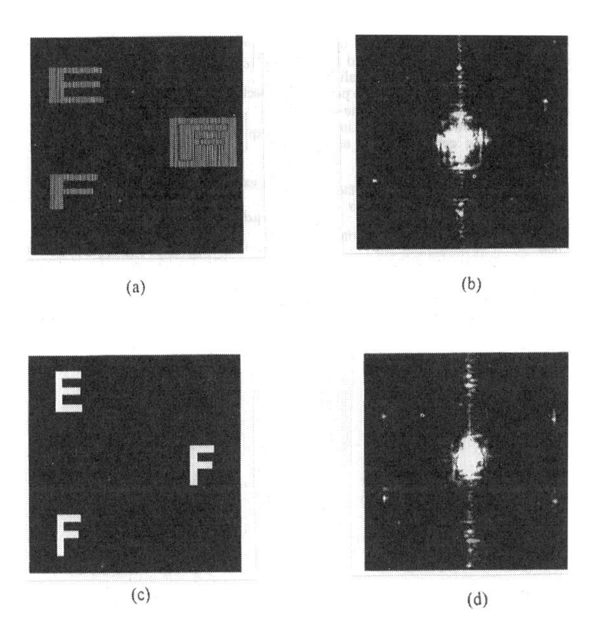

Fig. 7.19. Position-encoding JTC. (a) Position-encoded inputs, (b) Output correlation distribution. Conventional JTC. (c) Inputs, (d) Output correlation distribution.

as the cross-correlation peaks. For comparison, without using position encoding, Fig. 7.19c is presented to a CJTC, and the corresponding output correlation distribution is shown in Fig. 7.19d, in which it fails to differentiate the letters. Thus, we see that position encoding can readily improve *pattern discriminability*.

7.2.4. PHASE-REPRESENTATION JOINT-TRANSFORM DETECTION

Due to the availability of phase-modulating SLMs, phase-encoded inputs to a JTC are relatively convenient to implement. The CJTC offers the avoidance of filter synthesis and critical system alignment, but it suffers from poor detection efficiency for targets embedded in clutter, bright background noise, and multiple targets. A phase-encoded input can alleviate some of these shortcomings. Phase representation of an intensity $f(x, y)$ can be written as

$$pf(x, y) = e^{iT[f(x,y)]}, \tag{7.9}$$

where $T[\,]$ represents a monotonic real-to-phase transformation. By denoting $G_{\min}$ and $G_{\max}$ as the lowest and highest gray levels of an intensity object, the phase transformation can be written as

$$T[f(x, y)] = \frac{f(x, y) - G_{\min}}{G_{\max} - G_{\min}} 2\pi. \tag{7.10}$$

Applied to the JTC, the phase representation of the input can be written as $pf(x + a, y) + pr(x - a, y)$, where $pr(x, y) = e^{i\varphi_r(x,y)}$ represents the phase reference function. The corresponding output cross-correlation peak intensities is

$$C(\pm 2a, 0) = \left| \iint pf(x, y)pr^*(x, y)\, dx\, dy \right|^2 = |A|^2 \qquad 7.11)$$

for $\varphi_r(x, y) = T[f(x, y)]$, and A is a constant proportional to the size of the reference function.

The phase representation JTC (PJTC) is indeed an optimum correlator, regardless of nonzero-mean noise; that is,

$$\text{SNR} \leqslant \frac{1}{4K\pi^2} \iint E[|PF_n(p, q)|^2]\, dp\, dq. \tag{7.12}$$

The equality holds if and only if $\phi_r(x, y) = T[f(x, y)]$, where $E[\,]$ denotes the ensemble average and $F_n(p, q)$ is the noise spectrum. Thus, the PJTC is an optimal filtering system, independent of the mean value of the additive noise.

A CJTC often loses its pattern discriminability whenever a false target is similar to the reference function or the object is heavily embedded in a noisy background. For comparison, let the two images of an M60 tank and a T72 tank be embedded in a noisy background, as shown in Fig. 7.20a, where an M60 tank is used as the reference target. The corresponding output correlation distributions are shown in Figs. 7.20b and c, in which we see that CJTC fails to detect the M60 tank.

7.2.5. ITERATIVE JOINT-TRANSFORM DETECTION

Hybrid JTC is known for its simplicity of operation and real-time pattern recognition. By exploiting the flexibility of the computer, iterative operation is capable of improving its performance. In other words, it is rather convenient to feedback the output data into the input for further improvement of the operation.

Here we illustrate the iterative feedback to a composite filtering (as an example) to improve detection accuracy. Let us assume that the input scene to a JTC contains N reference patterns and a target $0(x, y)$ to be detected, as

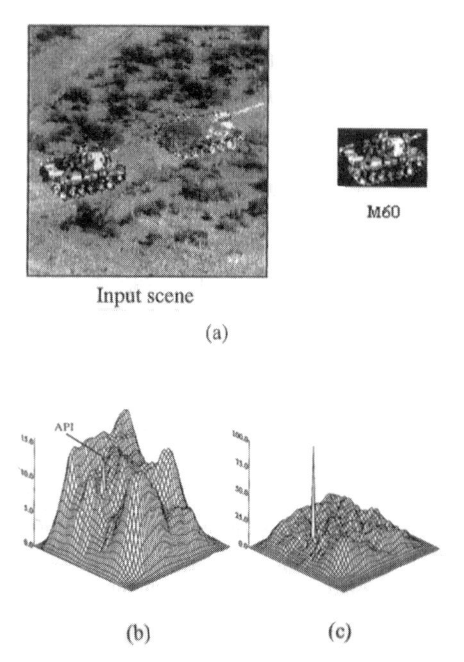

Fig. 7.20. (a) Input objects to the JTC. Output correlation results, (b) obtained from the CJTC, (c) obtained from PJTC.

written by

$$t(x, y) = r(x, y) + 0(x, y), \tag{7.13}$$

where

$$r(x, y) = \sum_{i=1}^{N} r_1(x - a_i, y - b_i)w(i),$$

and $r_i(x, y)$ is assumed the individual reference pattern centered at (a_i, b_i), and $w(i)$ represents an iterative feedback weight coefficient for each $r_i(x, y)$ with an initial weight value equal to 1. By joint-transforming $t(x, y)$, the corresponding JTPS is given by

$$\text{JTPS} = |R(p, q)|^2 + |0(p, q)|^2 + R^*(p, q)0(p, q) + R(p, q)0^*(p, q).$$

In the feedback iteration, JTPS is nonlinearized and inverse-transformed to obtain the output correlation, for which the feedback coefficient $w'(i)$ was calculated by multiplying the preceding $w(i)$ with the normalized correlation peak intensity $p_i/p_{\max}$, where p_i is the correlation peak intensity for $r_i(x, y)$ together with the target $0(x, y)$, and $p_{\max} = \max\{p_i, i = 1, \ldots, N\}$ is the highest peak intensity among them. By properly choosing a certain nonlinear function to the JTPS, the monostable convergence of this iteration can be achieved.

The proposed filtered iteration not only suggests that the weight coefficients $w(i)$ be fed back to the reference patterns but also implies that the feedback coefficients can be used to weight the composite filter. Since the feature of a reference pattern can be enhanced with an intensity compensation filter (ICF); i.e., the reciprocal of the reference power spectrum, the composite filter can be obtained by linearly combining all the ICFs contributed by each reference pattern; that is,

$$H(p, q) = \sum_{i=1}^{N} \frac{K_i}{T\{R_i|(p, q)|^2\}_{t_i}} \tag{7.14}$$

where T denotes the thresholding operation with t_i as the threshold value, and K_i is the iterative feedback weight coefficient for the filter.

To determine t_i, we assume that only one reference pattern [assuming $r_i(x, y)$] is present and the target $0(x, y)$ is equal to $r_i(x, y)$. t_i can then be adjusted until the sharpness of the output autocorrelation profile meets a certain requirement; e.g., the optimum sharpness under noisy constraint. Another way of determining t_i is to solve the following equation:

$$\sigma^2(t_i)/\mu^2(t_i) = C \tag{7.15}$$

where C is a user-provided parameter, $\mu(t_i)$ and $\sigma(t_i)$ are the mean and the standard deviation of $|R_i(p, q)|^2/T\{|R_i(p, q)|^2\}_{t_i}$; i.e., the cross-power spectrum of the reference and the object (now equal to the reference function) compensated with the ICF. $\sigma^2(t_i)/\mu^2(t_i)$ is inversely proportional to the ratio of peak intensity to the average background intensity, whereas the range for parameter C is usually chosen from 0.01 to 1. K_i is initialized such that all the ICF have the same average intensity. The new weight coefficient K_i' is equal to the preceding coefficient K_i times the normalized correlation peak intensity $p_i/p_{\max}$. The iteration process can be summarized as follows:

1. Compute all the ICF.
2. Initialize the reference and filter weight coefficients.
3. Compute the composite filter (combine all the weighted ICF).
4. Forward the Fourier transform of the target and reference functions.
5. Capture the JTPS and multiply it by the composite filter.
6. Inverse the Fourier transform of the filtered JTPS.
7. Analyze the correlation output for decision making.

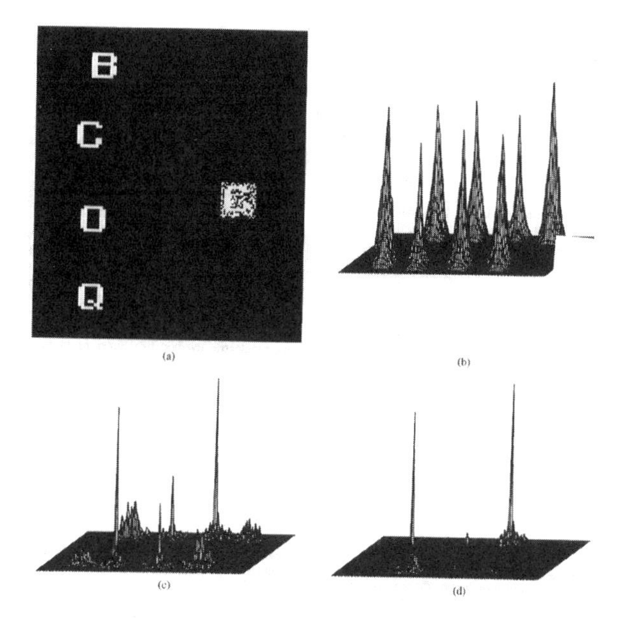

Fig. 7.21. (a) Reference patterns with a noisy input. (b) Obtained from the CJTC, (c) Output correlation after first iteration, (d) Output correlation after second iteration.

8. If the correlation output requirement is not met, update weight coefficients for reference and filters and repeat steps 3 to 7, and so forth.

Let us assume the input to a nonzero JTC represents a set of letters; "B", "C", "O," and "Q" are the reference targets with a noisy target "C," as shown in Fig. 7.21a. The corresponding output correlation using a CJTC is shown in Fig. 7.21b, in which we see that CJTC fails to detect the target. Figures 7.21c and 7.21d show the results obtained with the first and second iterations, respectively, in which we see that detection of letter "C" is rather visible.

By exploiting the flexibility operation of computers, signal detection reliability can be improved by continuous iteration.

7.3. POLYCHROMATIC PATTERN RECOGNITION

There are essentially two approaches to polychromatic correlation detection; one uses the VLC and the other uses JTP. We will describe these two approaches independently.

7.3.1. DETECTION WITH TEMPORAL FOURIER-DOMAIN FILTERS

The technique is similar to exploiting the coherence content from an incoherent source for partially coherent processing, as described in Sec. 2.8. Let us consider that a conventional VLC is illuminated by a red, green, and blue (RGB) coherent plane wave. A color image transparency in contact with a sinusoidal grating is assumed at the input plane. The spectral distribution at the Fourier domain can be written as

$$S(\alpha, \beta; \lambda) = \sum_{n=r}^{g,b} S_n(\alpha, \beta; \lambda_n) + \sum_{n=r}^{g,b} S_n\left(\alpha \pm \frac{f\lambda_n}{2\pi} p_0, \beta\right), \tag{7.16}$$

where (α, β) denotes the spatial coordinate at the Fourier plane, f is the focal length of the transform lens, and $S_r(\alpha, \beta)$, $S_g(\alpha, \beta)$, and $S_b(\alpha, \beta)$ are the corresponding color Fourier spectra. It is trivial to see that the red, green, and blue color spectra are scattering diffracted along the α axis. Since the transparency is spatial frequency limited, the RGB spectra are physically separated in the Fourier domain by using a sufficiently high sampling frequency p_0.

A set of RGB matched filters (called *temporal Fourier holograms*) can be synthesized by this interferometric technique of Vander Lugt, as written by

$$\begin{aligned} H_n(\alpha, \beta) = K_1 + K &\left|S_n\left(\alpha - \frac{f\lambda_n}{2\pi} p_0, \beta\right)\right| \\ &\cdot \cos\left[\frac{f\lambda_n}{2\pi}\alpha x_0 + \phi_n\left(\alpha - \frac{f\lambda_n}{2\pi} p_0, \beta\right)\right], \end{aligned} \tag{7.17}$$

where x_0 is an arbitrary carrier spatial frequency, Ks are appropriate proportionality constants, and $S_n(\alpha, \beta) = |S_n(x, \beta)| \exp[i\phi_n(\alpha, \beta)]$, are the primary color image spectra. If these temporal matched spatial filters are inserted in the Fourier domain, as illustrated in Fig. 7.22, the complex light distribution at the output plane of the VLC would be

$$\begin{aligned} g(x, y) = \sum_{n=r}^{g,b} [&s_n(x, y) \exp(ip_0 x) \\ &+ s_n(x, y) \exp(ip_0 x) * s_n(x - x_0, y) \exp(ip_0 x) \\ &+ s_n(x, y) \exp(ip_0 x) * s_n(-x + x_0, y) \exp(ip_0 x)], \end{aligned} \tag{7.18}$$

in which the last term represents the RGB image correlations that will be superimposely diffracted at $x = x_0$.

For experimental demonstration, a color image transparency of a campus

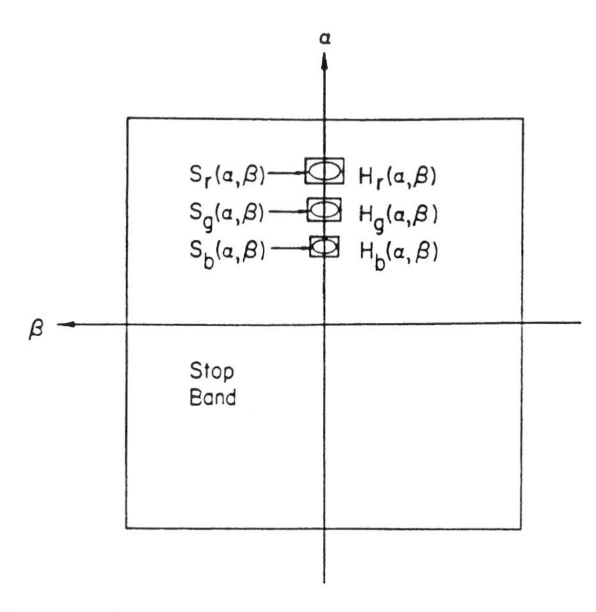

Fig. 7.22. Multicolor Fourier-domain filtering, H_r, H_g, and H_b are red, green, and blue color-matched spatial filters.

street scene (in gray level) is shown in Fig. 7.23a. Figure 7.23b shows the polychromatic detection, in which a blue correlation spot representing the location of the blue van and a red spot representing the stop sign are observed.

7.3.2. DETECTION WITH SPATIAL-DOMAIN FILTERS

Liquid crystal–TV (LCTV) has been widely used in optical processing. By exploiting the inherent polychromatic structure of the LCTV pixels, polychromatic target detection can be developed in a JTC. The LCTV we consider is a color LCTV (Seiko Model LVD-202) with 6 lines/mm resolution, which can be used as the spatial-carrier frequency. The enlarged liquid crystal–pixel array is shown in Fig. 7.24a. If the LCTV panel is illuminated by coherent red, blue, and green light, the dispersed spectral distribution at the Fourier domain is shown in Fig. 7.24b. For simplicity, let us consider a red reference signal $f_r(x, y)$, with blue, green, and red object images displayed on the LCTV panel, as written by

$$t(x, y) = f_r(x, y+c; \lambda_r) + f_r(x, y-c; \lambda_r) + f_b(x, y-d; \lambda_b) + f_g(x, y-e; \lambda_g), \tag{7.19}$$

(a)

(b)

Fig. 7.23. (a) A black-and-white picture of a color object transparency. The stop sign is red and the van is blue. (b) Corresponding correlation detections. In reality, the correlation spot for the van is blue and for the stop sign, red.

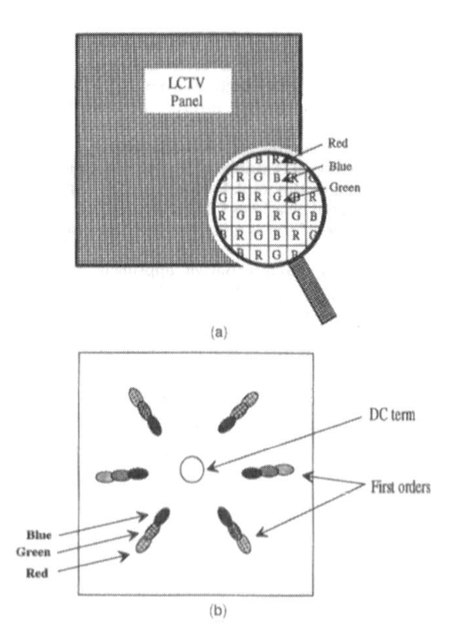

Fig. 7.24. (a) Color pixel array in a color LCD panel. (b) Spectral distribution of the LCTV panel.

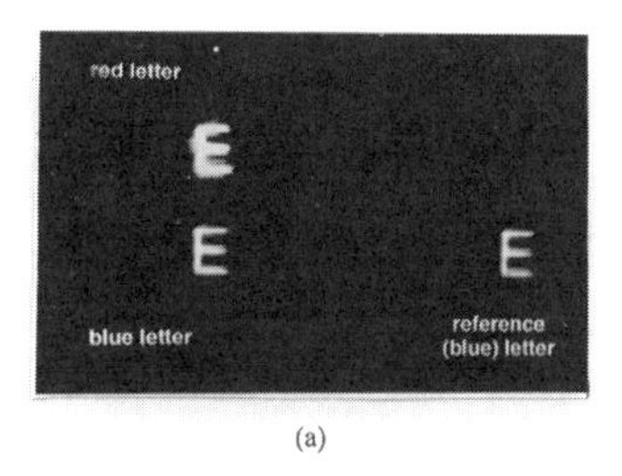

(a)

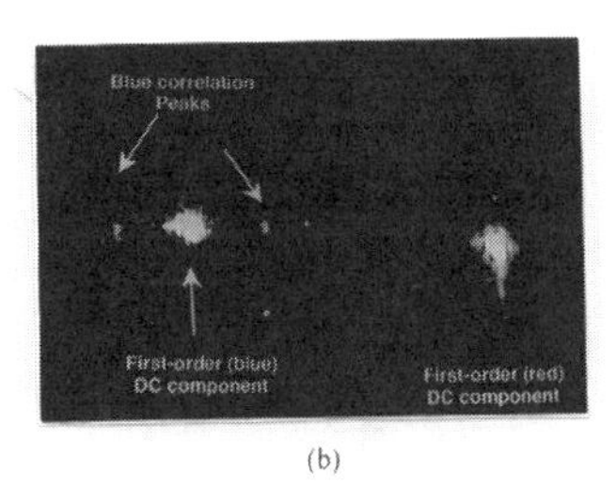

(b)

Fig. 7.25. (a) Input reference blue E (right) and object images (blue E, lower left; red E, upper left). (b) Output distribution.

where $t(x, y)$ represents the amplitude transmittance of the LCTV panel and c, d, and e are arbitrary constants. The corresponding JTPS can be shown as

$$\begin{aligned} |F(p, q)|^2 &= 2|F_r(p_r, q_r)|^2 + |F_b(p_b, q_b)|^2 + |F_g(p_g, q_g)|^2 \\ &\quad + F_r^2(p_r, q_r; \lambda_r) \exp(icq_r) + F_r^2(p_r, q_r; \lambda_r) \exp(-icq_r) \\ &= A + 2|F_r|^2 \cos(cq_r), \end{aligned} \tag{7.20}$$

in which we see that fringe modulation occurs only if the spectral content of the target signal matches the spatial content of the reference target. By displaying the polychromatic JTPS back on the LCTV panel for the correlation operation, the output light field would be

$$g(x, y; \lambda_r) = A_f + R_r(x - 2c) + R_r(x + 2c), \tag{7.21}$$

where a pair of red correlation peaks diffracted at the output plane occur at $x = \pm 2c$, and A_f represents the zero-order diffraction. For demonstration, we provide a black-and-white photo of the input functions as shown in Fig. 7.25a, in which a blue letter E is used as the reference target. The other letters in red and blue are used as input targets. The color JTPS was captured by a color

CCD and then sent back to the LCTV panel for correlation operation. Figure 7.25b shows the output distribution, in which two distinctive blue spots can easily be detected.

7.4. TARGET TRACKING

One interesting feature of using a hybrid-optical JTC is *adaptivity*; for example, the reference scene and the dynamic input scene can be continuously updated.

7.4.1. AUTONOMOUS TRACKING

By continually updating two sequential input scenes which display side-by-side at the input domain of a JTC, it is possible to track a moving target. Let us consider the two sequential scenes that are displayed with a video frame grabber on the input SLM, with the previous and the current frames positioned in the upper and lower half of the SLM. This is represented mathematically by

$$f_{t-1}\left(x-x_{t-1}, y-y_{t-1}-\frac{\alpha}{2}\right) \quad \text{and} \quad f_t\left(x-x_{t-1}-\delta x, y-y_{t-1}-\delta y+\frac{\alpha}{2}\right), \tag{7.22}$$

where 2α is the height of the display unit, t and $t-1$ represent the current and previous time frames, and $(\delta x, \delta y)$ is the relative translation of the target. The complex light field at the Fourier domain can be shown to be

$$\begin{aligned} T(u, v) = {} & F_{t-1}(u, v) \exp\left\{-i2\pi\left[ux_{t-1} + v\left(y_{t-1} + \frac{\alpha}{2}\right)\right]\right\} + F_1(u, v) \\ & \cdot \exp\left\{-i2\pi\left[u(x_{t-1} + \delta x) + v\left(y_{t-1} + \delta y - \frac{\alpha}{2}\right)\right]\right\}. \end{aligned} \tag{7.23}$$

If the detected JTPS is sent back to the SLM for the correlation operation, the output light distribution can be written as

$$\begin{aligned} C(x, y) = {} & R_{t,t}(x, y) + R_{t-1,t-1}(x, y) + R_{t,t-1}(x - \delta x, y + \delta y - \alpha) \\ & + R_{t-1,t}(x - \delta x, y - \delta y + \alpha), \end{aligned} \tag{7.24}$$

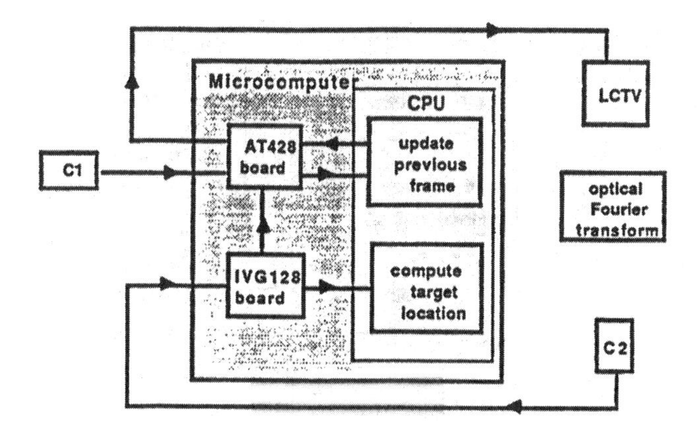

Fig. 7.26. Optical–digital interface diagram.

where

$$R_{m,n}(x, y) = \iint f_m(u, v) f_n^*(u - x, v - y)\, du\, dv$$

represents the correlations of f_m and f_n, which are diffracted at $x_1 = \delta x$, $y_1 = (\delta y - \alpha)$, and $x_2 = \delta x$, $y_2 = (-\delta y + \alpha)$, respectively. If the angular and scale tolerances of the JTC are approximately $\pm 5^\circ$ and $\pm 10\%$, and the motion of the target is relatively slow as compared to the processing cycle of the correlator, then f_{t-1} correlates strongly with f_t. Two high-intensity correlation peaks are diffracted into the output plane to locations given by

$$x_t = (x_{t-1} + x_1), \qquad y_t = (y_{t-1} + y_1 + \alpha). \tag{7.25}$$

A simple C language program can then be used to evaluate the target position. A block diagram of the system configuration is shown in Fig. 7.26. In spite of hardware limitations, the system constructed ran at approximately 1.2 cycle/s. It is possible for the hybrid system to run at half the video frame frequency, which is $\frac{1}{2} \times 30 = 15$ cycles/s, if specialized support hardware is used.

To demonstrate the performance of tracking, we assume a situation where a camera mounted on a moving space vehicle is focused on a fixed target on the ground for automatic navigation. As the space vehicle approaches the target, the detected scene changes continuously, the target size appears larger, and the scene's orientation and shape change due to the motion of the vehicle. Using computer-aided design graphics, a 3-D treelike model was created as a simulated target on the ground. Nine image sequences, simulating the changing

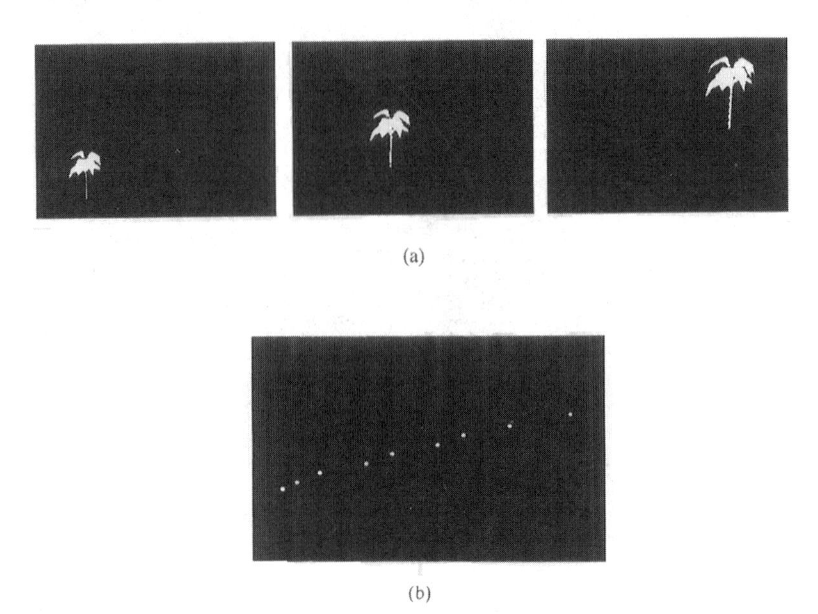

Fig. 7.27. (a) A sequence of nine images are recorded, simulating the exact scenes as captured by a camera mounted on a moving space vehicle. A total of nine images were taken but only frames 1, 5, and 9 are shown. (b) Tracked positions of the ground target as seen from the vehicle's coordinate frame.

scene viewed by a moving space vehicle, are shown in part in Fig. 7.27a. The JTC tracking system has little difficulty in correlating targets from different frames, even though the target in the first and the last frame look different. Figure 7.27b shows the tracked locations of the target as seen from the vehicle.

7.4.2. DATA ASSOCIATION TRACKING

Previously we discussed autonomous tracking for a single target by continuously updating the sequential frames on the input SLM of a JTC. We assumed the processing cycle of the system is sufficiently short; then a target will look much the same after a few sequential frames, and hence strong correlation signals will be generated at the output plane. The position measurements of the correlation peaks can be interpreted as the average velocities of the targets within a tracking system. Based on these measurements, the dynamic states of the targets can be updated. Since multiple correlation peaks are generated in a multitarget tracking problem, a data association algorithm can be employed to associate each correlation peak to the correct paths of the targets' motion.

The adaptive property of this system enables the tracking of targets under change in orientation, scale, and perspective during the course of their movement.

Let $f^n(x, y)$, $n = 1$ to N, be the targets in an image sequence, each of which are moving independently. Two sequential scenes are displayed on the input plane of a JTC such that the previous frame $(k - 1)$ and the current frame (k) are positioned in the upper and lower halves of the SLM, respectively. The corresponding image function is given by

$$T(x, y) = \sum_{n=1}^{N} f_{k-1}^{n}\left(x - x_{k-1}^{n}, y - y_{k-1}^{n} - \frac{a}{2}\right) + \sum_{n=1}^{N} f_{k-1}^{n}\left(x - x_{k-1}^{n} - \delta x_{k-1}^{n}, y - y_{k-1}^{n} - \delta y_{k}^{n} + \frac{a}{2}\right), \tag{7.26}$$

where $2a$ is the height of the active aperture of the SLM, (x_{k-1}^n, y_{k-1}^n) are the locations of the targets in the $k - 1$ frame, and $(\delta x_k^n, \delta y_k^n)$ are the relative translation of the targets from the $k - 1$ frame to the k frame. The corresponding JTPS is then introduced as modulation at the input plane SLM for the correlation detection cycle. The output complex light distribution is given by

$$\begin{aligned} C(x, y) = &\sum_{n=1}^{N}\sum_{m=1}^{N} R_{k-1,k-1}^{m,n}(x + x_{k-1}^{m} - x_{k-1}^{n}, y + y_{k-1}^{m} - y_{k-1}^{n}) \\ &+ \sum_{n=1}^{N}\sum_{m=1}^{N} R_{k,k}^{m,n}(x + x_{k-1}^{m} - \delta x_{k}^{m} - x_{k-1}^{n} + \delta x_{k}^{n}, y + y_{k-1}^{m} - \delta y_{k}^{m} - y_{k-1}^{n} + \delta y_{k}^{n}) \\ &+ \sum_{n=1}^{N}\sum_{m=1}^{N} R_{k-1,k}^{m,n}(x + x_{k-1}^{m} - x_{k-1}^{n} + \delta x_{k}^{n}, y + y_{k-1}^{m} - y_{k-1}^{n} + \delta y_{k}^{n} + a) \\ &+ \sum_{n=1}^{N}\sum_{m=1}^{N} R_{k,k-1}^{m,n}(x + x_{k-1}^{m} - \delta x_{k}^{m} - x_{k-1}^{n}, y + y_{k-1}^{m} - \delta y_{k}^{m} - y_{k-1}^{n} - a), \end{aligned} \tag{7.27}$$

where

$$R_{k,k}^{m,n}(x - \alpha, y - \beta) = f_k^m(x, y) \otimes f_k^n(x, y) * \delta(x - \alpha, y - \beta)$$

is the correlation function between f_k^m and f_k^n located at (α, β).

Note that the autocorrelation functions $R_{k,k}^{n,n}$ and $R_{k-1,k-1}^{n,n}$ are located at the origin of the output plane. The correlation between the same target in the k and the $k - 1$ frames, given as $R_{k,k-1}^{n,n}$, is diffracted around $(\delta x_k^n, \delta x_k^n + a)$ and

$(-\delta x_k^n, -\delta y_k^n - a)$. The rest of the terms ($R_{k,k-1}^{m,n}$ and $R_{k,k}^{m,n}$ for $m \neq n$) represent the cross-correlation functions between target m and target n. In this multiple-target tracking problem, we have assumed that the targets in any single input frame do not look alike, so the cross-correlation functions $R_{k,k-1}^{m,n}$ and $R_{k,k}^{m,n}$ generate much weaker correlation signals than that generated by the function $R_{n,k-1}^{n,n}$. Furthermore, the size and the gray level of the targets are also assumed to be relatively congruous with each other, so the intensities of the autocorrelation peaks generated from $R_{k,k-1}^{n,n}$, $n = 1$ to N, are not significantly different.

By translating the coordinate origin from the optical axis to $(0, a)$, the correlation peaks that are generated from the same target in the $k-1$ and the k frames are then located at $(\delta x_k^n, \delta y_k^n)$ in the new coordinate system. If these correlation peaks are associated with the proper target motions, the locations of the targets in the k frame are given by

$$x_k^n = x_{k-1}^n + \delta x_k^n, \qquad y_k^n = y_{k-1}^n + \delta y_k^n, \qquad n = 1 \text{ to } N. \tag{7.28}$$

The preceding equation shows that the new locations of the targets in the k frame can be updated based on their locations in the previous frame and the position measurements of the correlation peaks. It is apparent that the k and $k+1$ frames can be sent to the input plane in the next tracking cycle, so that multiple targets can be tracked simultaneously on a near real-time basis.

Note that the position of the correlation peak in the new coordinate system represents the average velocity of a target during the sampling interval δt; i.e.,

$$\bar{\dot{x}} = \delta x/\delta t, \qquad \bar{\dot{y}} = \delta y/\delta t. \tag{7.29}$$

Therefore, with a constant sampling interval and assuming that the sensor is not skewing or panning, the new coordinate system represents the plane of average velocity. For example, targets moving with *constant velocity* produce stationary correlation peaks, targets moving with *constant acceleration* generate correlation peaks that are located at equally separated intervals on a straight line, and correlation peaks located at the origin of the new coordinate plane correspond to *stationary* objects or background scene. In our demonstration, multiple targets are traveling at different velocities and in different directions; thus, the correlation peaks generated are located at different positions at the velocity plane.

In general, the motion of the targets can be represented by dynamic models, which are governed by well-known laws of physics. Unpredictable changes in target motions, commonly called *maneuvers*, can also be treated as gradual changes of motion parameters, if the sampling interval is sufficiently short compared with the maneuver time. Therefore, it is not difficult to associate the measurements in the velocity plane with the targets, based on the past history of the targets. Any prior knowledge of the targets before tracking will also be helpful in this association process. For example, if one target is known to travel much faster than the others, its correlation peak is more likely to be located farther away from the origin in the velocity plane than those of slower ones.

In the following, a parameter-based data association algorithm using a Kalman filtering model is described. Assume that the sampling interval δt is sufficiently short that the dynamic parameters (velocity, acceleration, angular velocity, etc.) of the targets are fairly consistent within a few sequential frames. Thus, given the dynamic parameters of a target in the k frame, the parameters in the next frame should be rather predictable. Let $z(k)$ be the measurement at the k frame and $\hat{z}(k|k-1)$ be the predicted measurement in the k frame, based upon the information evaluated up to the $k-1$ frame. Then the innovation (or measurement residue), defined as

$$v(k) = z(k) - \hat{z}(k|k-1), \tag{7.30}$$

can be used to evaluate the likelihood of association between $z(k)$ and the target under consideration. In stochastic models, one would evaluate the normalized squared distance D from the measured $z(k)$ to the current track,

$$D - v'S^{-1}v \tag{7.31}$$

where v' is the transpose of v and S is the innovation covariance of the target, which is basically the variance of the estimated states. The values of $\hat{z}(k|k-1)$, S, and the like can be evaluated by applying a standard Kalman filter to the dynamic model. (The use of Kalman filtering in stochastic modeling is a well-known subject on its own and will not be discussed here.)

A data association process can be carried out as follows:

Step 1: At the $k-1$ frame, the dynamic parameters of the N targets are determined at track 1 to N.

Step 2: At the k frame, N new measurements are made, given as $a, b, \ldots, N$. The normalized square distances are then computed:

$$D_{1a} = v'_{1a}S_1^{-1}v_{1a}, \qquad D_{1b} = v'_{1b}S_1^{-1}v_{1b}, \ldots, \text{etc.},$$

and similarly for $D_{2a}, D_{2b}, \ldots, D_{3a}, \ldots$, and so on.

Step 3: The most likely association is given by choosing the possible combination of Ds that yields the minimum sum.

To initiate the tracker, all the measurements in the first few tracking cycles are used to set up multiple potential tracks. Tracks that have inconsistent dynamic parameters are dropped until only a single track is assigned to each

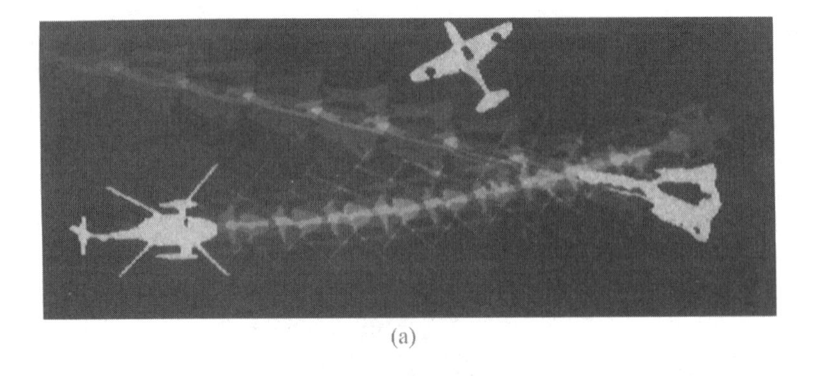

(a)

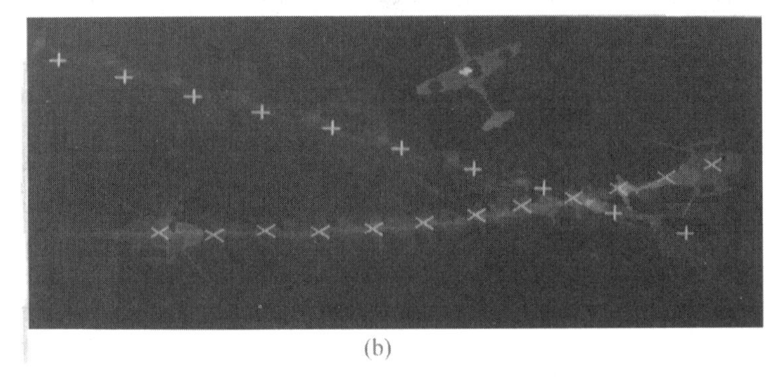

(b)

Fig. 7.28. (a) Motion sequence of three targets: helicopter and jet, moving from bottom to top; airplane, stationary. (b) Final results obtained after 11 tracking cycles. The actual locations of the targets at each of the tracking cycles are also given for comparison. The initial locations of the paths were assigned at the correct positions in this figure through manual interaction.

target. Recall that the measurement in each cycle is the average velocity of a target during the sampling period. Therefore, dynamic parameters can be assigned quickly in a few cycles; only two or three cycles are needed to determine targets with constant velocity or constant acceleration. The tracking scheme discussed so far, however, does not determine the initial positions of the targets in the first frame nor identify the measurements to the targets due to the nature of this adaptive correlation scheme. Therefore, an initial acquisition scheme is needed to perform this task before the start of the adaptive tracker. This can be done by using prestored reference images located at fixed positions at the image plane. (Further analysis of the initial target acquisition technique is beyond the scope of this section.)

To demonstrate the ability of the tracking algorithm, a motion sequence of three targets was generated, as shown in Fig. 7.28a, for which a target tracking model can then be set up for the continuous operation of the JTC. A computer program has been written to control the joint-transform process as well as the data association process. To initiate the coordinate, the system was first set to run with a stationary object as input, and the position of the correlation peak thus generated was assigned as the origin of the velocity plane coordinate. The tracking program was then set to run autonomously with the motion sequence. Figure 7.28b shows the actual location of the targets at each step of the tracking cycle. A relatively simple hybrid JTC is capable of tracking multiple targets rather accurately.

7.5. PATTERN RECOGNITION USING COMPOSITE FILTERING

Classical spatial matched filters are sensitive to rotational and scale variances. A score of approaches to developing composite-distortion-invariant filters are available. Among them, the synthetic discriminant function (SDF) filter (as described in Sec. 2.6.4) has played a central role for 3-D target detection. The original idea of SDF can be viewed as a linear combination of classical matched filters, where the coefficients of the linear combination are designed to yield equal correlation peaks for each of the distorted patterns. Since the dynamic range of an SDF is large, it is difficult to implement with currently available SLMs. On the other hand, the bipolar filter has the advantage of a limited dynamic range requirement and can be easily implemented with commercially available SLMs. Since the bipolar filter has a uniform transmittance function, it has the advantage of being light efficient. Several attempts have been made to construct bipolar SDF filters. However, binarization of SOF filters is not the best approach, since there is no guarantee that the SDF will be valid. Nevertheless, iterative approaches to optimize the bipolar SDF have been reported.

On the other hand, a *simulated annealing* (SA) algorithm (Sec. 2.6.5) as applied to the design of a bipolar filter is relatively new. A bipolar filter, as we will discuss, is in fact a *spatial-domain* filter, and can be directly implemented on an input phase-modulating SLM in a JTC. To demonstrate the performance of an (SA) bipolar composite filter (BCF), sets of out-of-plane–oriented T72 and M60 tanks, shown in Fig. 7.29a, have been used as target and antitarget training sets. The constructed BCF, using the SA algorithm, to detect a T72 tank is shown in Fig. 7.29c. If the input scene to the JTC shown in Fig. 7.30a is used (a BCF for detecting the T72 tank is located on the right-hand side), the output correlation distribution can be captured, as shown in Fig. 7.30b. We see that targets (T72 tanks) can be indeed detected from the noisy terrain, although T72 and M60 tanks are very similar.

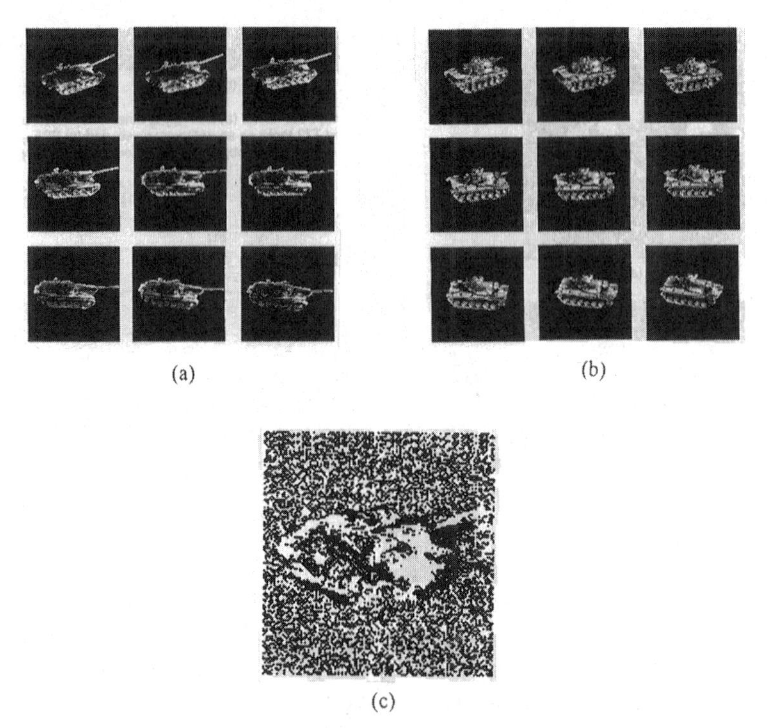

Fig. 7.29. (a) Out-of-plane rotation training images: M60 tanks and T72 tanks. (b) A bipolar composite filter to detect a T72 tank.

7.5.1. PERFORMANCE CAPACITY

Strictly speaking, an $n \times n$ pixel array bipolar spatial filter has 2^{n^2} images. However, in practice, a bipolar spatial filter may not memorize such a great number of training images. The reality is that some of the pixel elements would not be utilized in the synthesis. Thus, the actual capacity of a composite filter is to accommodate the largest possible number of training images, by which the effective pixels are efficiently used.

To look at the performance capacity of a composite filter, we focus our attention on correlation peak intensity detection. By setting the traditional 3 dB criterion, the performance capacity of SABCF can be defined as given by:

$$\text{PC (number of training images)} = \text{minimum of} \begin{cases} I_{\text{target}} \geqslant 0.5 I_{\text{target}-\max} \text{ or,} \\ \text{DR} \geqslant 1.3, \end{cases} \tag{7.32}$$

(a)

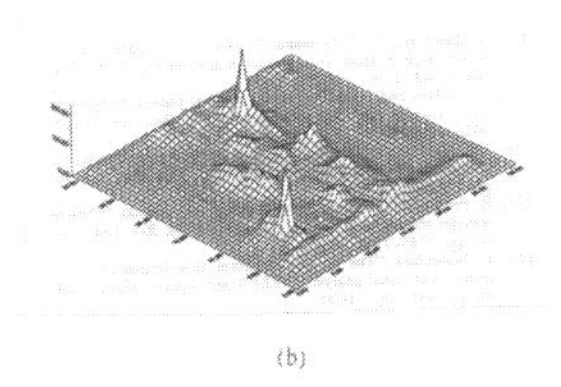

(b)

Fig. 7.30. (a) Input scene. (b) Output correlation distribution.

where

$$\mathrm{DR} = \frac{I_{\text{target}-\min}}{I_{\text{antitarget}-\max}} \tag{7.33}$$

is known as the *target discrimination ratio*, $I_{\text{target}-\min}$ is the minimum correlation peak intensity of the target set, and $I_{\text{antitarget}-\max}$ is the maximum correlation peak intensity of the antitarget set. The reason for using the combined criteria is that, although the correlation peak intensity (CPI) is primarily used to evaluate the capacity of the composite filters, we also have to consider the discrimination ability of these filters; that is, the DR should also be used as a *criterion* to evaluate the capacity of the filters. Thus, we see that these two criteria will be simultaneously used for the evaluation of performance capacity.

To investigate performance capacity, a set of 32×32, 64×64, and 128×128 pixel-element SABCFs are synthesized using the same target and antitarget images. This set of SABCFs is constructed with and without the antitarget set, respectively. By implementing this set of SABCFs respective to the JTC, normalized correlation peak intensities (CPIs) as a function of number of training images have been plotted. We have seen that the CPIs montonically

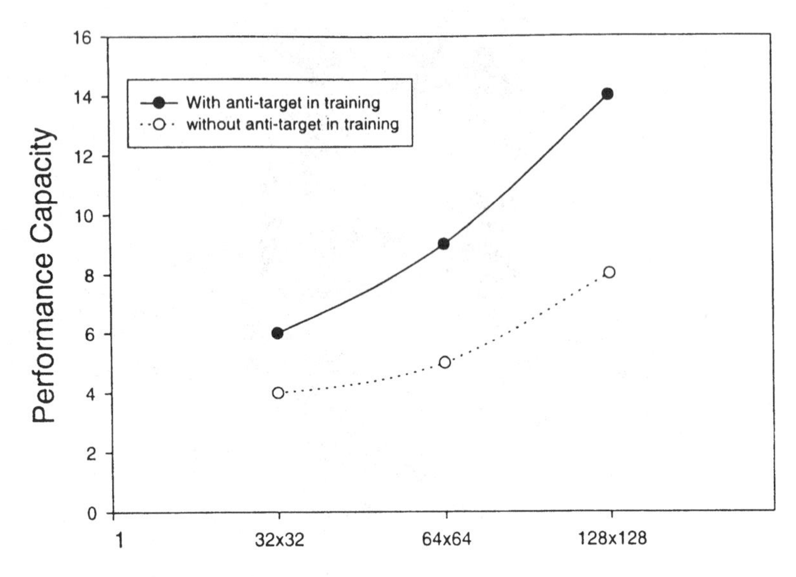

Fig. 7.31. Performance capacity as a function of pixel element.

decrease as the number of training images increases. We have also seen that the filters synthesis using both target and antitarget sets have higher detection performance, as compared with those filters not using the antitarget set. As anticipated, we have also seen that the set using a higher pixel element offers higher performance capacity.

Similarly, the target discrimination ratio (DR) (i.e., the *reliability* of detection) as a function of training images can also be obtained, in which we see that higher pixel-element SABCFs perform better. To conclude our observations, a plot of performance capacity is provided in Fig. 7.31. We see that the capacity increases as the number of pixel elements increases, and the ones that used both target and antitarget sets offer higher performance capacity.

7.5.2. QUANTIZATION PERFORMANCE

Limited dynamic range of a SA spatial-domain composite filter is one of the merits of practical implementation in a JTC; simplicity and realism. We now demonstrate the performance of quantized composite filters (QCFs), as applied to a JTC. Instead of synthesizing the composite filter into bipolar form, we have synthesized the filters into $2N + 1$ quantized gray levels. The SA algorithm for this quantized-level filter synthesis can be briefly described as follows:

1. Determine the effective and noneffective pixels of the training sets.

2. Noneffective pixels are randomly assigned with equal probability within the interval $[-N, N]$, where $n - 0, \pm 1, \pm 2, \ldots, \pm N$, and $2N + 1$ is the number of quantized gray levels.
3. Assign the desired correlation profiles for the target and the antitarget sets (i.e., W^t and W^a).
4. Set a desired allowable error value (e.g., a 1% error rate is used).
5. Assume the initial QCF has $2N + 1$ randomly distributed gray levels, and the initial system temperature is T. For example, the initial kT for the desired correlation energy is given in step 3.
6. Calculate the initial system energy E, which can be determined by its mean-square-error analysis between the desired and the actual correlation profiles, as given by

$$E = \sum_{m=1}^{N} \left\{ \iint [(0^t_m(x, y) - W^t(x, y))^2 + (0^a_m(x, y) - W^a(x, y))^2] \, dx \, dy \right\},$$

   where 0^t_m and 0^a_m are the actual target (auto) and the antitarget (cross) correlation profiles, where the subscript m represents the mth training image.
7. By perturbing one of the effective pixels, a new system energy can be calculated. If the new system energy decreases (e.g., $\Delta E < 0$), the perturbed pixel is unconditionally accepted; otherwise, the acceptance is based on the Boltzmann probability distribution.
8. Repeat the preceding steps for each of the effective pixels until the system energy is stable.
9. Then decrease the system temperature to a lower value and repeat steps 7 and 8, and so forth.
10. By repeating steps 7 to 9, a global minimum energy state of the system (i.e., the filter) will eventually be established. The system reaches the stable condition as the number of iterations increases beyond $4 \times M$, where M is the total number of pixels within the QCF.

Notice that the QCF is a spatial-domain filter; the effective pixels are basically the image pixels of the training sets. We have assumed that the noneffective pixels are randomly distributed within the interval $[-N, N]$, so they would not actually affect the output correlation peak intensity. This makes the QCF immune to background disturbances. To further optimize the noise performance, a bright background can be added into the antitarget training sets for SA QCF synthesis.

One of the major objectives in designing a QCF is to improve discriminability against the similar targets; namely, the antitarget set. The pixel value of the QCF is intentionally designed to have low-quantization levels, which can be actually implemented in a commercially available SLM. One important aspect of pattern recognition is *detectability*, which can be defined as the correlation peak intensity–to–sidelobe ratio, as given by

$$\text{detectability} \triangleq \frac{I_{\text{peak}}}{I_{\text{sidelobe}}}. \tag{7.34}$$

The other aspect of pattern recognition is *accuracy of detection*, which can be determined by the sharpness of the correlation profile, such as

$$\text{accuracy} \triangleq \frac{I_{\text{peak}}}{I_0}, \tag{7.35}$$

where

$$I_0 = \iint |R(x, y)|^2 \, dx \, dy, \tag{7.36}$$

and $R(x, y)$ is the output correlation distribution. The major objective of using a SA composite filter is to improve *discriminability* against similar targets. By using the conventional Fisher ratio (FR),

$$\text{FR} \triangleq \frac{|E\{I_1(0)\} - E\{I_2(0)\}|^2}{[\text{Var}\{I_1(0)\} + \text{Var}\{I_2(0)\}]^{1/2}}, \tag{7.37}$$

discriminability among similar targets can be differentiated, where I_1 and I_2 are the correlation and the cross-correlation peak intensities, respectively. In other words, the higher the FR would be, the higher the target discriminability. Another major objective for design of an SA filter is *reliability* of the target detection, which can be defined by the following discrimination ratio:

$$\text{reliability} = \text{DR} \triangleq \frac{\min_{i \in \pi_1} \{I_i\}}{\max_{j \in \pi_2} \{I_j\}}, \tag{7.38}$$

where I_i and I_j denote the correlation and the cross-correlation peak intensities, π_1 and π_2 represent the target and the antitarget sets, respectively, and min and max are the minimum and the maximum values that belong to the corresponding set. In other words, the higher the DR, the higher the *reliability* of target detection.

One of the advantages of using SAQCF is that the filters are *positive real* function filters, unlike the bipolar filter, which can be directly implemented in an amplitude-modulated SLM. By using the same training sets as shown in Fig. 7.29, several QCFs, for $N = 3, 5, \ldots, 17$ quantized levels, have been syn-

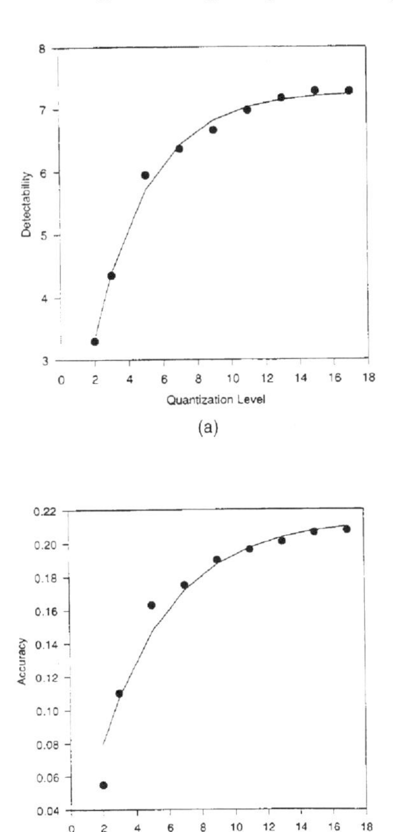

Fig. 7.32. (a) Detectability and (b) accuracy as a function of quantization level.

thesized. We have subsequently implemented this set of QCFs in a JTC, by which the *detectability* and *accuracy of detection* as a function of quantized levels are plotted in Figs. 7.32a and b, respectively. We see that both detectability and accuracy improve as the quantization level increases. However, they slow down somewhat when the quantization level is higher than 10 levels. We have also plotted the *discriminability* and the *reliability* of detections, as shown in Fig. 7.33. We see that both discriminability and reliability improve as quantization level increases, but discriminability levels off as $N > 10$.

Since Fisher ratio (FR) is affected by background noise, we have added 21 random noise levels into the test targets. In other words, the noisy background is formed by independently adding a zero-mean Gaussian noise to each pixel element of the test target. The FR for different quantization levels as a function of noise standard deviation is plotted in Fig. 7.34. Note that target discrimina-

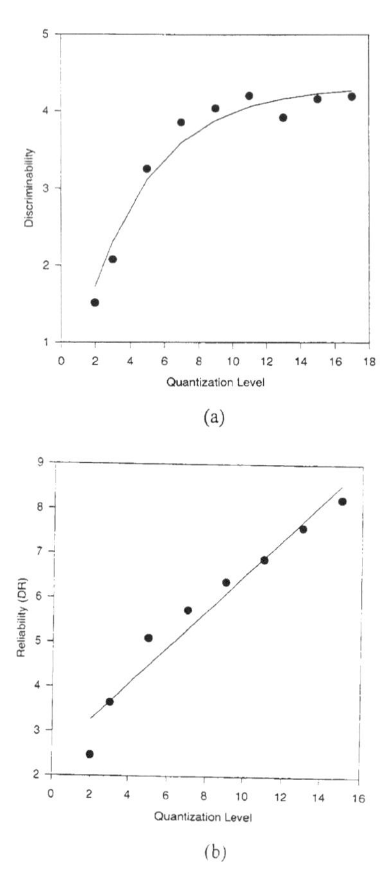

Fig. 7.33. (a) Discriminability and (b) reliability as a function of quantization level.

bility performs better for higher quantization levels, and it decreases rather rapidly as the standard deviation of the noise increases. Also note that, as the quantization level increases, the improvement of the FR tends to level off, as shown in Fig. 7.33a.

7.6. PATTERN CLASSIFICATION

Artificial neural pattern classifiers can be roughly classified into four groups: *global* discriminant classifiers, *local* discriminant classifiers, *nearest neighbor* classifiers, and *rule-forming* classifiers. In this section, we focus on nearest neighbor classifiers (NNCs) that perform classifications based on the distance

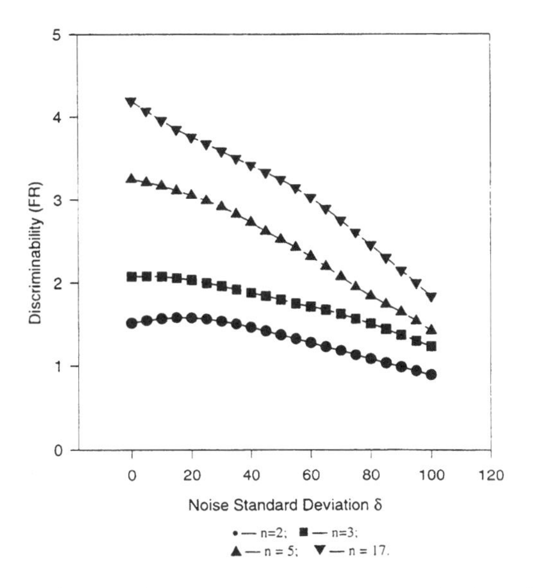

Fig. 7.34. Discriminability as a function of noise standard deviation for different gray levels.

among an unknown input and previously stored exemplars. They train extremely fast, but require a large amount of computation time on a serial processor for classifications, and also require large amounts of memory. While the memory requirement might be alleviated by rapid development of VLSI technologies, the calculation would still be limited by the bottleneck of serial processing. On the other hand, by taking advantage of free-space interconnectivity and parallel processing of optics, a hybrid JTC can be used as an NNC.

7.6.1. NEAREST NEIGHBOR CLASSIFIERS

A typical NNC is presented in Fig. 7.35, in which the first layer is the inner-product layer and the second layer is the maxnet layer.

Inner-Product Layer

Let $\{w_m(x), m = 0, 1, \ldots, M-1, x = 0, 1, \ldots, N-1\}$ be the interconnection weight matrix (IWM) of the inner-product layer, where M and N are the numbers of stored exemplars and input neurons, respectively. When an unknown input $u(x)$ is presented to the inner-product layer, the output will be

$$y_m = \sum_{x=0}^{N-1} w_m(x)u(x), \qquad m = 0, 1, \ldots, M-1. \tag{7.39}$$

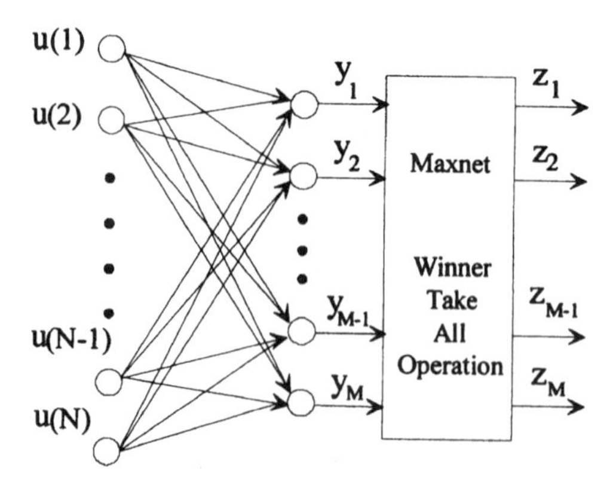

Fig. 7.35. A nearest neighbor classifier (NNC).

The preceding equation can be written as

$$y_m(\alpha) = \sum_{x=0}^{N-1} w_m(\alpha + x)u(x), \qquad m = 0, 1, \ldots, M-1, \tag{7.40}$$

which yields the correlation process as given by

$$y_m(x) = w_m(x) \otimes u(x), \qquad m = 0, 1, \ldots, M-1, \tag{7.41}$$

where $\otimes$ represents the correlation operation. The correlation peak tells which pattern the input belongs to and how much it has shifted relative to the trained pattern. While correlation-based processing may not be attractive for electronic processors, it is rather a simple matter for optical correlators.

MAXNET

After the inner-product operation, a *winner-take-all* operation is needed to locate the hidden node with the maximum inner product. A winner-take-all network using the adaptive thresholding maxnet (AT-maxnet) is shown in Fig. 7.36, in which the output of the maxnet is a function of an adaptive threshold, as given by

$$z_i(t) = f(y_i) = \begin{cases} 1, & y_i - \theta(t) > 0, \\ 0, & y_i - \theta(t) \leqslant 0, \end{cases} \qquad i = 0, 1, \ldots, M-1, \tag{7.42}$$

where $\theta(t)$ denotes an adaptive threshold value at the the tth iteration.

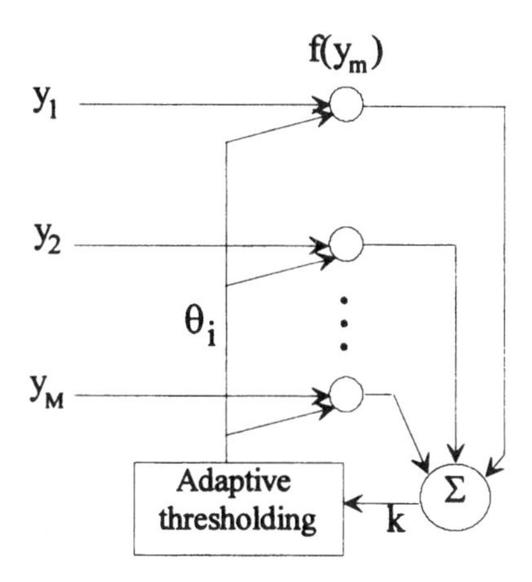

Fig. 7.36. The AT-maxnet.

The threshold $\theta(t)$ is controlled by the number of nonzero output nodes. Underthresholding would give rise to multiple nonzero nodes, while overthresholding would reject all the nodes. Therefore, the maximum hidden node can be extracted by simply adjusting the threshold value $\theta(t)$ in each iteration. Let us denote θ_{min} and θ_{max} to be the rejection and the maximum threshold levels, respectively. By denoting $\theta(t)$ the threshold level at the tth iteration, and k the number of nonzero output nodes, as given by

$$k = \sum_{m=0}^{M-1} z_m(t), \tag{7.43}$$

the adaptive thresholding strategy can be formed as follows:

1. Set $\theta(0) = \theta_{\text{min}}$.
   (a) Terminate the process if $k = 0$, which means the input pattern does not belong to any of the stored exemplars.
   (b) Terminate the process if $k = 1$, which means that the maximum node has been found.
   (c) Set $t = 1$ and go to step 2 if $k > 1$.
2. Set

   $$\delta\theta(t) = \frac{\theta_{\text{max}} - \theta_{\text{min}}}{2^t};$$

   then the threshold value for the tth iteration is

   $$\theta(t) = \begin{cases} \theta(t-1) + \delta\theta(t), & k > 1, \\ \theta(t-1) - \delta\theta(t), & k = 0 \end{cases}$$

3. Terminate the iteration process when $k=1$ or $t=\log_2[(\theta_{max}-\theta_{min})/\Delta\theta]$, where t is the iteration number, and $\Delta\theta$ is the minimum value that the AT-maxnet can resolve.

Notice that in the AT-maxnet, only a maximum number of

$$\log_2[(\theta_{max} - \theta_{min})/\Delta\theta]$$

iterations is needed to make a decision, which is independent of the number of stored exemplars. For example, given $\theta_{max} = 256$, $\theta_{min} = 0$, and $\Delta\delta = 1$, less than eight iterations are needed to search for the maximum node. Furthermore, the AT-maxnet can tell when multiple maximum nodes occur by checking whether k is greater than one or not, after $\log_2[(\theta_{max} - \theta_{min})/\Delta\theta]$ iterations, while the conventional maxnet cannot.

7.6.2. OPTICAL IMPLEMENTATION

It has been shown in a preceding section that transforming real-valued functions to phase-only functions offers higher light efficiency and better pattern discriminability in a JTC. Let

$$\{v_m(x), m = 0, 1, \ldots, M - 1, x = 0, 1, \ldots, N\}$$

be a set of exemplars. The phase representation of $v_m(x)$ is given by

$$T[v_m(x)] = \frac{v_m(x) - G_{min}}{G_{min} - G_{max}} 2\pi,$$

where $T[\cdot]$ represents a gray-level–to-phase transformation operator, $G_{min} = \min_{m,s}[v_m(x)\}$, and $G_{max} = \max_{m,x}\{v_m(x)\}$. The phase-transformed IWM of the inner-product layer can be written as

$$w_m(x) = \exp\{iT[v_m(x)]\}, \qquad m = 1, 2, \ldots, M, \qquad x = 1, 2, \ldots, N.$$

Thus, the inner product between $w_m(x)$ and the phase representation of the input $u(x)$ can be expressed as

$$pc_m = \sum_{x=1}^{N} \exp\{iT[v_m(x)] - iT[u(x)]\}.$$

To introduce the shift-invariance property, the preceding equation can be

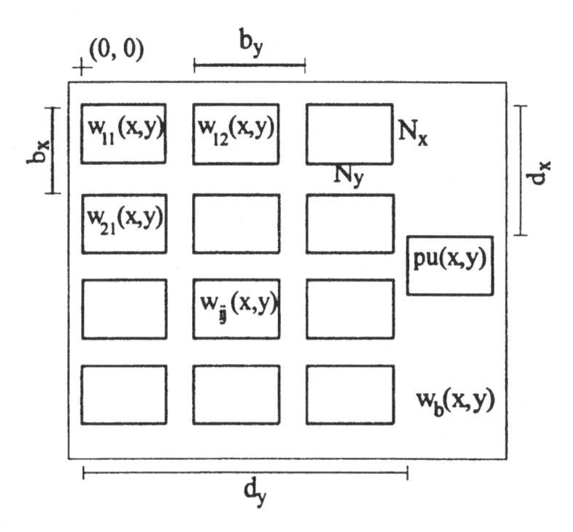

Fig. 7.37. An input arrangement for the JTC–NNC.

written in a correlation form as

$$pc_m(\alpha) = \sum_{x=1}^{N} \exp\{iT[v_m(x+\alpha)] - iT[u(x)]\}. \tag{7.44}$$

The correlation between the phase-transformed IWM and input pattern $u(x)$ can be easily realized by using a phase-modulated SLM. Notice that when binary input exemplars are used, this phase-transformed NNC is actually the *Hamming net*. In other words, the Hamming net is a special case of the proposed *phase-transformed* NNC.

Let the size of the stored exemplars be $N_x \times N_y$. The joint input to the JTC is illustrated in Fig. 7.37, in which a set of $M_x \times M_y$ exemplars are displayed on the left-hand side. The corresponding phase representation can be written as

$$f(x, y) = \sum_{m_x=0}^{M_x-1} \sum_{m_y=0}^{M_y-1} w_{m_x m_y}(x + m_x b_x, y + m_y b_y) \exp(ip\pi + iq\pi) w_b(x, y)$$
$$+ pu(x - d_x, y - d_y), \tag{7.45}$$

where (b_x, b_y) and (d_x, d_y) are the dimensions shown in the figure and $pu(x, y) \exp\{iT[u(x, y)]\}$.

The corresponding nonzero-order JTPS can be shown as

$$\text{NJTPS} = 4\left|\sum_{m_x=0}^{M_x-1} \sum_{m_y=0}^{M_y-1} W_{m_x m_y}(p, q) PU(p, q)\right| \cos[\delta\theta_{wu}(p, q)]$$
$$+ 4\left|\sum_{m_x=0}^{M_x-1} \sum_{m_y=0}^{M_y-1} W_b(p - \pi, q - \pi) PU(p, q)\right| \cos[\delta\theta_{bu}(p, q)], \tag{7.46}$$

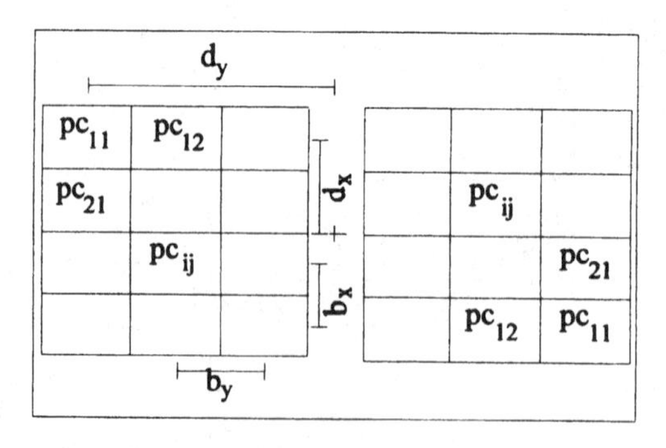

Fig. 7.38. The output correlation distribution from the JTC–NNC.

where

$$\delta\theta_{wu}(p, q) = \theta_{m_x m_y}(p, q) - \theta_u(p, q) + p(b_x m_x + d_x) + q(b_y m_y + d_y),$$

$$\delta\theta_{bu}(p, q) = \theta_b(p - \pi, q - \pi) - \theta_u(p, q) + pd_x + qd_y,$$

and θ_s are the corresponding phase components. Thus the output correlation between $u(x)$ and $\{w_{m_x m_y}(x, y)\}$ can be written as

$$pc(x, y) = \sum_{m_x=0}^{M_x-1} \sum_{m_y=0}^{M_y-1} w_{m_x m_y}(x + m_x b_x, y + m_y b_y) \otimes pu^*(x - d_x, y - d_y)$$

$$+ w_b(x, y) \exp(ip\pi + iq\pi) \otimes pu^*(x - d_x, y - d_y). \tag{7.47}$$

Since $w_{m_x m_y}(x, y)$ and $w_b(x, y)$ are physically separated, the cross-correlation between $w_b(x, y)$ and $pu(x, y)$ can be ignored. As shown in Fig. 7.38, the correlation between the input pattern and each exemplar will fall in a specific area with the size of $N_x \times N_y$. The highest correlation peak intensity within that area represents a match with this stored exemplar. The shift of the input pattern cannot exceed $N_x \times N_y$; otherwise, ambiguity occurs. Nevertheless, the full shift invariance is usually not required in 2-D classified applications. For instance, in character recognition, the shifts are usually introduced by noise in the segmentation process. Thus, the number of pixels shifted is often quite limited.

For demonstration, the input to the phase-transform JTC is shown in Fig. 7.39a, in which the Normal Times New Roman "2" is a segmented character to be classified. By using the nonzero order JTC, the corresponding output correlation distribution is shown in Fig. 7.39b. The autocorrelation peak intensity has been measured to about 10 times higher than the maximum

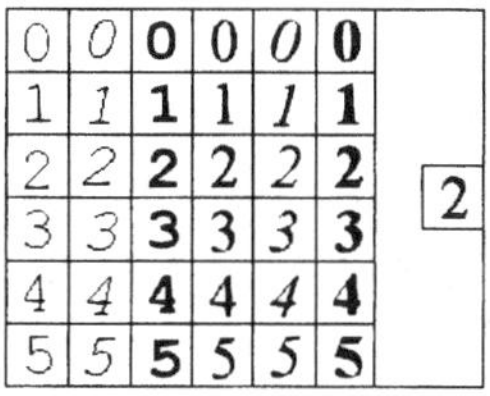

(a)

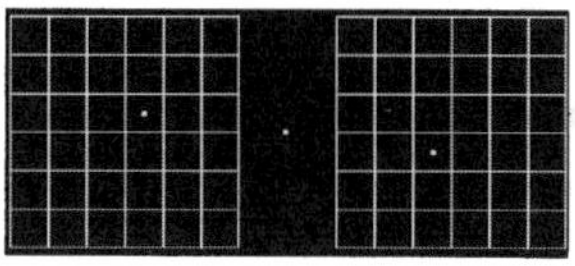

(b)

Fig. 7.39. (a) Input to the JTC–NNC, where a normal Times New Roman "2" is used as the input character. (b) Output correlation distribution.

cross-correlation peak intensity. The grids provided show the region of interest for each character.

7.7. PATTERN RECOGNITION WITH PHOTOREFRACTIVE OPTICS

Recent advances in photorefractive (PR) materials have stimulated interest in phase-conjugate correlators for pattern recognition. Although SLMs can be used to display complex spatial filters, current state-of-the-art SLMs are low-resolution and low-capacity devices. On the other hand, PR materials offer real-time recording, high resolution, and massive storage capacity; all desirable traits for multiplexed matched filter synthesis.

7.7.1. DETECTION BY PHASE CONJUGATION

In most optical correlators the input objects are required to be free from any phase distortion. The input objects are usually generated by an SLM. However, most SLMs introduce phase distortion, which severely degrade detection performance. Other problems are that the autocorrelation distribution is too broad and the cross-correlation intensity is too high. These result in low accuracy and reliability of detection. To improve performance, the input target can be pre-encoded, for which the inherent nonlinearities in phase conjunction can be used for the encoding process. Let us consider a *four-wave*

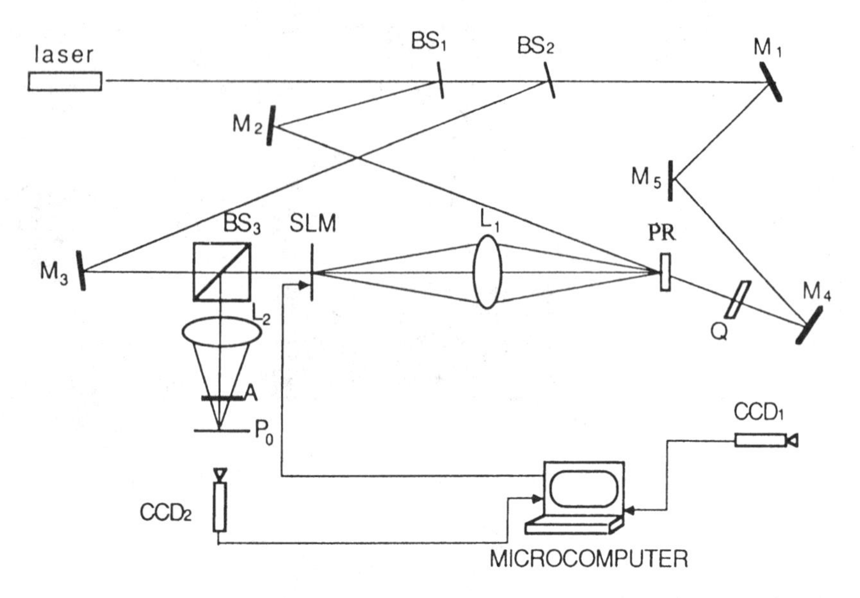

Fig. 7.40. Experimental setup. *BS*, beamsplitter; *M*, mirror; *Q*, quarter-wave plate; *A*, analyzer; L_1, imaging lens; L_2, transform lens; P_o, output plane.

mixing JTC, as shown in Fig. 7.40, in which the object beam is applied at the input of a JTC. This configuration has two major advantages:

1. The phase distortion due to the input SLM can be compensated for by the conjugated wavefront.
2. The modulation produced by the phase conjugation can be used to improve the accuracy of detection.

Since the readout beam is adjusted beyond the coherent length with respect to the writing beams; the setup is essentially a *two-wave mixing* configuration. Notice that the expanded collimated beam is divided into three paths. The first one is directed toward the PR crystal serving as the reference beam; the second beam is used to illuminate input objects O_1 and O_2, which are imaged onto the crystal; and the third beam (directed by M1, M5, and M4) serves as the readout beam. Thus we see that the reconstructed beam from the crystal is imaged back to the input objects for phase removal. After passing through the input objects, the beam is then jointly transformed in the output plane, P_o via a beamsplitter, BS_3. We notice that a half-wave plate, Q, is used to rotate the polarization of the readout beam and an analyzer, A, is used to reduce the light scattered from the optical elements in the system.

It is trivial the object beam can be written as

$$O(x, y)\exp[i\phi(x, y)] = O_1(x-b, y)\exp[i\phi_1(x-b, y)] + O_2(x+b, y)\exp[i\phi_2(x+b, y)], \tag{7.48}$$

where O_1, O_2, ϕ_1, and ϕ_2 are the amplitude and phase distortion of the input objects, respectively; b is the mean separation of the two input objects; and O_1, O_2 are assumed positive real. At the image plane (i.e., the crystal), the object beam is given by

$$O\left(\frac{x}{M}, \frac{y}{M}\right)\exp\left[i\phi\left(\frac{x}{M}, \frac{y}{M}\right)\right]\exp\left[ik\frac{x^2+y^2}{2L}\right], \tag{7.49}$$

where M represents the magnification factor of the imaging system; $L = s - f$, where s is the image distance and f is the focal length of lens L_1; and $k = 2\pi/\lambda$, where λ is the wavelength of the light source.

Thus, the reconstructed beam emerging from the crystal is given by

$$O'\left(\frac{x}{M}, \frac{x}{M}\right)\exp\left[-i\phi\left(\frac{x}{M}, \frac{y}{M}\right)\right]\exp\left[-ik\frac{x^2+y^2}{2L}\right]\exp\left[i\theta\left(\frac{x}{M}, \frac{y}{M}\right)\right]. \tag{7.50}$$

Notice that this expression does not represent the exact phase conjugation of Eq. (7.49). The amplitude distribution of $O'(x, y)$ deviates somewhat from, $O(x, y)$, which is primarily due to the nonlinearity of the crystal. However, a phase shift $\theta(x, y)$ between the phase conjugation and the reconstructed beam is also introduced, which is independent on the intensity ratio of the object beam and the reference beam. The reconstructed beam, imaged back on the object plane, can be written as

$$O'(x, y)\exp[-i\phi(x, y)]\exp[i\theta(x, y)].$$

After passing through the input objects, the complex light distribution becomes

$$O(x, y)O'(x, y)\exp[i\theta(x, y)], \tag{7.51}$$

which removes the phase distortion $\phi(x, y)$ of the input objects. By joint-transforming the preceding equation, a phase-distortion–free JTPS can be captured by the CCD_2.

For demonstration, a pair of input objects, shown in Fig. 7.41a, has been added with random phase noise (e.g., shower glass). Figure 7.41b shows the

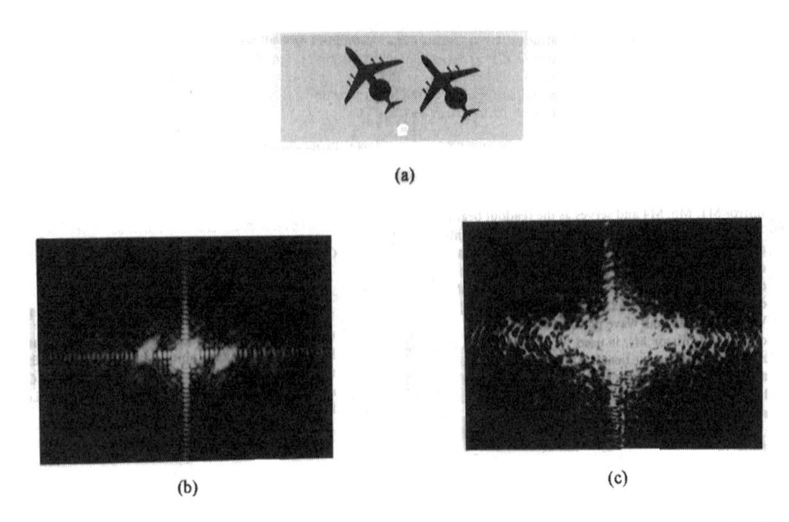

Fig. 7.41. (a) Input objects. (b) With phase compensation. (c) Obtained from CJTC.

output correlation spots as obtained with the preceding phase-conjugation detection. For comparison, the result obtained without using phase compensation is shown in Fig. 7.41c; it fails to detect the target object.

7.7.2. WAVELENGTH-MULTIPLEXED MATCHED FILTERING

We discussed Bragg diffraction-limited correlators in Chapter 2, in which we showed that a reflection-type thick-crystal filter performs better. Added to high wavelength selectivity, a wavelength-multiplexed matched filter can be realized.

For convenience, we recall this reflection-type correlator (RC) as depicted in Fig. 7.42, in which q_1 and q_2 represent the input scene and reference target, respectively. Besides the Bragg limitation, the PR crystal also imposes a *focal-depth limitation*, as given by

$$\delta \leqslant 2\lambda \left(\frac{f}{b}\right)^2, \tag{7.52}$$

where b is the width of the reference target, f is the focal length, and λ is the illuminating wavelength. It is apparent that focal depth within the crystal should be limited by

$$\delta' \leqslant 2\lambda\eta \left(\frac{f'}{b}\right)^2, \tag{7.53}$$

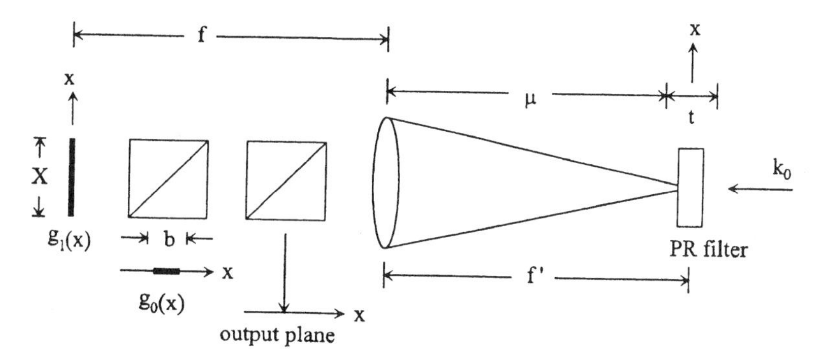

Fig. 7.42. A PR-based RC. $g_0(x)$, reference target; $g_1(x)$, input scene; k_0, reference beam.

where η is the refractive index of the PR crystal, and f' is the apparent (or actual) focal length, which is given by

$$f' = u + \delta' \approx f + 2\lambda(\eta - 1)\left(\frac{f}{b}\right)^2. \tag{7.54}$$

Thus, the allowable thickness of the PR crystal filter is limited by

$$t = 2\delta' \leqslant 4\lambda\eta\left(\frac{f}{b}\right)^2, \tag{7.55}$$

in which we assume that the thick-crystal filter is centered at the apparent focal length f'. By using the coupled-wave theory, the output correlation peak intensity can be shown as

$$R(s)|_{\mathrm{RC}} = \left|\int_{-b/2}^{b_2} |g_0(x)|^2 \operatorname{sinc}\left[\frac{\pi t}{\eta\lambda}\frac{S(x_0 - S)}{f^2}\right] dx\right|^2. \tag{7.56}$$

The width of the sinc factor is given by

$$W = \frac{2\eta\lambda f^2}{st}. \tag{7.57}$$

To maintain a high correlation peak intensity, the target width should be adequately small, such as

$$\frac{b}{2} + s \leqslant W,$$

for which we have

$$s \leqslant \frac{4\eta\lambda f^2}{bt} - \frac{2s^2}{b}. \tag{7.58}$$

It can be written as

$$s \leqslant \left[\frac{2\eta\lambda f^2}{t} + \frac{b^2}{16}\right]^{1/2} - \frac{b}{4}, \tag{7.59}$$

in which the *shift tolerance* of the target is inversely proportional to the square root of the thickness of the PR filter. Since the reflection filter has very high angular selectivities, the target-shift limitation is somewhat sensitive to the target translation. In other words, by making the thickness, t, sufficiently small, the shift tolerance in an RC can be made sufficiently high. Nevertheless, reducing the thickness of the PR filter also decreases storage capacity.

To demonstrate the shift tolerance of the RC, a 1 cm^3 Ce:Fe doped $LiNbO_3$ crystal has been used for the reflection-type matched-filter synthesis. By tuning the wavelength at $\Delta\lambda = 0.1$ nm per step with 10 sec per exposure, a set of wavelength-multiplexed reflection-type matched filters is synthesized. After constructing the PR filter, an input target can be used to reconstruct the reference beam by simply scanning the wavelength of the light source. If the input target is matched with one of the recorded matched filters, a sharp correlation peak can be detected by a CCD camera. Thus, we see that multitarget detection can be done by tuning the wavelength of the light source so that the spectral contents of the multiplexed filter can be exploited.

For shift-invariant demonstration, a set of English letters, shown in Fig. 7.43, are used as the training targets for the wavelength-multiplexed matched filter synthesis. Notice that each letter is centered on the optical axis and recorded separately during the construction of the PR filter. The corresponding

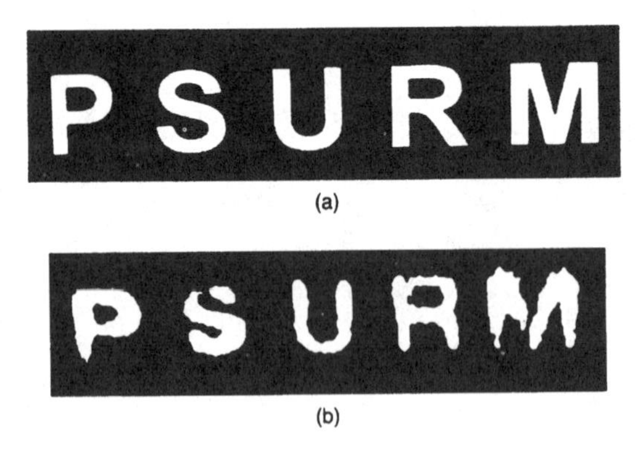

Fig. 7.43. (a) The input training objects and (b) the corresponding restructed images: $\lambda = 670.0$ nm, 670.1 nm, 670.2 nm, 670.3 nm, and 670.4.

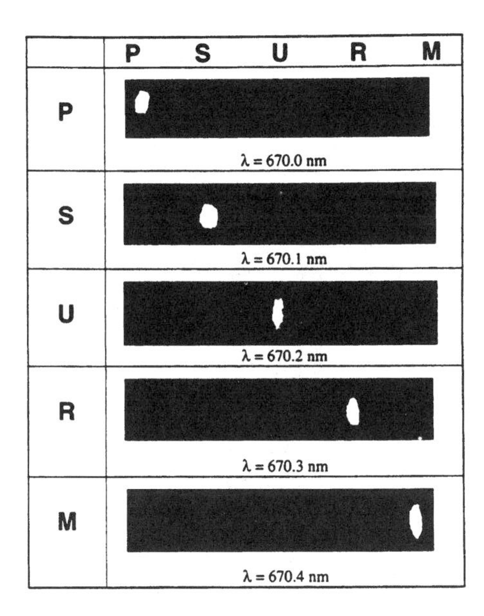

Fig. 7.44. Output correlation peak intensities show shift invariance and the spectral content.

reconstructed holographic images obtained with the specific spectral line are shown in Fig. 7.43b. To test the shift invariance of the PR filter, the five input letters are inserted at the input plane at the same time. Notice that these letters have been shifted with respect to the recorded PR filter. The corresponding correlation peak intensities for each spectral line are detected by a CCD camera. By properly thresholding these detected peak intensities, a set of correlation peaks corresponding to the recorded spectral contents is obtained, as depicted in Fig. 7.44. From these results, we see that the PR–based RC can indeed preserve the *shift invariance* by using a wavelength-multiplexed reflection-type PR filter.

7.7.3. WAVELET MATCHED FILTERING

Wavelet transform (WT) analysis is a viable alternative to Fourier transforms for optical pattern recognition. The WT has been utilized for multiresolution image analysis and in optical implementations. The 2-D WT of a signal is given by

$$\begin{aligned} w(a_x, a_y, b_x, b_y) &= \mathrm{WT}\{s(x, y)\} \\ &= \iint s(x, y) h^*_{ab}(x, y)\, dx\, dy, \end{aligned} \tag{7.60}$$

where the superasterisk represents the complex conjugate of the wavelet, which is defined as

$$h_{ab}(x, y) = \frac{1}{\sqrt{a_x a_y}} h\left(\frac{x - b_x}{a_x}, \frac{y - b_y}{a_y}\right). \tag{7.61}$$

This function is obtained by translating and dilating the *analyzing wavelet function* $h(x, y)$. Thus, the WT can be written as

$$\begin{aligned} w(a_x, a_y, b_x, b_y) &= \frac{1}{\sqrt{a_x a_y}} \iint s(x + b_x, y + b_y) h^*\left(\frac{x}{a_x}, \frac{y}{a_y}\right) dx\, dy \\ &= \frac{1}{\sqrt{a_x a_y}} s(x, y) \otimes h^*\left(\frac{x}{a_x}, \frac{y}{a_y}\right), \end{aligned} \tag{7.62}$$

which is essentially the cross-correlation between the signal and the dilated analyzing wavelet. Furthermore, $h(x/a_x, y/a_y)$ can be interpreted as a bandpass filter governed by the (a_x, a_y) dilation. Thus, both the dominant frequency and the bandwidth of the filter can be adjusted by changing the dilation of $h(x/a_x, y/a_y)$. In other words, the WT is essentially a filtered version of the input signal $s_i(x, y)$. The correlation of two WTs with respect to the input signal is an estimation of the similarity between the two signals. Due to the inherent local feature selection characteristic of the WT, the wavelet matched filter (WMF) generally offers a higher discriminability than the conventional matched filter (CMF). Although implementing the WMF in conventional VLC and JTC has been reported, the WMF can be synthesized with a PR–based Fourier-domain correlator.

We first illustrate WMF construction. If we let the WT target signal $s(x, y)$ be

$$w(a_x, a_y, b_x, b_y) = \frac{1}{\sqrt{a_x a_y}} s(x, y) \otimes h^*\left(\frac{x}{a_x}, \frac{y}{a_y}\right), \tag{7.63}$$

the corresponding Fourier-domain representation is

$$W(p, q) = \frac{1}{\sqrt{a_x a_y}} S(p, q) H_a^*(p, q). \tag{7.64}$$

Similarly, the Fourier-domain representation of the WT reference signal

$w_r(a_x, a_y, b_x, b_y)$ can be written as

$$W_r(p, q) = \frac{1}{\sqrt{a_x a_y}} S_r(p, q) H_a^*(p, q). \tag{7.65}$$

It is trivial that a conventional WT matched field (CMFW) is

$$\mathrm{CMF}_w(p, q) = W_r^*(p, q). \tag{7.66}$$

If the wavelet signal $w(a_x, a_y, b_x, b_y)$ is inserted in the FDP, the complex light field behind the CMF_w can be written as

$$W(p, q) W_r^*(p, q) = S_t(p, q) \mathrm{WMF}(p, q), \tag{7.67}$$

where

$$\mathrm{WMF}(p, q) = (1/a_x a_y) S_t^*(p, q) |H_a(p, q)|^2 \tag{7.68}$$

is defined as the *wavelet matched filter*. Thus, we see that WMF can be synthesized by conventional interferometric technique with appropriate spectra wavelet modulus $|H_a(p, q)|^2$.

The generation of WMFs is the key issue in the construction of a PR–based WT correlator, as shown in Fig. 7.45. The scaled moduli of the wavelet spectra can be sequentially generated by the SLM located at the back focal plane of lens L1; the emerging light field can be synchronously imaged onto the crystal by the imaging lens L2. Thus, a large number of WMFs can be recorded in the PR crystal as a set of reflection holograms by means of angular multiplexing. The light intensity within the crystal can be written as

$$I = \sum_{n=1}^{N} |S_r |H_n|^2 + R_n|^2, \tag{7.69}$$

where S_r and H_n are the Fourier transforms of the reference signal and the dilated analyzing wavelets, respectively; R_n is the reference beam; and N is the total number of wavelets. If a target signal s is inserted at the input plane, as shown in Fig. 7.45b, the reflected reconstructed light field from the crystal will be

$$O(p, q) = S_t\left(|S_r|^2 \sum_{n=1}^{N} |H|^4 + \sum_{n=1}^{N} |R_n|^2\right) + S_t S_r^* \sum_{n=1}^{N} |H|^2 R_n + S_t^* S_r \sum_{n=1}^{N} |H|^2 R_n^*. \tag{7.70}$$

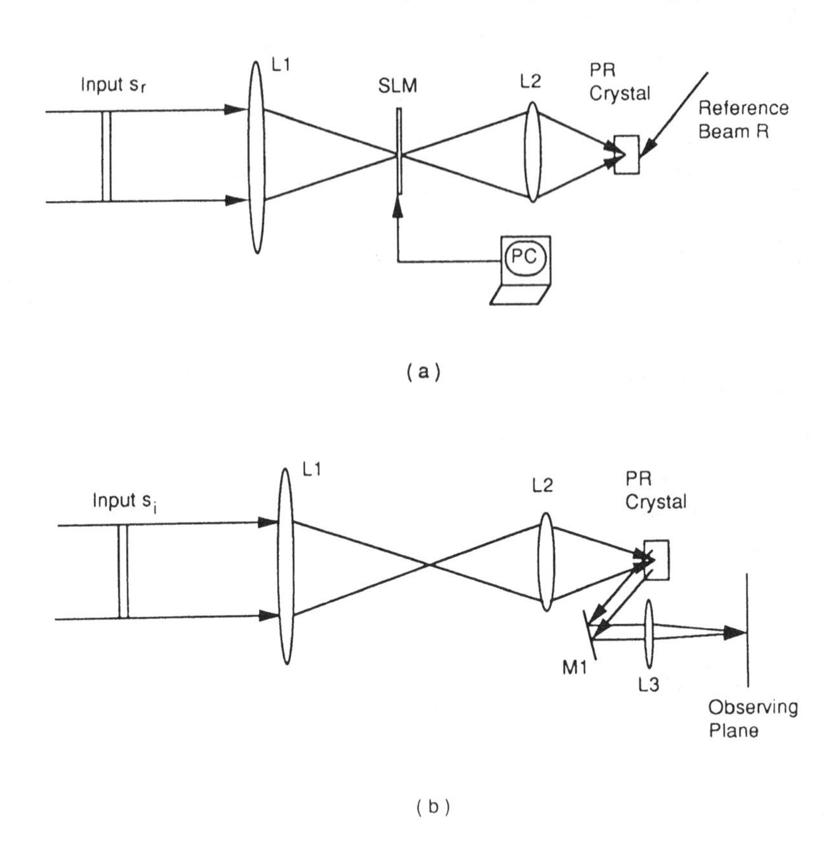

Fig. 7.45. An optical wavelet correlator using a PR crystal. (a) Reflection-type WMFs synthesis, (b) Correlation with a PR WMF.

Thus, by inverse Fourier transforming this equation, the correlation between the w_t and w_r can be obtained from the second term. If the input signal s is identical to the reference signal s_r, an array of autocorrelation peaks can be observed at the back focal plane of lens L3.

To demonstrate feasibility, a fingerprint, shown in Fig. 7.46a, is used for synthesis of the WMF. Since the typical fingerprints are circular striated-type patterns, a Mexican-hat–type analyzing wavelet is used. Inserting the fingerprint transparence at the input plane of the PR–based correlator gives the output correlation distribution, shown in Fig. 7.46b. We see that a sharp correlation peak can be detected. For comparison, the output correlation distribution as obtained by a conventional matched filter is depicted in Fig. 7.46c, in which we see that the WMF performs better.

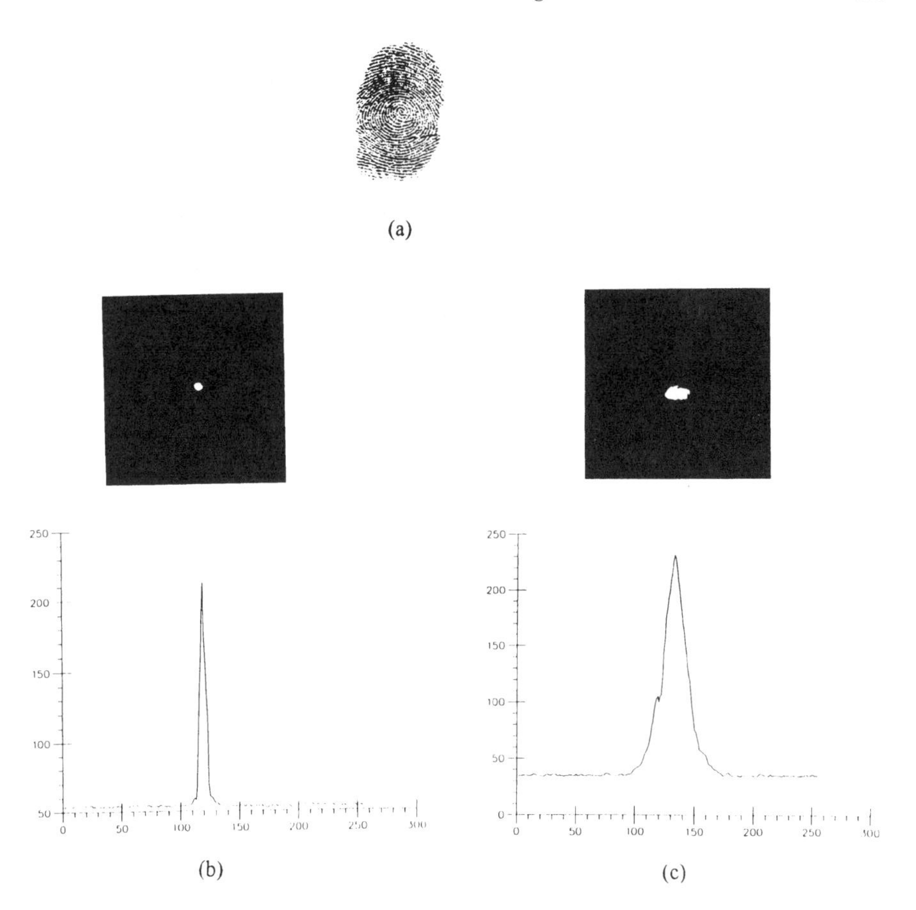

Fig. 7.46. (a) Input objects. Output correlation distributions (a) obtained from a WMF ($a_x = 0.05$, $a_y = 0.05$), and (b) using a CMF.

7.8. NEURAL PATTERN RECOGNITION

Strictly speaking, there are generally two kinds of neural networks (NNs), *supervised* and *unsupervised*. In supervised learning, a teacher is required to supply the NN with both input data and the desired output data, such as training exemplars (e.g., references). The network has to be taught when to learn and when to process information, but it cannot do both at the same time. In unsupervised learning, the NN is given input data but no desired output data; instead, after each trial or series of trials, it is given an evaluation rule that evaluates its performance. It can learn an unknown input during the process, and it presents more or less the nature of human self-learning ability.

Hopfield, Perceptron, error-driven back propagation, Boltzman machine, and interpattern association (IPA) are some of the well-known supervised-learning NNs. Models such as adaptive resonance theory, neocognitron, Madline, and Kohonen self-organizing feature map models are among the best-known unsupervised-learning NNs. In the following subsection, we first discuss a couple of supervised learnings.

7.8.1. RECOGNITION BY SUPERVISED LEARNING

The Hopfield and interpattern association NNs, presented in Sec. 2.9, are typical examples of supervised learning models. For example, if an NN has no pre-encoded memory matrix (e.g., IWM), the network is not capable of retreiving the input pattern. Instead of illustrating a great number of supervised NNs, we will now describe a heteroassociation NN that uses the interpattern association algorithm.

The strategy is to supervise the NN learning, so that the NN is capable of translating a set of patterns into another set. For example, we let patterns A, B, and C, located in a pattern space, to be translated into A′, B′, and C′, respectively, as presented by the Venn diagrams in Fig. 7.47. Then a hetero-association NN can be constructed by a simple *logic* function, such as

$$
\begin{aligned}
I &= A \wedge \overline{(B \vee C)} & I' &= A' \wedge \overline{(B' \vee C')} \\
II &= B \wedge \overline{(A \vee C)} & II' &= B' \wedge \overline{(A' \vee C')} \\
III &= C \wedge \overline{(A \vee B)} & III' &= C' \wedge \overline{(A' \vee B')} \\
IV &= (A \wedge B) \wedge \overline{C} & IV' &= (A' \wedge B') \wedge \overline{C'} \\
V &= (B \wedge C) \wedge \overline{B} & V' &= (B' \wedge C') \wedge \overline{A'} \\
VI &= (C \wedge A) \wedge \overline{B} & VI' &= (C' \wedge A') \wedge \overline{B'} \\
VII &= (A \wedge B \wedge C) \wedge \overline{\varnothing} & VII' &= (A' \wedge B' \wedge C') \wedge \overline{\varnothing}
\end{aligned}
$$

where $\wedge$, $\vee$ and $\overline{}$ stand for the logic AND, OR, and NOT operations, respectively, and $\varnothing$ denotes the empty set.

For simplicity, we assume that A, B, C, A', B', and C' are the input–output pattern training sets, as shown in Fig. 7.48a. Pixel 1 is the common pixel of patterns A, B, and C; pixel 2 is the common between patterns A and B; pixel 3 is the common between A and C; and pixel 4 represents the special feature of pattern C. Likewise, from the output pattern, pixel 4 is the special pixel of pattern B', and so on. A single-layer neural network can therefore be constructed, as shown in Fig. 7.48b. Notice that the second output neuron representing the common pixel of patterns A', B', and C' has positive interconnections from all the input neurons. The fourth output neuron, a special pixel of B', is

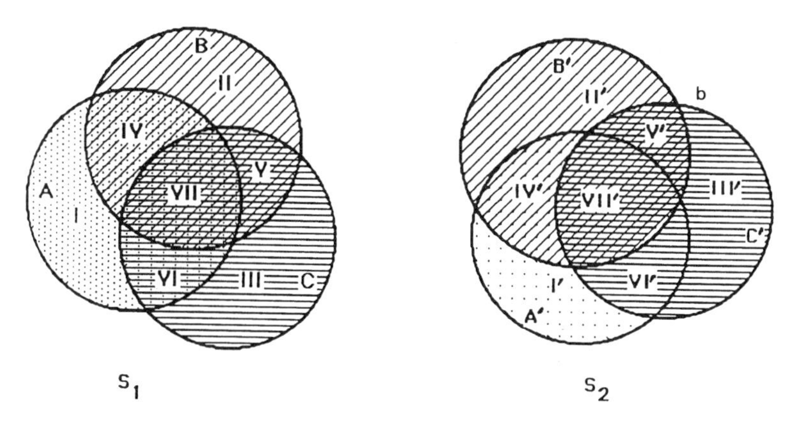

Fig. 7.47. Heteroassociation. S_1 a and S_2 are the input and output pattern spaces.

subjected to *inhibition* from both input neurons 3 and 4, and so on. The corresponding heteroassociation memory matrix is shown in Fig. 7.48c, for which 1, 0. −1 represents the excitory, null, and inhibitory connections. The memory matrix is partitioned into four blocks, and each block represents the corresponding output neurons. For illustration, we show a supervised hetero-association NN, as obtained from the optical NN of Fig. 7.7. A set of Arabic

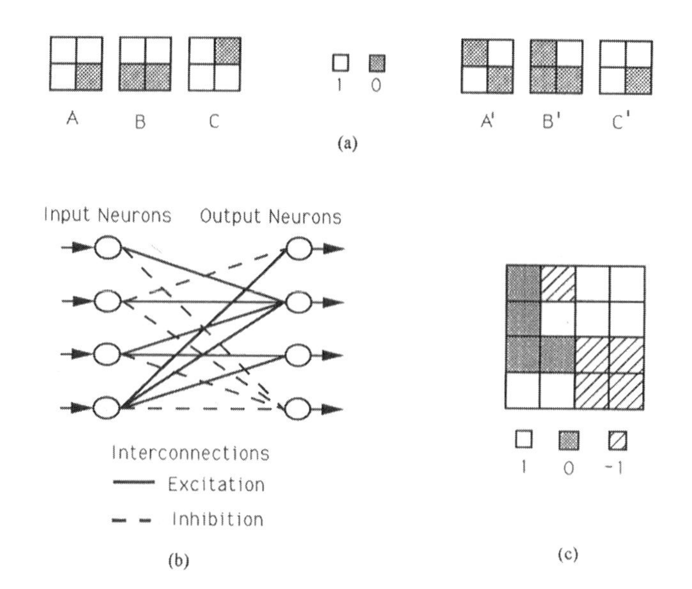

Fig. 7.48. Construction of an heteroasociative IWM (HA IWM). (a) Input–output training sets, (b) A tristate NN, (c) Heteroassociation IWM.

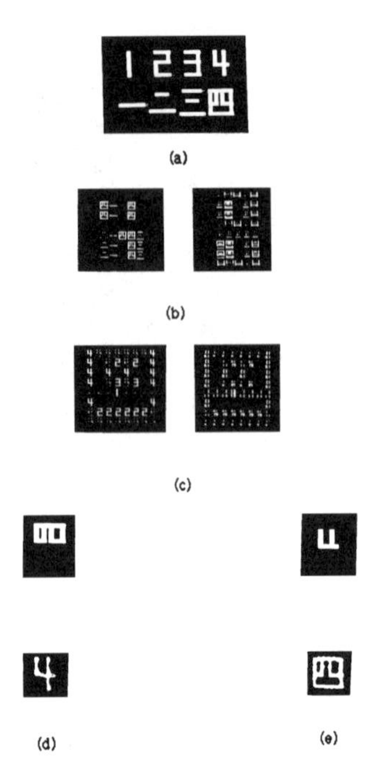

Fig. 7.49. (a) Arabic-Chinese numerics training sets. (b) Positive (left sides) and negative (right sides) parts of the heteroassociation IWM for Chinese to Arabic numerics translation, (c) Positive (left sides) and negative (right sides) parts of the HA IWM for Arabic to Chinese, (d), (e), Partial input patterns to the corresponding input patterns to the corresponding output translations.

and Chinese numeric numbers are presented to the NN. Their heteroassociation memory matrices for translating Arabic to Chinese numeric numbers and Chinese to Arabic numeric numbers are shown in Fig. 7.49. The corresponding translated results from Chinese to Arabic and Arabic to Chinese numerics for partial input patterns are shown in the figure. Thus, we see that NN is inherently capable of retrieving noisy or distorted patterns. In other words, as long as the presented input pattern contains the *main feature* of the supervised training patterns, the artificial NN has the ability of retrieving the pattern (in our case the translated pattern), as does a human brain.

7.8.2. RECOGNITION BY UNSUPERVISED LEARNING

In contrast with supervised learning, unsupervised learning (also called self-learning) means that the students learn by themselves, relying on some simple learning rules and past experiences. For an artificial NN, only the input data are provided, while the desired output result is not. After a single trial or series of trials, an evaluation rule (previously provided to the NN) is used to evaluate the performance of the network. Thus, we see that the network is capable of adapting and categorizing the unknown objects. This kind of self-organizing process is, in fact, a representation of the self-learning capability of a human brain.

To simplify our discussion, we will discuss here only the Kohonen's *self-organizing feature map*. Kohonen's model is the best-known unsupervised learning model; it is capable of performing statistical pattern recognition and classification, and it can be modified for optical implementation.

Knowledge representation in human brains is generally at different levels of abstraction and assumes the form of a *feature map*. Kohonen's model suggests a simple learning rule by continuously adjusting the interconnection weights between input and output neurons based on the matching score between the input and the memory matrix. A single-layer NN consists of $N \times N$ input and $M \times M$ output neurons, as shown in Fig. 7.50. Assume that a set of 2-D vectors (i.e., input patterns) are sequentially presented to the NN, as given by

$$x(t) = \{x_{ij}(t), i, j = 1, 2, \ldots, N\}, \tag{7.71}$$

where t represents the time index and (i, j) specifies the position of the input neuron. Thus, the output vectors of the NN can be expressed as the weighted

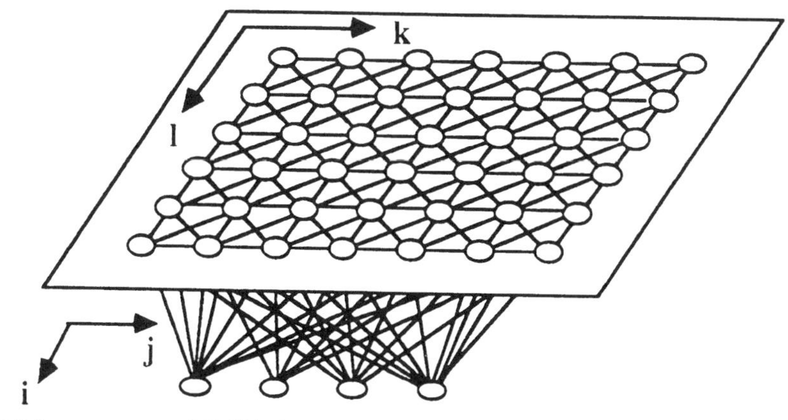

Fig. 7.50. A single-layer Kohonen NN.

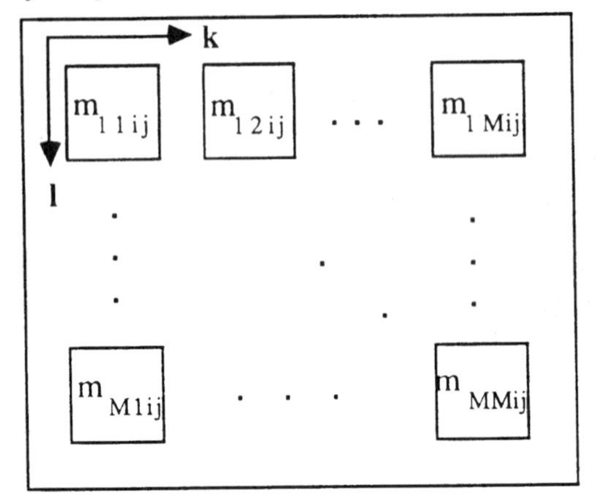

Fig. 7.51. Memory vectors in the memory matrix space.

sum of the input vectors,

$$y_{lk}(t) = \sum_{i=1}^{N} \sum_{j=1}^{N} m_{lkij}(t) x_{ij}(t), \qquad l, k = 1, 2, \ldots, M, \tag{7.72}$$

where $y_{lk}(t)$ represents the state of the (l, k)th neuron in the output space and $m_{lkij}(t)$ is the interconnection weight between the (i, j)th input and the (l, k)th output neurons. The preceding equation can also be written in matrix form, as given by

$$y_{lk}(t) = \mathbf{m}_{lk}(t)\mathbf{x}(t),$$

where $\mathbf{m}_{lk}(t)$ can be expanded in an array of 2-D submatrices, as shown in Fig. 7.51. Each submatrix can be written in the form

$$y_{lk}(t) = \begin{bmatrix} m_{lk11}(t) & m_{lk12}(t) & \cdots & m_{lk1N}(t) \\ m_{lk21}(t) & m_{lk21}(t) & \cdots & m_{lk2N}(t) \\ \vdots & \vdots & & \vdots \\ m_{lkN1}(t) & m_{lkN2}(t) & \cdots & m_{lkNN}(t) \end{bmatrix}, \tag{7.73}$$

where the elements in each submatrix represent the associative weight factors from each of the input neurons to one output neuron.

Notice that the Kohenen model does not, in general, specify the desired output results. Instead, a matching criterion is defined to find the best match between the input vector and the memory vectors and then determine the best matching output node. The optimum matching score d_c can be defined as

$$d_c(t) \triangleq \|\mathbf{x}(t) - \mathbf{m}_c(t)\| = \min_{l,k} \{d_{lk}(t)\} = \min_{l,k} \{\|\mathbf{x}(t) - \mathbf{m}_{lk}(t)\|\}, \quad (7.74)$$

where $c = (l, k)^*$ represents the node in the output vector space at which a best match occurred. $\| \ \|$ denotes the Euclidean distance operator.

After obtaining an optimum matching position c, a neighborhood $N_c(t)$ around the node c is further defined for IWM modification, as shown in Fig. 7.52. Notice that the memory matrix space is equivalent to the output space, in which each submatrix corresponds to an output node. As time progresses, the neighborhood $N_c(t)$ will slowly reduce to the neighborhood that consists of only the selected memory vector $\mathbf{m}_c$, as illustrated in the figure.

Furthermore, a simple algorithm is used to update the weight factors in the neighborhood topology, by which the similarity between the stored memory matrix $\mathbf{m}_{lk}(t)$ and the input vector $\mathbf{x}(t)$ increases. Notice that the input vectors can be either binary or analog patterns, while the memory vectors are updated in an analog incremental process.

Prior to the optical implementation, a few words must be mentioned. Referring to the optical NN of Fig. 7.7, LCTV1 is used to generate the IWM and LCTV2 is for displaying the input patterns. They form a matrix–vector product operation (i.e., the interconnection part). The output result emerging from LCTV2 is then picked up by a CCD array detector for maxnet operation, which can be obtained by microcomputer processing. Notice that the data flow in the optical system is controlled primarily by the computer. For instance, the

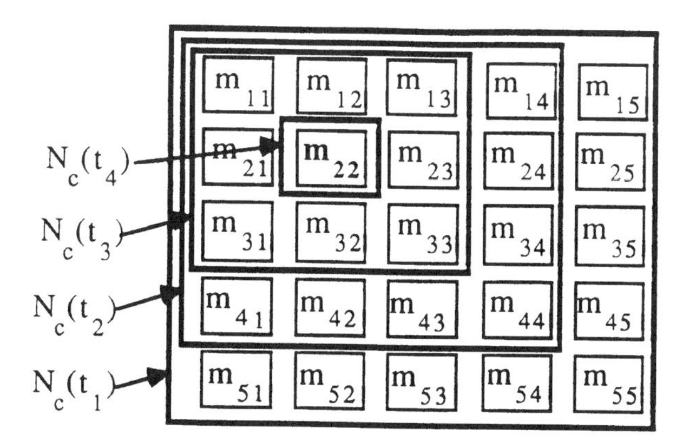

Fig. 7.52. Neighborhood selection in the memory matrix space using Kohonen's self-learning algorithm: the initial neighboring region is large. As time proceeds, the region shrinks until it reduces to one memory submatrix.

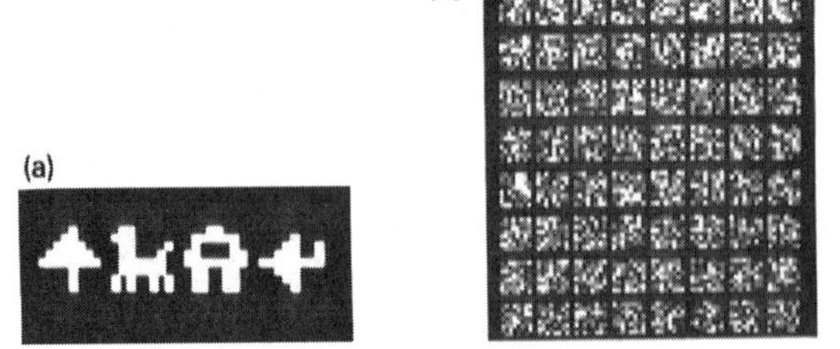

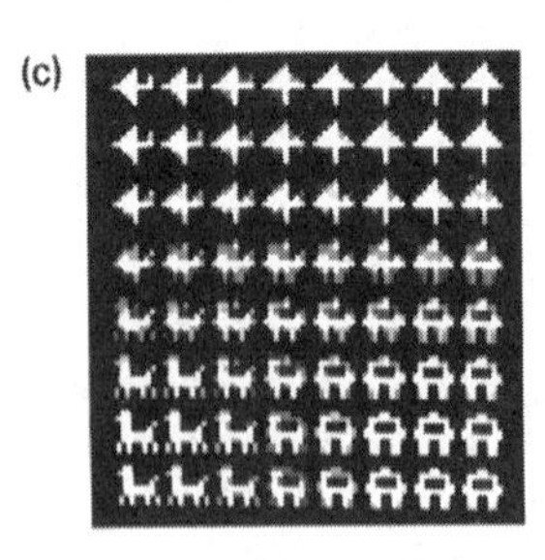

Fig. 7.53. (a) Input patterns. (b) Initial memory matrix space with random noise. (c) Final memory matrix representing a feature map.

IWMs and the input patterns can be written onto LCTV1 and LCTV2, and the computer can also make *decisions* based on the output results of the NN. Thus, we see that the illustrated optical NN is indeed a programmable and adaptive network.

For demonstration, four 8×8 pixel binary patterns (i.e., a tree, a dog, a house, and an airplane), shown in Fig. 7.53a, are sequentially presented to the optical NN. Figure 7.53b shows the initial memory matrix, which contains 50% random binary pixel elements. The output pattern picked up by the CCD camera must be normalized and the location of the maximum output intensity can be identified by using the maxnet algorithm. The memory submatrices in the neighborhood of the maximum output spot are adjusted based on the adaptation rule of the Kohonon model. In other words, changes of the memory vectors can be controlled by the learning rate. Thus, we see that the updated memory matrix can be displayed on LCTV1 for the next iteration, and so on. After some 100 sequential iterations, the memory matrix is converged into a *feature map*, shown in Fig. 7.53c. The centers of the feature patterns are located at (1,8), (7,1), (7,7), and (1,1), respectively, in the 8×8 memory matrix. In this sense, we see that the NN eventually learned these patterns after a series of encounters. It is, in fact, similar to one human early life experience: when in grade school, our teachers ferociously forced us to memorize the multiplication table; at that time we did not have the vaguest idea of the axiom of multiplication, but we learned it!

7.8.3. POLYCHROMATIC NEURAL NETWORKS

One interesting feature of optical pattern recognition is the exploitation of spectral content. There are two major operations in NN, the *learning phase* and the *recognition phase*. In the learning phase, the interconnection weights among the neurons are decided by the network algorithm, which can be implemented by a microcomputer. In the recognition phase, the NN receives an external pattern and then iterates the interconnective operation until a match with the stored pattern is obtained. The iterative equation for a two-dimensional NN is given by

$$U_{lk}(n+1) = f\left(\sum_{i=1}^{N}\sum_{j=1}^{N} T_{lkij}U_{ij}(n)\right), \tag{7.75}$$

where U_{lk} and U_{ij} represent the 2D pattern vectors, T_{lkij} is a 4D IWM, and f denotes a nonlinear operation, which is usually a sigmoid function for gray-level patterns and a thresholding operator for binary patterns.

A polychromatic neural network (PNN) is shown in Fig. 7.54, where two LCTVs are tightly cascaded for displaying the input pattern and the IWM, respectively. To avoid the moire' fringes resulting from the LCTVs, a fine-layer diffuser (e.g., Scotch tape) is inserted between them. To match the physical size of the IWM, the input pattern is enlarged so the input pattern pixel is the same size as the submatrix of the IWM. This is illustrated in Fig. 7.55. The summation of the input pattern pixels with the IWM submatrices can be obtained with a lenslet array by imaging the transmitted submatrices on the CCD array detector. By properly thresholding the array of detected signals, the result can be fed back to LCTV1 for the next iteration, and so on.

The liquid crystal panels we used happen to be color LCTVs, for which the color pixel distribution is depicted in Fig. 7.56. Every third neighboring RGB pixel element is normally addressed as one pattern pixel, called a *triad*. Although each pixel element transmits primary colors, a wide spectral content can be produced within each triad. If we denote the light intensity of the pixel element within a triad as $I_R(x, y)$, $I_G(x, y)$, and $I_B(x, y)$, the color image intensity produced by the LCTV is

$$I(x, y) = I_R(x, y) + I_G(x, y) + I_B(x, y). \tag{7.76}$$

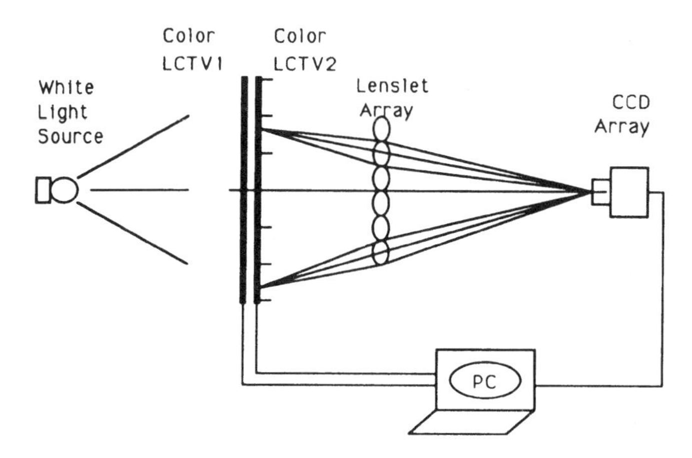

Fig. 7.54. A polychromatic NN using cascaded color LCTVs.

(a)

U_{11}	U_{12}	……	U_{1N}
U_{21}	U_{22}	……	U_{2N}
……	……	……	……
U_{N1}	U_{N2}	……	U_{NN}

(b)

T_{lk11}	T_{lk12}		T_{lk1N}
T_{lk21}	T_{lk22}		T_{lk2N}
T_{lkN1}	T_{lkN2}		T_{lkNN}

Fig. 7.55. Display formats. (a) For input pattern, (b) For the IWM.

A block diagram of the polychromatic neural network (PNN) algorithm is illustrated in Fig. 7.57. A set of reference color patterns is stored in the NN; then each pattern is decomposed into three primary color patterns, which are used as the basic training sets. For the learning phase, three primary color IWMs should be independently constructed, allowing a multicolor IWM to be displayed on LCTV2. If a color pattern is fed into LCTV1, the polychromatic iterative equation is

$$U_{lk}(n+1)=f\left(\sum_{i=1}^{N}\sum_{j=1}^{N}\{[T_{lkij}]_R[U_{ij}(n)]_R+[T_{lkij}]_G[U_{ij}(n)]_G+[T_{lkij}]_B[U_{ij}(n)]_B\}\right). \tag{7.78}$$

To demonstrate the PNN operation, we used a heteroassociation polychromatic training set, as shown in Fig. 7.58. By implementing the heteroassociation model, as described in Sec. 7.8.1, a multicolor IWM is converged in the network. When a red color A is presented to the PNN, a corresponding red Chinese character is translated at the output end. Similarly, if a blue Japanese Katakana is presented, then a blue color A will be observed. Thus, the PNN can indeed exploit the spectral content for pattern recognition.

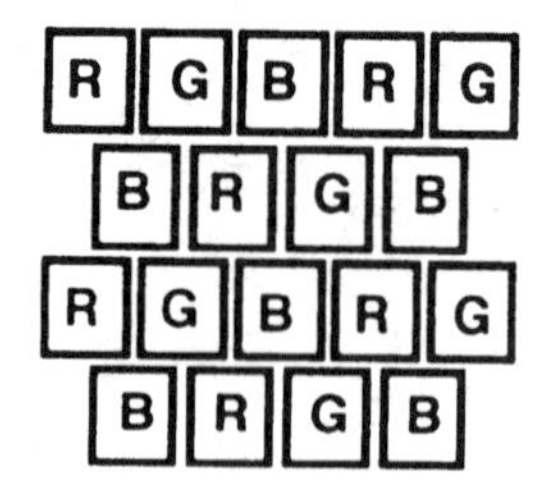

Fig. 7.56. Pixel structure of the color LCTV.

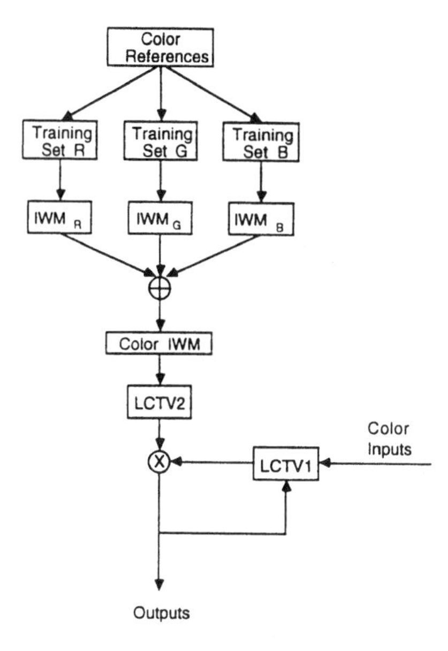

Fig. 7.57. Block diagram representation of the polychromatic NN algorithm.

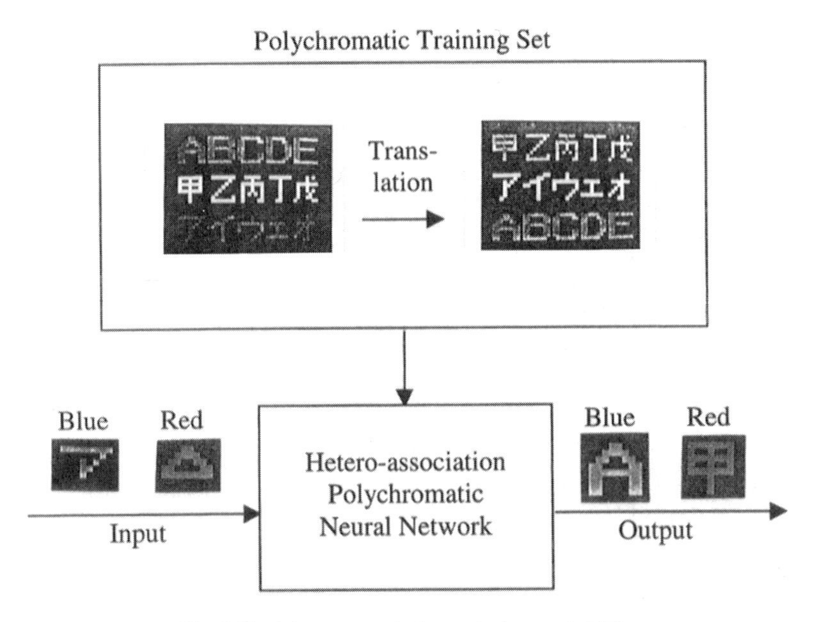

Fig. 7.58. A heteroassociation polychromatic NN.

REFERENCES

7.1 E. Abbe, 1873, "Beitrage zur Theorie des Mikroskops und der mikroskopischen Wahnehmung," *Archiv fur Mikroskopische Anatomie*, vol. 9, 413.

7.2 A. Vander Lugt, 1964, "Signal Detection by Complex Spatial Filtering," IEEE *Trans. Inform. Theory*, vol. IT-10, 139.

7.3 J. J. Hopfield, 1982, "Neural Network and Physical System with Emergent Collective Computational Abilities," *Proc. Natl. Acad. Sci.*, vol. 79, 2554.

7.4 T. Kohonen, *Self-Organization and Associative Memory*, Springer-Verlag, Berlin/New York, 1984.

7.5 F. T. S. Yu and T. H. Chao, 1983, "Color Signal Correlation Detection by Matched Spatial Filtering," *Appl. Phys.*, vol. B32.

7.6 D. Psaltis and N. Farhat, 1985, "Optical Information Processing Based on an Associative-Memory Model of Neural Nets with Thresholding and Feedback," *Opt. Lett.*, vol. 10, 98.

7.7 T. Lu et al., 1990, "Self-organizing optical neural network for unsupervised learning," *Opt. Eng.*, vol. 29, 1107.

7.8 E. C. Tam et al., 1990, "Data Association Multiple Target Tracking Using Phase-Mostly Liquid Crystal Television," *Opt. Eng.*, vol. 29, 1114.

7.9 M. S. Kim and C. C. Guest, 1990, "Simulated Annealing Algorithm for Binary-Phase–Only Filters in Pattern Classification," *Appl. Opt.*, vol. 29, 1203.

7.10 X. Yang, et al., 1990, "Compact Optical Neural Network Using Cascaded Liquid Crystal Televisions," *Appl. Opt.*, vol. 29, 5223.

7.11 F. T. S. Yu et al., 1991, "Wavelength-Multiplexed Reflection-Type Matched Spatial Filtering Using $LiNbO_3$," *Opt. Commun.*, vol. 81, 343.

7.12 F. T. S. Yu and S. Jutamulia, *Optical Signal Processing, Computing, and Neural Networks*, Wiley, New York, 1992.

7.13 F. T. S. Yu et al., 1992, "Polychromatic Neural Networks," *Opt. Commun.*, vol. 88, 81.

7.14 F. Cheng, et al., 1993, "Removal of Intraclass Association in Joint-Transform Power Spectrum," *Opt. Commun*, vol. 99, 7.

7.15 S. Jutamulia, "*Optical Correlator*," in *SPIE Milestone Series*, vol. 76, SPIE Opt. Eng. Press, Bellingham, Wash., 1993.

7.16 M. Wen, et al., 1993, "Wavelet Matched Filtering Using a Photorefractive Crystal," *Opt. Commun.*, vol. 99, 325.

7.17 P. Yeh, *Introduction to Photorefractive Nonlinear Optics*, Wiley, New York, 1993.

7.18 S. Yin et al., 1993, "Wavelength-Multiplexed Holographic Construction Using a Ce:Fe: doped $LiNbO_3$ Crystal with a Tunable Visible Light Diode Laser," *Opt. Commun.*, vol. 101, 317.

7.19 F. T. S. Yu and F. Cheng, 1994, "Multireference Detection Using Iterative Joint-Transform Operation with Composite Filtering," *Opt. Mem. and Neural Net.*, vol. 3, 34.

7.20 F. T. S. Yu, et al., 1995, "Application of Position Encoding to a Complex Joint-Transform Correlator," *Appl. Opt.*, vol. 34, 1386.

7.21. F. T. S. Yu and S. Yin, 1995, "Bragg Diffraction-Limited Photorefractive Crystal-Based Correlators," *Opt. Eng.*, vol. 34, 2225.

7.22 G. Lu, et al., 1995, "Phase-Encoded Input Joint-Transform Correlator with Improved Pattern Discriminability," *Opt. Lett.*, vol. 20, 1307.

7.23 G. Lu and F. T. S. Yu, 1996, "Pattern Classification Using a Joint-Transform Correlator–Based Nearest Neighbor Classifier," *Opt. Eng.*, vol. 35, 2162.

7.24 S. Yin et al., 1996, "Design of a Bipolar Composite Filter Using Simulated Annealing Algorithm," *Opt. Lett.*, vol. 20, 1409.

7.25 F. T. S. Yu and D. A. Gregory, 1996, "Optical Pattern Recognition: Architectures and Techniques," *IEEE Proc.*, vol. 84, 733.

7.26 P. Purwosumarto and F. T. S. Yu, 1997, "Robustness of Joint-Transform Correlator versus Vander Lugt Correlator," *Opt. Eng.*, vol 36, 2775.
7.27 G. Lu, et al., 1997, "Implementation of a Nonzero-Order Joint-Transform Correlator by Use of Phase-Shifting Techniques," *Appl. Opt.*, vol. 36, 470.
7.28 C. T. Li, et al., 1998, "A Nonzero-Order Joint-Transform Correlator," *Opt. Eng.*, vol. 37, 58.
7.29 F. T. S Yu et al., 1998, "Performance Capacity of a Simulated Annealing Bipolar Composite Filter (SABCF)," *Opt. Commun.*, vol. 154, 19.
7.30 F. T. S. Yu and S. Jutamulia, *Optical Pattern Recognition*, Cambridge University Press, Cambridge, UK, 1998.
7.31 F. T. S. Yu and S. Yin, *Optical Pattern Recognition*, in SPIE Milestone Series, vol. 156, SPIE Opt. Eng. Press, Bellingham, Wash., 1999.
7.32 M. S. Alam, *Optical Pattern Recognition Using Joint Transform Correlator*, in SPIE Milestone Series, vol. 157, SIE Opt. Eng. Press, Bellingham, Wash., 1999.

EXERCISES

7.1 Suppose that the resolution of the SLMs is given by 100 lines/mm and the CCD array detector is given as 20 lines/mm. Referring to the FPD of Fig. 7.5, compute the space–bandwidth product of the hybrid–optical processor.

7.2 Assume that the spatial frequency content of the SLM1 is given as 10 lines/mm. Calculate the space–bandwidth requirement for the SLM2 for a Vander Lugt spatial filter. Assume that the focal length of the transform lens is 400 mm.

7.3 Assume that K number of input objects are displayed on the input LCTV panel of a JTC, shown in Fig. 7.6, as given by

$$f(x, y) = \sum_{k=1}^{K} f_k(x - a_k, y - b_k),$$

where (a_k, b_k) denotes the locations of the input objects. The LCTV has an inherent grating structure so that its amplitude transmittance can be written as

$$t(x, y) = f(x, y)g(x, y),$$

where $g(x, y)$ represents a two-dimensional grating function of the LCTV.
(a) Compute the output complex light distribution.
(b) Identify the output auto- and cross-correlation terms of part (a).
(c) To ensure nonoverlapping distributions, determine the required locations of the input functions.

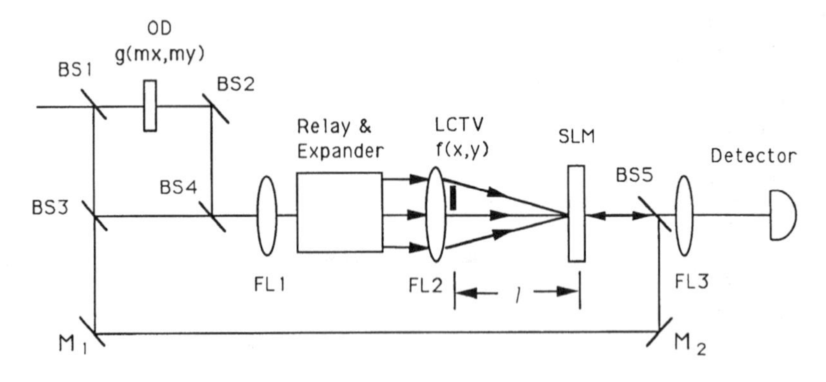

Fig. 7.59.

7.4 To avoid overlapping cross-correlation distributions, the symmetric locations of the object functions on the LCTV should be broken; for example

$$f(x, y) = f_1(x - a, y - b) + f_2(x, y - b) + f_3(x - a, y) + f_4(x, y + b).$$

Determine the output cross-correlation distributions.

7.5 The optical disk–based joint-transform correlator (ODJTC) of Fig. 7.8 can be represented by the block diagram of Fig. 7.59, where FL represents the Fourier-transform system and PBS is the polarized beam splitter. Let $f(x, y)$ and $g(mx, my)$ be the input and the magnified OD reference function, respectively. Assume that the 2-f Fourier-transform system is used in ODJTC
(a) Determine the relationship between the focal lengths FL1 and FL2.
(b) Assuming that the space bandwidth of the input and the reference OD functions are identical, compute the requirement of the Fourier-transform lenses.
(c) Letting the space–bandwidth product of part (b) be W, determine the required incident angle θ formed by the input and the reference beams. (*Hint*: Consider the focal depth of the Fourier-transform lens.)

7.6 One of the basic constraints of the ODJTC is the continuous movement of the OD reference functions. For simplicity, assume that the motion of the OD reference is restricted to the meridian plane perpendicular to the optical axis of FL1 of Fig. 7.8.
(a) Calculate the transmittance of the OD–based JTPS.
(b) Assuming that the allowable tolerance of the shifted fringes is limited to 1/10 cycle, calculate the allowable displacement of the OD during the operation cycle.

7.7 Consider the ODJTC as depicted in Fig. 7.59. Assume that the input function is located behind the joint-transform lens FL2 and also that the focal length of FL2 is f_2.
 (a) Calculate the Fourier spectrum of $f(x, y)$ at the SLM.
 (b) To match the quadratic-phase factors of the input and reference OD function, determine the location of the OD.
 (c) Compute the position tolerance of the OD. Assume that $m = 10$, SBP $= 200$, $b = 20$ mm, and $\lambda = 0.5\ \mu$m.

7.8 Note that line and continuous tone patterns can be encoded onto the optical disk in the pit array, thus allowing the enclosed OD patterns to be addressed in parallel. Assume that the encoding is in binary form.
 (a) Describe two encoding techniques for the OD images.
 (b) Are the techniques in part (a) suitable for both binary and gray-scale image patterns? Explain.

7.9 Referring to the ODJTC of Fig. 7.60, the duty cycle of the LCTV is 33 ms and the response time of the SLM is about 10 to 155 μm. Suppose that the disk size is 120 mm in diameter and that it is rotating at 1000 rpm. Further assume that the OD pattern size is about 0.5×0.5 mm^2 with 0.01 mm spacings, and that the patterns are recorded between radii of 30 and 55 mm within the disk. Calculate the average operating speed of the system.

7.10 If we use a 3-in color LCTV with 450×220 pixel elements and assume that the independent addressable pixels are 2×2 color pixel elements:
 (a) Evaluate the number of fully interconnected neurons that can be built using this LCTV.
 (b) If the number of output neurons is assumed to be 10, what would the input resolution requirement be?

7.11 Let the size of a 450×2220–pixel LCTV panel be 6×4 cm^2. If the LCTV is used to display the memory matrix (IWM) in an optical neuron network:
 (c) Compute the number of lenses required in the lenslet array for the fully interconnected network.

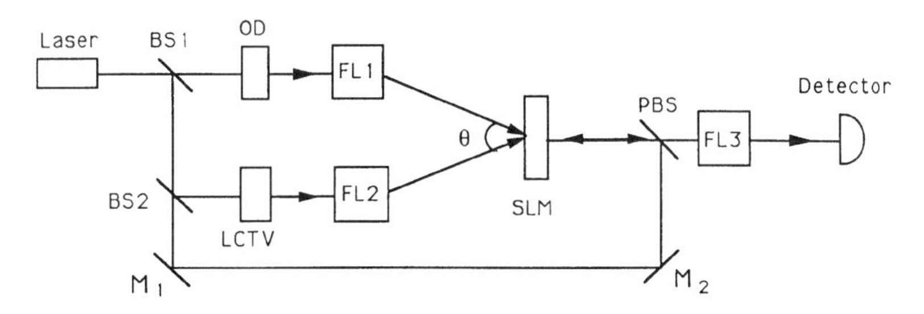

Fig. 7.60.

(d) If the focal length of these lenses is 50 mm, compute the required diameter of the lenses to match an input pattern of $2 \times 2\ \text{cm}^2$.

(e) Assuming that the diffraction angle of the diffuser is 18°, and the transmittance of the LCTVs is 10 percent, estimate the power budget of the ONN.

7.12 Refer to the misalignment limitations for the VLC and JTC as discussed in Sec. 7.1.4.

(a) Evaluate the transversal (filter) tolerance for these correlators.

(b) Sketch the correlation performance as a function of shifted filter plane.

7.13 Write an implicit correlation-detection equation as function of longitudinal misalignment for the VLC and JTC, respectively. Sketch the correlation performances for these correlators.

7.14 Consider a quasi-Fourier transformer as shown in Fig. 7.61, in which b is the width of an input object.

(a) Calculate the corresponding focal depth of the processor.

(b) Evaluate the allowable object displacement.

(c) Comment on your result.

7.15 Refer to the unconventional JTC of Fig. 7.62. Evaluate the allowable focal depth δ and incident angle θ, respectively, where b is the spatial width of the input objects.

7.16 Besides the trivial advantages of improving detection efficiency and SLM pixel utilization, show that the intrapattern correlations (which produce false alarms) can be eliminated by a nonzero-order JTC.

7.17 Show that a nonzero order JTC can also be achieved by using phase-shifting technique. In other words, by changing the sign of the reference function, a nonzero-order JTPS can be obtained by subtraction. Compare with the nonzero-order JTC in Sec. 7.2.2.

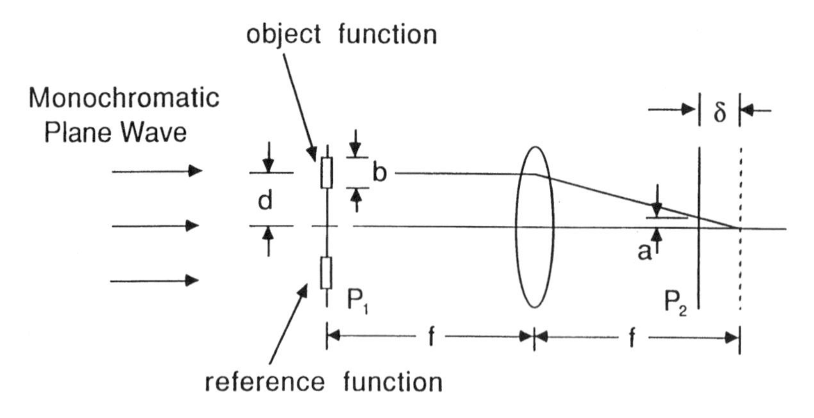

Fig. 7.61.

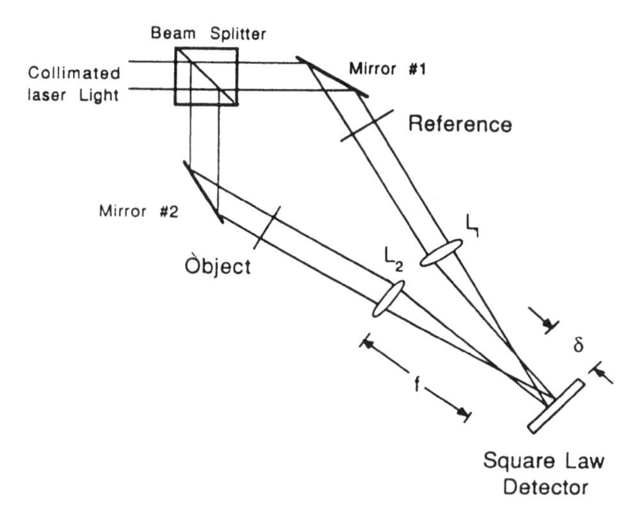

Fig. 7.62.

7.18 (a) State the advantages of using a nonzero-order JTC as compared with the conventional JTC.
(b) What is the price paid to gain these advantages?

7.19 Pattern discriminability is probably one of the most important criterion used for optical pattern recognition, which is defined by the signal-to-clutter ratio as given by

$$\mathrm{SCR} = \frac{\mathrm{API}}{\mathrm{CPI}},$$

where API is the autocorrelation peak intensity and CPI is the maximum cross-correlation peak intensity. Show that phase-encoded JTC can indeed improve the SCR (i.e., SCR > 1).

7.20 Most of the SLMs are amplitude-modulated devices. To improve the performance of pattern detection, one may use the position-encoded JTC. Illustrate that the position-encoded JTC can indeed improve pattern discriminability.

7.21 One of the major advantages of using a hybrid JTC is adaptivity, by which iterative processing can be performed. Conventional JTC suffers strong background noise, which produces low diffraction efficiency. We assume two targets are embedded in random bright background terrain, which produces poor correlation peak intensities at the output plane. Show that higher correlation peak intensities can be produced by continuous JTC–iterative processing.

7.22 Using a colinear RGB coherent illumination:
(a) Sketch a Fourier-domain polychromatic processor by which the multitarget set can be detected.
(b) If temporal spatial filters are being encoded with the Vander Lugt interferometric technique, sketch and explain the procedure for how those filters can be synthesized.
(c) By proper implementation of the filters of part (b) in the Fourier domain of part (a), show that polychromatic target detection can be indeed obtained at the output plane.

7.23 To synthesize a set of temporal holograms (i.e., matched filters) for polychromatic pattern detection in a VLC, one can use spectral dispersion with an input-phase grating. Assume the spatial-frequency limit of the polychromatic target is 10 hz per min, focal length of the transform lenses is 300 mm, and wavelengths of the RGB light source are 600 nm, 550 nm, and 450 nm, respectively.
(a) Calculate the sampling frequency requirement of the phase grating.
(b) Sketch the interferometric setup for the filter synthesis.
(c) What would be the spatial-carrier frequency of the set of matched filters?
(d) Evaluate the output polychromatic correlation distribution.

7.24 Assume a color LCTV is used at the input plane of a VLC; the pixel structure is shown in Fig. 7.24.
(a) Sketch a diagram to show that a set of RGB matched filters can be interferometrically synthesized at the Fourier domain.
(b) Evaluate the recorded matched filters.
(c) Evaluate the output polychromatic correlation distribution.

7.25 Refer to LCTV panel of Fig. 7.24.
(a) Show that isolated RGB–JTPS can be captured by a color CCD array detector.
(b) If one replicates the RGB–JTPS in the LCTV panel for correlation operation, show that high efficient polychromatic correlation detection can be obtained.
(c) If we assume that the RGB–JTPS is replicated into $N \times N$ arrays, what would be its correlation efficiency?
(d) Is it possible to use partial-spatial–coherent RGB light for JTC correlation detection of the replicated JTPS arrays of part (c)? What would be the spatial-coherent requirement of the light source?
(e) Comment on the correlation peak intensity and noise performance by using a strictly coherent and a partial-spatial–coherent source.

7.26 Refer to the autonomous target tracking of Sec. 7.4.1. The location of the correlation peak as a function of the target's translation in the x direction is plotted in Fig. 7.63(a). Considering the limited resolution of the array detector, find the minimum speed of the target that can be properly

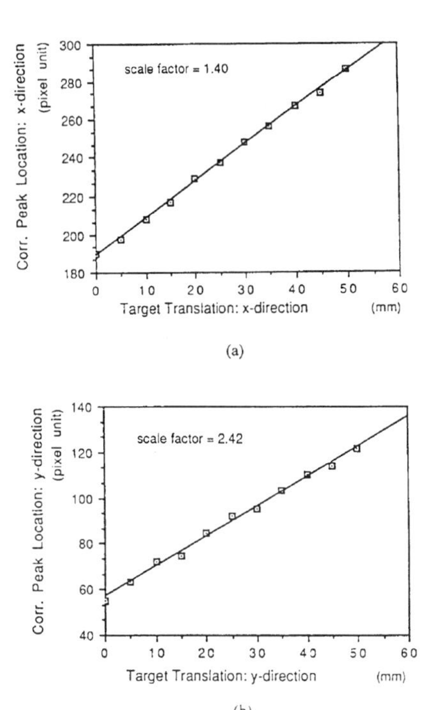

Fig. 7.63.

tracked using the JTC of Fig. 7.6, if the duty cycle of the JTC is 0.05 sec.

7.27 Refer to the role of the correlation peak position as a function of target motion in Fig. 7.63.
   (a) Determine the aspect ratio of the image array.
   (b) Assuming that the target is revolving in a circular path and the system is set to run autonomously, calculate the tracking errors after five revolutions.

7.28 Refer to Sec. 7.4, and consider the dynamic state of the target at the kth frame as given by

$$\mathbf{x}(k) = \begin{pmatrix} x(k) \\ \hline \dot{x}(k) \end{pmatrix}.$$

The velocity and the acceleration of the target can be handled by the JTC system of Fig. 7.6. Describe a Kalman filtering procedure so that multitarget tracking can be performed by the proposed hybrid–optical JTC.

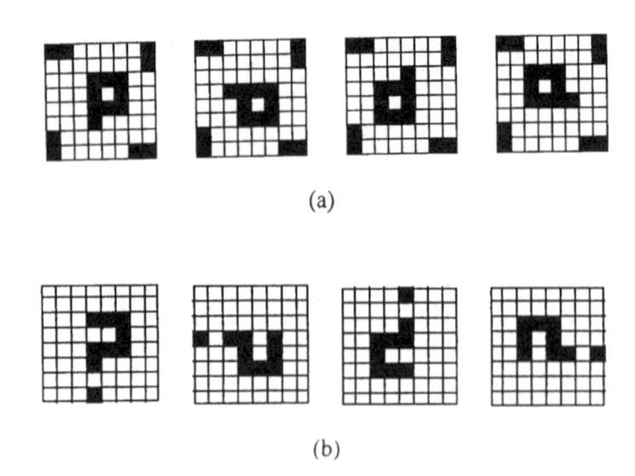

(a)

(b)

Fig. 7.64.

7.29 A target training set is provided in Fig. 7.64(a).
  (a) Using a simulated annealing algorithm, synthesize a bipolar composite filter (SABCF).
  (b) Evaluate (by computer simulation) the normalized correlation peak intensity (NCPI) as a function of training targets.
  (c) Plot the NCPI and the discrimination ratio (DR) as a function of training targets.

7.30 Repeat the preceding exercise for a 16×16 pixel frame, and compare your result with the 8×8 pixel frame.

7.31 If the antitarget set of Fig. 7.64(b) is included for the synthesis,
  (a) Evaluate the SABCF.
  (b) Plot the NCPI and DR as a function of training targets.
  (c) In view of the preceding results, comment on the performance capacity for using both target and antitarget sets.

7.32 Refer to the target and antitarget sets of Figs. 7.64(a) and (b).
  (a) Synthesize a 3-level SA composite filter.
  (b) Plot the NCPI and DR as a function of training targets.
  (c) Compare part (b) with the results obtained in the preceding exercise.

7.33 To demonstrate the shift-invariant property of the JTC–NNC in Sec. 7.6, we have shifted the input "2" upward and left by 8×8 pixel, is shown in Fig. 7.65. Assume that the SLM has 640×480 pixels, and each letter is fitted within a 32×32 pixel window.
  (a) Show (by computer simulation) the output correlation as obtained from a binarized JTPS.
  (b) What would be the allowable shift of "2" without causing any possible classified error?

7.34 To investigate the noise performance of the JTC–NNC of Sec. 7.6, assume that the set of input characters in the preceding exercise is embedded in

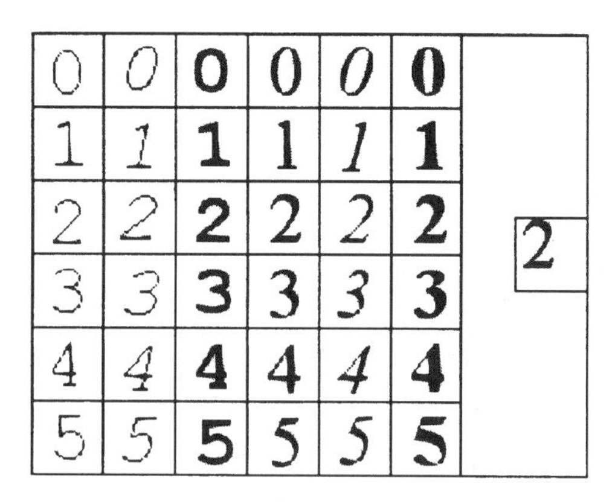

Fig. 7.65.

an additive white Gaussian noise with a standard deviation of $\sigma = 60$, and $\sigma = 100$, as shown in Fig. 7.66.

(a) Show (by computer simulation) the output correlation distributions for both cases.

(b) By defining the signal-to-clutter difference (SCD) for the pattern classification criterion,

$$\text{SCD} = \text{API} - \text{CPI},$$

where API and CPI are the auto- and cross-correlation peak intensities,

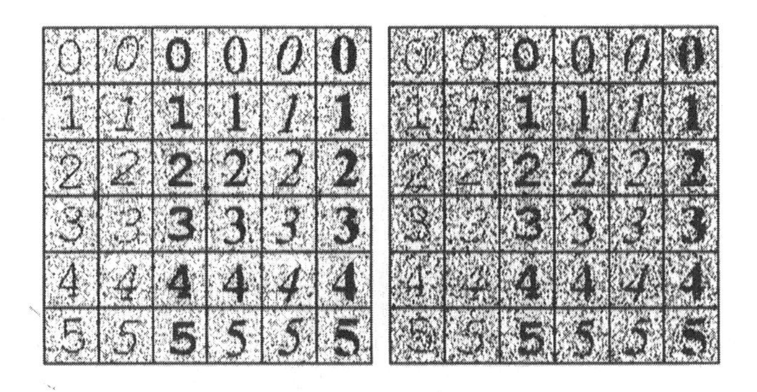

(a) (b)

Fig. 7.66.

plot SCD as function of character number for $\sigma = 60$ and $\sigma = 100$, respectively.

(c) Comment on the results obtained in part (b).

7.35 Assume that a 10-mm–thick $LiNbO_3$ photorefractive crystal is used for a transmission-type hologram, and that the wavelength of the light source $\lambda = 514$ nm and the writing angle is about 45°.

(a) Compute the allowable reading angular deviation.

(b) Repeat part (a) for a reflection-type hologram.

(c) Draw a vector diagram to represent the reading direction for the two-grating structure in a photorefractive crystal using the same reading wavelength. Assume the same amplitude is used and that the angular separation is denoted by $\Delta\theta$.

7.36. Refer to the spatial division scheme shown in Fig. 7.67 where (a) shows the recording setup and (b) shows the reading setup. Assume that the SLM has 256 × 256 pixel arrays and that the pixel size at P (the interconnection pattern) is about 50 μm. Calculate the minimum angular separation between the object illuminator beams B and B'. Notice that the illuminating wavelength is assumed $\lambda = 480$ nm.

7.37 Refer to the reflection-type matched filtering in Sec. 7.7. State the major advantages of using this type of photorefractive crystal filter. Are there any major disadvantages? Explain.

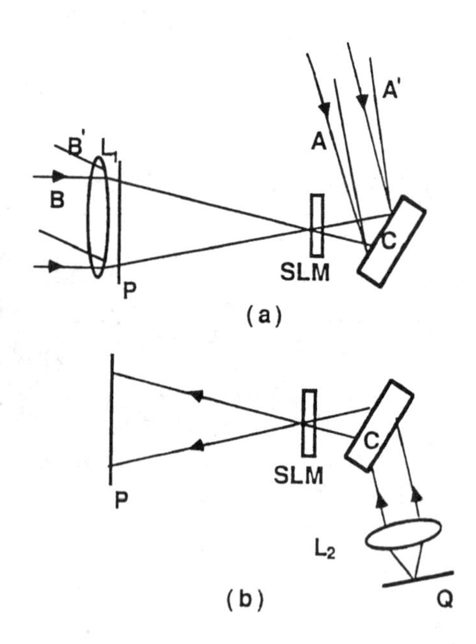

Fig. 7.67.

7.38 Refer to the compact joint-transform correlator of Fig. 7.40. Assume that a 10-mm–thick $LiNbO_3$ crystal is used for the joint-transform filter and that the illuminating wavelength is $\lambda = 500$ nm. If the object separation is about $2h = 10$ mm and the focal length of the transform lens is 300 mm, calculate the maximum allowable width of the object.

7.39 Assume that the distribution of a target signal is Gaussianly distributed as given by

$$h(x, y) = \frac{1}{\sqrt{2\pi}\,\sigma} e^{-1/2(r/\sigma)^2}, \quad x^2 + y^2 = r^2.$$

(a) Evaluate the corresponding wavelet matched filter (WMF).
(b) Evaluate the output correlation distribution, if one uses $a_x = a_y = \sigma$, and $a_x = a_y = 2\sigma$, respectively.
(c) If the input target is embedded with an additive white Gaussian noise $N = \sigma/2$, calculate the output signal-to-noise ratios for part (b).

7.40 A set of English letters {BPRFEHL} is to be stored in an 8×8 neuron network.
(a) Construct a Hopfield memory matrix.
(b) Construct an IPA memory matrix.
(c) Show that the Hopfield model becomes unstable, but that the IPA model can recall the letters with unique certainty.
(d) What is the minimum number of letters in this set that can be stored using the Hopfield model?

7.41 Estimate the minimum number of patterns that can be stored in an $N \times N$ neuron Hopfield network:
(a) For 8×8 neurons.
(b) For 64×64 neurons.

7.42 (a) Repeat the preceding exercise for the IPA model.
(b) Compare the results with those obtained from the Hopfield model.

7.43 Considering the input–output training set of Fig. 7.68 and using the IPA model,
(a) Develop a one-level heteroassociation neuron network.
(b) Construct a heteroassociation IWM.
(c) Estimate the number of storage patterns for a 4×4 neuron heteroassociation network.

7.44 (a) Develop a computer program using the IPA model for a heteroassociation neural network.
(b) If the input–output patterns of Arabic numerals {1, 2, 3} are to be translated into Roman numerals {I, II, III}, construct a heteroassociation IWM for 8×8 neurons.
(c) Construct a heteroassociation IWM to translate {I, II, III} as {1, 2, 3}.

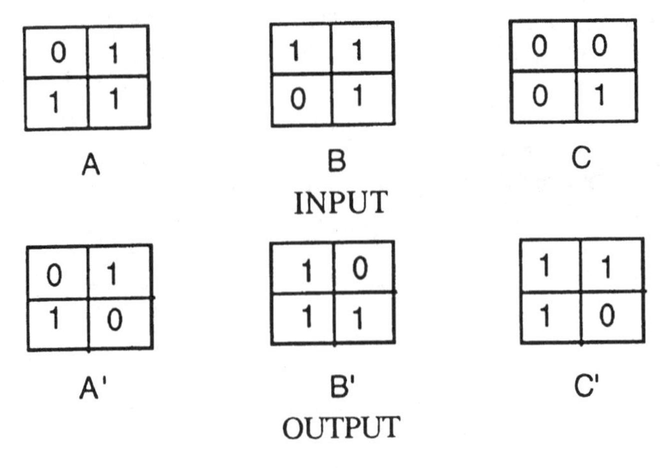

Fig. 7.68.

(d) Show that the partial input of an Arabic numeral will produce the correct Roman numeric translation.

7.45 (a) What are the major differences in operation between the Kohonen and the Hopfield neural nets?
(b) Describe the effects and the strategies for selecting the neighborhood $N_c(t)$ in a Kohonen neural net.

7.46 Assume that the initial memory matrix in a Kohonen neural net is distributed by the 50% random noise shown in Fig. 7.53(b). A training set of Roman letters $\{A, B, C, X\}$ is to be stored in the self-organizing neural net, in which a linear learning speed is used, where $a(0) = 0.02$ and $\alpha = 0.00005$.
(a) Simulate the memory matrix space after 200 iterations.
(b) Repeat part (a) for 400 iterations.
(c) Compare the results of parts (a) and (b).

7.47 Refer to the polychromatic neural network of Fig., 7.58.
(a) Simulate a polychromatic heteroassociation IWM for pattern translations.
(b) Show (by simulation) if a partial input of "red" A will result in the correct translation of a "red" Chinese character, as shown in this figure.

Chapter 8 Information Storage with Optics

Suganda Jutamulia[1] and Xiangyang Yang[2]
[1]BLUE SKY RESEARCH, SAN JOSE, CA
[2]SUN MICROSYSTEMS INC., MCLEAN, VA

8.1. DIGITAL INFORMATION STORAGE

Energy and mass are the fundamental physical quantities. Time and space are the fundamental physical dimensions. The pattern of the distribution of energy over time or space generates information. By this definition, any object continuously generates new information. Energy is transmitted from a source and finally is detected by a detector. The transmitted energy of desired information is known as the signal. The detected energy from unwanted information is called noise. Human beings are the ultimate detector for detecting information and translating it to meaningful abstract knowledge. Information storage involves a medium that stores the spatial energy pattern at a given time, and represents the stored pattern at a later time. A spatial-energy pattern can always be converted into a temporal pattern by a scanning process.

As was mentioned by Yu [1], light is not only part of the stream of energy that supports life, but also provides us with an important source of information. Information storage with optics, or simply *optical storage*, as we use this term throughout this chapter, is the medium that is able to store the pattern of light intensity. Upon stimulation, the medium presents back a pattern of light intensity identical to the previously stored pattern.

As mentioned above, information is a pattern of fluctuating energy. The signal is the desired information that propagates from one site to another and, therefore, is dynamic. However, data, which is the formatted information, is static. If the fluctuation of the energy is measured continuously and is represented by a real number, it is an analog signal. On the contrary, if it is measured in a discrete way and is represented by an integer, it is a digital signal. In other words, a digital signal is sampled and quantized.

A signal is meaningful only after it is detected. The accuracy of the interpretation of a received signal depends on the ability of the detector to resolve two positions and also two values. In the context of optical information processing and storage, energy is in the form of light intensity and the detector is a photodetector that converts light intensity into an electric signal. The resolving power for a given position is determined by the pixel size of the detector, while the resolving power for intensity is determined by the sensitivity of the detector. By taking both the smallest separable distance and the smallest separable intensity as units, any analog signal is indeed converted into a digital signal because it is unavoidably sampled and quantized by the detector.

Nevertheless, what is commonly called a digital signal is not the quantized signal but the binarized signal. Quantized values are represented by integers such as $0, 1, 2, 3, \ldots$; however, binarized values are represented by binary numbers; i.e., 0 and 1 only. Binary numbers use a base of 2 while decimal numbers use a base of 10. For example, decimal 1 is 1 in binary, 2 is 10, 3 is 11, 4 is 100, 5 is 101, etc. The length of a binary number is represented in bits, which stands for a binary unit. Thus, 1 is one bit, 10 and 11 are two bits, 100 and 101 are three bits, etc. Since there are only two states, 0 or 1, for binary data, it is easy to restore the distorted binary data to the correct data by applying a threshold at 0.5. Provided that information is stored in a binary form in optical storage, the amount of stored information can be represented in bits. For illustration, if a tiny photographic film can store just a picture of a checkerboard with 64 black-and-white squares, the storage capacity is 64 bits.

8.2. UPPER LIMIT OF OPTICAL STORAGE DENSITY

It is well known from diffraction theory [2, 3] that a lens can focus light to a spot that is limited by the diffraction. This spot is sometimes called an Airy disk, which has a central bright spot surrounded with ring fringes. The diameter of the central bright spot of the Airy disk is

$$Q = 2.44 \frac{f}{D} \lambda, \tag{8.1}$$

where f and D are the focal length and the diameter of the lens, and λ is the wavelength of light. In practice, one may take an approximation

$$Q \approx \lambda. \tag{8.2}$$

For further simplification, one may consider that an area of λ^2 is required to store a bit optically. Thus, the upper limit of storage capacity for a 2-D medium is

$$SC_{2D} = \frac{\text{Area}}{\lambda^2}. \tag{8.3}$$

By considering the third dimension, one may similarly infer that a volume of λ^3 is required to store a bit in a bulk optical memory, and the upper limit of the storage capacity for 3-D medium becomes

$$SC_{3D} = \frac{\text{Volume}}{\lambda^3}. \tag{8.4}$$

Instead of an image, we may alternatively record its Fourier-transform hologram as the memory. The original image would appear when the hologram is illuminated with the same reference beam as it was recorded. A hologram has a total area of $a \times a$, and the size of its smallest element is $\lambda \times \lambda$. The image is holographically formed by a Fourier-transform lens. Recalling Fourier analysis [2, 3, 4], the smallest element $\lambda \times \lambda$ will determine the size of the image that is proportional to $1/\lambda \times 1/\lambda$, and the size of the hologram $a \times a$ will determine the size of the smallest element of the image, which is proportional to $1/a \times 1/a$. Correspondingly, the storage capacity of a hologram is

$$SC_H = \frac{a^2}{\lambda^2}. \tag{8.5}$$

The storage capacity of the image is

$$SC_I = \frac{1/\lambda^2}{1/a^2} = \frac{a^2}{\lambda^2}. \tag{8.6}$$

The storage density of 2-D medium is the same; that is, $1/\lambda^2$, regardless of whether there is a direct bit pattern or a hologram. However, as we will find in a later section, the density of near field optical storage is higher than the limit set by the diffraction, since no diffraction occurs in near field optics.

When a bulk holographic medium is used to make a volume hologram, it can be multiplexed by n holograms. The maximum multiplexing is

$$n_{\max} = \frac{d}{\lambda}, \tag{8.7}$$

where d is the thickness of the volume hologram. By combining Eqs. (8.5) and (8.7), the storage capacity of a volume hologram can be calculated as follows:

$$SC_{VH} = \frac{a^2/\lambda^2}{d/\lambda} = \frac{\text{Volume}}{\lambda^3}. \qquad (8.8)$$

Referring to Eqs. (8.8) and (8.4), the upper limit of the storage density for a 3-D medium is $1/\lambda^3$ regardless of whether we record directly the bit pattern plane by plane in the 3-D storage or we record and multiplex each individual volume hologram. This principle was first formulated by van Heerden in 1963 [5, 6].

8.3. OPTICAL STORAGE MEDIA

Prior to discussing the architecture of optical storage, this section overviews commonly used materials for optical storage. Only the basic storage mechanism of each material is discussed in this section.

8.3.1. PHOTOGRAPHIC FILM

Photographic, or silver halide film is the most popular medium for storing an image. A photographic film or plate is generally composed of a base made of transparent acetate film or a glass plate, and a layer of photographic emulsion. The emulsion consists of a large number of tiny photosensitive silver halide grains which are suspended more or less uniformly in a supporting gelatin. When the photographic emulsion is exposed to light, some of the silver halide grains absorb optical energy and undergo a complex physical change; that is, grains that absorb sufficient light energy are immediately reduced, forming silver atoms. The aggregate of silver atoms in the silver halide grain is called a development center. The reduction to silver is completed by the chemical process of development. Any silver halide grain containing an aggregate of at least four silver atoms is entirely reduced to silver during development. The grains that have not been exposed or that have not absorbed sufficient optical energy will remain unchanged. If the developed film is then subjected to a chemical fixing process, the unexposed silver halide grains are removed, leaving only the metallic silver particles in the gelatin. These remaining metallic grains are largely opaque at optical frequencies, so the transmittance of the developed film depends on their density.

Silver halide emulsions can be used to produce phase as well as amplitude modulation [7]. In order to produce phase modulation, it is necessary to obtain the final image in the form of grains of a transparent dielectric compound (such as silver halide) instead of the opaque metallic silver. This can be achieved by bleaching, following the development process, to remove the silver image from the emulsion. The remaining silver halide is then desensitized. The desensitized silver halide modulates the phase of light passing through it.

8.3.2. DICHROMATED GELATIN

Gelatin is traditionally made from the parts of cows that people do not eat [8]. If one takes all the gristle, hooves, and bone and boils them for a long time, they end up as glue. If this glue is further refined, the final product is gelatin. Gelatin can absorb a very large amount of water and still remain rigid; in other words, it swells in water. Gelatin film is not intrinsically sensitive to light. Nevertheless, through chemical sensitization, usually by adding ammonium dichromate ($([NH_4)]_2Cr_2O_7)$, it is possible to induce changes in the gelatin that make it light sensitive.

The mechanism of the photochemical process which occurs when the dichromated gelatin is exposed to light is not well understood [7, 8, 9]. However, it is generally accepted that absorption of light is by the Cr^{6+} ion, and that reduction occurs to Cr^{3+} ions. As a result of reaction with the gelatin, the gelatin molecular chains in the regions exposed to more light have more cross-linking. These regions swell less when immersed in water, and, if rapidly dehydrated by exposure to alcohol, differential strains are produced between regions of maximum and minimum swelling. These strains modify the local index of refraction.

Gelatin film can be prepared by dip-coating and doctor-blading on a substrate. Since a photographic plate contains gelatin in the emulsion, an alternative approach is to remove the silver halide from the emulsion of a commercial photographic plate; e.g., Kodak 649F. As mentioned in the previous subsection, the silver halide can be removed by washing the unexposed photographic plate in fixer and then water.

8.3.3. PHOTOPOLYMERS

Polymerization is a process in which two or more molecules of the same substance unite to give a molecule (polymer) with the same composition as the original substance (monomer), but with a molecular weight which is an integral multiple of the original. Of the many methods available for polymerizing a substance, one important method involves the use of light; hence the term *photopolymer* [7, 10]. In a photopolymer, the required state of polymerization can be accomplished either through direct interaction of light with the polymer itself, or through an intermediary such as a photosensitizer.

A commercially available photopolymer contains a monomer and a photosensitizer incorporated with a polymeric binder to form a soft film [11]. During exposure, partial polymerization of the monomer occurs to an extent that is dependent upon the local intensity of the recording radiation. Since the distribution of the remaining monomer is not uniform, diffusion of monomer molecules takes place during and after exposure to regions of low concentration. The polymer molecules effectively do not move. After polymerization and diffusion are completed, diffraction efficiency can be considerably increased by a uniform postexposure using a fluorescent light. The postexposure may also desensitize the photosensitizer. The refractive index modulation is produced by the variations in polymer concentration.

8.3.4. PHOTORESISTS

Photoresists are organic photosensitive materials that are commonly used in photolithography for the fabrication of integrated circuit (IC) chips. There are two types of photoresists, negative and positive. After exposure to light, negative photoresists become insoluable in solvent due to polymerization or other processes. The areas which were not exposed to light are washed away, leaving a pattern of the exposed image. On the other hand, positive photoresists become soluble in the solvent because of depolymerization or other processes due to the action of light. An image or holographic interference pattern is recorded as a surface relief pattern of the photoresist layer. The resulting surface relief patterns enable preparation of reflection holograms through aluminum coating and mass duplication of holograms through embossing [8, 12]. The developed photoresist can also be used uncoated as a phase hologram [12].

8.3.5. THERMOPLASTIC FILM

Thermoplastic film, also known as a photoplastic device, involves the surface deformation of a transparent layer such that the phase of the light beam passing through the layer will be modulated. The essential elements in the thermoplastic process are the creation of an electric field pattern inside the thermoplastic layer, which is a copy of the incident optical pattern, and the pulse heating of the thermoplastic layer, which causes the thermoplastic to be molded according to the electric field pattern [13].

The device is composed of a glass substrate that is coated with a transparent conducting layer (tin oxide or indium oxide), on the top of which is a layer of photoconducting material, followed by a layer of thermoplastic. For the photoconductor, poly-n-vinyl carbazole (PVK) sensitized with trinitro-fluorenone (TNF) can be used with an ester resin thermoplastic [14].

Before exposure, the thermoplastic layer is charged by a corona device while the transparent conducting layer on the glass substrate is grounded. The positive ions from the ionized air are deposited on the surface of the thermo-plastic layer. The voltage is capacitively divided between the thermoplastic and photoconducting layers. After the charging process, the thermoplastic layer can be exposed to the signal light. The illuminated region displaces the charge from the photoplastic interface to the transparent conducting layer through the photoconducting layer, which in turn reduces the surface potential of the outer surface of the thermoplastic. The thermoplastic layer is then recharged to the original surface potential. The electric field across the thermoplastic increases in the illuminated area as a result of the second charging, although the surface potential is now uniform. The thermoplastic is developed by raising its temperature to the softening point and then rapidly reducing it to room temperature. Surface deformation caused by the electrostatic force then produces a phase recording of the intensity of the input light. This recording can be erased by raising the temperature of the thermoplastic above the melting point, thus causing the surface tension of the thermoplastic to flatten the surface deformation and erase the recording.

8.3.6. PHOTOREFRACTIVE MATERIALS

Holographic storage in lithium niobate ($LiNbO_3$), which is a photorefractive crystal, was first demonstrated by Chen et al., in 1968 [15]. Photorefractive materials are crystals that contain electron traps. Exposure of photorefractive materials to light excites electrons from traps to the conduction band, where electrons can freely move. The free electrons diffuse thermally or drift under applied or internal electric fields and become retrapped, preferably in regions of low-intensity light. The space charge buildup continues until the field completely cancels the effect of diffusion and drift; i.e., the current is zero throughout. The resulting charge distribution modifies the refractive index of the material by the linear electro-optic effect.

The electron traps are generated by intrinsic defects and residual impurities. The sensitivity of the crystal can be dramatically increased by adding iron to the crystal lattice. For example, Fe_2O_3 is added when a crystal of lithium niobate ($LiNbO_3$) is grown by means of the Czochralski process. Iron enters the crystal lattice as Fe^{2+} and Fe^{3+} ions. These two species of iron ions are initially evenly distributed throughout the crystal. When Fe^{2+} is exposed to the input light, an electron is excited to the conduction band of the crystal lattice, and Fe^{2+} becomes Fe^{3+}. Migration of electrons occurs in the conduction band of the crystal until they become trapped once more by Fe^{3+} ions in the low-intensity regions.

8.3.7. PHOTOCHROMIC MATERIALS

Photochromic materials have two distinct stable states A and B called color centers, since they are associated with absorption bands λ_1 and λ_2, respectively [16]. When illuminated with light of λ_1, the material undergoes a transition from stable state A to B. After the transition is completed, the material is no longer sensitive to λ_1. The absorption band of the material has shifted to λ_2. Thus, optical data can be written using light of λ_1 and read out using light of λ_2. However, when the material is illuminated with light of λ_2, the process reverses and the material reverts to its stable state A from B. The reversal may also occur naturally in the dark or it may be caused by heating.

Inorganic photochromic materials are often insulators or semiconductors with a large energy gap between the valence and conduction bands. The presence of imperfections in the crystal lattice or impurities cause the appearance of additional localized energy levels, A and B, within the bandgap. When the material is exposed to light of λ_1, electrons are excited from level A into the conduction band and then captured by electron traps in level B. The trapped electrons can be returned to level A by irradiating the material with light of λ_2, which is the energy required to excite an electron from level B into the conduction band. Since $\lambda_2 > \lambda_1$, in the thermally stable state, traps A will be occupied in preference to traps B.

Holograms may be recorded in photochromic materials either using λ_1 or λ_2. The recorded hologram can be read using the same wavelength λ_1 or λ_2. The reading process is rather complex, and involves refractive index modulation, which accompanies absorption modulation [17]. In organic photochromic materials, the effect generally involves a change of molecular structure, such as cis-trans photoisomerization. Isomers are compounds possessing the same composition and the same molecular weight, but differing in their physical or chemical properties. Cis-trans isomerism is due to different arrangments of dissimilar atoms or groups attached to two atoms joined by a double bond. While the change in refractive index of photochromic materials is usually small, photochromism in cis-trans isomers of stilbene and other organic materials can produce relatively large refractive index changes [18].

8.3.8. ELECTRON-TRAPPING MATERIALS

Electron-trapping or ET materials [19] are similar to photochromics in the sense that they both have two reversible stable states A and B that can be switched by the absorption of light of λ_1 and λ_2. In contrast to photochromics, the stimulated emission of λ_3 from ET material is utilized as output instead of the absorption of the reading light of λ_2. An example of an ET material is SrS:Eu,Sm. The mechanism of light emission is as follows: Both the ground and excited states of each impurity (Eu and Sm) exist within the bandgap of the host material (SrS). Blue writing light ($\lambda_1 = 488$ nm) excites an electron in the ground state of Eu^{2+} into its excited state. Some of the electrons at this higher energy level of the Eu^{2+} tunnel to the Sm^{3+} ions, where they remain trapped until stimulated by an infrared reading light of λ_2. The Sm^{2+} ions so formed are thermally stable deep traps of about 1.1 to 1.2 eV. Such a material has to be heated to about 250°C before the electrons are freed. Upon stimulation with an infrared reading light ($\lambda_2 = 1064$ nm), trapped electrons then tunnel back to the Eu ions, resulting in the characteristic Eu^{2+} emission ($\lambda_3 = 615$ nm) when the electrons return to the ground state.

In addition to ET materials [19] that were originally developed as infrared light sensors, an ET material using Eu-doped potassium chloride (KCl:Eu) has also been reported [20]. The specimen of KCl:Eu was first irradiated with ultraviolet light at about 240 nm. When the specimen was stimulated with visible light of 560 nm, intense emission with a peak at about 420 nm was observed.

8.3.9. TWO-PHOTON–ABSORPTION MATERIALS

The significance of two-photon absorption in optical storage has been known for some time [21]. However, recent application of two-photon–absorption materials to 3-D optical storage is usually associated with the work of Parthenopoulos and Rentzepis [22]. Two-photon–absorption materials work in a very similar way to ET materials. They require a write beam to record the data, and a read beam to stimulate the excited parts of the material to radiate luminescence. However, instead of only one photon, two-photon–absorption materials require two photons for the transition from the ground state to the stable excited state, and also another pair of photons to stimulate the excited state to return to the ground state. These two photons could have the same wavelength, although photons with different wavelengths are preferred. The two-photon–absorption phenomenon is generally known as photon gating, which means using a photon to control the behavior of another photon [23].

An example of a two-photon–absorption material is spirobenzopyran [22]. The mechanism of two-photon absorption is as follows: The molecule is initially at the S1 state. By simultaneously absorbing a photon at 532 nm and another photon at 1064 nm, the molecule is first excited to the S2 state and then stays at the stable S3 state. Upon absorption of two photons at the reading wavelength of 1064 nm, the molecule is excited to the S4 state, which is unstable. The excited molecule immediately evolves to the stable state S3 with emission at about 700 nm. In this mechanism the written data can be read without erasure. However, the real mechanism may involve other effects, since the written data has been observed to be partially erased when it is read out [24]. To completely erase the written data, the two-photon–absorption material must be heated to about 50°C or irradiated with infrared light. By raising the temperature, the molecules in the S3 state revert to the original state S1. Note that ET materials also demonstrate two-photon absorption [25].

8.3.10. BACTERIORHODOPSIN

Bacteriorhodopsin is a biological photochromic material [26]. It is the light-harvesting protein in the purple membrane of a microorganism called *Halobacterium halobium.* This bacterium grows in salt marshes where the salt concentration is roughly six times that of seawater. Bacteriorhodopsin film can be made by drying isolated purple membrane patches onto a glass substrate or embedding them into a polymer. The use of bacteriorhodopsin-based media is not restricted to the wild-type protein. A set of biochemical and genetic tools has been developed to greatly modify the properties of the protein.

Bacteriorhodopsin is the key to halobacterial photosynthesis because it acts as a light-driven proton pump converting light energy into chemical energy. The initial state of bacteriorhodopsin is called the B state. After absorption of a photon at 570 nm, the B state changes to the J state, followed by relaxation to the K and L states. Finally, the L state transforms into the M state by releasing a proton. The M state is a stable state, which can be driven back to the B state by absorption of a photon at 412 nm and capturing a proton. Reversing the M state to the B state can also be performed by a thermal process. The B and M states are the two distinct stable photochromic states in bacteriorhodopsin. Instead of absorption of a photon at 570 nm and 412 nm for each transition between the two states, the simultaneous absorption of two photons at 1140 nm and two photons at 820 nm can also stimulate the transition [27].

8.3.11. PHOTOCHEMICAL HOLE BURNING

In photochromism, the absorption of light at λ_1 changes the absorption coefficient at λ_2, and vice versa. The material has only two absorption bands at λ_1 and λ_2, of which only one can be in the activated condition. On the other hand, in photochemical hole burning, a large number of absorption bands, theoretically as large as from λ_1 to λ_{1000}, can exist at the same time and the same position [28, 29].

The absorption frequencies of dye molecules (impurities) embedded in a solid optical matrix are generally shifted due to their interaction with the surrounding matrix. The absorption lines of the dye molecules give rise to a broad continuous inhomogeneous optical absorption band. The dye molecules undergo photochemical reactions when they are irradiated by light of a certain frequency. After the material sufficiently absorbs optical energy at λ_1, the material can absorb no more light at λ_1. In other words, the λ_1 absorption band is bleached out, so that a spectral hole appears in the broad inhomogeneous absorption band. The spectral hole can be read using light of the same wavelength λ_1. Similar to two-photon absorption materials, two photons at λ_2 and λ_3 may be used at the same time for writing, while a photon at λ_1 is used for reading [23]. At the present time, the main difficulty of these materials, organic or inorganic, is that they must be kept at a very low temperature (< 100 K).

8.3.12. MAGNETO-OPTIC MATERIALS

Magneto-optic (MO) materials store binary information as upward and downward magnetization [30, 31]. The most commonly used MO medium is a thin film of a manganese bismuth (MnBi) alloy. The recorded data is read out using a linearly polarized laser beam, which undergoes a small rotation because of the Faraday or Kerr effect. The polarization of the beam is rotated to the left or right, depending on whether the magnetization is upward or downward. In the transmissive Faraday effect, the rotation angle is proportional to the film thickness. In reflective readout, there is a rotation of the polarization through the Kerr effect. This Kerr rotation is a sensitive function of the incidence angle. The Kerr rotation seldom exceeds 1°. In principle, the intensity modulation can be obtained by applying an analyzer.

Ferromagnetism is a property of some substances that are able to retain magnetization when the external magnetizing field is removed. In ferromagnetic materials, the magnetic moments of their atoms are aligned in the same direction. When ferromagnetic materials are heated over a temperature called the Curie point, atoms start a random motion due to the temperature. The magnetic moments are in random orientation, and the materials become paramagnetic. To write binary data onto MnBi film, the temperature of the MnBi medium at the data spot is raised in excess of the Curie point of the material (180°C to 360°C). The spot is heated by a focused laser beam. During cooling from the Curie point, the magnetization of the spot can be determined by an applied external magnetic field.

At room temperature, the MO medium is resistant to changes in magnetization. The reverse magnetic field required to reduce the magnetization of the recording material is called coercivity. The coercivity of MO film at room temperature is quite high. Some materials, such as phosphorus-doped cobalt

(Co[P]) exhibit a strong temperature dependence of the coercivity. At a temperature of about 150°C, the coercivity is decreased by a factor of 3 from that at room temperature. Thus, binary data can be written with a suitable field-applied coincident with the laser heating pulse that raises the temperature of the heated spot to 150°C (which is lower than the Curie point). Once cooled, the data will not switch because of the surrounding magnetic field, and thus only the area heated above 150°C is affected.

8.3.13. PHASE-CHANGE MATERIALS

Phase-change recording uses differences of reflected intensity to distinguish recorded binary data [30, 32]. The principle underlying optical recording using phase-change materials is the controlled, reversible switching of a spot between two states, usually the amorphous and crystalline states. In contrast to liquid and gaseous states, solids are bodies having constant shape and volume. However, a distinction is made between crystalline and amorphous solids. In a crystalline solid, the constituent atoms are in a periodically repeated arrangement, whereas the atoms are in a random pattern in the amorphous state. The stored data on the phase-change thin film can be read by passing a light beam through the thin film. The amorphous state is transparent while the crystalline state is opaque. On the other hand, the data can also be read by detecting the reflected light. The reflectivity of the crystalline state can be four times that of the amorphous state.

A distinctive class of amorphous solids are glasses, which are amorphous solids obtained by cooling from a melt. Upon slowly cooling below the melting temperature, T_m, the liquid freezes and becomes a crystalline solid. However, if the liquid is rapidly cooled from the melting temperature down to the glass transition temperature, T_g, crystallization cannot occur. Instead, an amorphous solid or glass is formed. In principle, to switch from the amorphous state to the crystalline state, the spot is heated by a laser beam to a temperature well above T_g, but just below T_m. The material then has sufficient time to crystallize when it cools to room temperature through T_g. To revert from the crystalline state to the amorphous state, the spot is heated by a laser beam to a temperature above T_m. In this case, glass will be formed provided that after the light pulse ends the material is cooled rapidly from T_m to T_g to prevent crystallization.

8.4. BIT-PATTERN OPTICAL STORAGE

This and successive sections will overview architectures for optical storage. Information to be stored and later recalled is a string of bits. Thus, the information is 1-D, which is related to electronic information processing systems such as computers and other communication equipment. However, this 1-D information is formatted in 2-D or 3-D in an optical storage system in order to achieve the maximum storage density allowed. A straightforward method for storing the optical information represented by a bit pattern is to record the bit pattern directly.

8.4.1. OPTICAL TAPE

A string of bits can be recorded on a tape. The drawback of a tape system is that the data cannot be quickly accessed at random. The tape must be wound from the beginning before the desired data can be read. Optical tape is traditionally applied to movie film to store the movie's sound track.

8.4.2. OPTICAL DISK

The most successful optical storage medium is the optical disk. Figure 8.1 shows the basic structure of an optical disk. The recorded signal is encoded in the length of the pit and the spacing of pits along the track. The distance between two adjacent tracks (track pitch) is 1.6 μm. The width of a pit is equal to a recording spot size of 0.5 to 0.7 μm. The light source used in an optical disk system is usually a GaAlAs semiconductor laser diode emitting at a wavelength of 0.78 to 0.83 μm. The spot size of the readout beam is determined by the numerical aperture (NA) of the objective lens. Typically, $\lambda/\text{NA} = 1.55$ is chosen, so the effective diameter of the readout spot is approximately 1 μm. The spot size is larger than the width of a pit, but a single readout spot does not cover two tracks.

8.4.2.1. *Read-Only Optical Disk*

For a read-only optical disk, such as a compact disk for music recording (CD) or a read-only memory compact disk for computers (CD-ROM), the

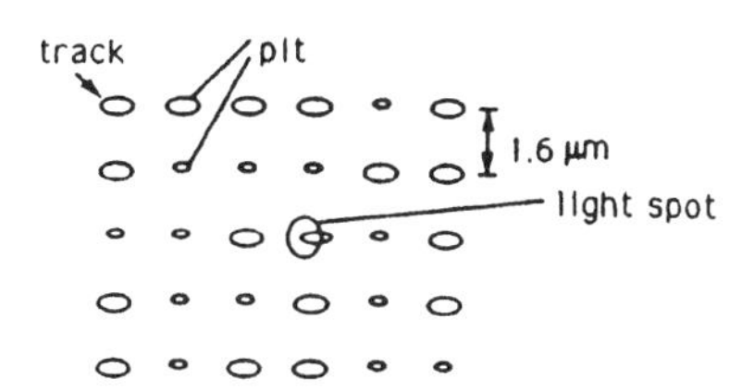

Fig. 8.1. Basic structure of an optical disk.

recorded data cannot be changed after the disk is manufactured. The process of optical pickup is as follows: When there is no pit, the light will be fully reflected to the detector. When there is a pit, the focused beam will cover both pit and the surrounding land. Both pit and land are coated with high-reflectivity material such as aluminum (Al), so light is reflected from both the pit and the land. The depth of the pit is made such that the phase difference between the reflected light from a pit and the land is π. Consequently, there is destructive interference at the detector, and less light is detected. The pit depth is typically 0.13 μm.

8.4.2.2. *WORM Optical Disk*

A WORM (write once, read many times) disk consists of either a polycarbonate or hardened glass substrate and a recording layer made of a highly reflective substance (dye polymer or tellurium alloy). As with other optical disks, the recording layer is covered by clear plastic to protect the recording medium. WORM disk systems use a laser beam to record data sequentially. A write beam burns a hole (pit) in the recording medium to produce a change in its reflectivity. In contrast to the previously discussed read-only optical disk or CD, the WORM optical disk modulates directly the reflected intensity of the readout beam.

8.4.2.3. *Magneto-optic Disk*

An erasable and rewritable optical disk is best represented by a magneto-optic (MO) disk. The MO disk makes use of MO recording material that at room temperature is resistant to changes in magnetization. The magnetization of the material can be changed by placing it in a magnetic field and heating it to its Curie point, about 180°. To erase and to write new data onto a MO disk, the heat of the write laser beam brings the recording material to the Curie point. A bias magnet then reverses the magnetization of the heated area that represents a bit. A low-power linearly polarized laser beam can be used to read the data on the MO disk. According to the Kerr magneto-optic effect, the polarization of the reflected readout beam will be rotated to the left or right, depending on whether the magnetization of the recording material is upward or downward. At the present time the typical rotation is less than 1°.

8.4.3. MULTILAYER OPTICAL DISK

While an optical disk is a 2-D optical storage device, a multilayer optical disk is in the class of 3-D optical storage. A new structure for an optical disk

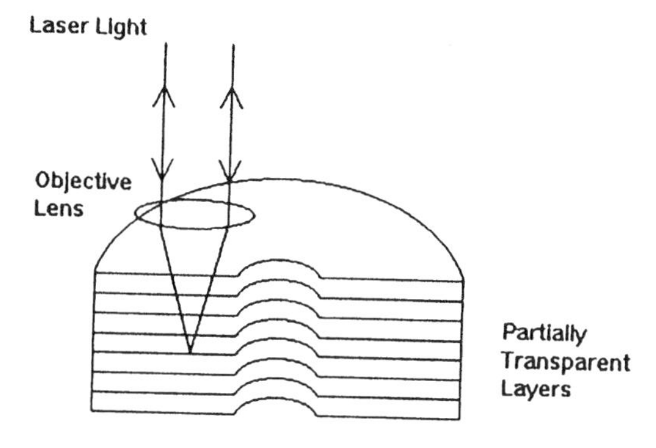

Fig. 8.2. Multilayer optical disk.

having up to six layers for a read-only disk and up to four layers for a WORM disk is also available [33]. There reportedly appear to be no fundamental hurdles for read-only disks with 10 to 20 layers. Several formatted thin-disk substrates are physically stacked with spacers into a single multilayer optical disk. To change from one layer to another, the optical pickup refocuses the objective lens to the new layer. The layers are separated by at least 100 μm so that other data surfaces are well out of focus. With the layers about 100 μm apart, the laser beam could spread over 10,000 times more area on a layer adjacent to the layer in focus, depending on the NA of the focusing lens.

Each disk layer must be partially transparent so that the laser can penetrate all the layers in a stack. Each surface, however, must also be reflective enough so the data can be read. On average, the surface would reflect only 5% of the incident light. For comparison, a read-only optical disk reflects about 95% of the incident light. The electronic part of the system should be modified to amplify the output signal of the photodetector to a level compatible with standard optical drives. At normal optical disk storage densities, a 10-layer disk would store 6.5 gigabytes of information (a single layer stores 650 megabytes). A multilayer optical disk is schematically shown in Fig. 8.2.

8.4.4. PHOTON-GATING 3-D OPTICAL STORAGE

The objective of 3-D storage is to arrange and store a string of bits in a 3-D structure. A book is a good analogy for 3-D storage. The 1-D bit pattern is first arranged in a 2-D format called page memory. The stack of page memories forms a 3-D storage medium. While we may freely open a book to select a page to read, we cannot mechanically open a 3-D optical storage medium to select a certain page or layer. A first approach to 3-D storage—a multilayer optical disk—was discussed in the previous subsection. Depending on the separation between the two layers, the laser beam diameter on the adjacent layer may significantly increase and the intensity may drop accordingly. This can be easily applied to an optical disk, because both writing and reading are based on a sequential mechanism. There is only one laser spot at a time. When a page memory is written or read in parallel, there are many bright laser spots at the same time. On the adjacent layers, although an individual laser spot is blurred, there is a possibility that those blurred laser spots overlap and the superimposed intensity is as high as a single focused spot.

In order to be able to select a page, a switch is required to activate only a layer at a time. The switch could be a photon-gating process in which the writing and reading of the optical memory requires two photons at the same site and the same time. By using a laser beam (first photon) to image the page memory onto the selected layer and providing another laser beam (second photon) to the selected layer only, we can selectively write or read a layer.

This scheme is best suitable for two-photon–absorption materials [22, 24, 25]. The principle of this architecture is schematically shown in Fig. 8.3. Note that the 3-D storage is a bulk transparent media made from a two-photon–absorption material lacking a delineated layered structure. The activated layer is formed optically by illuminating the bulk medium from the side with a sheet of light. The sheet of light is formed by focusing a collimated light using a cylindrical lens. An experimental demonstration with one activated layer in the bulk medium was performed recently by Hunter et al. [24]. Alternatively, the stored data can be read sequentially spot by spot instead of a parallel reading of the whole layer. This will eliminate the difficulty of forming a uniform sheet of light. The sequential reading architecture can also

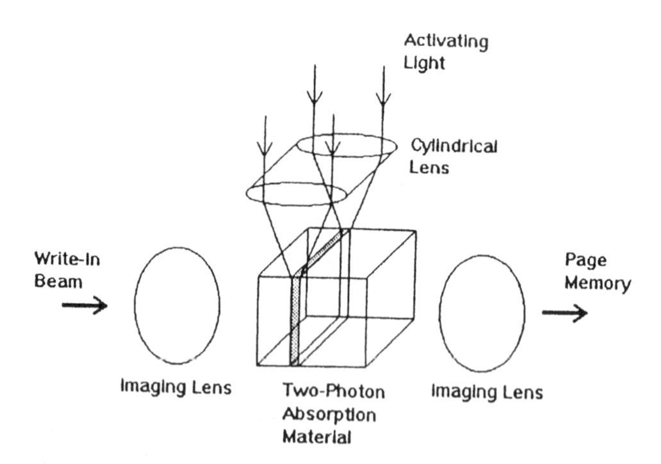

Fig. 8.3. Architecture of 3-D optical storage using two-photon–absorption materials.

be applied to bacteriorhodopsin [27]. However, instead of a luminescence output, a generated photovoltage is detected as the output signal from the 3-D storage device [27].

8.4.5. STACKED-LAYER 3-D OPTICAL STORAGE

In contrast to photon-gating 3-D storage that uses a bulk medium without a delineated layered structure, stacked-layer 3-D optical storage really consists of a stack of layers. This architecture can be represented by 3-D storage based on ET materials [34].

The concept that a stack of ET layers can form a 3-D optical storage device is schematically depicted in Fig. 8.4. The formatted 2-D page memory can be written by properly imaging the page composer onto a specific layer with blue laser light at 488 nm. Similar to multilayer disks [33], it had previously been calculated [34] that the diameter of the out-of-focus spot on the neighboring layer would be 200 times larger and the intensity would be reduced by 40,000 times. However, when 1000×1000 bits are written in parallel, the out-of-focus spot may overlap and superimpose. The superimposed spots may generate serious cross-talk noise.

This cross-talk problem can be overcome by using the coding and decoding technique given in Figs. 8.5(a) and (b), respectively. Since bits 1 and 0 consist of the same number of bright and dark pixels, the blurred 1 and 0 coded

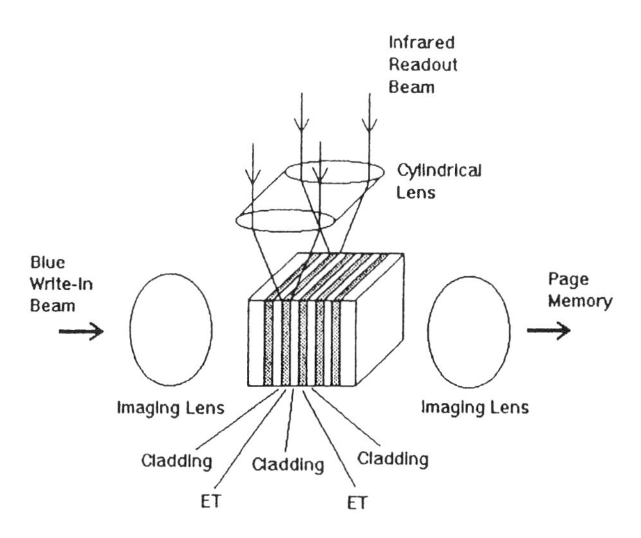

Fig. 8.4. Architecture of 3-D optical storage using ET materials.

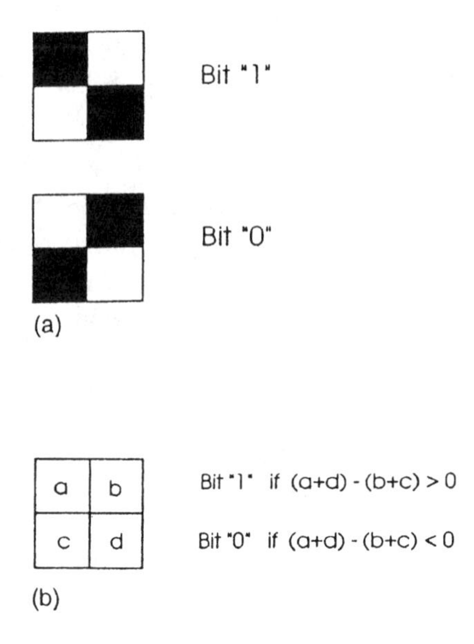

Fig. 8.5. (a) Coding and (b) decoding of bit "1" and bit "0" in 3-D optical storage using ET materials.

pattern contributes uniform intensity on other layers except the layer in focus.

To read the page memory stored in a specific layer, a sheet of 1064-nm infrared light is addressed to the ET layer from the side of the stacked-layer storage. The individual 4–10–μm–thick ET thin films together with transparent cladding layers form a slab waveguide. Since the infrared reading light is launched into the edge of the ET thin film, the infrared reading light will propagate and be trapped inside the waveguide. The orange luminescent emission (615 nm) corresponding to the written page memory at that layer will be produced as a result of the infrared stimulation. Since the luminescent light is emitted in all directions, part of it will transmit through all layers and arrive on an array detector.

A 3-D optical storage device consisting of five layers of ET thin films has already been demonstrated experimentally [35]. Figure 8.6(a) shows the encoded binary input patterns of five ET layers using the coding method shown in Fig. 8.5. Figure 8.6(b) shows the output patterns of five ET layers following the reading method described in the previous paragraph. The direct output patterns suffer from cross talk. However, the decoded binary outputs can be corrected as shown in Fig. 8.6(c). There is also a possibility of deactivating the ET thin film by applying an electric field across it.

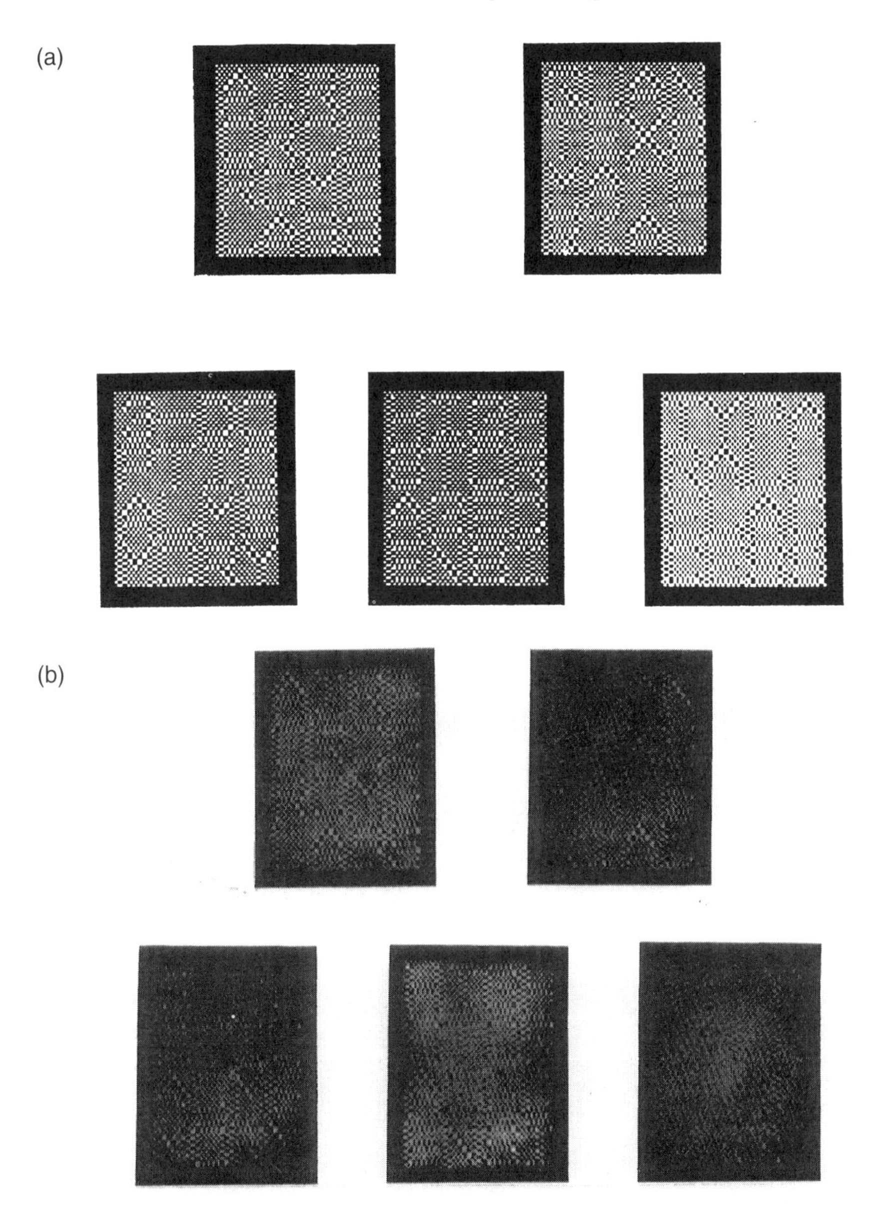

Fig. 8.6. Experimental results of 3-D optical storage using ET materials. (a) Encoded binary input patterns for five layers, (b) Retrieved output from each of the five ET layers, (c) Successfully decoded binary output patterns from (b).

Fig. 6. Continued.

8.4.6. PHOTOCHEMICAL HOLE-BURNING 3-D STORAGE

In photochemical hole-burning optical storage [23, 28, 29], the third dimension is frequency instead of depth or thickness. Thus, the storage may look like an ordinary single-layer optical disk or photographic plate, but the storage capacity can be as high as 1000 times that of single-layer disks. Currently, the media should be kept at a very low temperature, typically less than 100 K. This makes it impractical at present. The application of photochemical hole burning to optical storage is somewhat like the change from black-and-white to color photography. Consequently, low-cost and stable semiconductor lasers would be needed that must be tunable over a fairly large frequency range and must operate in single mode. Lasers fulfilling these requirements are still under development. Nevertheless, holography in photochemical hole-burning media has also been proposed recently [36].

8.5. HOLOGRAPHIC OPTICAL STORAGE

In optical signal processing [2, 4], it is widely known that an image can be recorded either directly on a photographic plate or in a hologram. The hologram records the interference pattern generated by the image and a reference beam. The hologram can then be reconstructed to produce the original image. This section reviews techniques to store a bit pattern in a hologram.

In the previous section, we learned that a 3-D storage in principle consists of multilayers whether there is a delineated layered structure (ET material) or not (two-photon absorption). When a specific layer is selected to be written or read, other layers must be effectively deactivated. The signal light interacts with only the selected layer and passes all other layers unaffected. This is analogous to selecting a page while reading a book.

In holographic 3-D optical storage, the input 2-D page memory is coded and distributed to the entire storage space. If we consider that the storage consists of layers, the page memory is stored in every layer with different coding. When the stored page memory is going to be read, every layer is read, and no layer is deactivated. With proper decoding, the stored page memory can be reconstructed. Instead of reading a book, this is analogous to imaging a cross section of the human body by employing computer tomography. However, no computer is needed to reconstruct a volume hologram.

8.5.1. PRINCIPLE OF HOLOGRAPHY

Holography was originally invented by Gabor [37] in 1948 to improve electron microscope images. Holography is a technique to record and to reconstruct wavefronts. Consider that we are in a room looking at an object, such as a flower, through a window. The light from the flower must transmit through the window to form an image on our retinas. If we could record the light at the window plane, then by reproducing the recorded light we would be able to see the flower, although there is no window. This cannot be done by traditional photographic techniques, because the recording medium is sensitive only to energy or light intensity and is not sensitive to phase. The light from the flower at the window plane is represented by its wavefront that has both amplitude and phase. The artificial window can only be made by recording and reconstructing the wavefront at the window plane. The wavefront can be recorded and later reconstructed by recording the interference pattern of the object wavefront and a reference wavefront [37]. Holography has been improved and made practical by Leith and Upatnieks [38], who introduced the concept of carrier frequency in 1962.

The principle of holography can be described briefly as follows. The recorded intensity on a hologram is

$$|O + R|^2 = |O|^2 + |R|^2 + OR^* + O^*R, \tag{8.9}$$

where O and R are complex amplitudes of object and reference beams, respectively, on the hologram plane. The symbol * indicates a complex conjugate. The amplitude transmittance of the hologram (which is a developed photographic plate) is proportional to the recorded intensity. When the hologram is illuminated by a reconstructing laser beam, which is the same as the reference beam when it was recorded, the wavefront of light transmitted through the hologram is given by

$$|O + R|^2 R = |O|^2 R + |R|^2 R + O|R|^2 + O^* R^2. \tag{8.10}$$

The third term is of particular interest. If R represents a plane wave, $|R|^2$ is constant across the photographic plate, and the third term is proportional to the complex amplitude O that is the original wavefront of the object on the hologram plane. It is important to note that following the scheme introduced by Leith and Upatnieks [38], propagation directions of the first, second, and fourth terms of Eq. (8.10) are separated from the propagation direction of the wavefront represented by the third term. Thus, in principle, the artificial window discussed previously can be realized using a hologram.

8.5.2. PLANE HOLOGRAPHIC STORAGE

Basic engineering concepts of holographic storage were introduced in the 1960s [39, 40, 41], following the publication of van Heerden's seminal papers [5, 6]. For bit-pattern storage such as conventional high-density microfiche, even a small dust particle on the film can create a missing portion on the record, and the missing information can never be recovered. However, when using holograms for high-density recording, a scratch or dust on the film will not destroy information but merely causes a slight increase in the noise of the reconstructed image, so that no particular portion of the recording is lost [42]. Consider that the information to be recorded is a string of bits, this string of bits being first arranged in a 2-D format called page memory. It is advantageous to record the Fourier-transform hologram of the page memory because the minimum space bandwidth is then required and the information about any one bit of the page memory is spread over the hologram plane [43].

In the simplest optical system, the page memory is displayed on a page composer which is a spatial-light modulator. A collimated coherent beam is modulated by the spatial-light modulator. The modulated light then passes through a lens that performs the Fourier transform of the page memory on the focal plane of the lens. A holographic medium records the interference pattern of a reference beam and the Fourier transform of the page memory on the focal plane. If the same reference beam is incident onto the recorded hologram, the Fourier transform of the page memory will be produced. The page memory can be reconstructed by passing its Fourier transform through another lens.

Equation (8.10) shows that only the same reference beam as was employed when the hologram was recorded will reconstruct the object. This characteristic provides multiplexing capability. A number of holograms can be recorded successively with reference beams having different incident angles on the same holographic plate. A specific angular reference beam would reconstruct only the object that was recorded with it at a certain position. Note that other objects are reconstructed at shifted positions. The multiplexing hologram can also be produced with reference beams having specific wavefronts [44]. The wavefront is generated by passing a plane wave through a phase-only spatial-light modulator. In fact, this phase modulator can also generate a wavefront similar to that of an oblique plane-wave reference beam.

Since the page memory must be displayed on a spatial-light modulator, and the reconstructed image of the page memory must be read by an array detector, the size of the page memory is restricted by the state of the art of the spatial-light modulator and the array detector. The size of the page memory is also restricted by the size of lenses and other optical elements used in the system. Based on today's technology, it should be realistic to construct a page memory that is 1000×1000 in size. Assuming the wavelength of light, λ, is approximately 1 μm, a page memory in principle could be stored in an area of 1 mm^2 either in a bit pattern or holographically.

The capacity of plane holographic storage might be larger than 10^6 bits, while the string of bits arranged in a page memory has only 10^6 bits. Consequently, many page memories can be stored in a plane hologram. As mentioned previously, one may apply angular multiplexing technique to superimpose a number of holograms on the same plate. However, larger optics and higher laser power are required to cover the whole holographic plate at one time. It is more practical to record each page memory in a tiny subhologram. Subholograms form an array on the holographic plate. A laser beam deflector is able to select a specific subhologram. In other words, a page memory is retrieved by addressing the read beam to a selected subhologram using the deflector. Figure 8.7 depicts the schematic diagram for plane holographic storage. Needless to say, the object beam in the recording is also deflected accordingly to form the subhologram array on the plate.

The experiment using an erasable magneto-optic MnBi thin film to record a 8×8 bit page memory was demonstrated by Rajchman in 1970 [45]. The MnBi thin film modulates the polarization of the read beam based on the Faraday effect and the Kerr effect in the transmissive and reflective modes, respectively. Note that the polarization modulation is actually phase modulation in two polarization directions. Thus, a MnBi thin film acts as a phase hologram and no analyzer is required in the reconstruction.

Instead of using beam deflectors, subholograms may be recorded on a moving media [46]. In fact, holographic disks consisting of 1-D subholograms have been experimentally demonstrated. The prototypes of a WORM holographic disk using a photographic plate and an erasable holographic disk using

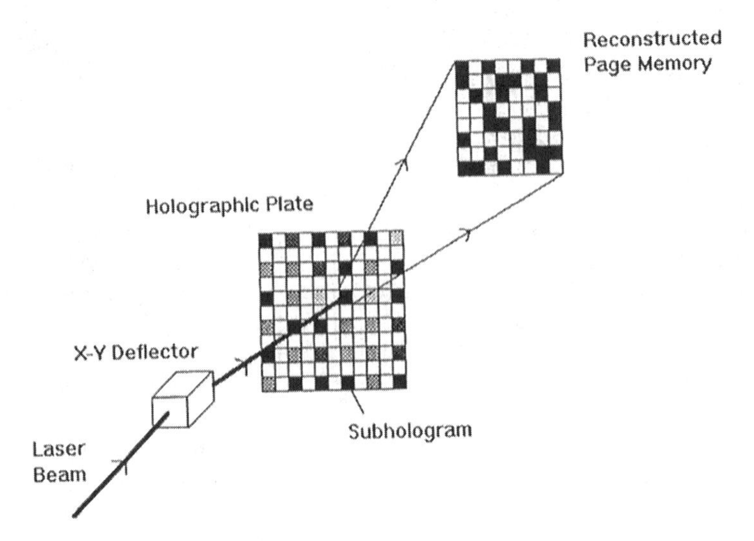

Fig. 8.7. Reference beam being addressed by a deflector to a subhologram on a holographic plane to retrieve a page memory.

a photorefractive layer were built by Kubota et al. in 1980 [47] and Mikaelian et al. in 1992 [48], respectively.

8.5.3. STACKED HOLOGRAMS FOR 3-D OPTICAL STORAGE

Several plane holograms can be stacked together in a layered structure to form a 3-D optical storage medium if it is possible to selectively activate any layer. In photon-gating 3-D storage using two-photon–absorption materials, the gating photon activates the layer. In stacked-layer 3-D storage using ET materials, the stimulating infrared that propagates in the thin film activates the layer. In stacked hologram 3-D storage, an electric signal can activate the selected layer.

8.5.3.1. *Stack of SBN Layers*

Strontium barium niobate ($Sr_xBa_{(1 \times x)}Nb_2O_6$), which is commonly abbreviated SBN, is a photorefractive material that can be used for generating a hologram. Thaxter and Kestigian demonstrated in 1974 that the sensitivity of SBN in hologram recording and the diffraction efficiency in hologram reconstruction could be controlled by an applied electric field [49]. The sensitivity of a pure SBN crystal ($Sr_{0.75}Ba_{0.25}Nb_2O_6$) is quite low for recording a hologram. However, when an external electric field of about 5×10^5 V/m is applied across the crystal in the proper direction, the sensitivity is significantly increased and enables the recording of a hologram. It is believed that this applied field aids the rate of photoionized charge separation by providing a drift field. The drift current attained under conditions of an applied field is significantly larger than the average diffusion current driven solely by the spatial variation of the photoionized charge density.

The recorded hologram can be made latent by applying an electric pulse that generates an electric field of about 10^5 V/m that is opposite to the field used for enhancing the recording sensitivity during recording. In the latent state, the diffraction efficiency of the reconstructed hologram is very low. The applied field causes a change of the charge distribution such that the hologram no longer diffracts the read beam. The change remains after the applied field is removed. However, the original charge distribution is not lost. The effect of the original charge is only suppressed and can be returned and even enhanced by applying an electric field of about 5×10^5 V/m opposite to the suppressing field and in the same direction as the field for enhancing the recording. Therefore, a stack of transparent SBN holographic layers with separated electric switches can be used as a 3-D optical storage device. Note that electrodes are at the edges of the SBN layer.

8.5.3.2. *Stack of PVA Layers*

Some organic photochromic materials, including azo-dye–doped polyvinyl alcohol (PVA), are sensitive to the polarization of light [50]. A hologram can be recorded and reconstructed using only light with certain polarization. Recently, a 3-D storage device formed by a stack of PVA layers and polarization rotators has been proposed by Chen and Brady [51].

In this 3-D storage, the PVA layer is used to store the hologram. Each PVA layer is covered by a polarization rotator that consists of a liquid crystal layer sandwiched between a pair of transparent electrodes. It is well known [2] that a liquid crystal polarization rotator can rotate the polarization of the incident light by 90° when no voltage is applied to the electrodes (on state). If voltage is applied, the liquid crystal polarization rotator will not affect the incident light (off state).

Consider that the incident light polarization is 0° and the PVA layer requires 90° polarization for activation. When the liquid crystal polarization rotators are in the off state, the incident light passes through the polarization rotators and the PVA layers unaffected. To activate a selected PVA layer, the liquid crystal polarization rotator just before it is turned on so as to rotate the light polarization by 90°. After passing the selected PVA layer, the next liquid crystal polarization rotator is also set on and rotates the polarization by 90°

once again, to return the light polarization to 0°. Therefore, the reconstructed light is transmitted through the rest of layers unaffected.

8.5.4. VOLUME HOLOGRAPHIC 3-D OPTICAL STORAGE

We have discussed a hologram that records the wavefront from a flower using a window analogy. Through the window we can see not only a flower but many other objects as well. The window can transmit wavefronts from an infinite number of objects, since photons represented by wavefronts do not interact with each other. However, only a limited number of wavefronts can be recorded on a hologram because the wavefronts must be recorded in interference patterns, and only a limited number of patterns can be recorded depending on the resolution and dynamic range of the hologram. Obviously, the larger the hologram is, the larger the storage capacity will be.

In contrast to conventional imaging that has to store the light pattern on a plane, the holographic technique transforms a 2-D pattern into a 3-D interference pattern for recording [5, 6, 52, 53]. The recorded 3-D interference pattern, which is known as a volume hologram, can be transformed back to the original 2-D object pattern in the reconstruction process. Consequently, the storage capacity can be increased significantly by recording the interference pattern in a 3-D medium. Since a volume hologram acquires the properties of a 3-D diffraction grating, the reconstruction is subject to Bragg's condition. The reconstruction beam with the same wavelength must be at the same angle as the reference beam in the recording. Contrary to a plane hologram that can be reconstructed from almost any angle, the reconstruction beam of a volume hologram will fail to produce the object wavefront at any angle except the recording angle.

Holographic storage techniques usually apply the same beam for writing and reading, or different wavelengths of light for writing and reading provided that Bragg's condition is satisfied, regardless of whether it is a plane or volume hologram. Therefore, there is almost no difference in the architectures of plane or volume holographic storage.

It is interesting to note that in ordinary applications, a plane hologram is used to produce the 3-D display of an object. On the other hand, in digital optical storage applications, a volume hologram is used to record and display a 2-D page memory.

The first prototype of 3-D holographic optical storage with an ingenious engineering design was demonstrated by D'Auria et al. in 1974 [54]. It was based on the architecture shown in Fig. 8.7. However, each tiny plane subhologram was replaced by a volume subhologram employing Fe–doped lithium niobate. With an additional deflector (not shown in Fig. 8.7), their ingenious design enabled the reference beam (which was also the read beam) to arrive at the tiny volume subhologram with a rotatable angle. Thus each tiny subhologram could be angular-multiplexed with many page memories. The area of the subhologram was 5×5 mm^2, and its thickness was 3 mm. Ten 8×8 bit page memories were multiplexed in a subhologram.

The cubic subholograms mentioned above can be replaced with SBN fibers [55, 56]. The diameter of the fibers is approximately 1 mm and they are about 4 mm long. Thirty to fifty page memories can be multiplexed and stored in an SBN fiber. Hundreds of fibers are arranged in a near-fiber-touching-fiber configuration, much like in a conventional microchannel plate. A merit of this approach is that a high-quality SBN crystal fiber is easier to grow than a bulk crystal. On the other hand, with constant progress of enabling technologies, Mok et al. successfully demonstrated the storage of 500 holograms of images consisting of 320×220 pixels in a 1 cm $\times$ 1 cm $\times$ 1 cm Fe–doped lithiun niobate crystal [57] and then 5000 holograms in a 2 cm $\times$ 1.5 cm $\times$ 1 cm Fe–doped lithium niobate crystal [58].

Recently, 3-D holographic optical storage was demonstrated using both wavefront and angular multiplexing [59]. Wavefront multiplexing is achieved by passing the reference beam through a phase-only spatial-light modulator that displays a specific phase pattern. The computer-generated phase patterns for multiplexing are orthogonal to each other. Angular multiplexing is provided by tilting the hologram. Thirty-six orthogonal phase patterns were generated using a computer based on a specific algorithm. The hologram was recorded at three angles. A total of 108 holograms were successfully recorded, six of which are shown in Fig. 8.8.

8.6. NEAR FIELD OPTICAL STORAGE

The resolution of an image displayed on a computer monitor is restricted by the size and number of pixels that constitute the display. For example, a liquid crystal display may have about 1 million pixels (1000×1000); each pixel is about 10 μm $\times$ 10 μm. Similarly, images we see in the real world are also limited by the size and number of light-sensitive cells in our retinas. We have about 120 million photoreceptor rod cells of 2 μm in diameter that are highly sensitive in dim light but insensitive to colors, and about 6 million photoreceptor cone cells of 6 μm in diameter sensitive to colors but insensitive at low light levels. If the image detail cannot be resolved by photoreceptor cells in the retina, to clearly see the image, it must be magnified optically before entering our eyes. This is normally done using an optical microscope.

However, there is a natural limitation that a microscope cannot magnify an image arbitrarily and infinitely. The resolution limit of a lens is determined by

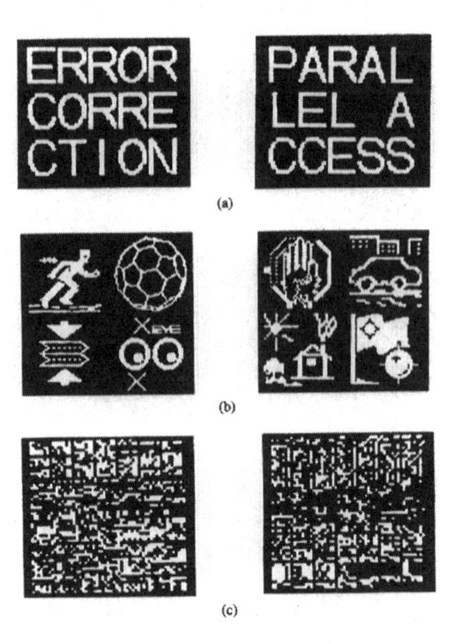

Fig. 8.8. Experimental results of wavefront and angular multiplexing. Two outputs from 36 patterns stored in (a) the first, (b) the second, and (c) the third angular position.

the Rayleigh criteria, as follows.

$$\frac{Q}{2} = 1.22\frac{f}{D}\lambda, \tag{8.11}$$

where Q is the diameter of the central spot of the Airy disk as given in Eq. (8.1), $Q/2$ is the Rayleigh resolution limit, f is the focal length of lens, D is the diameter of lens, and λ is the wavelength of light. We cannot keep increasing D to decrease $Q/2$ to zero, because, in practice, f and D of a lens are not independent. When f is made small, D will be limited by the curvature of the lens surface, which is, in turn, determined by f.

The Rayleigh criteria are based on the diffraction pattern of a circular aperture, which is commonly known as the Airy disk. $Q/2$ is actually the radius of the first-zero ring of the Airy disk. We must not consider that the highest spatial frequency passing through the optical system is $(Q/2)^{-1}$ or $D/1.22f\lambda$, which could be larger than $1/\lambda$. Spatial frequencies higher than $1/\lambda$ cannot propagate far, as will be shown in the following paragraph.

We first look at the optical field $f(x, y, z)$ provided that $f(x, y, 0)$ is known. The optical field $f(x, y, z)$ satisfies the wave equation as follows [4]:

$$(\nabla^2 + k^2)f(x, y, z) = 0. \tag{8.12}$$

Substituting $f(x, y, z)$ with its Fourier components yields

$$(\nabla^2 + k^2)\left[\iint F(u, v, z)\, e^{2\pi i(ux+vy)}\, du\, dv\right] = 0. \tag{8.13}$$

Notice that (x, y) is the two-dimensional space domain, and (u, v) is the two-dimensional spatial-frequency domain. Changing the order of integration yields

$$\iint \left[(2\pi iu)^2 + (2\pi iv)^2 + \frac{\partial^2}{\partial z^2} + k^2\right] F(u, v, z) e^{2\pi i(ux+vy)}\, du\, dv = 0. \tag{8.14}$$

Since the Fourier transform (or inverse Fourier transform) of zero is also zero, one obtains

$$\left[(2\pi iu)^2 + (2\pi iv)^2 + \frac{\partial^2}{\partial z^2} + k^2\right] F(u, v, z) = 0. \tag{8.15}$$

Remember $k = 2\pi/\lambda$; one finally gets a wave equation for $F(u, v, z)$ as follows:

$$\frac{\partial^2}{\partial z^2} F(u, v, z) + k^2[1 - \lambda^2(u^2 + v^2)]F(u, v, z) = 0. \tag{8.16}$$

The solution is

$$F(u, v, z) = F(u, v, 0)e^{izk\sqrt{1-\lambda^2(u^2+v^2)}}. \tag{8.17}$$

Notice that

$$f(x, y, z) \neq f(x, y, 0)e^{izk\sqrt{1-\lambda^2(u^2+v^2)}}. \tag{8.18}$$

Equation (8.17) indicates that $F(u, v, z)$ is a propagating wave, if

$$(u^2 + v^2) < \frac{1}{\lambda^2}. \tag{8.19}$$

However, for

$$(u^2 + v^2) > \frac{1}{\lambda^2}, \tag{8.20}$$

$F(u, v, z)$ is no longer a propagating wave, but is an evanescent wave. Thus, spatial frequencies higher than $1/\lambda$ decay in a very short distance, and never reach the image plane.

To collect the information carried by spatial frequencies higher than $1/\lambda$, the detector must be placed very close to the object, before the evanescent wave disappears. To transfer the high-frequency information over a significant distance, the evanescent wave must be converted into a propagating wave. This is done by placing a small aperture in the near field, where evanescent waves are present. The small aperture, which is smaller than λ, will convert the evanescent wave to a propagating wave by scattering. In practice, the small aperture is formed by a tapered fiber, and the light passing the aperture is propagating in the fiber to the detector. Furthermore, the aperture can be scanned to form a two-dimensional image. The image resolution now depends on the size of the aperture and the distance from the aperture to object.

Although the concept of near field optics can be traced back to 1928 [60], it was not demonstrated until the mid–1980s [61, 62], when the technologies for fabricating small apertures and for regulating the distance between the aperture and the object, and the scanning process became mature enough.

According to the Babinet principle [3], the diffraction effect generated by a very small aperture is the same as that of a needle tip (a needle tip is the negative image of a small aperture). Thus, instead of a very small aperture, a needle tip can also be brought in close proximity to the object. The light scattered by the tip is the same as the light collected through the aperture.

Near field optical storage typically uses an aperture probe, which is a tapered fiber [63]. The fiber is first placed into tension and then heated by a pulsed CO_2 laser, causing the fiber to stretch in the heated region and cleave, leaving a conically shaped fiber with a 20-nm aperture at the end. Because the fiber is narrowed at the end, it no longer works as a waveguide, and to keep the light inside the fiber, it is necessary to aluminize the outside of the conically shaped region. This is done by obliquely depositing aluminum on the side walls of the fiber, while avoiding covering the 20-nm aperture at the end. Betzig et al. [64] first demonstrated near field magneto-optical recording in a Co/Pt multilayer film using 20-nm–aperture tapered fiber. A 20×20 array of 60-nm domains (magnetic spots) written on 120-nm center-to-center spacing was demonstrated. This corresponds to a storage density of 45 Gbit/in^2. The wavelengths of writing and reading light were 515 nm and 488 nm, respectively. Consequently, the storage density of near field optical storage can go beyond the upper limit of storage density dictated by diffraction optics.

The drawback of tapered fiber is that most of the power is lost through heat dissipation in the tapered region of the fiber, because it is no longer a waveguide. Typically, output powers of up to 50 nW out of 100-nm aperture-tapered fiber at 514.5 nm can be obtained for 10 mW power input to the fiber. Thus the efficiency is 5×10^{-3}. To overcome this problem, the facet of a laser diode was metal-coated and then a small aperture (250 nm) was created at the metal coating [65]. The small-apertured laser diode directly delivers optical power to the near field substrate with unity efficiency. Light not passing through the aperture will be recycled in the laser diode and no energy is wasted.

8.7. CONCLUDING REMARKS

This chapter gave a brief overview of the basic mechanisms of materials that have been employed to store optical information or light patterns. The fundamental architectures for 3-D optical storage were discussed in detail. The architecture for 2-D bit pattern optical storage and the 2-D holographic storage were described. Novel near field optical storage that can store a bit in an area smaller that the wavelength of light was also presented. The aim of this chapter has been to present a broad survey of information storage with optics. The main objective of the research in optical storage is to realize an ideal device that can store a bit in an area of λ^2 or smaller, or a volume of λ^3 or smaller.

REFERENCES

8.1 F. T. S. Yu, *Optics and Information Theory*, R. E. Krieger, Malabar, Florida, 1984.

8.2 F. T. S. Yu and S. Jutamulia, *Optical Signal Processing, Computing, and Neural Networks*, Wiley-Interscience, New York, 1992.

8.3 E. Hecht and A. Zajac, *Optics*, Addison-Wesley, Reading, Mass., 1974.

8.4 J. W. Goodman, *Introduction to Fourier Optics*, McGraw-Hill, New York, 1968.

8.5 P. J. van Heerden, 1963, "A New Optical Method of Storing and Retrieving Information," *Appl. Opt.* 2, 387–392.

8.6 P. J. van Heerden, 1963, "Theory of Optical Information Storage in Solids," *Appl. Opt.* 2, 393–400.

8.7 L. Solymar and D. J. Cooke, *Volume Holography and Volume Gratings*, Academic Press, London, 1981.

8.8 G. Saxby, *Practical Display Holography*, Prentice Hall, New York, 1988, pp. 273–280.

8.9 J. R. Magariños and D. J. Coleman, 1985, "Holographic Mirrors," *Opt. Eng.* 24, 769–780.

8.10 S. V. Pappu, 1990, "Holographic Memories: A Critical Review," *Int. J. Optoelect.* 5, 251–292.

8.11 B. L. Booth, 1972, "Photopolymer Material for Holography," *Appl. Opt.* 11, 2994–2995.

8.12 J. W. Gladden and R. D. Leighty, "Recording Media," in *Handbook of Optical Holography*, H. J. Caulfield, Ed., Academic Press, San Diego, 1979, pp. 277–298.

8.13 T. C. Lee, 1974, "Holographic Recording on Thermoplastic Films," *Appl. Opt.* 13, 888–895.

8.14 L. H. Lin and H. L. Beaucamp, 1970, "Write-Read-Erase In Situ Optical Memory Thermoplastic Hologram," *Appl. Opt.* 9, 2088.

8.15 F. S. Chen, J. T. LaMacchia, and D. B. Fraser, 1968, "Holographic Storage in Lithium Niobate," *Appl. Phys. Lett.* 13, 223–225.

8.16 G. Jackson, 1969, "The Properties of Photochromic Materials," *Opt. Acta* 16, 1–16.

8.17 W. J. Tomlinson, 1975, "Volume Holograms in Photochromic Materials," *Appl. Opt.* 14, 2456–2467.
8.18 W. J. Tomlinson, E. A. Chandross, R. L. Fork, C. A. Pryde, and A. A. Lamola, 1972, "Reversible Photodimerization: A New Type of Photochromism" *Appl. Opt.* 11, 533–548.
8.19 S. Jutamulia, G. M. Storti, J. Lindmayer, and W. Seiderman, 1990, "Use of Electron-Trapping Materials in Optical Signal Processing. 1: Parallel Boolean Logics," *Appl. Opt.* 29, 4806–4811.
8.20 H. Nanto, K. Murayama, T. Usuda, F. Endo, Y. Hirai, S. Taniguchi, and N. Takeuchi, 1993, "Laser-Stimulated Transparent KCl:Eu Crystals for Erasable and Rewritable Optical Memory Utilizing Photostimulated Luminescence," *J. Appl. Phys.* 74, 1445–1447.
8.21 D. Von Der Linde and A. M. Glass, 1976, "Multiphoton Process for Optical Storage in Pyroelectrics," *Ferroelectrics* 10, 5–8.
8.22 D. M. Parthenopoulos and P. M. Rentzepis, 1989, "Three-Dimensional Optical Storage Memory," *Science* 245, 843–845.
8.23 W. E. Moerner, 1989, "Photon-Gated Persistent Spectral Hole Burning," *Japan. J. Appl. Phys.* (28 Suppl. 28-3), 221–227.
8.24 S. Hunter, C. Solomon, S. Esener, J. E. Ford, A. S. Dvornikov, and P. M. Rentzepis, 1994, "Three-Dimensional Optical Image Storage by Two-Photon Recording," *Opt. Mem. Neur. Netw.* 3, 151–166.
8.25 A. S. Dvornikov and P. M. Rentzepis, 1992, "Materials and Methods for 3D Optical Storage Memory," in *Photonics for Computers, Neural Networks, and Memories*, W. J. Miceli, J. A. Neff, and S. T. Kowel, Eds., *Proc. SPIE* 1773, 390–400.
8.26 N. Hampp, R. Thoma, D. Oesterhelt, and C. Bräuchle, 1992, "Biological Photochrome Bacteriorhodopsin and Its Genetic Variant Asp96 → Asn as Media for Optical Pattern Recognition," *Appl. Opt.* 31, 1834–1841.
8.27 R. P. Birge, November 1992, "Protein-Based Optical Computing and Memories," *Computer* 25(11), 56–67.
8.28 W. E. Moerner, W. Lenth, and G. C. Bjorklund, "Frequency-Domain Optical Storage and Other Applications of Persistent Spectral Hole Burning," in *Persistent Spectral Hole-Burning*, W. E. Moerner, Ed., Springer-Verlag, New York, 1988, pp. 251–307.
8.29 R. Ao, L. Kümmerl, M. Scherl, and D. Haarer, 1994, "Recent Advances in Frequency-Selective Optical Memories," *Opt. Mem. Neur. Netw.* 3.
8.30 G. R. Knight, "Interface Devices and Memory Materials," in *Optical Information Processing*, S. H. Lee, Ed., Springer-Verlag, New York, 1981, pp. 111–179.
8.31 D. Chen, 1979, "Magnetic Materials for Optical Recording," *Appl. Opt.* 13, 767–778.
8.32 K. A. Rubin and M. Chen, 1989, "Progress and Issues of Phase-Change Erasable Optical Recording Media," *Thin Solid Films*, 181, 129–139.
8.33 K. Rubin, H. Rosen, T. Strand, W. Imaino, and W. Tang, 1994, "Multilayer Volumetric Storage," in *Optical Data Storage, 1994 Tech. Digest Series* Vol. 10, OSA 104.
8.34 S. Jutamulia, G. M. Storti, W. Seiderman, and J. Lindmayer, 1990, "Erasable Optical 3D Memory Using Novel Electron Trapping, (ET) Materials," in *Optical Data Storage Technologies*, S.-J. Chua and J. C. McCallum, Eds., *Proc. SPIE* 1401, 113–118.
8.35 X. Yang, C. Y. Wrigley, and J. Lindmayer, "Three-Dimensional Optical Memory Based on Stacked-Layer Electron Trapping Thin Films," *Opt. Mem. Neur. Netw.* 3, 135–149.
8.36 A. Rebane, S. Bernet, A. Renn, and U. Wild, 1991, "Holography in Frequency-Selective Media: Hologram Phase and Causality," *Opt. Commun.* 86, 7–13, 1991.
8.37 D. Gabor, 1948, "A New Microscope Principle," *Nature* 161, 777–778.
8.38 E. N. Leith and J. Upatnieks, 1962, "Reconstructed Wavefronts and Communication Theory," *J. Opt. Soc. Am.* 52, 1123–1130.
8.39 V. A. Vitols, 1966, "Hologram Memory for Storing Digital Data," *IBM Tech. Discl. Bull.* 8, 1581–1583.
8.40 F. M. Smith and L. E. Gallaher, 1967, "Design Considerations for Semipermanent Optical Memory," *Bell Syst. Tech. J.* 46, 1267–1278.

8.41 L. K. Anderson, 1968, "Holographic Optical Memory for Bulk Data Storage," *Bell Lab. Rec.* 46, 319–325.
8.42 K. Iizuka, *Engineering Optics*, Springer-Verlag, New York, 1985.
8.43 C. B. Burckhardt, 1970, "Use of a Random-Phase Mask for the Recording of Fourier Transform Holograms of Data Masks," *Appl. Opt.* 9, 695–700.
8.44 L. Domash, Y.-M. Chen, M. Snowbell, and C. Gozewski, 1994, "Switchable Holograms and Approaches to Storage Multiplexing," in *Photonics for Processors, Neural Networks, and Memories II*, J. L. Horner, B. Javidi, S. T. Kowel, Eds., *Proc. SPIE* 2297, 415–424.
8.45 J. A. Rajchman, 1970, "Promise of Optical Memories," *J. Appl. Phys.* 41, 1376–1383.
8.46 H. Kiemle, 1974, "Considerations on Holographic Memories in the Gigabyte Region," *Appl. Opt.* 13, 803–807.
8.47 K. Kubota, Y. Ono, M. Kondo, S. Sugama, N. Nishida, and M. Sakaguchi, 1980, "Holographic Disk with High Data Transfer Rate: Its Application to an Audio Response Memory," *Appl. Opt.* 19, 944–951.
8.48 A. L. Mikaelian, E. H. Gulanyan, B. S. Kretlov, V. A. Semichev, and L. V. Molchanova, 1992, "Superposition of 1-D Holograms in Disk Memory System," *Opt. Mem. Neur. Netw.* 1, 7–14.
8.49 J. B. Thaxter and M. Kestigian, "Unique Properties of SBN and Their Use in a Layered Optical Memory," *Appl. Opt.* 13, 913–924.
8.50 T. Todorov, L. Nikolova, and N. Tomova, 1984, "Polarization Holography. 1: A New High-Efficiency Organic Material with Reversible Photoinduced Birefringence," *Appl. Opt.* 23, 4309–4312.
8.51 A. G. Chen and D. J. Brady, 1994, "Electrically Controlled Multiple Hologram Storage," *Opt. Mem. Neur. Netw.* 3, 129–133.
8.52 Y. N. Denisyuk, "Photographic Reconstruction of the Optical Properties of an Object in Its Own Scattered Radiation Field," *Sov. Phys. Dokl.* 7, 543–545.
8.53 E. N. Leith, A. Kozma, J. Upatnieks, J. Marks, and N. Massey, "Holographic Data Storage in Three-Dimensional Media," *Appl. Opt.* 5, 1303–1311.
8.54 L. d'Auria, J. P. Huignard, C. Slezak, and E. Spitz, "Experimental Holographic Read–Write Memory Using 3-D Storage," *Appl. Opt.* 13, 808–818.
8.55 L. Hesselink and J. Wilde, 1991, "Recent Advances in Holographic Data Storage in SBN," in *Soviet-Chinese Joint Seminar*, *Proc. SPIE* 1731, 74–79.
8.56 F. T. S. Yu and S. Yin, "Storage Dynamics of Photorefractive, (PR) Fiber Holograms," in *Optical Storage and Retrieval*, F. T. S. Yu and S. Jutamulia, Eds., Marcel Dekker, New York, 1996, Chapter 3.
57 F. H. Mok, M. C. Tackitt, and H. M. Stoll, 1991, "Storage of 500 High-Resolution Holograms in a $LiNbO_3$ Crystal," *Opt. Lett.* 16, 605–607.
8.58 F. H. Mok, 1993, "Angle-Multiplexed Storage of 5000 Holograms in Lithium Niobate," *Opt. Lett.* 18, 915–917.
8.59 X. Yang and S. Jutamulia, 1999, "Three-Dimensional Photorefractive Memory Based on Phase-Code and Rotation Multiplexing," *Proc. IEEE*, 87, 1941–1955.
8.60. E. H. Synge, 1928, "A Suggested Method for Extending Microscopic Resolution into Ultramicroscopic Region," *Phil. Mag.* 6, 356–362.
8.61. D. W. Pohl, W. Denk, M. Lanz, 1984, "Optical Stethoscopy: Image Recording with Resolution $\lambda/20$," *Appl. Phys. Lett.* 44, 651–653.
8.62. A. Lewis, M. Issacson, A. Harootunian, and A. Murray, 1984, "Development of a 500 Å Spatial-Resolution Light Microscopy. I. Light Is Efficiently Transmitted Through $\lambda/16$ Diameter Aperture," *Ultramicroscopy*, 13, 227–232.
8.63. M. H. Kryder, 1995, "Near-field Optical Recording: An Approach to 100 Gbit/in^2 Recording," *Optoelectronics — Devices and Technologies*, 10, 297–302.
8.64 E. Betzig, J. K. Trautman, R. Wolfe, E. M. Gyorgy, P. L. Finn, M. H. Kryder, and C.-H. Chang, 1992, "Near-Field Magneto-optics and High-Density Data Storage," *Appl. Phys. Lett.* 61, 142–144.

8.65 A. Partovi, D. Peale, M. Wuttig, C. A. Murray, G. Zydzik, L. Hopkins, K. Baldwin, W. S. Hobson, J. Wynn, J. Lopata, L. Dhar, R. Chichester, and J. H.-J. Yeh, 1999, "High-Power Laser Source for Near-Field Optics and Its Application to High-Density Optical Data Storage," *Appl. Phys. Lett.* 75, 1515–1517.

EXERCISES

8.1 (a) How are binary numbers 1 and 0 represented optically in intensity, amplitude, phase, and polarization?
(b) How are intensity, amplitude, phase, and polarization representing binary numbers stored in optical storage?

8.2 In a camera, the focal length of the lens is 5 cm and the diameter of the stop is 2 cm. What is the diameter of the central spot of the Airy disk when the camera is illuminated with a collimated beam from a sodium lamp with $\lambda = 589$ nm?

8.3 (a) How many bits can be stored in an optical storage of A4 paper size (21.5 cm $\times$ 27.5 cm) using a laser diode that emits light at HeNe wavelength of 633 nm?
(b) If the capacity of a CD-ROM is 650 Mbytes, the A4 paper size optical storage is equivalent to how many CD-ROMs?

8.4 (a) Using the same laser diode, how many bits can be stored in a 3-D optical storage having 1 cm^3 volume?
(b) To how many CD-ROMs is it equivalent?

8.5 (a) The resolution of a photographic plate is 1000 lp/mm. What is the size of the photosensitive silver halide grain?
(b) If our eyes can see a spot as small as 0.2 mm, at what magnification we will see the grain structure of the photographic plate?

8.6 (a) Can we use a dichromated gelatin film for photography? Explain why.
(b) Can we use a dichromated gelatin film for optical storage? If no, explain why not; if yes, explain how.

8.7 (a) In photopolymers, what information of light is stored: Intensity, amplitude, phase, or polarization of light?
(b) In the same photopolymers, is the information of light stored as the intensity transmission, amplitude transmission, phase modulation, or polarization rotation property of the material?

8.8 (a) Do photorcsists modulate the phase of light due to their local thickness or local refractive index?
(b) Why can photoresists coated with aluminum modulate reflected light?

8.9 What is the advantage of thermoplastic film as compared with dichromated gelatin, photopolymer, and photoresist films?

8.10 (a) A photorefractive crystal modulates the phase of light due to its local refractive index. Is the local refractive index proportional to the intensity, amplitude, phase, or polarization of the exposing light?
(b) Does the recording process in a photorefractive crystal involve developing and fixing steps such as that in a photographic plate?

8.11 A pattern A is recorded into a photochromic material using λ_1, and then read out using λ_2. Show that the output pattern at λ_2 is $(1 - A)$.

8.12 A pattern A is first recorded into an ET material using blue light ($\lambda_1 = 488$ nm). When a pattern B is projected into the ET material using infrared light ($\lambda_2 = 1064$ nm), the ET material emits orange-color light ($\lambda_3 = 615$ nm) with a pattern C. What is C in terms of A and B?

8.13 (a) An ET material is first exposed to uniform illumination of $\lambda_1 = 488$ nm, then to a pattern A of $\lambda_2 = 1064$ nm, and finally to a pattern B of $\lambda_2 = 1064$ nm, to produce an emission of pattern C at $\lambda_3 = 615$ nm. What is C in terms of A and B?
(b) An ET material is first exposed to uniform illumination of $\lambda_1 = 488$ nm, then to patterns A and B successively at the same wavelength $\lambda_2 = 1064$ nm, and finally to uniform illumination of $\lambda_2 = 1064$ nm, to produce an emission of pattern C at $\lambda_3 = 615$ nm. What is C in terms of A and B?

8.14 Explain why a two-photon–absorption material can be read without erasure while the readout of an ET material will erase the stored information.

8.15 Which of the following statements is correct?
(a) Bacteriorhodopsin emits light at 570 nm upon light stimulation at 412 nm.
(b) Bacteriorhodopsin modulates the amplitude of light passing through it.
(c) Bacteriorhodopsin modulates the phase of light passing through it.
(d) Bacteriorhodopsin modulates the frequency of light passing through it.

8.16 After bleaching out, spectral hole-burning material can be read at the beached-out band λ_1. Is the readout at λ_1 erasing the stored information?

8.17 (a) Do we need polarized light for writing information into magneto-optic material? Explain why or why not.
(b) Do we need polarized light for reading out the stored information from magneto-optic material? Explain why or why not.

8.18 Does phase-change material modulate the intensity, phase, or polarization of the readout light?

8.19 The following figure shows the basic pickup optics for a CD system (LD: laser diode; PBS: polarizing beam splitter; $\lambda/4$: quarter-wave plate). Assuming that the laser emits linearly polarized light, show that in principle, we can collect the reflected light 100%.

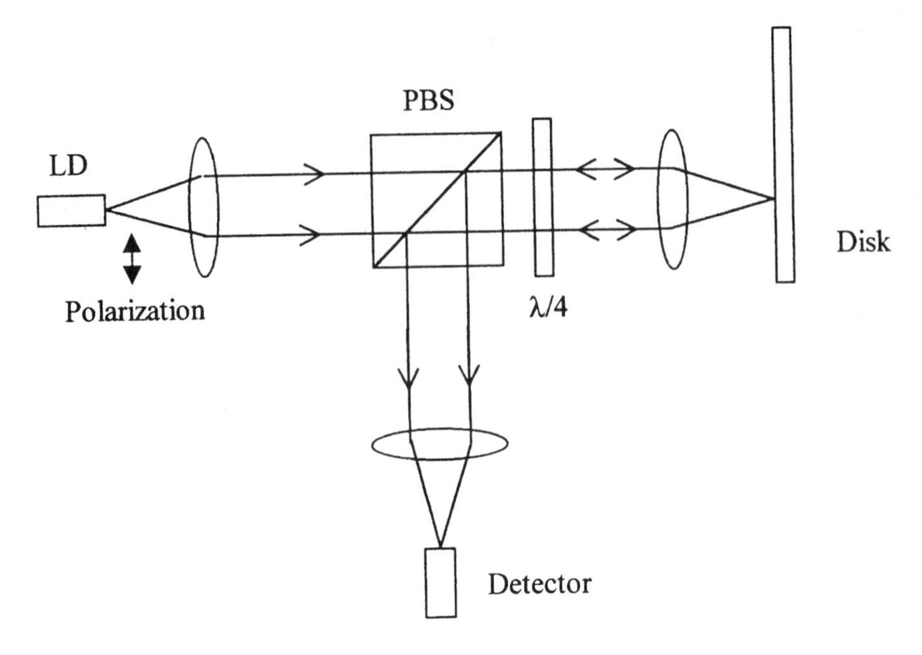

8.20 (a) Why does the size of the readout beam have to be larger than the width of the pit on a CD?
(b) On the other hand, why does the beam size have to be smaller than the size of the pit on a WORM and an MO disk?

8.21 The following figure shows the surface structure of the CD. Determine the pit depth d, in terms of the wavelength of light λ, and the refractive index of the clear coating n. If $\lambda = 0.78\ \mu m$, and $n = 1.5$, what is d?

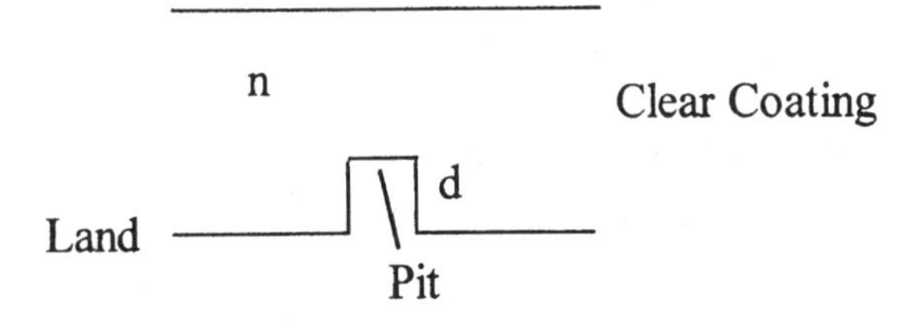

8.22 If bit 1 at a MO disk rotates polarization of light by 1°, and bit 0 does not rotate polarization of light, what are the output intensities of bit 1 and bit 0 by placing a polarization at 0° and 90° relative to the original polarization of readout beam? (*Hint*: Use Malus's law.)

8.23 A laser beam is focused onto a layer of a multilayer optical storage forming a spot with 1-μm diameter as shown in the following figure. If the layer separation is 50 μm and the NA of the focusing lens is 0.45, what is the approximate diameter of the beam at the adjacent layer? What is the approximate ratio of the area at the focused layer to the adjacent layer?

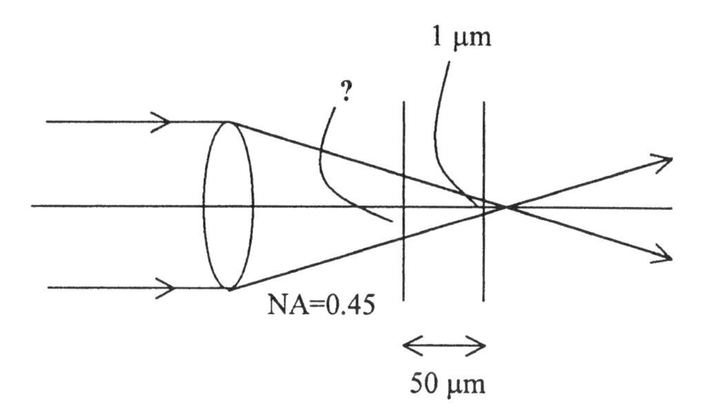

8.24 To read a page memory in photon-gating 3-D optical storage, the sheet of light is formed by focusing a collimated light using a cylindrical lens, as shown in the following figure.

(a) Show that the 3-D storage capacity is limited by the depth of focus of the focusing cylindrical lens.

(b) If the depth of focus is αd, where d is the layer thickness, and the 2-D storage density is 1 μm^{-2}, what is the maximum 3-D storage capacity of a cube of $1 \times 1 \times 1\ mm^3$?

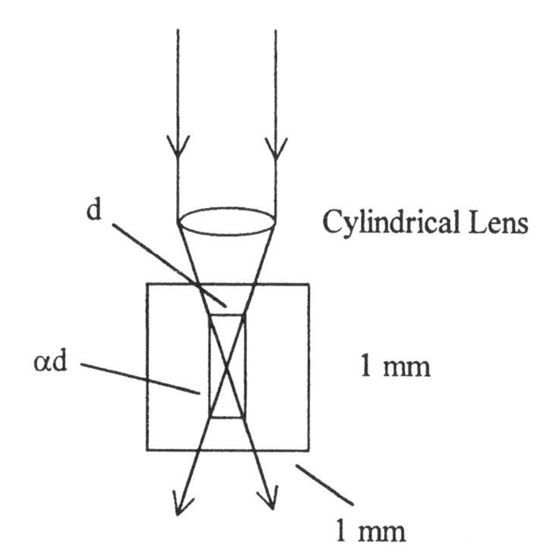

8.25 Based on Exercises 8.23 and 8.24, explain why for photon-gating 3-D optical storage we prefer high NA for the write-in and readout spherical lenses, and low NA for the activation cylindrical lens.

8.26 For stacked-layer 3-D optical storage using ET materials, show that the NA of the addressing cylindrical lens is

$$\mathrm{NA} = (n_2^2 - n_1^2)^{1/2},$$

where n_1 is the refractive index of the ET thin film, and n_2 is the refractive index of the cladding layer.

8.27 Why is a photoplastic device good for a hologram but not appropriate for photographic uses?

8.28 A thin-phase grating can be made using a photoplastic device. The thin-phase grating can be expressed as

$$t(x, y) = \exp[i\alpha \cos(2\pi by)] = \sum i^n J_n(\alpha) \exp(in2\pi by),$$

where J_n is the Bessel function of the first kind and the nth order, and α is a constant. Show that the maximum efficiency is 33.9%. (*Hint*: Use a Bessel function table.)

8.29 The amplitude transmittance of a grating recorded on a photographic plate can be expressed graphically in the following figure. What is the diffraction efficiency when the photographic plate is illuminated by a plane wave?

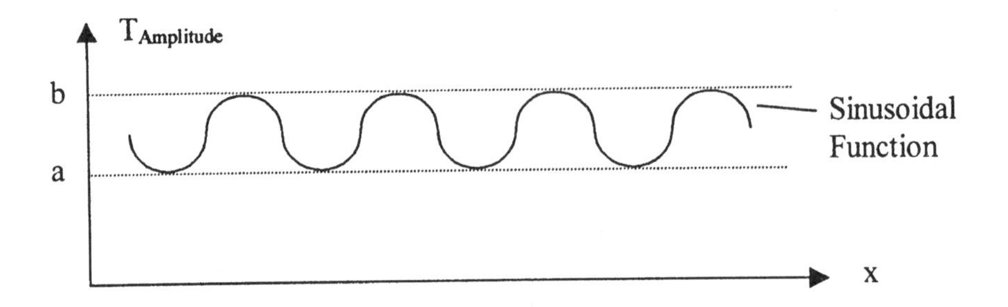

8.30 What is the optical gain of ET thin film? Assume the quantum efficiencies of ET thin film to blue and infrared exposures are 10% and 5%, respectively.

8.31 Can we employ an ET thin film to record and reconstruct a hologram? Explain why or why not.

8.32 We want to write a hologram into a photorefractive crystal that has a sensitivity of 200 nJ/μm^2 using an Ar laser with 10 mW output at 488 nm. Assume all the laser light can be utilized and focused onto an area of 0.5×0.5 mm^2. How long is the required exposure time?

8.33 Assume that a photorefractive crystal is employed as holographic storage that stores a 1000×1000 bit pattern for a computer. The stored pattern is read out by a laser beam. The reconstructed holographic image is detected by a CCD with the same 1000×1000 pixels, as shown in the following figure. Each pixel is $10 \times 10\ \mu m^2$, and the sensitivity is 10^{-6} W/cm^2. If the diffraction efficiency of the hologram is 10%, what is the minimum laser power required to read the memory?

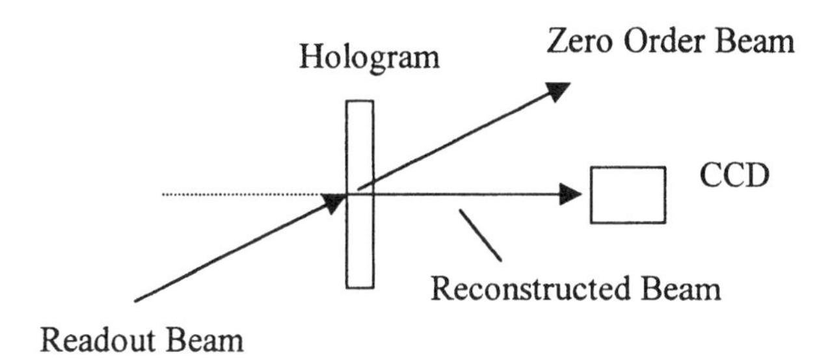

8.34 We have a sinusoidal pattern on a transmissive plate with period $\lambda/4$ and unity contrast. The plate is illuminated with a uniform light. What is the contrast at the plane $\lambda/100$, $\lambda/10$, $\lambda/2$, λ, and 10λ away from the plate?

8.35 Repeat Exercise 8.34 for a sinusoidal pattern with period $\lambda/20$.

Chapter 9 Computing with Optics

[1]Guoqiang Li and [2]Mohammad S. Alam
[1]Shanghai Institute of Optics and Fine Mechanics, Chinese Academy of Sciences, China
[2]The University of South Alabama, Mobile, Alabama

In this chapter, we review the recent advancements in parallel optoelectronic signed-digit computing. We begin by discussing signed-digit number systems, including modified signed-digit, trinary signed-digit, quaternary signed-digit, and negabinary signed-digit representations. Then, we cover the parallel arithmetic algorithms and architectures, fast-conversion algorithms between the signed and unsigned number systems, encoding, and sample experimental verification via optoelectronic implementation. The parallelism of the algorithms and the architectures are well mapped by efficiently utilizing the inherent parallelism of optics. By exploiting the redundancy of the signed-digit number systems, addition and subtraction can be completed in fixed steps independent of the length of the operands. The table lookup, content-addressable–memory, symbolic substitution, and parallel logic array approaches to optical computing can utilize the benefit of optics in parallel information processing and may lead to powerful and general purpose optoelectronic computing systems. Digital optical systems can be built by cascading two-dimensional planar arrays interconnected in free space. The proposed algorithms and architectures show great promise, especially with the development of optical interconnection, optoelectronic devices, and optical storage technology.

9.1. INTRODUCTION

The ever-increasing demand for high-speed computation and efficient processing of large amounts of data in a variety of applications have motivated the advancement of large-scale digital computers. Conventional computers suffer from several communication problems such as the Von Neumann interconnection bottleneck, limited bandwidth, and clock skew. The bandwidth and clock skew limit processing speed and add design complexity to the system. Communication bottlenecks at the architectural, bus, and chip levels usually come from the utilization of time multiplexing to compensate for the inability of electronics to implement huge interconnections in parallel. The sequential nature of data transport and processing prevents us from building high-performance computing systems. To overcome these bottlenecks, single-instruction multiple-data (SIMD)–based and multiple-instruction multiple-data (MIMD)–based architectures have been investigated. However, such architectures still suffer from interconnection bottlenecks. Another promising approach is to incorporate the attractive features of optics with those of digital computers [1–7], yielding a hybrid optoelectronic computer system.

Optics has several advantages over electronics, including ultrahigh processing speed, large bandwidth, massive parallelism, and noninterfering propagation of light rays. As a result, the application of optics has been widely investigated for arithmetic and logical processing, image processing, neural networks, data storage, interconnection networks, and ultrafast system bus. From the viewpoint of data representation, currently available optical processors can be classified into two categories, analog and digital; from the operational viewpoint, they are classified into numeric and nonnumeric processors. In earlier periods, optics was used to process only analog signals. However, in the last two decades, tremendous advances in nonlinear optoelectronic logic and switching devices have invigorated the research in digital optics.

Digital computing has excellent features including flexibility in computation, easy implementation, minimal effects of noise, and lower requirement of devices to identify two states of signal. Various optoelectronic systems have been demonstrated, including the pipelined array processor using symmetric self-electro-optic–effect devices (S-SEED) [8], optical cellular logic image processor [9], and the programmable digital optical computer using smart pixel arrays [10, 11]. Digital optical circuits can be constructed by cascading two-dimensional (2D) planar arrays of logic gates using free-space interconnections. These programmable logic arrays can be used to implement various complex functions, such as arithmetic and logical operations in constant time.

Recent advances in the algorithms, architectures, and optoelectronic systems that exploit the benefits of optics have led to the development of high-performance parallel computing systems. Since optical interconnections and neural networks will be discussed elsewhere in this book, this chapter focuses on parallel optical architectures and algorithms suitable for optical implementation. Parallel optical logic and architectures will be presented in Sec. 9.2, followed by a review of various number systems and arithmetic operations in Sec. 9.3. Section 9.4 is focused on parallel signed-digit algorithms, and number conversion is discussed in Sec. 9.5. Several optical implementation examples are given in Sec. 9.6, and a summary is presented in Sec. 9.7.

9.2. PARALLEL OPTICAL LOGIC AND ARCHITECTURES

A typical optical logic system is shown in Fig. 9.1. The input data are optically encoded and sent to the switching element for processing. The optical output can be obtained with a decoder. A logic operation itself is nonlinear. It can be performed by linear encoding with nonlinear optical devices, or nonlinear encoding with linear optical elements, or nonlinear encoding with nonlinear optical switches. The optical system has the inherent feature of performing 2D space-invariant logic in parallel so that SIMD processing can be easily realized. By programming the logic array, any combinational logic operations can be achieved. If the encoder, the switching element, or the decoder is space variant, we can perform space-variant logic, or perform multiple logic operations on the 2D data in parallel, or perform different logic operations in different areas. This corresponds to MIMD–processing based systems.

9.2.1. OPTICAL LOGIC

Nonlinear optical switching elements are array devices, which include the Pockels readout optical modulator (PROM) [12] based on the Pockels effect, liquid crystal switch [13], twisted nematic liquid crystal device [14], variable grating liquid crystal device [15], liquid crystal light valve (LCLV) [16], microchannel spatial-light modulator (SLM) [17], liquid crystal televisions

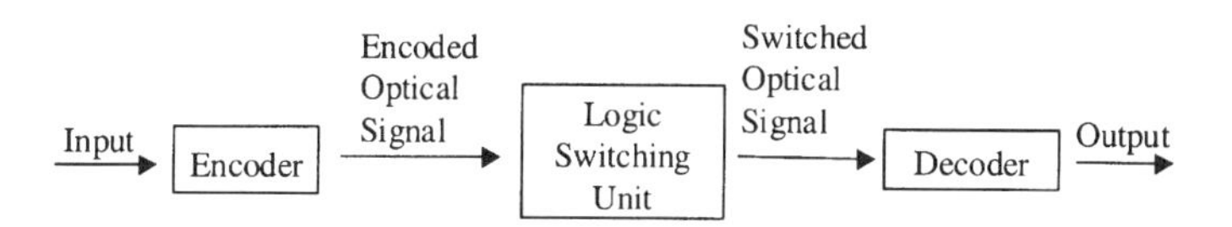

Fig. 9.1. Schematic diagram of an optical logic switching system.

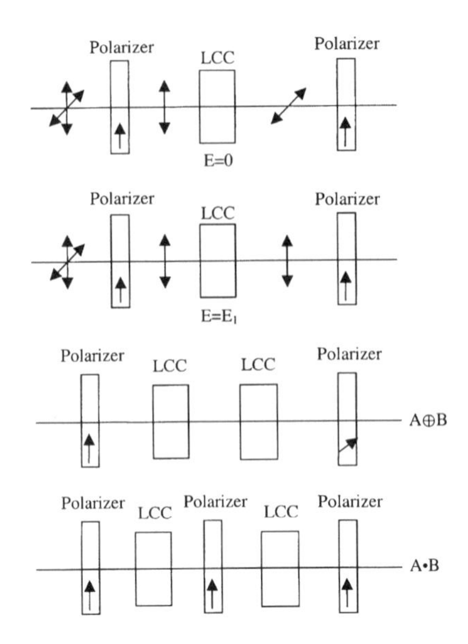

Fig. 9.2. Liquid crystal electro-optic switch for logic operations [13].

[18], magneto-optic SLM [19, 20], electron-trapping device [21, 22, 23], integrated optical waveguide [24, 25], etc. Other optical logic implementations utilize optoelectronic devices and optical fibers [26], two-beam coupling effect in photorefractive crystals [27], interference using 2D array of diffractive optical elements [28], and liquid valve and holographic elements [29]. Nonlinear devices based on multiple–quantum-well and optical bistability include S-SEED [8, 30], bistable etalons by absorbing transmission (BEAT) [31], and the vertical surface transmission electrophotonic device (VSTEP) [32].

For example, the optical logic gate using the liquid crystal electro-optic switch [13] is shown in Fig. 9.2. Polarized light entering a liquid crystal cell in the absence of any applied voltage is twisted by 90° on exit. An electrical field E_1 applied across the liquid crystal cell causes the polarized light to go through without being twisted. This property permits the liquid crystal cell to be used as an electro-optic switch. All sixteen logic functions can be realized by using various configurations of polarizers and liquid crystal cells. The XOR and AND operations are shown in Fig. 9.2. Figure 9.3 shows the basic operational principle of a magneto-optic SLM (MOSLM) [19, 20]. Two MOSLM arrays are aligned in series and switched individually. Each MOSLM provides two linearly polarized output states at 0° or 10°. Therefore, three-level polarization output (0°, 10°, or 20°) can be obtained from two cascaded MOSLMs. XOR and AND operations can be easily performed. Operations of the other logic functions have been presented in [19].

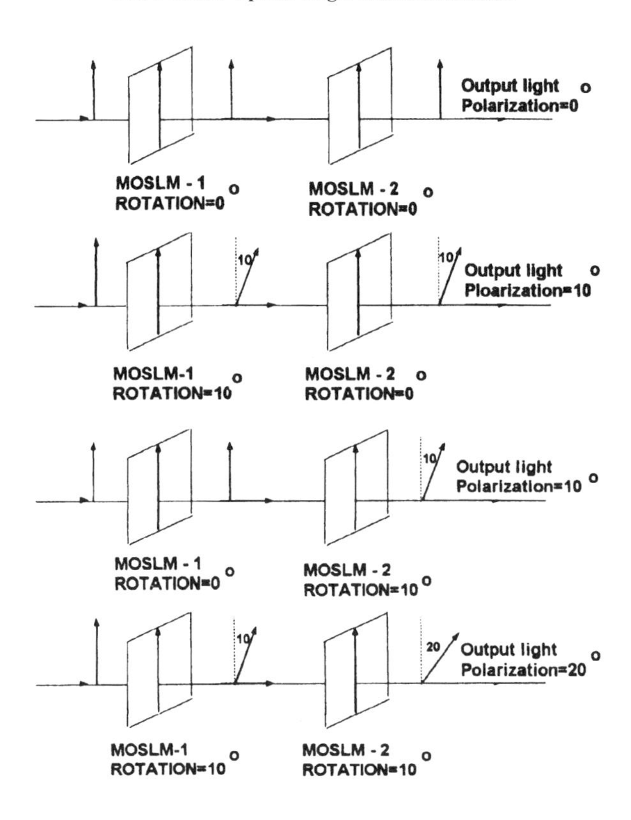

Fig. 9.3. Multiplicative configuration of optical logic gate using magneto-optic SLM [20].

For logic operations with nonlinear encoding and linear systems, polarization encoding was used in imaging systems [33], and theta modulation was applied in linear filtering systems [34] where the image pixels are represented by grating structures of different angles. Later, scattering [35], anisotropic scattering [36], and polarization encodings [37] were also employed in linear filtering systems. In addition, binary data can also be encoded with different grating frequencies [38] or with orthogonal gratings [39].

Spatial encoding has been extensively used in various optical logic systems. For example, a parallel optical logic gate array implemented by an optical shadow-casting technique [40] is shown in Fig. 9.4. It is designed for the 2D parallel logic operations of the corresponding pixels of images A and B (Fig. 9.4[c]). In the architecture of Fig. 9.4, dual-rail spatial encoding (Fig. 9.4[a]) is used; i.e., a binary value is represented by a bright–dark pattern. Such spatial encodings can be realized with a birefringent crystal and an SLM [41]. The two images are first separately encoded into spatial patterns and then overlap-

(a)

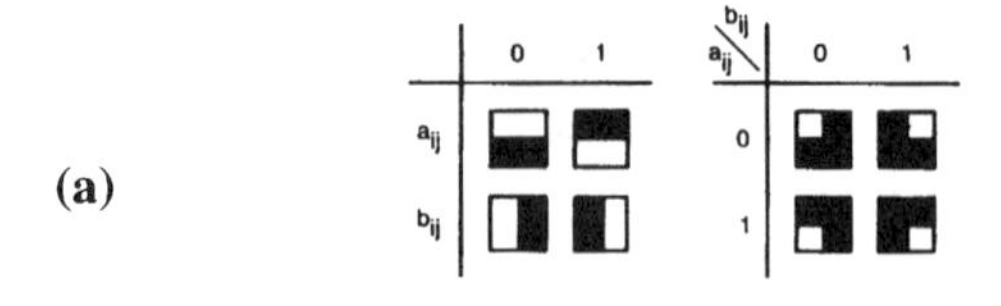

(b)

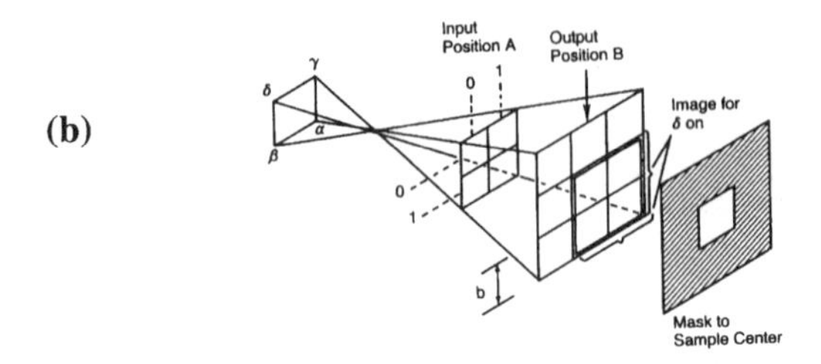

(c)

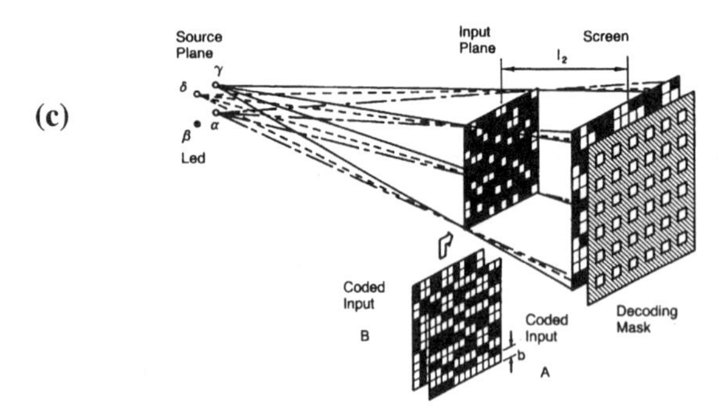

Fig. 9.4. Optical shadow casting for parallel optical logic [40]. (a) Encoding (left) and overlap of the input, (b) Operation of one cell, (c) Parallel operation on 2D images.

ped to compose a coded image as the input. The coded image is illuminated by a point-source array and projected onto the screen. The sampled image obtained from the decoding mask contains the logical output of all the coded pixels. The optical shadow-casting technique offers parallel processing capability with simple optical setup as well as programming capability by appropriately configuring the point-source array. Using the encoding scheme and a 2×2 LED source array (Fig. 9.4[b]), each coded pixel having four subcells can be expanded on the screen to a pixel having a pattern of 3×3 subcells. The decoding mask contains open windows at the center of each subcell. The detected signal can be represented by

$$F = \alpha \cdot (a \cdot b) + \beta \cdot (a \cdot \bar{b}) + \gamma \cdot (\bar{a} \cdot b) + \delta \cdot (\bar{a} \cdot b). \tag{9.1}$$

By controlling the ON or OFF state of each LED, all logical operations can be obtained. Each switching state of the LED array corresponds to one logical operation. This technique has been extended to perform logical operations of multiple variables [42, 43]. Shadow-casting with a source array is equivalent to the generation of multiple copies of the coded image and an overlap of the multiple images. Therefore, with spatial encoding, parallel optical logic operations can also be performed by multiple imaging using a lenslet array and decoding mask [44]. The output may have the same format as the encoded input, thus making the system cascadable. On this basis, a compact logic module using prisms and polarization elements can be implemented [45].

Parallel logic operations can also be performed by use of nonlinear encoding and nonlinear devices, such as polarization encoding and two-beam coupling in a photorefractice crystal [46], birefringence spatial encoding and field-enhanced effect of BSO crystal [47], hybrid intensity-polarization encoding and liquid crystal device [48].

9.2.2. SPACE-VARIANT OPTICAL LOGIC

The simplest way to perform space-variant logic operations is to utilize the spatial encoding-decoding method [49]. By appropriately encoding and overlapping the 2D input data, all logical operations can be obtained by decoding at prescribed subcells of the output pixel. The decoding mask can be space variant so that space-variant logic operations can be performed in parallel. This scheme has been extended to perform space-variant signed logic operations in arithmetic operations [50]. To reduce the number of decoding masks, several subcells of a coded pixel can be used for decoding in parallel [51]. By using additional polarization encoding in an optical shadow-casting system [52], a higher degree of freedom can be obtained, and logical operations on multiple variables can be processed with a smaller encoding area. Various logical operations can be obtained by changing the decoding mask. This method can also be used for parallel multiple logical operations on binary and multiple-valued variables by incorporating a space-variant polarized decoding mask [53, 54, 55]. Space-variant operations can also be realized by using multiple source arrays [56] and by using different encoding in different areas of the output pixel [57].

9.2.3. PROGRAMMABLE LOGIC ARRAY

A programmable logic array is a structure for implementing an arbitrary general-purpose logic operation [58]. It consists of a set of AND gates and a set of OR gates. With the help of coding of the input signal, various combinations of the coded signals can be generated to provide a set of output functions. The combination is specified by the connection pattern in the AND and OR gate arrays. In optics, AND logic is readily performed when light is transmitted through the modulators representing the inputs, and an OR operation is performed by collecting the light from several inputs to a common detector. An optical programmable logic array can be constructed by using AND and OR logic arrays interconnected by optics. The logic arrays are programmed by enabling or disabling certain interconnections in the arrays, and any logic functions can be implemented in the form of sum of products. Such an optical programmable logic array processor can be constructed by using 2D SEED arrays interconnected by a multistage interconnection network such as Banyan, crossover, or perfect shuffle, as shown in Fig. 9.5. All products are formulated by several stages of AND gate arrays and sums of the particular products are generated by several stages of OR gate arrays. Binary masks are used to select different optical paths among the gate arrays. Feedback paths are imaged back to the system with a vertical shift so that each row is imaged onto a different part of the masks at each iteration. However, this implementation requires a large number of gates. To alleviate this problem, space-invariant optical interconnections may be used.

The AND operation can also be realized by NOT and NOR operations. For example, using De Morgan's law, the logic function operation $F = AB + C\bar{D}$

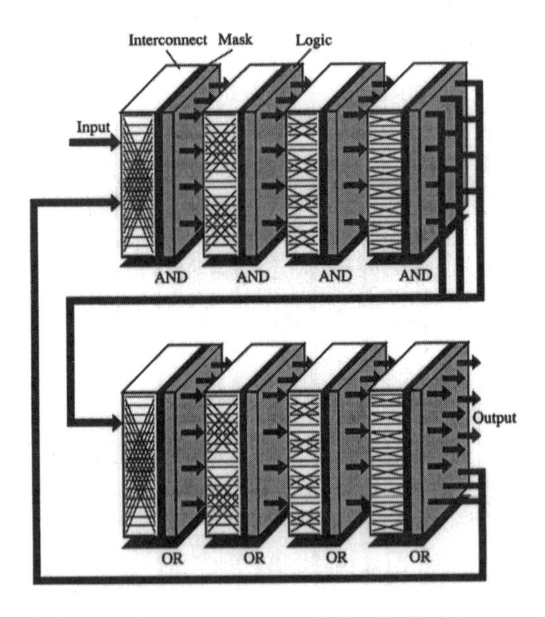

Fig. 9.5. A digital optical computing system composed of optically interconnected programmable logic arrays [58].

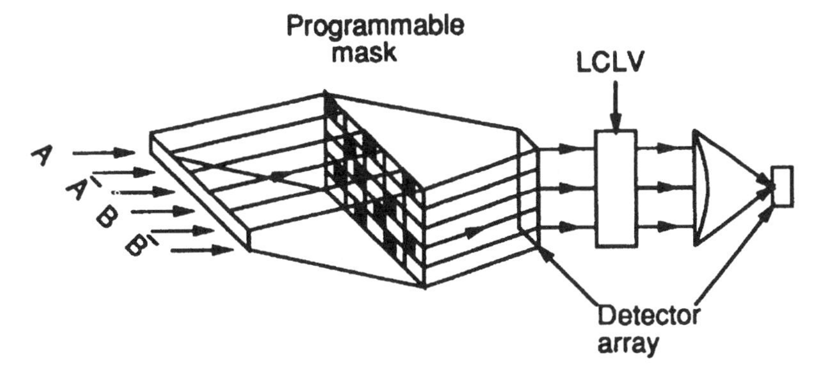

Fig. 9.6. Free-space programmable logic array based on a crossbar switch [7, 11].

can be expressed as

$$F = \overline{\bar{A} + \bar{B}} + \overline{\bar{C} + D}. \tag{9.2}$$

Thus, the AND operation can be implemented by inverting the input first, then performing the OR operation, and finally inverting the ORed outputs. The final sum of product is obtained by performing the OR operation on all the product terms. Therefore, a sum of products can be achieved by a NOT-OR-NOT-OR sequence. Correspondingly, an optical programmable logic array can be constructed based on a crossbar interconnection network, as shown in Fig. 9.6 [7, 11]. The switch SLM can be dynamically programmed to implement various operations. In this architecture, a LCLV is used for NOT operation, and all the products are collected in the detector using a cylindrical lens. A control unit constructed with free-space holographically interconnected 1-nanosecond–latency optoelectronic NOR gates is presented in [59]. Figure 9.7 shows a crossover interconnected OR–NOR gate array for a serial adder [58].

Recently, a logic array processing module using an electron-trapping (ET) device has been proposed (Fig. 9.8) [23]. The ET device has the advantages of high resolution, nanosecond response time, and high sensitivity. The ET device fabricated can be made from Eu- and Sm-doped CaS. Both ground and excited states of the impurities exist within the bandgap of the wide-bandgap host material. Short-wavelength visible light, such as blue, excites electrons from the ground state of Eu which are then transferred over to Sm. The electrons can remain in the ground state of Sm for a long time. But subsequent exposure to infrared light excites the trapped electrons to the excited state of Sm. Then the electrons transfer to the excited state of Eu and emit red light while returning

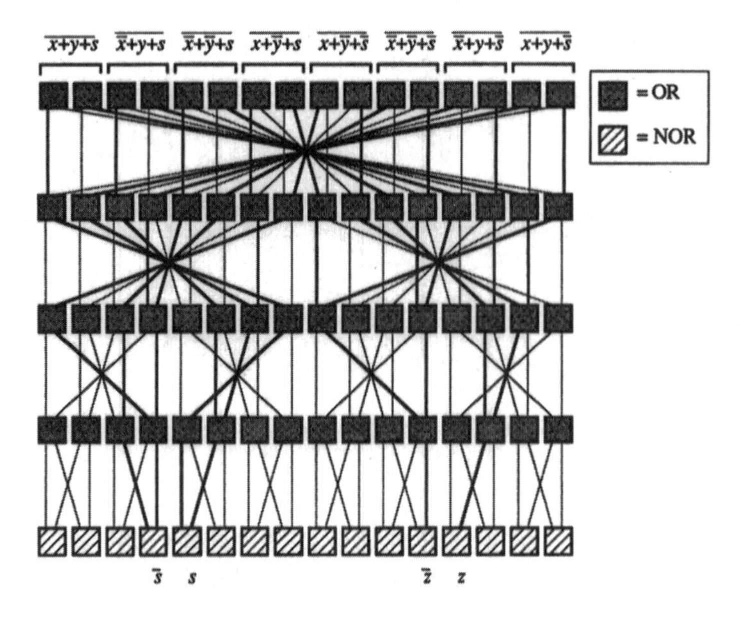

Fig. 9.7. Mapping terms to state (s) and output (z) functions for a dual-rail serial adder using crossover interconnects and OR-NOR logic.

to the ground state of Eu. The ET device possesses the inherent capability of performing OR and NOT functions. The OR operations on two-dimensional arrays are done in parallel by sequentially writing the binary arrays to be operated onto the device with enough energy to cause the device to saturate. When the device is read with infrared light, the output luminescence shows the

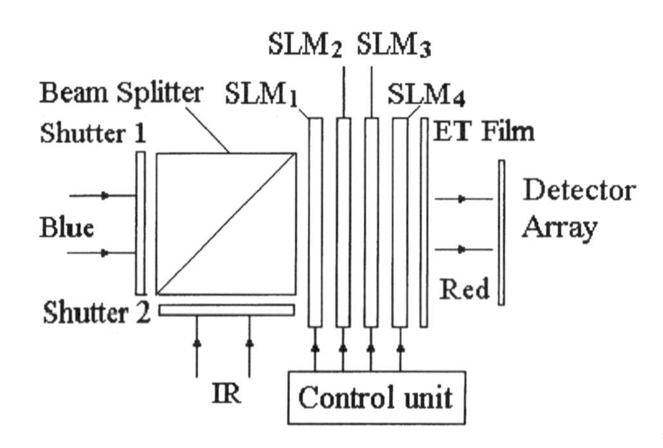

Fig. 9.8. Optical configuration for parallel logic operations using an electron-trapping device [23].

result. The NOT operations on the binary array A are performed (a) by uniform blue illumination and (b) with infrared illumination. The operational procedure for the remaining 14 binary logic functions of two variables has been reported [60]. With this module, all the intermediate results are stored in the ET device itself. This feature eliminates the feedback and optoelectronic and electro-optic conversion operations for complex logical operations.

9.2.4. PARALLEL ARRAY LOGIC

A cellular logic processor is well suited for optics because of its 2D data arrangement and parallel processing capability [61]. A cellular processor is generally composed of an n-dimensional interconnected cell structure. The state of a cell is represented by a function of the state of the neighboring cells. Both array logic and symbolic substitution can be considered as special cases of cellular logic.

Optical parallel array logic [62] is an extension of optical shadow-casting logic. In optical shadow-casting, encoding of the images and configuration of the source array correspond to the input encoding and combination of the coded signals in programmable logic array, respectively. Thus, optical shadow-casting algorithms can be easily modified to handle cellular logic operations. Figure 9.9 shows a schematic diagram of procedures of optical array logic, which consists of coding, discrete correlation, sampling, and decoding. Discrete

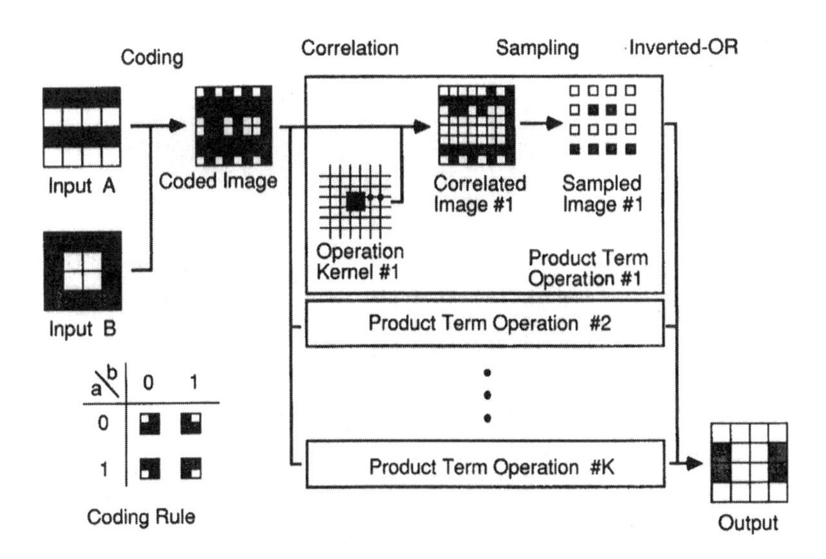

Fig. 9.9. Optical array logic processing [62].

correlation is a cross-correlation between a coded image and an operation kernel. The operational kernel is regarded as identical to the switching pattern of the source array in shadow-casting. To implement arbitrary neighborhood logical operation, multiple discrete correlations are usually required. Here, decoding may include logical inversion and sum operations. The logical neighborhood operations can be described by the following equation:

$$c_{i,j} = \sum_{k=1}^{K} \prod_{m=-L}^{L} \prod_{n=-L}^{L} f_{m,n,k}(a_{i+m,j+n}, b_{i+m,j+n}) \tag{9.3}$$

where $f(a, b)$ is a two-variable binary logic function operating on the $(i + m)$th and $(j + n)$th pixels of matrices A and B, L specifies the size of neighborhood area $(2L + 1) \times (2L + 1)$, and K corresponds to the number of discrete correlations. Each product term corresponds to an operational kernel and the logical sum of the product terms is equivalent to a combination of the operational kernels. With the powerful programming ability of optical array logic, it has been used for numerical processing, imaging processing, database management, and similar computation-intensive applications. With advances in optoelectronic devices and integration technology, the high-speed vertical-cavity surface-emitting laser (VCSEL) array can be used as an image emitter, the CMOS photodetector array can be used for detection, and ferroelectric liquid crystal SLM can be used for dynamic filters specifying the correlation kernels [62].

9.2.5. SYMBOLIC SUBSTITUTION

Symbolic substitution is a 2D parallel pattern transformation logic [63–69]. The logic functions are defined by substitution rules. A substitution rule consists of a search pattern (the left-hand side of the rule) and a scribe pattern (the right-hand side of the rule). Actually, the substitution rules are the spatial representations of a logical truth table. For example, with the dual-rail spatial encodings for 1 and 0 shown in Fig. 9.10(a), the symbolic representation of the substitution rules for binary half-addition is depicted in Fig. 9.10(b), where the left-hand side of the rule shows the bit pair to be operated and the right-hand side the intermediate sum (bottom) and carry (top). In truth-table look-up algorithms, each of the reduced minterms can be treated as a search pattern while the corresponding output can be treated as a scribe pattern. Since the substitution rules can be designed arbitrarily and each cell of the output pattern can be used for different functions, symbolic substitution can be employed as a versatile tool for space-invariant and space-variant logic operations, arithmetic operations, and image processing operations. It is

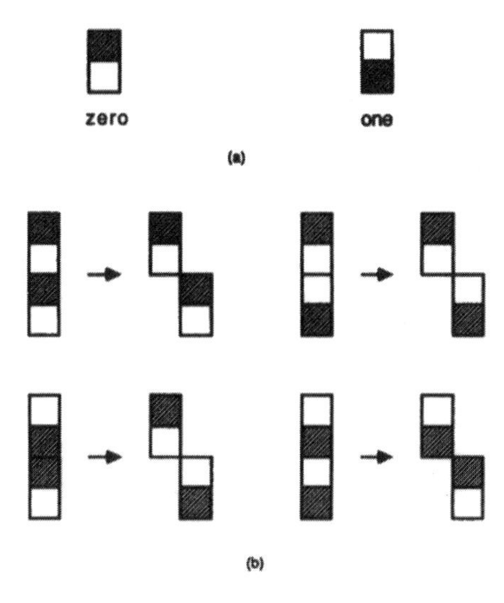

Fig. 9.10. (a) Dual-rail spatial encoding for 1 and 0, (b) Symbolic substitution rules for binary half-addition.

realized in two stages: recognition and substitution. In the recognition stage, all the occurrences of the search pattern in the 2-D input plane are marked in parallel. In the substitution stage, the scribe pattern is substituted in all the locations where the search pattern is found. In its operation, the recognition usually comprises the procedures of replicating, shifting, superimposing of the input pattern and thresholding [64]. The substitution phase is functionally similar to the recognition phase except for thresholding. Symbolic substitution can be performed by coherent correlator architectures, shown in Fig. 9.11. The first correlator (P1–P3) performs the recognition phase and the second

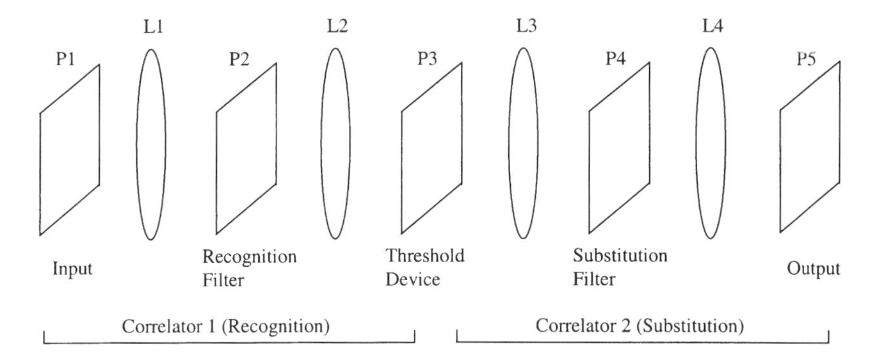

Fig. 9.11. Basic cascaded correlator architecture for symbolic substitution [70].

correlator (P3–P5) performs the substitution phase. In Fig. 9.11, plane P1 is Fourier-transformed by lens L1 onto plane P2, where a matched spatial filter that represents the Fourier transform of the encoded search pattern is placed. Thus, at P2 the product of the Fourier transform of the input and that of the search pattern is formed. This product is further Fourier-transformed by lens L2 to produce the correlation intensities at P3. The peak intensities indicate the presence of the search pattern. The correlation peaks can be thresholded by using an optically addressed SLM. In the second correlator (P3–P5), the Fourier transform of the substitution pattern is put at P4 and a substitution pattern will appear at each location that corresponds to the occurrence of the search pattern. This implements a single symbolic substitution rule. To perform multiple substitution rules in parallel, multichannel correlator architectures can be used.

Recently a novel approach which combines recognition and substitution into a single step has been proposed [70]. In this technique, symbolic substitution is formulated as a matrix–vector (M–V) multiplication for each pair of input recognition digits (the input vector $\boldsymbol{x}$) and the associated pair of output substitution digits (the output vector $\boldsymbol{y}$). The M–V multiplication is written as $\boldsymbol{y} = \boldsymbol{M}\boldsymbol{x}$, where $\boldsymbol{y}$ is the $L \times 1$ output vector (substitution pattern), $\boldsymbol{x}$ is the $K \times 1$ input vector (recognition pattern), and M is an unknown $K \times K$ recall matrix. This M–V equation is solved for the matrix $\boldsymbol{M}$ that satisfies all the possible input and output digit pairs of a truth table. The operation can be implemented by a single stage correlator. (An example will be shown in Sec. 9.6.1.)

9.2.6. CONTENT-ADDRESSABLE MEMORY

All arithmetic algorithms can be implemented either by logic operations or by truth-table look-up (shown in the following Section). For the truth-table look-up implementation, parallel architectures such as location-addressable memory (LAM) [71], content-addressable memory (CAM) [72, 73, 74], or symbolic substitution can be used. It has been pointed out that a CAM is more efficient than a LAM in terms of processing speed because a CAM implements the truth-table look-up by directly addressing its content rather than its location. In this scheme, the digit combinations of the input numbers are compared with the predetermined reference digits for generating the corresponding outputs. Therefore, the main objective in CAM implementation is to minimize the number of minterms and computational steps while ensuring a minimum number of variables in a minterm. For example, for different addition algorithms, one can choose the optimum algorithm in terms of the speed–minterm product. Both coherent [74] and incoherent CAM architectures have been proposed [72, 73]. Examples of incoherent CAM implementations will be shown in Sec. 9.6.2.

9.3. NUMBER SYSTEMS AND BASIC OPERATIONS

Various number systems and corresponding algorithms suitable for optical implementation have been investigated. Theses number systems can be categorized as binary and nonbinary. In the binary category, number systems with positive radix such as 2's complement and that with negative radix, named negabinary, were studied. The nonbinary systems include mixed radix, residue, and signed-digit number representations. In this section, we briefly discuss the basic operations of binary number systems and review the operations of nonbinary number systems.

9.3.1. OPERATIONS WITH BINARY NUMBER SYSTEMS

9.3.1.1. *Positive Binary Number System*

In conventional binary number systems with radix 2, three representations; namely, sign magnitude, 1's complement, and 2's complement, were used to encode a bipolar number. In signed-magnitude representation, the most significant bit (MSB) is identified as the sign of the number, 0 for positive and 1 for negative numbers. In each step of the operations, the MSB needs to be dealt with separately. Therefore, it is not suitable for parallel computation. In 1's complement representation, a negative number is formed by a bit-by-bit complement of the corresponding positive number. This produces two possible representations for 0, which is not desired. In 2's complement representation, a negative number is formed by adding 1 to its 1's complement representation. This technique eliminates the two representations of zero otherwise encountered with 1's complement representation, and a single addition or subtraction can be performed without any special concern for signs. Thus, 2's complement representation is widely used in digital computing. A 2's complement number may be represented by the following equation:

$$X = -x_{N-1}2^{N-1} + \sum_{i=0}^{N-2} x_i 2^i, x_i \in \{0, 1\}, \tag{9.4}$$

where x_{N-1} is the sign bit. A fractional number can be represented with negative indices.

Addition is the most fundamental arithmetic operation. Addition of two N-bit numbers A and B can be performed by using half-adders or a full-adder in N iterations. The half-adder is a two-input/two-output logical circuit. The truth table corresponding to its two inputs a_i and b_i, and the outputs,

Table 9.1

Truth Table for Binary Half-Addition

a_i	b_i	c_{i+1}	s_i
0	0	0	0
0	1	0	1
1	0	0	1
1	1	1	0

intermediate sum s_i and carry c_{i+1} to the next higher position, is shown in Table 9.1. The logical functions are found to be

$$\begin{aligned} s_i &= a_i \oplus b_i, \\ c_{i+1} &= a_i b_i. \end{aligned} \tag{9.5}$$

With N half-adder array, the addition can be completed in N iteration. The full-adder is a three-input/two-output logical circuit. At the ith bit position, it accepts two bits of the operands, a_i and b_i, and a carry c_i from the lower bit position and generates a sum bit s_i and carry c_{i+1}. The truth table is shown in Table 9.2. The logic equations are

$$\begin{aligned} s_i &= a_i \oplus b_i \oplus c_i, \\ c_{i+1} &= a_i b_i + a_i c_i + b_i c_i = a_i b_i + (a_i \oplus b_i) c_i. \end{aligned} \tag{9.6}$$

It takes N periods for a carry to propagate from the least significant bit (LSB) to MSB. This approach is called ripple-carry addition.

Table 9.2

Truth Table for Binary Full-Addition

a_i	b_i	c_i	c_{i+1}	s_i
0	0	0	0	0
0	0	1	0	1
0	1	0	0	1
0	1	1	1	0
1	0	0	0	1
1	0	1	1	0
1	1	0	1	0
1	1	1	1	1

For optical implementation, all the two-input logic gates described above can be used for the half-adder. The full adder can be constructed by two two-input logic gate or by a three-input logic gate. In addition, half-adders based on wavefront superposition [75] and threshold logic [76] have been proposed. Full-adders using spatial encoding and phase conjugation [77], SLM and polarizing elements [78], liquid crystal television [79], and photorefractive four-wave mixing [80] have also been suggested. A bit-slice digital optical computing prototype using directional coupler and optical fiber delay line has been demonstrated [81]. With free-space optical interconnection, a pipelined addition module based on symbolic substitution was proposed [82].

The conventional binary number system has a strong interdigit dependency on carry propagation, and serial addition does not fully employ the parallelism of optics. In order to reduce the time delay caused by the serial carry propagation, several approaches of advancing carries have been proposed. In the first approach [83, 84], the carry bits at all positions are generated *a priori* by the carry look-ahead technique so that the addition can be performed in three steps. To illustrate the principle of carry look-ahead addition, we introduce two auxiliary functions:

$$\begin{aligned} G_i &= a_i b_i, \\ P_i &= a_i \oplus b_i, \end{aligned} \tag{9.7}$$

where G and P are called the carry generation and propagation functions respectively. If G_i is true, it means a carry will be generated at the ith position. If P_i is true, it means that a carry c_i from the lower position will propagate to the higher position. Then Eq. (9.6) can be written as

$$\begin{aligned} s_i &= P_i \oplus c_i, \\ c_{i+1} &= G_i + P_i c_i. \end{aligned} \tag{9.8}$$

Consider a 4-bit module; we have

$$\begin{aligned} c_1 &= G_0 + P_0 c_0, \\ c_2 &= G_1 + P_1 c_1 = G_1 + G_0 P_1 + P_1 P_0 c_0, \\ c_3 &= G_2 + P_2 c_2 = G_2 + G_1 P_2 + G_0 P_1 P_2 + P_2 P_1 P_0 c_0. \end{aligned} \tag{9.9}$$

If we define G_M and P_M to be the carry generation and propagation functions of the module, the output carry of the module is then

$$c_{\text{out}} = G_M + P_M c_0, \tag{9.10}$$

in which

$$G_M = G_3 + P_3G_2 + P_3P_2G_1 + P_3P_2P_1G_0,$$
$$P_M = P_3P_2P_1P_0.$$

P_M is true when a carry enters and passes through the module, and G_M is true when the carry generated to the highest position (c_4) comes from the module itself. In addition, the carry bits at all positions can be obtained in parallel with P_i and G_i, and the sum bits can be obtained in parallel according to Eq. (9.8). Several optoelectronic implementations [85–88] have been proposed, including use of nonholographic optoelectronic content-addressable memory [88].

Another scheme to speed up the operation is the higher-order modular approach [43, 89, 90], which deals with several bit pairs in parallel; however, the operation time is still dependent on the length of the bit string. The third scheme for high-speed addition is the combinational logic approach with AND-OR-NOT operations [91]. All the output bits are represented by combinational logic and are obtained in parallel. The matrix–vector architecture can be used for optical implementation. These techniques are mainly suitable for shorter bit strings, and become extremely complex with larger bit strings. Some other special number systems such as the modified trinary number system [92] and redundant binary number system [93] were also used to represent positive numbers.

9.3.1.2. *Negabinary Number System*

Negabinary is one of the positional number systems with the fixed base -2 [42, 55, 94, 95]. In this system, any analog quantity X, has one and only one representation of a $(N + M)$–bit string $x_{N-1} \ldots x_1 x_0 x_{-1} \ldots x_{-M}$:

$$X = \sum_{i=-M}^{N-1} a_i(-2)^i, \quad a_i \in \{0, 1\}, \tag{9.11}$$

in which the digits with nonnegative subscripts constitute the integral part, and those with negative subscripts constitute the fractional part. The representation is unique; i.e., there is a one-to-one correspondence between X and x_i. Note that the digits carry not only magnitude but also polarity information. The signs of the weight assigned to each digit change alternately. When i is even, it is positive; when i is odd, it is negative. This property enables us to represent any bipolar number without a sign bit. It is this property that gives an advantage to performing negabinary arithmetic not at word level but at digit level. If a is positive then the most significant 1 must be in the even-indexed position, while the highest 1 being an odd index digit implies a negative. Moreover, the largest positive X is one having all its even index bits equal to 1 and odd index bits 0, i.e., ...1010101. The smallest negative value a is ...0101010. For even N, the representable number range is $[2(1 - 2^N)/3,$

$(2^N - 1)/3]$, while for odd N, the representable number range is $[2(1 - 2^{N-1})/3,$ $(2^{N+1} - 1)/3]$. Although the interval is asymmetrical, using N negabinary bits can accommodate 2^N integers. The negabinary complement operation is discussed in [108].

9.3.1.2.1. *Half-Addition and Half-Subtraction*

Because of the special characteristic of the negabinary number system, it is helpful to use symbolic substitution for negabinary arithmetic operations [55]. Assume A and B are two negabinary numbers to be operated, and a_i and b_i are their ith bits respectively.

Half-addition: For the 1-bit addition of a_i and b_i, there are four combinations; namely, ($(a_i b_i) = (0, 0), (0, 1), (1, 0)$, and $(1, 1)$. It is obviously seen that the first three combinations are carry free and the sums of these pairs are 0, 1, and 1, respectively. However, the fourth combination of bits (1, 1) will generate a carry-out. As described above, the signs of adjacent bits in negabinary are different, so the carry produced at one bit position should be subtracted from the sum of the next higher bit. It is therefore termed a negative carry (marked c_{i+1}^-. This can be concluded easily from the identical equation:

$$(1 + 1)(-2)^i = 0(-2)^i - 1(-2)^{i+1}. \tag{9.12}$$

The truth table is shown in Table 9.3.

Half-subtraction: When two bits a_i and b_i are taken for subtraction, three of the four combinations; namely, $(a_i, b_i) = (0, 0), (1, 0), (1, 1)$, do not yield any carry, and their sums equal to 0, 1, 0, respectively. (For the sake of unity, difference of two operands is also called sum.) Nevertheless, the combination (0, 1) will generate a positive carry-out (defined as $c_{i+1}+$), with the sum being 1. This is due to the fact that

$$(0 - 1)(-2)^i = 1(-2)^i + 1(-2)^{i+1}. \tag{9.13}$$

The truth table is shown in Table 9.4.

Table 9.3

Truth Table for Negabinary Half-Addition [55]

a_i	b_i	c_{i+1}^-	s_i
0	0	0	0
0	1	0	1
1	0	0	1
1	1	1	0

Table 9.4

Truth Table for Negabinary Half-Subtraction [55]

a_i	b_i	c^+_{i+1}	s_i
0	0	0	0
0	1	1	1
1	0	0	1
1	1	0	0

9.3.1.2.2. *Negabinary Symbolic Addition/Subtraction*

Consider the symbolic substitution rules necessary for negabinary addition. In the first iteration, the corresponding digits of a_i and b_i are added in parallel. So the truth table of half-addition is available for each bit position. The four substitution rules can be obtained from Table 9.3 and are illustrated in Fig. 9.12(a). The two digits are placed one over the other, and each column including a_i and b_i is replaced by a new two-column pattern including the sum s_i and the negative carry c^-_{i+1} to the next higher bit position. The numerical values rather than their spatial encoding are used. This operation will generate a sum string $s^{(1)}$ and a negative carry string $c^{-(1)}$ (the number in the parenthesis defines the iteration time).

In the second iteration, the two digits to be added in each column come from $s_i^{(1)}$ and $c_i^{-(1)}$. In fact, the addition between the sum $s_i^{(1)}$ and the negative carry $c_i^{-(1)}$ is a negabinary half-subtraction. According to Table 9.4, the four substitution rules are listed in Fig. 9.12(b). Consequently, another sum string $s^{(2)}$ and a positive carry string $c^{+(2)}$ are produced.

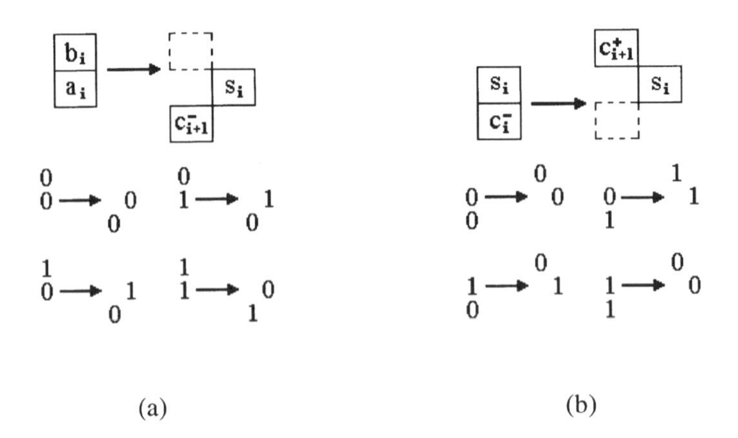

Fig. 9.12. Substitution rules for negabinary (a) half-addition, and (b) half-subtraction.

For the third iteration, it is easily seen that the two digits to be added in a column originate from $s_i^{(2)}$ and $c_i^{+(2)}$. The half-addition rules in Fig. 9.12 can be applied, resulting in the sum string $s_i^{(3)}$ together with the possible negative carry $c_i^{-(3)}$. As analog to the second iteration, in the fourth iteration the half-subtraction symbolic substitution rules are used. Therefore, the half-addition and half-subtraction rules are used alternately, and for odd-time and even-time iterations, respectively.

In reference to negabinary addition, one notices readily that for negabinary subtraction, the half-subtraction rules should be employed first and then the half-addition rules. This procedure will continue until no positive and negative carries occur.

9.3.1.2.3. *Unified Symbolic Arithmetic for Addition and Subtraction*

Generally speaking, both addition and subtraction can be performed easily by using two sets of substitution rules, one for half-addition and the other for half-subtraction. It is possible to combine the two sets of rules into one by introducing both positive and negative carries at the same time [55]. If we define the left-hand side of the rules containing the pixels $s_i^{(j-1)}$, $c_i^{+(j-1)}$ and $c_i^{-(j-1)}$ and and the right-hand side of the rules containing $s_i^{(j)}$, $c_{i+1}^{+(j)}$ and $c_{i+1}^{-(j)}$, the two sets of double-in–double-out rules can be merged into one set of triple-in–triple-out substitution rules. Since the condition that both c_i^+ and c_i^- equal to 1 does not happen, there are only six combinations for the negabinary triple-in variables. The unified arithmetic truth table and the corresponding symbolic substitution rules are shown in Table 9.5 and Fig. 9.13, respectively.

The initial values of the three digit strings $s_i^{(0)}$, $c_i^{+(0)}$, $c_i^{-(0)}$ should be assigned before the addition and subtraction are performed. For addition, the augend a can be viewed as the initial sum, and the addend b can be viewed as the initial positive carry; while for subtraction, the minuend a can also be taken as the original sum, and the subtrahend b can be taken as the original negative carry. Thus, for addition and subtraction we can write the following conditions.

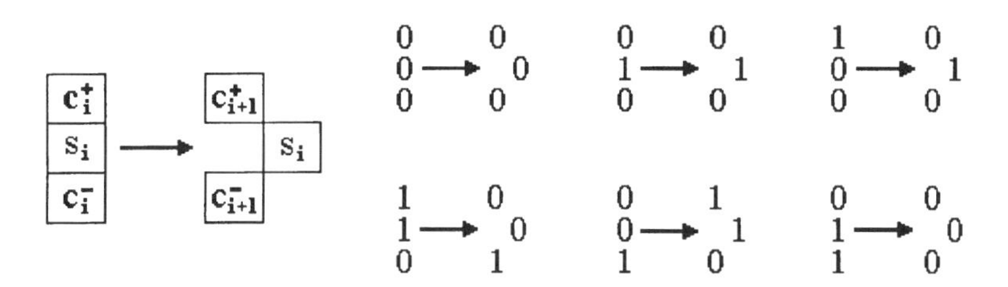

Fig. 9.13. Substitution rules for unified negabinary arithmetic [55].

Table 9.5

Truth Table for Unified Negabinary Addition/Subtraction [55]

c_i^+	c_i^-	s_i	c_{i+1}^+	c_{i+1}^-	s_i
0	0	0	0	0	0
0	0	1	0	0	1
1	0	0	0	0	1
1	0	1	0	1	0
0	1	0	1	0	1
0	1	1	0	0	0

For addition:

$$
\begin{aligned}
s_i^{(0)} &= a_i, \\
c_i^{+(0)} &= b_i, \\
c_i^{-(0)} &= 0, \quad (i = -M, \ldots, -1, 0, 1, \ldots, N-1).
\end{aligned}
\tag{9.14a}
$$

For subtraction:

$$
\begin{aligned}
s_i^{(0)} &= a_i, \\
c_i^{+(0)} &= 0, \\
c_i^{-(0)} &= b_i, \quad (i = -M, \ldots, -1, 0, 1, \ldots, N-1).
\end{aligned}
\tag{9.14b}
$$

After the initialization, both addition and subtraction of two numbers can be carried out in parallel by the successive use of the same substitution rules listed above. At each iteration, at least 1 bit of the final result for a pair of operands can be obtained. Therefore, a unified symbolic arithmetic for addition and subtraction has been established.

For clarity and without the loss of generality, addition of the 4-bit integral numbers $A = (3)_{10} = (0111)_{-2}$ and $B = (-6)_{10} = (1110)_{-2}$, and subtraction of the numbers $A = (-3)_{10} = (1101)_{-2}$ and $B = (-9)_{10} = (1011)_{-2}$ are illustrated below:

```
     -6 1 1 1 0     0 0 0 0       0 1
 +)   3 0 1 1 1       1 0 0 1       0 1 0 1      1 1 0 1 (=-3_10)
        0 0 0 0     0 1 1 0       0 0
                (1)           (2)           (3)
               --->          --->          --->
        0 0 0 0     0 0 1 0       0 0            1
     -3 1 1 0 1       0 1 1 0       0 0 1 0        1 0 1 0 --(4)--> 1 1 0 1 0
 -)  -9 1 0 1 1     0 0 0 0       0 1            0                  (=6_10)
```

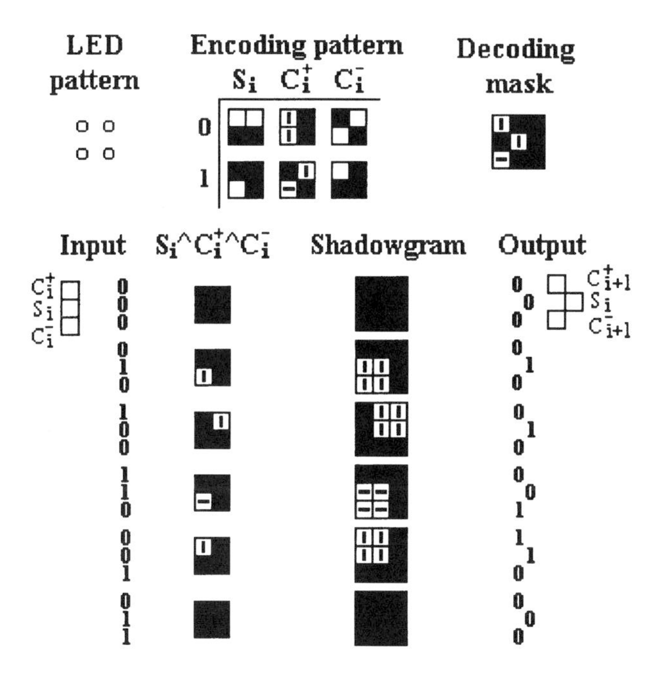

Fig. 9.14. Polarization optical shadow-casting design for a unified negabinary arithmetic unit [55]. The input c_i^+ (positive carry), c_i^- (negative carry), and s_i (sum bit) generate the output c_{i+1}^+ (positive carry to the next bit position), c_{i+1}^- (negative carry to the next bit position), and the new sum bit s_i.

Using the unified symbolic substitution rules in Fig. 9.13, the addition and subtraction are executed in parallel. From Eq. (9.11), it is understood that the range of representable 4-bit integral numbers expands from -10 to 5; i.e., $[-10, 5]$. In the addition example, there is no overflow. However, in the subtraction example, a positive carry is generated out of the MSB at the third step. As described earlier, the arithmetic operation is performed at the digit level, so this does not create any problem. One can continue to utilize the symbolic substitution rules for the final result (the fourth step). By using spatial and polarization encoding and space-variant decoding technique, a polarization optical shadow-casting system has been designed to perform negabinary symbolic addition and subtraction [55]. The encoding and decoding patterns are shown in Fig. 9.14.

Negabinary addition and subtraction can also be performed directly by binary logic operations. According to Table 9.5, the binary logical expressions for a full-adder and full-subtractor can be obtained as shown in the following equations.

For addition:

$$
\begin{aligned}
s_i &= a_i \oplus b_i \oplus c_i^+ \oplus c_i^-, \\
c_{i+1}^+ &= \overline{a_i}\,\overline{b_i}\,\overline{c_i^+}\,c_i^-, \\
c_{i+1}^- &= a_i b_i \overline{c_i^-} + (a_i \oplus b_i) c_i^+ \overline{c_i^-};
\end{aligned}
\tag{9.15}
$$

For subtraction:

$$
\begin{aligned}
d_i &= a_i \oplus b_i \oplus c_i^+ \oplus c_i^- \\
c_{i+1}^+ &= \overline{a_i} b_i \overline{c_i^+} + (a_i b_i + \overline{a_i}\,\overline{b_i}) \overline{c_i^+}\,c_i^-, \\
c_{i+1}^- &= a_i \overline{b_i} c_i^+ \overline{c_i^-}.
\end{aligned}
\tag{9.16}
$$

9.3.1.2.4. Higher-Order Negabinary Addition

In the above-mentioned 1-bit negabinary addition, the addition of two 1s generates a negative carry to the next higher bit position. If the carry generation is limited to be positive, twin carries will be generated to the next two higher bit positions according to the following identity [42, 95]:

$$
1 \cdot (-2)^i + 1 \cdot (-2)^i = 0 \cdot (-2)^i + 1 \cdot (-2)^{i+1} + 1 \cdot (-2)^{i+2}. \tag{9.17}
$$

The logical function is given by

$$
\begin{aligned}
s_i &= a_i \oplus b_i, \\
c_{i+2} &= c_{i+1} = a_i b_i.
\end{aligned}
\tag{9.18}
$$

The nature of the twin-carry generation in negabinary is different from other number system algorithms and matches well the representation of higher-order symbolic addition.

Consider a 2-bit–wide module with the inputs $a_{i+1}a_i$ and $b_{i+1}b_i$ (Fig. 9.15[a]). The outputs contain a 2-bit sum $s_{i+1}s_i$ and a 2-bit carry $c_{i+3}c_{i+2}$. The output–input relation can be represented by a higher-order substitution rule, as shown in the right side of the figure. The operands $a_{i+1}a_i$ are placed over the other operands $b_{i+1}b_i$, and in the output pattern, the lower row corresponds to the sum $s_{i+1}s_i$ and the upper row corresponds to the carry $c_{i+3}c_{i+2}$. There are 16 total combinations for the four input bits; the 16 substitution rules are listed in Fig. 9.15(b). With dual-rail spatial encoding, the number of reference patterns can be reduced [95]. An optical implementation based on

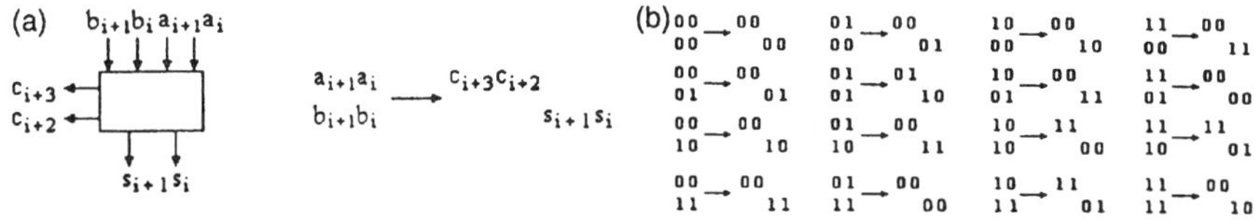

Fig. 9.15. Negabinary higher-order addition [95]. (a) Basic module, (b) Substitution rules.

an incoherent correlator has been shown. The logical function can be written as

$$\begin{aligned} s_i &= a_i \oplus b_i, \\ s_{i+1} &= a_{i+1} \oplus b_{i+1} \oplus (a_i b_i), \\ c_{i+2} &= a_i b_i \bar{a}_{i+1} \bar{b}_{i+1} + a_{i+1} b_{i+1} \overline{a_i b_i}, \\ c_{i+3} &= a_{i+1} b_{i+1} \overline{a_i b_i}. \end{aligned} \tag{9.19}$$

As an example, multiple additions can be performed using shadow-casting logic with spatial encoding and a 4 × 4 source array [42]. Note that in some cases, the twin-carry mechanism leads to a nonending sequence of carries to the left, but once the carry passes out the MSB of the operands, the remaining result is correct. For example, this occurs for the addition of 11 + 11, with the final sum being 10.

9.3.2. *Operations with Nonbinary Number Systems*

For operands encoded in nonredundant positional number systems such as the positive binary [96–101], 2's complement [102–105], signed-magnitude [106], negabinary [94, 107, 108, 109], and imaginary [110, 111] systems, intermediate results of addition and multiplication can be obtained in parallel based on the mixed-radix representation. In this approach, addition is done digit-wise, and multiplication is done through the well-known technique called digital multiplication via convolution (DMAC). Since convolution can exploit the advantages of optics in parallel two-dimensional data processing, inner-product and outer-product DMAC algorithms have been proposed and implemented through various optical systems. However, this cannot guarantee carry-free computation where the burden of carry propagation is transferred to electronic postprocessing requiring fast conversion from the mixed radix to its original binary format [112, 113, 114].

In the residue number system approach [115–120], a set of relatively prime numbers $(m_{N-1}, \ldots, m_1, m_0)$ called moduli are chosen to represent integers. An integer X is represented by a digit string $x_{N-1} \ldots x_1 x_0$ of which each digit x_i is the residue of X divided by m_i:

$$x_i = X \text{ modulo } m_i = |a|m_i, \quad (i = 0, 1, \ldots, N-1). \tag{9.20}$$

The range of the numbers to be represented is determined by the product of the moduli:

$$0 \leqslant X \leqslant M, \quad M = \prod m_i. \tag{9.21}$$

It can be directly utilized for carry-free addition, subtraction, and multiplication, because the operations on different moduli can be performed in parallel:

$$p_i = |P|m_i = |a * b|m_i = |a_i * b_i|m_i, \tag{9.22}$$

where $*$ represents an addition or a multiplication, and P is the sum or product. Subtraction can be transformed to addition by complementing the subtrahend with respect to the moduli. However, this technique also suffers from several disadvantages. For example, since different moduli operations are performed for different digits, it is not suitable for SIMD space-invariant global optical operations and the processing complexity is asymmetrically distributed. Moreover, decoding of the final result from residue to binary or decimal is rather complicated, and the increased time delay may offset almost all of the time saved in eliminating carries.

The signed-digit number system allows us to implement parallel arithmetic by using redundancy. The modified signed-digit (MSD) number system [70, 71, 72, 74, 121–145] has been widely studied. It confines the carry propagation to two adjacent digit positions and thus permits parallel carry-free addition and borrow-free subtraction of two arbitrary length numbers in constant time. Higher radix signed-digit number systems [73, 146–160] permit higher information storage density, less complexity, fewer components, and fewer cascaded gates and operations. Consequently, various parallel algorithms involving single as well as multiple computational steps have been proposed for MSD, trinary signed-digit (TSD) [146–154], and quaternary signed-digit (QSD) [73, 155–160] number systems. Among these algorithms, a two-step recoding technique [138, 139, 140, 149, 152, 153, 154, 156, 157, 160] has been recently introduced in which the operands are first recoded so that the second step operation is carry free. Moreover, negabinary number system [55, 94, 95, 108, 109, 160, 161], using -2 as the base, possess the particular superiority of uniquely representing both positive and negative numbers without a sign bit. Recently, a two-step parallel negabinary signed-digit (NSD) arithmetic has been proposed [160]. The negabinary numbers can be directly input for operation since negabinary is a subset of NSD.

The proposed signed-digit algorithms can be operated on all operand digits in parallel. This parallelism allows the aforementioned algorithms to be efficiently mapped on the parallel cellular optical computing architecture [1–7]. Consequently, these algorithms have been integrated in the parallel architectures via symbolic substitution, logic array processing, and matrix–matrix multiplication, all of which are powerful techniques for two-dimensional array data processing. For optoelectronic implementation, truth-table–based content-addressable memory (CAM), shared CAM, coherent and incoherent optical correlators, and nonlinear logic gates have been employed. In these systems, the signed digits can be encoded by spatial position, intensity, and polarization states.

Clever utilization of parallel optical cellular array architectures may lead to an even more powerful optoelectronic computing system. Spatial-light modulators (SLMs) with high processing speed, high resolution, and high contrast are being developed for processing the input, intermediate operands, and output operands. In the near future, the ultrahigh processing speed, parallelism, massive data rates from optical memory, and superior processing capabilities of optoelectronic technology can be fused to produce the next generation of optoelectronic computers.

9.4. PARALLEL SIGNED-DIGIT ARITHMETIC

9.4.1. GENERALIZED SIGNED-DIGIT NUMBER SYSTEMS

The aforementioned MSD, TSD, and QSD representations are all subsets of the previously defined signed-digit number system where the radix r can be greater than 2. With the introduction of the NSD number system, it is necessary to define a generalized signed-digit (GSD) number representation in which the radix r can be either positive or negative. In the GSD system, a decimal number X can be defined as

$$X = \sum_{i=-M}^{N-1} x_i r^i, \; x_i \in \{-a, \ldots, -1, 0, 1, \ldots, a\}, \tag{9.23}$$

where a is a positive integer. In this system, the number of elements of the digit set is usually greater than $|r|$ resulting in redundant representation; that is, there is more than one representation for each number. The redundancy also depends on the selection of a and the minimum and maximum values of a are given by

$$a_{\min} = \begin{cases} \dfrac{|r|+1}{2}, & |r| \text{ odd} \\ \dfrac{|r|}{2}, & |r| \text{ even}, \end{cases} \tag{9.24}$$

$$a_{\max} = |r| - 1.$$

For the GSD system $|r| \geqslant 4$, this gives the digit sets

$$\{-a_{\min}, \ldots, -1, 0, 1, \ldots, a_{\min}\} \qquad \text{and} \qquad \{-a_{\max}, \ldots, -1, 0, 1, \ldots, a_{\max}\},$$

respectively. All number systems can be expressed in terms of the GSD notation, as discussed in the following subsections.

9.4.1.1. *MSD Number System* [36–59]

For MSD numbers, $r = 2$, $a = 1$, and the digit set is $\{\bar{1}, 0, 1\}$, where the overbar indicates the logical complement; i.e., $\bar{1} = -1$. The redundant binary representation was originally proposed by Avizienis [121] and introduced to optics community by Drake et al. [71]. For example, the decimal number 5 can be represented in MSD by $(1\bar{1}01)_{MSD}$, or $(101)_{MSD}$, or $(10\bar{1}\bar{1})_{MSD}$, or $(1\bar{1}01\bar{1})_{MSD}$, etc. The negative value of an MSD positive number can be obtained by complementing each digit of that number. The complement of 1 is $\bar{1}$ and the complement of $\bar{1}$ is 1, while the complement of 0 is 0. For example, the decimal number -5 can be represented as $(\bar{1}10\bar{1})_{MSD}$. Thus, subtraction can be performed by complementing the subtrahend and then applying the addition operation.

9.4.1.2. *TSD Number System* [60–68]

In this case, $r = 3$, $a = 2$, and the digit set is $\{\bar{2}, \bar{1}, 0, 1, 2\}$. The degree of redundancy usually increases with the increase of radix. For illustration, consider the following TSD numbers and their corresponding binary and decimal representations:

$$(22)_3 = (1000)_2 = (8)_{10},$$

$$(222222222222222)_3 = (1101\ \ 1010\ \ 1111\ \ 0010\ \ 0110\ \ 1010)_2 = (14348906)_{10}.$$

In the first example, we observe that at most we need 4 bits to represent a

2-digit TSD number. In the second example, we need a maximum of 24 bits to represent a 15-digit TSD number. In the two examples cited, the TSD number system requires 50 to 37.5% fewer digits than the binary number system, which may result in substantial saving of memory space.

9.4.1.3. *QSD Number System* [69–75]

The QSD is an element of the GSD number system with $r = 4$, $a = 3$ (higher redundancy), and the digit set is $\{\bar{3}, \bar{2}, \bar{1}, 0, 1, 2, 3\}$. For instance, one of the above decimal numbers can be encoded in QSD as

$$(312233021222)_{\text{QSD}} = (1101\ \ 1010\ \ 1111\ \ 0010\ \ 0110\ \ 1010)_2 = (14348906)_{10},$$

which requires fewer digits compared to the binary, MSD, or TSD number systems. For minimum redundancy [156] with $a=2$, the digit set is $\{\bar{2}, \bar{1}, 0, 1, 2\}$.

9.4.1.4. *NSD Number System* [76]

In contrast to the aforementioned signed-digit number systems, the NSD representation uses a negative radix -2, with the digit set $\{\bar{1}, 0, 1\}$. The only difference between NSD and MSD lies in the radix. The advantage of the NSD system is that negabinary numbers can be directly used for parallel NSD arithmetic operations, because the NSD number system can uniquely represent a bipolar number without a sign and it is a subset of GSD number system. For example, the decimal number $(-486)_{10}$ can be represented by $(1001101110)_{-2}$.

Since the advent of optical signed-digit computing in 1986, a large number of parallel algorithms and architectures have been proposed. In the following, the basic arithmetic operations of addition, subtraction, multiplication, and division are summarized in terms of the signed-digit number systems used. All the algorithms of multiplication and division are developed on the basis of addition or subtraction.

9.4.2. MSD ARITHMETIC

MSD arithmetic operations, such as addition, are carried out by generating transfer and weight digits which are also known as the intermediate sum and carry digits, respectively. Assume x_i and y_i are the digits of the two operands X and Y at the ith digit position. Addition of the two digits at each position complies with the following basic rules:

$$x_i + y_i = 2c_{i+1} + s_i, \tag{9.25}$$

where c_{i+1} and s_i are the intermediate carry digit and sum digit, respectively, and they are also usually written as the transfer digit t_{i+1} and the weight digit w_i, respectively. By exploiting the redundant representations of the sum $1+0$ and $\overline{1}+0$ and using the property that $1+\overline{1}=0$, a carry can be absorbed within two digits of its left neighbor. According to the number of steps needed to yield the final sum, the algorithms for MSD addition can be classified into three main categories: the three-step approach, two-step approach, and one-step approach. According to the digit sets to which the intermediate carry and sum belong, the algorithms can be classified into the digit-set–nonrestricted approach and the digit-set–restricted approach. In the former approach, the intermediate carry and sum are selected from the entire MSD set $\{\overline{1}, 0, 1\}$, while in the latter approach, the intermediate carry and sum digits are restricted to the sets $\{\overline{1}, 0\}$ and $\{0, 1\}$, respectively. Subtraction can be implemented by first digit-by-digit complementing the subtrahend and then performing addition:

$$X - Y = X + \overline{Y} = \sum_{i=0}^{N-1} (x_i + \overline{y_i}). \tag{9.26}$$

The truth tables and the minimized minterms for the subtraction operation can be easily obtained by following the same procedure used for addition.

9.4.2.1. *Digit-Set–Unrestricted Addition/Subtraction*

9.4.2.1.1. Three-Step MSD Addition/Subtraction

In this category [70, 71, 122–130], the first stage generates a transfer and a weight string in parallel. These weight and transfer digits are then used to generate a second set of transfer and weight digits. The first two operations eliminate the occurrences of the (1, 1) and $(\overline{1}, \overline{1})$ pairs at the same digit position, so that the third-stage digitwise addition of the weight and transfer digits becomes carry free. Therefore, the addition of two arbitrary MSD numbers can be completed with three steps in constant time, independent of the operand length.

MSD addition can be realized by an array of trinary logic elements, as shown in Fig. 9.16, which depicts the addition of two five-digit MSD numbers A and B. In Fig. 9.16, logic elements T and W are used in the first stage while logic elements T^1 and W^1 are used in the second stage. The final sum S is obtained in the third stage by applying the logic element T. The truth tables for the above-mentioned four logic functions are shown in Table 9.6. Bocker et al. [122] effectively transformed these truth tables into symbolic substitution rules. Thus, a total of $3^3 = 27$ rules corresponding to the nine input combinations at each stage are required. A closer look at the truth tables corresponding

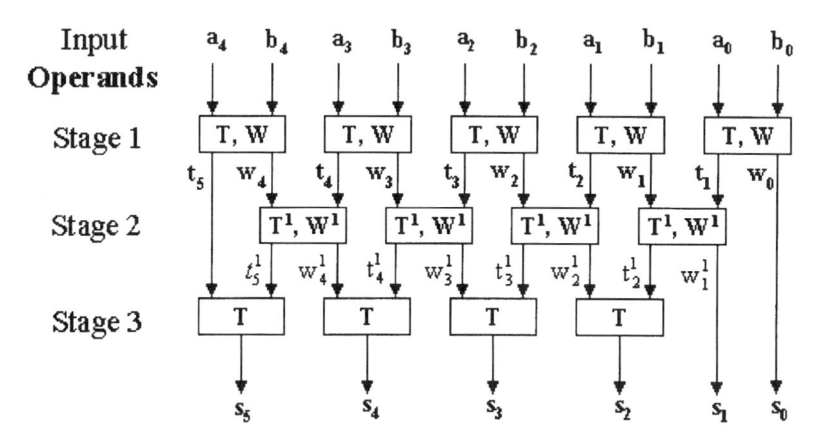

Fig. 9.16. Three-step MSD adder array for the addition of two five-digit numbers [71].

to the second and third stages reveals that five entries are identical. Therefore, the substitution rules can be reduced to 17 [123]. Recently, Zhou et al. [130] also studied multi-input three-stage MSD addition.

The truth tables can also be realized by logic operation [127, 128, 129]. W is the complement of W^1 while T is the logical OR of T^1 and W^1. Therefore, it is possible to achieve MSD addition by realizing only the T^1 and W^1 logical elements. Looking at the logic element W^1, the output W^1 is actually an XOR operation with the sign of the digit remaining unchanged. On the other hand, the truth table of logic element T^1 shows that T^1 corresponds to a logical AND operation in binary logic, with the sign of digits retained in the output.

Table 9.6

Truth Tables for the Three Logic Elements Employed in the Three-Step MSD Adder [71]

T

1	1	1	0
0	1	0	$\bar{1}$
$\bar{1}$	0	$\bar{1}$	$\bar{1}$
	1	0	$\bar{1}$

W

1	0	$\bar{1}$	0
0	$\bar{1}$	0	1
$\bar{1}$	0	1	0
	1	0	$\bar{1}$

T^1

1	1	0	0
0	0	0	0
$\bar{1}$	0	0	$\bar{1}$
	1	0	$\bar{1}$

W^1

1	0	1	0
0	1	0	$\bar{1}$
$\bar{1}$	0	$\bar{1}$	0
	1	0	$\bar{1}$

As a result, the logical operations required in MSD addition can be considered as the signed version of that in traditional binary addition.

9.4.2.1.2. *Two-Step MSD Addition/Subtraction*

The previous algorithm used two steps to guarantee that the third-step addition is carry free. The first two steps can be combined into a single step operation. For example, Li and Eichman [74] suggested the conditional symbolic substitution rules, where in addition to the digit pairs $a_i b_i$ to be substituted, a lower-level digit pair $a_{i-1} b_{i-1}$ is used as the reference digits. In the second step, the transfer and weight digits are added to yield carry-free addition. The data flow diagram for this two-step addition is shown in Fig. 9.17. To implement the two-step algorithm with a content-addressable memory (CAM), the truth tables are first classified in terms of the nonzero outputs 1 and $\bar{1}$. The minterms for the nonzero outputs are then logically minimized using a suitable minimization technique such as the Karnaugh map or the Quine-McClusky's algorithm [6]. The minimized minterms are listed in Table 9.7 where, in each four-variable minterm, the two first- (second) column digits denote the variables $a_i b_i (a_{i-1} b_{i-1})$, respectively. The symbol d implies a complete don't-care digit for $\bar{1}$, 0, and 1, whereas the symbol d_{xy} implies a partial don't-care for digits x and y.

Example: $10\bar{1}\bar{1}1010\bar{1}01\bar{1}(1433_{10}) + 1\bar{1}\bar{1}100\bar{1}10110(758_{10})$

Stage 1: $1\bar{1}\bar{1}01000011\bar{1}\phi\ (T)$
$\phi0100\bar{1}001\bar{1}\bar{1}01\ (W)$

Stage 2: $1\bar{1}001\bar{1}00100\bar{1}1(2191_{10})$,

where the symbol ϕ indicates a padded zero.

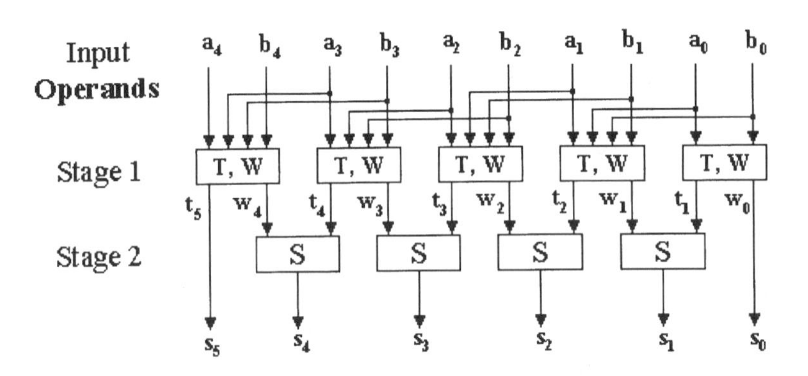

Fig. 9.17. Five-digit MSD addition configuration in two steps.

Table 9.7

Reduced Minterms for the MSD Addition [74]. The d Denotes a Do Not Care Digit

Logic	Output	Minterms $\begin{bmatrix} a_i & a_{i-1} \\ b_i & b_{i-1} \end{bmatrix}$	Logic	Output	Minterms $\begin{bmatrix} a_i & a_{i-1} \\ b_i & b_{i-1} \end{bmatrix}$
T	1	$1\mathrm{d} \quad 1\mathrm{d}_{01} \quad 0\mathrm{d}_{01}$	W	1	$1\bar{1} \quad 1\mathrm{d} \quad 0\bar{1} \quad 0\mathrm{d} \quad \bar{1}\mathrm{d}_{0\bar{1}} \quad 0\mathrm{d}_{0\bar{1}}$
		$1\mathrm{d}, \quad 0\mathrm{d}_{01}, \quad 1\mathrm{d}_{01}$			$0\mathrm{d}, \quad 0\bar{1}, \quad 1\mathrm{d}, \quad 1\bar{1}, \quad 0\mathrm{d}_{0\bar{1}}, \quad \bar{1}\mathrm{d}_{0\bar{1}}$
	$\bar{1}$	$\bar{1}\mathrm{d} \quad \bar{1}\mathrm{d}_{0\bar{1}} \quad 0\mathrm{d}_{0\bar{1}}$		$\bar{1}$	$\bar{1}1 \quad \bar{1}\mathrm{d} \quad 01 \quad 0\mathrm{d} \quad 1\mathrm{d}_{01} \quad 0\mathrm{d}_{01}$
		$\bar{1}\mathrm{d}, \quad 0\mathrm{d}_{0\bar{1}}, \quad \bar{1}\mathrm{d}_{0\bar{1}}$			$0\mathrm{d}, \quad 01, \quad \bar{1}\mathrm{d}, \quad \bar{1}1, \quad 0\mathrm{d}_{01}, \quad 1\mathrm{d}_{01}$
S	1	$1 \quad 0$			
		$0, \quad 1$			
	$\bar{1}$	$\bar{1} \quad 0$			
		$0, \quad \bar{1}$			

Later on, Cherri and Karim [131, 132] modified the above-mentioned two-step conditional arithmetic. The truth table can be arranged into two halves with each half having identical outputs. The two halves are separated by judging whether the less significant digit pair of MSD is binary or not. On this basis, a reference bit g_{i-1} is introduced such that when the less significant digit pair is binary, $g_{i-1} = 1$; otherwise, $g_{i-1}1 = 0$. The modified truth table is shown in Table 9.8 where only three digits instead of four are needed as inputs to each of the T and W elements. The second-step operation is an addition operation, as discussed in the previous section. The logically minimized minterms for generating the 1 and $\bar{1}$ outputs for this two-step addition technique is shown in Table 9.9.

Recently, the above algorithm has been further modified and implemented through binary logic operations [134]. With the programming of the reference digits, addition and subtraction can be performed by the same logic operations. With reference to Table 9.8, the truth table for subtraction can be obtained (Table 9.10). However, the reference bit g_{i-1} is true when a_{i-1} and the complement of b_{i-1} are binary; otherwise, $g_{i-1} = 0$. To avoid the logic

Table 9.8

Modified Conditional Truth Table for the First Step MSD Addition [131]

		t_{i+1}	w_i	t_{i+1}	w_i	t_{i+1}	w_i	t_{i+1}	w_i	t_{i+1}	w_i	t_{i+1}	w_i
g_{i-1}	0	1	0	0	1	0	0	0	0	$\bar{1}$	1	$\bar{1}$	0
	1	1	0	1	$\bar{1}$	0	0	0	0	0	$\bar{1}$	$\bar{1}$	0
$a_i b_i (g_i)$		11(1)		01(1)/10(1)		00(1)		$1\bar{1}(0)/\bar{1}1(0)$		$\bar{1}0(0)/0\bar{1}(0)$		$\bar{1}\bar{1}(0)$	

Table 9.9

Reduced Minterms for MSD Addition Rules Shown in Table 9.3 [131]

Function	Output	Minterms ($a_ib_iF_i$)	Function	Output	Minterms ($a_ib_iF_i$)
T	1	$d_{01}11$, $1d_{01}1$, $11d_{01}$	W	1	$0d_{1\bar{1}}0$, $d_{1\bar{1}}00$
	$\bar{1}$	$d_{0\bar{1}}\bar{1}0$, $\bar{1}d_{0\bar{1}}0$, $\bar{1}\bar{1}d_{01}$		$\bar{1}$	$0d_{1\bar{1}}1$, $d_{1\bar{1}}01$
S	1	01, 10			
	$\bar{1}$	$0\bar{1}$, $\bar{1}0$			

operations with signed digits, a closer look at the truth tables reveals that it is necessary to introduce another reference bit, h_i. For addition, h_i is true if both a_i and b_i are negative. For subtraction, h_i is true if both a_i and the complement of b_i are negative. A key feature of this definition is that we only need to consider the unsigned binary values of a_i and b_i, independent of their signs. This greatly simplifies the logical operations. Another advantage of this definition is that both addition and subtraction operations can be expressed by the same binary logic equations. In the first step, the required binary logic expressions at ith($i = 0, 1, \ldots, N - 1$) digit position for the two arithmetic operations are given by:

for t_i,

$$\text{output "1"}: a_ib_ig_i + (a_i \oplus b_i)g_ig_{i-1}, \tag{9.27}$$

$$\text{output "}\bar{1}\text{"}: h_i + (a_i \oplus b_i)\bar{g}_i\bar{g}_{i-1}. \tag{9.28}$$

for w_i,

$$\text{output "1"}: (a_i \oplus b_i)\bar{g}_{i-1}, \tag{9.29}$$

$$\text{output "}\bar{1}\text{"}: (a_i \oplus b_i)g_{i-1}. \tag{9.30}$$

where $g_{-1} = 1$. In the second step, the transfer and weight digits are added at

Table 9.10

Modified Conditional Truth Table for the First Step MSD Subtraction [134]

		t_{i+1}	w_i	t_{i+1}	w_i	t_{i+1}	w_i	t_{i+1}	w_i	t_{i+1}	w_i	t_{i+1}	w_i
g_{i-1}	0	0	0	0	0	0	1	$\bar{1}$	1	1	0	$\bar{1}$	0
	1	0	0	0	0	1	$\bar{1}$	0	$\bar{1}$	1	0	$\bar{1}$	0
$a_ib_i(g_i)$		$\bar{1}\bar{1}(0)/11(0)$		00(1)		$0\bar{1}(1)/10(1)$		$\bar{1}0(0)/01(0)$		$1\bar{1}(1)$		$\bar{1}1(0)$	

Table 9.11

Truth Table for the Second Step MSD Addition/Subtraction [134]

t_i	w_i	(g'_i)	s_i/d_i
0	0	(1)	0
0	1	(1)	1
1	0		
0	$\bar{1}$	(0)	$\bar{1}$
$\bar{1}$	0		
1	$\bar{1}$	(0)	0
$\bar{1}$	1		

each position in parallel to yield the final sum S or the difference D of the two operands without carries. The truth table is shown in Table 9.11. The logic functions for the final sum S and D can be obtained by defining a binary reference bit g'_i which is true for positive t_i and w_i:

for s_i and d_i,

$$\text{output "1": } (t_i \oplus w_i)g'_i, \tag{9.31}$$

$$\text{output "}\bar{1}\text{": } (t_i \oplus w_i)\overline{g'_i}. \tag{9.32}$$

The logic operations required for the 1 and $\bar{1}$ outputs of S are the same as those for the $\bar{1}$ and 1 outputs of W in the first step, respectively.

Therefore, both addition and subtraction can be performed in parallel by the same binary logic operations. This offers the advantage that through data programming we can perform space-variant arithmetic computation by space-invariant logic operations. The data flow diagram of the scheme is shown in Fig. 9.18. Additionally, in comparison with the approach [135] (to be described below) in which the MSD arithmetic was performed with binary logic via separating the positive and negative digits, the adoption of the reference bits yields simpler logical expressions, thus reducing the system complexity.

For example, consider the following illustrations for addition and subtraction using the proposed algorithm:

Addition: A1 + B1	Subtraction: A2 − B2
A1: $10\bar{1}\bar{1}11\bar{1}\bar{1}$ (89_{10})	A2: $\bar{1}011\bar{1}\bar{1}10$ (-90_{10})
B1: $11\bar{1}011\bar{1}0$ (170_{10})	B2: $10\bar{1}01111$ (111_{10})

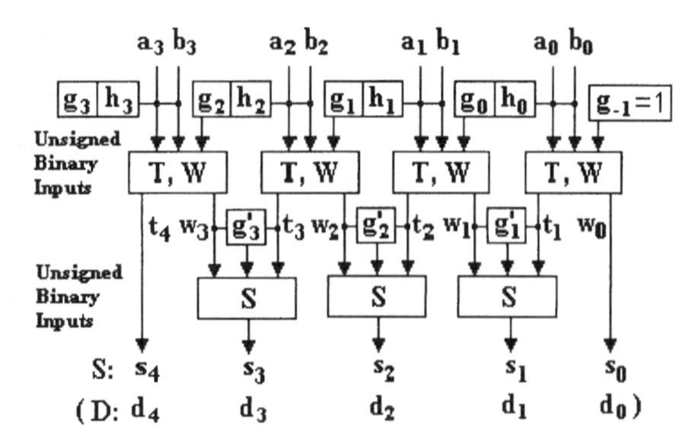

Fig. 9.18. Data flow diagram for the two-step MSD addition/subtraction implemented by binary logic operations [134].

Binary input:	
A1: 10111111	A2: 10111110
B1: 11101110	B2: 10101111
G1: 11001100	G2: 01110000
H1: 00100010	H2: 10001100
T1: $10\bar{1}010\bar{1}0\phi$	T2: $\bar{1}010\bar{1}\bar{1}00\phi$
W1: $\phi 010\bar{1}000\bar{1}$	W2: $\phi 0001000\bar{1}$
G1′: 1101 01000	G2′: 011100110
S: $10000\bar{1}\bar{1}0\bar{1}$ (259_{10})	D: $\bar{1}0100\bar{1}00\bar{1}$ (-201_{10})

By mimicking the binary-coded ternary representation [135], a MSD digit can be represented by a pair of bits; i.e., $1_{\text{MSD}} = [1, 0]$, $0_{\text{MSD}} = [0, 0] = [1, 1]$, $\bar{1}_{\text{MSD}} = [0, 1]$. So MSD numbers can be encoded as $A = [A_1, A_2]$, $B = [B_1, B_2]$, and their sum as $S = [S_1, S_2]$. Then

$$S = [S_1, S_2] = [A_1, A_2] + [B_1, B_2] = \{[A_1, A_2] + [B_1, 0]\} + [0, \mathrm{B}_2]. \tag{9.33}$$

Therefore, an MSD addition can be decomposed into a two-stage operation [136]. In the first stage, the partial sum $Z = [Z1, Z2] = [A1, A2] + [B1, 0]$ is computed. Thc sccond-stage operation can be transformed into that of the first stage by an exchange operation as shown in the following equation:

$$S = [S1, S2] = Ex\{[Z2, Z1] + [B2, 0]\}, \quad \text{where } Ex[x, y] = [y, x]. \tag{9.34}$$

By analyzing the truth table, the ith digit of Z1 and Z2 can be expressed by binary logic operations:

$$
\begin{aligned}
Z1_i &= A1_{i-1}\overline{A2}_{i-1} + \overline{A1_{i-1} \oplus A2_{i-1}}\, B1_{i-1}, \\
Z2_i &= (A1_i \oplus A2_i) \oplus B1_i,
\end{aligned}
\tag{9.35}
$$

where $\oplus$ indicates the Exclusive-OR logic operation. In reality, this scheme requires more than two steps.

9.4.2.1.3. Recoded Two-Step MSD Addition/Subtraction

Another important approach for two-step addition is based on the recoding of the MSD numbers (RMSD), which was proposed by Parhami [137] and introduced to optical computing by Awwal [138] in 1992. If there are no two consecutive 1s or $\bar{1}$s in an MSD number, then it is possible to perform carry-propagation–free addition. Accordingly, by exploiting the redundancy of the MSD number system, we can derive a truth table for recoding the input binary or MSD number to achieve limited carry-free addition (see column u_i of Table 9.12). The recoding algorithm converts a given MSD number X into

Table 9.12

Recording Truth Table for MSD Numbers [138, 139] where u_i, v_i, and w_i Correspond to the RMSD, the First and the Second SRMSD Algorithms, Respectively

$x_i x_{-1} x_{i-2} x_{i-3}$	u_i	v_i	w_i	$x_i x_{i-1} x_{i-2} x_{i-3}$	u_i	v_i	w_i
$\bar{1}\bar{1}\bar{1}d$	0	0	0	$111d$	0	0	0
$\bar{1}\bar{1}0\bar{1}$	0	0	0	1101	0	0	0
$\bar{1}\bar{1}01$	1	1	1	$110\bar{1}$	$\bar{1}$	$\bar{1}$	$\bar{1}$
$\bar{1}\bar{1}1d$	1	1	1	$11\bar{1}d$	$\bar{1}$	$\bar{1}$	$\bar{1}$
$\bar{1}\bar{1}00$	1	0	1	1100	0	0	$\bar{1}$
$\bar{1}0\bar{1}d$	1	1	1	$101d$	$\bar{1}$	$\bar{1}$	$\bar{1}$
$\bar{1}00d$	$\bar{1}$	1	$\bar{1}$	$100d$	$\bar{1}$	$\bar{1}$	1
$\bar{1}01d$	$\bar{1}$	$\bar{1}$	$\bar{1}$	$10\bar{1}d$	1	1	1
$\bar{1}1\bar{1}d$	$\bar{1}$	$\bar{1}$	$\bar{1}$	$1\bar{1}1d$	1	1	1
$\bar{1}10\bar{1}$	$\bar{1}$	$\bar{1}$	$\bar{1}$	$1\bar{1}01$	1	1	1
$\bar{1}100$	0	0	$\bar{1}$	$1\bar{1}00$	1	0	1
$\bar{1}101$	0	0	0	$1\bar{1}0\bar{1}$	0	0	0
$\bar{1}11d$	0	0	0	$1\bar{1}\bar{1}d$	0	0	0
$0\bar{1}\bar{1}d$	$\bar{1}$	$\bar{1}$	$\bar{1}$	$011d$	1	1	1
$0\bar{1}0\bar{1}$	$\bar{1}$	$\bar{1}$	$\bar{1}$	0101	1	1	1
$0\bar{1}00$	0	$\bar{1}$	0	0100	1	1	0
$0\bar{1}01$	0	0	0	$010\bar{1}$	0	0	0
$0\bar{1}1d$	0	0	0	$01\bar{1}d$	0	0	0
$00dd$	0	0	0				

Table 9.13

Reduced Minterms for the Recording Rules Shown in Table 9.12 [138, 139]

Operation	Method	Output	Minterms
	RMSD(U)	1	$d_{1\bar{1}}\bar{1}0d_{01}$, $d_{\bar{1}\bar{1}}\bar{1}1d$, $010d_{01}$, $011d$, $d_{0\bar{1}}0\bar{1}d$
		$\bar{1}$	$d_{1\bar{1}}0d_{01}d$, $d_{1\bar{1}}1\bar{1}d$, $0\bar{1}\bar{1}d$, $0\bar{1}0\bar{1}$, $d_{1\bar{1}}10\bar{1}$
Stage 1:	SRMSD(V)	1	$10\bar{1}d$, $011d$, $\bar{1}0d_{0\bar{1}}$,d, $010d_{01}$, $d_{1\bar{1}}\bar{1}1d$, $d_{1\bar{1}}\bar{1}01$
Recoding		$\bar{1}$	$\bar{1}01d$, $0\bar{1}\bar{1}d$, $10d_{01}d$, $0\bar{1}0d_{0\bar{1}}$, $d_{1\bar{1}}1\bar{1}d$, $d_{1\bar{1}}10\bar{1}$
	SRMSD(W)	1	0101, $011d$, $d_{1\bar{1}}\bar{1}0$, d_{10}, $d_{1\bar{1}}0\bar{1}d$, $d_{0\bar{1}}\bar{1}1d$, $10d_{0\bar{1}}d$
		$\bar{1}$	$0\bar{1}0\bar{1}$, $0\bar{1}\bar{1}d$, $d_{1\bar{1}}10d_{\bar{1}0}$, $d_{1\bar{1}}01d$, $d_{01}1\bar{1}d$, $\bar{1}0d_{01}d$
Stage 2:		1	$\bar{1}\bar{1}11$, 0011, $01d0$, $01d_{01}\bar{1}$, $100d$, $10\bar{1}d_{01}$
Addition		$\bar{1}$	$11\bar{1}\bar{1}$, $00\bar{1}\bar{1}$, $0\bar{1}d0$, $0\bar{1}d_{0\bar{1}}1$, $\bar{1}00d$, $\bar{1}01d_{0\bar{1}}$

Y with no consecutive $\bar{1}$s. Then the MSD number Y is reconverted to eliminate consecutive 1s in such a way that the reduction of 1s does not recreate an adjacent $\bar{1}$ digit. As a result, any number X is encoded into a number Z so that $z_i \times z_{-1} \neq 1$. It is necessary to pad one zero preceding the most significant digit and three trailing zeros after the least significant digit. An N-digit MSD number when recoded yields an $(N + 1)$–digit RMSD number. After recoding, the two numbers A and B can be added directly in the second step without generating any carry. The ith digit s_i of the sum depends on a_i, b_i, a_{i-1}, and b_{i-1}. After logical minimization, 10 minterms are required in the recoding stage as shown in Table 9.13. In the addition stage, the minterms for 1 and $\bar{1}$ are complements of each other. If the spatial encoding for a 1 is made equivalent to a 180°-rotated version of $\bar{1}$, then only 6 minterms are needed in the second step. Thus, the two-step recoded MSD arithmetic requires a total of 16 minterms.

A closer look at the entries of the u_i column of Table 9.12 reveals that the recoding table does not have a symmetrical complementary relationship between the entries that produce an output of 1 or $\bar{1}$. Thus, two symmetrically recoded MSD (SRMSD) schemes (columns of v_i and w_i of Table 9.12) were proposed [139]. The entries as well as their corresponding outputs are exactly digit-by-digit complement to each other. The logically minimized minterms are listed in Table 9.13, which shows that both SRMSD schemes requires only 6 reduced minterms. The addition table is the same as as the one shown in [136]. Therefore, the two-step SRMSD carry-free addition requires a total of 12 minterms.

Recently, the two-step higher-order MSD addition has been studied [140]. In the first step, the two-digit addend $a_i a_{i-1}$ and augend $b_i b_{i-1}$ as well as the reference digits $a_{i-1} b_{i-2}$ output a two-digit intermediate sum $s_i s_{i-1}$ and an intermediate carry to the $(i + 1)$th position. The number of minterms for s_i, s_{i-1}, and c_i is 38 (19 minterms for 1 and 19 minterms for $\bar{1}$), 16 (8 + 8), and 22 (11 + 11), respectively.

9.4.2.1.4. *One-Step MSD Addition/Subtraction*

As seen in the three-step arithmetic, the ith output sum digit depends on three adjacent digit pairs; i.e., a_i, b_i, a_{i-1}, b_{i-1}, a_{i-2}, and b_{i-2}. Mirsalehi and Gaylord [141] developed a truth table with the above variables as the inputs and s_i as the output. Since each variable has three possible values ($\bar{1}$, 0, and 1), the table contains $3^6 = 729$ entries. There are 183 input patterns that produce an output 1, and 183 patterns that produce an output $\bar{1}$. By logical minimization, the number of patterns for each case is reduced from 183 to 28. Thus, the output digit can be obtained by comparing the input patterns with a total of 56 reduced minterms. Note that a reference pattern that produces a $\bar{1}$ output is a digit-by-digit complement of the corresponding minterm producing a 1. The reduced minterms for generating the output 1 are shown in Table 9.14.

Recently, redundant bit representation was used as a mathematical description of the possible digit combinations [142]. Based on this representation and the two-step arithmetic, a single-step algorithm was derived which requires 14 and 20 minterms for generating the 1 and $\bar{1}$ outputs, respectively. Furthermore, by classifying the three neighboring digit pairs into 10 groups, another one-step algorithm was proposed where it is critical to determine the appropriate group to which the digit pairs belong. Another one-step addition algorithm [143] was investigated in which the operands are in binary while the result is obtained in MSD. In this case, 16 terms are necessary which can be reduced to 12 minterms. This special case [143] is of limited use since MSD–to-binary conversion is needed for addition of multiple numbers.

9.4.2.2. *Digit-Set–Restricted MSD Addition/Subtraction*

To derive a parallel MSD algorithm, it must be guaranteed that at each digit position, the two digit combinations of (1, 1) and ($\bar{1}$, $\bar{1}$) never occur. This can be also realized by limiting the intermediate carry and sum digits to the sets $\{\bar{1}, 0\}$ and $\{0, 1\}$, respectively [144]. However, according to Eq. (9.25), in this case the sum $x_i + y_i$ can contain the values from the set $\{-2, -1, 0, 1, 2\}$ while the result of operation $2c_{i-1} + s_i$ can contain the values from the set

Table 9.14

Reduced Minterms that Produce a 1 in the Output Digit of the MSD Adder [141]

$d_{\bar{1}1}1dd_{\bar{1}1}1d$, $01d01d$, $d_{\bar{1}1}d_{01}1d_{\bar{1}1}1d$, $0d_{01}d_{01}01d_{01}$, $\bar{1}\bar{1}$ $d01d$, $0d_{01}101d$

$d_{\bar{1}1}1dd_{\bar{1}1}d_{01}1$ $d_{\bar{1}1}\bar{1}d_{01}001$ $\bar{1}0d00d$ $01d0d1$ $01d_{01}0d_{01}d_{01}$ $d_{\bar{1}1}\bar{1}100d_{01}$

$\bar{1}1d0\bar{1}d$ $0110d_{01}d$ $0\bar{1}d_{01}d_{\bar{1}1}01$ $d_{\bar{1}1}0d_{01}0\bar{1}1$ $0\bar{1}d\bar{1}1d$ $0d_{01}d011$

$0\bar{1}1d_{\bar{1}1}0d_{01}$ $d_{\bar{1}1}010\bar{1}d_{01}$ $01d\bar{1}\bar{1}d$ $d_{\bar{1}1}11d_{\bar{1}1}d_{01}d$ $00d_{01}d_{\bar{1}1}\bar{1}1$

$d_{\bar{1}1}d_{01}d_{01}d_{\bar{1}1}1d_{01}$ $00d\bar{1}0d$ $d_{\bar{1}1}d_{01}dd_{\bar{1}1}11$ $001d_{\bar{1}1}\bar{1}d_{01}$ $d_{\bar{1}1}1d_{01}d_{\bar{1}1}d_{01}d_{01}$

$\{-2, -1, 0, 1\}$. Obviously, the operation $2c_{i+1} + s_i$ cannot produce the value 2. Some modifications are thus necessary to be made in Eq. (9.25). In order to ensure that each formed intermediate carry digit is nonpositive and each formed intermediate sum digit is nonnegative, a reference digit r_i is employed to transfer quantities from one digit position of the input operand digits to another position to keep the identity. For any digit position i, there exists an output value, which is equal to $2r_{i+1}$, to the next higher order digit position and an incoming value, which is equal to r_i, from the next lower order digit position. The sum of the input operand digits at the ith position can be represented in terms of the reference digit values mentioned above and the values of the intermediate carry and sum digits. So a new algebraic equation is written as

$$x_i + y_i = 2(c_{i+1} + r_{i+1}) + s_i - r_i. \tag{9.36}$$

The reference digit r_i may be either nonrestricted, belonging to the digit set $\{\bar{1}, 0, 1\}$, or restricted to a smaller set $\{0, 1\}$. The addition algorithm based on this scheme consists of three steps for generating the reference digits, the intermediate sum and carry, and the final sum, respectively. The three-step algorithm with nonrestricted reference digits is developed in Sec. 9.3.2.2.1 and that with digit-set–restricted reference digits in Sec. 9.3.2.2.2. Through the proposed encoding scheme, the algorithms can be completed within two steps, presented in Sec. 9.3.2.2.3.

9.4.2.2.1. Algorithms with Nonrestricted Reference Digits

The first step is to generate the reference digits, which are of the set $\{\bar{1}, 0, 1\}$. The reference digit at the ith position depends only upon the addend and augend digits at the $(i-1)$th digit position. It should be mentioned that for a fixed input digit pair (x_i, y_i), the reference digit r_{i+1} has a unique value, otherwise there maybe no solutions for c_{i+1} and s_i to satisfy their digit-set constraints. For example, if (x_i, y_i) is $(1, 1)$, then the reference digit r_{i+1} must be 1. Provided that r_{i+1} and r_i are $\bar{1}$ and 0, respectively, from Eq. (4.36), we can deduce that c_{i+1} and s_i satisfy the equation $2c_{i+1} + s_i = 4$. There are no solutions for c_{i+1} and s_i since c_{i+1} is of the set $\{\bar{1}, 0\}$ and s_i is of the set $\{0, 1\}$. If r_{i+1} and r_i are 0 and 1, respectively, the same contradiction will occur. Hence, the reference digit r_{i+1} based on the digit pair $(1, 1)$ cannot bc $\bar{1}$ or 0. Similarly, the reference digits generated by other fixed-input digit pairs also have the corresponding unique values. Table 9.15 shows the computation rules for the reference digit r_{i+1} based on the operand digit pair (x_i, y_i). The addend and augend digits are classified into six groups since they can be interchanged. After this step, a reference word has been obtained.

Table 9.15

Truth Table for the Nonrestricted Reference Digits in the Three-Step Digit-Set–Restricted MSD Addition [144]

x_i	y_i	r_{i+1}
$\bar{1}$	$\bar{1}$	$\bar{1}$
$\bar{1}$	0	0
0	$\bar{1}$	
0	0	0
1	$\bar{1}$	0
$\bar{1}$	1	
1	0	1
0	1	
1	1	1

The second step is to generate the digit-set–restricted intermediate carry and sum digits based on the reference digits. At each digit position, according to the digit pair (x_i, y_i) and the determined reference digits r_{i+1} and r_i, the unique solutions of c_{i+1} and s_i will be calculated. For example, if the pair is $(\bar{1}, \bar{1})$, r_{i+1} and r_i are $\bar{1}$ and $\bar{1}$, respectively, we can produce an equation: $2c_{i+1} + s_i = -1$ from Eq. (9.36). So c_{i+1} and s_i must be $\bar{1}$ and 1, respectively, to satisfy this equation. Actually, the formation of the intermediate carry and sum is dependent on two consecutive input digit pairs. Table 9.16 is the truth table for this step.

The last step is to generate the final result by adding the intermediate carry and sum words digit-by-digit in parallel. The principle is proven below. The summation of two N-digit MSD numbers X and Y can be expressed as

$$X + Y = \sum_{i=0}^{N-1} 2^i(x_i + y_i). \tag{9.37}$$

A substitution of Eq. (9.36) into Eq. (9.37) leads to

$$\begin{aligned} X + Y &= \sum_{i=0}^{N-1} 2^i[2(c_{i+1} + r_{i+1}) + s_i - r_i] \\ &= \sum_{i=0}^{N-1} c_{i+1}2^{i+1} + \sum_{i=0}^{N-1} s_i 2^i + \sum_{i=0}^{N-1} r_{i+1}2^{i+1} - \sum_{i=0}^{N-1} r_i 2^i. \end{aligned} \tag{9.38}$$

The first term of Eq. (9.38), which is equivalent to $c_N \ldots c_2 c_1 0$, can be rewritten

Table 9.16

Truth Table for the Three-Step Digit-Set–Restricted MSD Addition Based on the Nonrestricted Reference Digits: the Second Step Rules [144]

x_i	y_i	r_i	c_{i+1}	s_i
$\bar{1}$	$\bar{1}$	$\bar{1}$	$\bar{1}$	1
		0	0	0
		1	0	1
$\bar{1}$	0	$\bar{1}$	$\bar{1}$	0
or		0	$\bar{1}$	1
0	$\bar{1}$	1	0	0
0	0	$\bar{1}$	$\bar{1}$	1
		0	0	0
		1	0	1
1	$\bar{1}$	$\bar{1}$	$\bar{1}$	1
or		0	0	0
$\bar{1}$	1	1	0	1
1	0	$\bar{1}$	$\bar{1}$	0
or		0	$\bar{1}$	1
0	1	1	0	0
1	1	$\bar{1}$	$\bar{1}$	1
		0	0	0
		1	0	1

as $\sum_{i=0}^{N} 2^i c_i$ since $c_0 = 0$. The second term $\sum_{i=0}^{N-1} 2^i s_i$ can be expressed as

$$\sum_{i=0}^{N-1} 2^i s_i = \sum_{i=0}^{N} 2^i s_i - 2^N s_N. \tag{9.39}$$

The third term of Eq. (9.38) is equivalent to the word $r_N \ldots r_2 r_1 0$ and the fourth term is equivalent to the word $r_{N-1} \ldots r_2 r_1 r_0$, where $r_0 = 0$. Therefore, the difference of the third and fourth terms is

$$\sum_{i=0}^{N-1} 2^{i+1} r_{i+1} - \sum_{i=0}^{N-1} 2^i r_i = 2^N r_N. \tag{9.40}$$

As a result, Eq. (9.38) becomes

$$X + Y = \sum_{i=0}^{N} 2^i (c_i + s_i) + 2^N (r_N - s_N). \tag{9.41}$$

Note that $x_N = 0$ and $y_N = 0$. The reference digit r_N has three possible states;

$\bar{1}$, 0, and 1. In the following, we discuss these three cases based on the computation rule for the input digit pair (0,0) in Table 9.16.

Case I — $r_N = \bar{1}$: there exist $c_{N+1} = \bar{1}$, $s_N = 1$. The second term of Eq. (9.41) is written as

$$\begin{aligned} 2^N(r_N - s_N) &= 2^N(\bar{1} - 1) \\ &= 2^{N+1}c_{N+1}. \end{aligned} \tag{9.42}$$

Case II — $r_N = 0$: there exist $c_{N+1} = 0$, $s_N = 0$. The second term of Eq. (9.41) is written as

$$2^N(r_N - s_N) = 2^{N+1}c_{N+1}. \tag{9.43}$$

Case III — $r_N = 1$: there exist $c_{N+1} = 0$, $s_N = 1$. The second term of Eq. (9.41) is written as

$$2^N(r_N - s_N) = 2^{N+1}c_{N+1}. \tag{9.44}$$

Consequently, Eq. (9.14) can be simplified as

$$\begin{aligned} X + Y &= \sum_{i=0}^{N} 2^i(c_i + s_i) + 2^{N+1}c_{N+1} \\ &= \sum_{i=0}^{N+1} 2^i c_i + \sum_{i=0}^{N+1} 2^i s_i, \end{aligned} \tag{9.45}$$

where $s_{N+1} = 0$. The first and second terms represent the $(N + 2)$–bit intermediate carry word C and sum word S, respectively. That is, Eq. (9.45) can be rewritten as

$$X + Y = C + S. \tag{9.46}$$

From Eq. (9.46), we conclude that the sum of two N-digit MSD words is equal to the sum of the $(N + 2)$–bit intermediate carry and sum words, regardless of the reference word.

Figure 9.19 is the diagram of the three-step MSD adder based on this scheme. Using the Karnaugh map or the Quine-McCluskey method, the corresponding reduced-logic minterms in each step are summarized in Table 9.17. The first two steps can be combined into a single step. An example is shown below for addition of two 8-digit MSD numbers. It is necessary to pad one zero preceding the most significant digit of the operand digits, as shown

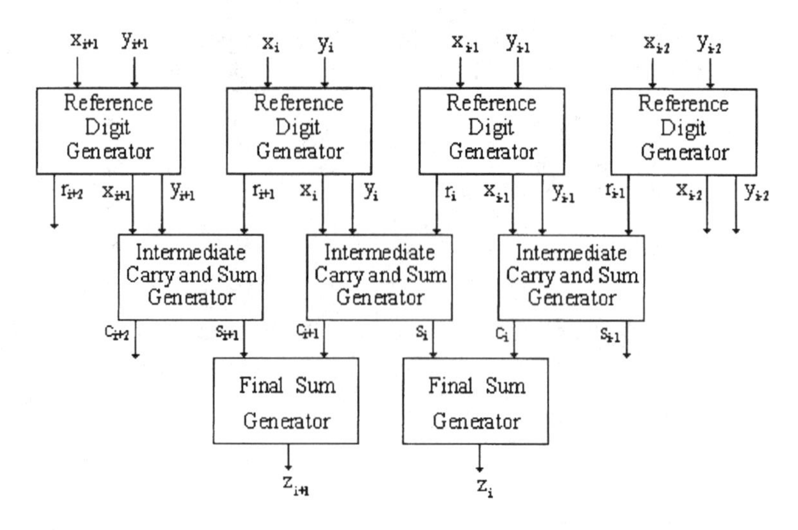

Fig. 9.19. Diagram of the three-step digit-set–restricted MSD adder/subtractor [144].

in the following example:

$$
\begin{aligned}
&(X)\ \phi\bar{1}01\bar{1}\bar{1}011 = (-117)_{10}\\
&(Y)\ \phi1\bar{1}001\bar{1}01 = (69)_{10}\\
\text{step 1:}\ &(R)\ 00100011\phi\\
\text{step 2:}\ &(C)\ 000\bar{1}\bar{1}0000\phi\\
&(S)\ \phi000110000\\
\text{step 3:}\ &(Z)\ 000\bar{1}010000 = (-48)_{10}.
\end{aligned}
$$

Here the symbol ϕ denotes a padded zero.

Table 9.17

Reduced Truth Table for the Three-Step Digit-Set–Restricted MSD Addition with Nonrestricted Reference Digits [144]

Step order	Output function	Output literal	Input function	Minterms
Step 1	r_{i+1}	$\bar{1}$	$x_i y_i$	$\bar{1}\bar{1}$
		1		$1d_{01}$, 01
Step 2	c_{i+1}	$\bar{1}$	$x_i y_i r_i$	$d_{\bar{1}1}00$, $0d_{\bar{1}1}0$, $dd\bar{1}$
	s_i	1		$d_{\bar{1}1}00$, $0d_{\bar{1}1}0$, $00d_{\bar{1}1}$, $d_{\bar{1}1}d_{\bar{1}1}d_{\bar{1}1}$
Step 3	z_i	$\bar{1}$	$c_i s_i$	$\bar{1}0$
		1		01

The key feature of this algorithm lies in the utilization of the reference digits. Their function is to determine the values transferred from one digit position of the operand words to another digit position so that the digits of the formed intermediate carry and sum words are all restricted to a smaller set and their sum is equal to the sum of the operand words.

9.4.2.2.2. Algorithms with Digit-Set–Restricted Reference Digits

In the above-mentioned algorithm, the reference digits can contain the value from the set $\{\bar{1}, 0, 1\}$ while the intermediate carry and sum digits are all digit-set restricted. For the purpose of simplifying the algorithm, the reference digits can also be restricted to the digit set $\{0, 1\}$. The merit of this scheme is that each of the intermediate carry and sum digits and the reference digits has only two states; i.e., $\bar{1}$ and 0 or 0 and 1. Therefore, only a single binary bit may be required for their representations independent of their signs. This makes it possible to realize associated operations by binary logic [134]. Tables 9.18 and 9.19 are the truth tables for generating the digit-set–restricted reference digits, intermediate carry, and sum digits, respectively. The reduced minterms for this three-step scheme are listed in Table 9.20. Similarly, the first two steps can be combined, as shown in Table 9.21. Furthermore, the three-step operations can be completed in a single step. The logic minterms for the one-step digit-set–restricted MSD addition are shown in Table 9.22.

Now we discuss Eq. (9.41) again under the condition of digit-set–restricted reference digits. According to the computation rule of digit pair (0, 0) in Table 9.19, r_N and s_N are equal to each other and Eq. (9.41) is thus rewritten as

$$X + Y = \sum_{i=0}^{N} 2^i c_i + \sum_{i=0}^{N} 2^i s_i. \tag{9.47}$$

Table 9.18

Truth table for the Three-Step MSD Digit-Set–Restricted Addition Based on the Digit-Set–Restricted Reference Digits: the First Step Rules [144]

x_i	y_i	r_{i+1}
$\bar{1}$	$\bar{1}$	0
$\bar{1}$	0	0
0	$\bar{1}$	
0	0	0
1	$\bar{1}$	1
$\bar{1}$	1	
1	0	1
0	1	
1	1	1

Table 9.19

Truth Table for the Three-Step Digit-Set–Restricted MSD Addition Based on the Digit-Set–Restricted Reference Digits: the Second Step Rules [144]

x_i	y_i	r_i	c_{i+1}	s_i
$\bar{1}$	$\bar{1}$	0	$\bar{1}$	0
		1	$\bar{1}$	1
$\bar{1}$	0	0	$\bar{1}$	1
0	$\bar{1}$	1	0	0
0	0	0	0	0
		1	0	1
1	$\bar{1}$	0	$\bar{1}$	0
$\bar{1}$	1	1	$\bar{1}$	1
1	0	0	$\bar{1}$	1
0	1	1	0	0
1	1	0	0	0
		1	0	1

Table 9.20

Reduced Truth Table for the Three-Step Digit-Set–Restricted MSD Addition with Digit-Set–Restricted Reference Digits [144]

Step order	Output function	Output literal	Input function	Minterms
Step 1	r_{i+1}	1	$x_i y_i$	1d, d1
Step 2	c_{i+1}	$\bar{1}$	$x_i y_i r_i$	$0d_{\bar{1}1}0$, $d_{\bar{1}1}00$, $\bar{1}d_{\bar{1}1}d_{01}$, $1\bar{1}d_{01}$
	s_i	1		$d_{\bar{1}1}00$, $0d_{\bar{1}1}0$, 001, $d_{\bar{1}1}d_{\bar{1}1}1$
Step 3	z_i	$\bar{1}$	$c_i s_i$	$\bar{1}0$
		1		01

Table 9.21

Reduced Truth Table for Digit-Set–Restricted MSD Addition by Combining the First Two Steps of Table 9.20 into One [144]

Output literal	Minterms $x_i y_i x_{i-1} y_{i-1}$
$c_{i+1} = \bar{1}$	$d_{\bar{1}1}\bar{1}dd$, $\bar{1}d_{\bar{1}1}dd$, $d_{\bar{1}1}0d_{\bar{1}0}d_{\bar{1}0}$, $0d_{\bar{1}1}d_{\bar{1}0}d_{\bar{1}0}$
$s_i = 1$	001d, 00d1, $d_{\bar{1}1}d_{\bar{1}1}1d$, $d_{\bar{1}1}d_{\bar{1}1}d1$, $d_{\bar{1}1}0d_{\bar{1}0}d_{\bar{1}0}$, $0d_{\bar{1}1}d_{\bar{1}0}d_{\bar{1}0}$

Table 9.22

Reduced Truth Table for the One-Step Digit-Set–Restricted MSD Addition with Digit-Set–Restricted Reference Digits

Output literal	Minterms $x_i y_i x_{i-1} y_{i-1} x_{i-2} y_{i-2}$
$z_i = \bar{1}$	$00\bar{1}\bar{1}dd$, $d_{\bar{1}1}d_{\bar{1}1}\bar{1}\bar{1}dd$, $00\bar{1}0d_{\bar{1}0}d_{\bar{1}0}$, $000\bar{1}d_{\bar{1}0}d_{\bar{1}0}$, $d_{\bar{1}1}d_{\bar{1}1}\bar{1}0d_{\bar{1}0}d_{\bar{1}0}$, $d_{\bar{1}1}d_{\bar{1}1}0\bar{1}d_{\bar{1}0}d_{\bar{1}0}$, $d_{\bar{1}1}01\bar{1}dd$, $d_{\bar{1}1}0\bar{1}1dd$, $0d_{\bar{1}1}1\bar{1}dd$, $0d_{\bar{1}1}\bar{1}1dd$, $d_{\bar{1}1}010d_{\bar{1}0}d_{\bar{1}0}$, $d_{\bar{1}1}001d_{\bar{1}0}d_{\bar{1}0}$, $0d_{\bar{1}1}10d_{\bar{1}0}d_{\bar{1}0}$, $0d_{\bar{1}1}01d_{\bar{1}0}d_{\bar{1}0}$
$z_i = 1$	$00101d$, $0010d1$, $00011d$, $0001d1$, $d_{\bar{1}1}d_{\bar{1}1}101d$, $d_{\bar{1}1}d_{\bar{1}1}10d1$, $d_{\bar{1}1}d_{\bar{1}1}011d$, $d_{\bar{1}1}d_{\bar{1}1}01d1$, $0011dd$, $d_{\bar{1}1}d_{\bar{1}1}11dd$, $d_{\bar{1}1}0\bar{1}01d$, $d_{\bar{1}1}00\bar{1}1d$, $d_{\bar{1}1}0\bar{1}0d1$, $d_{\bar{1}1}00\bar{1}d1$, $0d_{\bar{1}1}\bar{1}01d$, $0d_{\bar{1}1}0\bar{1}1d$, $0d_{\bar{1}1}\bar{1}0d1$, $0d_{\bar{1}1}0\bar{1}d1$, $d_{\bar{1}1}000dd$, $0d_{\bar{1}1}00dd$

Equation (9.47) implies that the summation of two N-digit MSD numbers X and Y is equal to the summation of $(N+1)$–bit intermediate carry word C and sum word S, which is unlike the algorithm with the nonrestricted reference digits mentioned above. As an example, let us consider the three-step addition of the following two 8-digit MSD numbers.

$$
\begin{aligned}
(X)\ & \phi 101\bar{1}\bar{1}\bar{1}01 = (133)_{10} \\
(Y)\ & \phi 1\bar{1}\bar{1}01\bar{1}11 = (39)_{10} \\
\text{step 1: } (R)\ & 10101011\phi \\
\text{step 2: } (C)\ & 00\bar{1}0\bar{1}\bar{1}00\phi \\
(S)\ & 100000100 \\
\text{step 3: } (Z)\ & 10\bar{1}0\bar{1}\bar{1}100 = (172)_{10}.
\end{aligned}
$$

Table 9.18 shows that the digit pairs $(\bar{1}, \bar{1})$, $(\bar{1}, 0)$, $(0, \bar{1})$, and $(0, 0)$ determine the reference digits valued 0 while the pairs $(1, \bar{1})$, $(\bar{1}, 1)$, $(1, 0)$, $(0, 1)$, and $(1, 1)$ determine the reference digits valued 1. From Table 9.19, we notice that some digit pairs; e.g., pairs $(0, 0)$ and $(1, 1)$, generate the identical outputs. It is possible and important to design an appropriate encoding scheme so that the algorithm can be simplified and implemented by binary logic functions. Three binary bits (u_i, v_i, w_i) can be used to code the input digit pairs (x_i, y_i), as shown in Table 9.23. The digit combinations generating the same intermediate carry and sum in terms of determined reference digit have the same u_i and v_i. For instance, the (u_i, v_i) of the combinations $(\bar{1}, \bar{1})$, $(1, \bar{1})$, and $(\bar{1}, 1)$ are all $(1, 0)$. The third encoding digit w_i of one digit pair is equal to the reference digit generated by this pair as described below:

$$r_{i+1} = w_i. \tag{9.48}$$

Table 9.23

Encoding Scheme and Computation Rules to be Used for the Digit-Set–Restricted MSD Addition by Logic Operation [144]

x_i	y_i	u_i	v_i	w_i	r_{i+1}	r_i	c_{i+1}	s_i
$\bar{1}$	$\bar{1}$	1	0	0	0	0	$\bar{1}$	0
						1	$\bar{1}$	1
$\bar{1}$	0	0	0	0	0	0	$\bar{1}$	1
0	$\bar{1}$					1	0	0
0	0	0	0	0	0	0	0	0
						1	0	1
1	$\bar{1}$	1	0	1	1	0	$\bar{1}$	0
$\bar{1}$	1					1	$\bar{1}$	1
1	0	0	1	1	1	0	$\bar{1}$	1
0	1					1	0	0
1	1	0	0	1	1	0	0	0
						1	0	1

So the encoding contains the information of the input digit combinations and the corresponding reference digits. From Table 9.23, by using a Karnaugh map to minimize the logical minterms for the nonzero outputs, the binary logic expressions for the intermediate carry c_{i+1} and sum s_i can be obtained:

$$c_{i+1} = u_i + v_i\overline{r_i}, \tag{9.49}$$

$$s_i = \overline{v_i}r_i + v_i\overline{r_i}. \tag{9.50}$$

It should be mentioned that for the intermediate carry digits, which are restricted to the set $\{\bar{1}, 0\}$, the true logic denotes negative digit $\bar{1}$. Inserting Eq. (9.48) in Eqs. (9.49), and (9.50) results in

$$c_{i+1} = u_i + v_i\bar{w}_{i-1}, \tag{9.51}$$

$$s_i = \overline{v_i}w_{i-1} + v_i\bar{w}_{i-1}. \tag{9.52}$$

The first and the second steps are thus combined into a single step since from the encoding of the input operand pairs, the intermediate carry and sum digits can be obtained directly through binary logic operations on the inputs. We employ variable Z^+ to denote all of the occurrences of positive digit 1 in the final sum string Z by a 1 symbol and Z^- to denote all of the occurrences of negative digit $\bar{1}$ in Z by a 1 symbol. The final result Z can thus be yielded by combination of Z^+ and Z^-. The binary logic expressions for z_i^+ and z_i^- are

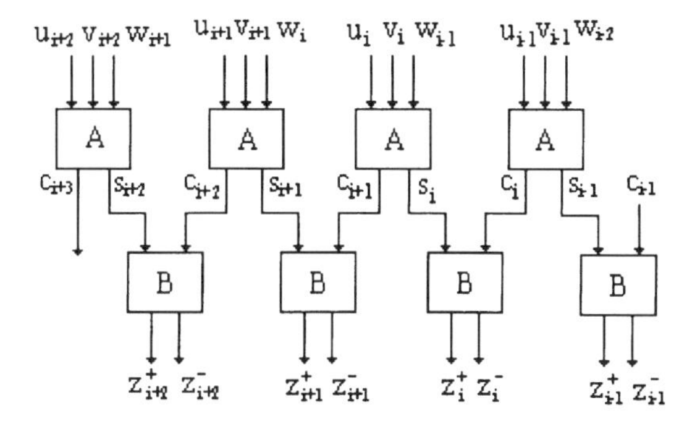

Fig. 9.20. Tree structure for the two-step digit-set–restricted MSD addition/subtraction [144]. The functional block A represents the computation rules generated from Eqs. (9.51) and (9.52) while the functional block B represents the computation rules generated from Eqs. (9.53) and (9.54).

listed as follows:

$$z_i^- = c_i \overline{s_i}, \tag{9.53}$$

$$z_i^+ = s_i \overline{c_i}. \tag{9.54}$$

Figure 9.20 shows the tree structure for the two-step digit-set–restricted MSD addition. The functional block A represents the computation rules generated from Eqs. (9.51) and (9.52), while the functional block B represents the computation rules generated from Eqs. (9.53) and (9.54). An example showing the application of the aforementioned technique is illustrated below.

$$
\begin{array}{lll}
 & (X1) & \phi 11\bar{1}0\bar{1}100 = (156)_{10} \\
 & (Y1) & \phi 0100\bar{1}\bar{1}1\bar{1} = (53)_{10} \\
 & (U) & 000001100 \\
 & (V) & 010100011 \\
 & (W) & 011000110 \\
 & & \\
\text{step 1:} & (C) & 00101111\phi \\
 & (S) & 100101111 \\
 & & \\
\text{step 2:} & (Z^-) & 001010000 \\
 & (Z^+) & 100100001 \\
 & & \\
 & (Z): & 10\bar{1}1\bar{1}0001 = (209)_{10}
\end{array}
$$

9.4.2.3. *Multiplication*

Multiplication of two N-digit MSD numbers $A = a_{N-1}a_{N-2}\ldots a_0$ and $B = b_{N-1}b_{N-2}\ldots b_0$ generates a $2N$–digit product $P = p_{2N-1}p_{2N-2}\ldots p_0$. The procedure consists of two steps: generate the partial products $P^{(i)}$ and add them [123, 126]. The product is formed as

$$P = AB = \sum_{i=0}^{N-1} P^{(i)} = \sum_{i=0}^{N-1} Ab_i 2^i. \tag{9.55}$$

Each partial product $P^{(i)}$ is formulated by multiplying the multiplicand A by the ith digit b_i of the multiplier B and shifting by i digit positions, which corresponds to the weight 2^i. The partial product is a shifted version of $\bar{A}$, or 0, or A, depending on whether the value of b_i is $\bar{1}$, or 0, or 1. The traditional multiplication algorithm checks each digit of the multiplier B from the least significant digit b_0 to the most significant digit b_{N-1}. To speed up the multiplication, the following two measures must be pursued.

- Generate all partial products in parallel. This can be accomplished via symbolic substitution, truth-table look-up, and outer-product, crossbar, and perfect-shuffle interconnection networks.
- Add all partial products pairwise by means of an adder tree. With a total of N partial products at the leaves of the tree, the summation process takes $\log_2 N$ steps. At each step j, we perform $N/2^j$ MSD additions in parallel:

$$P_{i,j} = P_{2i-2,j-1} + P_{2i-1,j-1}, \quad \text{for } i = 1, 2, \ldots, N/2^j, j = 1, 2, \ldots, \log_2 N. \tag{9.56}$$

The final product is produced at the root of the binary tree. The partial products are generated in constant time while the addition of partial products using the tree structure requires $\log_2 N$ iterations. Therefore, the multiplication of two N-digit MSD numbers can be carried out in $O(\log_2 N)$ time. The following numerical example illustrates the multiplication process between the numbers $A = 10\bar{1}\bar{1} = 5_{10}$ and $B = 10\bar{1}1 = 7_{10}$.

Step 1: Generation of the partial products

$$p^{(0)}\text{: Ab}_0 \times 2^0 = 00010\bar{1}\bar{1}$$

$$p^{(1)}\text{: Ab}_1 \times 2^1 = 00\bar{1}0110$$

$$p^{(2)}\text{: Ab}_2 \times 2^2 = 0000000$$

$$p^{(3)}\text{: Ab}_3 \times 2^3 = 10\bar{1}\bar{1}000$$

Step 2: Summation of the partial products

$$\left.\begin{array}{l}\left.\begin{array}{l}00010\bar{1}\bar{1}\\00\bar{1}0110\end{array}\right\rangle \rightarrow 00000\bar{1}\bar{1}1\\[2ex]\left.\begin{array}{l}0000000\\10\bar{1}\bar{1}000\end{array}\right\rangle \rightarrow 1\bar{1}\bar{1}01000\end{array}\right\rangle \rightarrow 0010010\bar{1} = 35_{10}.$$

9.4.2.4. *Division*

Conventional storing and nonstoring division methods require knowledge of the sign of the partial remainder for exact selection of the quotient digits. However, in MSD the sign of a partial remainder is not always readily available. Therefore, we must seek an effective division algorithm which can overcome this difficulty and use the parallel addition and multiplication arithmetic operations developed in the previous sections. To this end, two different approaches have been proposed: the convergence approach [123] and the quotient-selected approach [145].

9.4.2.4.1. Convergence Division

A parallel MSD division algorithm which meets the aforementioned requirements is based on a convergence approach [123]. Consider a dividend X and a divisor D in normalized form, $1/2 \leqslant |X| < |D| < 1$. We want to compute the quotient $Q = X/D$ without a remainder. The algorithm utilizes a sequence of multiply factors $m_0, m_1, \ldots, m_n$ so that $D \times (\Pi_{i=0}^{i=n} m_i)$ converges to 1 within an acceptable error criterion. Initially, we set $X_0 = X$ and $D_0 = D$. The algorithm repeats the following recursions:

$$X_{i+1} = X_i \times m_i, \qquad D_{i+1} = D_i \times m_i, \tag{9.57}$$

so that for all n

$$D \times \left(\prod_{i=0}^{n} m_i\right) \rightarrow 1, \qquad Q = X \times \left(\prod_{i=0}^{n} m_i\right). \tag{9.58}$$

The effectiveness of this convergence method relies on the ease of computing the multiply factors m_i using only the MSD addition and multiplication. It is found that for D_{i+1} to converge quadratically to 1, the factors m_i should be chosen according to the following equation:

$$m_i = 2 - D_i, 0 < D_i < 2. \tag{9.59}$$

It is just the 2's complement of the previous denominator D_i. In the MSD system, the subtraction can be computed in constant time by complement and addition. Since the convergence is quadratic, the accumulated denominator length is doubled after each iteration. Thus, for a desired quotient of length n, the maximum number of iterations needed is $\log_2 n$. In each iteration, three operations are involved; i.e., two MSD multiplication and an MSD subtraction. If the binary-tree architecture is employed to perform a multiplication, a total of $\log_{2n} n$ steps are required and up to $m/2$ additions are needed in each step.

9.4.2.4.2. *Quotient-Selected Division*

MSD division based on the conventional add/subtract-and-shift-left principle is realized by shifting the partial remainder, generating a quotient digit, and then performing an addition or subtraction operation to produce the next partial remainder. It generates one quotient digit per iteration from the most significant digit toward the least significant digit. The number of required iterations is n. The conventional algorithm requires knowledge of the sign of the partial remainder for selecting a quotient digit, and it is necessary to check all digits of the partial remainder. Recently a novel quotient-selected MSD division algorithm [145] has been proposed in which we only need to know the range rather than the exact value of the partial remainder. The selection function is realized by checking only several of its most significant digits.

The algorithm based on the add/subtract-and-shift-left principle to compute the division X/D relies on the following equations:

$$w_i = 2w_{i-1} - q_i D, \tag{9.60}$$

$$Q_i = \sum_{i=0}^{n} 2^{-i} q_i, \qquad Q_0 = 0, \tag{9.61}$$

where i is the iteration step, w_i is the partial remainder at step i having the form of $w_{i,0} \cdot w_{i,1} \cdot w_{i,2} \ldots w_{i,n-1}$, q_i is the quotient digit, and Q_i is the value of the quotient after the ith iteration.

In any signed-digit–based division scheme with quotient digits selected from the set $\{-s, \ldots, 0, \ldots, s\}$, the range of the partial remainder in the $(i-1)$th iteration is $-\eta D < w_{i-1} < \eta D$, where $\eta = s/(r-1)$ is called the index of redundancy. For MSD numbers with radix $r = 2$, the quotient digits belong to the set $\{\bar{1}, 0, 1\}$ and thus $\eta = 1$. So the partial remainder w_{i-1} should be restricted to the range $(-D, D)$ and the range of $2w_{i-1}$ is $(-2D, 2D)$. To guarantee $w_i \in (-D, D)$, the rules determining the three possible values of q_i are listed as follows:

$$\begin{aligned} &\text{if } -2D < 2w_{i-1} < 0, && \text{then } q_i = \bar{1} \\ &\text{if } -D < 2w_{i-1} < D, && \text{then } q_i = 0 \\ &\text{if } 0 < 2w_{i-1} < 2D, && \text{then } q_i = 1. \end{aligned} \tag{9.62}$$

From Eq. (9.62), one can see that there exists one overlap region $(-D, 0)$ for the ranges of $2w_{i-1}$ for $q_i = \bar{1}$ and $q_i = 0$, and another overlap region $(0, D)$ for $q_i = 0$ and $q_i = 1$. In these overlap regions, selection of either of the quotient digits is valid due to the redundant nature of $2w_{i-1}$. Figure 9.21 plots the relationship of the shifted partial remainder $2w_{i-1}$ versus the divisor D. It clearly shows the change of the range that $2w_{i-1}$ can represent with the normalized D varied from 1/2 to 1. The shaded regions 1 and 2 are the overlap regions where more than one quotient digit may be selected. If we can find a constant value C_1 (C_2), for all the values of $D \in [1/2, 1)$, within the overlap region, then $2w_{i-1}$ in any iteration can be compared with this constant value to select either $q_i = 1(0)$ if $2w_{i-1} > C_1(C_2)$, or $q_i = 0(\bar{1})$ otherwise. Such a constant should be independent of the value of the divisor D so that the quotient selection will only depend on $2w_{i-1}$ in any iteration. From Fig. 9.21, one can observe that C_1 is independent of D only when it lies in the range

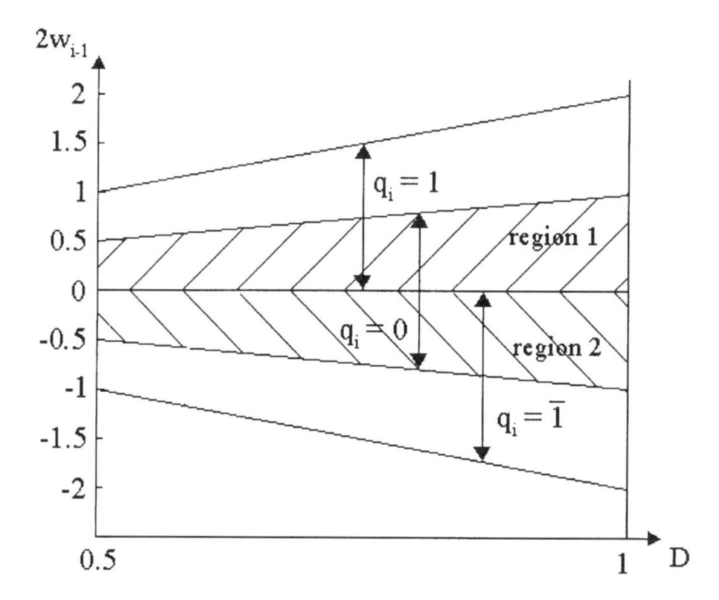

Fig. 9.21. Shifted partial remainder versus divisor plot for quotient-selected MSD division [145].

(0, 0.5]. Similarly, the constant C_2 should be in the range [−0.5, 0). Therefore a number of values in the above ranges can be selected for C_1 and C_2. Note that the constants should also result in a simple quotient selection function. To compare $2w_{i-1}$ with C_1 or C_2, one can approximate the result according to several of their most significant digits instead of performing a subtraction operation. With the observation of several most significant digits, the quotient selection is thus independent of the length of the divisor operand. These most significant digits should not only guarantee the correct choice of the quotient digits but also be as few as possible to simplify the physical implementation. Here three most significant digits $w_{i,0} \cdot w_{i,1} w_{i,2}$ of partial remainder are used for quotient selection since they are sufficient to represent a constant value in range (0, 0.5] or [−0.5, 0). For example, $C_1 = (0.10)_{\text{MSD}} = (1.\bar{1}0)_{\text{MSD}} = (0.5)_{10}$ is selected as the reference constant value for selection of $q_i = 0$ and $q_i = 1$, while constant value $C_2 = (0.\bar{1}0)_{\text{MSD}} = (\bar{1}.10)_{\text{MSD}} = (-0.5)_{10}$ is selected for $q_i = 0$ and $q_i = \bar{1}$. If $w_{i,0} w_{i,1} w_{i,2}$ is equal to or greater than $(0.10)_{\text{MSD}}$ or $(1.\bar{1}0)_{\text{MSD}}$, then the quotient digit is set to 1; if $w_{i,0} w_{i,1} w_{i,2}$ is equal to or lower than $(0.\bar{1}0)_{\text{MSD}}$ or $(\bar{1}.10)_{\text{MSD}}$, then the quotient digit is set to −1; otherwise, the quotient digit is set to 0. The selection rules are listed in Table 9.24. As an example, we compute the MSD division with $X = (0.1011\bar{1}01)_{\text{MSD}} = (0.6640625)_{10}$ and $D = (0.1110\bar{1}\bar{1}1)_{\text{MSD}} = (0.8359375)_{10}$. The partial remainder and the quotient digit in each iteration are shown in Table 9.25. After eight iterations, the division results are generated with $Q = (0.1100110)_{\text{MSD}} = (0.796875)_{10}$.

Using the Karnaugh map or Quine-McCluskey method, the reduced logic minterms for outputs 1 and $\bar{1}$ of the quotient digits can be obtained. We used symbol d_{xy} to imply a partial don't-care for digits x and y and symbol d to imply a complete don't-care for digits $\bar{1}$, 0, and 1. As a result, to determine the

Table 9.24

Truth Table for the Quotient Digit q_i According to $w_{i,0} w_{i,1} w_{i,2}$ of the Partial Remainder [145]

$w_{i,0}$	$w_{i,1}$	$w_{i,2}$	q_{i+1}	$w_{i,0}$	$w_{i,1}$	$w_{i,2}$	q_{i+1}
1	0	1	1	0	0	$\bar{1}$	0
1	0	0	1	0	$\bar{1}$	1	0
1	0	$\bar{1}$	1	0	$\bar{1}$	0	$\bar{1}$
1	$\bar{1}$	1	1	0	$\bar{1}$	$\bar{1}$	$\bar{1}$
1	$\bar{1}$	0	1	$\bar{1}$	1	1	0
1	$\bar{1}$	$\bar{1}$	0	$\bar{1}$	1	0	$\bar{1}$
0	1	1	1	$\bar{1}$	1	$\bar{1}$	$\bar{1}$
0	1	0	1	$\bar{1}$	0	1	$\bar{1}$
0	1	$\bar{1}$	0	$\bar{1}$	0	0	$\bar{1}$
0	0	1	0	$\bar{1}$	0	$\bar{1}$	$\bar{1}$
0	0	0	0				

Table 9.25

Example of the Quotient-Selected MSD Division, [145] where
$X = (0.1011\bar{1}01)_{\mathrm{MSD}} = (0.6640625)_{10}$ and $D = (0.1110\bar{1}\bar{1}1)_{\mathrm{MSD}} = (0.8359375)_{10}$

Iteration	Partial remainder	Quotient digit	Quotient word
1	$w_0 = 0.1011\bar{1}01$	$q_1 = 1$	$Q_1 = 0.1$
2	$w_1 = 2w_0 - D = 0.100000\bar{1}$	$q_2 = 1$	$Q_2 = 0.11$
3	$w_2 = 2w_1 - D = 0.001010\bar{1}$	$q_3 = 0$	$Q_3 = 0.110$
4	$w_3 = 2w_2 = 0.01010\bar{1}0$	$q_4 = 0$	$Q_4 = 0.1100$
5	$w_4 = 2w_3 = 0.1010\bar{1}00$	$q_5 = 1$	$Q_5 = 0.11001$
6	$w_5 = 2w_4 - D = 0.10\bar{1}\bar{1}11\bar{1}$	$q_6 = 1$	$Q_6 = 0.110011$
7	$w_6 = 2w_5 - D = 0.00\bar{1}000\bar{1}$	$q_7 = 0$	$Q_7 = 0.1100110$
8	$w_7 = 2w_6 = 0.0\bar{1}000\bar{1}0$		

quotient digits, one can use three minterms $10d$, $1\bar{1}d_{01}$, and $01d_{01}$ for output 1, and three minterms $\bar{1}0d$, $\bar{1}1d_{0\bar{1}}$, and $0_{\bar{1}}d_{01}$ for output $\bar{1}$.

If we use dual-rail spatial encoding S^*S^{**} to encode an MSD digit, the quotient-digit selection can be done by binary logic operations. The code used here is $S^*S^{**} = 01$ for $1, S^*S^{**} = 00$ for 0, and $S^*S^{**} = 10$ for $\bar{1}$. Therefore, a quotient digit q_i can be obtained from a six-variable string $w^*_{i,0}w^{**}_{i,0}w^*_{i,1}w^{**}_{i,1}w^*_{i,2}w^{**}_{i,2}$, as shown in Table 9.26 for $q_i = 1$ and Table 9.27 for $q_i = \bar{1}$. Using the Karnaugh map, we can derive the minimized logical expressions for outputs 1 and $\bar{1}$ of the quotient digits as shown below.

$$\text{for output 1:}\quad w^{**}_{i,0}\overline{w^*_{i,1}} + w^{**}_{i,0}\overline{w^*_{i,2}} + \overline{w^*_{i,0}}w^{**}_{i,1}\overline{w^*_{i,2}}, \tag{9.63}$$

$$\text{for output } \bar{1}:\quad w^*_{i,0}\overline{w^{**}_{i,1}} + w^*_{i,0}\overline{w^{**}_{i,2}} + \overline{w^{**}_{i,0}}w^*_{i,1}\overline{w^{**}_{i,2}}, \tag{9.64}$$

Table 9.26

Encodings of $w_{i,0}w_{i,1}w_{i,2}$ Generating a Quotient Digit 1 [145]

$w_{i,0}$	$w_{i,1}$	$w_{i,2}$	$w^*_{i,0}$	$w^{**}_{i,0}$	$w^*_{i,1}$	$w^{**}_{i,1}$	$w^*_{i,2}$	$w^{**}_{i,2}$
1	0	1	0	1	0	0	0	1
1	0	0	0	1	0	0	0	0
1	0	$\bar{1}$	0	1	0	0	1	0
1	$\bar{1}$	1	0	1	1	0	0	1
1	$\bar{1}$	0	0	1	1	0	0	0
0	1	1	0	0	0	1	0	1
0	1	0	0	0	0	1	0	0

Table 9.27

Encodings of $w_{i,0}w_{i,1}w_{i,2}$ Generating a Quotient Digit $\bar{1}$

$w_{i,0}$	$w_{i,1}$	$w_{i,2}$	$w^{*}_{i,0}$	$w^{**}_{i,0}$	$w^{*}_{i,1}$	$w^{**}_{i,1}$	$w^{*}_{i,2}$	$w^{**}_{i,2}$
$\bar{1}$	0	$\bar{1}$	1	0	0	0	1	0
$\bar{1}$	0	0	1	0	0	0	0	0
$\bar{1}$	0	1	1	0	0	0	0	1
$\bar{1}$	1	$\bar{1}$	1	0	0	1	1	0
$\bar{1}$	1	0	1	0	0	1	0	0
0	$\bar{1}$	$\bar{1}$	0	0	1	0	1	0
0	$\bar{1}$	0	0	0	1	0	0	0

9.4.3. TSD ARITHMETIC

9.4.3.1. *Addition*

A two-step TSD algorithm [146] was designed by mapping the two digits a_i and b_i into an intermediate sum and an intermediate carry so that the ith intermediate sum and the $(i - 1)$th intermediate carry do not generate a carry. The digits $a_{i-1}b_{i-1}$ from the next lower order position are used as the reference. Both the intermediate sum and the intermediate carry belong to the set $\{\bar{2}, \bar{1}, 0, 1, 2\}$, and a total of 28 minterms are required. Later on, the higher-order TSD operation [147] was developed, which involves the operand digits $a_i a_{i-1}$ and $b_i b_{i-1}$ and the reference digits $a_{i-2}b_{i-2}$ to yield a two-digit intermediate sum $s_i s_{i-1}$ and an intermediate carry. In this technique, the truth table becomes relatively large and more minterms are needed for implementation. Recently, a two-digit trinary full-adder [148] was designed with $a_i a_{i-1} b_i b_{i-1}$ as the inputs. In this technique, the intermediate carry and one of the intermediate sum digits s_{i-1} are restricted to $\{\bar{1}, 0, 1\}$, thus ensuring that the second-step addition is carry free.

In a TSD number system, a carry will be generated for the addition of the digit combinations $\bar{2}\bar{2}$, $\bar{2}\bar{1}$, $\bar{1}\bar{2}$, 12, 21, and 22. For carry-free addition, the aforementioned digit combinations must be eliminated from the augend and the addend. By exploiting the redundancy of the TSD numbers, a recoding truth table [149] has been developed to eliminate the occurrence of these digit combinations. An N-digit TSD number $A = a_{N-1}a_{N-2} \ldots a_1 a_0$ is recoded into an $(N + 1)$–digit TSD number $C = c_N c_{N-1} \ldots c_1 c_0$ such that C and A are numerically equal and the recoded TSD output c_i does not include the 2 and $\bar{2}$ literals. Thus, this recoding operation maps the TSD set $\{\bar{2}, \bar{1}, 0, 1, 2\}$ to a smaller set $\{\bar{1}, 0, 1\}$. This recoding naturally guarantees that the second-step addition of the recoded numbers is carry free. The recoded digit c_i depends on the current TSD value a_i and three lower-order digits $a_{i-1}a_{i-2}a_{i-3}$ and the

Table 9.28

Reduced Minterms for TSD Recoding [149] where the Minterms Generating a $\bar{1}$ Output are Digit-by-Digit Complement of that Generating a 1 Output

Output	Minterms $a_i a_{i-1} a_{i-2} a_{i-3}$
1	$d_{1\bar{2}}\bar{2}2d_{\bar{1}012}$, $d_{1\bar{2}}\bar{1}\bar{1}d_{\bar{1}012}$, $012d_{\bar{1}012}$, $02\bar{1}d_{\bar{1}012}$, $d_{\bar{1}2}\bar{2}1d_{10\bar{1}\bar{2}}$, $d_{1\bar{2}}11d_{10\bar{1}\bar{2}}$, $d_{12}\bar{1}\bar{2}d_{10\bar{1}\bar{2}}$, $d_{1\bar{2}}2\bar{2}d_{10\bar{1}\bar{2}}$, $d_{1\bar{2}}\bar{1}d_{012}d$, $d_{\bar{1}\bar{2}}1d_{0\bar{1}\bar{2}}d$, $d_{\bar{1}2}\bar{2}d_{0\bar{1}\bar{2}}d$, $02d_{012}d$, $d_{1\bar{2}}\bar{2}12$, $d_{1\bar{2}}\bar{1}\bar{2}2$, $d_{1\bar{2}}12\bar{2}$, $d_{1\bar{2}}2\bar{1}\bar{2}$, $d_{\bar{1}2}\bar{2}2\bar{2}$, $d_{\bar{1}2}\bar{1}\bar{1}\bar{2}$, $d_{1\bar{2}}0dd$, $02\bar{2}2$, 0112

recoding operation can be performed at all digit positions in parallel. To generate the 1 and $\bar{1}$ outputs of c_i, 42 four-variable ($a_i a_{i-1} a_{i-2} a_{i-3}$) minterms are required. Notice that the minterms for generating the 1 and $\bar{1}$ outputs are exact complements of each other. Table 9.28 lists the 21 minterms required for generating the 1 output in the recoding step. In the second step, 6 minterms are required for generating the sum output. In the second step, only 2 minterms, 10 and 01 ($\bar{1}0$ and $0\bar{1}$) are required for generating the 1 ($\bar{1}$) output, and only 1 minterm 11 ($\bar{1}\bar{1}$) is needed for generating the 2 ($\bar{2}$) output of s_i. In addition, the minterms for $s_i = 1(2)$ are digit-by-digit complement of the minterms for $s_i = \bar{1}(\bar{2})$. Therefore, the algorithm requires 21 minterms for generating the 1 output in the recoding step, and 3 minterms for generating the 1 and 2 outputs of s_i in the addition step.

The two-step TSD addition [150] was further simplified by restricting the ith intermediate sum s_i and the $(i + 1)$th intermediate carry c_{i+1} values to the set $\{\bar{1}, 0, 1\}$. Each sum of the two TSD digits from the set $\{\bar{2}, \bar{1}, 0, 1, 2\}$ has at least one representation that satisfies the requirement. The reduced minterms for this operation are listed in Table 9.29. Also note that the minterms for generating the 1 output of $s_i(c_{i+1})$ are digit-by-digit complement of the corresponding minterms for the $\bar{1}$ minterms. Thus, only three minterms are required for generating the 1 output of s_i and three minterms are required for

Table 9.29

Reduced Minterms for the First Step Operation of the Simplified Nonrecoded TSD Addition [150]

Output	Minterms $a_i b_i$
$S_i = 1$	$d_{\bar{1}2}d_{\bar{1}2}$, $d_{\bar{2}1}0$, $0d_{\bar{2}1}$
$S_i = \bar{1}$	$d_{1\bar{2}}d_{1\bar{2}}$, $d_{2\bar{1}}0$, $0d_{2\bar{1}}$
$C_{i+1} = 1$	02, 20, $d_{12}d_{12}$
$C_{i+1} = \bar{1}$	$0\bar{2}$, $\bar{2}0$, $d_{\bar{1}\bar{2}}d_{\bar{1}\bar{2}}$

yielding the 1 output of c_{i+1}. For the second-step addition, the minterms are the same as those used in the addition of recoded TSD numbers. Comparing the nonrecoding technique [150] to all of the previously reported TSD algorithms which require a large number of four- or six-variable minterms; this nonrecoding scheme requires only two-variable minterms and the number of minterms has been drastically reduced.

The simplified two-step nonrecoded TSD algorithm can be combined to a single-step operation [152], but inclusion of the lower-order digits $a_{i-1}b_{i-1}$ together with $a_i b_i$ makes the truth table more complex and more minterms are needed (21, 21, 9, and 9 minterms for 1, $\bar{1}$, $\bar{2}$, and 2, respectively). The two-step recoded and one-step nonrecoded addition algorithms were also investigated based on the redundant bit representation [153]. However, in this later case, the truth table becomes even more complex, thus complicating its practical implementation.

Algorithms for TSD subtraction can be derived directly by methods similar to TSD addition. Alternatively, we can perform TSD subtraction by complement and addition operations.

9.4.3.2. *Multiplication*

Because the TSD digits belongs to the set $\{\bar{2}, \bar{1}, 0, 1, 2\}$, the partial product formation during the multiplication of two digits may generate a carry. If the operands are recoded first, such carry generation can be eliminated. Hence TSD multiplication [153] can be classified into four cases according to whether the multiplicand A and the multiplier B are recoded or nonrecoded. The difference lies in how to generate the partial product as shown below:

(i) *Recoded TSD Multiplier and Multiplicand* If both the multiplier and multiplicand are recoded, all digits of a_i and b_j belongs to the set $\{\bar{1}\ 0, 1\}$. Then the partial products can be generated in parallel as in the MSD multiplication; i.e., a partial product is the complement of the multiplicand A, or 0, or the multiplicand itself, depending on whether the multiplier digit b_j is $\bar{1}$, 0, or 1.

(ii) *Multiplication with Only a Recoded Multiplier* If the multiplier B is recoded to the digit set $\{\bar{1}, 0, 1\}$, the product of b_j and a_i is still carry free, and we can get all the partial products in parallel. The a_i digit as well as the partial-product digit belongs to the set $\{\bar{2}, \bar{1}, 0, 1, 2\}$. A partial product is $\bar{A}$ or 0, or A depending on whether the multiplier digit b_j is $\bar{1}$, or 0, or 1.

(iii) *Nonrecoded Multiplication with Carries* If the multiplicand and the multiplier are not recoded, carries may be transmitted to the next higher order position when a_i and b_j are either 2 or $\bar{2}$. The carry c_{i+1} and the product p_i

				a_3	a_2	a_1	a_0	$= A =$ Multiplicand
				b_3	b_2	b_1	b_0	$= B =$ Multiplier
∅	∅	∅	c_{03}	c_{02}	c_{01}	c_{00}	∅	Partial Products and Partial Carries
∅	∅	∅	∅	p_{03}	p_{02}	p_{01}	p_{00}	
∅	∅	c_{13}	c_{12}	c_{11}	c_{10}	∅	∅	
∅	∅	∅	p_{13}	p_{12}	p_{11}	p_{10}	∅	
∅	c_{23}	c_{22}	c_{21}	c_{20}	∅	∅	∅	
∅	∅	p_{23}	p_{22}	p_{21}	p_{20}	∅	∅	
c_{33}	c_{32}	c_{31}	c_{30}	∅	∅	∅	∅	
∅	p_{33}	p_{32}	p_{31}	p_{30}	∅	∅	∅	

Fig. 9.22. Four-digit nonrecoded TSD multiplication [153]. p and c are for the partial product and the carry, respectively.

can be limited to the set $\{\bar{1}, 0, 1\}$. When A is multiplied by b_i, both the partial product and partial carry can be arranged as separate numbers, as shown in Fig. 9.22. As a result, we have N partial-carry words, in addition to N partial-product words to be added.

(iv) *Nonrecoded Multiplication without Carries* As described above, when two TSD digits are multiplied, the carry c_{i+1} and the partial product p_i can be limited to the set $\{\bar{1}, 0, 1\}$. Therefore, partial-carry generation can be avoided by considering the product of two consecutive digits of the multiplicand $A(a_i a_{i-1})$ and one digit of the multiplier $B(b_j)$. The truth table can be obtained directly from the multiplication of two digits.

For each of the above-mentioned cases, when all partial products are generated in parallel, they can be added in a tree structure using the nonrecoding addition technique. But for the third case, an extra addition step is required.

9.4.3.3. *Division*

Based on the TSD addition and multiplication algorithms, recently TSD division through the convergence approach [154] has been studied. Table 9.30 shows an illustration of the TSD division algorithm using $X = (0.\bar{1}\bar{2})_3 = (0.5555)_{10}$ and $Y = (0.20)_3 = (0.6666)_{10}$.

Table 9.30

An Example for TSD Convergence Division with $X = (0.\bar{1}\bar{2})_3$ and $D = (0.20)_3$ [154]

Iteration number	Multiplication factor	Accumulated numerator	Accumulated denominator
0	$m_0 = 2 - Y_0 = (1.1000)_3$ $= (1.3333\ldots)_{10}$	$X_1 = X_0 m_0 = (0.\bar{2}0\bar{1}00)_3$ $= (-0.7407\ldots)_{10}$	$Y_1 = Y_0 m_0 = (0.2200)_3$ $= (0.8888\ldots)_{10}$
1	$m_1 = 2 - Y_1 = (1.0100)_3$ $= (1.1111\ldots)_{10}$	$X_2 = X_1 m_1 = (0.\bar{2}\bar{1}\bar{1}0\bar{2})_3$ $= (-0.8230\ldots)_{10}$	$Y_2 = Y_1 m_1 = (0.2222)_3$ $= (0.9876\ldots)_{10}$
2	$m_2 = 2 - Y_2 = (1.0001)_3$ $= (1.0124\ldots)_{10}$	$X_3 = X_2 m_2 = (0.\bar{2}\bar{1}\bar{1}\bar{1}\bar{1}\ldots)_3$ $= (-0.8333\ldots)_{10}$	$Y_3 = Y_2 m_2 = (0.2222222)_3$ $= (0.9998\ldots)_{10}$

From the above example, it is evident that the proposed technique yields the quotient just after three iterations; i.e.,

$$Q = X_3 = (0.\bar{2}\bar{1}\bar{1}\bar{1}\bar{1}\ldots)_3 = (-0.8333\ldots)_{10}.$$

Each iteration of the TSD division algorithm involves three operations — a subtraction operation (or a complement and addition operation) for calculating the multiplication factors m_i and two TSD multiplication operations for calculating the numerator and the denominator. The subtraction (or complement plus addition) and multiplication operations required for each step of the division algorithm can be performed in constant time using TSD subtraction (addition) and multiplication algorithms developed in the previous sections. A block diagram for the TSD division process is shown in Fig. 9.23, which can also be applied to MSD division.

9.4.4. QSD ARITHMETIC

9.4.4.1. *Addition*

For QSD addition, three two-step techniques have been suggested. One [155] is based on checking two pairs of operand digits and one pair of reference digits from the lower-order digit position, requiring a total of 2518 six-variable operation rules. The second approach [156, 157] is realized by recoding the input numbers to restrict the digit range to a smaller set $\{\bar{2}, \bar{1}, 0, 1, 2\}$ before performing addition. In this scheme, the number of minterms can be significantly reduced. In addition, the number of variables involved in each minterm is reduced from six to four. Alam and his colleagues [156] used 38 minterms for recoding and 66 minterms for addition while Cherri [157] used the same number of minterms for recoding and 64 minterms

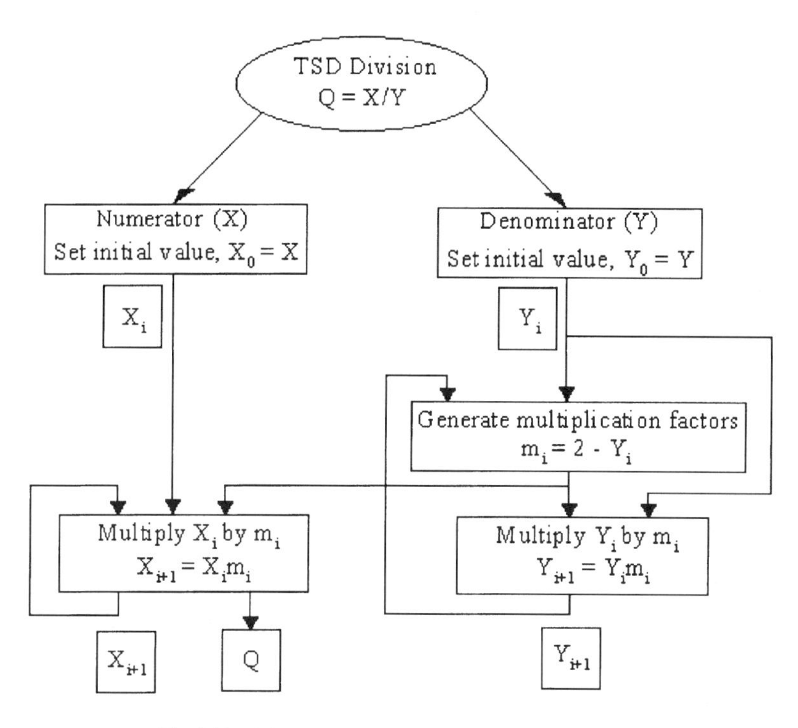

Fig. 9.23. A block diagram for TSD division [154].

for addition. In the recoding (addition) step of both schemes, the minterms for generating the outputs 1 and 2 (1, 2, and 3) are the exact complement of those used for generating the outputs $\bar{1}$ and $\bar{2}$ ($\bar{1}$, $\bar{2}$, and $\bar{3}$). Consequently, the former required 19 (33) minterms for the outputs 1 and 2 (1, 2, and 3) in recoding (addition), as shown in Table 9.31, while the latter required 19 (32) minterms for the outputs 1 and 2 (1, 2, and 3) in recoding (addition) as shown in Table 9.32.

Table 9.31

Reduced Truth Table for QSD Recoding [156] where the Minterms for the $\bar{1}$ and $\bar{2}$ Outputs are Digit-by-Digit Complement of that Generating the 1 and 2 Outputs, Respectively

Output literal	Minterms $x_i x_{i-1} x_{i-2} x_{i-3}$
2	$d_{\bar{3}1}223$ $d_{\bar{3}1}23d$ $d_{\bar{3}1}3dd$ $2\bar{2}2d_{\bar{2}\bar{1}0123}$ $2\bar{2}d_{\bar{1}0123}d$ $2d_{\bar{1}01}dd$ $22d_{\bar{3}\bar{2}\bar{1}01}d$ $222d_{\bar{3}\bar{2}\bar{1}012}$
1	$d_{\bar{3}1}\bar{2}\bar{2}d_{\bar{2}\bar{1}0123}$ $d_{\bar{3}1}\bar{2}d_{\bar{1}0123}d$ $d_{\bar{3}1}d_{\bar{1}01}dd$ $d_{\bar{3}1}2d_{\bar{3}\bar{2}\bar{1}01}d$ $d_{\bar{3}1}22d_{\bar{3}\bar{2}\bar{1}012}$ $d_{\bar{2}2}\bar{3}dd$ $d_{\bar{2}2}\bar{2}\bar{3}d$ $d_{\bar{2}2}\bar{2}\bar{2}\bar{3}$ 0223 023d 03dd

Table 9.32

Reduced Truth Table for the Second Step Operation of the Recoded QSD Addition [156]

Output literal	*Minterms* $a_i b_i a_{i-1} b_{i-1}$
3	1122 $12\bar{2}d_{\bar{1}012}$ $12d_{\bar{1}012}d$ $21\bar{2}d_{\bar{1}012}$ $21d_{\bar{1}012}d$ $22\bar{2}\bar{2}$
2	0122 $02\bar{2}d_{\bar{1}012}$ $02d_{\bar{1}012}d$ 1022 $112d_{\bar{1}012}$ $11d_{\bar{1}01}d$ $112d_{\bar{2}\bar{1}01}$ $12\bar{2}\bar{2}$ $20\bar{2}d_{\bar{1}012}$ $20d_{\bar{1}012}d$ $21\bar{2}\bar{2}$
1	$\bar{1}122$ $\bar{1}2\bar{2}d_{\bar{1}012}$ $\bar{1}2d_{\bar{1}012}d$ 0022 $01\bar{2}d_{\bar{1}012}$ $01d_{\bar{1}01}d$ $012d_{\bar{2}\bar{1}01}$ $02\bar{2}\bar{2}$ $1\bar{1}22$ $10\bar{2}d_{\bar{1}012}$ $10d_{\bar{1}01}d$ $102d_{\bar{2}\bar{1}01}$ $11\bar{2}\bar{2}$ $2\bar{1}\bar{2}d_{\bar{1}012}$ $2\bar{1}d_{\bar{1}012}$ $20\bar{2}\bar{2}$

Since these algorithms are usually implemented by optical symbolic substitution or content-addressable memory where the input numbers and the minterms are spatially encoded and compared, the optical system can be simplified by the use of fewer rules and fewer variables in a rule. For this purpose, the number of minterms may be cut down further. Li et al. [73] have proposed a simplified two-step QSD arithmetic by exploiting the redundant representation of two single-digit sums. In this scheme, addition of two numbers is realized by adding all digit pairs at each position in parallel. The addition of a_i and b_i will yield an intermediate sum (s_i) and an intermediate carry (c_{i+1}) for the next higher order digit position. If s_i and c_{i+1} generated in the first step are restricted to the sets $\{\bar{2}, \bar{1}, 0, 1, 2\}$ and $\{\bar{1}, 0, 1\}$, respectively, then the addition of s_i and c_i in the second step becomes carry free. Consequently, in the first step only 20 two-variable minterms are needed to produce the intermediate sum and the intermediate carry, and in the second step only 12 two-variable minterms are required to generate the final sum. By taking the complementary relationship into account, only 10 (6) minterms are required for the outputs of 1 and 2 (1, 2, and 3) in the first (second)-step addition. These minterms are shown in Table 9.33. Comparing Li's [73] technique with the other QSD techniques [155–158], the information content of the minterms is significantly reduced, which alleviates system complexity. For illustration, consider the following numerical example for four-digit addition:

Augend A:	$\bar{3}2\bar{2}3$ (decimal -165)
Addend B:	$12\bar{1}2$ (decimal 94)
Intermediate sum S:	$\phi\bar{2}011$
Intermediate carry C:	$01\bar{1}1\phi$
Final sum S:	$\bar{1}\bar{1}21$ (decimal -71)

Table 9.33

Reduced Truth Table for the Simplified Two-Step Nonrecoded QSD Addition [73]

Step number	Output literal	Minterms $a_i b_i$
1	$s_i = 2$	$d_{\bar{3}1}d_{\bar{3}1}$, $\bar{2}0$, $0\bar{2}$
	$s_i = 1$	$d_{3\bar{1}}d_{2\bar{2}}$, $d_{2\bar{2}}d_{3\bar{1}}$, $0d_{\bar{3}1}$, $d_{\bar{3}1}0$
	$c_{i+1} = 1$	$3d_{3210}$, $d_{3210}3$, $d_{12}d_{12}$
2	$s_i = 3$	21
	$s_i = 2$	20, 11
	$s_i = 1$	$2\bar{1}$, 10, 01

The minterms for QSD subtraction can be derived in a similar fashion. Alternatively, we can complement the subtrahend and then apply the addition truth table.

9.4.4.2. *Multiplication*

Usually the multiplication of two N-digit QSD numbers A and B produces a $2N$-digit product $P = p_{2N-1}p_{2N-2}\cdots p_1 p_0$. For fast multiplication [160], we need to generate all N partial products in parallel and add them in a tree structure. Each partial product $p^{(i)}$ $(i = 0, 1, \ldots, N-1)$ is formed by multiplying the multiplicand A and the ith digit of the multiplier B and shifting i digits to the left; i.e., $p^{(i)} = Ab_i 4^i$. Thus, the product P can be computed from the summation

$$P = AB = \sum_{i=0}^{N-1} p^{(i)} = \sum_{i=0}^{N-1} Ab_i 4^i = \sum_{i=0}^{2N-1} p_i 4^i. \tag{9.65}$$

The difficulty lies in how to generate all the partial products in parallel for higher-radix multiplication. In QSD multiplication with the digit set $\{\bar{3}, \bar{2}, \bar{1}, 0, 1, 2, 3\}$, carries may be transmitted to the next higher digit positions when both digits a_j and b_i to be multiplied take the values $\bar{3}$, $\bar{2}$, 2, 3. Notice that the digits of the product whose weights are larger than 4^{i+j} are called carries herein. The carry propagation prevents us getting the partial products Ab_i directly and sequential additions are required. To overcome this problem and lessen complexity, three methods can be chosen:

- Recode the operands to map the digit set from $\{\bar{3}, \bar{2}, \bar{1}, 0, 1, 2, 3\}$ to a smaller one $\{\bar{2}, \bar{1}, 0, 1, 2\}$. The multiplication can thus be simplified. The product can be represented by two digits (partial carry and partial product), with the partial-product digit limited to $\{\bar{2}, \bar{1}, 0, 1, 2\}$ and the partial-carry digit limited to $\{\bar{1}, 0, 1\}$. This property guarantees that

Table 9.34

Reduced Truth Table for the Recorded QSD Multiplication [160]

Output literal	Minterms $a_j a_{i-1} b_i$	
3	122	$\bar{1}\bar{2}\bar{2}$
2	$1\mathrm{d}_{\bar{1}01}2 \quad 2\mathrm{d}1$	$\bar{1}\mathrm{d}_{10\bar{1}}\bar{2} \quad \bar{2}\mathrm{d}\bar{1}$
1	$1\mathrm{d}1 \quad \mathrm{d}_{20}\bar{2}\bar{2} \quad 1\bar{2}2$	$\bar{1}\mathrm{d}\bar{1} \quad \mathrm{d}_{\bar{2}0}22 \quad \bar{1}2\bar{2}$
$\bar{1}$	$1\mathrm{d}\bar{1} \quad \mathrm{d}_{20}\bar{2}2 \quad 1\bar{2}\bar{2}$	$\bar{1}\mathrm{d}1 \quad \mathrm{d}_{\bar{2}0}2\bar{2} \quad \bar{1}22$
$\bar{2}$	$1\mathrm{d}_{\bar{1}01}\bar{2} \quad 2\mathrm{d}\bar{1}$	$\bar{1}1\mathrm{d}_{10\bar{1}}2 \quad \bar{2}\mathrm{d}1$
$\bar{3}$	$12\bar{2}$	$\bar{1}\bar{2}2$

addition of the products $b_i a_0 4^0 + b_i a_1 a_j 4^1 + \cdots + b_i a_{n-1} 4^{N-1}$ is carry free if the partial product is permitted to be represented in the original format. Hence after recoding, partial products can be formed in two steps by digitwise multiplication and addition. Furthermore, by the above analysis, we can conclude that the value of the $(i+j)$th digit $p^{(i)}_{i+j}$ of the ith partial product ($p^{(i)} = Ab_i 4^i$) of the recoded numbers is dependent on the values of $a_j a_{j-1}$ and b_i. Therefore, by examining every two consecutive digits of A and b_i in parallel, we can generate the ith partial product in a single step. In this way, all partial products can be produced simultaneously. The reduced minterms for parallel multiplication are summarized in Table 9.34. Note that the minterms at the right side are digit-by-digit complement of those at the left side. The partial products are represented in the original QSD format. When employing this table, it is necessary to pad one zero trailing the least significant digit of the multiplicand. From the recoding truth table (Table 9.31), one can deduce that the most significant digit of any recoded number is restricted to $\{\bar{1}, 0, 1\}$. According to the multiplication truth table (Table 9.34), when $a_j a_{j-1} = 0\bar{1}$ or 01, and $b_i = \bar{2}, \bar{1}, 1$, or 2, the resultant product is always 0. Therefore, we do not need to pad a zero preceding the most significant digit of the multiplicand, and the partial product Ab_i has the same number of digits as recoded A.

- Recode only the multiplier B into the small set $\{\bar{2}, \bar{1}, 0, 1, 2\}$. In this case, when a_i is multiplied by b_j, the partial product digit is limited to $\{\bar{2}, \bar{1}, 0, 1, 2\}$ and the partial carry digit to $\{\bar{1}, 0, 1\}$. Similarly, we can form all the partial products in parallel by examining every pair of consecutive digits of A and b_i at the same time.
- When performing nonrecoded QSD multiplication, both the partial-product digit and the partial-carry digit of $a_i \times b_1$ belong to the set $\{\bar{2}, \bar{1}, 0, 1, 2\}$; thus, it is impossible for us to obtain the result of Ab_i as a single partial product in parallel by checking two consecutive digits of the multiplicand $A(a_i a_{i-1})$ and one digit of the multiplier $B(b_j)$. The solution

to this problem is to arrange the partial products and the partial carries as separate numbers, as shown in Fig. 9.22.

When all the partial products are formed in parallel, we can use the simplified nonrecoded QSD addition algorithm to sum them in a tree structure. However, the third scheme just discussed is simpler than alternate techniques to generate the partial products, although it requires an additional step to add the partial products compared to the other schemes.

9.4.5. NSD ARITHMETIC OPERATIONS

In [50], a two-step parallel negabinary addition and subtraction algorithm is proposed, in which the sum of two negabinary numbers is represented in NSD form. As shown in Table 9.35, in the first step the transfer and weight digits are generated by signed AND and XOR operations. In the second step, the final sum is obtained by signed XOR operation. The algorithm can be performed by spatial encoding and space-variant decoding technique. Later the two-step algorithm was combined into a single step [161] and implemented with space-polarization encoding and decoding. For multiple addition and multiplication, Li et al. [162] developed a generalized algorithm for NSD arithmetic. As discussed in the following sections, the former algorithm is a special case of generalized NSD arithmetic.

In NSD arithmetic, the digits of the operands belong to the set $\{\bar{1}, 0, 1\}$. For a digitwise operation, there are nine possible combinations; i.e., $(a_i, b_i) = (1, 1)$, $(1, 0)$, $(1, \bar{1})$, $(0, 1)$, $(0, 0)$, $(0, \bar{1})$, $(\bar{1}, 1)$, $(\bar{1}, 0)$, and $(\bar{1}, \bar{1})$. For addition, at each position an intermediate sum (defined by a weight function, W) $w_i \in \{\bar{1}, 0, 1\}$ and an intermediate carry (defined by a transfer function, T) $t_i \in \{\bar{1}, 0, 1\}$ will be produced, satisfying the condition $a_i + b_i = (-2)t_{i+1} + w_i$. The digit combinations $(1, 1)$ and $(\bar{1}, \bar{1})$ generate a $\bar{1}$ and a 1 carry, respectively, while the combinations $(0, 0)$, $(\bar{1}, 1)$, and $(1, \bar{1})$ do not generate any carry. The remaining combinations, however, can yield two different results based on the redundancy of NSD: $0 + 1 = 1 + 0 = 01$ or $\bar{1}\bar{1}$, and $0 + \bar{1} = \bar{1} + 0 = 0\bar{1}$ or 11. For subtrac-

Table 9.35

Truth Table for Logic Functions in Two-Step Addition and One-Step Subtraction [50]

T (Signed AND)

a_i \ b_i	0	1
0	0	0
1	0	$\bar{1}$

W (XOR)

a_i \ b_i	0	1
0	0	1
1	1	0

S (Signed XOR)

a_i \ b_i	0	$\bar{1}$
0	0	$\bar{1}$
1	1	0

D (Signed XOR)

a_i \ b_i	0	1
0	0	$\bar{1}$
1	1	0

Table 9.36

Truth Table for the First Step Operation in NSD Addition [162]

a_i	b_i	$(g_i\ h_i)$	$a_{-1}b_{-1}$	(f_i)	T_{i+1}	w_i
1	1	(1 0)	Don't care		$\bar{1}$	0
1	0	(1 0)	Both are nonnegative	(1)	0	1
0	1					
			Otherwise	(0)	$\bar{1}$	$\bar{1}$
0	0	(1 0)	Don't care		0	0
1	$\bar{1}$	(0 0)	Don't care		0	0
$\bar{1}$	1					
0	$\bar{1}$	(0 0)	Both are nonnegative	(1)	1	1
$\bar{1}$	0					
			Otherwise	(0)	0	$\bar{1}$
$\bar{1}$	$\bar{1}$	(0 1)	Don't care		1	0

tion, on the other hand, the condition $a_i - b_i = (-2)t_{i+1} + w_i$ should be satisfied. The combinations $(1, \bar{1})$ and $(\bar{1}, 1)$ generate a $\bar{1}$ and a 1 carry, respectively, while the combinations (0, 0), (1, 1), and $(\bar{1}, \bar{1})$ do not generate a carry. The remaining combinations, however, have two different redundant results: $0 - 1 = \bar{1} - 0 = 0\bar{1}$ or $\bar{1}1$, and $0 - \bar{1} = 1 - 0 = 01$ or $1\bar{1}$. By exploiting this redundancy, it is possible to perform addition (subtraction) in two steps.

Addition is considered first. To avoid generating a carry in the second step, the first step should guarantee that identical nonzero digits t_i and w_i do not occur at the same position; i.e., both w_i and t_i are neither 1s nor $\bar{1}$s. The optimized rules are shown in Table 9.36, in which the augend and the addend digits are classified into six groups since they can be interchanged. For the second and the fifth groups, the two halves having two different outputs can be distinguished by examining whether both a_{i-1} and b_{i-1} are binary. It is helpful to introduce a reference bit f_i, 1 for binary and 0 for the other cases. With this reference bit, the number of variables in a minterm can be reduced from four to three. In the second step, carry-free addition at each position results in the final sum. The corresponding rules are shown in Table 9.37.

Subtraction can be performed by addition after complementing the subtrahend, and this requires three steps. The direct implementation can be performed in two steps. The first step transforms the subtraction into addition, generating the transfer t_{i+1} and the weight w_i at all positions. The substitution rules for subtraction in the first step are listed in Table 9.38. There are nine rows in the table since the minuend and the subtrahend cannot be interchanged. The reference bit f_i is true if both a_{i-1} and the complement of b_{i-1} are nonnegative. These rules also ascertain that the second-step addition is carry free, according to Table 9.37.

Table 9.37

Truth Table for the Second Step Operation in NSD Addition and Subtraction [162]

t_i	w_i	(g_i)	s_i
0	0	1	0
0	1	(1)	1
1	0		
0	$\bar{1}$	(0)	$\bar{1}$
$\bar{1}$	0		
1	$\bar{1}$	(0)	0
$\bar{1}$	1		

The above algorithms for two-step addition and subtraction can be mapped to the architecture of symbolic substitution or content-addressable memory (CAM). The logical minterms for the nonzero output are summarized in Table 9.39. Alternatively, one can implement the involved arithmetic through logic operations. We define the operations with binary logic, but signed digits are included. A closer look at the truth tables reveals that it is necessary to introduce two additional reference bits, g_i and h_i. For addition, g_i is true if both

Table 9.38

Truth Table for the First Step Operation in NSD Subtraction [162] where $\hat{b}_{i-1}$ Indicates the Complement of b_{i-1}

a_i	b_i	$(g_i\ h_i)$	$a_{i-1}b_{i-1}$	(f_i)	t_{i+1}	w_i
1	1	(0 0)	Don't care		0	0
1	0	(1 0)	a_{i-1} and $\hat{b}_{i-1}$ are nonnegative	(1)	0	1
			Otherwise	(0)	$\bar{1}$	$\bar{1}$
1	$\bar{1}$	(1 0)	Don't care		$\bar{1}$	0
0	1	(0 0)	a_{i-1} and $\hat{b}_{i-1}$ are nonnegative	(1)	1	1
			Otherwise	(0)	0	$\bar{1}$
0	0	(1 0)	Don't care		0	0
0	$\bar{1}$	(1 0)	a_{i-1} and $\hat{b}_{i-1}$ are nonnegative	(1)	0	1
			Otherwise	(0)	$\bar{1}$	$\bar{1}$
$\bar{1}$	1	(0 1)	Don't care		1	0
$\bar{1}$	0	(0 0)	a_{i-1} and $\hat{b}_{i-1}$ are nonnegative	(1)	1	1
			Otherwise	(0)	0	$\bar{1}$
$\bar{1}$	$\bar{1}$	(0 0)	Don't care		0	0

Table 9.39

Reduced Minterms for the Two-Step NSD Addition and Subtraction [162] where d_{xy} Indicates a Partial Don't-Care Digit

Step number	Operation	Function	Output digit	Logical terms $(a_i b_i f_i)$
	Addition	t_{i+1}	1	$\bar{1}\bar{1}d_{10}, 0\bar{1}1, \bar{1}01$
			$\bar{1}$	$11d_{10}, 100, 010$
First step		w_i	1	$d_{1\bar{1}}01, 0d_{1\bar{1}}1$
			$\bar{1}$	$d_{1\bar{1}}00, 0d_{1\bar{1}}0$
		t_{i+1}	1	$\bar{1}1d_{01}, 011, \bar{1}01$
	Subtraction		$\bar{1}$	$1\bar{1}d_{01}, 100, 0\bar{1}0$
		w_i	1	$d_{1\bar{1}}01, 0d_{1\bar{1}}1$
			$\bar{1}$	$d_{1\bar{1}}00, 0d_{1\bar{1}}0$
Second step	Addition and subtraction	s_i	1	$01, 10$
			$\bar{1}$	$0\bar{1}, \bar{1}0$

a_i and b_i are positive while h_i is true if both a_i and b_i are negative. For subtraction, g_i is true if both a_i and the complement of b_i are positive while h_i is true if both a_i and the complement of b_i are negative. For the reference bits, we only need to consider the unsigned binary values of a_i and b_i, independent of their signs. This greatly simplifies the logical operations. Another advantage of this definition is that both addition and subtraction operations can be expressed by the same logic equations. In the first step, the required binary logic expressions for the two arithmetic operations are given by:

for t_i,

$$\textit{output}\ \text{“1”: } h_i + (a_i \oplus b_i)\overline{g_i} f_i, \tag{9.66}$$

$$\text{output “}\bar{1}\text{”: } (a_i b_i + (a_i \oplus b_i)\overline{f_i}) g_i; \tag{9.67}$$

for w_i,

$$\textit{output}\ \text{“1”: } (a_i \oplus b_i) f_i, \tag{9.68}$$

$$\text{output “}\bar{1}\text{”: } (a_i \oplus b_i)\overline{f_i}, \qquad (f_i = g_{i-1}). \tag{9.69}$$

In the second step, the logic operations for the final sum S and D are the same as W in the first step by defining a binary reference bit g_i which is true for positive t_i and w_i:

for s_i and d_i,

$$\textit{output}\ \text{“1”: } (t_i \oplus w_i) g_i, \tag{9.70}$$

$$\text{output “}\bar{1}\text{”: } (t_i \oplus w_i)\overline{g_i}. \tag{9.71}$$

Therefore, both addition and subtraction can be performed in parallel using the same logic operations. This offers the advantage that through data programming we can perform space-variant arithmetic computation by space-invariant logic operations [55]. As described above, the adoption of reference bits yields simpler logical expressions, thus reducing system complexity in comparison with the previous approach [136] where MSD arithmetic is performed with binary logic by separating the positive and negative digits. For example, consider the following illustrations for addition and subtraction using the proposed algorithm:

Addition		Subtraction	
$A1$:	$1\bar{1}101\bar{1}0\bar{1}$ (-237)	$A2$:	$10\bar{1}0101\bar{1}$ (-107)
$+B1$:	$0\bar{1}0\bar{1}101\bar{1}$ (-91)	$-B2$:	$\bar{1}\bar{1}0100\bar{1}\bar{1}$ (83)
a_i:	11101101		10101011
b_i:	01011011		11010011
f_i:	01010101		10011101
g_i:	10101010		11001110
h_i:	01000001		00000001
t_i:	$\bar{1}1\bar{1}1\bar{1}1\bar{1}10$		$\bar{1}\bar{1}0100\bar{1}10$
w_i:	$0\bar{1}0\bar{1}101\bar{1}0$		$00\bar{1}\bar{1}11000$
s_i:	$\bar{1}0\bar{1}001000$ (-328)	d_i:	$\bar{1}\bar{1}\bar{1}011\bar{1}10$ (-190)

Multiplication of two N-digit NSD numbers A and B is done by adding the N partial products. The operation can be performed by adopting the same procedure as the MSD multiplication. However, if the numbers to be multiplied are in negabinary, the partial product may be 0 or the multiplicand A itself, depending on whether the multiplier bit value is b_i, 0, or 1, respectively.

9.5. CONVERSION BETWEEN DIFFERENT NUMBER SYSTEMS

Before using the above-mentioned parallel MSD, TSD, QSD, or NSD algorithms, the operands should be uniquely encoded in 2's complement, 3's complement, 4's complement, or negabinary, respectively. These algorithms also yield the results in MSD, TSD, QSD, or NSD which should be converted back to 2's complement [70, 123, 163], 3's complement, 4's complement, or negabinary [162], respectively. The whole operational procedures for MSD and NSD arithmetic are shown in Fig. 9.24. In the aforementioned parallel algorithms, the addition/subtraction time is independent of the operand length. For multiplication, the computation time is proportional to $O(\log_2 N)$. To enhance overall performance, the remaining task is to develop a fast-conver-

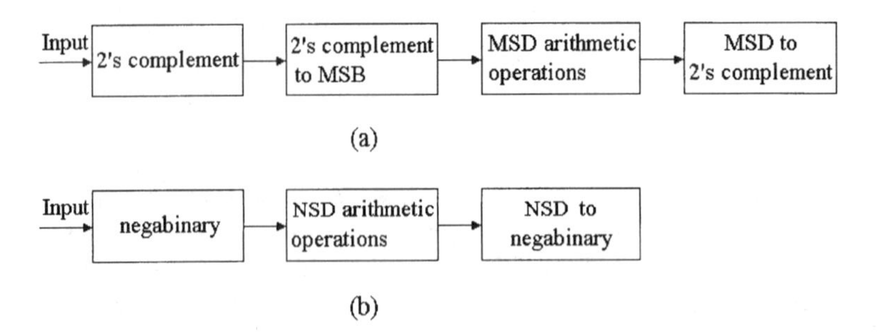

Fig. 9.24. The operation procedures for MSD and NSD number systems [162].

sion algorithm among these number systems. The speed of the conventional conversion algorithms [70, 123] is limited by ripple-carry propagation. To compensate, we present carry-lookahead–mode conversion algorithms among the associated number systems [162, 163].

9.5.1. CONVERSION BETWEEN SIGNED-DIGIT AND COMPLEMENT NUMBER SYSTEMS

To perform MSD addition, the bipolar operands must be encoded in 2's complement. An N-bit 2's complement number X has the same form as Eq. (9.4). Therefore, the conversion of a 2's complement number to an MSD equivalent can be done by changing the MSB x_{N-1} to $\bar{1}$ if the x_{N-1} equals to 1, while keeping other bits unchanged.

Conventionally, the conversion from MSD to 2's complement is carried out by separating an MSD number into positive and negative parts and then performing subtraction:

$$Z = X^{+} - X^{-} \tag{9.72}$$

where

$$X^{+} = \sum_{x_i = 1} x_i \times 2^i, \qquad X^{-} = \sum_{x_i = -1} x_i \times 2^i. \tag{9.73}$$

The operation can be done serially by a 2's complement adder, and an N-digit MSD number is converted to an $(N + 1)$–digit 2's complement representation.

Recently a carry-lookahead conversion algorithm [163] has been introduced and implemented optically. In this algorithm, the 2's complement equivalent of X can be written as

$$Z = \sum_{x_i = \pm 1} |x_i| \times 2^i - \sum_{x_i = -1} |x_i| \times 2^{i+1}. \tag{9.74}$$

Let us define a binary variable C_i corresponding to the ith digit position. At the current MSD digit position i, if there is at least one, $\bar{1}$ to the right of this position and no 1 between the $\bar{1}$s and the current position, then $C_i = 1$; otherwise, $C_i = 0$. Note that $C_0 = 0$. Furthermore, each digit x_i is represented by two binary bits (S_i, D_i); i.e., $0 = (0, 0)$, $1 = (0, 1)$, and $\bar{1} = (1, 1)$. For each position i, if the input MSD digits S_i and D_i, and C_i are known, we can obtain the binary output B_i and the output variable C_{i+1} according to the following binary logic expressions:

$$B_i = D_i \oplus C_i \tag{9.75}$$

$$C_{i+1} = S_i + \bar{D}_i C_1, \qquad C_0 = 0. \tag{9.76}$$

Therefore, at a given stage, a carry is generated if S_i is true, and a stage propagates an input carry to the higher-level position if D_i is false. On this basis, we can derive a fast-conversion algorithm in carry-lookahead format. If we replace $\bar{D}_i$ with P_i and extend it to a four-digit block, we can write the following equations:

$$C_1 = S_0 + P_0 C_0 \tag{9.77}$$

$$C_2 = S_1 + P_1 C_1 = S_1 + P_1 S_0 + P_1 P_0 C_0 \tag{9.78}$$

$$C_3 = S_2 + P_2 C_2 = S_2 + P_2 S_1 + P_2 P_1 S_0 + P_2 P_1 P_0 C_0 \tag{9.79}$$

$$C_4 = S_3 + P_3 C_3 = S_3 + P_3 S_2 + P_3 P_2 S_1 + P_3 P_2 P_1 S_0 + P_3 P_2 P_1 P_0 C_0. \tag{9.80}$$

Equation (9.80) can be rewritten as

$$C_4 = S'_0 + P'_0 C_0, \tag{9.81}$$

where

$$S'_0 = S_3 + P_3 S_2 + P_3 P_2 S_1 + P_3 P_2 P_1 S_0$$

$$P'_0 = P_3 P_2 P_1 P_0.$$

In general,

$$C_{4i+4} = S'_i + P'_i C_{4i}, \tag{9.82}$$

where

$$S'_i = S_{4i+3} + P_{4i+3} S_{4i+2} + P_{4i+2} P_{4i+2} S_{4i+1} + P_{4i+3} P_{4i+2} P_{4i+1} S_{4i}$$

$$P'_i = P_{4i+3} P_{4i+2} P_{4i+1} P_{4i}.$$

The logic in Eq. (9.82) determines the carry status by looking across a wide word. In each block, the output indicates that a carry has been generated within that block or a low-order carry would be propagated through that block.

For TSD arithmetic operations, the input numbers are first encoded in 3's complement, as shown in the following equation:

$$X = -x_{N-1} \times 3^{N-1} + \sum_{i=0}^{N-2} x_i \times 3^i, \quad x_i \in \{0, 1, 2\}. \tag{9.83}$$

It is evident that conversion form 3's complement to TSD is straightforward by changing the most significant digit x_{N-1} to 0, $\bar{1}$, or $\bar{2}$, if x_{N-1} is 0, 1, or 2, while the remaining digits need not be changed. The conversion from TSD to 3's complement can also be completed in the carry-lookahead mode. Similarly, a 4's complement number takes the form

$$X = -x_{N-1} \times 4^{N-1} + \sum_{i=0}^{N-2} x_i \times 4^i, \quad x_i \in \{0, 1, 2, 3\}. \tag{9.84}$$

It is also possible to derive a carry-lookahead conversion algorithm from QSD to the 4's complement number systems.

9.5.2. CONVERSION BETWEEN NSD AND NEGABINARY NUMBER SYSTEMS

Since negabinary can uniquely encode any positive and negative numbers without a sign digit and negabinary is a subset of NSD, the input operands encoded in negabinary can be directly used for NSD operations without any conversion. We only need to focus on developing a fast-conversion algorithm from NSD to the normal negabinary [162]. Consider an NSD number X, represented by

$$X = \sum_{i=0}^{N-1} x_i(-2)^i, \quad x_i \in \{\bar{1}, 0, 1\}. \tag{9.85}$$

Since $\bar{1} = 11$, Eq. (9.85) can be rewritten as

$$\begin{aligned} X &= \sum_{x_i=1} x_i(-2)^i + \sum_{x_i=\bar{1}} x_i(-2)^i \\ &= \sum_{i=0}^{N-1} |x_i|(-2)^i + \sum_{x_i=\bar{1}} |x_i|(-2)^{i+1} \\ &= A + B. \end{aligned} \tag{9.86}$$

Thus, conversion of a number X from NSD to negabinary is transformed to the addition of two negabinary numbers A and B, where the bit a_i $(i = 0, 1, \ldots, N-1)$ is the unsigned binary value of each digit x_i, and B is formed by putting 1s to the left of $\bar{1}$s in x_i and padding 0s to the other bit positions. Let the negabinary form of X be Z. Notice that for the same number of digits, NSD can represent a larger integer value than the normal negabinary. An N-digit NSD representation can have a value in the range $[-(2^N-1), 2^N-1]$. Thus an N-digit NSD number X should be converted to an $(N+2)$–digit negabinary number Z.

In negabinary addition, either a positive or a negative carry (denoted by c_{i+1}^{+} and c_{i+1}^{-}, respectively) may be generated from the ith digit position to the $(i+1)$th digit position. The speed at which negabinary numbers A and B can be added is restricted by the time taken by the carry to propagate serially from the LSB to the MSB. To speed up the conversion process, the carry-lookahead principle can be employed.

In the carry-lookahead technique, the carries at all bit positions of each block must be generated simultaneously. Thus, the logical expression for carry generation and propagation must be expressed as a function of the operand bits. It is usually in the form of a sum of product, which can be extended from the LSB to the MSB. For a negabinary full-adder, the carries to the next higher bit position can written as

$$c_{i+1}^{+} = \overline{a_i + b_i}, \tag{9.87}$$

$$\begin{aligned} c_{i+1}^{-} &= a_i b_i (c_i^{+} + \overline{c_i^{-}}) + \overline{a_i}\,\overline{b_i}(c_i^{+} + \overline{c_i^{-}}) + c_i^{+}\overline{c_i^{-}} \\ &= \overline{a_i \oplus b_i} c_i^{+} + (\overline{a_i \oplus b_i} + c_i^{+})\overline{c_i^{-}}, \end{aligned} \tag{9.88}$$

and

$$\overline{c_{i+1}^{-}} = (a_i \oplus b_i)\overline{c_i^{+}} + (a_i \oplus b_i + \overline{c_i^{+}})c_i^{-}. \tag{9.89}$$

Let us define two auxiliary functions as

$$G_i = \overline{a_i + b_i}, \tag{9.90}$$

and

$$P_i = \overline{a_i \oplus b_i}, \tag{9.91}$$

then we can derive

$$c_0^{+} = 0, \qquad c_0^{-} = 0, \tag{9.92}$$

$$c_{i+1}^{+} = \overline{a_i + b_i} = G_i, \quad (i = 0, 1, 2, \ldots, N + 1) \tag{9.93}$$

$$c_1^{-} = P_0, \tag{9.94}$$

$$c_2^{-} = P_1 G_0 + (P_1 + G_0)\overline{P_0}, \tag{9.95}$$

$$c_3^{-} = P_2 G_1 + (P_2 + G_1)\overline{P_1}\,\overline{G_0} + (P_2 + G_1)(\overline{P_1} + \overline{G_0})P_0, \tag{9.96}$$

$$\begin{aligned} c_4^{-} &= P_3 G_2 + (P_3 + G_2)\overline{P_2}\,\overline{G_1} + (P_3 + G_2)(\overline{P_2} + \overline{G_1})P_1 G_0 \\ &\quad + (P_3 + G_2)(\overline{P_2} + \overline{G_1})(P_1 + G_0)\overline{P_0}. \end{aligned} \tag{9.97}$$

If we rewrite c_4^{-} as

$$c_4^{-} = U_0 + V_0[\overline{P_0 c_0^{+}} + (\overline{P_0} + \overline{c_0^{+}})c_0^{-}], \tag{9.98}$$

where

$$U_0 = P_3 G_2 + (P_3 + G_2)\overline{P_2}\,\overline{G_1} + (P_3 + G_2)(\overline{P_2} + \overline{G_1})P_1 G_0,$$

$$V_0 = (P_3 + G_2)(\overline{P_2} + \overline{G_1})(P_1 + G_0).$$

It is valid to apply the aforementioned methodology to higher bit positions yielding

$$c_{4i+4}^{-} = U_i + V_i[\overline{P_{4i} c_{4i}^{+}} + (\overline{P_{4i}} + \overline{c_{4i}^{+}})c_{4i}^{-}], \quad i = 0, 1, 2. \tag{9.99}$$

Therefore, with a four-bit module, it is possible to generate the carries in the lookahead mode for any length of operands. For long operands, a multistage carry-lookahead architecture can be used where all carries can be generated in parallel. Finally, the negabinary equivalence of x is given by

$$z_i = a_i \oplus b_i \oplus c_i^{+} \oplus c_i^{-}. \tag{9.100}$$

The above operation is performed bitwise in parallel, as shown in the following example:

Input:	$\bar{1}\bar{1}\bar{1}011\bar{1}10$ (−190)
a_i:	111011110
b_i:	1110001000
P_i:	0110101001
G_i:	0000100001
c_i^{+}:	00001000010
c_i^{-}:	00101010010
Output:	01101000110 (−190).

It is obvious that the carry-lookahead addition used in the number conversion process is also applicable for the fast addition of two numbers. When only two numbers are added, it is preferable to use carry-lookahead addition directly because two-step carry-free addition requires number conversion from NSD to the normal negabinary, which takes the same time as carry-lookahead addition.

9.6. OPTICAL IMPLEMENTATION

The parallel algorithms and architectures described above have been implemented optically in different ways. In this section, three examples of arithmetic-logic operations are shown for implementations of symbolic substitution, CAM, and logic array processing. Symbolic substitution is realized through M–V multiplication and is usually achieved with coherent correlators. An incoherent optical correlator–based shared-CAM (SCAM) is used for implementing simplified two-step QSD addition [73], and an optical logic array processor using an electron-trapping device is demonstrated for parallel implementation of two-step NSD addition and subtraction [162]. The other arithmetic-logic operations can be implemented similarly.

9.6.1. SYMBOLIC SUBSTITUTION IMPLEMENTED BY MATRIX–VECTOR OPERATION

To show how symbolic substitution can be realized by matrix–vector multiplication (Sec. 9.2.5), an example is illustrated here. Consider the second-step substitution rules in the recoded or nonrecoded TSD addition (Sec. 9.4.3.1) [152]. The digits to be added are both from the set $\{\bar{1}, 0, 1\}$ and their sum belongs to the set $\{\bar{2}, \bar{1}, 0, 1, 2\}$. We use the symbols o, z, ob, t, and tb to represent the input pixel patterns for 1, 0, $\bar{1}$, 2, and $\bar{2}$, respectively. The nine possible input digit pairs to be added can be written as the columns of the input x. Then matrix X can be written as

$$X = \begin{bmatrix} o & o & o & z & z & ob & ob & ob & ob \\ o & z & ob & o & z & ob & o & z & ob \end{bmatrix}. \tag{9.101}$$

The corresponding nine output substitution patterns can be written as the columns of the matrix Y:

$$Y = [t \quad o \quad z \quad o \quad z \quad ob \quad z \quad ob \quad tb]. \tag{9.102}$$

To reduce the space–bandwidth product, and hence increase the throughput of a system, it is preferable to reduce the number of columns in the X matrix. Note that the digits to be added can be interchanged without affecting the final results. Thus, three columns in the X and Y matrices can be eliminated. We rewrite the input matrix X by superimposing (OR) all encoded digits in a column as:

$$X = [o + o \quad o + z \quad o + ob \quad z + z \quad z + ob \quad ob + ob], \tag{9.103}$$

where + denotes a logical OR operation. The corresponding output matrix Y becomes

$$Y = [t \quad o \quad z \quad z \quad ob \quad tb]. \tag{9.104}$$

An exact solution for the M–V equation exists if X is invertible. This implies that X must be a square matrix. In the previously reported techniques [70], six pixels are used to encode each TSD digit. To enhance the space–bandwidth product, here four pixels are used to encode a TSD digit, as shown in Fig. 9.25. Substituting the encoding patterns into the matrices X and Y, we obtain

$$X = \begin{bmatrix} 1 & 1 & 1 & 0 & 1 & 1 \\ 1 & 1 & 1 & 0 & 0 & 0 \\ 0 & 1 & 1 & 1 & 1 & 1 \\ 0 & 1 & 0 & 1 & 1 & 0 \end{bmatrix}, \quad Y = \begin{bmatrix} 0 & 1 & 0 & 0 & 1 & 0 \\ 1 & 1 & 0 & 0 & 0 & 0 \\ 0 & 0 & 1 & 1 & 1 & 1 \\ 0 & 0 & 1 & 1 & 0 & 0 \end{bmatrix}. \tag{9.105}$$

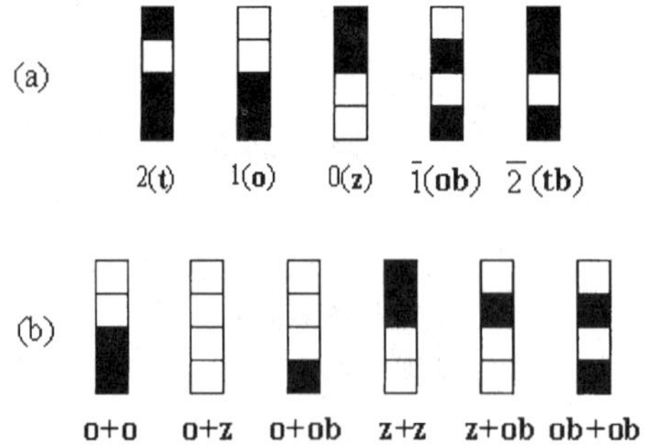

Fig. 9.25. Four-pixel TSD encoding for the digits in (a) the Y matrix and (b) the X matrix [152].

A nonexact pseudoinverse solution M of $Y = MX$ can be obtained as

$$M = \begin{bmatrix} 0.54 & -0.18 & -0.45 & 0.82 \\ 0.36 & 0.54 & -0.64 & 0.54 \\ 0.09 & -0.36 & 1.09 & -0.36 \\ -0.91 & 0.64 & 1.09 & -0.36 \end{bmatrix}. \quad (9.106)$$

The matrix M can be modulated by an amplitude modulator together with a phase modulator and it can be used as a matched spatial filter (MSF) to implement the symbolic substitution. Since M is not exact, a threshold operation is needed at the output plane. Now the Y matrix becomes

$$Y = MX = \begin{bmatrix} 0.36 & 0.73 & -0.09 & 0.36 & 0.91 & 0.09 \\ 0.91 & 0.82 & 0.27 & -0.09 & 0.27 & -0.27 \\ -0.27 & 0.45 & 0.82 & 0.73 & 0.82 & 1.18 \\ -0.27 & 0.45 & 0.82 & 0.73 & -0.18 & 0.18 \end{bmatrix}. \quad (9.107)$$

After thresholding at a value of 0.5, we obtain the correct output, given by

$$Y = \text{thresh}(MX) = \begin{bmatrix} 0 & 1 & 0 & 0 & 1 & 0 \\ 1 & 1 & 0 & 0 & 0 & 0 \\ 0 & 0 & 1 & 1 & 1 & 1 \\ 0 & 0 & 1 & 1 & 0 & 0 \end{bmatrix}. \quad (9.108)$$

It should be mentioned that the encoding of the TSD digits is not unique. Further improvement of the system is possible if we combine the $o + ob$ and $z + z$ digit pairs into a single pair. Other truth tables can be designed following this method [70, 126, 152].

9.6.2. SCAM–BASED INCOHERENT CORRELATOR FOR QSD ADDITION

Various holographic or nonholographic methods have been proposed to implement the truth-table look-up, or CAM. Through a holographic approach, a complete set of reduced minterms is used as the reference for generating an output digit. The number of minterms increases with the increase in the number of operand digits; hence, this technique is memory inefficient. In comparison, nonholographic schemes are simpler and more flexible. Recently, an efficient shared-CAM (SCAM) scheme has been explored [73, 95, 125, 156,

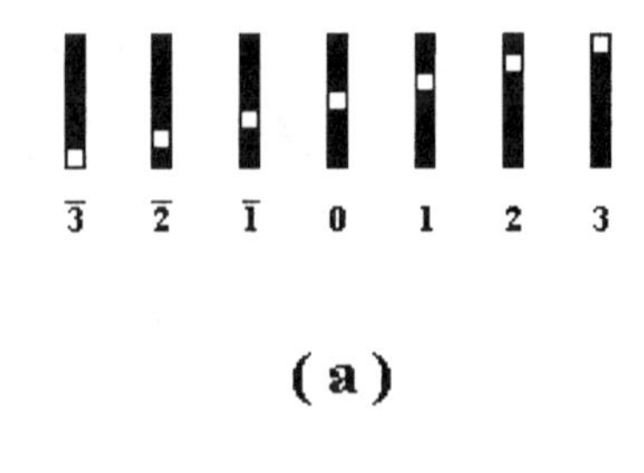

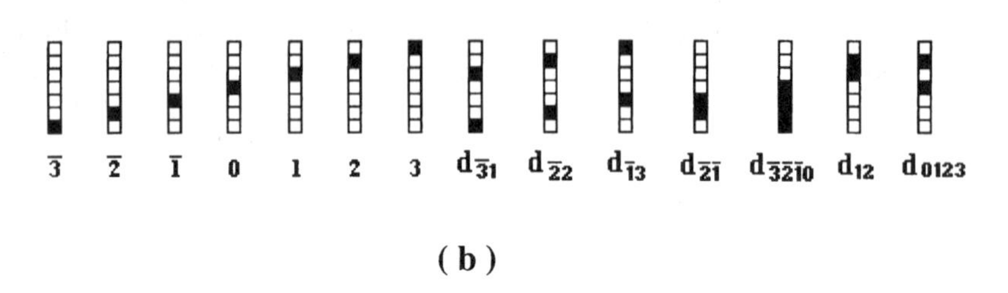

Fig. 9.26. Symmetric spatial encoding patterns for (a) the input digits and (b) storing the digits $\bar{3}$ through 3 and the partial don't-care digits [73].

160] where all the output digits can be produced by using only a single set of reference patterns. Here we use the incoherent correlator–based SCAM processor as the basic building module for implementing the simplified two-step QSD addition discussed in Sec. 9.4.4.1.

The input numbers and the minterms of the truth table should be spatially encoded to compare them and generate the nonzero outputs. The bright–dark encoding patterns for the input digits $\bar{3}$ through 3 are shown in Fig. 9.26(a). Each pattern is composed of seven pixels, one of which is bright. Note that the patterns for the complement digit pairs are symmetric; i.e., the encoding pattern for the digit d is a 180° rotated version of that for the digit $\bar{d}$.

The SCAM patterns are designed in such a way that when the input matches one of the stored minterms, no light will be transmitted to the output plane. This implies that a zero intensity value over a particular area corresponds to a valid output. Therefore, the SCAM patterns for storing the digits $\bar{3}$ through 3 are pixel-by-pixel complement of the individual input patterns of the same digit (Fig. 9.26[b]). Because the encoding patterns of the input are symmetric for complement digit pairs, the patterns for storing the complement digits are symmetric, too. The pattern for a partial don't-care digit should generate a dark output intensity for any of the possible input digits involved, so it can be designed by ANDing the individual storing patterns of these digits. The patterns for all of the partial don't-care digits of Table 9.33 are shown in Fig.

9.26(b). For example, the pattern for $d_{\bar{3}1}$ will block the light for input digits $\bar{3}$ and 1, and transmit light for other input digits.

To optically realize the pattern recognition among an input digit combination and the SCAM patterns, we can overlap the two encoded patterns and integrate the light intensity of the particular area. In the two addition steps, each minterm consists of 2 digits (a_i and b_i in the first step, S_i and C_i in the second step) and therefore consists of 14 transparent–opaque pixels. This operation is equivalent to performing vector inner product (VIP), where each vector has 14 binary elements. Each input digit combination ($a_i b_i$ or $S_i C_i$) is multiplied by all minterms that are responsible for generating the nonzero output. If the number of logically minimized minterms in a step is k, there will be k VIPs for each digit combination, and if one VIP is zero, it indicates a match. To accomplish this operation, an incoherent correlator–based optoelectronic SCAM processor, shown in Fig. 9.27(a), can be used. Two SLMs, SLM1, and SLM2, are utilized for encoding the input and storing the SCAM patterns, respectively. In SLM2, each minterm is stored in a column and one set of the minterms (the number being k) of a reduced truth table are displayed side by side. In SLM1, the different input digit combinations (assume J, $J = M + N$, where M and N are the numbers of digits for the fraction and integer parts of a QSD number, respectively) are encoded separately, and the separation between the two neighboring combinations is k times the pixel size p. All correlation outputs, including the desired VIPs, are produced in the back focal plane of the lens. Postprocessing electronics detects the intensity values. The operational principle of this unit will be illustrated next with an example.

As mentioned earlier, the minterms for outputs $\bar{1}$ and $\bar{2}$ ($\bar{1}$, $\bar{2}$, and $\bar{3}$) are digit-by-digit complement of the corresponding minterms for outputs 1 and 2 (1, 2, and 3), respectively, in step 1 (step 2). With symmetric spatial encoding, a pattern of $\bar{3}$, $\bar{2}$, $\bar{1}$, or 0 is a 180° rotated version of 3, 2, 1, or 0, respectively. Therefore, if we display the input pattern in one channel and form a digit-by-digit complement version of it in another channel by geometric optics, we can generate all outputs in the two channels using the same SCAM pattern, which only stores the minterms for the positive or negative outputs. Thus, it is possible to reduce the number of minterms by 50%; i.e., the number of minterms to be stored for steps 1 and 2 can be reduced from 20 and 12 to 10 and 6, respectively. The SCAM pattern for the two steps corresponding to the negative outputs of Table 9.33 are shown in Figs. 9.27(b) and 9.21(c), respectively. To obtain the positive outputs in parallel through another channel, we can either complement the input pattern digit-by-digit and use the same SCAM pattern, or complement the SCAM pattern digit-by-digit and use the same input pattern to perform the VIP operations.

For experimental verification, we consider the basic SCAM processor unit for adding two QSD numbers A and B. In the first step, using the same SCAM

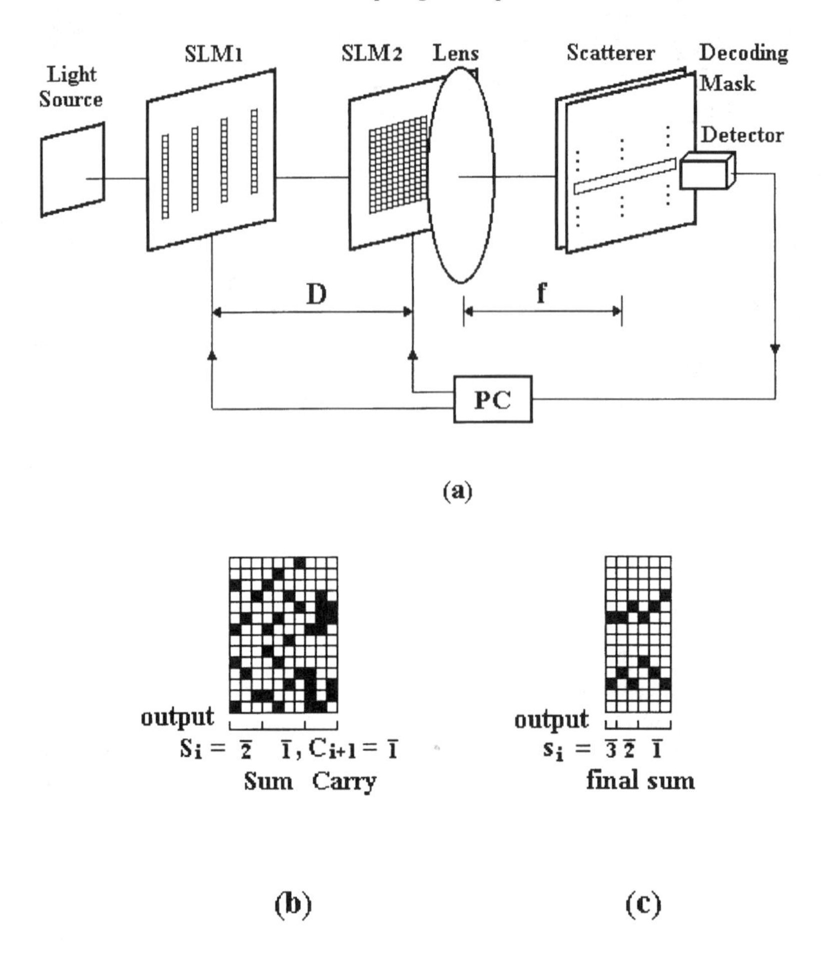

Fig. 9.27. (a) Incoherent correlator–based optoelectronic shared–CAM processor unit, (b) and (c), Shared–CAM patterns for simplified two-step QSD addition.

pattern shown in Fig. 9.27(b), two separate channels are employed to encode

$$(a_{N-1}b_{N-1}, a_{N-2}b_{N-2}, \ldots, a_1b_1, a_0b_0, a_{-1}b_{-1}, \ldots, a_{-M}b_{-M})$$

and

$$(\bar{a}_{N-1}\bar{b}_{N-1}, \bar{a}_{N-2}\bar{b}_{N-2}, \ldots, \bar{a}_1\bar{b}_1, \bar{a}_0\bar{b}_0, \bar{a}_{-1}\bar{b}_{-1}, \ldots, \bar{a}_{-M}\bar{b}_{-M})$$

for yielding the negative and positive outputs, respectively. In this example, $N = 4$ and $M = 0$. In channel 1, the digit combinations a_3b_3, a_2b_2, a_1b_1, a_0b_0

are spatially encoded as vectors in different columns, as shown in Fig. 9.28(a). Since there are 10 minterms stored in the SCAM ($k = 10$), the separation between the two adjacent vectors should be $10p$. In channel 2, the digit combinations $\bar{a}_3\bar{b}_3$, $\bar{a}_2\bar{b}_2$, $\bar{a}_1\bar{b}_1$, $\bar{a}_0\bar{b}_0$ are encoded following a similar procedure. Because the encoding of each digit combination comprises 14 pixels (2 digits—each encoded by 7 pixels) and the SCAM has 10 minterms, the correlation output between each digit combination and the SCAM pattern has a total of 27×10 pixels, as shown in Fig. 9.28(b) where the correlation of two 14-pixel vectors generated the 27-pixel vector. However, only the intensities of the 10 pixels in the central segment correspond to the exact VIP between the input digit combination and 10 minterms. Therefore, only these pixels, which are circumscribed in Fig. 9.28(b), need to be detected. Consequently, a decoding mask which is transparent in the center is used at the correlation plane [not shown in Fig. 9.28(b)]. Remember that the use of an extended light source is energy inefficient. To improve energy efficiency, an LED array or VCSEL array can be used [108–111, 125]. The desired negative (positive) output distribution for channel 1 (channel 2) is shown in Fig. 9.28(c), where a zero intensity value indicates a valid output. Note that the output digit indices are reversed compared to the input, due to the reverse imaging property of the lens. The intermediate results of S_0 and C_5 in this step are the final sum digits of s_0 and s_5, respectively, and they need not be put into operation in the second

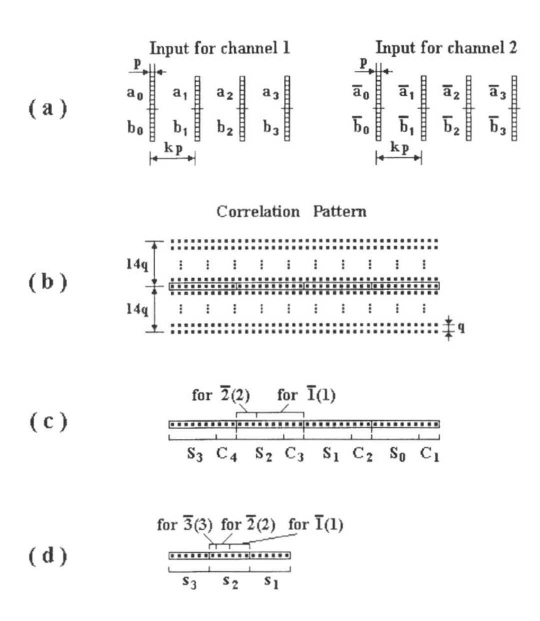

Fig. 9.28. Encoding and decoding [73]. (a) The input format; (b) and (c), Correlation pattern and the intermediate result after the first-step operation, (d) Final sum after the second-step addition.

The pixels in the central row:

The detected intensity values:

0221122112 2221122222 2122112110 2221221222

Output: $S_3 = \bar{2}$ $C_2 = \bar{1}$

(a)

The pixels in the central row:

The detected intensity values:

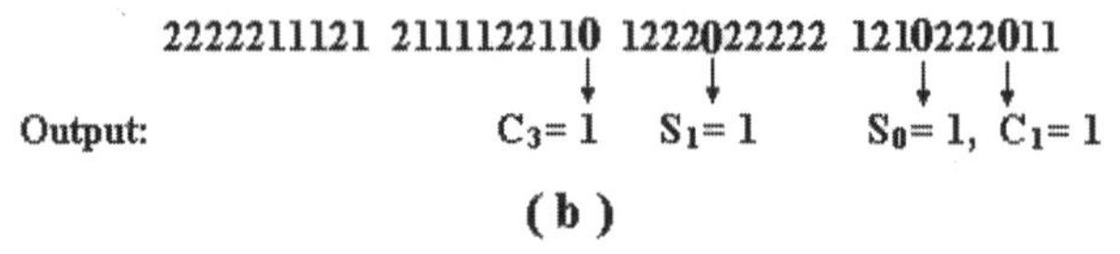

(b)

Fig. 9.29. Experimental results. (a) Central pixels of the correlation pattern of channel 1, the detected intensity values of these pixels, and the corresponding negative outputs, (b) Central pixels of the correlation pattern of channel 2, the detected intensity values, and the corresponding positive outputs, (c) and (d), Negative and positive final sum digits obtained in the second step.

step. In the second step, the intermediate results $S_3S_2S_1$ and $C_3C_2C_1$ are encoded in the same way as the inputs. Two channels of the SCAM processor are utilized and the SCAM pattern in Fig. 9.27(c) is used, which includes 6 minterms. The digit combinations S_3C_3, S_2C_2, and S_1C_1 and their complements are aligned in the same way as shown in Fig. 9.28(a). However, the separation between the two neighboring vectors is now $6p$ ($k = 6$ in this case). In the correlation output between each digit combination and the 6 minterms stored in the SCAM pattern, only the intensities of the central 6 pixels correspond to the desired VIPs. The negative or positive output distribution is illustrated in Fig. 9.28(d). With the operations similar to the first step, the final sum is obtained. Note that the detectors just need to detect whether there is an output at the individual positions and so thc threshold valuc can be selected between 0 and 1.

In the experiment, for example, the addition $\bar{3}2\bar{2}3 + 12\bar{1}2$ is verified. At first, the vectors $\bar{3}1$, 22, $\bar{2}\bar{1}$, and 32 are encoded as the input of channel 1, and their complemented counterparts $3\bar{1}$, $\bar{2}\bar{2}$, 21, and $\bar{3}\bar{2}$ are encoded as the input of

The pixels in the central row:

The detected intensity values:

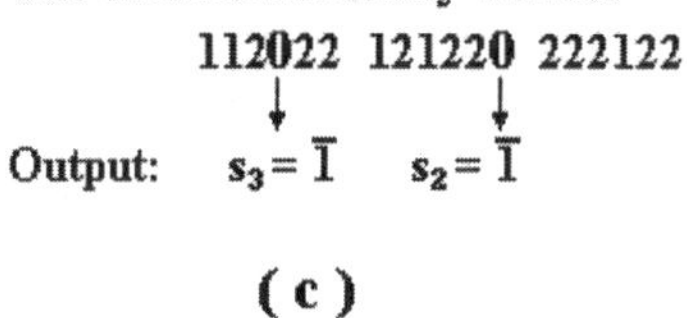

The pixels in the central row:

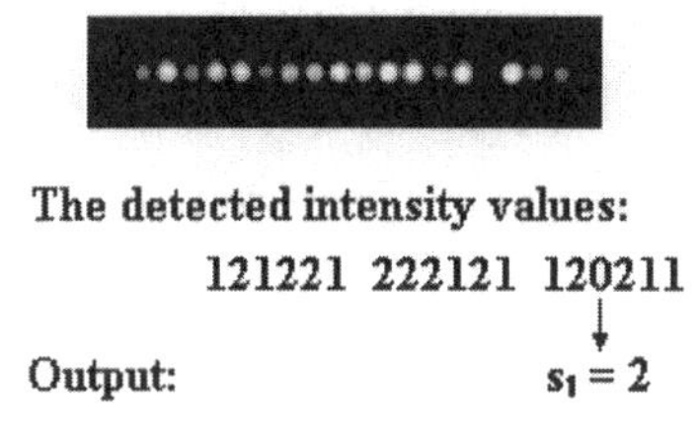

(d)

Fig. 9.29. Continued.

channel 2. The distance D between the two SLMs is 540 mm, and the focal length f of the lens is 240 mm. Figure 9.29(a) shows the pixels in the central row of the detected correlation pattern and the detected intensity values of these pixels. With reference to Fig. 9.28(c), we obtain the negative intermediate output $S_3 = \bar{2}$ and $C_2 = \bar{1}$ by identifying the zero intensities. Figure 9.29(b) shows the decoded result in channel 2 for the positive intermediate output. From Fig. 9.29(b), it is evident that $S_0 = S_1 = 1$ and $C_3 = C_1 = 1$. The intermediate result S_0 in this step is also the final sum digit of s_0; , i.e., $s_0 = 1$. Thus, the operands for the second-step addition become $S_3S_2S_1 = \bar{2}01$ and $C_3C_2C_1 = 1\bar{1}1$. The digit combinations $\bar{2}1$, $0\bar{1}$, and 11 and their complemented counterparts are encoded in channels 1 and 2, respectively. The experimental result for the negative and positive sum digits are obtained as shown in Figs. 9.29(c) and 9.29(d), respectively. Combining these results, we obtain the final sum $\bar{1}\bar{1}21$, which is consistent with the results obtained using numerical simulation.

9.6.3. OPTICAL LOGIC ARRAY PROCESSOR FOR PARALLEL NSD ARITHMETIC

In earlier sections, we discussed a set of NSD arithmetic operations. Addition and subtraction are performed at each digit position in parallel by the same logic operations, which are thus space invariant. This property is well suited to optical cellular architecture and, therefore, these algorithms can be effectively mapped for optical implementation. Since all algorithms can be executed via binary logic operations, we focus on a new implementation of parallel optical logic. As an example, we demonstrate the parallel execution of both addition and subtraction concurrently using the two-step carry-free algorithm.

Various optoelectronic systems have been suggested for implementing arithmetic-logic units. They use either linear encoding with nonlinear optical devices, or nonlinear encoding with linear optical elements. To perform space-variant operations, space-variant encoding and decoding must be used. In these systems, polarization encoding is usually employed and several subpixels are needed to represent a single digit, which increases the implementation difficulty and sacrifices the space–bandwidth product. Moreover, complex logic operations can be realized in a recursive fashion by decomposing them into simple ones. When the intermediate results are obtained, the optical signals are converted to electronic signals, stored in the controlling electronics, and then fed back as the input for the next iteration using SLMs. This operation is time consuming and greatly diminishes system performance.

Recently, a novel optical logic device using electron-trapping (ET) material has been reported, which has the advantages of high resolution, nanosecond response time, and high sensitivity. As a result, it has been used in 3D memory, neural networks, and optical computing systems. Here a novel method for implementing arbitrary arithmetic and logic functions without temporal latency is presented, and in principle, it allows the operations of any number of variables. Based on the above algorithms and space-variant operations, addition and subtraction can be easily realized in parallel. In this technique, only one pixel and binary intensity values are needed for encoding each digit.

The basic principle of the ET device for logic operations was described in Sec. 9.2.3. This technique can be extended to implement the combinational logic involved in negabinary arithmetic. For example, the operational procedure for the logic in the two-step addition/subtraction is shown in Table 9.40, where the overbar represents blue illumination and the underline represents infrared illumination. The symbol xy is just an overlap of x and y. Since the NOT operation can be realized either by negating the corresponding digits in the SLM or by the ET device, the operational principle can be classified into two categories, depending on whether the negation is included in the input. If the numbers are input without negation, it is seen from Table 9.40 that the

Table 9.40

Procedure for Performing the Binary Logic Operations Involved in the Two-Step NSD Addition/Subtraction [162] where $\overline{\overline{x}}$ Represents Input x with Blue Illumination, $\underline{\underline{x}}$ Represents Input x with Infrared Illumination, xy Implies the Overlap of x and y, and the Numbers can be Input with or without Negation

	Operation	Procedure
Functions	Input without negation	Input with negation
$(a_i \oplus b_i) f_i$	$\overline{\overline{a_i f_i}}\,\overline{\overline{b_i f_i}}\,\underline{\underline{a_i b_i f_i}}$	$\overline{\overline{a_i b_i f_i}}\,\overline{\overline{a_i \bar{b}_i f_i}}$
$(a_i \oplus b_i) \overline{f}_i$	$\overline{\overline{a}}_i\,\underline{\underline{b_i f_i}}\,\overline{\overline{b}}_i\,\underline{\underline{a_i b_i}}\,\underline{\underline{b_i f_i}}$	$\overline{\overline{\bar{a}_i b_i \bar{f}_i}}\,\overline{\overline{a_i \bar{b}_i \bar{f}_i}}$
$h_i + (a_i \oplus b_i)\overline{g}_i f_i$	$\overline{\overline{b_i f_i}}\,\overline{\overline{a_i f_i}}\,\underline{\underline{a_i b_i}}\,\underline{\underline{a_i g_i}}\,\underline{\underline{b_i g_i}}\,\overline{\overline{h}}_i$	$\overline{\overline{h}}_i\,\overline{\overline{\bar{a}_i b_i \bar{g}_i f_i}}\,\overline{\overline{a_i \bar{b}_i \bar{g}_i f_i}}$
$(a_i b_i + (a_i \oplus b_i)\bar{f}_i) g_i$	$\overline{\overline{b_i g_i}}\,\overline{\overline{a_i g_i}}\,\underline{\underline{a_i b_i}}\,\underline{\underline{a_i f_i}}\,\underline{\underline{b_i f_i}}\,\overline{\overline{g}}_i$	$\overline{\overline{a_i b_i g_i}}\,\overline{\overline{\bar{a}_i b_i \bar{f}_i g_i}}\,\overline{\overline{a_i \bar{b}_i \bar{f}_i g_i}}$

overlap involves a maximum of three variables. The optical system shown in Fig. 9.8 can be effectively employed for this implementation. Shutters 1 and 2 are used to control the exposure of blue and infrared light beams. SLMs 1, 2, and 3 are utilized to input the data arrays. When a complete arithmetic-logic function is performed, the final result is read with infrared and detected by a CCD. Since both addition and subtraction are realized by the same logic, the operands to be added or subtracted are arranged in two sectors for parallel processing of the two types of operations. The corresponding digits are matched to each other in the SLMs. For instance, to perform the logic $(a_i \oplus b_i) f_i$ on two-dimensional data arrays, first data arrays A and F enter the system through SLM_1 and SLM_3, respectively, with the blue illumination. Then, data arrays B and F are input through SLM_2 and SLM_3, respectively, also with blue illumination. Finally, the data arrays A, B, and F are overlapped with the infrared illumination. Other logic operations are performed in a similar fashion. Operations $(a_i \oplus b_i) \overline{f}_i$ on two-dimensional data arrays are performed by (a) input A with blue illumination, (b) input B with infrared illumination, (c) input F with infrared illumination, (d) input B with blue illumination, (e) input AB with infrared illumination, and (f) input BF with infrared illumination. Operations $h_i + (a_i \oplus b_i)\overline{g}_i f_i$ on two-dimensional arrays are performed by (a) input BF with blue illumination, (b) input AF with blue illumination, (c) input AB with infrared illumination, (d) input AG with infrared illumination, (e) input BG with infrared illumination, and (f) input H with blue illumination. Operations $(a_i b_i + (a_i \oplus b_i)\overline{f}_i) g_i$ on two-dimensional data arrays are performed by (a) input BG with blue illumination, (b) input

AG with blue illumination, (c) input *AB* with infrared illumination, (d) input *AF* with infrared illumination, (e) input *BF* with infrared illumination, and (f) input *G* with blue illumination. If the numbers are input with negation, the operation becomes more simple and fewer steps are required.

In fact, after each illumination, the intermediate result is automatically stored in the ET device. Therefore, it is not necessary to convert the intermediate result through the CCD and store it in the computer. Furthermore, only binary values are included both in the input and output. The CCD detector only needs to distinguish the binary states rather than act as a threshold device. This makes the system more tolerable to optical noise. Notice that the main computing hardware can be stacked together. Thus, it is possible to construct a compact negabinary computer for arithmetic and logical operations. Moreover, the system is simple since the signed-digit arithmetic is realized by binary logic. This leads to an efficient and inexpensive general-purpose optical computing system. The configuration used here is much simpler than that used for MSD computing operations [127, 128].

The experimental results verifying the numerical examples shown in Sec. 9.4.5 for $A2 + B2$ and $A2 - B2$ are shown in Fig. 9.30, where the input numbers are introduced with negation. In each photograph, the upper row corresponds the result of $A1 + B1$ and the lower row corresponds to the result of $A2 - B2$. Figures 9.30(a) and (b) show the outputs 1 and $\bar{1}$ of W, and the relative combination generates $0\bar{1}0\bar{1}101\bar{1}0$ and $0\bar{1}\bar{1}11000$ for W output. Figures 9.30(c) and (d) show the outputs 1 and $\bar{1}$ of T, and the relative combination generates $\bar{1}1\bar{1}1\bar{1}1\bar{1}10$ and $\bar{1}\bar{1}0100\bar{1}10$ for T output. Then the digits of T and W are used as input for the second-step addition. Figures 9.30(e) and (f) depict the outputs 1 and $\bar{1}$ of the sum and difference, respectively. A combination of the outputs generates the final sum $\bar{1}0\bar{1}001000$ and the final difference $\bar{1}\bar{1}\bar{1}011\bar{1}10$. Note that the optical signal output has the same form as the input encoding so that the result can be directly used as input for the next addition or subtraction stage.

9.7. SUMMARY

We have reviewed the parallel optical logic architectures and the various algorithms of different number systems including binary, negabinary, residue, modified signed-digit, trinary signed-digit, quaternary signed-digit, and negabinary signed-digit representations. Ultrafast parallel arithmetic algorithms, fast-conversion algorithms between the signed and unsigned number systems, encoding schemes, and architectures for implementation as well as experimental demonstration of sample arithmetic operations are also incorporated. The inherent parallelism of optics is a good match between the parallelism of the

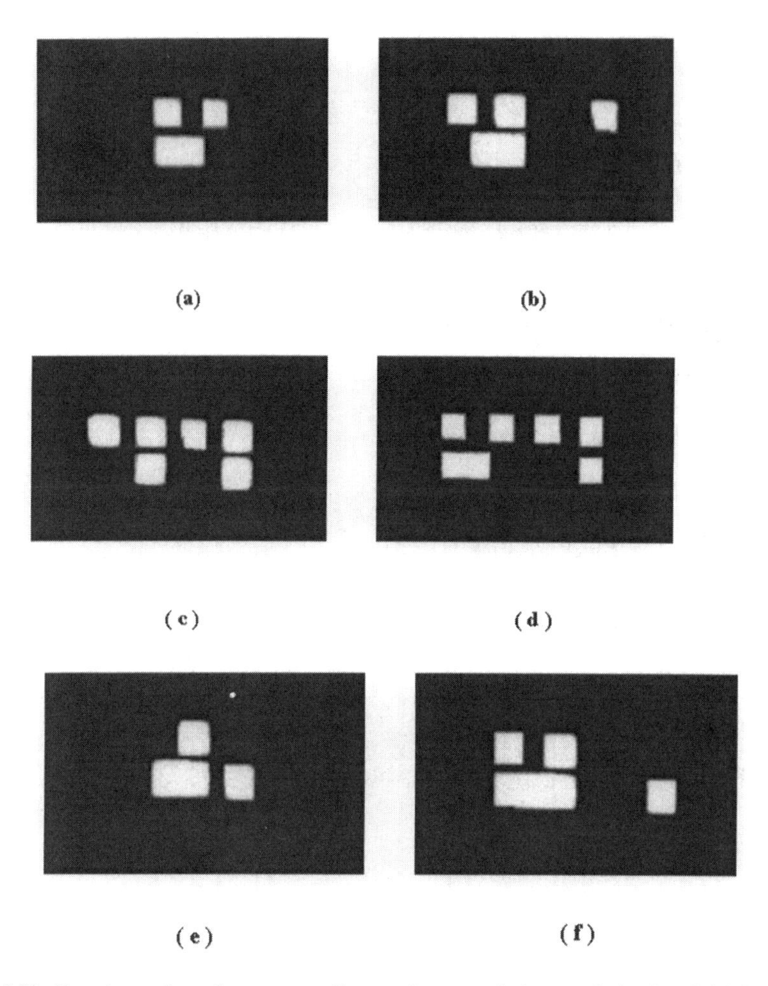

Fig. 9.30. Experimental result corresponding to the numerical example in Sec. 9.4.5 [162]. (a) Output 1 of W, (b) Output $\bar{1}$ of W. The relative combination generates the W result $0\bar{1}0\bar{1}101\bar{1}0$ and $0\bar{1}\bar{1}11000$. (c) Output 1 of T, (d) Output $\bar{1}$ of T. The relative combination generates the T result $\bar{1}1\bar{1}1\bar{1}1\bar{1}10$ and $\bar{1}\bar{1}0100\bar{1}10$. (e) Output of 1 of the sum and difference, (f) Output of $\bar{1}$ of the sum and difference. A combination generates the final $\bar{1}1\bar{1}001000$ and the final difference $\bar{1}\bar{1}\bar{1}011\bar{1}10$.

algorithms and the architectures proposed. By exploiting the redundancy of signed-digit number systems, addition, subtraction, multiplication, division, and other complex arithmetic operations can be completed in fixed steps independent of the operand length. The table look-up, content-addressable memory, symbolic substitution, and parallel logic array approaches to optical computing can effectively exploit the benefits of optics in parallel information processing, which may lead to extremely powerful and general-purpose computer systems. Currently, the impact of optics is mainly in the noninterfering interconnects at different levels. The ultrahigh processing speed and inherent parallelism of optics, huge data rates from optical storage, and the suitable advantages of microelectronics can be combined in the evolution of the next generation of computers. The performance of optoelectronic processor has already been analyzed in some references [70, 123]. With the rapid advancement of optoelectronic integrated circuits, photonic switching, optical interconnect, and optical storage, tremendous progress is expected to be seen in the development of optoelectronic computers in the near future.

REFERENCES

9.1 F. T. S. Yu and S. Jutamulia, *Optical Signal Processing, Computing, and Neural Networks*, John Wiley & Sons, Inc., New York, 1992.

9.2 Special issue on optical computing, 1984, *Proc. IEEE* 72, No. 7, 1984, *Proc. IEEE*, No. 11; 1994; Special issue on optical information processing, 1996, *Proc. IEEE*, No. 5 and 6.

9.3 W. T. Cathey, K. Wagner, and W. J. Miceli, 1989, "Digital Computing with Optics," *Proc. IEEE* 77, 1558–1572.

9.4 N. Streibl, K.-H. Brenner, A. Huang, J. Jahns, J. Jewell, A. W. Lohmann, D. A. B. Miller, M. Murdocca, M. E. Prise, and T. Sizer, 1989, "Digital Optics," *Proc. IEEE* 77, 1954–1969.

9.5 P. B. Berra, A. Ghafoor, M. Guizani, S. J. Marcinkowski, and P. A. Mitkas, 1989, "Optics and Supercomputing," *Proc. IEEE* 77, 1797–1815.

9.6 A. D. McAulay, *Optical Computer Architectures*, John Wiley & Sons, Inc., New York, 1991.

9.7 M. A. Karim and A. A. S. Awwal, *Optical Computing: An Introduction*, John Wiley & Sons, Inc., New York, 1992.

9.8 N. C. Craft and M. E. Prise, May 1990, "Processor Does Light Logic," *Laser Focus World*, 191–200.

9.9 F. A. P. Tooley and S. Wakelin, 1993, "Design of a Symmetric Self-Electro-optic-Effect-Device Cellular-Logic Image Processor," *Appl. Opt.* 32, 1850–1862.

9.10 V. P. Heuring, 1991, "Systems Considerations in Designing and Implementing a Bit Serial Optical Computer," *Opt. Eng.* 30, 1931–1935.

9.11 P. S. Guilfoyle, May 1993, "A Third-Generation Digital Optical Computer," *Photonics Spectra*, 116–124.

9.12 B. A. Horwitz and F. J. Corbett, 1978, "The PROM—Theory and Applications for the Pockels Readout Optical Modulator," *Opt. Eng.* 17, 353–364.

9.13 A. H. Khan and U. R. Nejib, 1987, "Optical Logic Gates Employing Liquid Crystal Optical Switches," *Appl. Opt.* 26, 270–273.

9.14 R. A. Athale and S. H. Lee, 1979, "Development of an Optical Parallel Logic Device and a Half-Adder Circuit for Digital Optical Processing," *Opt. Eng.* 18, 513–517.

9.15 P. Chavel, A. A. Sawchuk, T. C. Strand, A. R. Tanguay, and B. H. Soffer, 1980, "Optical Logic with Variable-Grating–Mode Liquid Crystal Devices," *Opt. Lett.* 5, 398–400.

9.16 M. T. Fatehi, K. C. Wasmundt, and S. A. Collins, 1981, "Optical Logic Gates Using Liquid Crystal Light Valve: Implementation and Application Example," *Appl. Opt.* 20, 2250–2256.

9.17 C. Warde, A. M. Weiss, A. D. Fisher, and J. I. Thackara, 1981, "Optical Information Processing Characteristics of the Microchannel Spatial Light Modulator," *Appl. Opt.* 20, 2066–2074.

9.18 F. T. S. Yu and S. Jutamulia, 1987, "Optical Parallel Logic Gates Using Inexpensive Liquid Crystal Televisions," *Opt. Lett.* 12, 1050–1052.

9.19 F. T. S. Yu, S. Jutamulia, and T. Lu, 1987, "Optical Parallel Logic Based on Magneto-optic Spatial Light Modulator," *Opt. Commun.* 63, 225–229.

9.20 Y.-D. Wu, D.-S. Shen, V. K. Bykovsky, J. Rosetti, and M. A. Fiddy, 1994, "Digital Optical Computing with Magneto-optic Spatial Light Modulators: A New and Efficient Multiplication Algorithm," *Appl. Opt.* 33, 7572–7578.

9.21 A. D. McAulay, 1988, "Logic and Arithmetic with Luminescent Rebroadcasting Devices," *Proc. SPIE* 936, 321–326.

9.22 S. Jutamulia, G. Storti, J. Lindmayer, and W. Seiderman, 1990, "Use of Electron–Trapping Materials in Optical Signal Processing I: Parallel Boolean Logic," *Appl. Opt.* 29, 4806–4811.

9.23 G. Li, F. Qian, H. Ruan, and L. Liu, 1999, "Compact Parallel Optical Modified Signed-Digit Arithmetic-Logic Array Processor by Using Electron-Trapping Device," *Appl. Opt.* 38, 5039–5045.

9.24 A. D. McAulay, 1999, "Optical Guided Wave Arithmetic," *Opt. Eng.* 38, 468–476; H. F. Taylor, 1978, "Guided Wave Electro-optic Devices for Logic and Computation," *Appl. Opt.* 17, 1493–1498.

9.25 V. G. Krasilenko, A. E. Volosovich, L. I. Konopaltseva, and O. K. Kolesnitsky, 1995, "Integrated Optical Multichannel Logic Elements," *Opt. Eng.* 34, 1049–1052.

9.26 D. H. Schaefer and J. P. Strong, 1979, "Tse Computers," *Proc. IEEE* 65, 129–138.

9.27 Y. Fainman, C. C. Guest, and S. H. Lee, 1986, "Optical Digital Logic Operations by Two-Beam Coupling in Photorefractive Material," *Appl. Opt.* 25, 1598–1603.

9.28 J. A. Davis, R. P. Tiangco, T. H. Barnes, and T. G. Haskell, 1995, "Programmable Optical Logic Systems Using Free-Space Optical Interconnections," *Appl. Opt.* 34, 1929–1937.

9.29 B. K. Jekins, A. A. Sawchuk, T. C. Strand, R. Forchheimer, and B. H. Soffer, 1984, "Sequential Optical Logic Implementation," *Appl. Opt.* 23, 3455–3464.

9.30 M. E. Prise, N. C. Craft, R. E. LaMarche, M. M. Downs, S. J. Walker, L. A. D'Asaro, and L. M. Chirovsky, 1990, "Module for Optical Logic Circuits Using Symmetric Self-Electro-optic Effect Devices," *Appl. Opt.* 29, 2164–2170.

9.31 R. G. A. Craig, G. S. Buller, F. A. P. Tooley, S. D. Smith, A. C. Walker, and B. S. Wherrett, 1990, "All-Optical Programmable Logic Gate," *Appl. Opt.* 29, 2148–2152.

9.32 T. Yanase, K. Kasahara, Y. Tashiro, S. Asada, K. Kubota, S. Kawai, M. Sugimoto, N. Hamao, and N. Takadou, *Optical Computing in Japan*, S. Ishihara, Ed., Nova Science Publishers, Commack, USA, 1989, pp. 475.

9.33 B. M. Watrasiewicz, 1975, "Optical Digital Computers," *Opt. Laser Technol.* 7, 213–215.

9.34 H. Bartelt, A. W. Lohmann, and E. E. Sicre, 1984, "Optical Logical Processing in Parallel with Theta Modulation," *J. Opt. Soc. Am.* A 1, 944–951; A. W. Lohmann and J. Weigelt, 1986, "Digital Optical Adder Based on Spatial Filtering," *Appl. Opt.* 25, 3047–3053.

9.35 A. W. Lohmann and J. Weigelt, 1984, "Optical Logic Processing Based on Scattering," *Opt. Commun.* 52, 255–258.

9.36 A. W. Lohmann and J. Weigelt, 1985, "Optical Logic by Anisotropic Scattering," *Opt. Commun.* 54, 81–86.

9.37 A. W. Lohmann and J. Weigelt, 1987, "Spatial Filtering Based on Polarization," *Appl. Opt.* 131–135.

9.38 X. Yang, K. Chin, and M. Wu, 1987, "Real-Time Optical Logic Processing," *Opt. Commun.* 64, 412–416.

9.39 A. A. Rizvi, S. U. Rehman, and M. S. Zubairy, 1988, "Binary Logic by Grating Structure," *J. Mod. Opt.* 35, 1591–1594.

9.40. J. Tanida and Y. Ichioka, 1983, "Optical Logic Array Processor Using Shadowgrams," *J. Opt. Soc. Am.* 73, 800–809.

9.41 J. Tanida, J. Nakagawa, and Y. Ichioka, 1989, "Local Variable Logic Operation Using Birefringent Optical Elements," *Appl. Opt.* 28, 3467–3473.

9.42 G. Li, L. Liu, M. He, and H. Cheng, 1997, "Optical Multi-input Parallel Logic Array Processor for Higher-Order Negabinary Addition," *Optik* 105, 93–98.
9.43 A. K. Datta and M. Seth, 1994, "Multi-input Optical Parallel Logic Processing with the Shadow-Casting Technique," *Appl. Opt.* 33, 8146–8152.
9.44 L. Liu and X. Liu, 1991, "Cascadable Binary Pattern Logic Processor Using Multiple Imaging," *Opt. Commun.* 82, 446–452.
9.45 Z. Zhang and L. Liu, 1992, "Solid-State Integrated Optical Parallel Dual-Rail Logic Gate Module," *Opt. Commun.* 91, 185–188.
9.46 W. Wu, S. Campbell, S. Zhou, and P. Yeh, 1993, Polarization-Encoded Optical Logic Operations in Photorefractive Media," *Opt. Lett.* 18, 1742–1744.
9.47 L. Liu and X. Liu, 1991, "Logic Gate Modules Using Opto-optical Birefringence Switching," *Opt. Lett.* 16, 1439–1441.
9.48 K. W. Wong, L. M. Cheng, and M. C. Poon, 1992, "Design of Digital-Optical Processors by Using Both Intensity and Polarization-Encoding Schemes," *Appl. Opt.* 31, 3225–3232.
9.49 T. Yatagai, 1986, "Optical Space-Variant Logic-Gate Array Based on Spatial-Encoding Technique," *Opt. Lett.* 11, 260–262.
9.50 G. Li, L. Liu, L. Shao, Y. Yin, and J. Hua, 1997, "Parallel Optical Negabinary Arithmetic Based on Logic Operations," *Appl. Opt.* 36, 1011–1016.
9.51 J. N. Roy and S. Mukhopadhyay, 1995, "A Minimization Scheme of Optical Space-Variant Logical Operations in a Combinational Architecture," *Opt. Commun.* 119, 499–504.
9.52 Y. Li, G. Eichmann, and R. R. Alfano, 1986, "Optical Computing Using Hybrid-Encoded Shadow Casting," *Appl. Opt.* 25, 2636–2638.
9.53 M. Karim, A. A. S. Awwal, and A. K. Cherri, 1987, "Polarization-Encoded Optical Shadow-Casting Logic Units: Design," *Appl. Opt.* 2720–2725.
9.54 A. A. S. Awwal and M. Karim, 1990, "Multiprocessor Design Using Polarization-Encoded Optical Shadow Casting," *Appl. Opt.* 29, 2107–2112; J. U. Ahmed and A. A. S. Awwal, 1992, "Polarization-Encoded Optical Shadow-Casting Arithmetic-Logic–Unit Design: Separate and Simultaneous Output Generation," *Appl. Opt.* 31, 5622–5631.
9.55 G. Li, L. Liu, and J. Hua, 1997, "Unified Optical Negabinary Arithmetic with Polarization-Encoded Optical Shadow Casting," *Opt. Laser Technol.* 29, 221–227.
9.56 S. Zhou, X. Yang, M. Wu, and K. Chin, 1987, "Triple-In Double-Out Optical Parallel Logic Processing System," *Opt. Lett.* 12, 968–970.
9.57 K. W. Wong and L. M. Cheng, 1994, "Space-Variant Optical Logic Operations Based on Operation-Dependent Encoding Method," *Appl. Opt.* 33, 2134–2139.
9.58 M. Murdocca, *A Digital Design Methodology for Optical Computing*, The MIT Press, Cambridge, 1990.
9.59 V. P. Heuring, L. H. Ji, R. J. Feuerstein, and V. N. Morozov, 1994, "Toward a Free-Space Parallel Optoelectronic Computer: A 300-MHz Optoelectronic Counter Using Holographic Interconnects," *Appl. Opt.* 33, 7579–7587.
9.60 H. Ruan, S. Chen, and F. Gan, 1998, "Parallel Optical Logic Processor and Bit Slice Full-Adder Using a Single Electron-Trapping Device," *Opt. Laser Technol.* 30.
9.61 T. Yatagai, 1986, "Cellular Logic Architecture for Optical Computers," *Appl. Opt.* 25, 1571–1577.
9.62 J. Tanida, T. Konishi, and Y. Ichioka, 1994, "P-OPALS: Pure Optical-Parallel Array Logic System," *Proc. IEEE* 82, 1668–1676.
9.63 A. Huang, "Parallel Algorithms for Optical Digital Computers," in *Technical Digest*, IEEE Tenth International Optical Computing Conference, S. Horvitz, Ed., IEEE Computer Society Press, Silver Spring, Maryland, 1983, pp. 13–17.
9.64 K.-H. Brenner, A. Huang, and N. Streibl, 1986, "Digital Optical Computing with Symbolic Substitution," *Appl. Opt.* 25, 3054–3060.
9.65 K.-H. Brenner, M. Kufner, and S. Kufner, 1990, "Highly Parallel Arithmetic Algorithms for a Digital Optical Processor Using Symbolic Substitution Logic," *Appl. Opt.* 29, 1610–1618.

9.66 G. Li and L. Liu, 1993, "Optical One-Rule Symbolic Substitution for Pattern Processing," *Opt. Commun.* 101, 170–174.

9.67 H.-I. Jeon, M. A. G. Abushagur, A. A. Sawchuk, and B. K. Jenkis, 1990, "Digital Optical Processor Based on Symbolic Substitution Using Holograph Matched Filtering," *Appl. Opt.* 29, 2113–2125.

9.68 S. D. Goodman and W. T. Rhodes, 1988, "Symbolic Substitution Applications for Image Processing," *Appl. Opt.* 27, 1708–1714.

9.69 A. K. Cherri, A. A. S. Awwal, and M. A. Karim, 1991, "Morphological Transformations Based on Optical Symbolic Substitution and Polarization-Encoded Optical Shadow-Casting System," *Opt. Commun.* 82, 441–445.

9.70 D. Casasent and P. Woodford, 1994, "Symbolic Substitution Modified Signed-Digit Optical Adder," *Appl. Opt.* 33, 1498–1506.

9.71 B. L. Drake, R. P. Bocker, M. E. Lasher, R. H. Patterson, and W. J. Miceli, 1986, "Photonic Computing Using the Modified Signed-Digit Number Representation," *Opt. Eng.* 25, 38–43.

9.72 Y. Li, D. H. Kim, A. Kostrzewski, and G. Eichmann, 1989, "Content-Addressable Memory-Based Single-Stage Optical Modified Signed-Digit Arithmetic," *Opt. Lett.* 14, 1254–1256.

9.73 G. Li, L. Liu, H. Cheng, and H. Jing, 1997, "Simplified Quaternary Signed-Digit Arithmetic and Its Optical Implementation," *Opt. Commun.* 137, 389–396.

9.74 Y. Li and G. Eichmann, 1987, "Conditional Symbolic Modified Signed-Digit Arithmetic Using Optical Content-Addressable Memory Logic Elements," *Appl. Opt.* 26, 2328–2333.

9.75. Y. Takaki and H. Ohzu, 1990, "Optical Half-Adder Using Wavefront Superposition," *Appl. Opt.* 29, 4351–4358.

9.76. P. P. Banerjee and A. Ghafoor, 1988, "Design of a Pipelined Optical Binary Processor," *Appl. Opt.* 27, 4766–4770.

9.77 Y. Li, M. Turner, P. Neos, R. Dorsinville, and R. R. Alfano, 1989, "Implementation of a Binary Optical Full-Adder Using a Venn Diagram and Optical Phase Conjugation," *Opt. Lett.* 14, 773–775.

9.78 S. Fukushima, T. Kurokawa, and H. Suzuki, 1990, "Optical Implementation of Parallel Digital Adder and Subtractor," *Appl. Opt.* 29, 2099–2106.

9.79 Y. Jin and F. T. S. Yu, 1988, "Optical Binary Using Liquid Crystal Television," *Opt. Commun.* 65, 11–16.

9.80 W. Wu, S. Campbell, and P. Yeh, 1995, "Implementation of an Optical Multiwavelength Full-Adder with a Polarization Encoding Scheme," *Opt. Lett.* 20, 79–81.

9.81 V. P. Heuring, H. F. Jordan, and J. P. Pratt, 1992, "Bit-Serial Architecture for Optical Computing," *Appl. Opt.* 31, 3213–3224.

9.82 K.-H. Brenner, W. Eckert, and C. Passon, 1994, "Demonstration of an Optical Pipeline Adder and Design Concepts for its Microintegration," *Opt. Laser Technol.* 26, 229–2237.

9.83 K. Hwang, *Computer Arithmetic: Principles, Architecture, and Design*, John Wiley & Sons, 1979.

9.84 E. Swartzlander, 1986, "Digital Optical Arithmetic," *Appl. Opt.* 25, 3021–3032.

9.85 B. Arazi, 1985, "An Electro-optical Adder," *Proc. IEEE* 73, 162–163.

9.86 J. B. McManus and R. S. Putnam, 1987, "Construction of an Optical Carry-Adder," *Appl. Opt.* 26, 1557–1562.

9.87 V. Chandran, T. F. Krile, and J. F. Walkup, 1986, "Optical Techniques for Real-Time Binary Multiplication," *Appl. Opt.* 25, 2272–2276.

9.88 A. Kostrzewski, D. H. Kim, Y. Li, and G. Eichmann, 1990, "Fast Hybrid Parallel Carry-Lookahead Adder," *Opt. Lett.* 15, 915–917.

9.89 S. P. Kozaitis, 1988, "Higher-Order Rules for Symbolic Substitution," *Opt. Commun.* 65, 339–342.

9.90 G. Eichman, A. Kostrzewski, D. H. Kim, and Y. Li, 1990, "Optical Higher-Order Symbolic Recognition," *Appl. Opt.* 29, 2135–2147.

9.91 P. S. Guilfoyle and W. J. Wiley, 1988, "Combinational Logic Based on Digital Optical Computing Architecture," *Appl. Opt.* 27, 1661–1673.
9.92 A. K. Datta, A. Basuary, and S. Mukhopadhyay, 1989, "Arithmetic Operations in Optical Computations Using a Modified Trinary Number System," *Opt. Lett.* 14, 426–428.
9.93 G. A. De Biase and A. Massini, 1993, "Redundant Binary Number Representation for an Inherently Parallel Arithmetic on Optical Computers," *Appl. Opt.* 32, 659–664.
9.94 G. Li, L. Liu, L. Shao, and Z. Wang, 1994, "Negabinary Arithmetic Algorithms for Digital Parallel Optical Computation," *Opt. Lett.* 19, 1337–1339.
9.95 G. Li, L. Liu, L. Shao, and Y. Yin, 1996, "Optical Negabinary Carry-Lookahead Addition with Higher-Order Substitution Rules," *Opt. Commun.* 129, 323–330.
9.96 D. Psaltis, D. Casasent, D. Neft, and M. Carlotto, 1980, "Accurate Numerical Computation by Optical Convolution," *Proc. SPIE* 232, 151–156.
9.97 P. S. Guilfoyle, 1984, "Systolic Acousto-optic Binary Convolver," *Opt. Eng.* 23, 20–25.
9.98 C. K. Gary, 1992, "Matrix–Vector Multiplication Using Digital Partitioning for More Accurate Optical Computing," *Appl. Opt.* 31, 6205–6211.
9.99 D. S. Kalivas, 1994, "Real-Time Optical Multiplication with High Accuracy," *Opt. Eng.* 33, 3427–3431.
9.100 D. S. Athale, W. C. Collins, and P. D. Stilwell, 1983, "High-Accuracy Matrix Multiplication with Outer-Product Optical Processor," *Appl. Opt.* 22, 368–370.
9.101 Y. Li, G. Eichmann, and R. R. Alfano, 1987, "Fast Parallel Optical Digital Multiplication," *Opt. Commun.* 64, 99–104.
9.102 R. P. Bocker, 1984, "Optical Digital RUBIC (Rapid Unbiased Bipolar Incoherent Calculator) Cube Processor," *Opt. Eng.* 23, 26–33.
9.103 B. K. Taylor and D. P. Casasent, 1986, "2's Complement Data Processing for Improved Encoded Matrix–Vector Processors," *Appl. Opt.* 25, 956–961.
9.104 G. Li and L. Liu, 1994, "Complex-Valued Matrix–Vector Multiplication Using 2's Complement Representation," *Opt. Commun.* 105, 161–166.
9.105 G. Li, L. Liu, L. Shao, and Y. Yin, 1995, "Modified Direct 2's Complement Parallel Array Multiplication Algorithm for Optical Complex Matrix Operation," *Appl. Opt.* 34, 1321–1328.
9.106 D. Casasent and B. K. Taylor, 1985, "Banded-Matrix High-Performance Algorithm and Architecture," *Appl. Opt.* 24, 1476–1480.
9.107 C. Perlee and D. Casasent, 1986, "Negative Base Encoding in Optical Linear Algebra Processors," *Appl. Opt.* 25, 168–169.
9.108 G. Li and L. Liu, 1994, "Negabinary Encoding for Optical Complex Matrix Operation," *Opt. Commun.* 113, 15–19.
9.109 L. Liu, G. Li, and Y. Yin, 1994, "Optical Complex Matrix–Vector Multiplication Using Negabinary Inner Products," *Opt. Lett.* 19, 1759–1761.
9.110 G. Li and L. Liu, 1994, "Bi-imaginary Number System for Optical Complex Calculations," *Opt. Commun.* 113, 25–30.
9.111 G. Li and L. Liu, 1995, "Simplified Optical Complex Multiplication Using Binary-Coded Quarter-Imaginary Number System," *Opt. Commun.* 122, 16–22.
9.112 M. A. Karim, 1991, "Smart Quasiserial Postprocessor for Optical Systolic Systems," *Appl. Opt.* 30, 910–911.
9.113 D. Psaltis and R. A. Athale, 1986, "High-Accuracy Computation with Linear Analog Optical Systems: A Critical Study," *Appl. Opt.* 25, 3071–3076.
9.114 E. J. Baranoski and D. P. Casasent, 1989, "High-Accuracy Optical Processors: A New Performance Comparison," *Appl. Opt.* 28, 5351–5357.
9.115 A. Huang, Y. Tsunoda, J. W. Goodman, and S. Ishihara, 1979, "Optical Computation Using Residue Arithmetic," *Appl. Opt.* 18, 149–162.
9.116 A. P. Goutzoulis, D. K. Davies, and E. C. Malarkey, 1987; "Prototype Position-Encoded Residue Look-Up Table Using Laser Diodes," *Opt. Commun.* 61, 302–308.

9.117 T. K. Gaylord, M. M. Mirsalehi, and C. C. Guest, 1985, "Optical Digital Truth-Table Look-Up Processing," *Opt. Eng.* 24, 48–58.
9.118 Y. Li, G. Eichmann, R. Dorsinville, and R. R. Alfano, 1988, "Demonstration of a Picosecond Optical-Phase Conjugation-Based Residue-Arithmetic Computation," *Opt. Lett.* 13, 178–180.
9.119 M. Seth, M. Ray, and A. Basuray, 1994, "Optical Implementation of Arithmetic Operations Using the Positional Residue System," *Opt. Eng.* 33, 541–547.
9.120 G. Li, L. Liu, B. Liu, F. Wang, and Y. Yin, 1996, "Digital Partitioning for Optical Negabinary Modular Multiplication," *Opt. & Laser Technol.* 28, 593–596.
9.121 A. Avizienis, 1961, "Signed-Digit Number Representations for Fast Parallel Arithmetic," *IRE Trans. Elect. Comput*, EC-10, 389–398.
9.122 R. P. Bocket, B. L. Drake, M. E. Lasher, and T. B. Henderson, 1986, "Modified Signed-Digit Addition and Subtraction Using Optical Symbolic Substitution, *Appl. Opt.* 25, 2456–2457.
9.123 K. Hwang and A. Louri, 1989, "Optical Multiplication and Division Using Modified Signed-Digit Symbolic Substitution," *Opt. Eng.* 28, 364–372.
9.124 S. Zhou, S. Campbell, and P. Yeh, 1992, "Modified Signed-Digit Optical Computing by Using Fanout Elements," *Opt. Lett.* 17, 1697–1699.
9.125 B. Ha and Y. Li, 1994, "Parallel Modified Signed-Digit Arithmetic Using an Optoelectronic Shared Content-Addressable–Memory Processor," *Appl. Opt.* 33, 3647–3662.
9.126 K. A. Ghoneim and D. Casasent, 1994, "High-Accuracy Pipelined Iterative-Tree Optical Multiplication," *Appl. Opt.* 33, 1517–1527.
9.127 K. W. Wong and L. M. Cheng, 1994, "Optical Modified Signed-Digit Addition Based on Binary Logical Operations," *Opt. Laser Technol.* 26, 213–217.
9.128 B. Wang, F. Yu, X. Liu, P. Gu, and J. Tang, 1996, "Optical Modified Signed-Digit Addition Module Based on Boolean Polarization-Encoded Logic Algebra," *Opt. Eng.* 35, 2989–2994.
9.129 F. Qian, G. Li, H. Ruan, L. Liu, 1999, "Modified Signed-Digit Addition by Using Binary Logic Operations and Its Optoelectronic Implementation," *Optics & Laser Technology* 31, 403–410.
9.130 S. Zhou, S. Campbell, W. Wu, P. Yeh, and H.-K. Liu, 1994, "Modified Signed-Digit Arithmetic for Multi-input Digital Optical Computing," *Appl. Opt.* 33, 1507–1516.
9.131 A. K. Cherri and M. A. Karim, 1988, "Modified Signed-Digit Arithmetic Using an Efficient Symbolic Substitution," *Appl. Opt.* 27, 3824–3827.
9.132 A. K. Cherri and M. A. Karim, 1989, "Symbolic Substitution–Based Glagged Arithmetic Unit Design Using Polarization-Encoded Optical Shadow-Casting System," *Opt. Commun.* 70, 455–461.
9.133 S. Zhou, S. Campbell, P. Yeh, and H.-K. Liu, 1995, "Two-Stage Modified Signed-Digit Optical Computing by Spatial Data Encoding and Polarization Multiplexing," *Appl. Opt.* 34, 793–802.
9.134 G. Li, F. Qian, H. Ruan, and L. Liu, 1999, "Compact Parallel Optical Modified Signed-Digit Arithmetic-Logic Array Processor by Using Electron-Trapping Device," *Appl. Opt.* 38, 5039–5045.
9.135 G. Eichmann, Y. Li, and R. R. Alfano, 1986, "Optical Binary-Coded Ternary Arithmetic and Logic," *Appl. Opt.* 25, 3113–3121.
9.136 Y. Wu, Z. Zhang, L. Liu, and Z. Wang, 1993, "Arithmetic Operation Using Binary-Encoding Modified Signed-Digit System," *Opt. Commun.* 100, 53–58.
9.137 B. Parhami, 1988, "Carry-Free Addition of Recoded Binary Signed-Digit Numbers," *IEEE Trans. Comput.* 37, 1470–1476.
9.138 A. A. S. Awwal, 1992, "Recoded Signed-Digit Binary Addition/Subtraction Using Opto-electronic Symboloic Substitution," *Appl. Opt.* 31, 3205–3208.
9.139 A. K. Cherri, 1994, "Symmetrically Recoded Modified Signed-Digit Optical Addition and Subtraction," *Appl. Opt.* 33, 4378–4382.

9.140 M. S. Alam, A. A. S. Awwal, and M. A. Karim, 1992, "Digital Optical Processing Based on Higher-Order Modified Signed-Digit Symbolic Substitution," *Appl. Opt.* 31, 2419–2425.
9.141 M. M. Mirsalehi and T. K. Gaylord, 1986, "Logical Minimization of Multilevel-Coded Functions," *Appl. Opt.* 25, 3078–3088.
9.142 H. Huang, M. Itoh, and T. Yatagai, 1994, "Modified Signed-Digit Arithmetic Based on Redundant Bit Representation," *Appl. Opt.* 33, 6146–6156.
9.143 S. Barua, 1991, "Single-Stage Optical Adder/Subtractor," *Opt. Eng.* 30, 265–270.
9.144 F. Qian, G. Li, H. Ruan, H. Jing, and L. Liu, 1999, "Two-Step Digit-Set–Restricted Modified Signed-Digit Addition–Subtraction Algorithm and Its Optoelectronic Implementation," *Appl. Opt.* 5621–5630.
9.145 F. Qian, G. Li, and M. Alam (2001), "Optoelectronic Quotient-Selected Modified Signed-Digit Division," *Opt. Eng.* 40, 275–282.
9.146 A. A. S. Awwal, M. N. Islam, and M. A. Karim, 1992, "Modified Signed Digit Trinary Arithmetic by Using Optical Symbolic Substitution," *Appl. Opt.* 31, 1687–1694.
9.147 M. S. Alam, M. A. Karim, A. A. S. Awwal, and J. J. Westerkamp, 1992, "Optical Processing Based on Conditional Higher-Order Trinary Modified Signed-Digit Symbolic Substitution," *Appl. Opt.* 31, 5614–5621.
9.148 J. U. Ahmed, A. A. S. Awwal, and M. Karim, 1994, "Two-Bit Trinary Full-Adder Design Based on Restricted Signed-Digit Numbers," *Opt. & Laser Technol.* 26, 225–228.
9.149 M. S. Alam, 1994, "Parallel Optical Computing Using Recoded Trinary Signed-Digit Numbers," *Appl. Opt.* 33, 4392–4397.
9.150 M. S. Alam, 1994, "Efficient Trinary Signed-Digit Symbolic Arithmetic," *Opt. Lett.* 19, 353–355.
9.151 M. M. Hossain, J. U. Ahmed, A. A. S. Awwal, and H. E. Michel, 1998, "Optical Implementation of an Efficient Modified Signed-Digit Trinary Addition," *Opt. & Laser Technol.* 30, 49–55.
9.152 A. K. Cherri, M. K. Habib, and M. S. Alam, 1998, "Optoelectronic Recoded and Nonrecoded Trinary Signed-Digit Adder That Uses Optical Correlation," *Appl. Opt.* 37, 2153–2163.
9.153 A. K. Cherri and M. S. Alam, 1998, "Recoded and Nonrecoded Trinary Signed-Digit Adders and Multipliers with Redundant-Bit Representation," *Appl. Opt.* 37, 4405–4418.
9.154 M. S. Alam (March 1999, in press), "Parallel Optoelectronic Trinary Signed-Digit Division," *Opt. Eng.* 38.
9.155 M. S. Alam, K. Jemili, and M. A. Karim, 1994, "Optical Higher-Order Quaternary Signed-Digit Arithmetic," *Opt. Eng.* 33, 3419–3426.
9.156 M. S. Alam, Y. Ahuja, A. K. Cherri, and A. Chatterjea, 1996, "Symmetrically Recoded Quaternary Signed-Digit Arithmetic Using a Shared Content-Addressable Memory," *Opt. Eng.* 35, 1141–1149.
9.157 A. K. Cherri, 1996, "High-Radix Signed-Digit Arithmetic Using Symbolic Substitution Computations," *Opt. Commun.* 128, 108–122.
9.158 A. K. Cherri, 1998, "Classified One-Step High-Radix Signed-Digit Arithmetic Unit," *Opt. Eng.* 37, 2324–2333.
9.159 K. Hwang and D. K. Panda, 1992, "High-Radix Symbolic Substitution and Superposition Techniques for Optical Matrix Algebraic Computations," *Opt. Eng.* 31, 2422–2433.
9.160 G. Li, L. Liu, H. Cheng, and X. Yan, 1998, "Parallel Optical Quaternary Signed-Digit Multiplication and Its Use for Matrix–Vector Operation," *Optik* 107, 165–172.
9.161 S. Zhang and M. A. Karim, 1998, "One-Step Optical Negabinary and Modified Signed-Digit Adder, *Opt. Laser Technol.* 30, 193–198.
9.162 G. Li, F. Qian, H. Ruan, and L. Liu, 1999, "Parallel Optical Negabinary Computing Algorithm and Optical Implementation," *Opt. Eng.* 38, 403–414.
9.163 F. Qian, G. Li, H. Ruan, and L. Liu, 2000, "Efficient Optical Lookahead Conversion from Modified Signed-Digit to 2's Complement Representation," *Optik* 113, 13–19.

EXERCISES

9.1 Design a free-space optical programmable logic array using a crossbar switch to implement the function $F(A, B, C, D) = AB + CD + DA$.

9.2 Design 16 logic gates of two variables using a liquid crystal electro-optic switch.

9.3 With reference to Fig. 9.4, design the light source patterns for all 16 logic operations.

9.4 Perform $(151)_{10} - (101)_{10}$ using residue arithmetic with the bases 5, 7, and 11.

9.5 (a) Show how 10010011 may be added to 01100101 by means of symbolic substitution.

(b) Derive higher-order symbolic substitution rules for 2-bit addition, and then perform the addition in part (a) again.

9.6 Perform MSD addition $11\bar{1}10\bar{1}01 + \bar{1}11\bar{1}\bar{1}\bar{1}1\bar{1}$ by binary logic operations.

9.7 Perform MSD addition $\bar{1}\bar{1}1\bar{1}010\bar{1} + 1\bar{1}\bar{1}1111\bar{1}1$ using the symmetrically recoded MSD algorithm.

9.8 Using the convergence MSD division algorithm, calculate the division of $X = (0.\bar{1}0)_{MSD} = (-0.5)_{10}$ by $Y = (0.11)_{MSD} = (0.75)_{10}$ with 16-digit precision.

9.9 Using the quotient-selected MSD division algorithm, calculate the division of $X = (0.101\bar{1}\bar{1}01)_{MSD}$ by $Y = (0.11\bar{1}0\bar{1}\bar{1}1)_{MSD}$.

9.10 Perform $(1401)_{10} - (1001)_{10}$ using two-step trinary signed-digit arithmetic.

9.11 Suppose the incoherent correlator–based content-addressable memory processor is used to implement two-step trinary signed-digit addition (Table 9.29). Design the spatial-encoding patterns for the trinary digits and the content-addressable memory patterns for recognition.

9.12 (a) Using the two-step quaternary signed-digit algorithm, calculate the sum $32\bar{2}\bar{3} + \bar{1}\bar{2}12$.

(b) Calculate the product $32\bar{2}\bar{3} \times \bar{1}\bar{2}12$.

9.13 Derive an algorithm for number conversion from the trinary signed-digit system to the 3's complement system.

9.14 Derive an algorithm for number conversion from the quaternary signed-digit system to the 4's complement system.

9.15 Perform negabinary signed-digit addition $\bar{1}1\bar{1}0\bar{1}1\bar{1}1 + \bar{1}00\bar{1}1\bar{1}1\bar{1}$.

9.16 Convert the MSD number $100\bar{1}001\bar{1}0\bar{1}\bar{1}1$ to 2's complement.

9.17 Convert the negabinary signed-digit number $\bar{1}1\bar{1}0\bar{1}11\bar{1}0$ to negabinary.

9.18 Design an optical system to perform matrix–vector operation for symbolic substitution.

9.19 Explain how optical shifting can be accomplished with a hologram, a polarizing beamsplitter, or a birefrigent crystal.

9.20 Design a 512×512 optical crossbar interconnection network.

Chapter 10 Sensing with Optics

Shizhuo Yin
THE PENNSYLVANIA STATE UNIVERSITY

10.1. INTRODUCTION

A light field is a high-frequency electromagnetic field. In general, a complex monochromatic light field, $\boldsymbol{E}(\boldsymbol{r}, t)$, can be written as

$$\boldsymbol{E}(\boldsymbol{r}, t) = \boldsymbol{A}(\boldsymbol{r}, t)e^{i(\omega t + \phi(\boldsymbol{r},t))}, \tag{10.1}$$

where $\boldsymbol{A}(\boldsymbol{r}, t)$ is the amplitude of the complex light field, ω is the angular frequency of the monochromatic light field, and $\phi(\boldsymbol{r}, t)$ is the phase of the complex field. Note that a broadband light source can be viewed as the summation of different-frequency monochromatic light fields. Thus, Eq. (10.1) can be treated as a general mathematical description of a light field. This equation also shows us the key parameters, which are used to describe the light fields, including:

1. The polarization–direction of the electric field.
2. The amplitude (i.e., $|\boldsymbol{A}(\boldsymbol{r}, t)|$) or intensity (i.e., $I(\boldsymbol{r}, t) = |\boldsymbol{A}(\boldsymbol{r}, t)|^2$).
3. The frequency (i.e., $f = \omega/2\pi$) or wavelength of the light field (i.e., λ).
4. The phase $\phi(\boldsymbol{r}, t)$.

All these parameters may be subject to changes due to external perturbations. Thus, by detecting the changes in these parameters, external perturbations can be detected or sensed. In other words, we can do sensing with optics.

In recent years, with the development and incorporation of optical fibers, sensing with optics has been largely focused on optical fiber sensors; this offers some potential advantages such as immunity to electromagnetic interference, a nonelectrical method of operation, small size and weight, low power, and relatively low cost. Therefore, this chapter will focus on fiber-optic sensors. In particular, since distributed fiber-optic sensors offer the unique advantage of multiple location measurements with only a single fiber, we will discuss distributed fiber-optic sensors in a separate section. Note that the distributed measurement capability of the fiber-optic sensor distinguishes it from other types of sensors, such as electronic sensors. This is one of the major benefits of employing fiber-optic sensors.

10.2. A BRIEF REVIEW OF TYPES OF FIBER-OPTIC SENSORS

As mentioned in Sec. 10.1, the key parameters of light fields include amplitude (or intensity), polarization, frequency, and phase. Thus, we categorize types of fiber sensors based on their sensing parameters. For example, if the sensing parameter is based on a change in light-field intensity, this type of sensor will belong to the category of intensity-based fiber-optic sensors.

10.2.1. INTENSITY-BASED FIBER-OPTIC SENSORS

There are many transduction mechanisms which can result in a change in light intensity when light passes through an optical fiber, so intensity-based fiber optic sensors can be used. These mechanisms may include:

- Microbending loss
- Breakage
- Fiber-to-fiber coupling
- Modified cladding
- Reflectance
- Absorption
- Attenuation
- Molecular scattering
- Molecular effects
- Evanescent fields [1]

For the purpose of illustration, several kinds of widely used intensity-type fiber-optic sensors are briefly summarized in the following subsections.

10.2.1.1. *Intensity-Type Fiber-Optic Sensors Using Microbending*

When a fiber is bent, there may be losses due to this bending. This localized bending is called microbending. The output light intensity is proportional to the amount of microbending. Therefore, by detecting changes in output light intensity, the amount of microbending can be measured so a fiber-optic sensor can be used.

Figure 10.1 illustrates a microbending-based displacement sensor [2, 3, 4]. The light from a light source is coupled into an optical fiber. As the deformer moves closer to the fiber, radiation losses increase due to microbending. Thus, the detected transmitted light decreases. Besides displacement measurement, several other parameters such as strain, pressure, force, and position can also be mechanically coupled to displacement of this microbending device so that these parameters can also be measured by the same fiber-sensor setup. It is reported that 0.25% linearity was achieved [5]. (Applications of microbend sensors for industrial pressure, acceleration, and strain sensors are described in detail in [6, 7].)

10.2.1.2. *Intensity-Type Fiber-Optic Sensors Using Reflection*

Figure 10.2. shows the basic principle of fiber-optic sensors using reflection [8]. Light travels along the fiber from left to right, leaves the fiber end, and incidents on a movable reflector. If the reflector moves closer to the fiber, most of the light can be reflected back into the fiber so that a high light intensity signal is detected. However, when the reflector moves farther away from the exit end of the fiber, less light is coupled back into the fiber, so a weak signal is detected. Therefore, the monotonic relationship between fiber–reflector distance, D, and returned light intensity can be used to measure displacement distance. To avoid the influence of the intensity fluctuation of the light source,

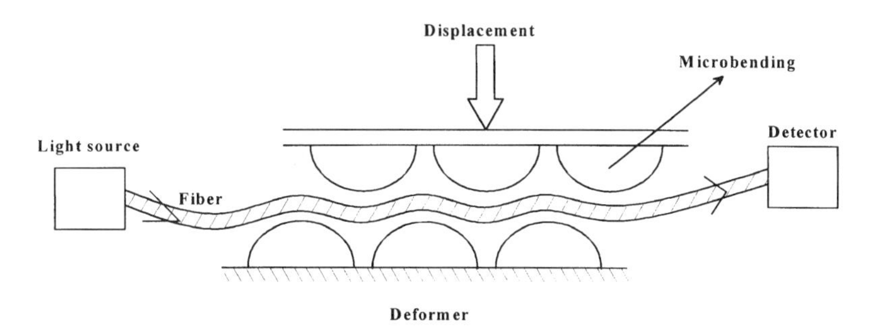

Fig. 10.1. Intensity-type fiber-optic sensor based on microbending.

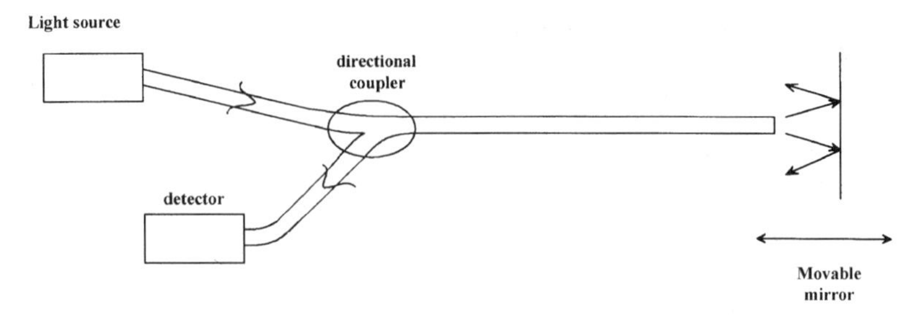

Fig. 10.2. Intensity-type fiber-optic sensor based on reflection.

a suitable reference signal is usually added in this type of intensity-based fiber-optic sensor.

10.2.1.3. *Intensity-Type Fiber-Optic Sensors Using Evanescent Wave Coupling*

The evanescent wave phenomenon comes from the fact that when light propagates along a single mode optical fiber, it is not totally confined to the core region but extends into the surrounding glass cladding region. The lightwave portion in the surrounding cladding region is called the evanescent wave. This phenomenon has been used to fabricate one of the most widely used fiber-optic components, the directional coupler [9, 10]. The coupling intensity between two fibers is a function of the distance between the two fiber cores. The closer the distance, the stronger the coupling will be. Figure 10.3 shows a fiber sensor based on this evanescent wave coupling concept. Light is launched into one of the fibers, and it propagates to a region where a second core is placed in close proximity so that part of the evanescent wave of the first fiber is within

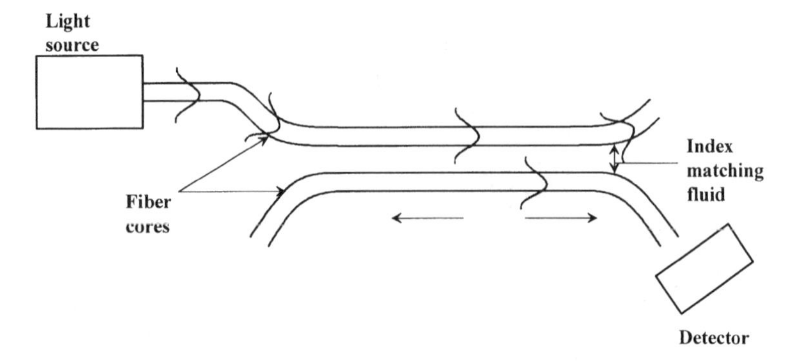

Fig. 10.3. Intensity-type fiber-optic sensor based on evanescent wave coupling.

the second fiber region. Thus, evanescent wave coupling occurs. The coupling coefficient is directly proportional to the separation distance between the two fibers. When an environmental effect such as a pressure, an acoustic wave, or a temperature change causes a change in the distance between two fibers, the result is a change in the coupling coefficient. Thus, the detected light intensity of the second fiber is also changed. Therefore, by monitoring the intensity change of the second fiber, the change in the environment can be sensed.

10.2.2. POLARIZATION-BASED FIBER-OPTIC SENSORS

10.2.2.1. *Mathematical Description of Polarization*

As mentioned previously, the light field is a vector. The direction of the electric field portion of the light field is defined as the polarization state of the light field. There are different types of polarization states of the light field, including linear, elliptical, and circular. For the linear polarization state, the direction of the electric field always keeps in the same line during light propagation. For the elliptical polarization state, the direction of the electric field changes during the light propagation. The end of the electric field vector forms an elliptical shape. That is why it is called elliptical polarized light.

Mathematically, the possible polarization states can be written as

$$\boldsymbol{E}_T = A_x \hat{x} + A_y e^{i\delta} \hat{y}, \tag{10.2}$$

where $\boldsymbol{E}_T$ represents the total electric field vector; A_x and A_y represent the components of the electric field in the x and y directions, respectively; δ represents the relative phase difference between x and y components; and $\hat{x}$ and $\hat{y}$ represent the unit vectors in the x and y directions, respectively. When $\delta = 0$, the polarization state is a linear polarization state. When $\delta = \pi/2$, the polarization state is a circular polarization state. In general, the polarization state is an elliptical polarization state, as illustrated in Fig. 10.4.

Since any optical detector can only detect light intensity directly, instead of using (A_x, A_y, δ) to describe a polarization state, a four-component Stokes vector representation is often used to describe a polarization state. The four components of this vector are given by

$$\begin{aligned} s_0 &= A_x^2 + A_y^2 \\ s_1 &= A_x^2 - A_y^2 \\ s_2 &= 2A_x A_y \cos\delta \\ s_3 &= 2A_x A_y \sin\delta. \end{aligned} \tag{10.3}$$

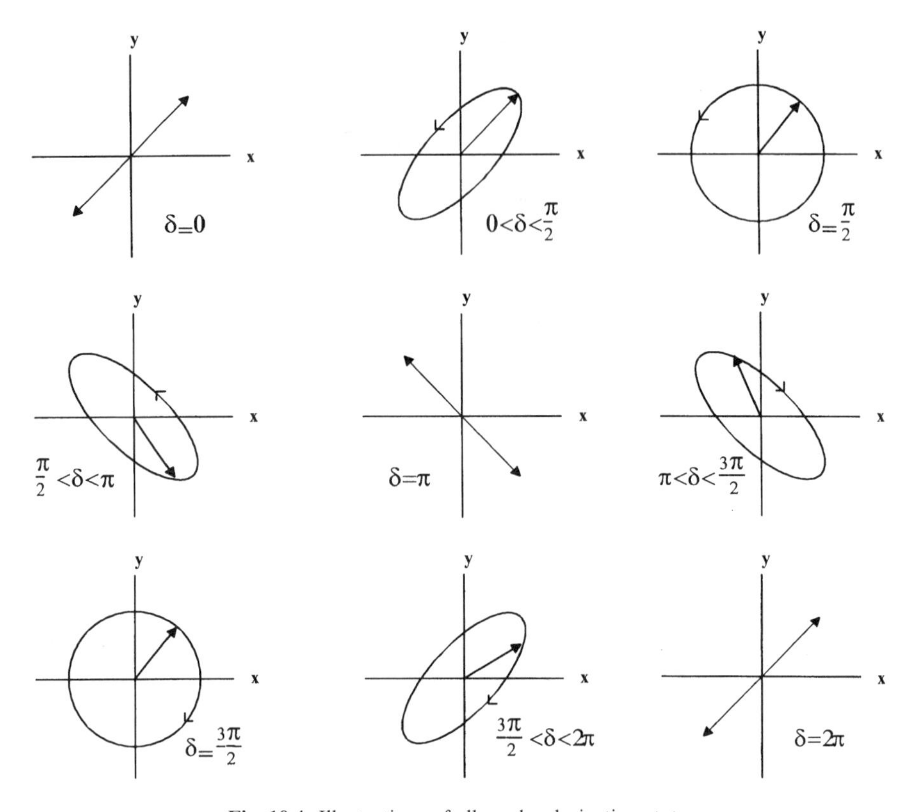

Fig. 10.4. Illustrations of allowed polarization states.

Obviously, there are only three independent components due to the relation $s_0^2 = s_1^2 + s_2^2 + s_3^2$. The beauty of the Stokes vector is that all the components of the vector can be measured in a relatively easy way. Figure 10.5 shows an experimental setup used to measure the Stokes vector, which includes a photodetector sensitive to the wavelength of interest, a linear polarizer, and a quarter-wave plate. Assume that the polarizer transmission axis is oriented at an angle θ relative to the $\hat{x}$ axis, while the fast axis of the wave plate is along $\hat{x}$. Six light beam intensities $(I_1, I_2, \ldots, I_6)$ are measured under different conditions:

1. I_1 is measured with the polarizer aligned in the $\hat{x}$ direction without a quarter-wave plate.
2. I_2 is measured with the polarizer aligned in the $\hat{y}$ direction without a quarter-wave plate.
3. I_3 is measured with the polarizer aligned in 45° relative to the $\hat{x}$ direction without a quarter-wave plate.

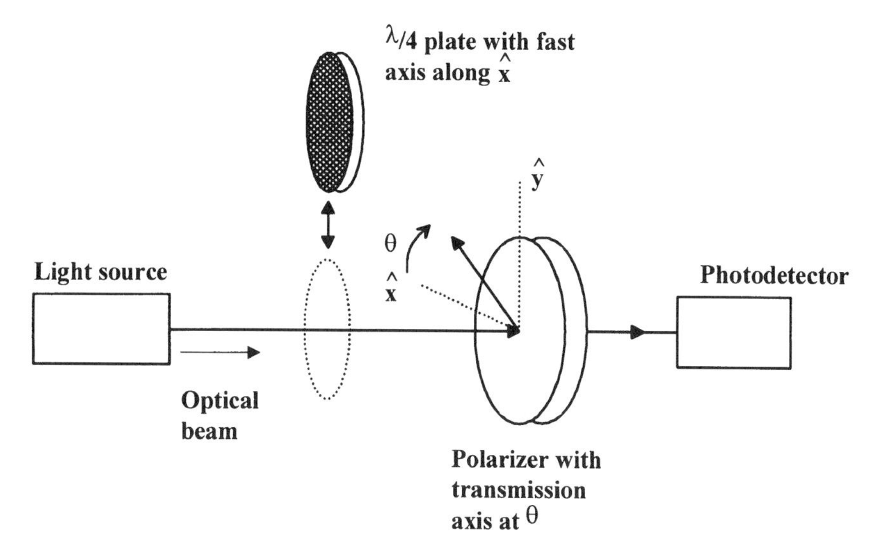

Fig. 10.5. Optical experimental setup to determine all four components of Stokes vector.

4. I_4 is measured with the polarizer aligned in 135° relative to the $\hat{x}$ direction without a quarter-wave plate.
5. I_5 is measured with the polarizer aligned in 90° relative to the $\hat{x}$ direction with a quarter-wave plate.
6. I_6 is measured with the polarizer aligned in 270° relative to the $\hat{x}$ direction with a quarter-wave plate.

The Stokes components can be derived from these measurements, as given by

$$\begin{aligned} s_0 &= I_1 + I_2 \\ s_1 &= I_1 - I_2 \\ s_3 &= I_3 - I_4 \\ s_4 &= I_5 - I_6. \end{aligned} \tag{10.4}$$

The physical meaning of Stokes vectors are as follows: s_0 represents the total intensity of the beam, s_1 represents the difference in intensities between the vertical and horizontal polarization components, s_2 represents the difference in intensities between $\pi/4$ and $3\pi/4$, and s_3 represents the difference between the right and left circular polarization components.

The possible polarization states of a light field can also be represented by a one-to-one mapping of these states to the surface of a sphere. This spherical

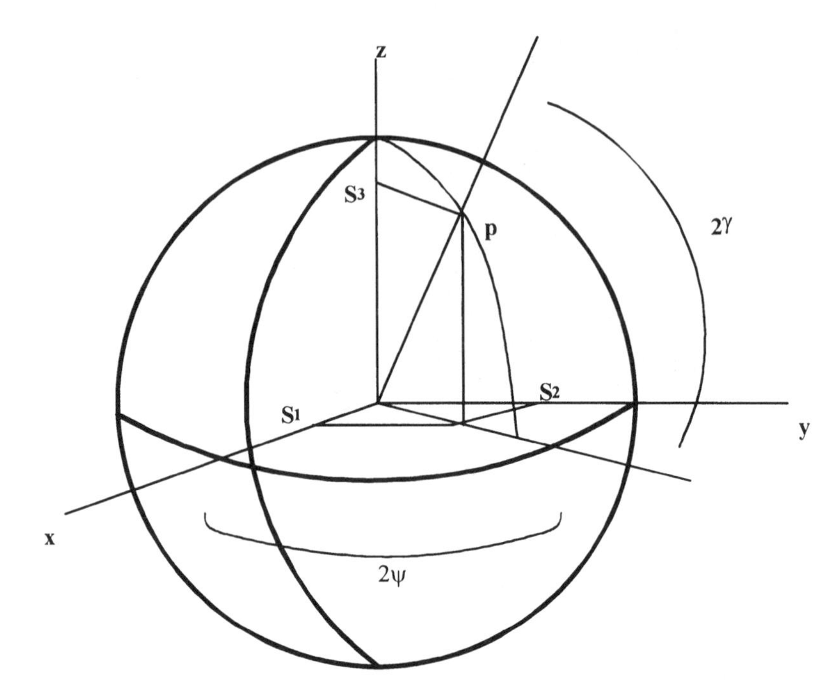

Fig. 10.6. Poincaré sphere.

representation of polarization is called the Poincaré sphere, as shown in Fig. 10.6. The radius of the sphere equals the s_0 component of the Stokes vector, and the Cartesian axes with origin at the center of the sphere are used to represent the s_1, s_2, and s_3 components of the vector. Thus, any point on this sphere may be projected on the axes to yield its corresponding Stokes vector components. Based on the Poincaré sphere, the change from one polarization state to another polarization state can be transformed as two rotations of the sphere; one through an appropriate 2ψ, followed by a second rotation of 2γ.

To quantitatively determine the polarization state transition from one polarization state to another polarization state, a polarization transformation matrix approach is used. The most widely used ones are the Mueller calculus and the Jones calculus [11].

In the Jones calculus, the polarization state is represented by the Jones vector, which is a two-component complex number

$$\boldsymbol{a} = \begin{bmatrix} A_x e^{i\delta_x} \\ Aye^{i\delta_y} \end{bmatrix}, \tag{10.5}$$

where A_x and A_y represent the amplitude for the x and y components, respectively, while δ_x and δ_y represent the phase for the x and y components, respectively. The transformation from one polarization state to another polarization state is represented by a 2×2 transformation matrix with complex components called the Jones matrix.

On the other hand, in the Mueller calculus, the polarization state is represented by the Stokes vector, $\boldsymbol{s} = (s_1, s_2, s_3, s_4)$ and transformation is represented by a 4×4 transformation matrix with real components called the Mueller matrix. The basic difference between the Jones calculus and the Mueller calculus is that in the Jones approach, all the elements can be complex numbers, while in the Mueller approach, all the elements are real numbers.

Table 10.1 illustrates the polarization state expression in terms of Jones vectors and Mueller vectors. Table 10.2 illustrates the most commonly used polarization state transformation matrices in terms of Jones matrices and Mueller matrices.

With the Jones and Mueller calculi, the output polarization state can be calculated from the input polarization state as well as the transformation matrix. Assume that the light passes through a series of optical elements consisting of polarizers and retarders. The output state can be obtained by sequentially multiplying the vector for the input state by the matrix representing each optical element in the order that it encounters them. Mathematically, this can be written as

$$\boldsymbol{I}_0 = M_f M_{f-1} \cdots M_1 \boldsymbol{I}_i, \tag{10.6}$$

Table 10.1

Polarization for Light Vector $\boldsymbol{E}$ with $E_x = A_x e^{i\delta_x}$, $E_y = A_y e^{i\delta_y}$, $\delta = \delta_x - \delta_y$, and $A^2 = A_x^2 + A_y^2$

Status of polarization	Jones vector	Stokes (or Mueller) vector
Linear polarization: transmission axis at angle θ to x axis	Jones: $\bar{a} = \begin{bmatrix} A_x e^{i\delta_x} \\ A_y e^{i\delta_y} \end{bmatrix}$	Mueller: $\bar{s} = \begin{bmatrix} A^2 \\ A^2 \cos 2\theta \\ A^2 \sin 2\theta \\ 0 \end{bmatrix}$
Right (+) and left (−) circular polarization: $Ax = Ay = A/\sqrt{2}$, $\delta = \pm\pi/2$	Jones: $\bar{a} = \begin{bmatrix} \dfrac{A}{\sqrt{2}\, e^{i\delta_x}} \\ \dfrac{A}{\sqrt{2}} e^{i(\delta_x \pm \pi/2)} \end{bmatrix}$	Mueller: $\bar{s} = \begin{bmatrix} A^2 \\ 0 \\ 0 \\ \pm A^2 \end{bmatrix}$
Elliptical polarization	Jones: $\bar{a} = \begin{bmatrix} A_x e^{i\delta_y} \\ A_y e^{i\delta_y} \end{bmatrix}$	Mueller: $s = \begin{bmatrix} A_x^2 + A_y^2 \\ A_x^2 - A_y^2 \\ 2A_x A_y \cos\delta \\ 2A_x A_y \sin\delta \end{bmatrix}$

Table 10.2

Jones and Mueller Matrices for Linear Polarizer, Linear Retarder, and Rotation Matrices

Type of component	Jones matrices	Mueller matrices
Linear polarizer: Transmission axis at angle θ to the axis in the x-y plane	$\begin{bmatrix} \cos^2\theta & \sin\theta\cos\theta \\ \sin\theta\cos\theta & \sin^2\theta \end{bmatrix}$	$\frac{1}{2}\begin{bmatrix} 1 & \cos 2\theta & \sin 2\theta & 0 \\ \cos 2\theta & \cos^2 2\theta & \sin 2\theta\cos 2\theta & 0 \\ \sin 2\theta & \sin 2\theta\cos 2\theta & \sin^2 2\theta & 0 \\ 0 & 0 & 0 & 0 \end{bmatrix}$
Linear retarder: Fast-axis angle θ to the x axis in the x-y plane, retardation $\delta = \delta_x - \delta_y$	$\begin{bmatrix} e^{i\delta}\cos^2\theta + \sin^2\theta & (e^{i\delta} - 1)\sin\theta\cos\theta \\ (e^{i\delta} - 1)\sin\theta\cos\theta & e^{i\delta}\sin^2\theta + \cos^2\theta \end{bmatrix}$	$\begin{bmatrix} 1 & 0 & 0 & 0 \\ 0 & \cos^2 2\theta + \sin^2 2\theta\cos\delta & (1-\cos\delta)\sin 2\theta\cos 2\theta & -\sin 2\theta\cos\delta \\ 0 & (1-\cos\delta)\sin 2\theta\cos 2\theta & \sin^2 2\theta + \cos^2 2\theta\cos\delta & \cos 2\theta\sin\delta \\ 0 & \sin 2\theta\sin\delta & -\cos 2\theta\sin\delta & \cos\delta \end{bmatrix}$
Rotation transform with rotation angle θ	$\begin{bmatrix} \cos\theta & \sin\theta \\ -\sin\theta & \cos\theta \end{bmatrix}$	$\begin{bmatrix} 1 & 0 & 0 & 0 \\ 0 & \cos 2\theta & \sin 2\theta & 0 \\ 0 & -\sin 2\theta & \cos 2\theta & 0 \\ 0 & 0 & 0 & 1 \end{bmatrix}$

where $\boldsymbol{I}_i$ represents the input polarization vector, M_n ($n = 1, 2, \ldots, f$) represents the nth transformation matrix, and $\boldsymbol{I}_0$ represents the output polarization vector.

To illustrate how to use above method, let us look at the following example.

Example 10.1. A unit intensity light beam linearly polarized in the x direction passes through a linear retarder with fast axis at angle θ, relative to the x axis and retarding amount, δ. Calculate the output polarization state in terms of θ and δ using Mueller calculus.

Solve: In this case,

$$\boldsymbol{I}_i = \begin{bmatrix} 1 \\ 1 \\ 0 \\ 0 \end{bmatrix},$$

$$M = \begin{bmatrix} 1 & 0 & 0 & 0 \\ 0 & \cos^2 2\theta + \sin^2 2\theta \cos\delta & (1-\cos\delta)\sin 2\theta \cos 2\theta & -\sin 2\theta \cos\delta \\ 0 & (1-\cos\delta)\sin 2\theta \cos 2\theta & \sin^2 2\theta + \cos^2 2\theta \cos\delta & \cos 2\theta \sin\delta \\ 0 & \sin 2\theta \sin\delta & -\cos 2\theta \sin\delta & \cos\delta \end{bmatrix}.$$

Thus,

$$\boldsymbol{I}_0 = M\mathbf{I}_i = \begin{bmatrix} 1 & 0 & 0 & 0 \\ 0 & \cos^2 2\theta + \sin^2 2\theta \cos\delta & (1-\cos\delta)\sin 2\theta \cos 2\theta & -\sin 2\theta \cos\delta \\ 0 & (1-\cos\delta)\sin 2\theta \cos 2\theta & \sin^2 2\theta + \cos^2 2\theta \cos\delta & \cos 2\theta \sin\delta \\ 0 & \sin 2\theta \sin\delta & -\cos 2\theta \sin\delta & \cos\delta \end{bmatrix}$$

$$\times \begin{bmatrix} 1 \\ 1 \\ 0 \\ 0 \end{bmatrix}$$

$$= \begin{bmatrix} 1 \\ \cos^2 2\theta + \sin^2 2\theta \cos\delta \\ (1-\cos\delta)\sin 2\theta \cos 2\theta \\ \sin 2\theta \sin\delta \end{bmatrix}$$

Thus, in general, the output becomes an elliptical polarization state.

10.2.2.2. *Polarization-Based Fiber-Optic Sensors*

Optical fiber is made of glass. The refractive index of the fiber can be changed due to the application of stress or strain. This phenomenon is called the photoelastic effect. In addition, in many cases, the stress or strain in different directions is different so that the induced refractive index change is also different in different directions. Thus, there is an induced phase difference among different polarization directions. In other words, under external perturbation, such as stress or strain, the optical fiber works like a linear retarder, as described in Sec. 10.2.2.1. Therefore, by detecting the change in the output polarization state, the external perturbation can be sensed.

Figure 10.7 shows the optical setup for the polarization-based fiber-optic sensor [12]. It is formed by polarizing the light from a light source via a polarizer that could be a length of polarization-preserving fiber. The polarized light is launched at 45° to the preferred axes of a length of birefringent polarization-preserving fiber. This section of fiber serves as a sensing fiber. As mentioned earlier, under external perturbation such as stress or strain, the phase difference between two polarization states is changed. Mathematically speaking, δ is a function of external perturbation; i.e., $\delta = \delta(p)$ where p represents the amount of possible perturbations. Then, based on Eq. (10.6), the output polarization state is changed according to the perturbation. Therefore, by analyzing the output polarization state by using an analyzer as the exit end of the fiber, as shown in Fig. 10.7, the external perturbation can be sensed. One of the major advantages of polarization-based fiber-optic sensors is the capability of optical common mode rejection.

To make the fiber-optic sensor practical, it must be sensitive to the phenomena it is designed to measure and insensitive to changes in other environmental parameters. For the strain or stress measurement, environmental temperature is an unwanted environmental parameter. For the polarization-based fiber-optic sensor, environmentally induced refractive index changes in the two polarization directions are almost the same. There is almost no induced phase difference between two polarization states; in other words, $\delta \approx 0$. Thus, environmental temperature fluctuation will not substantially deteriorate the performance of the sensor.

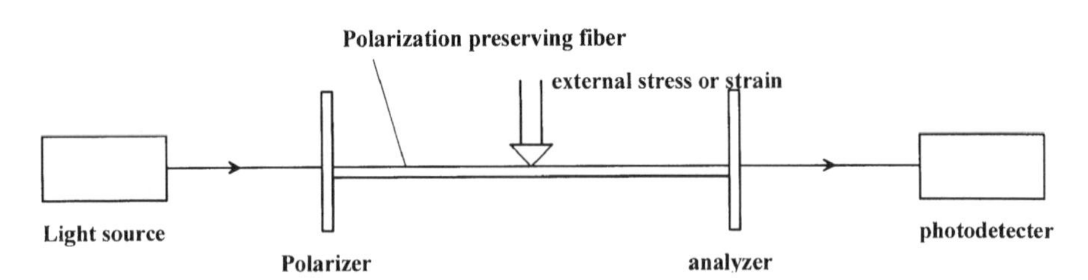

Fig. 10.7. Polarization-based fiber-optic sensor.

10.2.3. PHASE-BASED FIBER-OPTIC SENSORS

As described by Eq. (10.1), the phase of the light field, $\phi(\mathbf{r}, t)$ can also be changed by external perturbations so that the fiber-optic sensor can also be built based on the phase changes of the light field. The relationship between the phase change and the optical path change can be written as

$$\phi(\mathbf{r}, t) = \frac{2\pi}{\lambda} L(\mathbf{r}, t), \tag{10.7}$$

where λ is the light wavelength and $L(\mathbf{r}, t)$ represents the optical path change. Since the optical wavelength is very small, in the order of microns, a small change in the optical path may result in a large fluctuation in the phase change. Thus, in general, the phase-based fiber-optic sensor is more sensitive than the intensity-based fiber-optic sensor. Note that, since the optical detector cannot detect the optical phase directly, some types of interferometric techniques are exploited to implement phase-type fiber-optic sensors, as described in the following subsections.

10.2.3.1. *Fiber-Optic Sensors Based on the Mach–Zehnder Interferometer*

Figure 10.8 shows one kind of widely used fiber Mach–Zehnder interferometer–based fiber-optic sensor [13]. The interferometer consists of two arms, the sensing arm and the reference arm. The light coming from a coherent light source, such as from a distributed feedBack (DFB) semiconductor laser, is launched into the single mode fiber. The light is then split into two beams of nominal equal intensity by a 50/50 fiber-optic directional coupler, part being sent through the sensing fiber arm, the remainder through the reference arm. The output from these two fibers, after passing through the sensing and reference fiber coils, is recombined by the second fiber-optic directional coupler. Thus, an interference signal between the two beams is formed and detected by the photodetector.

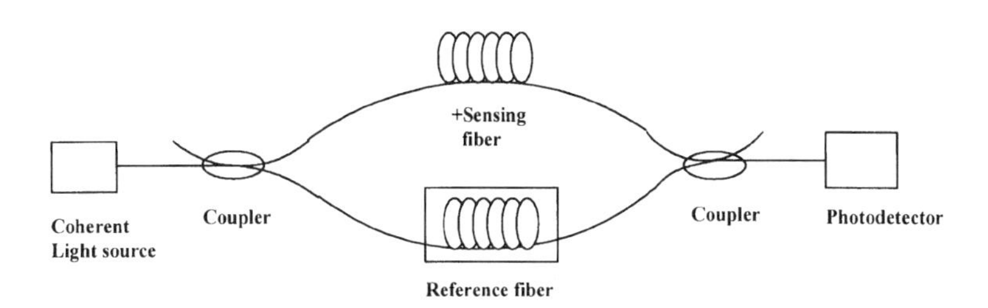

Fig. 10.8. Mach–Zehnder interferometer–based fiber-optic sensor.

For the purpose of simplicity, assume that the power coupling coefficients of the two couplers are 0.5 with no optical loss. In this case, the output light intensity can be written as [13]

$$I = I_0 \cos^2\left[\frac{\pi}{\lambda}(n_s L_s - n_r L_r)\right], \tag{10.8}$$

where I_0 represents a constant light intensity; L_s and L_r represent the sensing and reference fiber lengths, respectively; n_s and n_r represent the sensing and reference fiber refractive indices, respectively; and λ is the operating wavelength. Both n_s and L_s can change as a function of external perturbations such as stress or strain. Thus, by detecting the change in the output light intensity I, external perturbations can be sensed.

To minimize the influence from slowly changing environmental factors such as temperature and enhance the performance of the fiber sensor, in many cases the length of the reference arm is periodically modulated [14]. This can be realized by winding the reference arm fiber on a PZT drum. A sinusoidal electric signal is added on the PZT drum so that the diameter of the drum is periodically modulated by the sinusoidal electric signal, which in turn results in the periodic change in the reference arm fiber length.

Note that interferometric fiber sensors are usually constructed using conventional single mode optical fibers. Conventional single mode fiber can support two orthogonal polarization modes, owing to such effects as bending, the fiber becomes birefringent. This effect can result in a change in the interference fringe visibility. Thus, the signal-to-noise ratio of the sensing signal can be influenced by this effect [15]. To overcome polarization-induced effects in fiber interferometry, methods include (1) using polarization-preserving fiber throughout the entire sensor system; (2) actively controlling the input polarization state; and (3) employing polarization diversity techniques in which certain output polarization states are selected to avoid polarization fading [16, 17].

10.2.3.2. *Fiber-Optic Sensor Based on the Michelson Interferometer*

Figure 10.9 shows a kind of Michelson interferometer–based fiber-optic sensor. In this case, a single directional coupler is used for both splitting and recombining the light. The light traveling from the source is split into the sensing and reference arms. After traversing the length of the arms, the light is reflected back through the same arms by reflectors. The light is then recombined by the initial beam splitter.

There are similarities and differences between Michelson and Mach–Zehnder interferometers. In terms of similarities, the Michelson is often considered to be a folded Mach–Zehnder, and vice versa. Thus, from this argument one

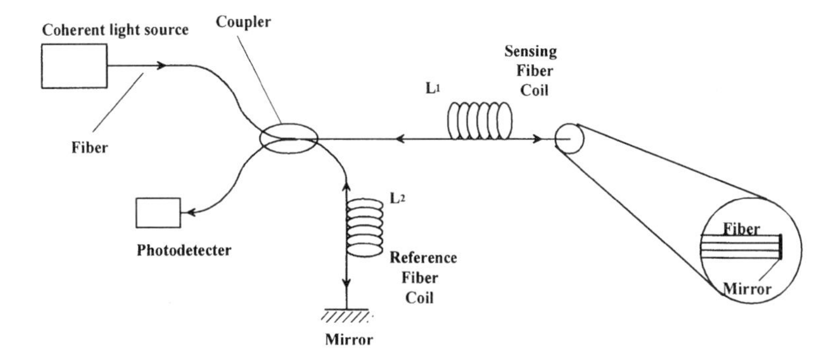

Fig. 10.9. Machelson interferometer–based fiber-optic sensor.

can see that the optical loss budget for both configurations is similar. The outputs, of course, have the same form as the Mach–Zehnder. In terms of differences, the Michelson configuration requires only one optical fiber coupler. Owing to the fact that the light passes through both the sensing and reference fibers twice, the optical phase shift per unit length of fiber is doubled. Thus, the Michelson intrinsically can have better sensitivity. From the practical point of view the physical configuration of the Michelson interferometer is sometimes somewhat easier to package, although this is obviously dependent on its application. Another clear advantage is that the sensor can be interrogated with only a single fiber between the source/detector module and the sensor. However, a good-quality reflection mirror is required for the Michelson interferometer. In addition, part of the light is fed back into the optical source due to the complementary output. This can be extremely troublesome for semiconductor diode laser sources. An optical isolator is needed to minimize this adverse effect.

10.2.3.3. *Fiber-Optic Sensors Based on the Fabry-Perot Interferometer*

The Fabry-Perot interferometer is a multiple-beam interferometer. Figure 10.10 shows a fiber-optic Fabry-Perot interferometer [18]. In this type of interferometer, due to the high reflectivity of the mirrors, the light bounces back and forth in the cavity many times, which increases the phase delay many times. The transmitted output intensity of the Fabry-Perot interferometer is given by

$$I(\phi) = \frac{A^2T^2}{(1-R)^2} \cdot \frac{1}{1 + \dfrac{4R}{(1-R)^2}\sin^2\phi} \tag{10.9}$$

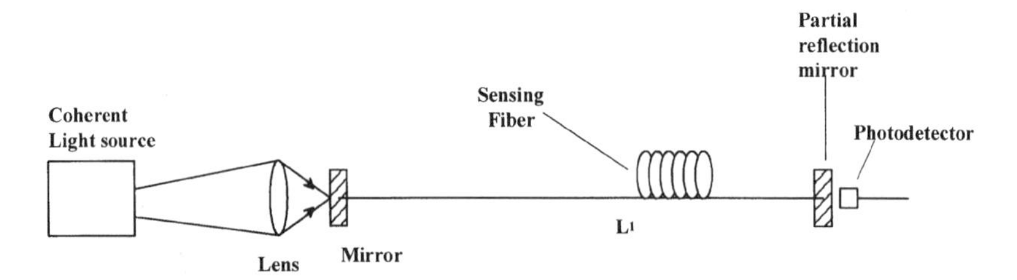

Fig. 10.10. Fabry-Perot interferometer–based fiber-optic sensor.

where T and R are the transmission and reflection coefficients of the mirrors, A is the amplitude of the input, and ϕ the total phase delay for a single transmit through the cavity (i.e., $2\pi nL/\lambda$, where n and L are the refractive index and the length of the cavity, respectively). Figure 10.11 shows the output intensity $I(\phi)$ as a function ϕ for different reflection coefficients. The higher the reflection coefficient, the sharper the interference peak will be. In other words, near the peak region, the output light intensity is very sensitive to a small change in the phase delay. Based on Eq. (10.9), it can be shown that the maximum sensitivity of the Fabry-Perot interferometer is proportional to the reflection coefficients, as given by [18]

$$\left.\frac{dI(\phi)}{d\phi}\right|_{\max} \propto \sqrt{F}, \tag{10.10}$$

where $F = 4R/(1 - R)^2$ is termed the coefficient of finesse. The larger the F number, the sharper (or finer) the interference peak will be. Fiber Fabry-Perot interferometers with cavity finesses of over 100 have been demonstrated [19]. Thus, the sensitivity of the fiber Fabry-Perot interferometer–based fiber sensor can be much higher then that of the Mach–Zehnder or Michelson interferometers.

However, the Fiber Fabry-Perot interferometer also suffers two major drawbacks: sensitivity to source coherence length and frequency jitter, and the complex shape of the function $I(\phi)$. For a long-cavity Fabry-Perot interferometer–based sensor, a long coherence length is required. Most semiconductor diode lasers have linewidths of a few tens of MHz; thus, the Fabry-Perot configuration is really incompatible with diode laser sources for high-sensitivity measurements.

As mentioned in the previous paragraph, the other difficulty with the Fabry-Perot interferometer is the shape of the rather complex transfer function $I(\phi)$. Although single sensors may be implemented with the active homodyne approach locking the interferometer to the maximum sensitive region (i.e.,

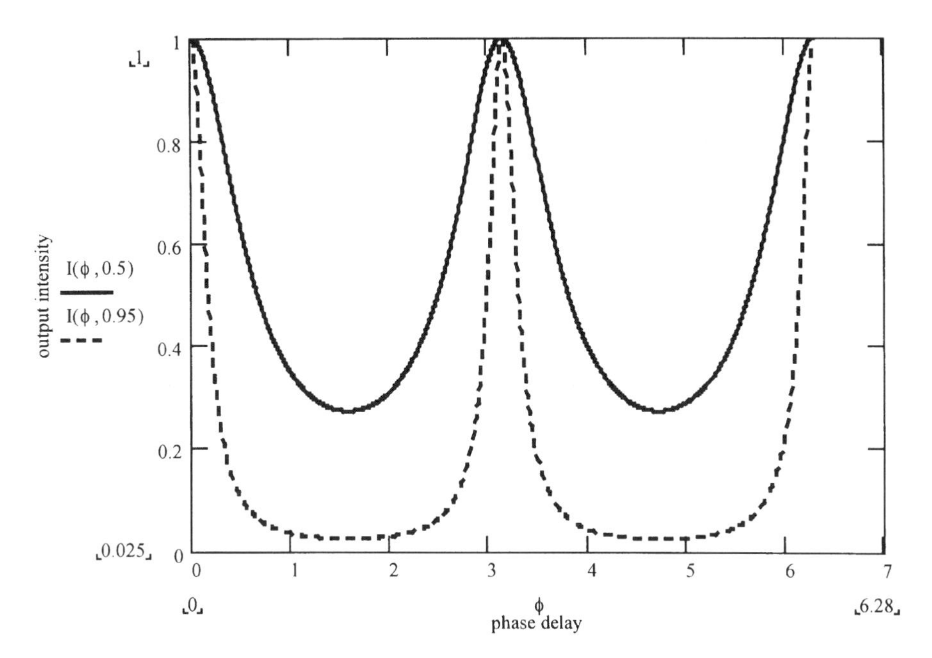

Fig. 10.11. Transfer function of fiber Fabry-Perot interferometer, $I(\phi)$ as a function of phase delay ϕ for difference reflection coefficients. Solid line: $R = 0.5$; dashed line: $R = 0.95$.

$dI(\phi)/d\phi|_{max}$), no passive techniques have been developed for this complex transfer function. In addition, at most values of static phase shift, the sensitivity of the Fabry-Perot interferometer is zero.

On the other hand, when finesse, F, is small, the sensitivity of the Fabry-Perot interferometer is not high. However, it offers a simple sensor configuration, which does not require fiber couplers and does not incur the difficulties associated with the multiple-beam approach outlined above.

10.2.4. FREQUENCY (OR WAVELENGTH)–BASED FIBER-OPTIC SENSORS

As described in Eq. (10.1), a light field is also a function of frequency (or wavelength). The emitted light frequency or wavelength may also be influenced by some types of perturbations. Thus, certain types of fiber-optic sensors can also be built based on frequency or wavelength changes.

As an example, let us look at a fluorescent light–based temperature sensor [20, 21]. The fluorescent material is used at the sensing head of the fiber sensor. Since the fluorescent lifetime of this material is temperature dependent, the

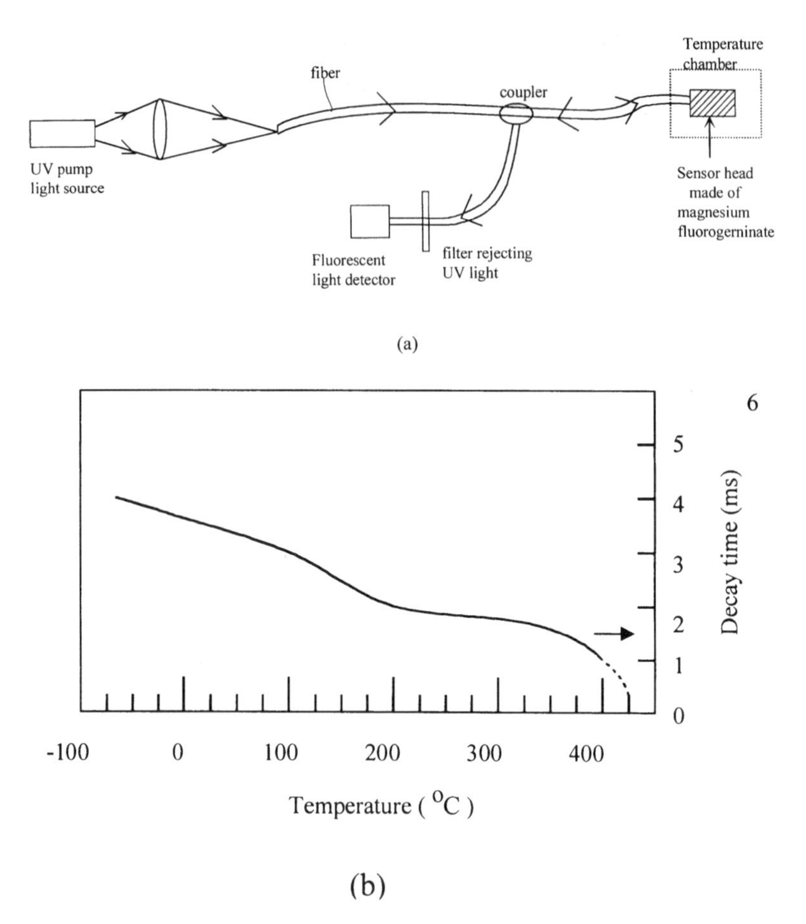

Fig. 10.12. Illustration of fluoroptic temperature sensor. (a) Schematic of device, (b) Decay-time of luminescent emission as a function of temperature.

changing of decay time with temperature can be used to sense the temperature. Figure 10.12 illustrates a fluoroptic temperature sensor. A UV lamp source is used as the means of excitation of the magnesium fluorogerminate material. The decay time characteristics of magenesium fluorogerminate is used to sense the measurand. To further enhance the temperature-sensing range, a combination of black-body radiation and fluorescence decay with internal cross-referencing to enable both high and low temperature operation was also developed, as shown in Fig. 10.13 [22]. Thus, the use of frequency-based fiber-optic sensors is a very effective approach for wide-range temperature sensing.

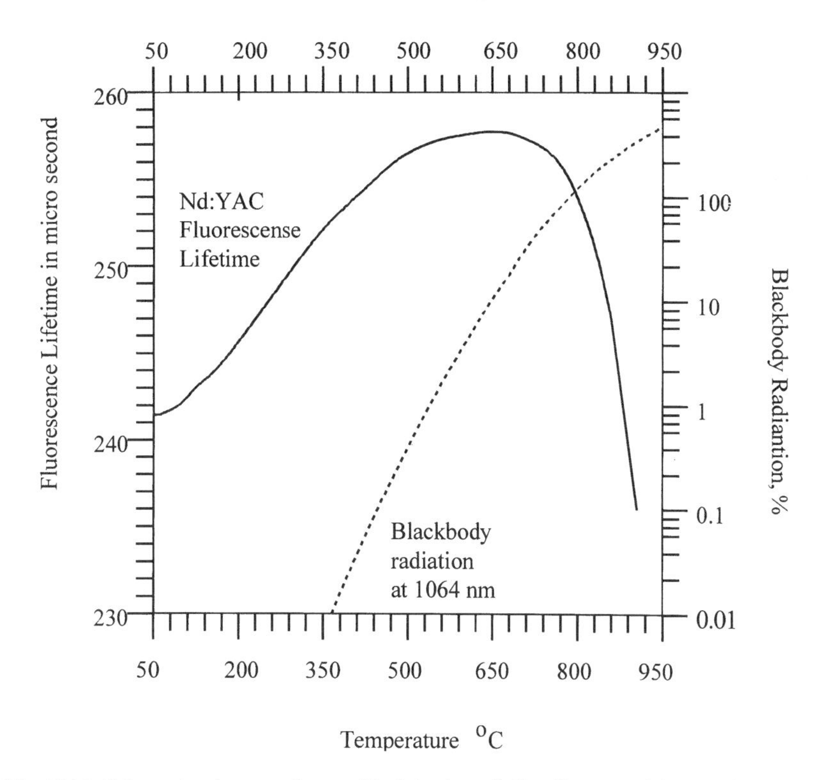

Fig. 10.13. Schematic of cross-reference black-body radiation/fluorescent temperature sensor.

10.3. DISTRIBUTED FIBER-OPTIC SENSORS

One of the most important features of fiber-optic sensors is their ability to implement distributed sensing. Multiple-point measurements can be achieved by using a single fiber, which provides light weight, compact size, and low cost.

There are basically two types of distributed fiber-optic sensors, intrinsic distributed sensors and quasi-distributed sensors. In this section, we briefly introduce the most commonly used distributed fiber-optic sensors.

10.3.1. INTRINSIC DISTRIBUTED FIBER-OPTIC SENSORS

Intrinsic distributed fiber-optic sensors are particular effective for use in applications where monitoring of a single measurand is required at a large number of points or continuously over the path of the fiber. These applications may include stress monitoring of large structures such as buildings, bridges, dams, storage tanks, aircraft, submarines, etc.; temperature profiling in electric power systems; leakage detection in pipelines; embedding sensors in composite materials to form smart structures; and optical fiber-quality testing and fiber-optic network line monitoring. Intrinsic distributed fiber-optic sensors may depend on different principles so that different types of intrinsic distributed fiber-optic sensors can be developed as described in the following subsections.

10.3.1.1. *Optical Time–Domain Reflectometry Based on Rayleigh Scattering*

One of the most popular intrinsic distributed fiber-optic sensors is optical time–domain reflectrometery based on Rayleigh scattering. When light is launched into an optical fiber, loss occurs due to Rayleigh scattering that arises as a result of random microscopic (less than wavelength) variations in the index of refraction of the fiber core. A fraction of the light that is scattered in a counterpropagation direction (i.e., 180° relative to the incident direction) is recaptured by the fiber aperture and returned toward the source. When a narrow optical pulse is launched in the fiber, by monitoring the variation of the Rayleigh backscattered signal intensity, the spatial variations in the fiber-scattering coefficient, or attenuation, can be determined. Since the scattering coefficient of a particular location reflects the local fiber status, by analyzing the reflection coefficient, the localized external perturbation or fiber status can be sensed. Thus, distributed sensing can be realized. Since this sensing technique is based on detecting the reflected signal intensity of the light pulse as a function of time, this technique is called optical time–domain reflectometry (OTDR). OTDR has been widely used to detect fault/imperfection location in fiber-optic communications [23, 24]. In terms of sensing applications, OTDR can be effectively used to detect localized measurand-induced variations in the loss or scattering coefficient of a continuous sensing fiber.

Figure 10.14 shows the basic configuration of OTDR. A short pulse of light from a laser (e.g., semiconductor laser or *Q*-switched YAG laser) is launched into the fiber that needs to be tested. The photodetector detects the Rayleigh backscattering light intensity as a function of time relative to the input pulse. If the fiber is homogeneous and is subject to a uniform environment, the backscattering intensity decays exponentially with time due to the intrinsic loss in the fiber. However, if at a particular location, there is a nonuniform environment (e.g., having a localized perturbation), the loss coefficient at that particular location will be different from the regular Rayleigh scattering coefficient. Thus, the backscattering intensity will not decay according to the exponential issue, as in other unperturbed locations. If we draw a curve using time as the horizontal axis and Log(P_s) (where P_s is the detected intensity) as

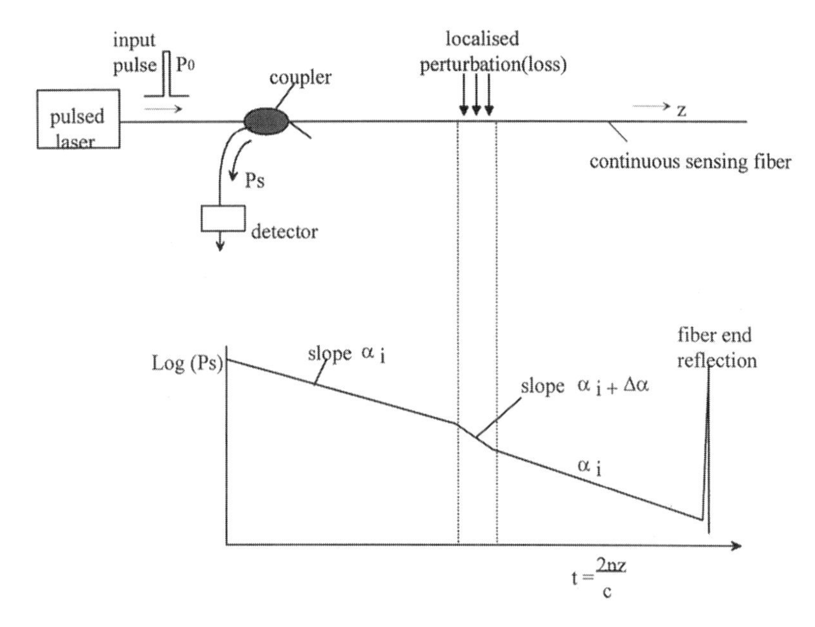

Fig. 10.14. Principle of optical time–domain reflectometry based on Rayleigh scattering.

the vertical axis, a sudden change in this curve is expected at the perturbed locations as shown in Fig. 10.14.

Mathematically, the detected scattering light intensity, P_s is given by

$$P_s(t) = P_0 r(z) e^{-\int_0^z 2\alpha(z)\,dz}, \tag{10.11}$$

where P_0 is a constant determined by the input pulse energy and the fiber-optic coupler power-splitting ratio, $z = tc/2n$ reflects the location of the launched pulse at time t (where c is the light speed in vacuum, and n is the refractive index of the fiber), $r(z)$ is the effective backscattering reflection coefficient per unit length that takes into account the Rayleigh backscattering coefficient and fiber numerical aperture, and $\alpha(z)$ is the attenuation coefficient. The slope of the logarithm of the detected signal is proportional to the loss coefficient $\alpha(z)$. The spatial resolution of an OTDR is the smallest distance between two scatterers that can be resolved. If the input pulse has a width τ, the spatial resolution is given by

$$\Delta Z_{\min} = \frac{c\tau}{2n}. \tag{10.12}$$

Example 10.2. An OTDR instrument has a pulse width of 10 ns; assume that the refractive of the fiber is $n = 1.5$. Calculate the spatial resolution of this OTDR.

Solve: Substituting $c = 3 \times 10^8$ m/s, $\tau = 10^{-8}$ s, and $n = 1.5$ into Eq. (10.12), we get $\Delta Z_{\min} = 1$ m.

From the above example, it can be seen that the spatial resolution OTDR is in the order of meter. This may not be high enough for certain distributed sensing requirements, such as smart structure monitoring. To increase the spatial resolution, one has to reduce the pulse width. However, the reduction in the pulse width may cause a decrease in the launched pulse energy so that the detected light signal also decreases. Thus, the detected signal-to-noise ratio may become poor. In particular, when long sensing range is required, a certain power level is needed to guarantee that the backscattering signal is detectable in the whole sensing range. Thus, there is a basic trade-off between spatial resolution and sensing range. The pulse width and pulse energy must be optimized based on their application requirements.

To enhance the capability of detecting a weak backscattering signal, several techniques were developed (e.g., coherent OTDR [25] and pseudorandom-coded OTDR [26]). In coherent OTDR, the weak returned backscatter signal is mixed with a strong coherent local oscillator optical signal to provide coherent amplification. In pseudorandom-coded OTDR, correlation techniques are used in conjunction with pseudorandom input sequences. Due to the use of multiple pulses, the pulse energy is increased without sacrificing the spatial resolution.

10.3.1.2. *Optical Time–Domain Reflectometry Based on Raman Scattering*

Rayleigh scattering is caused by density and composition fluctuations frozen into the material during the drawing process. This type of scattering is largely independent of ambient temperature provided that the thermo-optic coefficients of the fiber constituents are similar. To sense the ambient temperature distribution, optical time–domain reflectometry based on Raman scattering was developed.

Raman scattering involves the inelastic scattering of photons. The molecular vibrations of glass fiber (induced by incident light pulse) cause incident light to be scattered, and as a result, produce components in a broadband about the exciting (pump) wavelength comprising Stokes (λ_s, lower photon energy) and anti-Stokes (λ_a, higher photon energy) emissions [27]. Its usefulness for temperature sensing is that the intensity ratio between Stokes and anti-Stokes is temperature dependent, as given by

$$R_r = \left(\frac{\lambda_s}{\lambda_a}\right)^4 e^{-hc\bar{v}/kT}, \tag{10.13}$$

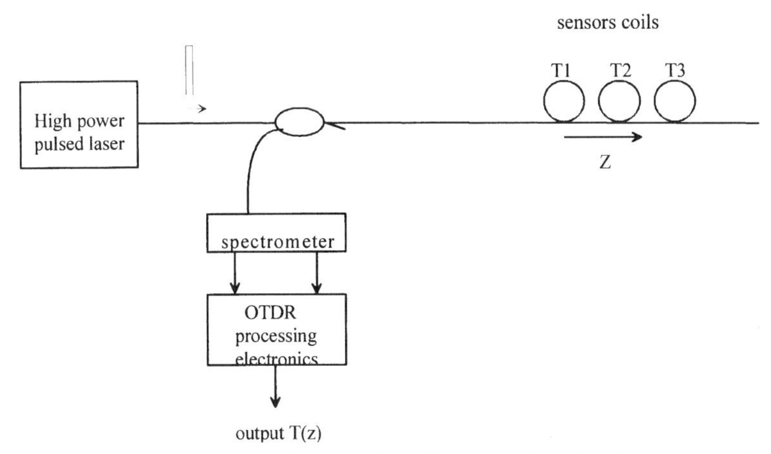

Fig. 10.15. Principle of optical time–domain reflectometry based on Raman scattering.

where R_r is the ratio of anti-Stokes to Stokes intensity in the backscattered light, h is the Planck's constant, $\bar{v}$ is the wave number separation from the pump wavelength, k is the Boltzmann's constant, and T is the absolute temperature. For example, for glass fiber, if the pump wavelength is at 514 nm, R_r has a value about 0.15 at room temperature with temperature dependent about 0.8%/°C in the range 0°C to 100°C.

Figure 10.15 illustrates OTDR temperature sensing based on Raman scattering. The basic configuration is similar to the Rayleigh scattering case, except that a spectrometer is added in front of the photodetector so that the Stokes and anti-Stokes spectral lines can be separately detected. Both the spectral line locations and intensities are recorded as a function of time. Then, the ratio between anti-Stokes to Stokes lines is calculated based on the measured data for different times. By substituting this $R_r(t)$ and wavelengths of Stokes and anti-Stokes lines into Eq. (10.13), the temperature as a function time, $T(t)$ can be obtained. Similar to the Rayleigh scattering case, there is also a one-to-one relationship between the spatial location, z, and the recording time, t, as given by $t = 2nz/c$. Thus, the temperature profile of the sensing fiber can be obtained.

The major difficulty of Raman OTDR is the low Raman scattering coefficient that is about three orders of magnitude weaker than that of the Rayleigh. Thus, high input power is needed to implement Raman-OTDR.

10.3.1.3. *Optical Time–Domain Reflectometry Based on Brillouin Scattering*

When an optical pulse launched into an optical fiber, stimulated Brilliouin scattering occurs from acoustic vibrations stimulated in the optical fiber [28].

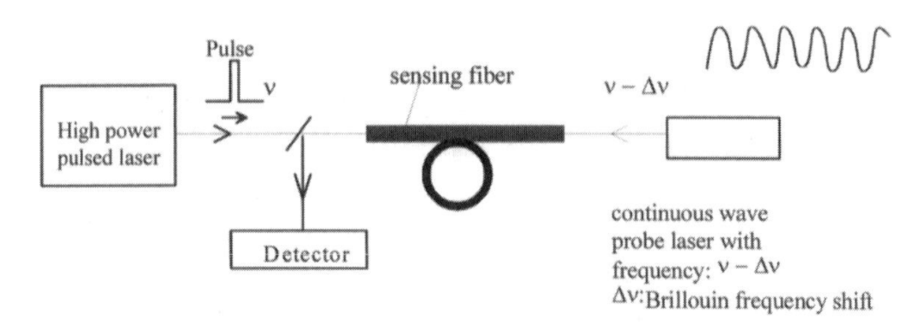

Fig. 10.16. Basic configuration of optical time–domain reflectometry based on Brillouin scattering.

The acoustic vibrations cause a counterpropagating wave that drains energy away from the forward-going input pulse. This counterpropagating scattering wave is called a Brillouin scattering wave. To satisfy the requirement of energy conservation, there is a frequency shift between the original light pulse frequency and the Brillouin scattering wave, which, in general, is on the order of tens of GHz.

Since the frequency shift of the Brillouin gain spectrum is sensitive to temperature and strain, it becomes very useful for building a fiber-optic sensor. In particular, the frequency shift depends on the magnitude of longitudinal strain, which comes from the fact that, under different longitudinal strain conditions, the acoustic wave frequency induced by the photon is different. Thus, the longitudinal strain distribution can be measured based on the Brillouin scattering effect [29]. Note that longitudinal strain distribution is difficult to measure using other techniques.

Figure 10.16 shows the configuration of Brillouin optical time–domain reflectometry. Since the counterpropagated Brillouin scattering signal is, in general, a very weak signal. To detect this weak signal, the coherent detection technique is used. To realize this coherent detection, the fiber is interrogated from one end by a continuous wave (CW) source and the other by a pulsed source, as shown in Fig. 10.16. The difference in wavelength between the two sources is carefully controlled to be equal to the Brillouin frequency shift of Brillouin scattering. Thus, there is an interaction between the Brillouin scattering light and CW light. The lower frequency wave (that can be CW or pulsed) is then amplified, which depends on the match between the frequency difference of the two sources and the local frequency shift. From the variation of the amplified signal, the local gain can thus be determined. Figure 10.17 illustrates the Brillouin gain at different longitudinal strain conditions. It can be seen that, indeed, there is a frequency shift under different longitudinal strain conditions. The solid curve represents no longitudinal strain and the dotted curve represents a 2.3×10^{-3} longitudinal strain. The frequency shift is about 119 MHz; that is detectable with current technology.

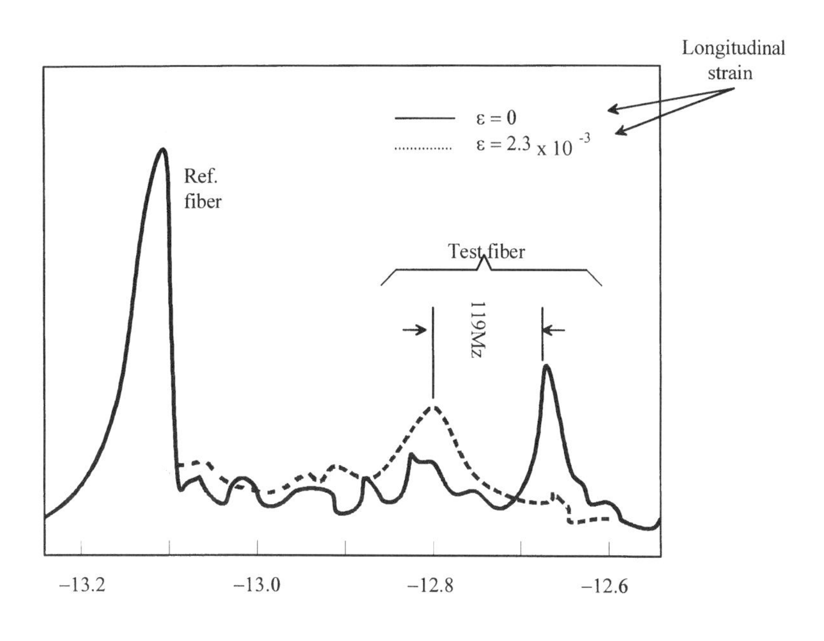

Fig. 10.17. An illustration of Brillouin scattering-frequency shift under different longitudinal strain conditions.

Since the Brillouin frequency shift is not much, to detect this frequency shift, a very stable, single-frequency, tunable laser is needed. Currently, 50-m spatial resolution with 0.01% strain resolution over >1 km fiber lengths is achieved by using a Nd:YAG ring laser, which is capable of delivering 1 mW power into the single mode fiber with remarkably narrow (5 kHz) spectra [29].

10.3.1.4. *Optical Frequency-Domain Reflectometry*

As discussed earlier, to obtain high spatial resolution, a very narrow light pulse is required, which results in a proportionally lower level of the backscattering signal and an increased receiver bandwidth requirement for detecting these pulses. Thus, a large increase in the noise level is expected so that only strong reflections can be detected in noise. To increase the spatial resolution without sacrificing backscattering signal intensity, optical frequency-domain reflectometry (OFDR) was developed [30, 31, 32].

Figure 10.18 shows the configuration of OFDR. A highly monochromatic light is coupled into a single mode fiber, and the optical frequency, ω, is modulated in a linear sweep. Assume that the power loss constant is α (due to

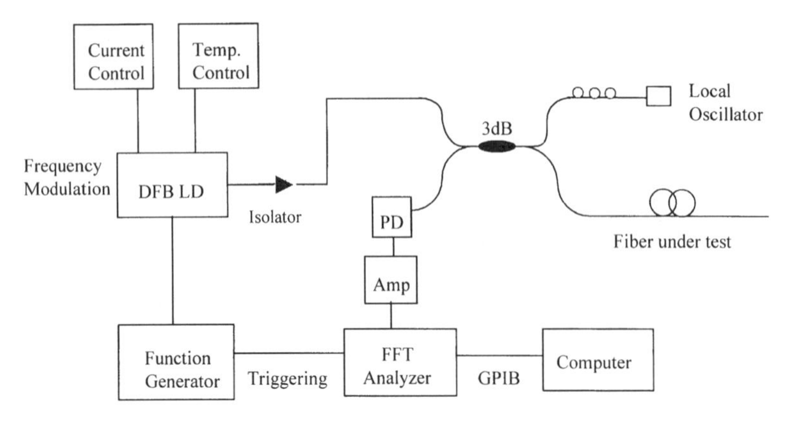

Fig. 10.18. An optical setup for optical frequency-domain reflectometry.

effects such as Rayleigh scattering) and there is only one perturbation at location, x, which results in a reflectivity $r(x)$. The returned light signal from the testing fiber can be written as

$$A(\beta) = A_0 r(x) e^{-\alpha x} e^{i2\beta x}, \tag{10.14}$$

where A_0 is a constant, and β is the propagation constant. For the purpose of simplicity, let

$$R(x) = A_0 r(x) e^{-\alpha x}. \tag{10.15}$$

Note that $R(x)$ is related to the disturbance distribution. Once this $R(x)$ is obtained, the disturbance along the fiber can be determined. Due to the introduction of $R(x)$, Eq. (10.15) can be rewritten as

$$A(\beta) = R(x) e^{i2\beta x}. \tag{10.16}$$

However, in real cases, the reflection may not come from one point. The reflection itself is a distribution. Thus, Eq. (10.16) needs to be written as the summation from the contributions of different locations. Mathematically, it can be written as

$$A(\beta) = \int_0^L R(x) e^{i2\beta x}\, dx, \tag{10.17}$$

where L is the total length of the testing fiber.

In the experiment, since the frequency is linearly tuned with the time, the propagation constant β is also linearly tuned with the time. Substituting $\beta = 2\pi nf/c$ (where c is the light speed in the vacuum), the reflected signal as a function of frequency, f, $A(f)$ can be written as

$$A(f) = \int_0^L R(x)e^{i(4\pi nfx)/c}\,dx. \tag{10.18}$$

Similarly, the reflected reference signal from the local oscillator $A_r(f)$ is

$$A_r(f) = A_1 e^{i(4\pi nfx_r)/c}, \tag{10.19}$$

where A_1 is a constant and x_r is the location of the reference point. In general, it is the exit ending point of the reference fiber. Then, via coherent detection, the interference term can be shown to be

$$I(f) = A(f)A_r^*(f) + A^*(f)A_r(f). \tag{10.20}$$

Substituting Eqs. (10.18) and (10.19) into Eq. (10.20), we get

$$I(f) = A_1 \int_0^L R(x)\cos\left[\frac{4\pi nf}{c}(x - x_r)\right]dx. \tag{10.21}$$

By taking the inverse Fourier transform of Eq. (10.21), one can obtain $R(x)$; i.e.,

$$R(x - x_r) = \int_0^{\Delta f} I(f)\cos\left[\frac{4\pi nf}{c}(x - x_r)\right]df, \tag{10.22}$$

where Δf is the total frequency tuning range. Since $R(x)$ is directly related to the strain and temperature distribution of the fiber, distributed sensing is achieved.

From Eqs. (10.21) and (10.22), it can shown that the spatial resolution of this sensing system is

$$\Delta x = \frac{c}{2n\Delta f} = \frac{\lambda^2}{2n\Delta\lambda}, \tag{10.23}$$

where Δf and $\Delta\lambda$ are the total frequency tuning range and wavelength tuning range, respectively. Similarly, it can also be shown that the sensing range, L, of this distributed fiber sensor is

$$L = \frac{c}{2n\delta f} = \frac{\lambda^2}{2n\delta\lambda}, \tag{10.24}$$

where δf and $\delta\lambda$ are the effective frequency resolution and wavelength resolution, respectively.

For the semiconductor laser, the frequency tuning range can be 10 GHz changing the driving current. Substituting this number, $n = 1.5$, and $c = 3 \times 10^8$ m/s into Eq. (10.23), we obtain $\Delta x = 1$ cm. Similarly, if the effective frequency resolution can be 1 kHz (this is possible with current diode laser technology), then, based on Eq. (10.24), the sensing range L can be as long as 10 km.

In OFDR, coherent detection is needed. Since conventional single mode fiber cannot hold the polarization state of the light propagating in the fiber, polarization-insensitive detection is expected. Figure 10.19 shows polarization-insensitive detection using the polarization-diversity receiver. Since the reference signal is equally divided between two detectors, squaring and summing the two photocurrents produces a signal independent of the unknown polarization angle θ. If the returning test signal is elliptically polarized, the only difference will be a phase shift in the fringe pattern on one of the detectors. This phase shift is not important since the envelope detector eliminates any phase information before the summing process. Therefore, the output signal is

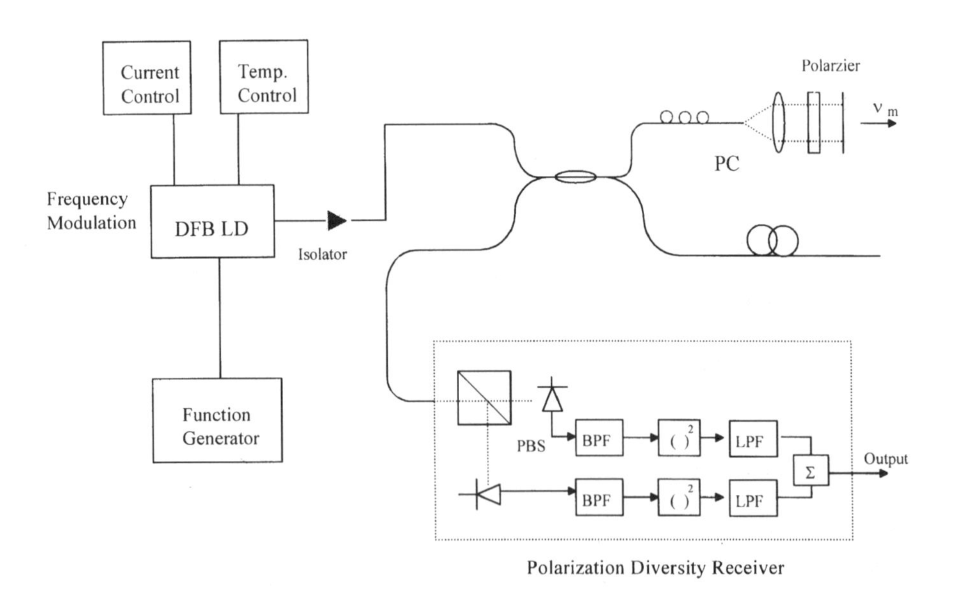

Fig. 10.19. An experimental setup for the implementation of a polarization-diversity receiver.

completely independent of the polarization state returning from the test arm.

Another critical issue of coherent OFDR is the suppression of phase noise. The high spatial resolution of OFDR depends on the light source having a large, phase-continuous, and linear tuning range. The fact is that no laser source is perfect in practice. Any laser source has phase noise and other problems in its output signal spectrum (the laser modulation problem mentioned above is an example). The phase noise of the laser limits two aspects of system performance. One is the distance over which measurements of discrete reflections can be made before incoherent mixing predominates, and the other is the dynamic range between the reflection signal of interest and the level of phase noise. Phase noise is the effect of quantum noise in a diode laser. This effect causes the phase of the laser output to change with time in a random fashion. It is this random phase fluctuation that determines the theoretical minimum line width of a diode laser [33].

To reduce the phase noise influence, first we need a quantitative analysis of the phase noise for a coherent OFDR system. For a linear frequency sweep of slope γ, the optical field $E(t)$ can be described by

$$E(t) = E_0 e^{j(\omega_0 t + \pi\gamma t^2 + \phi_t)}, \tag{10.25}$$

where ω_0 is the initial optical frequency and ϕ_t is the randomly fluctuating optical phase at time t The photocurrent $I(t)$ of the photodetector is proportional to the optical intensity incident on the photodetector, which is made up of the coherent mixing of the optical field from the source with a delayed version of that field, as given by

$$I(t) = |E(t) + \sqrt{R}\, E(t - \tau_0)|^2, \tag{10.26}$$

where $E(t)$ is the reflected back signal from the reference arm and $\sqrt{R}\, E(t - \tau_0)$ is the signal from the test fiber ($\tau = 2nx_0/c$). Combining Eqs. (10.25) and (10.26), we get

$$I(t) = E_0^2(1 + R + 2\sqrt{R}\cdot\cos(\omega_b t + \omega_0\tau_0 - \tfrac{1}{2}\omega_b\tau_0 + \phi_t - \phi_{t-\tau_0})), \tag{10.27}$$

where the beat frequency ω_b is given by $2\pi\gamma\tau_0$. The normalized autocorrelation function $R_1(T)$ of the photocurrent is given by

$$\begin{aligned} R_I &= \left(\frac{1}{E_0^2}\right)\langle I(t)I(t + T)\rangle \\ &= (1 + R)^2 + 2R\cos\omega_b T\langle\cos(\phi_{t+T} + \phi_{t-\tau_0} - \phi_{t-\tau_0+T} - \phi_t)\rangle \\ &\quad - 2R\sin\omega_b T\langle\sin(\phi_{t+T} + \phi_{t-\tau_0} - \phi_{t-\tau_0+T} - \phi_t)\rangle, \end{aligned} \tag{10.28}$$

where $\langle\,\rangle$ denotes a time average. Since the power spectral density (PSD) is the Fourier transform of the autocorrelation function, we can calculate a general expression for PSD. Considering the OFDR scheme in [33], one-sided PSD $S_I(f)$ can be denoted by

$$S_I(f) = (1 + R)^2\delta(f) + 2Re^{-(2\tau_0/\tau_c)}\delta(f - f_b) + \frac{2R\tau_c}{1+\pi^2\tau_c^2(f-f_b)^2}\left[1-e^{-(2\tau_0/\tau_c)}\left\{\cos 2\pi(f-f_b)\tau_0 + \frac{\sin 2\pi(f-f_b)\tau_0}{\pi\tau_c(f-f_b)}\right\}\right]. \tag{10.29}$$

There are three terms in Eq. (10.29). The first term is a delta function at *dc*, the second is a delta function at the beat frequency, and the third is a continuous function of frequency, strongly affected by the coherence time and the delay time. It represents the distribution of phase noise around the beat frequency. Note that, for the purpose of simplicity, we ignore the Rayleigh backscattering, the intensity noise of laser source, and the shot noise at the photodetector receiver here because their spectrum amplitudes are smaller compared with the third term in Eq. (10.29). To suppress the phase noise, one technique is to vary the length of the reference arm and then use a one-dimension smooth filter to find the desired beat spectrum peaks. In other words, we can just measure the Fresnel back-reflection heterodyne beat signal by varying the reference arm length.

As an example, let us look at an experimental result of OFDR. A 5 km-long single mode fiber dispenser is selected as the testing fiber. The strain distribution among different layers is measured by using the OFDR. Figure 10.20 shows the experimental results. The upper curve of Fig. 10.20 shows the detected interference signal between the reference arm and sensing arm. Clearly, the beating effect of frequency can be seen. The lower curve of Fig. 10.20 shows the strain distribution of the fiber achieved by taking the Fourier transform of the upper curve. It can be seen that a set of peaks appeared in the curve, which represents a sudden change in the strain among different layers.

10.3.2. QUASI-DISTRIBUTED FIBER-OPTIC SENSORS

When truly distributed sensing is difficult to realize, quasi-distributed fiber-optic sensor technique is used. In this technique, the measurand is not monitored continuously along the fiber path, but at a finite number of locations. This is accomplished either by sensitizing the fiber locally to a particular field of interest or by using extrinsic-type (bulk) sensing elements. By using quasi-distributed fiber-optic sensors, more measurands can be sensed.

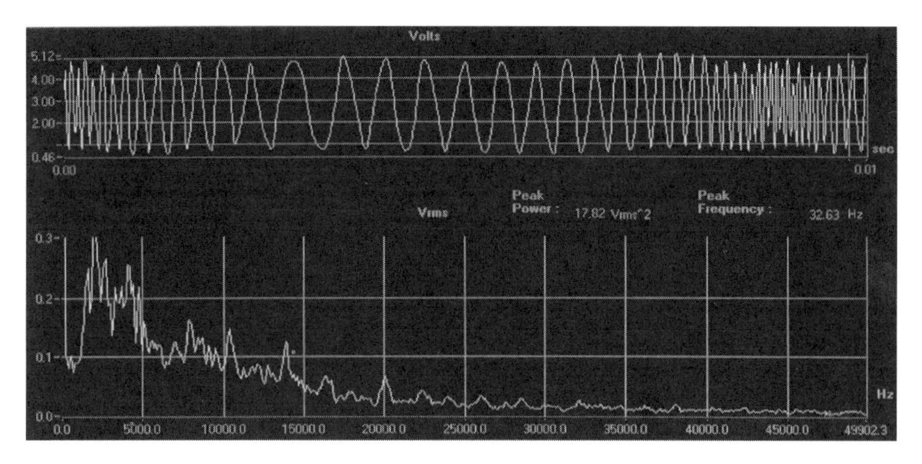

Fig. 10.20. An illustration of experimental results of optical frequency-domain reflectometry.

Theoretically speaking, by cascading a set of point sensors together (as described in Sec. 10.2), quasi-distributed sensing can be achieved. For example, when a series of reflectors are fabricated in the fiber, a quasi-distributed fiber-optic sensor based on discrete reflectors can be built. The OTDR technique can then be used to analyze the relative positions of these reflectors. Changes in these reflection signals can be used to sense changes in these discrete locations.

Although quasi-distributed fiber-optic sensors may be based on a variety of principles such as Fresnel reflection and cascaded interferometers, fiber Bragg grating-based quasi-distributed fiber-optic sensors have unique features including high sensitivity, high multiplexing capability (such as using wavelength division multiplexing), and cost effectiveness. Thus we will discuss quasi-distributed fiber-optic sensors based on fiber Bragg gratings in detail.

10.3.2.1. *Quasi-Distributed Fiber-Optic Sensors Based on Fiber Bragg Gratings*

10.3.2.1.1. Fiber Bragg Grating and Its Fabrication

The concept of fiber Bragg grating can be traced back to the discovery of the photosensitivity of germanium-doped silica fiber [34, 35]. It was found that when an argon ion laser was launched into the core of the fiber, under prolonged exposure, an increase in the fiber attenuation was observed. In addition, almost all of the incident radiation back-reflected out of the fiber. Spectral measurements confirmed that the increase in reflectivity was the result of a permanent refractive index grating being photoinduced in the fiber. This photoinduced permanent grating in germanium-doped silica fiber is called fiber Bragg grating. During that time, it was also found that the magnitude of the photoinduced refractive index depended on the square of the writing power at the argon ion wavelength (488 nm). This suggested a two-photo process as the possible mechanism of refractive index change.

Almost a decade later, in 1989, Metlz et al. [36] showed that a strong index of refraction change occurred when a germanium-doped fiber was exposed to direct, single-photon UV light close to 5 eV. This coincides with the absorption peak of a germania-related defect at a wavelength range of 240–250 nm. Irradiating the side of the optical fiber with a periodic pattern derived from the intersection of two coherent 244-nm beams in an interferometer resulted in a modulation of the core index of refraction, inducing a periodic grating. Changing the angle between the intersecting beams alters the spacing between the interference maxima; this sets the periodicity of the gratings, thus making possible reflectance at any wavelength. This makes the fiber Bragg grating have practical applications because the original approach was limited to the argon ion writing wavelength (488 nm), with very small wavelength changes induced by straining the fiber. Reliable mass-produced gratings can also be realized by using phase masks [37].

The principle of UV–induced refractive index in germanium-doped silica fiber may be explained as follows: Under UV light illumination, there are oxygen vacancies located at substitutional Ge sites, which results in ionized defect band bleaching, liberating an electron and creating a GeE' hole trap. Thus, the refractive index for the regions under UV exposure is different from the unexposed regions.

Figure 10.21 shows the basic configuration of fabricating Bragg gratings in photosensitive fiber using a phase mask. The phase mask is a diffractive optical element that can spatially modulate the UV writing beam. Phase masks may be formed either holographically or by electrobeam lithography. The phase mask is produced as a one-dimensional periodic surface-relief pattern, with period Λ_{pm} etched into fused silica. The profile of the phase is carefully designed so that when a UV beam is incident on the phase mask, the zero-order diffracted beam is suppressed (typically less than 3%) of the transmitted power. On the other hand, the diffracted plus and minus first orders are maximized. Thus, a near-field interference fringe pattern is produced. The period of this interference fringe pattern, Λ, is one-half that of the mask (i.e., $\Lambda = \Lambda_{pm}/2$). This fringe pattern photo-imprints a refractive index modulation in the core of a photosensitive fiber that is placed in contact with or close proximity to the phase mask. To increase the illumination efficiency, a cylindrical lens is generally used to focus the fringe pattern along the fiber core. Note that the KrF excimer laser is often used as the UV light source because it has the right output wavelength and output power.

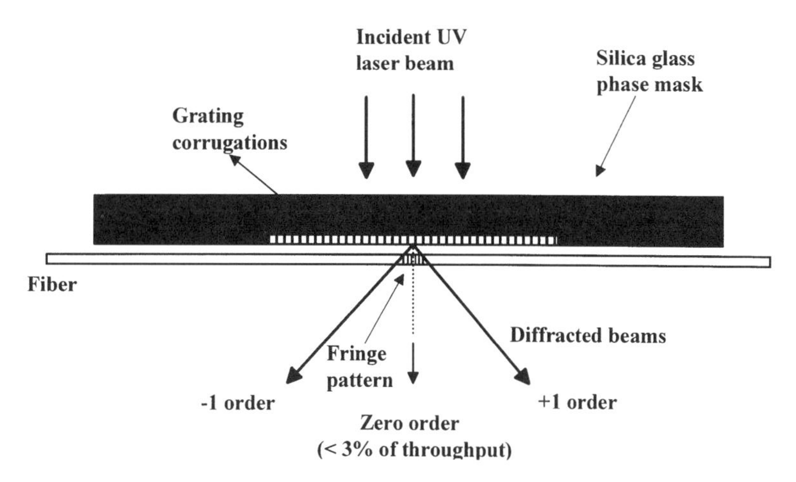

Fig. 10.21. Fabrication of fiber Bragg grating using the phase mask approach.

One of the most important properties of fiber Bragg grating is wavelength-selective reflection. Assume that a broadband light is coupled into a fiber with fiber Bragg grating inside. Light that has a wavelength matching the Bragg condition will be reflected back. Light that has a wavelength not matching the Bragg condition will be transmitted through the fiber, as shown in Fig. 10.22. Mathematically the Bragg condition is given by

$$\lambda_B = 2n_{eff}\Lambda, \tag{10.30}$$

where λ_B is the Bragg grating wavelength that will be reflected back from the Bragg grating, and n_{eff} is the effective refractive index of the fiber core at wavelength λ_B.

For the uniform Bragg grating, the reflectivity and spectral width of the fiber Bragg grating were quantitatively analyzed [38, 39]. The refractive index profile for the uniform Bragg grating with grating period, Λ, is given by

$$n(z) = n_0 + \Delta n \cos(2\pi z/\Lambda), \tag{10.31}$$

where Δn is the amplitude of the induced refractive index perturbation (typical values $10^{-5} \sim 10^{-3}$), n_0 is the average refractive index, and z is the distance along the fiber longitudinal axis. The reflectivity for this grating can be shown to be [38]

$$R(L, \lambda) = \frac{\Omega^2 \sinh^2(SL)}{\Delta\beta^2 \sinh^2(SL) + S^2 \cosh^2(SL)}, \tag{10.32}$$

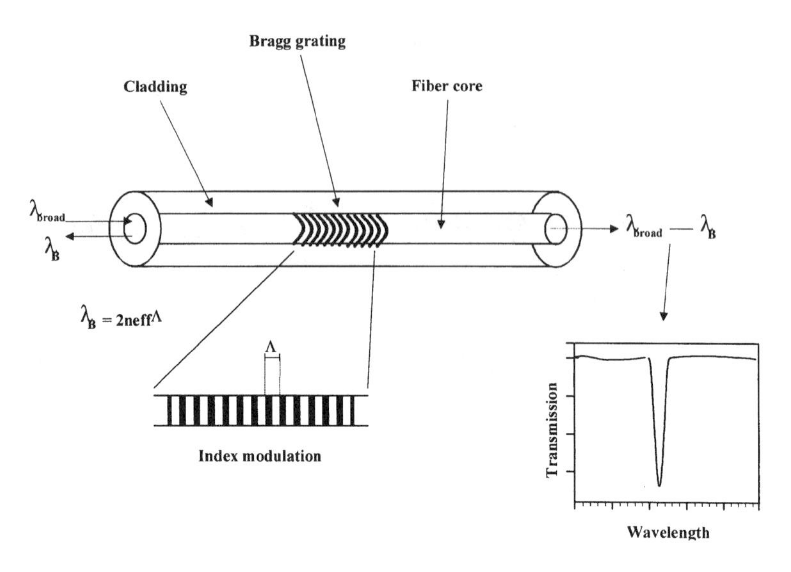

Fig. 10.22. Illustration of a uniform fiber Bragg grating and its wavelength-selective reflection property.

where $R(L, \lambda)$ is the reflectance as a function of wavelength, λ, and fiber length, L; Ω is the coupling coefficient; $\Delta\beta = \beta - \pi/\Lambda$ is the detuning wave vector; $\beta = 2\pi n_0/\lambda$ is the propagation constant; Λ is the Bragg grating period; and $S = \sqrt{\Omega^2 - \Delta\beta^2}$. For the single mode sinusoidal modulated grating as described by Eq. (10.14), the coupling constant, Ω, is given by

$$\Omega = \frac{\pi \Delta n}{\lambda} M_p, \tag{10.33}$$

where M_p is the fraction of the fiber mode power contained by the fiber core, which can be approximately expressed as

$$M_p = 1 - V^{-2}, \tag{10.34}$$

where V is the normalized frequency of the fiber and given by

$$V = \frac{2\pi a}{\lambda} \sqrt{n_{\text{co}}^2 - n_{\text{cl}}^2}$$

where a is the core radius, and n_{co} and n_{cl} are the core and cladding refractive indices, respectively. At the Bragg grating center wavelength, the detuning factor $\Delta\beta = 0$. Thus, the peak reflectivity of the Bragg grating is

$$R(L, \lambda_B) = \tanh^2(\Omega L). \tag{10.35}$$

From Eq. (10.35), it can be seen that the peak reflectivity increases as the refractive index modulation depth, Δn, and/or grating length L increases.

A general expression for the approximate full-width-half-maximum bandwidth of the grating is given by [39]

$$\Delta\lambda = \lambda_B \cdot q \sqrt{\left(\frac{\Delta n}{2n_0}\right)^2 + \left(\frac{\Lambda}{L}\right)^2}, \tag{10.36}$$

where q is a parameter that approximately equals 1 for strong gratings (with near 100% reflection) whereas $q \sim 0.5$ for weak gratings.

Equation (10.36) shows that, to achieve narrow spectral width, long grating length and small refractive index modulation need to be used.

As an example, a calculated reflection spectrum as a function of the wavelength is shown in Fig. 10.23. The following parameters are used during the calculation: $n_{co} = 1.45$, $n_{cl} = 1.445$, $a = 4\ \mu$m, and $\Lambda = 0.535\ \mu$m. The solid line corresponds to the stronger coupling case with $\Delta n = 10^{-3}$ and grating

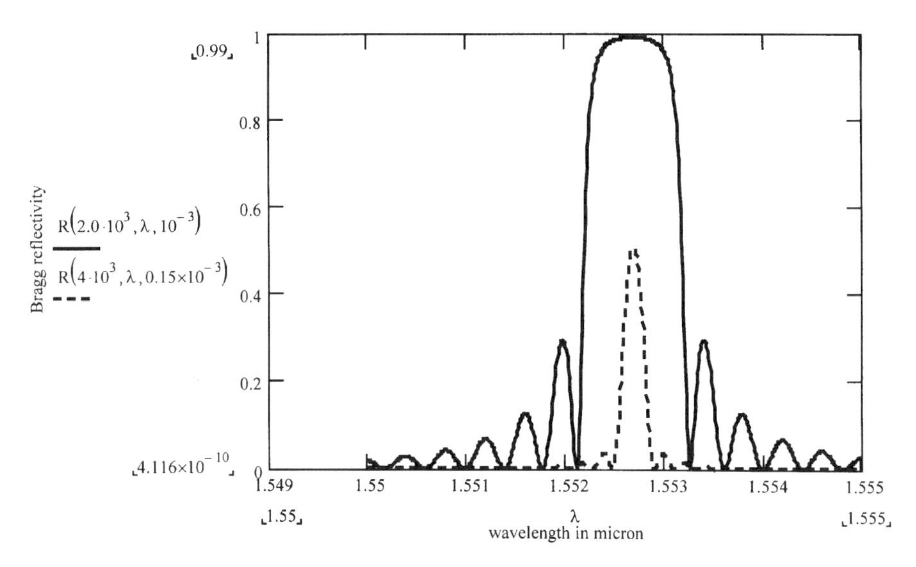

Fig. 10.23. Reflection spectrum of a Bragg grating as a function of wavelength with different coupling constant and length. Solid line: $\Delta n = 10^{-3}$ and $L = 2$ mm; dashed line: $\Delta n = 1.5 \times 10^{-4}$ and $L = 4$ mm.

length $L = 2$ mm. The dashed line corresponds to the weaker coupling case with $\Delta n = 1.5 \times 10^{-4}$ and grating length $L = 4$ mm. From this figure, it can be seen that a higher reflectivity is achieved for the stronger coupling case. However, a narrower spectrum width could be achieved with a weaker coupling case, which is consistent with the conclusion given by Eq. (10.36).

10.3.2.1.2. Fiber Bragg Grating–Based Point Sensor

From Eq. (10.30), we see that the reflected wavelength by Bragg grating, λ_B, is a function of both the effective refractive index of the fiber, n_{eff}, and the Bragg grating period, Λ. Since both n_{eff} and Λ may be affected by the change of strain and temperature on the fiber, λ_B is also sensitive to the change of strain and temperature on the fiber. Thus, by detecting the change in λ_B, the strain and temperature can be sensed.

By differentiating Eq. (10.30), the shift in the Bragg grating center wavelength, λ_B, due to strain and temperature changes can be expressed as

$$\Delta\lambda_B = 2\left(\Lambda\frac{\partial n_{eff}}{\partial L} + n_{eff}\frac{\partial\Lambda}{\partial T}\right)\Delta L + 2\left(\Lambda\frac{\partial n_{eff}}{\partial T} + n_{eff}\frac{\partial\Lambda}{\partial T}\right)\Delta T. \quad (10.37)$$

The first term in Eq. (10.37) represents the strain effect on an optical fiber. This corresponds to a change in the grating period and the strain-induced change in the refractive index due to the photoelastic effect. To make the calculation easier, the strain effected may be expressed as [40]

$$\Delta\lambda_B = \lambda_B(1 - P_e)\varepsilon, \quad (10.38)$$

where p_e is an effective strain-optic constant defined as

$$p_e = \frac{n_{eff}^2}{2}[p_{12} - v(p_{11} + p_{12})], \quad (10.39)$$

p_{11} and p_{12} are components of the strain-optic tensor, v is the Poisson's ratio, and ε is the applied strain. The typical values for germanosilicate optical fiber are $p_{11} = 0.113$, $p_{12} = 0.252$, $v = 0.16$, and $n_{eff} = 1.482$. Substituting these parameters, into Eqs. (10.38) and (10.39), the anticipated strain sensitivity at ~1550 nm is about 1.2-pm change when 1 $\mu\varepsilon$ (i.e., 10^{-6}) is applied to the Bragg grating.

The second term in Eq. (10.37) represents the effect of temperature on an optical fiber. A shift in the Bragg wavelength due to thermal expansion changes the grating period and the refractive index. Similar to the strain case, for silica fiber, this wavelength shift due to the temperature change may be expressed as [40]

$$\Delta\lambda_B = \lambda_B(\alpha_\Lambda + \alpha_n)\Delta T, \tag{10.40}$$

where $\alpha_\Lambda = (1/\Lambda)(\partial\Lambda/\partial T)$ is the thermal expansion coefficient and $\alpha_n = (1/n_{\text{eff}})(\partial n_{\text{eff}}/\partial T)$ represents the thermo-optic coefficient. For the fiber Bragg grating, $\alpha_\Lambda \approx 0.55 \times 10^{-6}/^\circ\text{C}$ and $\alpha_n \approx 8.6 \times 10^{-6}/^\circ\text{C}$. Thus, the thermo-optic effect is the dominant effect for the wavelength shift of the Bragg grating when there is a temperature change on the grating. Based on Eq. (10.40), the expected temperature sensitivity for a 1550-nm Bragg grating is approximately 13.7 pm/°C.

Figure 10.24 shows the conceptual configuration of the fiber Bragg grating point sensor. This sensor may be used as a strain or temperature sensor. The broadband light source, such as that coming from an erbium-doped fiber amplifier, is coupled into a single mode optical fiber with a fiber Bragg grating inside the fiber. Due to the existence of the grating, the wavelengths that match the Bragg condition will be reflected back and the other wavelengths will pass through. Then wavelength monitoring devices (such as a spectrometer) can be used to monitor the wavelength spectra for both the reflected signal and the transmitted signal, as shown in Fig. 10.24. The changes in the spectra can be used to detect the applied strain or temperature changes.

As an example, Figs. 10.25(a) and 10.25(b) show the experimental results of the peak wavelength as a function of applied strain and ambient temperature for standard germanosilicate optical fiber, respectively. From these data, it can be seen that a wavelength resolution of ~1 pm is required to resolve a

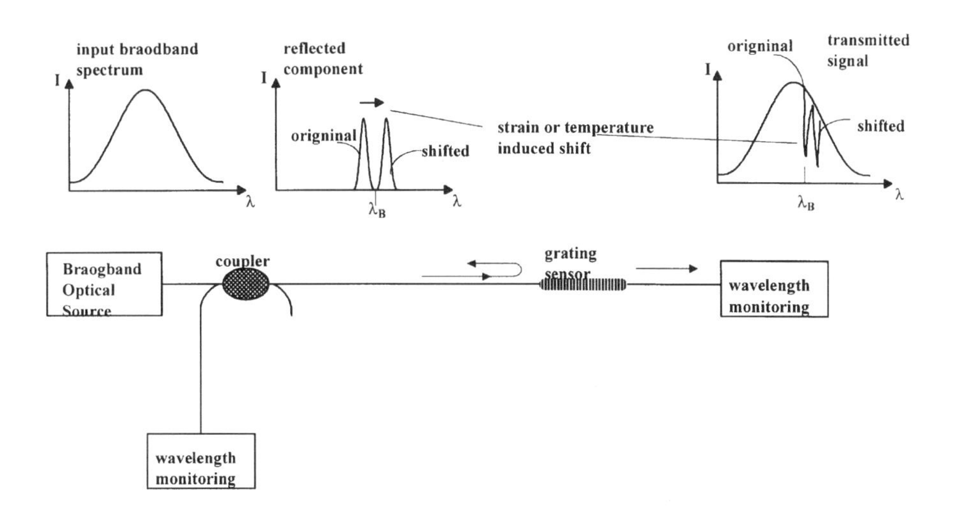

Fig. 10.24. An illustration of a fiber Bragg grating–based point sensor.

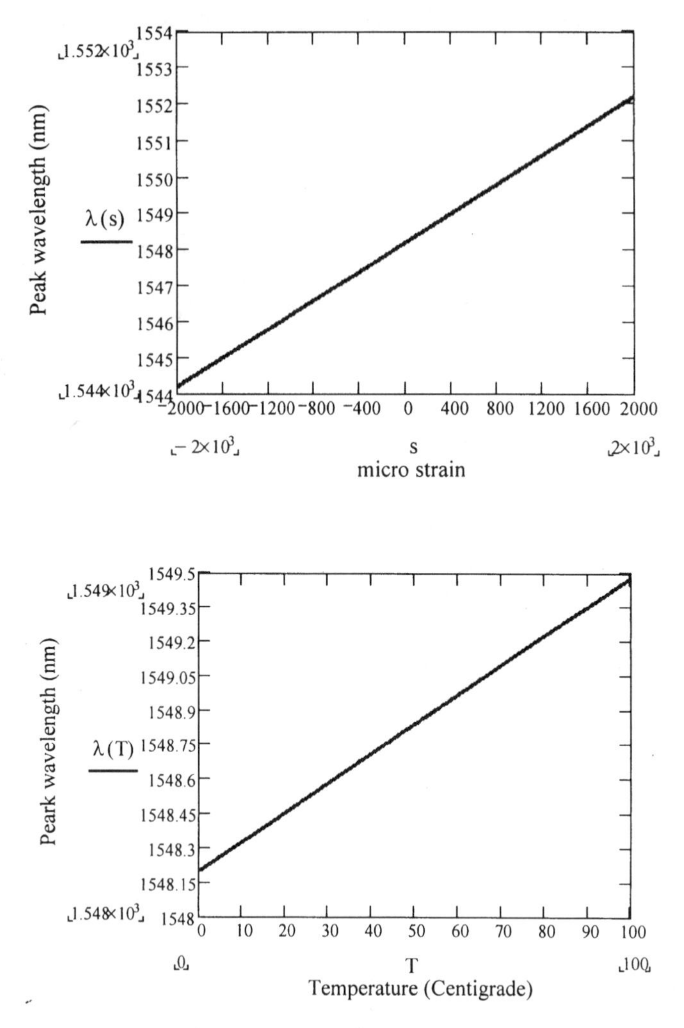

Fig. 10.25. Peak wavelength of fiber Bragg grating. (a) Under applied stress, (b) At different temperatures with zero applied strain.

temperature change of $\sim 0.1°C$, or strain change of 1 μ strain. To detect such a tiny wavelength shift, thc tcchniquc of using the unbalanced Mach Zehnder interferometer was proposed and implemented [41]. Figure 10.26 shows the basic configuration of applying an unbalanced interferometer to detect the wavelength shift of the Bragg grating induced by external perturbations. Due to the inherent wavelength dependence of the phase of an unbalanced interferometer on the input wavelength, shifts in Bragg wavelength are converted

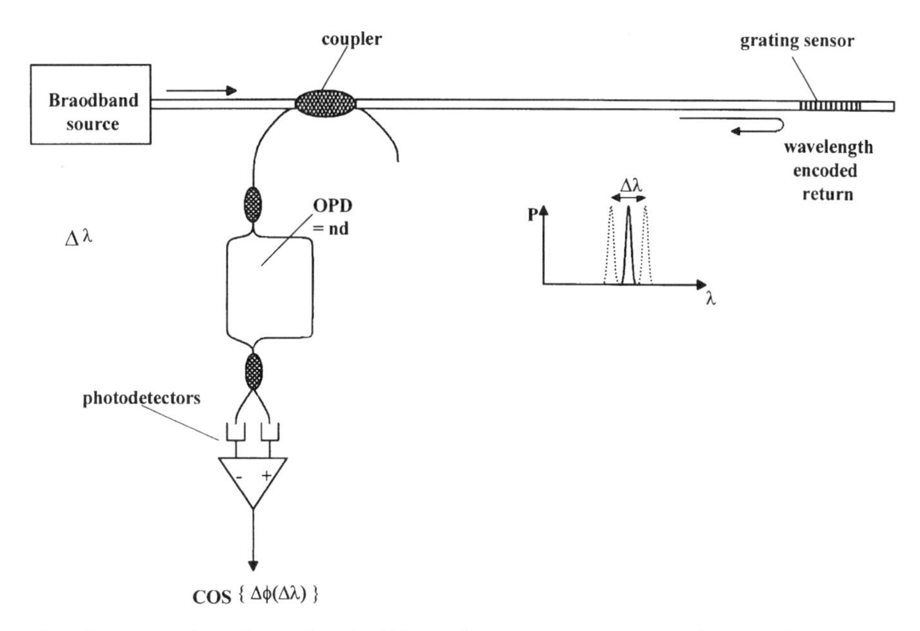

Fig. 10.26. Detection of wavelength shift of fiber Bragg grating by using unbalanced fiber Mach–Zehnder interferometer.

into phase shifts. Note that the interferometer path difference must be kept less than the effective coherent length of the light reflected from the grating. Mathematically, the dependence of the output phase change on the Bragg grating wavelength shift is given by

$$\Delta\phi = \frac{2\pi nd}{\lambda^2}\Delta\lambda, \tag{10.41}$$

where $\Delta\phi$ represents the phase change, λ is the operating wavelength, $\Delta\lambda$ is the wavelength shift, n is the refractive index of the fiber, and d is the length difference between the two arms of the Mach–Zehnder interferometer.

Example 10.3. Assume that an unbalanced interferometer has a path difference $d = 5$ mm; refractive index $n = 1.5$, operating wavelength $\lambda = 1550$ nm, and minimum phase resolution of the system is 0.01 radians. Calculate the minimum wavelength shift that can be detected by this unbalanced interferometer.

Solve: Based on Eq. (10.41), the minimum detectable wavelength shift is

$$\Delta\lambda_{\min} = \frac{\lambda^2}{2\pi nd}\Delta\phi_{\min} = \frac{1550^2\ \text{nm}^2}{2\times\pi\ \text{radians}\times 1.5\times 5\times 10^6\ \text{nm}}\times 0.01\ \text{radians} = 0.51\ \text{pm}.$$

Thus, such an unbalanced interferometer can detect tiny wavelength shift, which makes the built fiber sensor have good sensitivity.

10.3.2.1.3. Quasi-Distributed Fiber-Optic Sensors Based on Fiber Bragg Gratings

The major motivation of applying fiber Bragg gratings to fiber sensors is the capability of integrating a large number of fiber Bragg gratings in a single fiber so that quasi-distributed fiber sensing can be realized in a compact and cost-effective way [42]. Currently, with the rapid advent of optical communication networks, more than 100 wavelength channels can be put in a single fiber by using the wavelength-division multiplexing (WDM) technique [43]. Thus, if we assign one central wavelength for each grating, more than 100 sensors can be integrated into a single fiber. Furthermore, applying time-division multiplexing (TDM) to each wavelength channel creates a severalfold increase in the number of sensors that can be integrated. Therefore, a compact, cost-effective distributed fiber sensor can be built.

Figure 10.27 shows the configuration of combining WDM and TDM in quasi-distributed fiber-optic sensors based on fiber Bragg gratings. By launching a short pulse of light from the source, the reflections from the fiber Bragg grating will return to the detector at successively later times. The detection instrumentation is configured to respond to the reflected signals only during a selected window of time after the pulse is launched. Thus, a single WDM set of sensors is selected for detection.

As a practical example, Figure 10.28(a) shows the Stork Bridge in Winterthur, Switzerland, in which carbon fiber–reinforced polymer (CFRP) cables were used instead of the usual steel cables. Each CFRP cable is equipped with an array of seven Bragg gratings, as well as other types of regular sensors. Figure 10.28(b) shows the output from these Bragg grating sensors during daily variations of strain and temperature in one CFRP cable. BG1, BG4, BG6, and BG7 represent the outputs from the first, fourth, fifth, sixth, and seventh Bragg grating sensors. Results of the measurements are pretty consistent with the

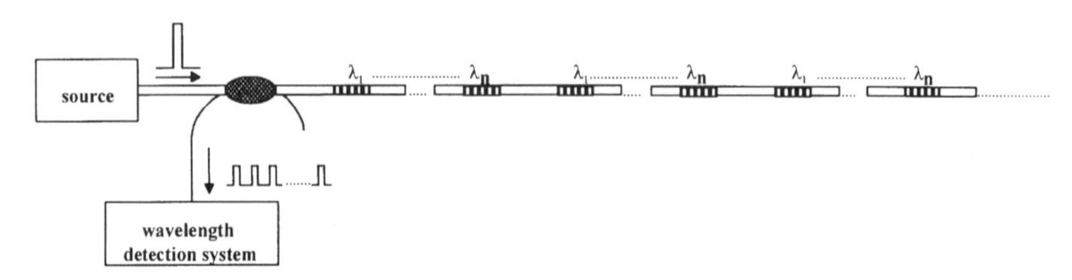

Fig. 10.27. A configuration of combining WDM/TDM in a fiber Bragg grating–based quasi-distributed fiber sensor.

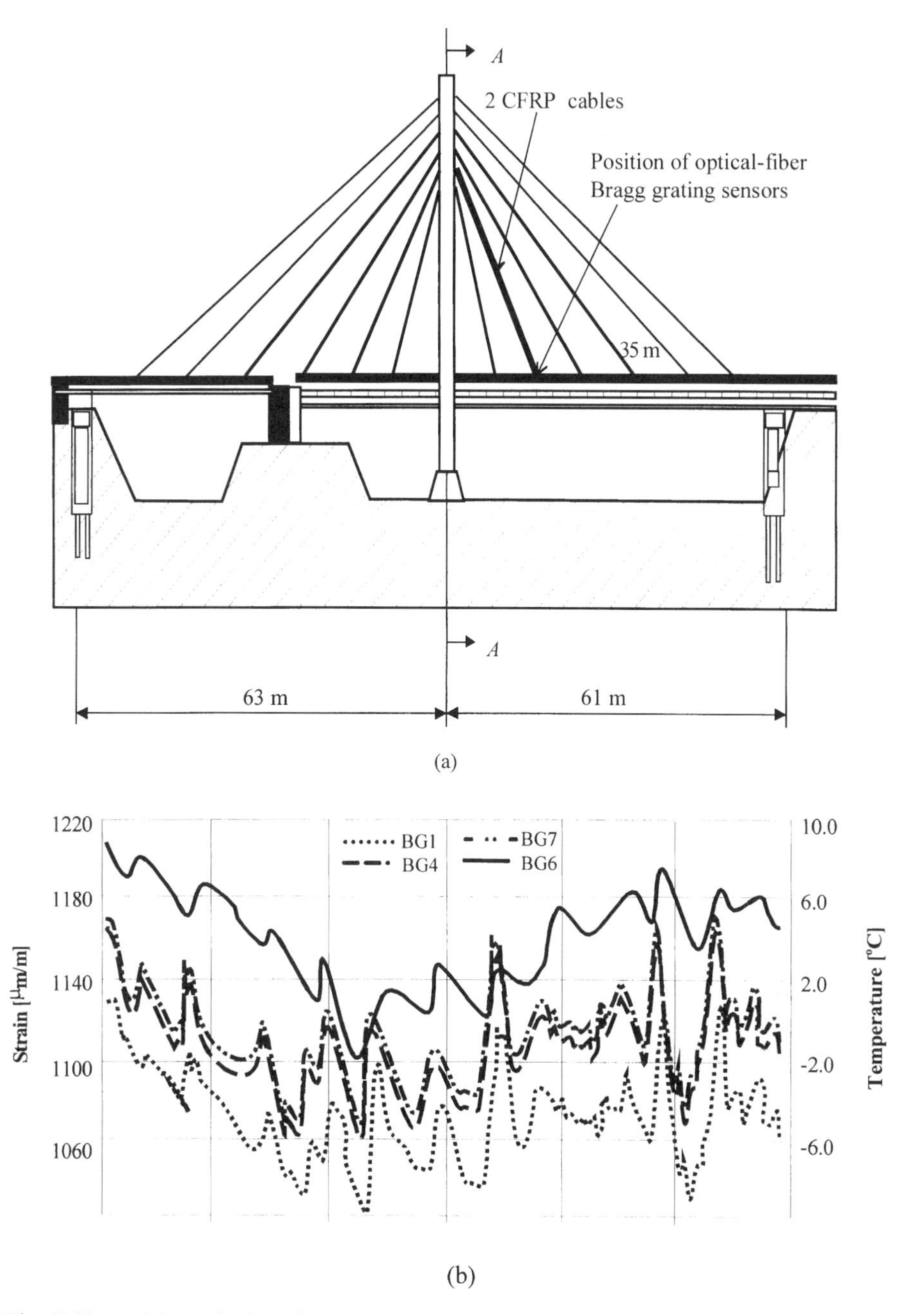

Fig. 10.28. (a) Schematic view of Stork Bridge in Winterthur, Switzerland. (b) A draft figure of recorded daily variations of strain and temperature in one CFRP cable by fiber Bragg grating sensors.

results obtained with conventional resistant strain gauges [44]. Thus, indeed, fiber Bragg grating–based quasi-distributed fiber sensors may be used in real-world sensing.

10.3.2.1.4. Multiple-Parameter Measurements Using Fiber Bragg Gratings Written in Highly Birefringent Polarization-Preserving Fibers

In real-world applications, strain and temperature perturbations may add to the fiber Bragg gratings simultaneously. To avoid the use of additional sensors to separate influence from strain and temperature changes, the techniques of using double Bragg gratings written at widely spaced wavelengths at the same location in a polarization-maintaining (PM) fiber were developed [45, 46]. In this case, the spectrum reflected from one location contains four peaks. Thus, in principle, one can determine axial strain, two components of transverse strain, and temperature change in that location based on these four peaks, as given by

$$\begin{bmatrix} \Delta\lambda_{1a} \\ \Delta\lambda_{2a} \\ \Delta\lambda_{1b} \\ \Delta\lambda_{2b} \end{bmatrix} = K \begin{bmatrix} \varepsilon_1 \\ \varepsilon_2 \\ \varepsilon_3 \\ \Delta_T \end{bmatrix}, \tag{10.42}$$

where a and b designate peak wavelength measurements for Bragg gratings written at wavelengths a and b, $\Delta\lambda_{1a}$ represents the wavelength shift from horizontal polarization direction at wavelength a, $\Delta\lambda_{2a}$ represents the wavelength shift from vertical polarization direction at wavelength a, $\Delta\lambda_{1b}$ represents the wavelength shift from horizontal polarization direction at wavelength b, $\Delta\lambda_{2b}$ represents the wavelength shift from vertical polarization direction at wavelength b, K is a 4×4 matrix the elements of which may be determined by separate experimental calibrations of sensor response to transverse strain, axial strain and temperature change, ε_1, ε_2, and ε_3 are axial, two transverse strains, respectively, and ΔT represents the temperature change. This technique is still under development and some promising results have been achieved [46].

10.4. SUMMARY

This chapter provided a brief summary on sensing with optics. First, it introduces the mathematical description of the light field, which includes amplitude, polarization, phase, and frequency. Since all these parameters of the light field may change with external perturbations, sensing may be performed by light field. Second, the application of each parameter to different types of fiber sensors are introduced. The advantages and limitations of each type of sensor are discussed. Finally, since distributed sensing capability is one of the most important features of fiber-optic sensors, distributed fiber-optic sensors were introduced in the third section of this chapter. In particular, fiber Bragg grating–based fiber sensors were discussed in detail due to their practicability for quasi-distributed fiber sensing that may be used in structure (e.g., bridges and dams) monitoring and for synthesizing smart composite materials. On the other hand, with the rapid advent of fiber-optic communication networks, intrinsic distributed fiber-optic sensors such as OTDR, and OFDR may become more and more important for real-time network monitoring and integrated optical circuits inspection.

REFERENCES

10.1 K. T. V. Grattan and B. T. Meggitt, *Optical Fiber Sensor Technology*, Chapman & Hall, New York, 1995.

10.2 F. Corke, F. Gillham, A. Hu, D. W. Stowe, and L. Sawyer, 1988, "Fiber-Optic Pressure Sensors Employing Reflective Diaphragram Techniques," *SPIE*, vol. 985, 164–171.

10.3 I. P. Giles, 1985, "Self-Compensating Technique for Remote Fiber-Optic Intensity Modulated Transducers," *SPIE*, 522, 233–239.

10.4 I. P. Giles, S. McNeill, and B. Sulshaw, 1985, "A Stable Remote Intensity-Based Optical Fiber Sensor," *J. Phys. E (GB)*, 18(6), 502–504.

10.5 C. M. Lawson and V. J. Tekippe, 1983, "Environmentally Insensitive Diaphragm Reflectance Pressure Sensor," *SPIE*, 412, 96–103.

10.6 J. W. Berthold, W. L. Ghering, and D. Varshneya, "Design and Characterization of a High-Temperature, Fiber-Optic Pressure Transducer," 1987, *IEEE J. Lightwave Technol.*, *LT*-5, 1–6.

10.7 D. R. Miers, D. Raj, and J. W. Berthold, 1987, "Design and Characterization Fiber-Optic Accelerometers," *SPIE*, 838, 314–317.

10.8 G. L. Mitchell, "Intensity-Based and Fabry-Perot Interferometer Sensors," in *Fiber-Optic Sensors*, E. Udd, Ed., pp. 139–140, John Wiley & Sons, Inc., New York, 1991.

10.9 R. A. Bergh, G. Kotler, and J. J. Shaw, 1980, "Single Mode Fiber-Optics Directional Coupler," *Electron. Lett.*, 16, 260–261.

10.10 P. D. McIntyre and A. W. Snyder, 1973, "Power Transfer between Optical Fibers," *J. Opt. Soc. Am.*, 63, 1518–1527.

10.11 P. S. Theocaris and E. E. Gdoutos, *Matrix Theory of Photoelasticity*, Springer-Verlag, New York, 1979.

10.12 B. W. Brennan, 1988, "Prototype Polarimetric-Based Fiber-Optic Strain Gauge," in *Review of Progress in Quantitative Nondestructive Evaluation*, Plenum Press, New York, 1988, 547–552.

10.13 C. D. Butter and G. B. Hocker, 1980, "Measurements of Small Phase Shifts Using a Single Mode Optical Fiber Interferometer," *Opt. Lett.*, 5, 139–141.

10.14 D. A. Jackson, R. G. Priest, A. Dandrige, and A. B. Tveten, 1980, "Elimination of Drift in a Single Mode Optical Fiber Interferometer Using a Piezoelectrically-Stretched Coiled Fiber," *Appl. Opt.* 19, 2926–2929.

10.15 D. W. Stowe, D. R. Moore, and R. G. Priest, 1982, "Polarization Fading in Fiber Interferometer Sensors," *IEEE. J. Quantum Electron*, OE-18, 1644–1647.
10.16 N. J. Frigo, A. Dandridge, and A. B. Tveten, 1984, "Technique for the Elimination of Polarization Fading in Fiber Interferometers," *Electron. Lett.* 20, 319–320.
10.17 K. H. Wanser and N. H. Safar, 1987, "Remote Polarization Control of Fiber-Optic Interferometers," *Opt. Lett.*, 12, 217–219.
10.18 S. J. Petuchowski, T. G. Giallorenzi, and S. K. Sheem, 1981, "A Sensitive Fiber-Optic Fabry-Perot Interferometer," *IEEE J. Quantum Electron*, OE-17, 2168–2170.
10.19 J. Stone, 1985, "Optical Fiber Fabry-Perot Interferometer with Finesse of 300," *Electron. Lett.*, 21, 504–505.
10.20 S. W. Allison, M. R. Coates, and M. B. Scudiero, 1987, "Remote Thermometry in Combustion Environment Using the Phosphor Technique," *SPIE*, 788, 90–99.
10.21 K. T. V. Grattan, A. W. Palmer, and F. A. S. Al-Ramadhan, 1990, "Optical-Fiber Sensing Techniques for Wide-Range Temperature Measurement," *Proceedings TEMPMEK0 90*, Helsinki, Finland, Finnish Society of Automatic Control, pp. 182–192.
10.22 Z. Zhang, K. T. V. Grattan, and A. W. Palmer, 1992, "Fiber-Optic Temperature Sensor Based on the Cross Referencing between Black-Body Radiation and Fluorescence Lifetime," *Rev. Sci. Instrum.*, 63, 3177–3181.
10.23 M. K. Barnoski and S. M. Jensen, 1976, "Fiber Waveguides: A Novel Technique for Investigating Attenuation Characteristics," *Appl. Opt.*, 15, 2112–2115.
10.24 D. Marcuse, *Principles of Optical Fiber Measurements*, Academic Press, New York, 1981.
10.25 R. I. MacDonald, 1981, "Frequency Domain Optical Reflectometer," *Appl. Opt.*, 20, 1840–1844.
10.26 J. K. A. Everard, 1987, "Novel Signal Techniques for Enhanced OTDR Sensors," *SPIE*, 798, 42–46.
10.27 J. P. Dakin, 1985, "Temperature-Distributed Measurement Using Raman Ratio Thermometry," *SPIE*, 566, 249–256.
10.28 N. Shibata, R. Waarts, and R. Braun, 1987, "Brillouin-Gain Spectra for Single Mode Fibers Having Pure-Silica, GeO_2–Doped and P_2O_5–Doped Cores," *Opt. Lett.*, 12, 269–271.
10.29 M. Tateda, 1990, "First Measurement of Strain Distribution Along Field-Installed Optical Fibers Using Brillouin Spectroscopy," *J. Lightwave Technol.*, 8, 1269–1272.
10.30 J. Nakayama, K. Lizuka, and J. Nielsen, 1987, "Optical Fiber Fault Locator by the Step Frequency Method," *Appl. Opt.*, 26, 440–443.
10.31 W. Eickhoff and R. Ulrich, 1981, "Optical Frequency-Domain Reflectometry in Single Mode Fiber," *Appl. Phys. Lett.*, vol. 39, 693–695.
10.32 F. T. S. Yu, J. Zhang, S. Yin, and P. B. Ruffin, "A Novel Distributed Fiber Sensor Based on the Fourier Spectrometer Technique," OSA Annual Meeting, MA1, p. 50, Dallas, Texas, 1994.
10.33 Shalini Venkatesh and Wayne V. Sorin, 1993, "Phase Noise Considerations in Coherent Optical FMCW Reflectometry," *J. Lightwave Technol.*, vol. 11, no. 10, 1694–1700.
10.34 K. O. Hill, Y. Fujii, D. C. Johnson, and B. S. Kawasaki, 1978, "Photosensitivity in Optical Fiber Waveguides: Application to Reflection Filter Fabrication," *Appl. Phys. Lett.*, vol. 32, 647–649.
10.35 B. S. Kawasaki, K. O. Hill, D. C. Johnson, and Y. Fujii, 1978, "Narrow-Band Bragg Reflectors in Optical Fibers," *Opt. Lett.*, vol. 3, 66–68.
10.36 G. Meltz, W. W. Moery, and W. H. Glenn, 1989, "Formation of Bragg Gratings in Optical Fibers by a Transverse Holographic Method," *Opt. Lett.*, vol. 14, 823–825.
10.37 K. O. Hill, et al., 1993, "Bragg Gratings Fabricated in Monomode Photosensitive Optical Fiber by UV Exposure through a Phase Mask," *Appl. Phys. Lett.*, vol. 62, 1035–1037.
10.38 D. K. Lam and B. K. Garside, 1981, "Characterization of Single Mode Optical Fiber Filters," *Appl. Opt.*, vol. 20, 440–445.

10.39 S. J. Russell, J. L. Archambault, and L. Reekie, October 1993, "Fiber Gratings," *Physics World*, 41–46.
10.40 G. Meltz and W. W. Morey, 1991, "Bragg Grating Formation and Gemanosilicate Fiber Photosensitivity," International Workshop on Photoinduced Self-Organization Effect in Optical Fiber, Quebec City, Quebec, May 10–11, *Proceedings of SPIE*, vol. 1516, 185–199.
10.41 A. D. Kersey, T. A. Berkoff, and W. W. Morey, 1992, "High-Resolution Fiber Bragg Grating Based Strain Sensor with Interferometric Wavelength Shift Detection," *Electron. Lett.*, vol. 28, 236–238.
10.42 A. D. Kersey, M. A. Davis, H. J. Patrick, M. LeBlanc, K. Koo, C. Askins, M. A. Putnam, and E. J. Friebele, 1997, "Fiber Grating Sensors," *J. Lightwave Technol.*, vol. 15, 1442–1463.
10.43 S. V. Kartalopoulos, *Introduction to DWDM Technology*, IEEE Press, New York, 2000.
10.44 U. Sennhauser, R. Bronnimann, P. Mauron, and P. M. Nellen, "Reliability of Optical-Fibers and Bragg Grating Sensors for Bridge Monitoring," in *Fiber-Optic Sensors for Construction Materials and Bridges*, Farhad Ansari, Ed., Technomic Publishing Co., Inc., Lancaster, 1998, pp. 117–128.
10.45 C. M. Lawrence, D. V. Nelson, and E. Udd, 1996, "Multiparameter Sensing with Fiber Bragg Gratings," *SPIE*, 2872, 24–31.
10.46 C. M. Lawrence, D. V. Nelson, E. Udd, and T. Bennett, 1999, "A Fiber-Optic Sensor for Transverse Strain Measurement," *Experi. Mech.*, vol. 38, 202–209.

EXERCISES

10.1 Write down a general expression for the light field including amplitude, phase, frequency, and polarization.

10.2 Assume that a microbending device is used to implement an intensity-based fiber-optic sensor. If the effective refractive index for the guided mode is $n_g = 1.45$ and the effective refractive index for the radiation mode is $n_r = 1.446$, and the operating wavelength is $\lambda = 1.55\ \mu$m, calculate the optimum period for the microbending device.

10.3 For evanescent wave–based fiber-optic sensors, what is the order of maximum separation between two fibers (as shown in Fig. 10.3) in terms of wavelength?

10.4 A unit-intensity light beam linearly polarized in the x direction passes through a linear retarder with fast axis at angle, $\theta = 45^\circ$ relative to the x axis and retarding amount, $\delta = \pi/2$. Calculate the output polarization state in terms of the Stokes vector.

10.5 For a Mach–Zehnder interferometer–based fiber-optic sensor, assume that the optical path difference between two arms is $0.1\ \mu$m, the operating wavelength is $\lambda = 1.55\ \mu$m, and the maximum output intensity of this sensor is $I_{\max} = 1$ mW. What is the output power under the current situation?

10.6 A fiber-optic sensor is based on a Fabry-Perot interferometer. Assume that the input amplitude is $A = 1$, the reflection coefficient $R = 0.95$, and transmission coefficient $T = 0.1$.

(a) Calculate the coefficient of finesse F.
(b) Draw the output light intensity as a function phase shift ϕ.
(c) Based on the curve arrived at in part (b), estimate the maximum relative percentage change in output light intensity if the phase, ϕ, shifts 0.1 radians.

10.7 Assume that the black-body radiation method is used to sense the temperature of the sun's surface. If the maximum light intensity happens at $\lambda = 550$ nm, estimate the temperature of the sun's surface based on black-body radiation.

10.8 Explain what is fluorescent light.

10.9 An optical time–domain reflectometer (OTDR) is used to detect the fatigue location in optics networks. Assume that the refractive index of the fiber is $n = 1.5$ and a reflection peak is detected at a time $t = 10\,\mu s$ after launching the pulse. Where is the location of the fatigue relative to the incident end of the fiber?

10.10 In Exercise (10.9), if the pulse width of OTDR is $\tau = 5$ ns, what is the best spatial resolution of this OTDR?

10.11 Distributed temperature measurement is realized by using Raman scattering. Assume that the pump wavelength is $\lambda = 514$ nm, the Stokes and anti-Stokes wavelength shift in a particular location in terms of wave number is $\mp 400\,\text{cm}^{-1}$, and the intensity ratio from the anti-Stokes line to the Stokes line is 0.167. Calculate the temperature of that location.

10.12 A tunable diode laser is used as the light source of an optical frequency-domain reflectometer. Assume that the total tuning range of the diode laser is $\Delta\lambda = 1$ nm, the frequency stability of the laser is $\delta v = 300$ kHz, and the refractive index of the fiber is $n = 1.5$. Estimate the maximum sensing range and the spatial resolution of this distributed fiber optic sensor.

10.13 A fiber Bragg grating is written by UV light. Assume that the maximum refractive index modulation of the grating is $\Delta n = 3 \times 10^{-4}$, the length of the grating is $L = 5$ mm, the operating wavelength of the grating is $\lambda = 1550$ nm, the effective refractive index of the fiber is $n = 1.5$, and 60% of light energy is inside the couple.
(a) What is the grating period?
(b) What is the maximum diffraction efficiency of this grating?
(c) What is the FWHM bandwidth of the grating?

10.14 A fiber Bragg grating is used to sense temperature change. Assume that the wavelength shift induced by temperature change can be written as $\Delta\lambda = \lambda \cdot \alpha \cdot \Delta T$, where α is the total thermocoefficient given by $\alpha = 10^{-5}/°\text{C}$ and the operating wavelength $\lambda = 1550$ nm. To achieve 0.1°C temperature resolution, what is the required wavelength resolution of the wavelength detector?

Chapter 11 | Information Display with Optics

[1]Ting-Chung Poon and [2]Taegeun Kim

[1]Optical Image Processing Laboratory
(www.ee.vt.edu/~oiplab/),
Bradley Department of Electrical and
Computer Engineering,
Virginia Polytechnic Institute and
State University (Virginia Tech),
Blacksburg, Virginia 24061, USA
[2]Department of Optical Engineering,
Sejong University,
98 Kunja-Dong, Kwangjin-ku,
Seoul, 143-747, Korea

11.1. INTRODUCTION

From conventional two-dimensional (2-D) projection display to 3-D holographic display, techniques using photographic films have been extremely well developed. Photographic films have shown good optical properties in modulating certain properties of optical wavefronts, like amplitude or phase. However, in spite of good optical properties, photographic films have a critical drawback in time delay. Since they require chemical processing, long time delay is required from information storage to display. In modern display applications, there are increased needs for a real-time device for optical information processing; such a device is often called a spatial light modulator (SLM). Examples of SLMs are acousto-optic modulators and electro-optic modulators. In this chapter, we will concentrate on these two types of modulators.

We first discuss 2-D information display using acousto-optic (AO) modulators in Sec. 11.2. Sections 11.2.1 to 11.2.3 discuss the rudiments of acousto-optics and some of its applications, such as intensity modulation and deflection of laser. In Sec. 11.2.4, we illustrate how laser TV display can be achieved using AO devices. Since the real world is 3-D, in which the human visual system can sense it, this makes a real-time 3-D display an ultimate goal in the area of information display. Holography has been considered a most likely candidate for this task. In Sec. 11.3, we discuss 3-D display. Within the section, the principles of holography are first discussed, which will then be followed by two modern holographic techniques called optical scanning holography and synthetic aperture holography. In Sec. 11.4, we discuss information display using electro-optic modulators. We provide background for the electro-optic effect and discuss two types of electro-optic SLMs: electrically addressed SLMs and optically addressed SLMs. The use of these types of SLMs for 3-D display will also be discussed. Finally, in Sec. 11.5, we make some concluding remarks.

11.2. INFORMATION DISPLAY USING ACOUSTO-OPTIC SPATIAL LIGHT MODULATORS

Acousto-optics deals with the interaction between light and sound. It can result in light beam deflection, amplitude modulation, phase modulation, frequency shifting, and spectrum analysis [1]. Devices and systems based on acousto-optic interaction have played and continue to play a major role in various types of optical information processing [2, 3]. In later sections, we cover acousto-optic interactions using a plane-wave scattering model and discuss how intensity and frequency modulation of a laser beam can be accomplished by acousto-optic interaction. Finally, we present laser television displays using these modulation effects.

11.2.1. THE ACOUSTO-OPTIC EFFECT

An acousto-optic modulator (AOM) or Bragg cell is a spatial light modulator that consists of an acoustic medium, such as glass, to which a piezoelectric transducer is bonded. When an electrical signal is applied to the transducer, a sound wave propagates through the acoustic medium, causing perturbations in the index of refraction proportional to the electrical excitation, which in turn modulates the laser beam traversing the acoustic medium.

There are a variety of ways to explain acousto-optic interaction [1–11]. An instinctive approach considers the interaction of sound and light as a collision of photons and phonons [7]. Basically, the conservation of energy and momentum laws are applied to the process of collision. If we denote the wave vectors of the incident plane wave of light, scattered or diffracted plane wave of light, and sound plane wave by $\vec{k}_0$, $\vec{k}_{+1}$, and $\vec{K}$, respectively, the conservation of momentum may be written as

$$\hbar\vec{k}_{+1} = \hbar\vec{k}_0 + \hbar\vec{K}, \tag{11.1}$$

where $\hbar = h/2\pi$ and h denotes Planck's constant. From Fig. 11.1(a) and the division of Eq. (11.1) by $\hbar$, it is apparent that the condition for wave matching

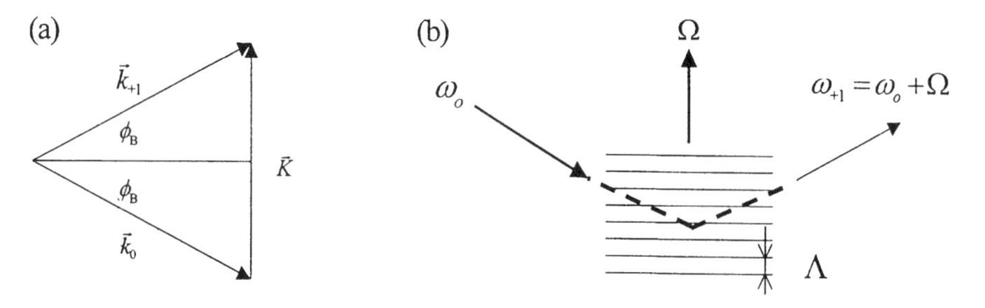

Fig. 11.1. Upshifted acousto-optic interaction.

to occur is expressed as

$$\vec{k}_{+1} = \vec{k}_0 + \vec{K}. \tag{11.2}$$

For all practical cases $|\vec{K}| \ll |\vec{k}_0|$, therefore the magnitude of $\vec{k}_{+1}$ is essentially equal to that of $\vec{k}_0$, and the wave vector triangle shown in Fig. 11.1(a) is nearly isosceles. Because we really have a traveling sound wave, the frequency of the diffracted plane wave is Doppler shifted by an amount equal to that of the sound frequency. The conservation of energy takes the form (after division by $\hbar$)

$$\omega_{+1} = \omega_0 + \Omega, \tag{11.3}$$

where ω_{+1}, ω_0, and Ω are the radian frequencies of the diffracted light, incident light, and sound, respectively. Since the frequency of the diffracted light is upshifted by an amount equal to the sound frequency, the interaction described above is called upshifted acousto-optic interaction. The situation is shown in Fig. 11.1(b). Now, suppose the direction of the incident light is changed such that it is incident at an angle opposite to that for upshifted interaction, as shown in Fig. 11.2. This will cause a downshift in the frequency of the diffracted light beam. Again, the conservation laws for momentum and energy can be

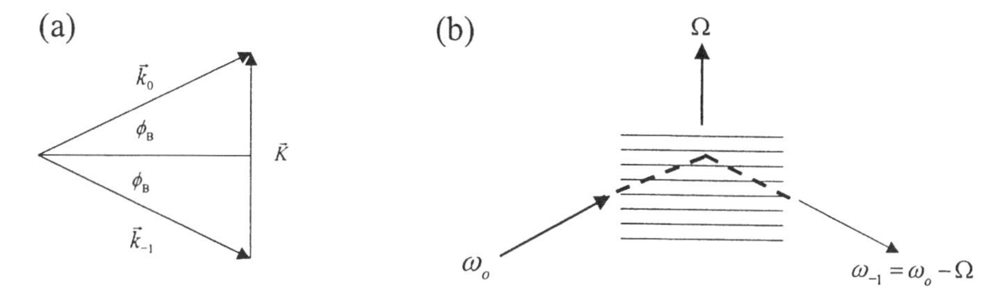

Fig. 11.2. Downshifted acousto-optic interaction.

applied, respectively, to obtain relationships similar to those expressed in Eqs. (11.2) and (11.3); i.e.,

$$\vec{k}_{-1} = \vec{k}_0 - \vec{K}, \tag{11.4}$$

and

$$\omega_{-1} = \omega_0 - \Omega, \tag{11.5}$$

where the -1 subscripts indicate the interaction is downshifted. Note that the closed triangles in Figs. 11.1(a) and 11.2(a) stipulate that there are certain critical angles of incidence for plane waves of light and sound to interact. The angle ϕ_B is called the Bragg angle, and it is given by

$$\sin(\phi_B) = \frac{K}{2k_0} = \frac{\lambda}{2\Lambda}, \tag{11.6}$$

where λ is the wavelength of light inside the acoustic medium, and Λ is the wavelength of sound. Note that the diffracted beams differ in direction by an angle equal to $2\phi_B$, as shown in Figs. 11.1 and 11.2.

In actual experiments, scattering happens even though the direction of incident light is not exactly at the Bragg angle. However, the maximum diffracted intensity occurs at the Bragg angle. The reason is that a finite length transducer does not produce ideal plane waves, and the sound waves actually spread as they propagate inside the acoustic medium. As the length of the transducer decreases, the sound column will act less and less like a single plane wave; in fact, it is now more appropriate to consider it an angular spectrum of plane waves.

For a transducer with an aperture L, sound waves spread out over an angle $\pm\Lambda/L$, as shown in Fig. 11.3. Considering the upshifted Bragg interaction and, referring to Fig. 11.3, we see that the $\vec{K}$-vector can be orientated through an angle $\pm\Lambda/L$ due to the spread of sound. In order to have only one diffracted order of light generated (i.e., $\vec{k}_{+1}$), we have to impose the condition that

$$\lambda/\Lambda \gg \Lambda/L,$$

or

$$L \gg \Lambda^2/\lambda. \tag{11.7}$$

This is because for $\vec{k}_{-1}$ to be generated, for example, a pertinent sound wave vector must lie along $\vec{K}'$; however, this either is not present, or is present in negligible amounts in the angular spectrum of sound, if condition in Eq. (11.7) is satisfied. If L satisfies this condition, the acousto-optic device is said to operate in the Bragg regime and the device is commonly known as a Bragg cell. Thus, physical reality dictates that a complete energy transfer between the two diffracted beams is impossible since there always exists more than two dif-

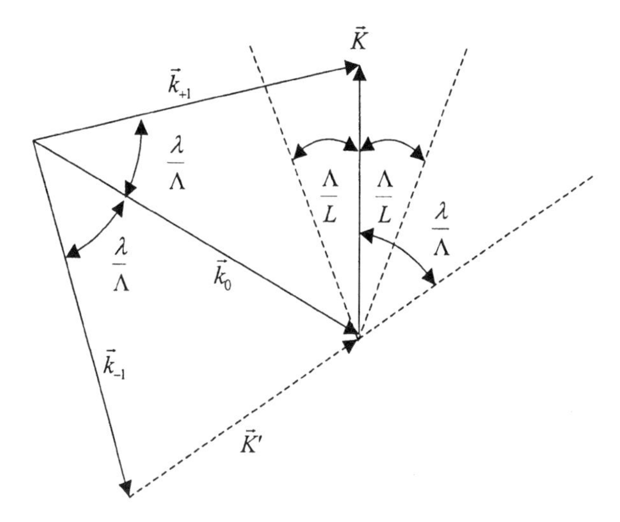

Fig. 11.3. Wave–vector diagram illustrating the condition for defining the Bragg regime.

fracted plane waves, $\tilde{E}_0$ and $\tilde{E}_1$, as shown in Fig. 11.4, where $\tilde{E}_n$ is the nth-order diffracted plane-wave complex amplitude at frequency $\omega_n = \omega_0 + n\Omega$, and is commonly called the nth-order light in acousto-optics.

In the ideal Bragg regime, only two diffracted plane waves exist. In contrast, the generation of multiple diffracted plane waves defines the so-called Raman-Nath regime. The Klein-Cook parameter [8],

$$Q = \frac{K^2 L}{k_0} = 2\pi \frac{\lambda L}{\Lambda^2},$$

has been defined to allow the proper classification of the acousto-optic device as a Bragg or Raman-Nath cell, where L is the length of the transducer which defines the so-called interaction length between the light and the sound. The Bragg regime is defined arbitrarily as the condition when the diffraction efficiency for the first-order diffracted plane wave; i.e., $n = 1$ for upshifted diffraction or $n = -1$ for downshifted diffraction, is 90%. Consequently, it can be shown that operation in the Bragg regime is defined by $Q \geqslant 7$ [8, 11]. Notice that for ideal Bragg diffraction, Q would have to be infinity (i.e., $L = \infty$).

Although the above discussion describes the necessary conditions for Bragg diffraction to occur, it does not predict how the acousto-optic interaction process affects the amplitude distribution among the various diffracted plane waves. We adopt the Korpel-Poon multiple–plane-wave scattering theory for

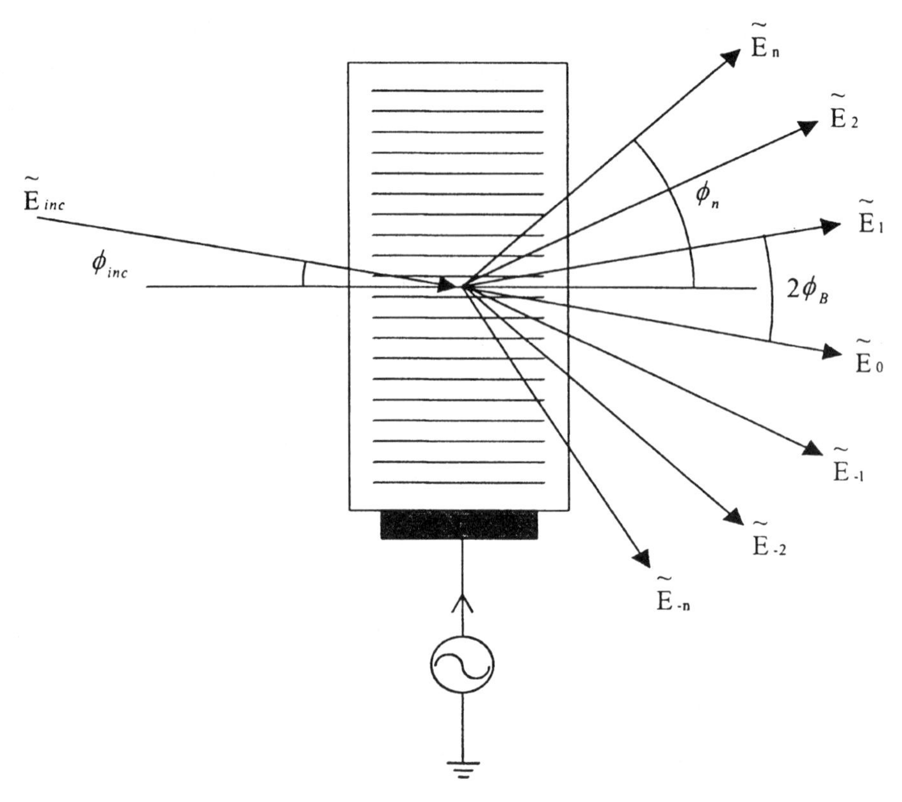

Fig. 11.4. Basic acousto-optic Bragg cell. $\tilde{E}_{\text{inc}}$ is the incident complex amplitude, and $\tilde{E}_n$ is the complex amplitude of the nth-order diffracted light at $\phi_n = \phi_{\text{inc}} + 2n\phi_B$ with frequency $\omega_n = \omega_0 + n\Omega$.

the investigation [10, 11, 12]. The theory represents the light and sound fields as plane-wave decompositions together with a multiple scattering for the interaction. The general formalism is applicable not only to hologram-type configurations but also to physically realistic sound fields subject to diffraction. For a typical rectangular sound column with plane-wave incidence, as shown in Fig. 11.5, the general 2-D multiple-scattering theory can be reduced to the following infinite coupled equations:

$$\begin{aligned}\frac{d\tilde{E}_n}{d\xi} = &-j\frac{\tilde{\alpha}}{2}\exp\left\{\frac{-jQ\xi}{2}\left[\frac{\phi_{\text{inc}}}{\phi_B} + (2n-1)\right]\right\}\tilde{E}_{n-1} \\ &-j\frac{\tilde{\alpha}^*}{2}\exp\left\{\frac{jQ\xi}{2}\left[\frac{\phi_{\text{inc}}}{\phi_B} + (2n+1)\right]\right\}\tilde{E}_{n+1}\end{aligned} \tag{11.8}$$

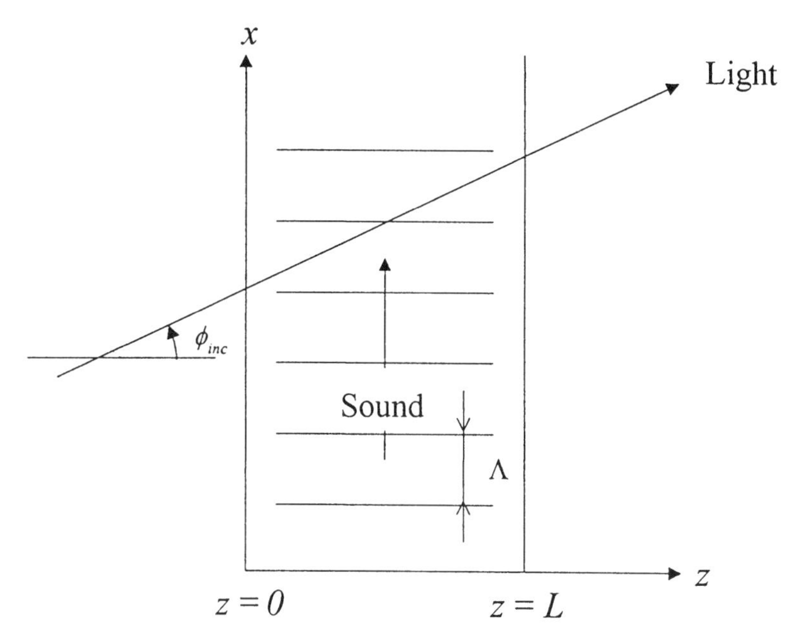

Fig. 11.5. Conventional 2-D sound column configuration.

with the boundary condition $\tilde{E}_n = \tilde{E}_{\text{inc}}\delta_{no}$ at $z < 0$, where δ_{no} is the Kronecker delta and $\tilde{E}_n$ is the complex amplitude of the nth-order plane wave of light in the direction $\phi_n = \phi_{\text{inc}} + 2n\phi_B$. ϕ_{inc} is the incident angle of the plane wave, $\tilde{E}_{\text{inc}}$. The other parameters in Eq. (11.8) are defined as follows: $\tilde{\alpha} = Ck_0AL/2$, where C represents the strain–optic coefficient of the medium, A is the complex amplitude of sound with the sound field $S(x, t) = Re[A \exp[j(\Omega t - Kx)]$, $\xi = L/z$ is the normalized distance inside the sound cell, and L is the width of the sound column. $\xi = 0$ signifies when a plane wave of light enters into the acousto-optic cell, and $\xi = 1$ denotes when a plane wave of light exits from the cell.

Physically, it is clear from Eq. (11.8) that the equation identifies the plane wave contributions to $\tilde{E}_n$ from neighboring orders, $\tilde{E}_{n-1}$ and $\tilde{E}_{n+1}$, with the phase terms indicating the degree of phase mismatch between the orders. The equation is a special case of general multiple-scattering theory valid for any sound field, not just a sound column. Note that the sign convention for ϕ_{inc} is counterclockwise positive; that is, $\phi_{\text{inc}} = -\phi_B$ signifies upshifted diffraction. For a given value of $\tilde{\alpha}$ and Q the solution to Eq. (11.8) represents the contributions to the nth-order plane wave of light, $\tilde{E}_n$, owing to the plane wave $\tilde{E}_{\text{inc}}$ incident at ϕ_{inc}.

For $\phi_{\text{inc}} = -(1 + \delta)\phi_B$, where δ represents the deviation of the incident plane wave away from the Bragg angle, and limiting ourselves to $\tilde{E}_0$ and $\tilde{E}_1$,

Eq. (11.8) is reduced to the following set of coupled differential equations:

$$\frac{d\tilde{E}_0}{d\xi} = -j\frac{\tilde{\alpha}^*}{2}\exp(-j\delta Q\xi/2)\tilde{E}_1,$$

and

$$\frac{d\tilde{E}_1}{d\xi} = -j\frac{\tilde{\alpha}}{2}\exp(j\delta Q\xi/2)\tilde{E}_0 \tag{11.9}$$

with the boundary conditions $\tilde{E}_0(\xi = 0) = \tilde{E}_{\text{inc}}$, and $\tilde{E}_1(\xi = 0) = 0$. Assuming $\tilde{\alpha}$ being real positive for simplicity; i.e., $\tilde{\alpha}^* = \tilde{\alpha} = Ck_0|A|L/2 = \alpha$, and α is called the peak phase delay of the sound through the medium, Eq. (11.9) may be solved analytically by a variety of techniques; the solutions are given by the well-known Phariseau formula [13]:

$$\tilde{E}_0(\xi) = \tilde{E}_{\text{inc}}\exp(-j\delta Q\xi/4)\left\{\cos[(\delta Q/4)^2 + (\alpha/2)^2]^{1/2}\xi + j\frac{\delta Q}{4}\frac{\sin[(\delta Q/4)^2 + (\alpha/2)^2]^{1/2}\xi}{[(\delta Q/4)^2 + (\alpha/2)^2]^{1/2}}\right\} \tag{11.10a}$$

$$\tilde{E}_1(\xi) = \tilde{E}_{\text{inc}}\exp(j\delta Q\xi/4)\left\{-j\frac{\alpha}{2}\frac{\sin[(\delta Q/4)^2 + (\alpha/2)^2]^{1/2}\xi}{[(\delta Q/4)^2 + (\alpha/2)^2]^{1/2}}\right\}. \tag{11.10b}$$

Equations (11.10a) and (11.10b) are similar to the standard two-wave solutions found by Aggarwal and adapted by Kogelnik to holography. More recently, it has been rederived with the Feynman diagram technique [14]. Equations (11.10a) and (11.10b) represent the plane-wave solutions that are due to oblique incidence, and by letting $\delta = 0$ we can reduce them to the following set of well-known solutions for ideal Bragg diffraction:

$$\tilde{E}_0(\xi) = \tilde{E}_{\text{inc}}\cos(\alpha\xi/2), \tag{11.11a}$$

$$\tilde{E}_1(\xi) = -j\tilde{E}_{\text{inc}}\sin(\alpha\xi/2). \tag{11.11b}$$

For a more general solution; i.e., assuming α^* is complex, Eq. (11.11b) becomes

$$\tilde{E}_1(\xi) = -j\frac{A}{|A|}\tilde{E}_{\text{inc}}\sin(\alpha\xi/2) \tag{11.11c}$$

with Eq. (11.11a) remaining the same.

11.2.2. INTENSITY MODULATION OF LASER

Because it is common to refer to light by its intensity (i.e., $I_n = |\tilde{E}_n|^2$, where I_n represents light intensity for the nth-order plane wave), we write the *zero*th-order light and the first-order light intensities, according to Eqs. (11.11a,b), as

$$I_0 = |\tilde{E}_0|^2 = I_{\text{inc}} \cos^2(\alpha\xi/2), \tag{11.12a}$$

and

$$I_1 = |\tilde{E}_1|^2 = I_{\text{inc}} \sin^2(\alpha\xi/2), \tag{11.12b}$$

where $I_{\text{inc}} = |\tilde{E}_{\text{inc}}|^2$ represents the intensity of the incident light. The value of ξ is set to unity to represent the intensities at the exit of the Bragg cell. Figure 11.6 shows a plot of the solutions given by Eq. (11.12) and illustrates complete energy transfer between the two coupled modes (located at $\alpha = \pi$) for ideal Bragg diffraction. Note that by changing the amplitude of the sound; i.e., through α, we can achieve intensity modulation of the diffracted beams.

In fact, one of the most popular applications for an acousto-optic Bragg cell is its ability to modulate laser light. Figure 11.6 shows the *zero*th- and first-order diffraction curves plotted as a function of α, where P represents the bias point necessary for linear operation. Figure 11.7 illustrates the relationship between the modulating signal, $m(t)$ and the intensity modulated output signal, $I(t)$. As shown in the figure, $m(t)$ is biased along the linear portion of the first-order diffraction curve.

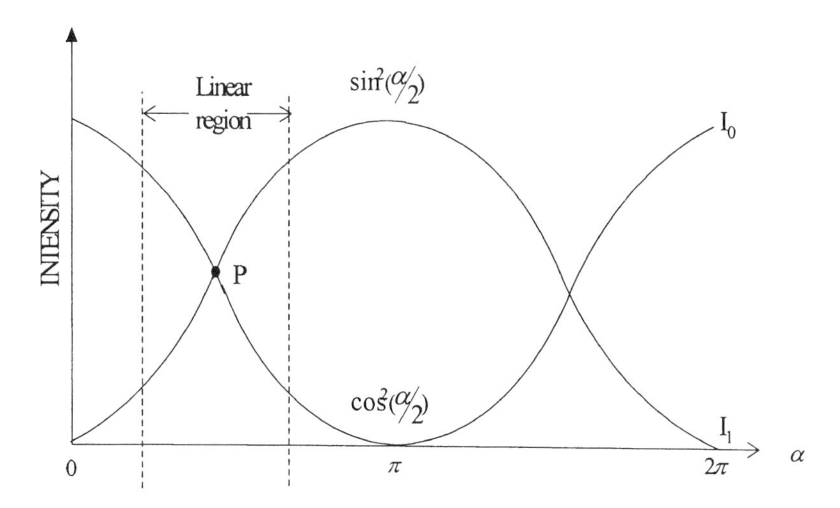

Fig. 11.6. Complete energy transfer at $\alpha = \pi$.

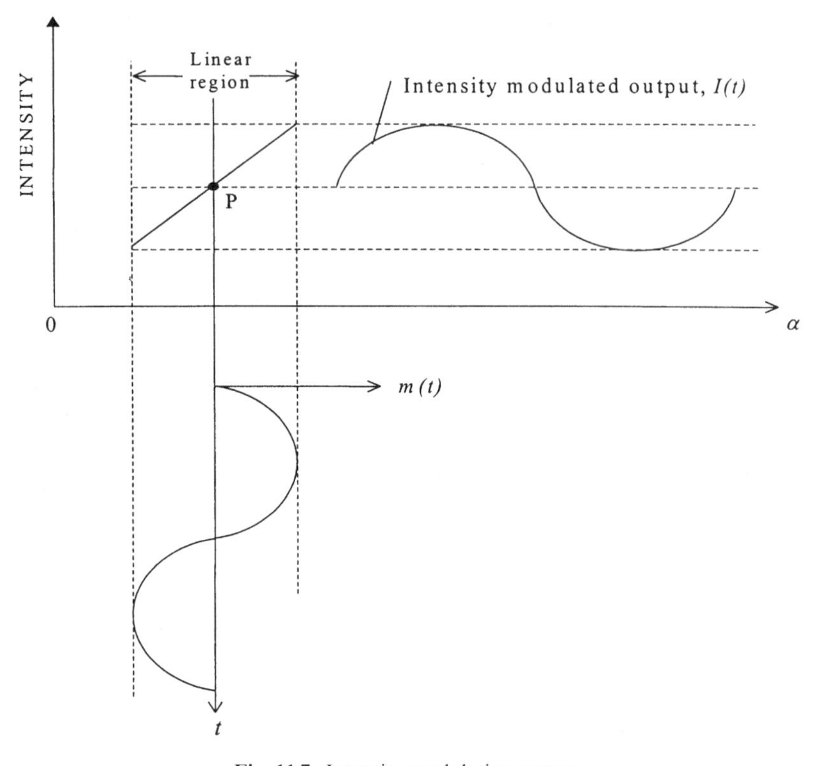

Fig. 11.7. Intensity modulation system.

It should be clear that intensity modulation can also be achieved by using the *zero*th-diffracted order. However, note that the two diffracted orders process information in the linear regions with opposite slopes. This suggests that any demodulated electrical signal received via the first diffracted order will be 180° out of phase with the electrical signal received via the *zero*th-diffracted order. This has been demonstrated experimentally [15].

For intensity modulation using acousto-optics devices, it is instructive to estimate the *modulation bandwidth*. The modulation bandwidth, B, can be defined as the difference between the highest and lowest sound frequencies at which the normalized diffracted intensity drops by 50%. To estimate this, we assume α is small such that $(\delta Q/4)^2 \gg (\alpha/2)^2$; i.e., under the case of weak inteaction. The normalized intensity is, therefore, given by

$$I_1 \bigg/ \left(\frac{\alpha}{2}\right)^2 = \frac{\sin^2(\delta Q/4)}{(\delta Q/4)^2}. \tag{11.13}$$

When $\delta Q/4 = 0.45\pi$, the normalized intensity drops to about one half. Using the definition of Q and knowing that $\Lambda = v_s/f_B$, we write

$$\delta Q/4 = \frac{\pi L \lambda f_B \delta f_B}{v_s^2} = 0.45\pi,$$

where f_B is the center frequency at which the Bragg condition is satisfied and v_s is the sound velocity in the acoustic medium. We then find that the bandwidth is

$$B = 2\delta f_B \approx \frac{1.8 v_s}{\lambda} \frac{\Lambda}{L}. \tag{11.14}$$

The situation is shown in Fig. 11.8.

It seems from the above equation that the bandwidth may be increased by reducing the interaction length L. However, there is a limit to this procedure. The limit is set by the angular spread of the incident light λ/D, where D is the width of the light. As L becomes smaller, the spread of the sound becomes larger. For efficient interaction, the angular spread of the sound must be approximately equal to that of the light; i.e., $\Lambda/L \simeq \lambda/D$ so that every plane wave of light can interact with the sound. Hence, Eq. (11.14) becomes

$$B \approx \frac{1.8 v_s}{D} \approx 1/\tau \tag{11.15}$$

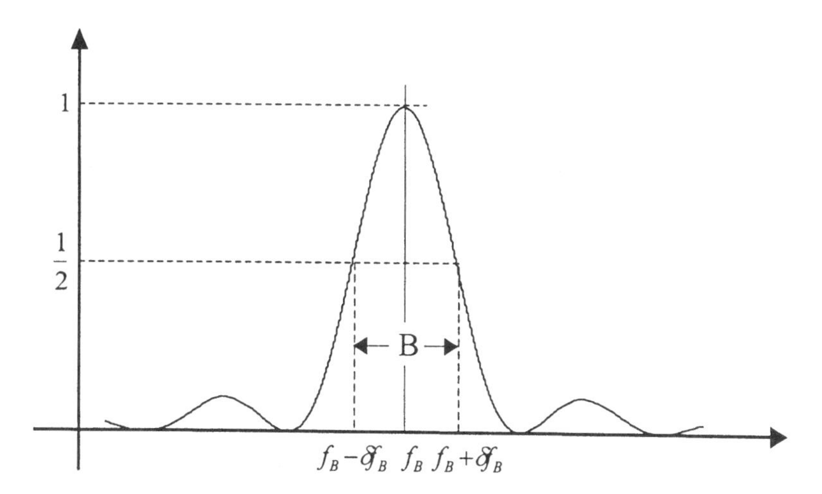

Fig. 11.8. Bandwith estimation.

by ignoring the factor 1.8 for simplicity. $\tau = D/v_s$ is the transit time for the sound wave to cross the light beam and the modulation bandwidth for the acousto-optic modulator is, therefore, the inverse of the sound propagation time through the light beam. In practice, the bandwidth can be increased by focusing the light inside the sound column. By special beam-steering techniques, it is possible to further increase the allowable bandwidth. The interested reader is referred to the existing literature [11].

11.2.3. DEFLECTION OF LASER

In contrast to intensity modulation, where the amplitude of the modulating signal is varied, the frequency of the modulating signal is changed for applications in laser deflection. We, therefore, have the so-called acousto-optic deflector (AOD), as shown in Fig. 11.9, where the acousto-optic device operates in the Bragg regime.

In Fig. 11.9, the angle between the first-order beam and the *zero*th-order beam is defined as the deflection angle ϕ_d. We can express the change in the deflection angle $\Delta\phi_d$ upon a change $\Delta\Omega$ of the sound frequency as

$$\Delta\phi_d = \Delta(2\phi_B) = \frac{1}{2\pi}\frac{\lambda}{v_s}\Delta\Omega. \tag{11.16}$$

Using a He-Ne laser ($\lambda \sim 0.6\,\mu$m), a change of sound frequency of 20 MHz

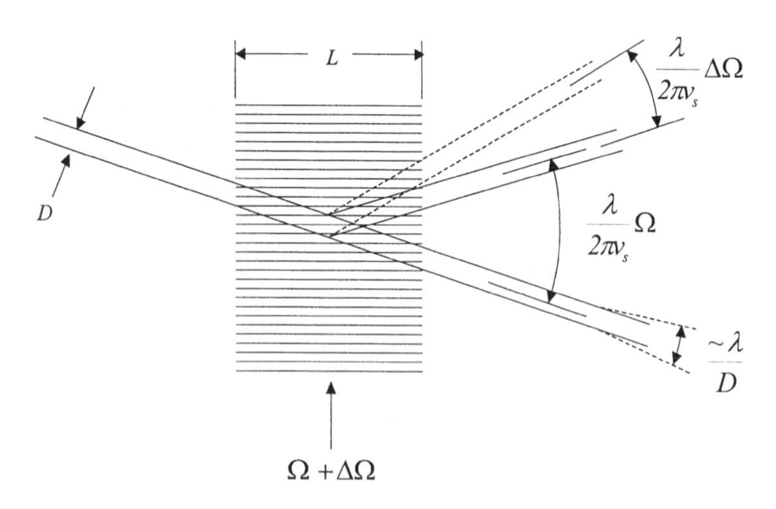

Fig. 11.9. Acousto-optic deflector.

around the center frequency of 40 MHz, and $v_s \sim 4 \times 10^3$ m/s for the velocity of sound in glass, a change in the deflection angle is $\Delta\phi_d \sim 3$ mrad. The number of resolvable angles N in such a device is determined by the ratio of the range of deflection angles $\Delta\phi_d$ to the angular spread of the scanning light beam. Since the angular spread of a beam of width D is of the order of λ/D, we then have

$$N = \frac{\Delta\phi_d}{\lambda/D} = \tau\frac{\Delta\Omega}{2\pi}. \tag{11.17}$$

With the previously calculated $\Delta\phi_d \sim 3$ mrad and using a light beam of width $D \sim 5$ mm, the number of resolvable angles $N \sim 25$ spot. One can, for example, increase the number of resolvable angles by increasing the transit time through the expansion of the incident laser beam along the direction of sound propagation.

11.2.4. LASER TV DISPLAY USING ACOUSTO-OPTIC DEVICES

We have shown in previous sections that Bragg cells can be used for intensity modulation and laser-beam scanning. In fact, Korpel et al. at Zenith Radio Corporation demonstrated the first laser TV display using two Bragg cells, one for intensity modulation and the other for laser beam deflection [16]. The system is shown in Fig. 11.10.

A low-power telescope compresses the He-Ne laser beam to a diameter suitable for the acousto-optic modulator (AOM), into which an amplitude-modulated video signal with a carrier of 41.5 MHz is fed (typically in AOMs,

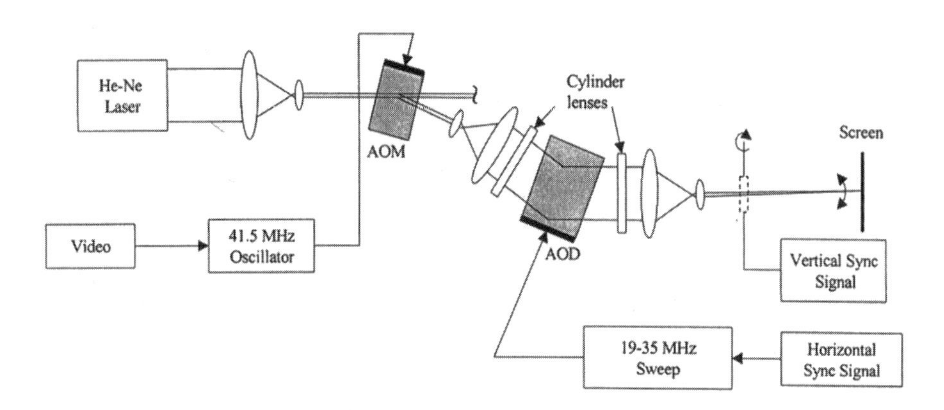

Fig. 11.10. 1966's Zenith TV laser display system.

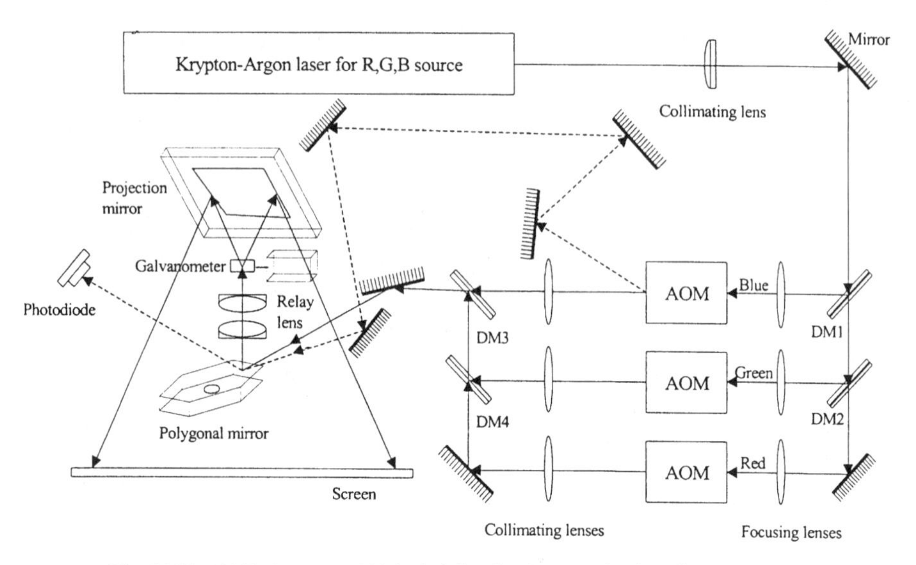

Fig. 11.11. 1998's Samsung 200-inch full color laser projection display system.

$L \gg d$, where d is the height of the acoustic medium and is often neglected). The diffracted light is then expanded and focused into a light wedge by a cylinder lens while passing through the acousto-optic deflector (AOD) and then restored to its original circular form. The purpose of forming the light wedge is to increase the number of resolvable angles along the horizontal scan line (see Eq. [11.17]). Indeed, the system reported has produced $N = 200$ with a sweep signal from 19 to 35 MHz. Finally, the deflection angle is magnified by a third telescope to bring the display screen closer. Vertical scanning is provided by a galvanometer through a vertical synchronizing signal, as shown in the figure. The laser TV display system reported just falls a little short of meeting commercial resolution standards. This idea of using acousto-optic devices for laser TV display, however, has recently been revitalized by Samsung in the pursuit of high-definition TV (HDTV). Figure 11.11 shows the Samsung's system [17].

A 4-watt Krypton-Argon laser is used to provide three primary colors of red (R), blue (B), and green (G). The dichroic mirrors (DM1 and DM2) separate the three colors into three light channels. The focusing lenses are used to generate smaller laser spots inside the three AOMs so as to create a larger modulation bandwidth for intensity modulation (see Eq. [11.15]). The three AOMs are driven by RGB video signals, which are derived from a composite video signal. The composite video also provides the composite sync signals for x–y scan synchronization. The three modulated lasers are combined by dichroic mirrors DM3 and DM4, and a mirror and then the combined lasers are scanned by a rotating polygonal mirror for horizontal lines and a galvanometer for vertical lines. Finally, the 2-D scanning image is projected by a projection mirror on a screen for 2-D display. Note that even though the polygonal scan mirror is synchronized by the horizontal sync signal derived from the composite video, it is still necessary to correct the jittering due to the mirror's fast rotating speed. The correction is done by monitoring a reflected beam from the polygonal mirror, where the reflected light is derived from the first order diffracted light of one of the AOMs, as shown in the figure (see the dotted line). A photodiode detecting the deviation of the position of the reflected light together with some kind of feedback circuit delivers a correcting signal to the polygonal mirror. It is interesting to point out that this kind of feedback mechanism to compensate for any joggling of the mirror is not necessary if an acousto-optic deflector for horizontal scanning is used, as in Zenith's experiment. In Fig. 11.12, we show a projected image obtained by the laser video projector developed by Samsung. The projection is an impressive $4\,\text{m} \times 3\,\text{m}$ display [18]. It is expected to apply the projector system to HDTV applications by developing acousto-optic modulators with a bandwidth of 30 MHz instead of the currently developed modulators of NTSC's 5-Mhz video bandwidth. The acousto-optic modulators to be developed must be able to sustain focused lasers of high power (for higher bandwidth) and at the same time have high diffraction efficiency; say, about 80% of light gets diffracted. Remember that these two requirements, higher bandwidth and higher diffraction efficiency, are conflicting when analyzed in the weak interaction regime

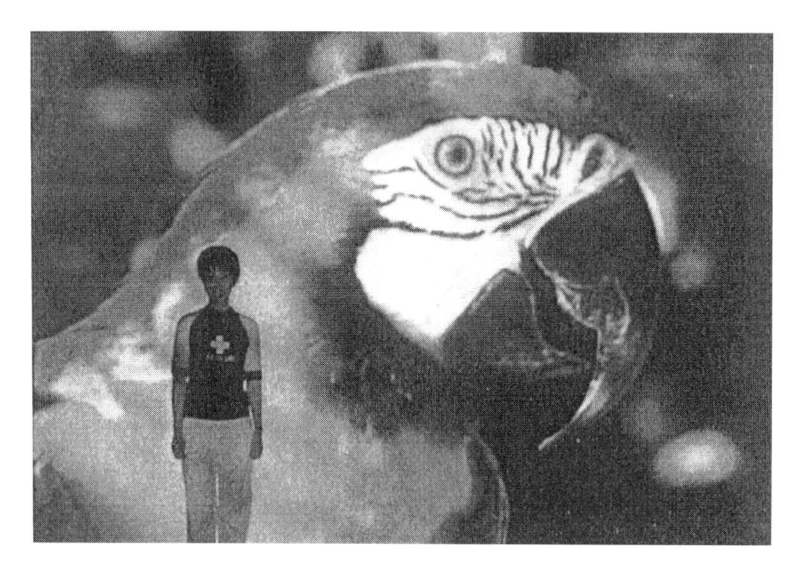

Fig. 11.12. Two hundred-inch full color laser projection display (courtesy of Samsung Advanced Institute of Technology [18]).

($\alpha \sim$ small). For the strong interaction regime, the situation is compounded by another well-known fact that the diffracted light-beam profile is distorted and hence tends to further lower the diffraction efficiency [11, 14]. (Interested readers are encouraged to explore this further.)

11.3. 3-D HOLOGRAPHIC DISPLAY

In this section, we first develop the principles of holography. We then discuss two modern and novel electronic holographic techniques: optical scanning holography and synthetic aperture holography. Acousto-optic modulators play an essential role in both of these techniques.

11.3.1. PRINCIPLES OF HOLOGRAPHY

The principle of holography can be explained by using a point object, since any 3-D object can be considered as a collection of points. Figure 11.13 shows

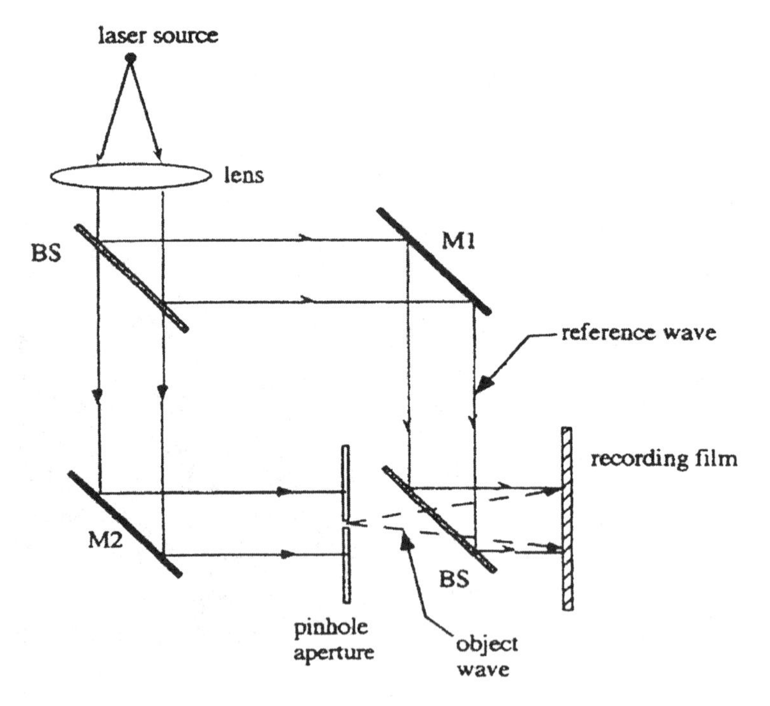

Fig. 11.13. On-axis holographic recording of a point-source object.

a collimated laser split into two plane waves and recombined through the use of two mirrors (M1 and M2) and two beam splitters (BS).

One plane wave is used to illuminate the pinhole aperture (our point object to be recorded), and the other illuminates the recording film directly. The plane wave that is scattered by the point object generates a diverging spherical wave. This diverging wave is known as an object wave in holography. The plane wave that directly illuminates the photographic plate is known as a reference wave. Let ψ_o represent the field distribution of the object wave on the plane of the recording film, and similarly let ψ_r represent the field distribution of the reference wave on the plane of the recording film. The film now records the intensity of the interference of the reference wave and the object wave; i.e., what is recorded is given by $|\psi_r + \psi_o|^2$ mathematically, provided the reference wave and the object wave are mutually coherent over the film. The coherency of the light waves is guaranteed by the use of a laser source. This kind of recording is commonly known as holographic recording.

It is well known in optics that if we describe the amplitude and phase of a light field in a plane, say $z = 0$, by a complex function $\psi(x, y)$, we can obtain for the light field a distance away, say $z = z_0$, according to Fresnel diffraction [19]

$$\psi(x, y; z_0) = \psi(x, y) * h(x, y; z = z_0), \tag{11.18}$$

where the symbol $*$ denotes convolution and $h(x, y; z)$ is called the free-space impulse response, given by

$$h(x, y; z) = \frac{jk_0}{2\pi z} \exp[-jk_0(x^2 + y^2)/2z]. \tag{11.19}$$

In Eq. (11.19), $k_0 = 2\pi/\lambda$, λ being the wavelength of the light field. Now, referring back to our point object example, let us model the point objcct at a distance z_0 from the recording film by an offset delta function; i.e., $\psi(x, y) = \delta(x - x_0, y - y_0)$. According to Eq. (11.18), the object wave arises from the point object on the film as given by

$$\psi_o = \psi(x, y; z_0) = \frac{jk_0}{2\pi z_0} \exp\{-jk_0[(x - x_0)^2 + (y - y_0)^2]/2z_0\}. \tag{11.20}$$

This object wave is a spherical wave. For the reference plane wave, the field distribution is constant on the film; say, $\psi_r = a$ (a constant) for simplicity.

Hence, the intensity distribution being recorded on the film is

$$
\begin{aligned}
t(x, y) &= |\psi_r + \psi_o|^2 \\
&= \left| a + \frac{jk_0}{2\pi z_0} \exp\{-jk_0[(x - x_0)^2 + (y - y_0)^2]/2z_0\} \right|^2 \qquad (11.21) \\
&= A + B \sin\left\{\frac{k_0}{2z_0}[(x - x_0)^2 + (y - y_0)^2]\right\},
\end{aligned}
$$

where

$$A = a^2 + \left(\frac{k_0}{2\pi z_0}\right)^2 \quad \text{and} \quad B = \frac{k_0}{\pi z_0}.$$

This expression is called a Fresnel zone plate [20], which is the hologram of a point object; we shall call it a point-object hologram. Note that the center of the zone plate specifies the location x_0 and y_0 of the point object and the spatial variation of the zone plate is governed by a sine function with a quadratic spatial dependence. Hence, the spatial rate of change of the phase of the zone plate; i.e., the fringe frequency, increases linearly with the spatial coordinates, x and y. The fringe frequency, therefore, depends on the depth parameter, z_0.

Figure 11.14(a) shows the hologram of a point object for $x_0 = y_0 = c < 0$, located a distance z_0 away from the recording film. In Fig. 11.14(b) we show the hologram of another point object $x_0 = y_0 = -c$, but now the point object is located away from the film at $z_1 = 2z_0$. Note that since $z_1 > z_0$ in this case, the fringe frequency on the hologram due to this point object varies slower than that of the point object which is located closer to the film. Indeed, we see that the fringe frequency contains the depth information, whereas the center of

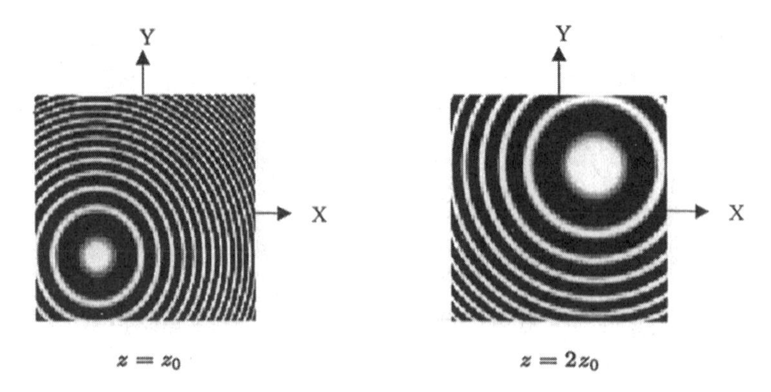

Fig. 11.14. Point-object holograms with different x, y, and z locations.

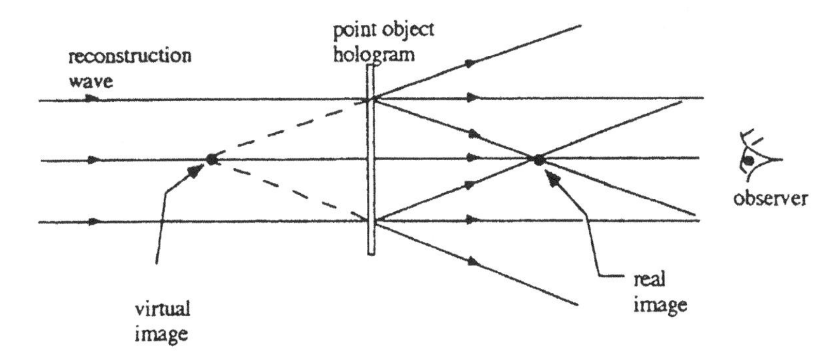

Fig. 11.15. Holographic reconstruction.

the zone defines the transverse location of the point object. For an arbitrary 3-D object, we can think of the object as a collection of points, and therefore we can envision that we have a collection of zones on the hologram, with each zone carrying the transverse location as well as the depth information of each individual point. In fact, a hologram has been considered as a type of Fresnel zone plate [21] and the holographic imaging process has been discussed in terms of zone plates [22].

So far, we have discussed the transformation of a point object to a zone plate on the hologram during holographic recording, and this corresponds to a coding process. In order to decode it, we need to obtain the point object back from the hologram. This can be done by simply illuminating the hologram with a reconstruction wave, as shown in Fig. 11.15. Figure 11.15 corresponds to the reconstruction of a hologram of the point object located on-axis; i.e., for the simple case where $x_0 = y_0 = 0$.

Note that in practice, the reconstruction wave usually is identical to the reference wave; therefore, we assume the reconstruction wave to have a field distribution on the plane of the hologram given by $\psi_{rc}(x, y) = a$. Hence, the field distribution of the transmitted wave immediately after the hologram is $\psi_{rc}t(x, y) = at(x, y)$ and the field at arbitrary distance of z away is according to Eq. (11.18), given by the evaluation of $at(x, y) * h(x, y; z)$. For the point-object hologram given by Eq. (11.21), we have, after expanding the sine term of the hologram $t(x, y)$,

$$t(x, y) = A + \frac{B}{2j}\left\{\exp\left(j\frac{k_0}{2z_0}[(x - x_0)^2 + (y - y_0)^2]\right) - \exp\left(-j\frac{k_0}{2z_0}[(x - x_0)^2 + (y - y_0)^2]\right)\right\} \quad (11.22)$$

Therefore, we will have three terms resulting from the illumination of the hologram by the reconstruction wave. These contributions, according to Fresnel diffraction upon illumination of the hologram; i.e., $at(x, y) * h(x, y; z_0)$, are as follows:

$$\text{First term:} \quad aA * h(x, y; z = z_0) = aA, \quad \text{(zero-order beam).} \tag{11.23a}$$

$$\text{Second term:} \quad \sim \exp\left(j\frac{k_0}{2z_0}[(x - x_0)^2 + (y - y_0)^2]\right) * h(x, y; z = z_0) \tag{11.23b}$$

$$\sim \delta(x - x_0, y - y_0), \quad \text{(real image).}$$

$$\text{Third term:} \quad \sim \exp\left(-j\frac{k_0}{2z_0}[(x - x_0)^2 + (y - y_0)^2]\right) * h(x, y; z = -z_0)$$

$$\sim \delta(x - x_0, y - y_0), \quad \text{(virtual image).} \tag{11.23c}$$

In the terminology of holography, the first term is the zero-order beam due to the bias in the hologram. The result of the second term is called the real image and the third term is the virtual image. Note that the real image and the virtual image are located at a distance z_0 in front and back of the hologram, respectively. This situation is shown in Fig. 11.15. The real image and the virtual image are called the twin images in on-axis holography.

Figure 11.16(a) shows the holographic recording of a 3-point object and Fig. 11.16(b) shows the reconstruction of the hologram. Note that the virtual image appears at the correct 3-D location of the original object, while the real image

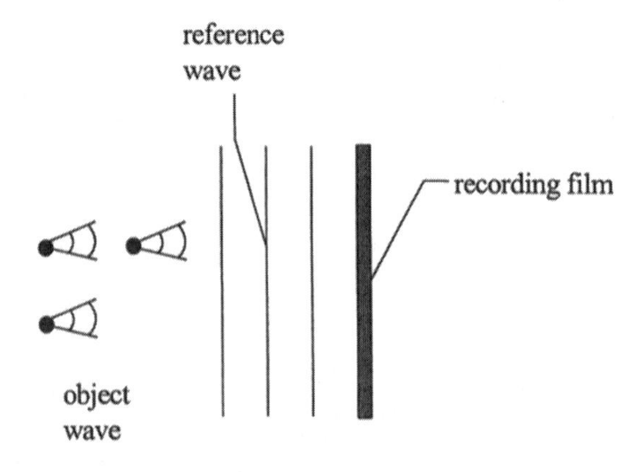

Fig. 11.16(a). Holographic recording.

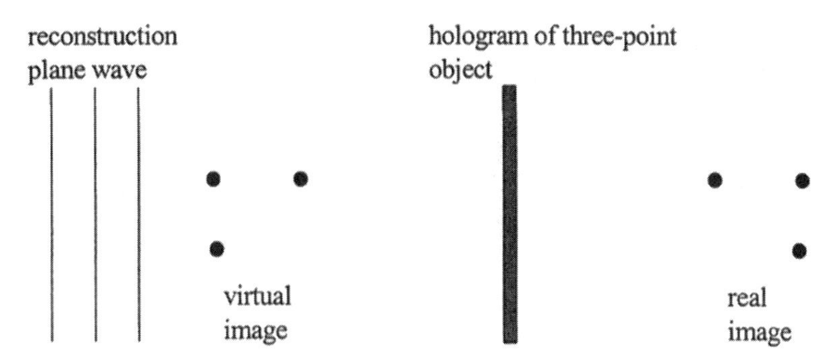

Fig. 11.16(b). Holographic reconstruction.

is the mirror image of the original object, with the axis of reflection on the plane of the hologram. Indeed, the original 3-D wavefront has been completely stored and now reconstructed. This is further explained by inspecting Eq. (11.21). By expanding Eq. (11.21), we have

$$t(x, y) = |\psi_r + \psi_o|^2 = |\psi_r|^2 + |\psi_o|^2 + \psi_r\psi_o^* + \psi_o\psi_r^*. \tag{11.24}$$

Note that the original object wavefront (the fourth term) ψ_o has been successfully recorded (amplitude and phase) on the hologram. When the hologram is reconstructed; i.e., the hologram is illuminated by ψ_{rc}, the transmitted wave immediately behind the hologram is

$$\psi_{rc}t(x, y) = \psi_{rc}[|\psi_r|^2 + |\psi_o|^2 + \psi_r\psi_o^* + \psi_o\psi_r^*].$$

Again, assuming $\psi_{rc} = \psi_r = a$, we have

$$\psi_{rc}t(x, y) = a[|\psi_r|^2 + |\psi_o|^2] + a^2\psi_o^* + a^2\psi_o.$$

The last term is identical to, within a constant multiplier, the object wave which was present on the plane of the hologram when the hologram was recorded. If we view the reconstructed object wave, we would see a virtual image of the object precisely at the location where the object was placed during recording with all the effects of parallax and depth. The third term is proportional to the complex conjugate of ψ_o and is responsible for the generation of the real image. Physically, it is the mirror image of the 3-D object, as discussed previously. Finally, the first two terms; i.e., the zero-order beam, are space-variant bias terms as ψ_o is a function of x and y in general. This would produce a space-variant background (noise) on the observation plane.

11.3.2. OPTICAL SCANNING HOLOGRAPHY

Optical scanning holography (OSH) is a novel holographic recording technique first suggested by Poon and Korpel [23], and subsequently formulated by Poon [24], in which holographic information of an object can be recorded using heterodyne 2-D optical scanning. Corresponding to the principle of holography, the technique also consists of two stages: the recording or coding stage, and the reconstruction or decoding stage. In the recording stage, the 3-D object is two-dimensionally scanned by the so-called time-dependent Fresnel zone plate (TDFZP) [24]. The TDFZP is created by the superposition of a plane wave and a spherical wave of different temporal frequencies. The situation is shown in Fig. 11.17.

The plane wave is generated at the output of beam expander BE1. Note that the frequency of the plane wave has been up-shifted by Ω through the acousto-optic frequency shifter (AOFS). The generation of a spherical wave is accomplished by lens L following beam expander BE2. After the focusing of the lens, we have the spherical wave. Beam splitter BS2 is used to combine the plane wave emerging from beam expander BE1 and the spherical wave. Thus, a time-dependent Fresnel zone plate, created by the interference of mutually coherent spherical and plane wavefronts, is formed on the 3-D volume or the 3-D object to be inspected. The intensity pattern of the combined beam is thus given by the following expression:

$$\left| A \exp[j(\omega_0 + \Omega)t] + B\left(\frac{jk_0}{2\pi z}\right) \exp[-jk_0(x^2+y^2)/2z] \exp(j\omega_0 t)\right|^2, \quad (11.25)$$

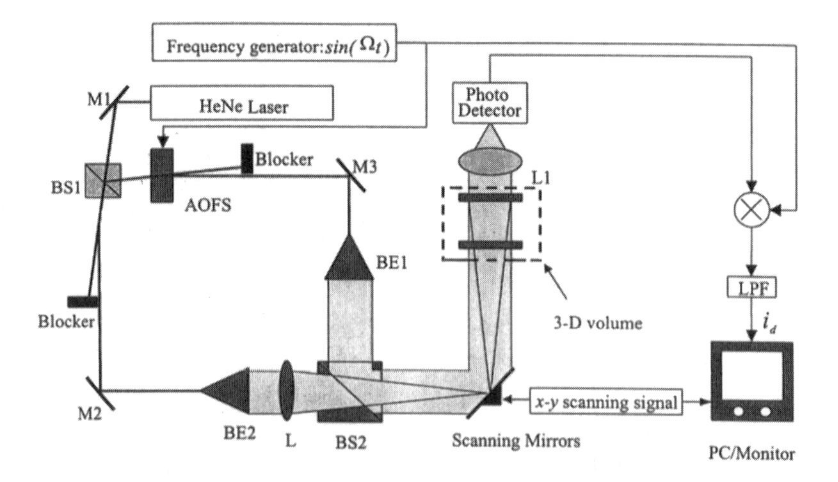

Fig. 11.17. Optical scanning holography. BS1, 2: beam splitters; AOFS: acousto-optic frequency shifter; M1, 2, 3: mirrors; BE1, 2: beam expanders; ⊗: electronic multipliers; LPF: low-pass filter.

where A and B are constants, representing the amplitude of the plane wave and the diverging spherical wave, respectively. The frequencies $\omega_0 + \Omega$ and ω_0 are the temporal frequencies of the plane wave and the spherical wave, respectively. The parameter z is a depth parameter measured away from the focal plane of lens L (which is on the surface of the 2-D scanning mirrors) toward the 3-D object. Equation (11.25) can be expanded and written as

$$I(x, y; \Omega) = A^2 + C^2 + 2AC \sin\left[\frac{k_0}{2z}(x^2 + y^2) + \Omega t\right] \tag{11.26}$$

where $C = Bk_0/2\pi z$. This is a familiar Fresnel zone plate (FZP) expression (see Eq. [11.21]) but has a time-dependent variable. We shall call it a time-dependent Fresnel zone plate (TDFZP). This TDFZP is now used to scan the 3-D object in two dimensions and the photodetector collects all the transmitted light. For the sake of explaining the concept, let us assume that a single off-axis point object $\delta(x - x_0, y - y_0)$ is located z_0 away from the focal plane of lens L. The scanning of the TDFZP on the point object will cause the photodetector to deliver a heterodyne-scanned current $i(x, y; z, t)$ given by

$$i_{\text{scan}}(x, y; z, t) \sim \sin\left\{\frac{k_0}{2z_0}[(x - x_0)^2 + (y - y_0)^2] + \Omega t\right\}. \tag{11.27}$$

After electronic multiplying with $\sin(\Omega t)$ and low-pass filtering, the scanned demodulated electrical signal i_d, as indicated in Fig. 11.17, is given by

$$i_d(x, y) \sim \cos\left\{\frac{k_0}{2z_0}[(x - x_0)^2 + (y - y_0)^2]\right\}, \tag{11.28}$$

where x and y in the above equation are determined by the motion of the scanning mechanism. Note that the electrical signal i_d contains the location (x_0, y_0) as well as the depth (z_0) information of the off-axis point object. If this scanned demodulated signal is stored in synchronization with the $x - y$ scan signals of the scanning mechanism, then what is stored is the 3-D information of the object or a hologram and the transmission function of the hologram is $t_\delta(x, y) \propto i_d(x, y)$. Hence $t_\delta(x, y)$ is the hologram of $\delta(x - x_0, y - y_0)$ when it is located z_0 from the focus of the spherical wave which is used to generate the scanning time-dependent Fresnel zone plate. This way of generating holographic information is nowadays commonly known as electronic holography [25].

11.3.3. SYNTHETIC APERTURE HOLOGRAPHY

This section discusses synthetic aperture holography (SAH), one promising technique for reconstructing computer-generated holographic information for real-time 3-D display. Synthetic aperture holography, proposed by Benton [26], showed the display of images 150 mm × 75 mm with a viewing angle of 30° [27]. The system is enough to be considered a real-time display for the human visual system [28].

In the holographic display, a large amount of information is required to display a large area with a wide viewing angle. In fact, for a given spatial resolution capability, f_0 of a spatial light modulator, it determines the viewing angle, θ. The situation is illustrated in Fig. 11.18 for on-axis Fresnel zone plate reconstruction.

For intensity pattern given of the form as bias $+\cos(k_0 x^2/2z_0)$, an instantaneous spatial frequency $f_{ins} = (1/2\pi)(d/dx)(k_0 x^2/2z_0) = x/\lambda z_0$. By setting $f_{inst} = f_0$, we solve for the size $x_{max} = \lambda z_0 f_0$ of the limiting aperture of the hologram, which determines the reconstructed image resolution. Defining the numerical aperture of the system as $NA = \sin\theta = x_{max}/z_0$, we have $NA = \lambda f_0$. Now, according to the Nyquist sampling theorem, the sampling interval $\Delta x \leqslant 1/2f_0$; hence, in terms of NA, we have $\Delta x \leqslant \lambda/2NA$. Assuming the size of the SLM is $l \times l$, the number of samples (or resolvable pixels) is $N = (l/\Delta x)^2$. In terms of NA, we have $N = (l \times 2NA/\lambda)^2$. Hence, for a full parallax 100 mm × 100 mm on-axis hologram that is presented on a SLM, $\lambda = 0.6\ \mu$m, and a viewing angle of 60°, i.e., $\theta = 30°$, the required number of resolvable pixels is about 6.7 billion on the SLM. Because the human visual system normally extracts depth information of objects through their horizontal parallax, vertical parallax can be eliminated in order to reduce the amount of information to be stored in the SLM. For 256 vertical lines, the number of pixels required is $256 \times (l/\Delta x) \sim 21$ million if vertical parallax is eliminated. This technique of information reduction is the well-known principle of rainbow holography invented by Benton

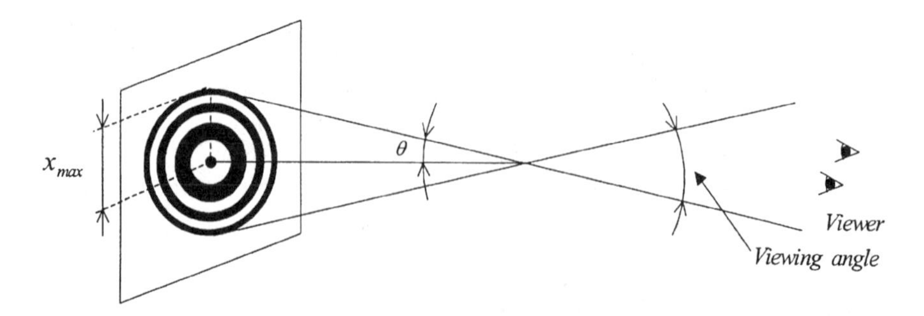

Fig. 11.18. Viewing angle for on-axis Fresnel-zone plate reconstruction.

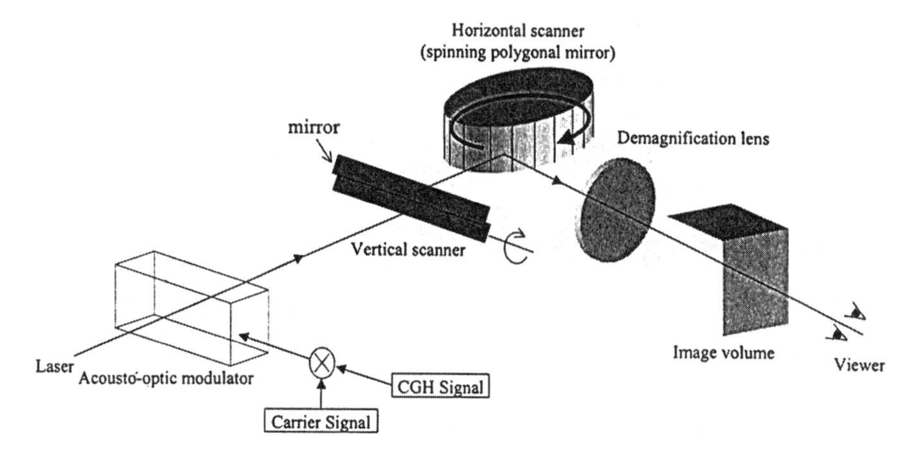

Fig. 11.19. Schematic view of synthetic aperture holography.

[29]. Figure 11.19 shows a schematic view of a system that is capable of displaying computer-generated holograms (CGHs) optically through the use of an acousto-optic modulator.

Let us first discuss the computation of a holographic pattern that will result in holograms only with horizontal parallax. In short, only object points along a vertical position of the object will contribute to the interference pattern along the same vertical position on the recording plane; i.e., we ignore the contribution of other object points that do not lie along the same vertical position. This type of calculation gives rise to holograms possessing horizontal parallax only. A full 2-D CGH is then computed simply by generating an array of these vertical lines for each value of y over the entire vertical extent of the object. These vertical lines are then fed to an acousto-optic modulator sequentially for display. We now examine how these lines are displayed by the AOM. Figure 11.20 shows the principle behind it.

The electrical signal at frequency Ω to the piezoelectric transducer is represented by the analytic signal $e(t)\exp(j\Omega t)$ and the real signal at the transducer is $Re[e(t)\exp(j\Omega t)]$. The analytic signal in the soundcell may then be written as $e(t - x/v_s)\exp[j(\Omega t - Kx)] \propto s(-x + v_s t)\exp[j(\Omega t - Kx)]$. The first-order diffracted light along x for weak scattering; i.e., $\alpha \ll 1$ and according to Eq. (11.11c), is written as

$$-j\tilde{E}_{\text{inc}} s(-x + v_s t)\exp[j(\omega_0 + \Omega)t - k\phi_B x)], \tag{11.29}$$

whereas the *zero*th-order light is $\tilde{E}_{\text{inc}}\exp[j(\omega_0 t + k\phi_B x)]$. We can now see that if the computer-generated holographic (CGH) signal after being multiplied by

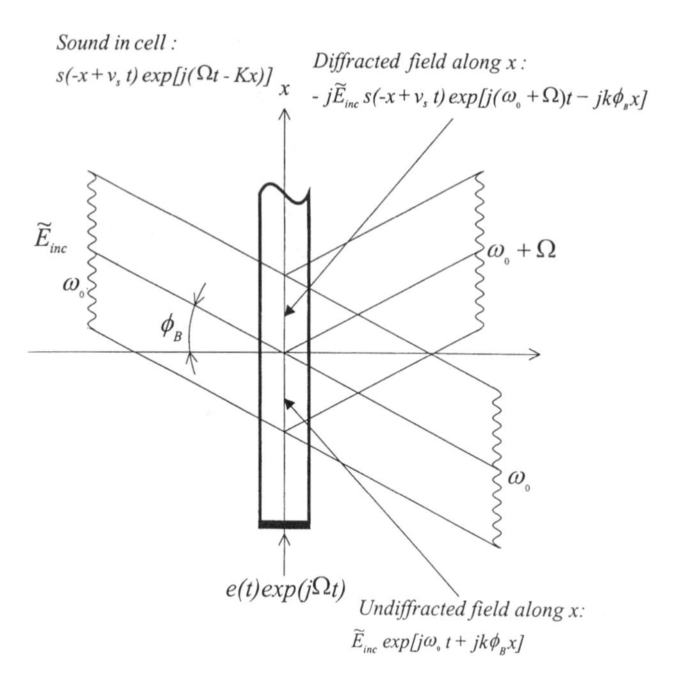

Fig. 11.20. Field distributions in acousto-optic diffraction.

the carrier at frequency Ω, is fed to the transducer; say, for example an on-axis holographic information of the form $e(t) \sim \sin(bt^2)$, the diffracted first-order light is then proportional to $\sin[b(-x + v_s t)^2]\exp[j(\omega_0 + \Omega)t - k\phi_B x)]$. It is a traveling focused spot with the focal length controlled by the parameter b along the direction of the first-order light, and thus the laser light that passes through the soundcell is diffracted according to the holographic information of one horizontal line of the 3-D image in general. Now, the spinning polygonal mirror scans the diffracted image with the opposite direction of the diffracted light's moving. This makes the diffracted image appear stationary [30]. This horizontal scan actually creates a virtual soundcell that is exactly as long as one horizontal line of the CGH signal. This situation is similar to synthetic aperture radar (SAR), where a small antenna is horizontally scanned to give an effective aperture equal to the whole scan line. Hence, this holographic display technique is called synthetic aperture holography. Each reconstruction of a 1-D hologram of each one horizontal line of the 3-D image is scanned onto the corresponding vertical location by the vertical scanner. When this vertical scanning is fast enough to trick the human visual system, a viewer can see a real-time 3-D reconstruction of the 3-D image. Since the diffracted angles are small, a demagnification lens is usually needed to magnify the angles in order to bring the viewing angle to a more acceptable value, as shown in Fig. 11.19.

11.4. INFORMATION DISPLAY USING ELECTRO-OPTIC SPATIAL LIGHT MODULATORS

In this section, we first discuss the electro-optic effect. We then present an electron-beam–addressed spatial light modulator and an optically addressed spatial light modulator as examples, which use the electro-optic effect to accomplish the important phase and intensity modulation.

11.4.1. THE ELECTRO-OPTIC EFFECT

Many SLMs are based on the electro-optic effect of crystals. This section presents the electro-optic effect through the use of a mathematical formalism known as Jones calculus. The refractive index of a birefringent crystal depends on the direction of the crystal. Specifically, uniaxial birefringent crystals, which are the most commonly used materials, have two different kinds of refractive indices. In them, two orthogonal axes have the same refractive index; these are called ordinary axes. The other orthogonal axis, which is called the extraordinary axis, has a different index. A uniaxial birefringent crystal is shown in Fig. 11.21. When light propagates along one of the ordinary axes ($z-$ axis) in the crystal and with the polarization in the $x-y$ plane as shown, the light experiences two different refractive indices.

The electric field, $\boldsymbol{E}$, that propagates along the z-axis is then decomposed into two components along the ordinary and extraordinary axes:

$$\boldsymbol{E} = \hat{x}A_{n_o}e^{j(w_0t-k_{n_o}z)} + \hat{y}A_{n_e}e^{j(w_0t-k_{n_e}z)}, \tag{11.30}$$

where $\hat{x}$ and $\hat{y}$ represent the unit vectors along the x- and y-axes, respectively, $k_{n_o} = (2\pi/\lambda_v)n_o$ and $k_{n_e} = (2\pi/\lambda_v)n_e$, respectively, represent the wavenumbers associated with the different indexes, λ_v is the wavelength of light in free space, and and n_o and n_e are the ordinary and extraordinary refractive indexes, respectively. The Jones vector for $\boldsymbol{E}$, which is composed of two components of linear polarization, is defined by:

$$\boldsymbol{J} = \begin{pmatrix} A_{n_o}e^{-jk_{n_o}z)} \\ A_{n_e}e^{-jk_{n_o}z)} \end{pmatrix}. \tag{11.31}$$

Note that the Jones vector represents the polarization state of the linearly polarized plane wave by means of two complex phasors. Referring to the coordinates of Fig. 11.21, we therefore see that the Jones vector for the incident light (at $z=0$) to the birefringent crystal is given by

$$\boldsymbol{J}_{\text{in}} = \begin{pmatrix} A_{n_o} \\ A_{n_e} \end{pmatrix}. \tag{11.32}$$

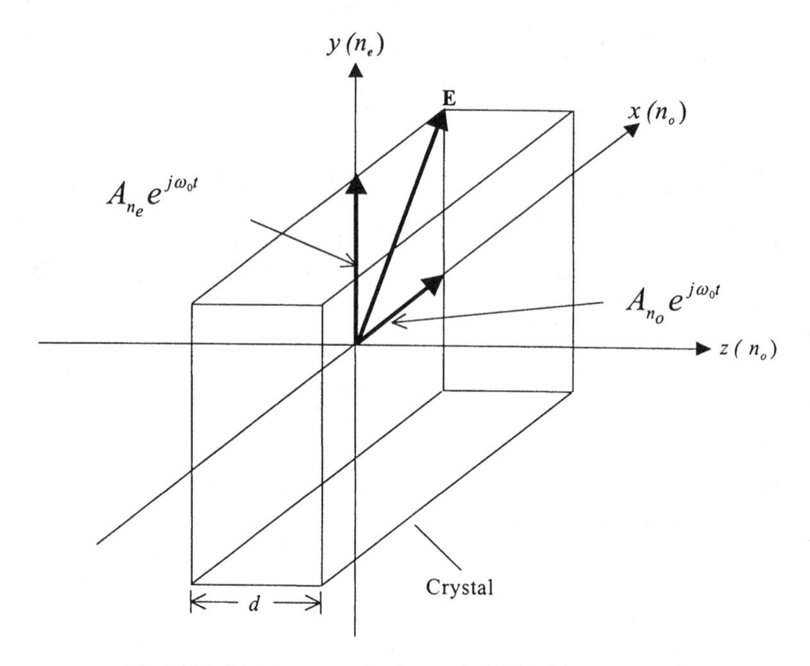

Fig. 11.21. Light propagation in a uniaxial birefringent crystal.

Each component of the incident light experiences different refractive indices when propagating along the crystal. The Jones vector for the emerging light from the crystal at $z = d$ is then given by:

$$\boldsymbol{J}_{\text{out}} = \begin{pmatrix} A_{n_o} e^{-jk_{n_o}d} \\ A_{n_e}{}^{-jk_{n_e}d} \end{pmatrix}. \tag{11.33}$$

Assuming that $n_o < n_e$; i.e., the x-axis is a fast axis and y-axis is a slow axis, we define the phase retardation, Δ, of the extraordinary axis component with respect to the ordinary axis component as

$$\Delta = (k_{n_e} - k_{n_o})d = \frac{2\pi}{\lambda_v}(n_e - n_o)d. \tag{11.34}$$

Expressing in terms of the phase retardation, the Jones vector for the emerging light is given by

$$\boldsymbol{J}_{\text{out}} = \begin{pmatrix} A_{n_o} \\ A_{n_e} e^{-j\Delta} \end{pmatrix}. \tag{11.35}$$

Now the transformation of one polarization state to another can be represented by a matrix called the Jones matrix; for instance, transformation from $\boldsymbol{J}_{\text{in}}$

to $\boldsymbol{J}_{\text{out}}$ may be realized by applying the matrix $\boldsymbol{M}_{pr}$ as follows:

$$\boldsymbol{J}_{\text{out}} = \boldsymbol{M}_{pr}(\Delta)\boldsymbol{J}_{\text{in}} = \begin{pmatrix} 1 & 0 \\ 0 & e^{-j\Delta} \end{pmatrix} \boldsymbol{J}_{\text{in}}, \tag{11.36}$$

where $\boldsymbol{M}_{pr}$ is the Jones matrix of the phase retarder.

The refractive indices of certain types of crystals can be changed according to the external applied electric field. This induces birefringence in the crystals and is called the electro-optic effect. Two commonly used crystals for SLMs, based on the electro-optic effect, are electro-optic crystals and liquid crystals. In electro-optic crystals, the applied electric field redistributes the bonded charges in the crystals. This causes a slight deformation of the crystal lattice. As a result of this, the refractive indices of the crystals are changed according to the applied electric field [31]. In the case of liquid crystals, an external electric field exerts torque on the electric dipole in each liquid crystal molecule. This causes the rotation of the dipole in the liquid crystal, which induces different phase retardation for the propagating light through the liquid crystal. For a detailed discussion of the electro-optic effect in liquid crystals, refer to [32].

Electro-optic phase retardation can be described mathematically through the use of the Jones matrix. For the linear (or *Pockels*) electro-optic effect in uniaxial crystals, the refractive indices are changed linearly proportional to the applied electric field and the Jones matrix for the effect, as given by:

$$\boldsymbol{M}_{\text{poc}}(\Delta) = \begin{pmatrix} 1 & 0 \\ 0 & e^{-j\Delta} \end{pmatrix} \tag{11.37}$$

with $\Delta = (2\pi/\lambda)\gamma E_z d$, where γ, E_z, and d represent the electro-optic coefficient, the magnitude of the applied electric field along the z-direction, and the length of the crystal, respectively. Note that in the Pockels effect, a change in the electric field along the z-axis induces phase retardation along the x- and y-axes [19, 31]. Figure 11.22 shows an intensity modulation system that is composed of an electro-optic crystal and two polarizers.

A polarizer is an optical device that allows light with a certain polarization state to pass through. The polarizers, shown in Fig. 11.22, allow polarization transmit through their polarization axes along the x and y directions, respectively. The Jones matrices of the two polarizers for the x and y directions are given by

$$\boldsymbol{M}_{\text{pol},x} = \begin{pmatrix} 1 & 0 \\ 0 & 0 \end{pmatrix} \tag{11.38}$$

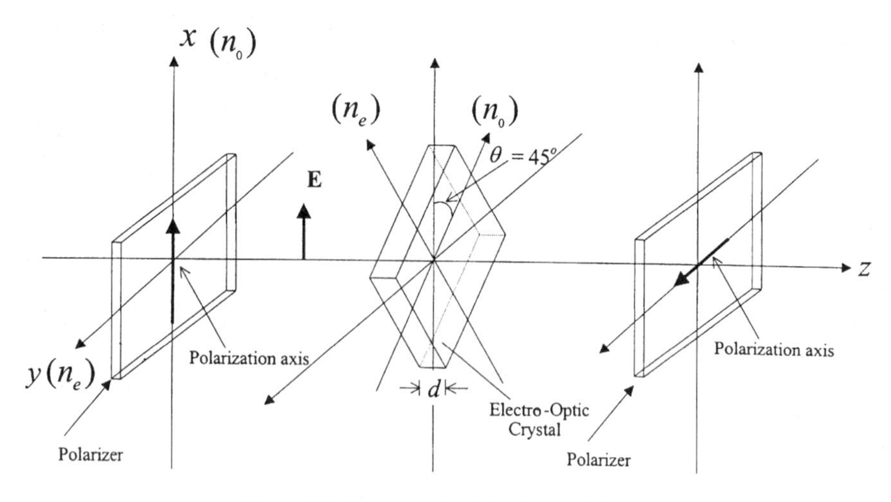

Fig. 11.22. Electro-optic intensity modulator.

and

$$\boldsymbol{M}_{\text{pol},y} = \begin{pmatrix} 0 & 0 \\ 0 & 1 \end{pmatrix}, \tag{11.39}$$

respectively. The two polarizers, aligned in such a way that their polarization axes are 90° with respect to each other, as shown in Fig. 11.22, are called crossed polarizers.

The light that passes through the polarizer can be described by the product of the Jones vector and the Jones matrix of the polarizer, while the rotation of a coordinate system can be described by a rotation matrix. The rotation matrix is given by

$$\boldsymbol{M}_{\text{rot}}(\theta) = \begin{pmatrix} \cos\theta & \sin\theta \\ -\sin\theta & \cos\theta \end{pmatrix}, \tag{11.40}$$

where θ represents an anticlockwise rotation of the original axes so as to form a new coordinate system. It can be shown that the Jones matrix of a polarizer with a polarization axis making an angle α with the x-axis is

$$\boldsymbol{M}_{\text{pol}}(\alpha) = \begin{pmatrix} \cos^2\alpha & \sin\alpha\cos\alpha \\ \sin\alpha\cos\alpha & \sin^2\alpha \end{pmatrix}. \tag{11.41}$$

The state of the output light through an optical system can be calculated by the successive multiplication of Jones matrices. As an example, with reference to Fig. 11.22, the state of the output light is given by

$$\boldsymbol{J}_{\text{out}} = \boldsymbol{M}_{\text{pol},y}\boldsymbol{M}_{\text{rot}}(\theta)\boldsymbol{M}_{\text{poc}}(\Delta)\boldsymbol{M}_{\text{rot}}(-\theta)\boldsymbol{J}_{\text{in}}, \tag{11.42}$$

where the negative sign in $\boldsymbol{M}_{\text{rot}}$ reflects the fact that the new axes along n_o and n_e have been rotated clockwise.

When the rotation angle between the x- and y-axes and the crystal axes is 45°, and the input light is polarized along the x-axis with its Jones vector given by

$$\boldsymbol{J}_{\text{in}} = \begin{pmatrix} E_x \\ 0 \end{pmatrix},$$

as shown in Fig. 11.22, the output light is then given by, according to Eq. (11.42),

$$\begin{aligned}\boldsymbol{J}_{\text{out}} &= \begin{pmatrix} 0 & 0 \\ 0 & 1 \end{pmatrix} \frac{1}{\sqrt{2}} \begin{pmatrix} 1 & 1 \\ -1 & 1 \end{pmatrix} \begin{pmatrix} 1 & 0 \\ 0 & e^{-j\Delta} \end{pmatrix} \frac{1}{\sqrt{2}} \begin{pmatrix} 1 & -1 \\ 1 & 1 \end{pmatrix} \begin{pmatrix} E_x \\ 0 \end{pmatrix} \\ &= \frac{E_x}{2} \begin{pmatrix} 0 \\ e^{-j\Delta} - 1 \end{pmatrix}. \end{aligned} \tag{11.43}$$

Thus the intensity of output light with respect to the input intensity is given by

$$\frac{I_{\text{out}}}{I_{\text{in}}} = \frac{|\boldsymbol{J}_{\text{out}}|^2}{|\boldsymbol{J}_{\text{in}}|^2} = \frac{1}{4}|e^{-j\Delta} - 1|^2 = \sin^2\left(\frac{\Delta}{2}\right) = \sin^2\left(\frac{\pi}{\lambda}\gamma E_z d\right). \tag{11.44}$$

Note that the intensity of output light can be controlled by the applied electric field and hence the system can be used for intensity modulation applications.

11.4.2. ELECTRICALLY ADDRESSED SPATIAL LIGHT MODULATORS

This section presents the electron-beam–addressed spatial light modulator (EBSLM) as an example of SLMs that are categorized as electrically addressed SLMs [33]. The EBSLM is a spatial light modulator which converts a serial video signal into a parallel coherent optical image. The basic structure of the EBSLM is illustrated in Fig. 11.23. It is composed of an electron-gun, deflection coils, an accelerating mesh electrode, and an electro-optic crystal coated with a reflecting mirror [34]. The controller basically controls the writing and erasing of data for the EBSLM.

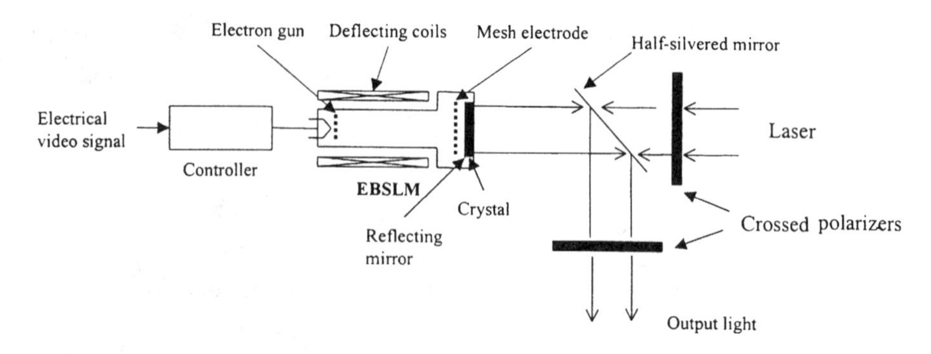

Fig. 11.23. Structure of electron-beam–addressed spatial light modulator (EBSLM) [34].

The major operation of an EBSLM is based on the electro-optic effect discussed in Sec. 11.4.1. We shall, however, briefly discuss its overall operation. First, according to electrical video signals, the electron beam is modulated through the electron gun and then the modulated electron beam is accelerated by the mesh electrode. The electron beam scans onto the surface of the crystal two-dimensionally by deflecting coils. As a result of 2-D scanning, electric charges are distributed onto the crystal's surface corresponding to the input video signals. The resulting charge distribution induces a spatially varying electric field within the electro-optic crystal This spatially induced electric field deforms the crystal as a result of the electro-optic effect discussed in Sec. 11.4.1.

To read out the resulting spatial distribution on the crystal by laser, a pair of crossed polarizers, as shown in Fig. 11.23, are used. This corresponds to the situation shown in Fig. 11.22, where the crystal is inserted between a pair of crossed polarizers for intensity modulation. In this way, a coherent spatial distribution of the output light would correspond to the 2-D scanned video information on the crystal. The typical performance specifications of commercially available EBSLM from Hamamatsu are spatial resolution ~ 10–20 lp/mm with maximum readout laser power $\sim 0.1\ \mathrm{W/cm^2}$ and active area $\sim 9.5 \times 12.7\ \mathrm{mm^2}$ [34].

A real-time 3-D holographic recording (based on optical scanning holography) and display using EBSLM was first demonstrated in the Optical Image Processing Laboratory at Virginia Tech [35, 36]. The recording stage of Fig. 11.24 is essentially the same as the one shown and discussed in Fig. 11.17, where $\Omega/2\pi$ is 40 MHz for the acousto-optic frequency shifter. The focal length of lens L is 17.5 cm. Collimated beams are about 1 cm. The scanning beams of the system scan the 3-D object by the movement of scanning mirrors according to the x–y scanning signal. The scanning signal is also used to synchronize the electron scanning beam inside EBSLM.

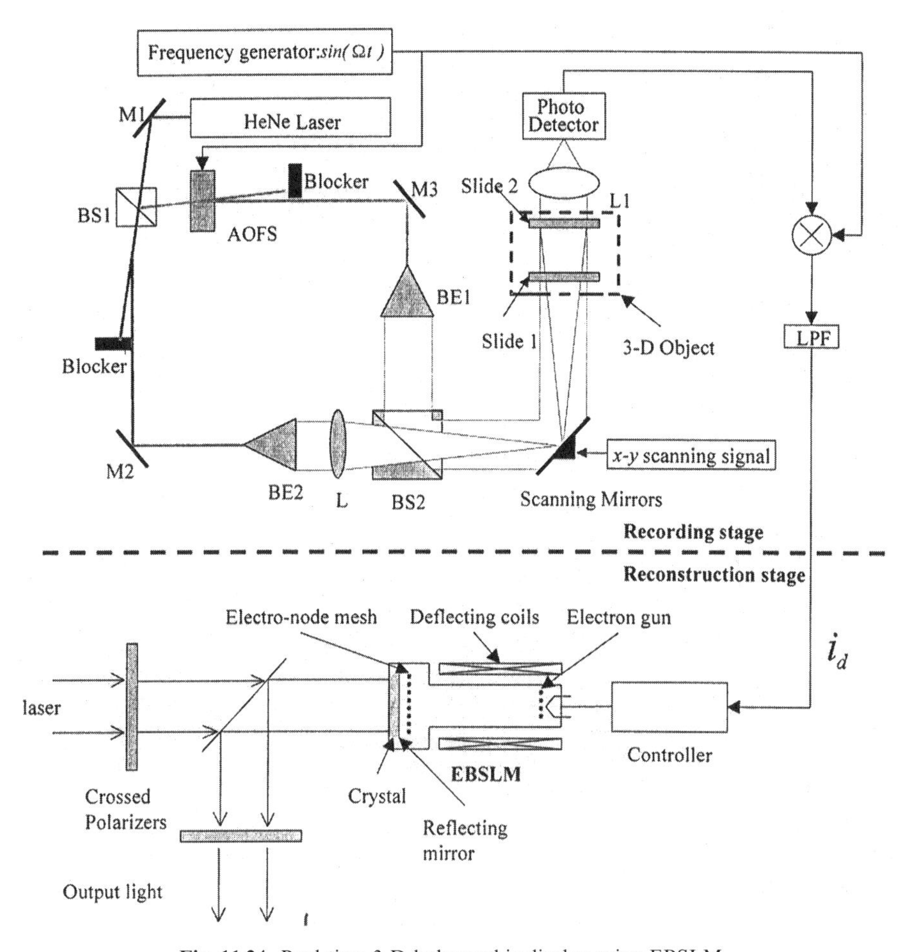

Fig. 11.24. Real-time 3-D holographic display using EBSLM.

As an example, we show a simple 3-D object consisting of two transparencies or slides, as shown in Fig. 11.24. The slides of a "V" and a "T", shown in Fig. 11.25, are located side by side but separated by a depth distance of about 15 cm, with the "V" located closer to the 2-D scanning mirror at a distance about 23 cm. Both the "V" and the "T" are approximately 0.5 cm × 0.5 cm, have a line width of about 100 μm, and are transmissive on an opaque background. The output current from the photodetector is demodulated and sent as scanned holographic data to the controller in the reconstruction stage and eventually the hologram of the scanned 3-D object is sent to EBSLM and displayed on the crystal for spatial modulation of light for reconstruction. The hologram of the 3-D object is shown in Fig. 11.26 [37, 38]. Figures 11.27(a),

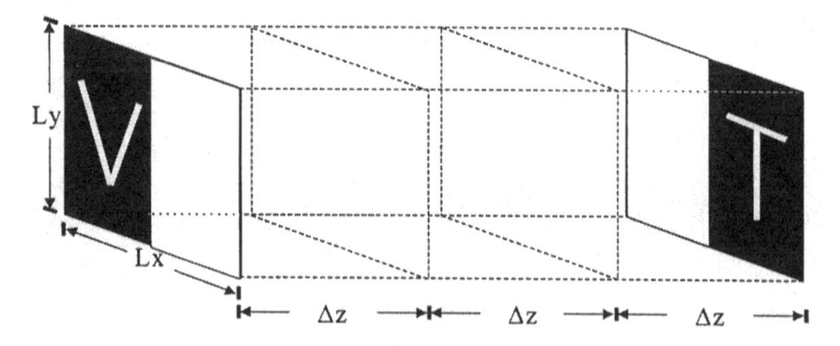

Fig. 11.25. A 3-D object that is composed of two transparencies.

(b), and (c) show the real-time reconstruction of the hologram using EBSLM at three different depths. We see that in Fig. 11.27(a), "V" is focused, whereas "T" is focused as shown in Fig. 11.27(c) at a different depth; hence, the resulting 3-D display.

11.4.3. OPTICALLY ADDRESSED SPATIAL LIGHT MODULATORS

The other category of SLMs is the optically addressed spatial light modulator, which converts incoherent spatial distribution of optical fields to the coherent distribution of optical fields through the use of liquid crystals (LCs). Liquid crystals are organic materials that process properties that are between fluids and solids [32]. They have fluidlike properties such as fluidity and elasticity and yet the arrangement of molecules within them exhibits structural orders such as anisotropicity. Most liquid crystal molecules can be visualized

Fig. 11.26. Hologram of the 3-D object shown in Fig. 11.25.

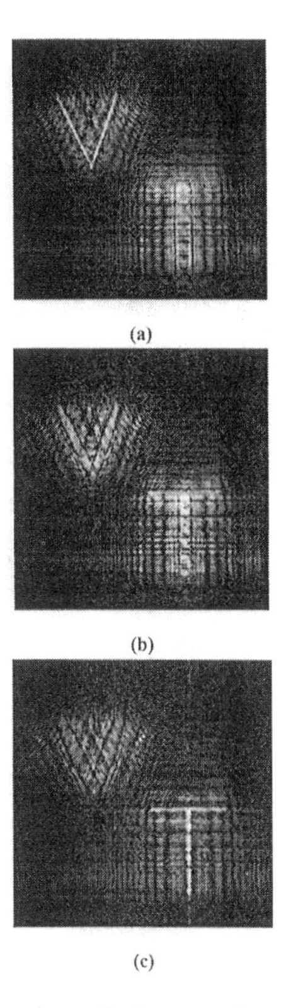

Fig. 11.27. Real-time optical reconstruction of hologram shown in Fig. 11.26 through three depths: (a) shows that "V" is in focus, while (c) shows that "T" is in focus; (b) reconstructs the hologram at a depth between "V" and "T."

as ellipsoids or have rodlike shapes and have a length-to-breadth ratio of approximately 10 with a length of several nanometers. Liquid crystals are classified in terms of their phases. In the so-called nematic phase of the crystals, the molecules tend to align themselves with their long axes parallel, but their molecular centers of gravities are random. The situation is shown in Fig. 11.28. Note that the vector $\hat{\boldsymbol{n}}$ denotes the macroscopic direction of the aligned nematic liquid crystal molecules and is called a director. The director can be reoriented by an external electric field. Because of the elongated shape of the molecules, individual molecules have an anisotropic index of refraction, and the materials in bulk form, therefore, are birefringent. Denoting $n_\perp$ and $n_\parallel$ as the respective refractive indices of the liquid crystal when an electric field is applied perpendicularly or parallelly to its director, the amount of birefringence is $n_\parallel - n_\perp$. Birefringence in the range of 0.2 or more is often found, which makes LCs highly useful.

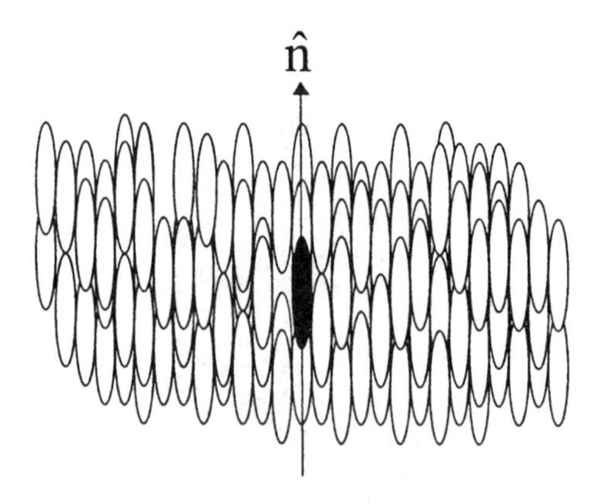

Fig. 11.28. Molecular ordering of nematic phase of liquid crystal.

For LC molecules without alignment, such as when contained in a bottle, their appearance is often milky as they scatter light due to the randomness of LC clusters. For realizing useful electro-optic effects, liquid crystals have to be aligned. The nematic liquid crystal cell is composed of two glass plates (or glass substrates) containing a thin liquid crystal layer (about 5 to 10 μm thick) inside, as shown in Fig. 11.29.

The two glass plates impose boundary conditions on the alignment of nematic liquid crystal molecules contained between them. Each glass plate is

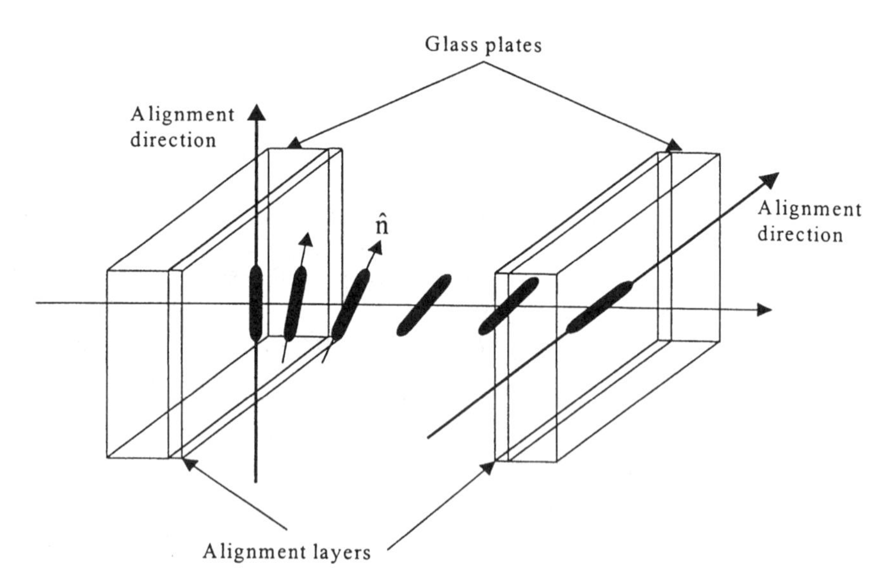

Fig. 11.29. Molecular arrangement in a twisted nematic liquid crystal cell.

coated with a thin electrically conductive but optically transparent metallic film (such as indium-tin-oxide, ITO) called the alignment layer; then the layer is rubbed with a fine cotton cloth in a unidirectional manner. Fine grooves about several nanometers wide are formed by rubbing and thus cause the liquid crystal molecules to lie parallel to the grooves. This rubbing method has been widely used for fabricating large-panel LC devices. High-quality alignment can be made by vacuum deposition of a fine silicon monoxide (SiO) layer to create microgroves onto the surface of the glass for aligning LC molecules. If each alignment layer is polished with different directions, the molecular orientation rotates helically about an axis normal to plates, such as the situation shown in Fig. 11.29. The configuration shown in Fig. 11.29 is called the twist alignment as the back glass plate is twisted at an angle with respect to the front plate. Hence, if the alignment directions between the two plates are 90°, we have the perpendicular alignment. If the alignment directions are parallel, the LC molecules are parallelly aligned, and we have parallel alignment [39].

The twisted nematic liquid can act as a polarization rotator under certain conditions. For example, if a x-plane polarized light is incident on the crystal cell, as shown in Fig. 11.30, the light will rotate its polarization in step with the twisted structure (i.e., align with the directors as the light propagates along the cell) and eventually will leave the cell with its polarization aligned along the y-direction [40].

We model the twisted nematic liquid crystal, which has width d, as a stack of N incremental layers of equal widths $\Delta z = d/N$. Each of the layers acts as a

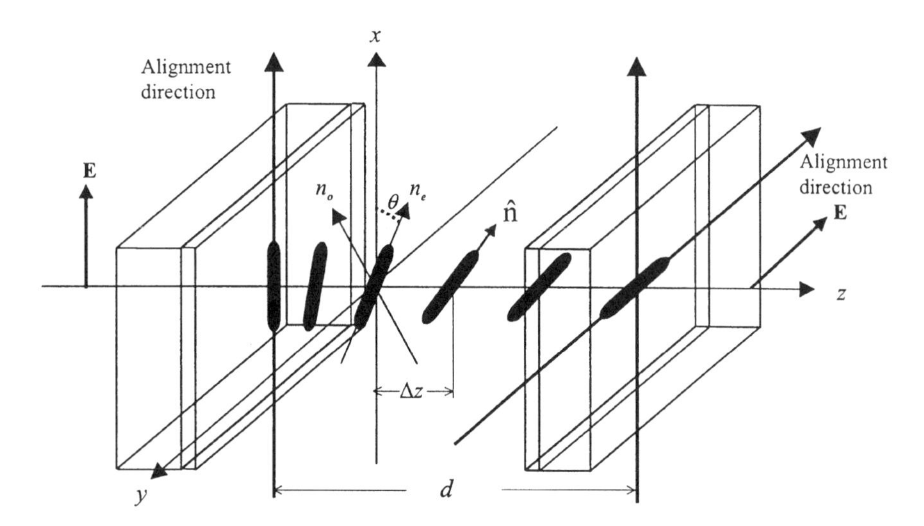

Fig. 11.30. Twisted nematic liquid crystal as a polarization rotator.

uniaxial crystal with n_e and n_o representing the refractive indices of the extraordinary axis and the ordinary axis of the crystal. The angle between the extraordinary axis and the x-axis is called the twist angle; assume that the twist angle varies linearly with the rotation of z. The rotation of each layer then becomes $\Delta\theta = \alpha\Delta z$, where α is the twist coefficient (degree per unit length). Let us say the mth layer that is located at $z = z_m = m\Delta z$, $m = 1, 2, \ldots, N$ is a wave retarder the extraordinary axis of which is rotated by an angle $\theta_m = m\Delta\theta$ with the reference x-axis. This mth layer is expressed by the successive multiplications of Jones matrices:

$$\boldsymbol{T}_m = \boldsymbol{M}_{\text{rot}}(\theta_m)\boldsymbol{M}_{pr}(\Delta z)\boldsymbol{M}_{\text{rot}}(-\theta_m), \tag{11.45}$$

where

$$\boldsymbol{M}_{pr}(\Delta z) = \begin{pmatrix} \exp(-jn_e k_v \Delta z) & 0 \\ 0 & \exp(-jn_o k_v \Delta z) \end{pmatrix}$$

represents the phase retardation of the mth layer, and k_v is the wavenumber of light in free space. The overall Jones matrix of the twisted nematic liquid crystal is expressed by the successive multiplication of the mth layer's Jones matrix:

$$\boldsymbol{T} = \prod_{m=1}^{N} \boldsymbol{T}_m = \prod_{m=1}^{N} \boldsymbol{M}_{\text{rot}}(\theta_m)\boldsymbol{M}_{pr}(\Delta z)\boldsymbol{M}_{\text{rot}}(-\theta_m). \tag{11.45}$$

Since $\boldsymbol{M}_{\text{rot}}(-\theta_m)\boldsymbol{M}_{\text{rot}}(\theta_{m-1}) = \boldsymbol{M}_{\text{rot}}(-\theta_m + \theta_{m-1}) = \boldsymbol{M}_{\text{rot}}(-\Delta\theta)$, we get

$$\begin{aligned} \boldsymbol{T} &= \boldsymbol{M}_{\text{rot}}(\theta_N)[\boldsymbol{M}_{pr}(\Delta z)\boldsymbol{M}_{\text{rot}}(-\Delta\theta)]^N \boldsymbol{M}_{\text{rot}}(-\theta_1) \\ &= \boldsymbol{M}_{\text{rot}}(\theta_N)\left[\exp(-j\phi\Delta z)\begin{pmatrix} \exp\left(-j\beta\dfrac{\Delta z}{2}\right) & 0 \\ 0 & \exp\left(j\beta\dfrac{\Delta x}{2}\right) \end{pmatrix}\right. \\ &\qquad \left. \times \begin{pmatrix} \cos\alpha\Delta z & -\sin\alpha\Delta z \\ \sin\alpha\Delta z & \cos\alpha\Delta z \end{pmatrix}\right]^N \boldsymbol{M}_{\text{rot}}(-\theta_1), \end{aligned} \tag{11.47}$$

where $\phi = (n_o + n_e)k_v/2$ and $\beta = (n_e - n_o)k_v$. Since $\alpha \ll \beta$ as is the case in practice, we can assume the rotation per Δz is small enough for rotation matrix,

$$\begin{pmatrix} \cos \alpha\Delta z & -\sin \alpha\Delta z \\ \sin \alpha\Delta z & \cos \alpha\Delta z \end{pmatrix}$$

to be considered a unit matrix. Therefore, the overall Jones matrix is approximately given by

$$\boldsymbol{T} \approx \boldsymbol{M}_{\text{rot}}(\alpha N\Delta z)\exp(-jN\phi\Delta z)\begin{pmatrix} \exp\left(-j\beta N\ \dfrac{\Delta z}{2}\right) & 0 \\ 0 & \exp\left(j\beta N\ \dfrac{\Delta z}{2}\right) \end{pmatrix}. \quad (11.48)$$

In the limit as $N \to \infty$, $\Delta z \to 0$, and $N\Delta z = d$,

$$\boldsymbol{T} \approx \exp(-jd\phi)\boldsymbol{M}_{\text{rot}}(\alpha d)\begin{pmatrix} \exp\left(-j\beta\dfrac{d}{2}\right) & 0 \\ 0 & \exp\left(j\beta\dfrac{d}{2}\right) \end{pmatrix}. \quad (11.49)$$

This Jones matrix represents a wave retarder of retardation βd, followed by a polarization rotation of angle αd. Thus, the polarization state of an incident

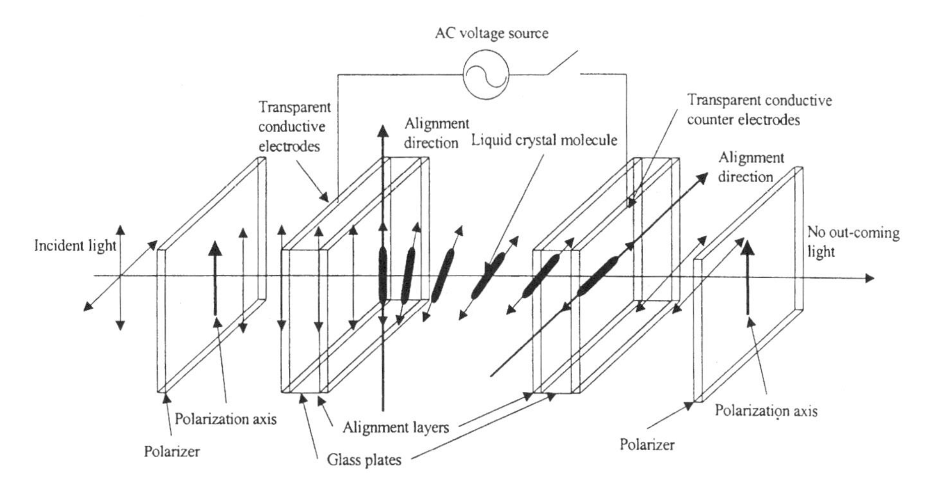

Fig. 11.31(a). Operation of the twisted nematic liquid crystal cell in the absence of an electric field.

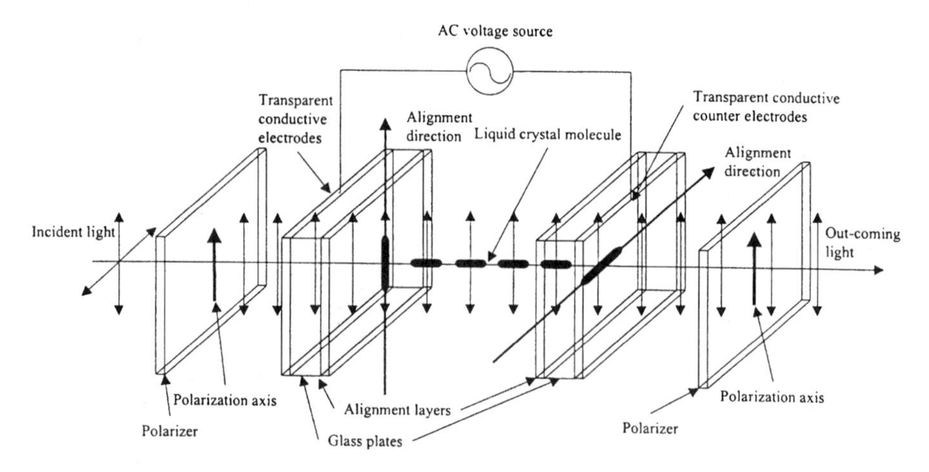

Fig. 11.31(b). Operation of the twisted nematic liquid crystal cell in the presence of an electric field.

light is changed as it propagates within the LC cell. This is shown graphically in Fig. 11.31(a).

So far we have not considered any external applied electric field, which can change the behavior of the cell. Indeed the application of an electric field across the cell can induce an electric dipole in each liquid molecule. For the usual electro-optic applications, liquid crystals for which $n_{\parallel} > n_{\perp}$ are selected; i.e., the dielectric constant of a molecule is larger in the direction of the long axis of the molecule than normal to that axis. The electric field can then induce dipoles having charges at opposite ends of the long direction of the molecule and the induced dipoles can cause the liquid crystal molecules to be aligned parallel to the direction of the applied electric field. The situation is shown in Fig. 11.31(b). Indeed, the system shown in Fig. 11.31 can be used as a liquid crystal light switch. When there is no electric field applied, the output light is blocked and when there is an applied field, the light will be transmitted. Because the long time exposure of electric fields with fixed directions can cause permanent chemical changes to liquid crystals, the cell is usually driven by AC voltages and the typical frequency range of AC voltages is around 1 KHz, with voltages of the order of 5 volts [41].

We shall now consider the so-called parallel-aligned liquid crystal (PAL) cell and discuss its operation in terms of phase modulation and intensity modulation. The PAL–SLM is shown in Fig. 11.32(a). In perpendicular-aligned LC modulators, the light will have output polarization changed after propagating through the SLM (see Fig. 11.31(a). The change in polarization is sometimes undesirable; for instance, if one wants to perform interference downstream of the optical system, linearly polarized waves of orthogonal polarizations will not interfere [42]. For PAL–SLM, however, it is clear that

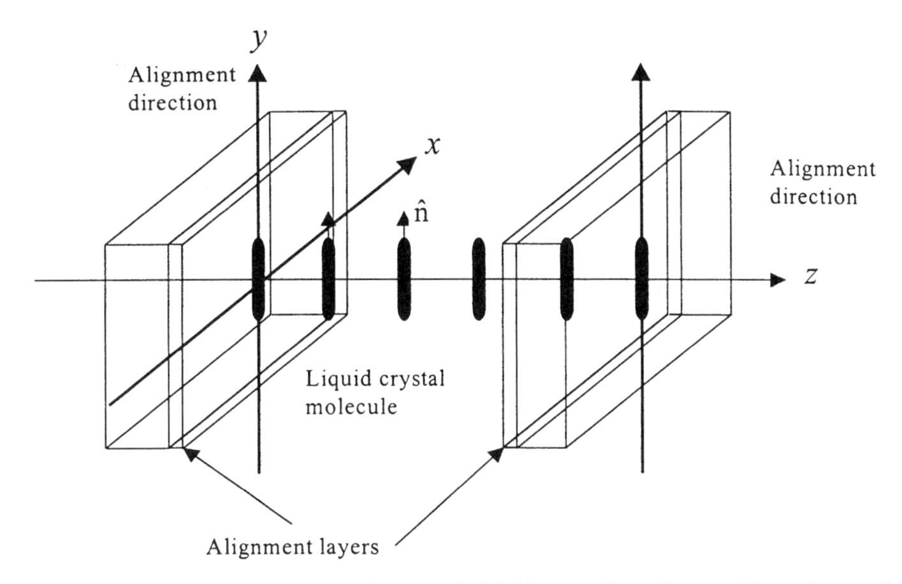

Fig. 11.32(a). PAL–SLM when there is no applied field; pure phase change without change of output polarization if the incident light is linearly polarized along the director's axis.

the phase change can be a pure phase modulation (without changing the polarization state). For example, consider the case of a linearly polarized light incident on a PAL–SLM with the light polarization parallel to the director's axis. The output light will experience a phase change and remains the same polarization at the exit of the SLM, as shown in Fig. 11.32(a).

For a voltage-controlled phase modulator, we look at the situation in Fig. 11.32(b), where the electric field is applied along the z-direction. The electric field tends to tilt the molecules as shown in the figure. When the applied electric field is strong enough, most of the molecules tilt, except those adjacent to the alignment layer due to the elastic forces at the surfaces. The tilt angle as a function of the voltage V (strictly speaking, it is the root-mean-square value of the applied voltage as the voltage is AC across the SLM) has been described by the following equation [40, 43]:

$$\theta = \begin{cases} 0, & V \leqslant V_{th} \\ \dfrac{\pi}{2} - 2\tan^{-1}\exp\left(-\dfrac{V - V_{th}}{V_o}\right), & V > V_{th}, \end{cases} \tag{11.50}$$

where V_{th} is a threshold voltage below which no tilting of the molecules occurs and V_0 is a constant, when $V - V_{th} = V_0$, $\theta \sim 50°$. For $V - V_{th} > V_0$, the angle θ keeps increasing with V, eventually reaching a saturation value of $\pi/2$ for

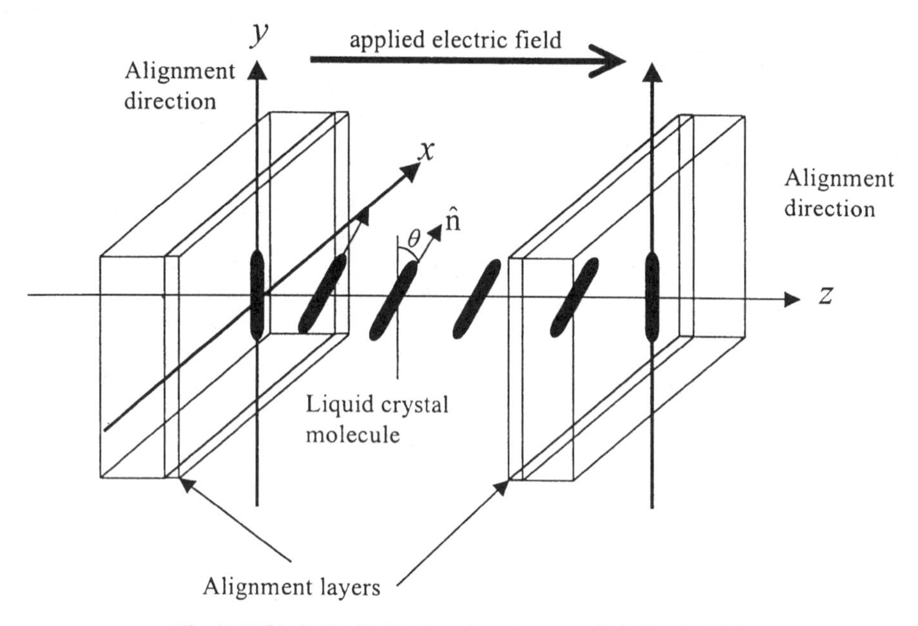

Fig. 11.32(b). PAL–SLM when there is an applied electric field.

large voltage. When the voltage is turned off, the molecules return to their original positions. Hence, the LC materials have memory, in a sense. Now, when the molecules are tilted and light is traveling along the z-direction, polarizations in the x- and y-directions will have refractive indices n_0 and $n(\theta)$, respectively, where

$$\frac{1}{n^2(\theta)} = \frac{\cos^2\theta}{n_e^2} + \frac{\sin^2\theta}{n_o^2}. \tag{11.51}$$

Figure 11.33 illustrates the cross section of the index ellipsoid. Since the ellipsoid is rotated clockwise by angle θ on the y–z coordinates under the influence of the applied electric field, we have modeled it conveniently with the propagation vector $\vec{k}_0$, rotated anticlockwise on the coordinates, as shown in Fig. 11.33. It shows that the value of the refractive index depends on the orientation of the molecules tilted. Under this situation, the phase retardation becomes $\Delta = 2\pi[n(\theta) - n_0]d/\lambda$. The retardation becomes $\Delta_{\max} = 2\pi[n_e - n_0]d/\lambda$, which is maximum when there is no tilt in the molecules (i.e., no applied electric field or applied voltage). Hence, the SLM can be used as a voltage-controlled phase modulator. The SLM can be used as an intensity modulator if suitably aligned between two crossed polarizers. In the next

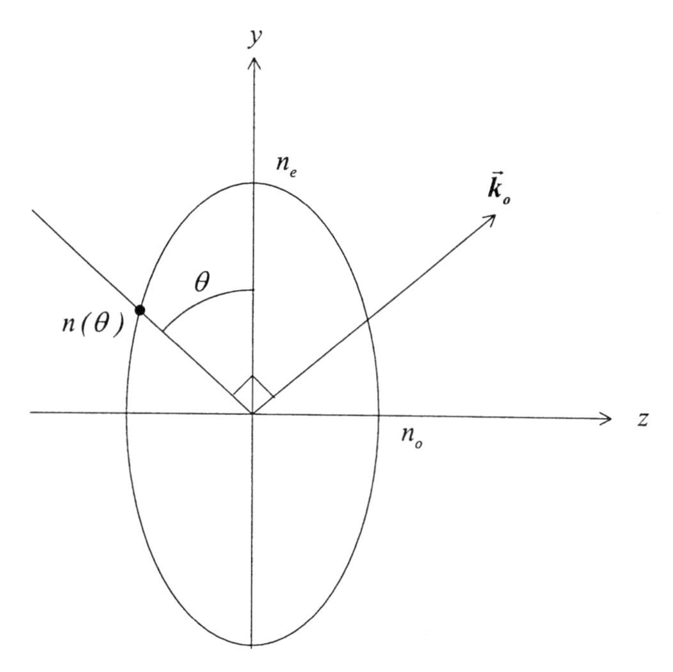

Fig. 11.33. Cross section of tilted index ellipsoid illustrating that the value of refractive index depends on the orientation of the tilt.

section, we will discuss a commercially available PAL–SLM operated in a reflective mode.

Figure 11.34 shows a reflective PAL–spatial light modulator (PAL–SLM) [41]. The SLM consists of two major sections separated by an optional light-blocking layer and a dielectric mirror: the photoconductive section on the left-hand side is for writing incoherent images on the SLM, and the nematic liquid crystal section on the right-hand side is for reading out images coherently. Hence, the SLM is considered an incoherent-to-coherent image converter. The photoconductor layer and the twisted nematic liquid crystal layer are sandwiched between transparent conductive electrodes across which AC voltages are applied. The function of the light-blocking layer is to isolate the photoconductive layer from the read-out light, which might otherwise write if the read-out light is strong enough. The glass plates are optical flats to achieve phase flatness in the device and the antireflecting layers are used to reduce multiple reflections in the flats so as not to spoil the spatial resolution of the device.

We now describe how the PAL–SLM works. Figure 11.35 shows a simplified AC electrical model for the PAL–SLM. Here the LC and the photoconductive layer are simply modeled as RC circuits, and the dielectric

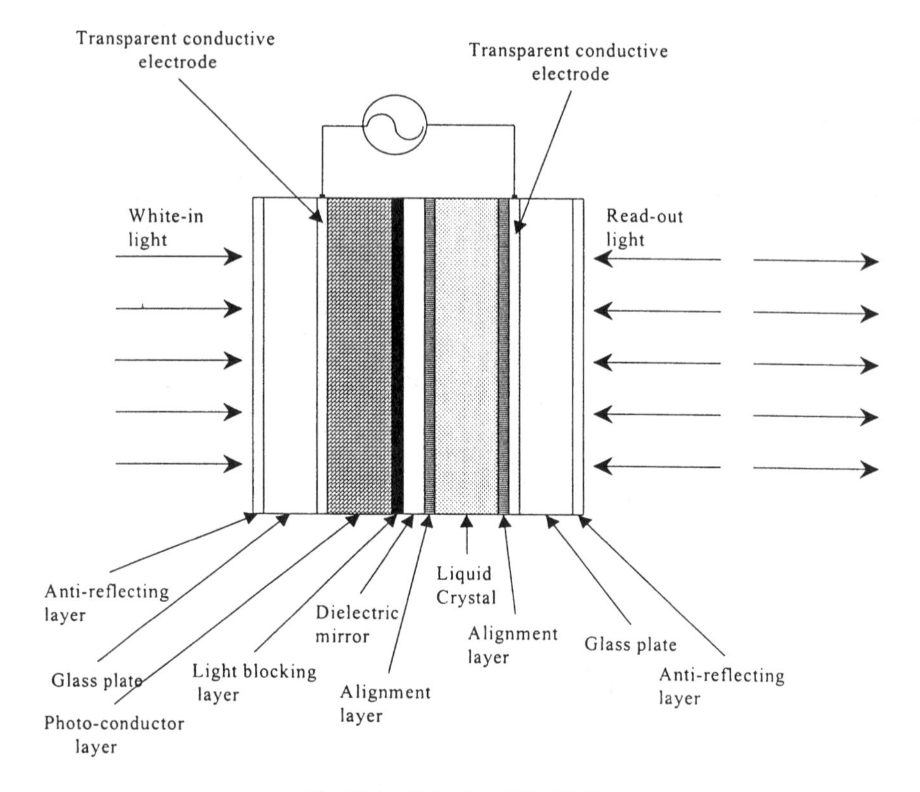

Fig. 11.34. Reflective PAL–SLM.

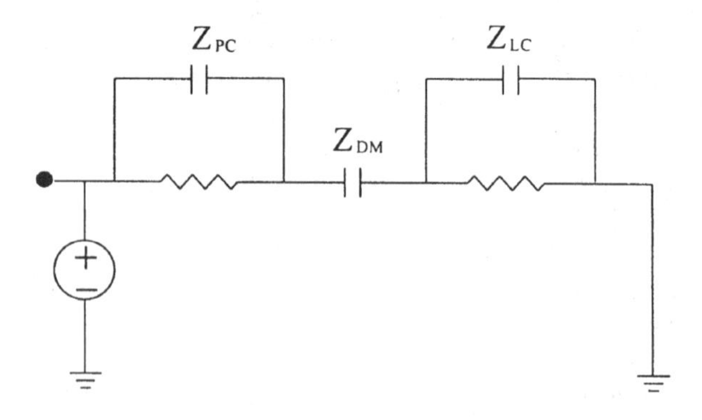

Fig. 11.35. Electrical model for the PAL–SLM. Z_{PC} and Z_{LC} are the impedances of the photoconductive layer and the liquid crystal layer, respectively. Z_{DM} is the impedance of the dielectric mirror.

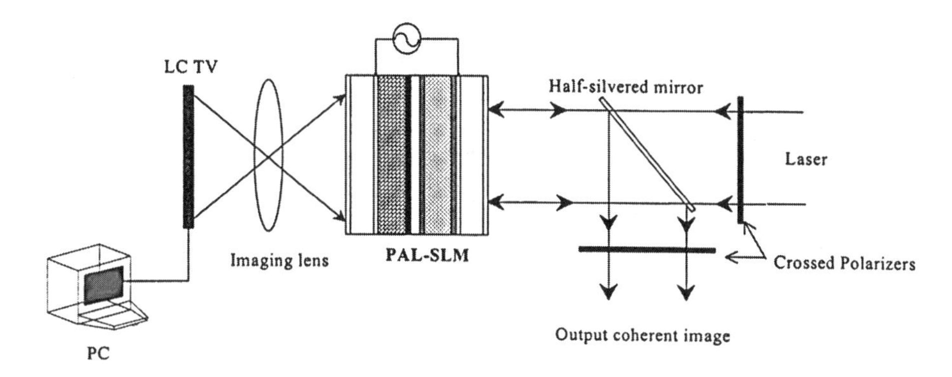

Fig. 11.36. Reflective PAL–SLM system for incoherent to coherent image conversion.

mirror is modeled as a capacitor with impedance Z_{DM}. In the PAL–SLM, the bias electric field within the system is obtained by an applied AC voltage between the electrodes, and the field is varied according to the change of the impedance of the photoconductive layer upon illumination by the write light. In areas where the layer is dark, its impedance is high, and small applied voltage is dropped across the LC layer; hence, the molecules will not tilt. However, when the layer is illuminated, its conductivity increases and therefore the impedance decreases and the applied voltage dropped across the liquid increases. The result leads to the tilting of the molecules and hence the phase of the read light is changed locally according to the intensity distribution of the write light or the input image upon the area of the photoconductive layer. Typical spatial resolution is 50 lp/mm with 18×18 mm^2 active area [41]. Figure 11.36 illustrates the use of PAL–SLM for holographic reconstruction. The hologram shown in Fig. 11.26 has been stored in a PC and is ouputted to a standard LC-TV. The hologram is subsequently imaged onto the photoconductive side of the SLM. A pair of crossed polarizers are aligned such that their polarization axes are 45° away from the directors of the LC molecules, such as the situation illustrated in Fig. 11.22 for electro-optic crystals. Results identical to those obtained in Fig. 11.27 have been observed [44].

11.5. CONCLUDING REMARKS

We have discussed 2-D and 3-D information display using optics. Displays using acousto-optic modulators have been emphasized in the majority of this chapter. These modulators indeed have been used for 2-D displays as well as for 3-D display applications. In 2-D displays, laser TVs have been discussed, whereas in 3-D applications, holographic information has been acquired with the help of AOMs, as in the case of optical scanning holography. In addition, 3-D displays using an AOM have been used in synthetic aperture holography. Spatial light modulators based on the use of the electro-optic effect have also been discussed. Specifically, electron-beam–addressed SLMs and parallel-aligned nematic liquid crystal SLMs have been emphasized and they have been used for 3-D holographic display as well. While applications involving SLMs are abundant [45], large high-resolution 2-D displays still present a challenging problem [14, 17] and large 3-D displays with sizable viewing angles remain formidable. This becomes one of the ultimate problems in modern optics, as current SLMs are not suitable for such applications due mainly to their limited spatial resolution and size.

REFERENCES

11.1 A. Korpel, 1981, "Acousto-optics—A Review of Fundamentals," *Proc. IEEE*, 69, 48–53.
11.2 P. K. Das and C. M. Decusatis, *Acousto-Optic Signal Processing: Fundamentals and Applications*, Artech House, Boston, 1991.
11.3 A. Vanderlugt, *Optical Signal Processing*, John Wiley and Sons, Inc., New York, 1992.
11.4 P. Debye and F. W. Sears, 1932, "On the Scattering of Light by Supersonic Waves," *Proc. Nat. Acad. Sci.*, 18, 409–414.
11.5 C. V. Raman and N .S. N. Nath, 1935; 1936, "The Diffraction of Light by High-Frequency Sound Waves," *Proc. Ind. Acad. Sci.*, 2, 406–420; 3, 75–84, 119–125, 259–465.
11.6 R. R. Aggarwal, 1950, "Diffraction of Light by Ultrasonic Waves (Deduction of Different Theories for the Generalized Theory of Raman and Nath)," *Proc. Ind. Acad. Sci. A*31, 417–426.
11.7 R. Adler, 1967, "Interaction between Light and Sound," *IEEE Spectrum*, 4(5), 42–54.
11.8 W. R. Klein and B. D. Cook, 1967, "Unified Approach to Ultrasonic Light Diffraction," *IEEE Trans. Sonics Ultrason.*, SU-14, 723–733.
11.9 H. Kogelnik, 1969, "Coupled Wave Theory for Thick Hologram Gratings," *Bell Sys. Tech. J.*, 48, 2909–2947.
11.10 A. Korpel and T.-C. Poon, 1980, "Explicit Formalism for Acousto-optic Multiple Pane-Wave Scattering," *J. Opt. Soc. Am.*, 70, 817–820.
11.11 A. Korpel, *Acousto-optics*, Marcel Dekker, Inc., 1988.
11.12 R. A. Appel and M. G. Somekh, 1993, "Series Solution for Two-Frequency Bragg Interaction Using the Korpel-Poon Multiple-Scattering Method," *J. Opt. Soc. Am. A*, 10, 466–476.
11.13 P. Phariseau, 1956, "On the Diffraction of Light by Progressive Supersonic Waves," *Proc. Indian Acad. Sci. Sect.*, A 44, 165–170.
11.14 M. R. Chatterjee, T.-C. Poon, and D. N. Sitter, 1990, "Transfer Function Formalism for Strong Acousto-optic Bragg Diffraction of Light Beams with Arbitrary Profiles," *Acustica*, 71, 81–92.
11.15 T.-C. Poon, M. D. McNeill, and D. J. Moore, 1997, "Modern Optical Signal Processing Experiments Demonstrating Intensity and Pulse-Width Modulation Using Acousto-optic Modulator," *American Journal of Physics*, 65, 917–925.

11.16 A. Korpel, R. Adler, P. Desmares, and W. Watson, 1966, "A Television Display Using Acoustic Deflection and Modulation of Coherent Light," *Proceedings of the IEEE*, 54, 1429–1437.
11.17 Y. Hwang, J. Lee, Y. Park, J. Park, S. Cha, and Y. Kim, 1998, "200-inch Full Color Laser Projection Display," *Proceedings of SPIE*, 3296, 116–125.
11.18 B. Choi, Samsung Advanced Institute of Technology, private communications.
11.19 P. P. Banerjee and T.-C. Poon, *Principles of Applied Optics*, R. D. Irwin, Boston, 1991.
11.20 J. B. De Velis and G. O. Reynolds, *Theory and Applications of Holography*, Addison-Wesley, Reading, Mass., 1967.
11.21 G. L. Rogers, 1950, "Gabor Diffraction Microscopy: The Hologram as a Generalized Zone Plate," *Nature*, 166, 237.
11.22 W. J. Siemens-Wapniarski and M. Parker Givens, 1967, "The Experimental Production of Synthetic Holograms," *App. Opt.*, 7, 535–538.
11.23 T.-C. Poon and A. Korpel, 1979, "Optical Transfer Function of an Acousto-optic Heterodyning Image Processor," *Opt. Lett.*, 4, 317–319.
11.24 T.-C. Poon, 1985, "Scanning Holography and Two-Dimensional Image Processing by Acousto-optic Two-Pupil Synthesis," *J. Opt. Soc. Am.*, 2, 521–527.
11.25 C. J. Kuo, ed., 1996, special issue on "Electronic Holography," *Opt. Eng.*, 35.
11.26 S. A. Benton, 1990, "Experiments in Holographic Video Imaging," in P. Greguss and T. J. Jeong, eds., SPIE Proc. paper #IS-8, 247–267.
11.27 P. St. Hilaire, 1995, "Scalable Optical Architecture for Electronic Holography," *Opt. Eng.*, 34, 2900–2911.
11.28 P. St. Hilaire, August 1997, "Holographic Video", *Optics & Photonics News*, 35–68.
11.29 S. A. Benton,"The Mathematical Optics of White Light Transmission Holograms," T. H. Jeong, ed., Proc. First International Symposium on Display Holography, Lake Forest College, July 1982.
11.30 L. M. Myers, 1936, "The Scophony System: an Analysis of Its Possibilities," *TV and Shortwave World*, 201–294.
11.31 A. Yariv and P. Yeh., *Optical Waves in Crystals*, John Wiley & Sons, New York, 1984.
11.32 E. Kaneko, *Liquid Crystal TV Displays: Principles and Application of Liquid Crystal Displays*, KTK Scientific Publishers, Tokyo, Japan, 1987.
11.33 K. Shinoda and Y. Suzuki, 1986, "Electron Beam–Addressed Spatial Light Modulator," *Proceedings of SPIE*, 613, 158–164.
11.34 Electron beam spatial light modulator (EBSLM), Model X3636, provided by Hamamatsu, Photonics K. K., Japan and Hamamatsu Corp., Bridgewater, New Jersey.
11.35 B. D. Duncan and T.-C. Poon, 1992, "Gaussian Beam Analysis of Optical Scanning Holography," *J. Opt. Soc. Am. A*, 9, 229–236.
11.36 T.-C. Poon, B. W. Schilling, M. H. Wu, K. Shinoda, and Y. Suzuki, 1993, "Real-Time Two-Dimensional Holographic Imaging Using an Electron Beam–Addressed Spatial Light Modulator," *Opt. Lett.*, 18, 63–65.
11.37 T.-C. Poon, K. Doh, B. Schilling, M. Wu, K. Shinoda, and Y. Suzuki, 1995, "Three-Dimensional Microscopy by Optical Scanning Holography," *Opt. Eng.*, 34, 1338–1344.
11.38 T.-C. Poon, K. Doh, B. Schilling, K. Shinoda, Y. Suzuki, and M. Wu, 1997, "Three-Dimensional Holographic Display Using an Electron Beam–Addressed Spatial Light-Modulator," *Optical Review*, 4, 567–571.
11.39 U. Efron, ed., *Spatial Light Modulator Technology*, Marcel Dekker, Inc., New York, 1995.
11.40 B. E. A. Saleh and M. C. Teich, *Fundamentals of Photonics*, John Wiley & Sons, New York, 1991.
11.41 Parallel Aligned Nematic Liquid Spatial Light Modulator (PAL–SLM), Model X5641 Series, Hamamatsu, Photonic K. K., Japan and Hamamatsu Corp., Bridgewater, New Jersey.

11.42 N. Mukohzaka, N. Yoshida, H. Toyoda, Y. Kobayashi, and T. Hara, 1994, "Diffraction Efficiency Analysis of a Parallel-Aligned Nematic Liquid Crystal Spatial Light Modulator," *App. Opt.*, 33, 2804–2811.
11.43 P.-G. de Gennes, *The Physical of Liquid Crystals*, Clarendon Press, Oxford, 1974.
11.44 T.-C. Poon, experiments conducted at Research Institute of Electronics, Shizuoka University, Hamamatsu, Japan, 1997.
11.45 F. T. S. Yu and S. Jutamulia, eds., *Optical Pattern Recognition*, Cambridge University Press, 1998.

EXERCISES

11.1 For the system of equations given in Eq. (11.9)
(a) Show the energy of conservation: $|\tilde{E}_0|^2 + |\tilde{E}_1|^2 = |\tilde{E}_{inc}|^2$.
(b) Solve for $\tilde{E}_0$ and $\tilde{E}_1$ for $\tilde{\alpha}^* = \tilde{\alpha} = \alpha$.

11.2 Verify that Eq. (11.11c) is the solution for Eq. (11.9) when $\delta = 0$.

11.3 Starting from Eq. (11.8) with $\tilde{\alpha}^* = \tilde{\alpha} = \alpha$, and assuming that $Q = 0$ for normal incidence (i.e., $\phi_{inc} = 0$), find the solutions under this situation. This situation is known as the ideal Raman-Nath regime for the acousto-optic modulator.

11.4 Design an acousto-optic intensity modulation system based on Raman-Nath diffraction. We want to use the *zero*th-order light for transmission of information. Sketch an optical system that would do the job.

11.5 Consider an acousto-optic deflector. If the interaction length is 10 mm and its center frequency is 150 Mhz, estimate the number of resolvable spots and the transit time of the deflector if the laser beam width is 2 mm and its wavelength is 0.6 μm. We assume that the sound velocity in the acoustic medium is 4000 m/s.

11.6 Verify the three equations given by Eq. (11.23).

11.7 A point-object hologram is given of the form as bias + $\cos(k_0 x^2/2z_0)$, where the wavenumber of the recording light is k_0 and z_0 is the location of the point away from the holographic film. Find the real image location of the point upon illumination by plane wave of light with wavenumber k_1.

11.8 For the hologram given by Exercise x11.7, we now assume that the recording film has a finite resolution limit in such a way that NA is given. Estimate the spot size of the reconstructed real image in terms of NA.

11.9 A three-point object given by $\delta(x, y; z - z_0) + \delta(x - x_0, y; z - z_0) + \delta(x, y; z - (z_0 + \Delta z_0))$ is scanned by the time-dependent Fresnel zone plate expressed by Eq. (11.26). Find an expression for its hologram, $t_{3\delta}(x, y)$.

11.10 Hologram scaling: After we have recorded the hologram of the three-point object, $t_{3\delta}(x, y)$, obtained in Exercise 11.9, we want to scale the hologram to $t_{3\delta}(Mx, My)$, where M is the scale factor. Find the location of the real-image reconstruction of the three-point object if the hologram is illuminated by a plane wave. Draw a figure to illustrate the reconstruction locations of the three points.

11.11 With reference to Exercises 11.9 and 11.10 and defining the lateral magnification M_{lat} as the ratio of the reconstructed lateral distance to the original lateral distance x_0, express M_{lat} in terms of M.

11.12 Defining the longitudinal magnification M_{long} as the ratio of the reconstructed longitudinal distance to the original longitudinal distance Δz_0 for hologram $t_{3\delta}(Mx, My)$, express M_{long} in terms of M.

11.13 Show that the Jones matrix of a polarizer with a polarization axis making an angle α with the x axis is given by Eq. (11.41).

11.14 A crystal called KDP (potassium dihydrogen phosphate) with width, $d = 3\,\text{cm}$ and proportional coefficient, $\gamma = 36.5\,\text{m/V}$, is used as the elecro-optic crystal in the electro-optic intensity modulator that is shown in Fig. 11.22. Calculate the required magnitude of the electric field and the applied voltage in order to completely block the outgoing light. Assume that the wavelength is $\lambda = 0.628\,\mu\text{m}$.

11.15 Plot Eq. (11.50) as a function of $(V - V_{th})/V_o$. Also plot the normalized phase retardation, $\Delta/\Delta_{\text{max}}$, as a function of $(V - V_{th})/V_o$ for $n_0 = 1.5$, $n_e = 1.6$ and $\lambda = 0.628\,\mu\text{m}$.

Chapter 12 Networking with Optics

Lintao Zhang, Guohua Xiao and Shudong Wu
AVANEX CORPORATION

Networking with optics includes optical networking based on time, on optical frequency/wavelength, on space, and on optical coding. While research has been ongoing in all these areas since fiber optics emerged as a dominant communication technology in the 1970s, up until now, only optical networks based on frequency/wavelength become a reality.

12.1. BACKGROUND

Wavelength division multiplexing (WDM) is the underlying technology for optical networking based on the wavelength of optical signal [1, 2, 3]. Among existing transmission media, optical fiber provides the most bandwidth, compared to copper wire, coaxial cable, and microwave [4]. Today, AllWave^{TM} fiber made by Lucent Technology has about 300 nm ($\sim$43 THz) of bandwidth usable for optical transmission. The two most common ways of multiplexing additional signals onto the huge bandwidth of fiber is by assigning different optical frequencies to each signal (frequency-division multiplexing or FDM), or by assigning different time slots to each signal (time-division multiplexing or TDM). Because wavelength is the reciprocal of frequency, WDM is logically equivalent to FDM. WDM is performed in the optical domain using a passive optical wavelength division multiplexer (WDM mux or for short, mux) and demultiplexer, whereas TDM is performed in the electrical domain using digital technology. Although there has been an ongoing battle over the choice of TDM or WDM technology since optical fiber became the undisputed choice of transmission media for high-speed digital communications in the late 1970s, TDM always ended up the winner because of simple economics: everything in the electrical domain was much cheaper to carry out than in the optical domain. WDM mux and demux were very expensive, compare to electrical TDM digital mux and demux. In long-distance communication systems, signals need to be regenerated along the way to clean up transmission loss, noises, and distortion. While regeneration, reshaping, and retiming (3R) in the electrical domain progressed from tens of megabits per second to multiple gigabits per second, optical 3R technology did not even exist. Therefore, fiber-optic communication systems rely on optical-to-electrical (O/E) conversion, electrical 3R for each individual channel, and electrical-to-optical (E/O) conversion to reach hundreds and thousands of kilometers.

The emergence of the Erbium-doped fiber amplifier (EDFA) and surging bandwidth demand in the late 1980s, though, finally pushed WDM onto center stage of optical communications [5, 6, 7]. EDFAs boost light signals in the optical domain, regardless of signal data rate and format. Meanwhile, progress in optical mux/demux and laser diode technologies have also reduced the cost of packing and unpacking optical signals by their wavelengths. The combination of EDFA and WDM technology makes it possible to transmit many optical channels at different wavelengths down one single optical fiber hundreds of kilometers without the need of O/E/O. Economics favors EDFA/WDM over electrical TDM as more capacities are demanded from transmission systems. TDM capacity is limited by high-speed electronics. It tops out at 40 gigabits/s at the moment, rather small compared to the >40 THz transmission bandwidth on optical fiber. WDM can pack as many top-speed TDM channels as optical mux/demux allows and boost them all using EDFA. It is no surprise that both terrestrial and undersea fiber-optic systems began to massively deploy EDFA and dense WDM (DWDM) technologies in the early 1990s when EDFA became commercially available. The difference between WDM and DWDM lies only in that in DWDM, wavelength difference (spacing) between channels is very small, from several nanometers to less than 1 nanometer, while traditionally, WDM implied multiplexing optical signals at 1.3 μm with optical signals at 1.55 μm. Figure 12.1 shows a typical DWDM system configuration. Optical signals are generated by transmitters emitting different wavelengths. Optical mux combines these signals together onto a single optical fiber. Boosting EDFA amplifies the composite WDM signals before launching them into the transmission fiber. Inline EDFA reboosts the composite signal to compensate the loss of fiber. At the receiving end, optical demux separates the composite WDM signals into separate channels, and optical receivers recover all the signals.

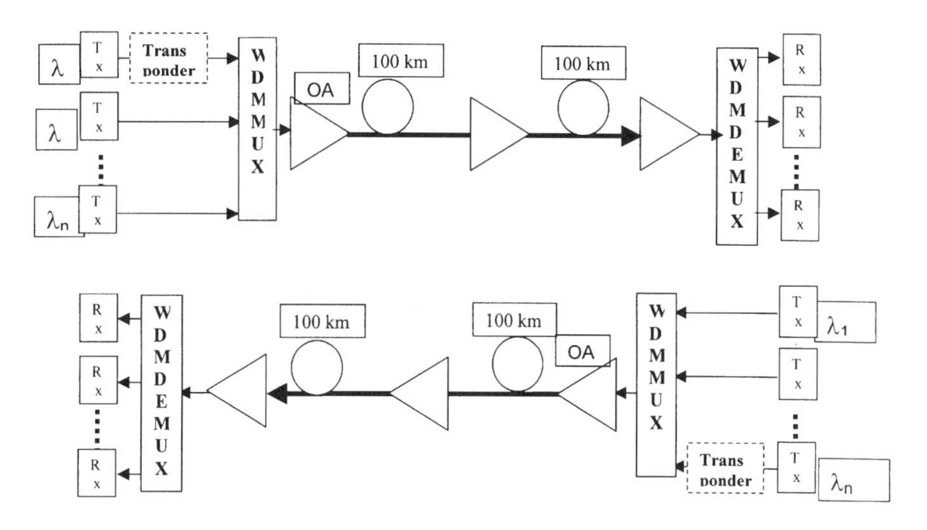

Fig. 12.1. DWDM transmission system configuration.

It took about 20 years for the data rate of TDM fiber-optic systems to increase from 44 Mb/s in 1976 to 10 Gb/s in 1996. While the next TDM hurdle, a 40–Gb/s commercial system, is expected to be surmounted by 2001, DWDM systems have evolved from less than 10 channels in 1995 to more than 100 channels in 2000, delivering >1 Tb/s capacity. Having secured its success in transmission capacity boosting, the next challenge for WDM is to help improve communication network architecture so that the unprecedented bandwidth of optical fiber can be accessed easily by data traffic.

By the end of the 20th century, there were two logical communication network infrastructures: a circuit-switched network that carries voice traffic, and a packet-switched network that carries data traffic. Physically, both circuit-switched and packet-switched networks run on the same telecommunication networks. The core network infrastructure is optimized for the circuit-switched voice traffic because voice traffic came first and there was much more voice traffic at the beginning. Voice transmission started from a single voice circuit per copper wire pair on open-wire pairs in 1910s. Voice signals were digitized, multiplexed, and transmitted from one central office to another through the use of channel banks based on TDM technology. T1, which stands for Trunk Level 1, is a digital transmission link that has a total signaling speed of 1.544 Mb/s consisting of 24 voice channels at 64 kb/s per channel plus multiplexing overhead. It is the standard for digital transmission in North America. T1 service has a variety of applications. These include the trunking of voice and data services from the customer's premises to the central office or long-distance point of presence (POP), Internet access, access to frame relay and ATM public data service networks, and transmission among office sites within or between local regions. Another technology that relies heavily on TDM is synchronous optical network (SONET), which defines a synchronous frame structure for synchronous transmission of multiplexed digital traffic over optical fiber. It is a set of standards defining the rates and formats for optical networks specified in ANSI T1.105, ANSI T1.106, and ANSI T1.117. A similar standard, synchronous digital hierarchy (SDH), has also been established in Europe by the International Telecommunication Union Telecommunication Standardization Sector (ITU-T). SONET equipment is generally used in North America and SDH equipment is generally used everywhere else in the world. Both SONET and SDH are based on a structure that has a basic frame and speed. The frame format used by SONET is the synchronous transport signal (STS), with STS-1 being the base level signal at 51.84 Mbps. A STS-1 frame can be carried in an OC-1 (optical carrier-1) signal. The frame format used by SDH is the synchronous transport module (STM), with STM-1 being the base level signal at 155.52Mbps. A STM-1 frame can be carried in an OC-3 signal. Both SONET and SDH have a hierarchy of signaling speeds. Multiple lower-level signals can be multiplexed together to form higher-level signals. For example, three STS-1 signals can be multiplexed together to form a STS-3 signal, and four STM-1 signals multiplexed together will form a STM-4 signal. SONET and SDH are technically comparable standards.

While both T1 and SONET/SDH still enjoy huge success in telecommunication networks decades after their introduction, they were designed to handle voice traffic, not data traffic. During the 1980s and 1990s, voice traffic grew at a linear rate while data traffic increased exponentially, mostly due to advances in computer technology, data communication, and the Internet. By 2000, telecommunication networks were carrying five times as much data traffic as voice traffic. However, the voice-centric core network architecture model has four layers for data traffic: IP (Internet Protocol) and other content-bearing traffic, ATM (asynchronous transfer mode) for traffic engineering, a SONET/SDH transport network, and DWDM for fiber capacity. DWDM has made available the huge bandwidth of optical fiber for transmission. But the circuit-switched core network infrastructure makes this bandwidth inaccessible for the huge data traffic at the IP layer, since there are two much slower layers of ATM and SONET/SDH in between. The fastest interface of commercial ATM switches only reaches 155 Mb/s (OC-3), much slower than current SONET gear with a top rate of 9.95328 Gb/s (OC-192). And the fastest bandwidth manager SONET ADM tops out at OC-192, much slower than the top DWDM trunk of 1 Tb/s. The existing four-layer approach has functional overlap among its layers, contains outdated functionality, and is too slow to scale, which makes it ineffective as the architecture for optical data networks.

Bandwidth demand for data traffic and the huge bandwidth capacity of DWDM, combined with advances in optical switch and optical filter technologies, brought forth huge interest in the commercial deployment of optical networking with optical wavelength by the end of the 20th century [8, 9]. As carriers deploy more and more DWDM in the networks, there is an increase demand for managing capacity at optical wavelength, forging a new optical layer. If this optical layer can perform many traditional network functions, such as add/drop multiplexing (ADM), cross-connect, signal restoration, and service provision in the optical domain, it will be much faster than SONET/SDH and ATM, since the latter two are all limited by electronics processing speed. Not only will the optical layer eliminate $>70\%$ of O/E/O conversion in today's network, but also it will enable a direct connection between the IP router and DWDM terminals, thus eliminating the bandwidth bottleneck SONET and ATM layers, and in the process making the huge bandwidth provided by DWDM readily available for data traffic.

An optical network, shown in Fig. 12.2(a), consists of a set of interconnected network nodes, shown in Fig. 12.2(b). Each network node has an optical amplifier for signal reboosting, optical cross-connect for wavelength routing, and DWDM terminals for transport to a local terabit router. In the next section we discuss details about the major elements in optical networks.

12.2. OPTICAL NETWORK ELEMENTS

12.2.1. OPTICAL FIBERS

Optical fibers are the transmission media for optical networks. They are basically tiny strands of glass consisting of core, cladding, and outside protection coatings. Proper material selections and the cylindrical waveguide design of the core and cladding will ensure that light energy is mostly confined inside the core while propagation occurs along the optical fiber. While there are many types of optical fibers being used in today's communication networks, they are all characterized by loss, dispersion, and nonlinearities. Loss in optical fiber attenuates optical signals, thus limiting the span distance. Current commercial transmission fiber has a loss specification of about 0.25 dB/km. Dispersion is defined as the group velocity delay difference among traveling optical signals. It is cataloged as chromatic dispersion, which includes dispersions caused by optical material and waveguides modal dispersion, which is dispersion caused by different propagation modes; and polarization mode dispersion (PMD), which is dispersion caused by different polarizations in fiber. Nonzero dispersion in optical fibers results in the widening and/or distortion of propagating optical pulses, causing receiving errors. Dispersion limit is bit rate dependent. Nonlinearities in optical fibers include four wave mixing (FWM), self-phase modulation (SPM), cross-phase modulation (XPM),

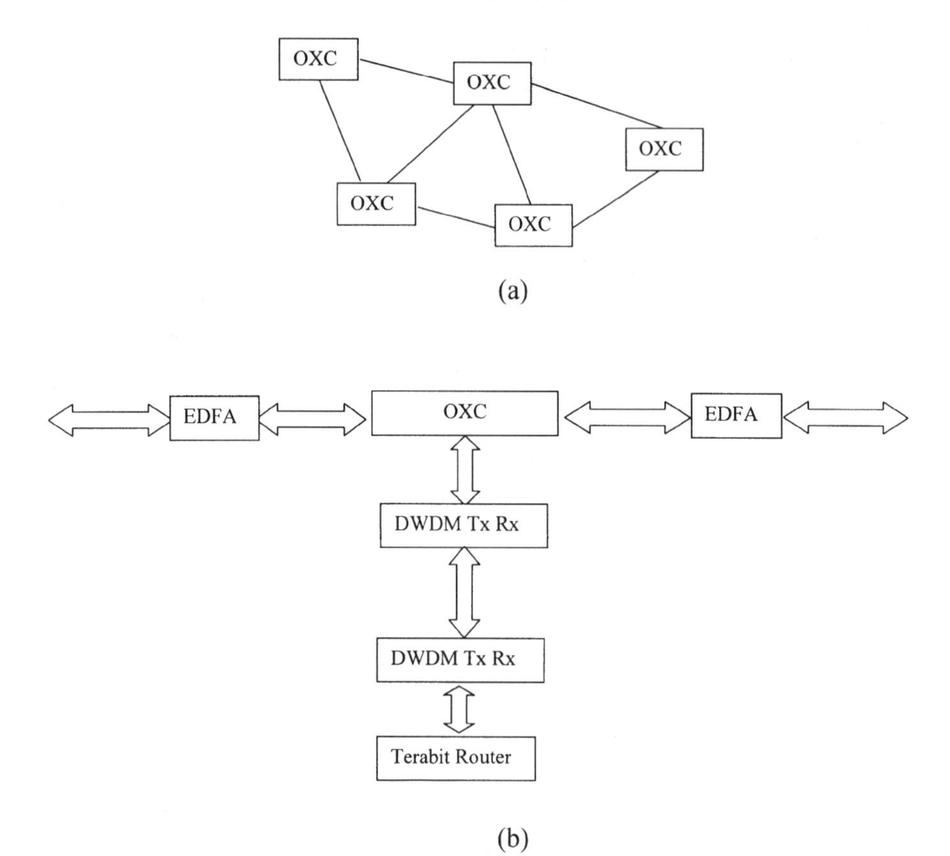

Fig. 12.2. Optical network physical architecture. (a) Network layout, (b) Node configuration.

stimulated Brillouin scattering (SBS), and stimulated Raman scattering (SRS). These cause optical signal distortion and/or noise, and need to be considered in system design.

Two types of optical fibers are commonly used in optical networks. The first is single mode fiber, which is a fiber with very small core, about 10 μm in diameter. As the name implies, single mode fiber only supports the transmission of one or two lowest-order–bound modes at the wavelength of interest. It is used for long-distance and high–bit-rate applications. Several types of single mode fiber are popularly used, including nondispersion-shifted fiber, also called standard single mode fiber (SSMF), which supports wavelengths in both the 1.3 μm region with no dispersion and in the 1.55 μm region with chromatic dispersion of about 17 ps/nm-km; dispersion-shifted fiber (DSF), which supports wavelengths in the 1.55 μm region with no dispersion; and nonzero dispersion–shifted fiber (NZ-DSF), such as LEAF (large effective area fiber) fiber made by Corning and TrueWave fiber by Lucent, with about ± 2 ps/nm-km dispersion in the 1.55 μm region. NZ-DSF fibers are designed mainly for high-speed long-distance transmission. Its nonzero local dispersion helps to combat the fiber nonlinearities induced by the high signal power required for long-haul transmission. The third generation of LEAF also aims at bringing down PMD, which is quite crucial for 40 Gb/s systems.

The second type of fiber is multimode fiber, which has a large core about 50 μm in diameter and thus can support the propagation of more than one propagation mode at the wavelength of interest. In addition to chromatic dispersion, this type of fiber leads to modal dispersion, which significantly limits the bandwidth–distance product in system design. It is mainly used for low speed and intraoffice applications.

12.2.2. OPTICAL AMPLIFIERS

The advent of semiconductor-laser–pumped, erbium-doped fiber amplifiers (EDFAs) accelerated greatly the pace of deployment of high-capacity lightwave systems, and, combined with DWDM technology, laid the foundation for optical networking.

Figure 12.3 is a simplified schematic of an EDFA showing the erbium-doped fiber pumped by a semiconductor pump laser using a pump/signal multiplexer [10]. In practice, additional components are usually added to EDFAs, including optical isolators to prevent the amplifier from oscillating due to spurious reflections, additional tap couplers to enable amplifier performance monitoring, and optical filters to compensate the gain spectral profile for DWDM applications. The critical component of the EDFA is the section of optical fiber

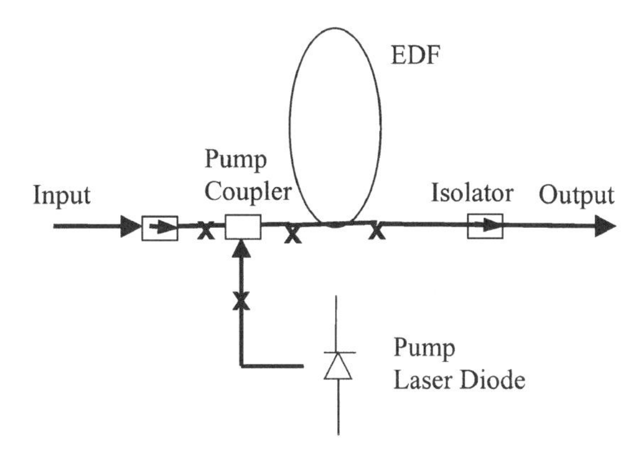

Fig. 12.3. Single-stage Erbium-doped fiber amplifier configuration.

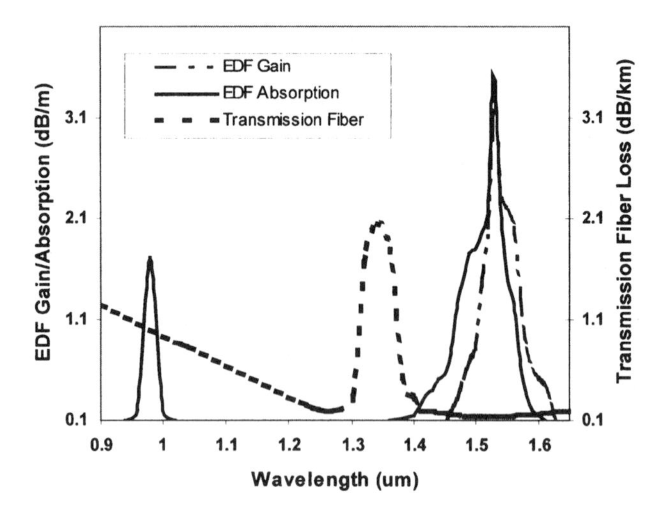

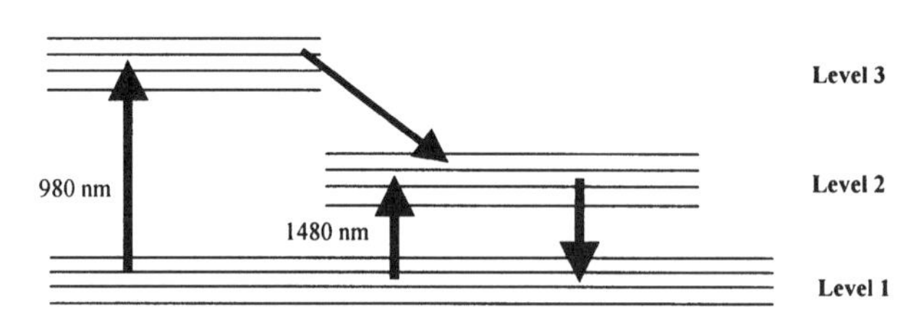

Fig. 12.4. Loss of transmission fiber, absorption/gain of Erbium-doped fiber. The bottom shows energy levels of Erbium-doped fiber.

that is slightly doped with erbium ions which, when properly pumped optically, becomes an amplifying medium in the 1520–1620 nm wavelength window. Erbium-doped fiber is fabricated by selective doping during fabrication to confine the <0.1% erbium to the single mode fiber core. Figure 12.4 compares the optical behavior of erbium to conventional optical fiber used for transmission [5]. The absorption in conventional fiber is very low, about 0.2 dB/km for commercial single mode fiber at the 1550 nm region. By contrast, an erbium concentration of 100 parts per million in the core causes absorption of 2 dB/m at 1530 nm. Absorption of pump light excites the erbium ions, which store the energy until, ideally, a signal photon stimulates its conversion into another signal photon. Figure 12.4 shows that the erbium fiber can be pumped at several wavelengths; the absorption bands at 980 and 1480 nm are the most efficient. As Fig. 12.4 shows, erbium ions are pumped up to an upper energy level by the absorption of pump light at 1480 nm or 980 nm. The transition to the ground state emits a photon and may be either spontaneous (the natural decay of the excited ion in the absence of any interactions) or stimulated (in the presence of photons possessing the transition energy, stimulated emission produces additional photons identical to the stimulating photons at a rate proportional to their flux). Signal photons in the EDFA stimulate depopulation of the excited state (the metastable level), which amplifies the signal. The long lifetime of the excited state, approximately 10 milliseconds, assures that, instead of emitting noise by spontaneous emission, most erbium ions will wait to amplify signals by stimulated emission.

Theoretical models based on the three-level-lasing system of Fig. 12.4 are quite successful in explaining the observed properties of EDFAs [10, 11]. The three levels are ground level, metastable level, and pump level. The rate equations describing the effects of the pump (P_p), signal (P_p), and ASE (P_p) power on the population densities are [11]

$$\begin{aligned}\frac{dN_1(z,t)}{dt} &= -\left[\frac{\sigma_{sa}\Gamma_s}{hv_sA}(P_s + P_a^+ + P_a^-) + \frac{\sigma_{pa}\Gamma_p}{hv_pA}(P_a^+ + P_a^-)\right]N_1 \\ &+ \left[\frac{\sigma_{se}\Gamma_s}{hv_sA}(P_s + P_a^+ + P_a^-) + A_{21}\right]N_2 + \frac{\sigma_{pe2}\Gamma_p}{hv_sA}(P_a^+ + P_a^-)N_2 \\ &+ \frac{\sigma_{pe}\Gamma_p}{hv_pA}(P_a^+ + P_a^-)N_3 \end{aligned} \tag{12.1}$$

$$\begin{aligned}\frac{dN_2(z,t)}{dt} &= \frac{\sigma_{sa}\Gamma_s}{hv_sA}(P_s + P_a^+ + P_a^-)N_1 - \left[\frac{\sigma_{se}\Gamma_s}{hv_sA}(P_s + P_a^+ + P_a^-) + A_{21}\right]N_2 \\ &- \frac{\sigma_{pe2}\Gamma_p}{hv_pA}(P_a^+ + P_a^-)N_2 + A_{32}N_3. \end{aligned} \tag{12.2}$$

By conservation, $N_3 = N_t - N_1 - N_2$, where N_1, N_2, and N_3 are the population densities of the ground level, metastable level, and pump level, respectively, and N_t is the total erbium ion density. The superscript + designates pump and ASE copropagating with the signal, and − designates when they counter-propagate to the signal. The absorption a and emission e cross sections of the pump p and signal s are $\sigma_{s,p;a,e,e2}$. (For example, σ_{pe2} represents the cross section for pump emission from the metastable level.) With pumping into the metastable level ($\lambda_p = 1450$–1500 nm), the amplifier behaves as a two-level system and $\sigma_{pe2} = \sigma_{pe}$. Pumping into other absorption bands (e.g., $\lambda_p = 980$ nm) have $\sigma_{pe2} = 0$. Other parameters are the fiber core area A, the signal-to-

Table 12.1

Summary of Material Parameters Applicable to Optical Amplifier

Parameter	Symbol	Typical value	Unit
Pump emission cross section	σ_{pe}	0.42×10^{-21}	cm^2
Pump absorption cross section	σ_{pa}	1.86×10^{-21}	cm^2
Signal emission cross section	σ_{se}	5.03×10^{-21}	cm^2
Signal absorption cross section	σ_{sa}	2.85×10^{-21}	cm^2
Amplifier homogeneous bandwith	Δv	3100	GHz
Radiative transition rate	A_{21}	100	s^{-1}
Nonradiative transition rate	A_{32}	10^9	s^{-1}
Fiber core area	A	12.6×10^{-8}	cm^2
Signal to core	Γ_s	0.4	
Pump to core	Γ_p	0.4	

core overlap Γ_s, and the pump-to-core overlap Γ_p. No other effects of the radial distribution of ions or the optical mode are included here, since the erbium ions are confined to the region of the optical mode's peak intensity and $\Gamma_{s,p}$ are small. The nonradiative transition rate from level 3 to 2 is A_{32} and the radiative transition rate from level 2 to level 1 is A_{21}. Table 12.1 summarizes the material parameters and typical fiber parameters applicable to fiber amplifiers.

The convective equations describing the spatial development of the pump, signal, and ASE in the fiber are

$$\frac{dP_p^{\pm}(z,t)}{dz} = \mp P_p^{\pm}\Gamma_p(\sigma_{pa}N_1 - \sigma_{pe2}N_2 - \sigma_{pe}N_3) \mp \alpha_p P_p^{\pm} \tag{12.3}$$

$$\frac{dP_s(z,t)}{dz} = P_s\Gamma_s(\sigma_{se}N_2 - \sigma_{sa}N_1) - \alpha_s P_s \tag{12.4}$$

$$\frac{dP_p^{\pm}(z,t)}{dz} = \pm P_a^{\pm}\Gamma_s(\sigma_{se}N_2 - \sigma_{sa}N_1) \pm 2\sigma_{se}N_2\Gamma_s h v_s \Delta v \mp \alpha_s P_a^{\pm}. \tag{12.5}$$

The second term in Eq. (12.5) is ASE power produced in the amplifier per unit length within the amplifier homogeneous bandwidth Δv for both polarization states. The loss term $\alpha_{s,p}$ represents internal loss of the amplifier.

Gain, output power, and noise figure are the most important characteristics of EDFAs for use in optical communication systems [5–12]. Gain is defined as the ratio of output power to input power. Saturation occurs when large signal power in the EDFA decreases gain, limiting the signal output power from the amplifier. This gain saturation results when the signal power grows large and causes stimulated emission at such a high rate that the inversion is decreased; that is, the number of excited erbium ions decreases substantially. The output signal power is limited only by available pump power. Heavily saturated amplifiers can convert pump photons to signal photons with efficiencies exceeding 90%.

An optical amplifier always degrades the signal by adding noise from amplified spontaneous noise (ASE), which arises from amplified light produced by spontaneous emission from erbium ions [13]. Noise figure is defined as the input signal-to-noise ratio (SNR) divided by the output signal-to-noise ratio, when the input signal is shot noise limited. Shot noise represents the deviation from ideal current, which is caused by random generation of discrete charge carriers. In semiconductor photodiodes, shot noise arises from random generation and recombination of the free electrons and holes. Shot noise–limited optical signal in essence means pure optical sine wave as any photodiode can tell. As in any communication system, noise is measured within certain bandwidths. So in practical terms, noise figure is defined as the ratio of input and output SNR's of ideal optoelectronic receivers in an infinitesimally optical bandwidth. As the optical bandwidth tends to zero, the contribution from spontaneous-spontaneous beat noise diminishes, leaving just the signal-spontaneous and signal shot noises. The noise figure then is obtained from measurements of the EDFA gain, ASE, and optical bandwidth of the optical spectrum analyzer:

$$\text{Noise figure} = \frac{N_{\text{EDFA}}}{h\upsilon G B_w} + \frac{1}{G} = \frac{N_{\text{out}} - N_{\text{in}} G}{h\upsilon G B_w} + \frac{1}{G} \tag{12.6}$$

where N_{EDFA} is the amplified spontaneous emission power generated by EDFA, N_{out} is the total output noise power from EDFA, N_{in} is the input noise power to EDFA, G is amplifier gain, B_w is the measurement bandwidth in Hz, υ is the signal frequency in Hz, and h is the Planck constant. The first term results from signal-spontaneous beat noise, and the 1/G term results from the amplified signal–shot noise contribution. The best noise figure occurs when the amplifier is operated in a regime where the active fiber (EDF) is completely inverted; i.e., all erbium ions are pumped to meta-state, corresponding to the quantum limit of 3 dB, for the 980 nm pump, or 4.1 dB for the 1480 pump. For optical amplifiers with gain in excess of 20 dB (100 times in linear scale), the shot noise contribution 1/G term can be neglected, and the noise figure becomes proportional to the ratio of EDFA ASE power to the gain. Therefore, for high-gain amplifiers, noise figure becomes a useful figure of merit for describing the quantity of ASE power generated in an amplifier.

Gain fluctuations in an optical amplifier will cause distortions to the optical signal it is boosting. EDFA is relatively immune to gain fluctuations, fortunately, because the spontaneous decay of the erbium-excited state has a long lifetime, and gain saturation is very slow. Typically, it takes 0.1 to 1 ms for gain compression to occur after a saturation signal is launched into the EDFA. These times are very long compared to the pulse period of the 6 μs (155 Mb/s, OC-3) or shorter in DWDM digital systems. EDFAs add little signal distortion or cross talk for point-to-point transmission systems. However, as DWDM networking becomes more and more attractive, the gain dynamics of EDFA invoke interest. In DWDM networks, optical channels are added and/or dropped along the system. When this happens, the number of channels into EDFA changes. Since EDFAs used in high–channel count systems are operated under saturation conditions already, any change of input channels will change the saturation condition, thus changing the gain. The most serious difficulty is the relatively big over- or undershooting spike in gain transient responses. These spikes may result in optical saturation or loss of signal (LOS) at the receiver. EDFA transient response depends on the level of EDF saturation, the percentage of input power change, and number of EDFAs in the chain. Research is still underway to overcome this problem in network design. Most solutions rely on fast electronics to suppress or alleviate at least the transient.

12.2.3. WAVELENGTH DIVISION MULTIPLEXER/DEMULTIPLEXER

Dense wavelength division multiplexing (DWDM) is a fiber-optic transmission technology which combines multiple optical channels at different light wavelengths and then transmits them over the same optical fiber. Using DWDM, theoretically hundreds and even thousands of separate wavelengths can be multiplexed into a single optical fiber. In a 100-channel DWDM system with each channel (wavelength) carrying 10 gigabits per second (billion bits per second), up to 1 terabits (trillion bits) can be delivered in a second by the optical fiber. This kind of capacity can impressively transmit 20 million simultaneous two-way phone calls or transmit the text from 300 years' worth of daily newspapers per second. Nevertheless, terabits can be easily consumed by one million families watching video on Web sites at the same time.

The implementation of a DWDM system requires some critical optical components. The DWDM multiplexer and demultiplexer are among the most important ones. Figure 12.5 demonstrates a point-to-point DWDM optical transmission system where the multiplexer combines N channels into a single optical fiber; the demultiplexer, on the other hand, separates the N channel light into individual channels detected by different receivers. Exponentially growing network traffic demands that many more channels be combined into limited usable bandwidth of the same single fiber. As a result, the channel spacing of DWDM systems becomes narrower and narrower so that more and

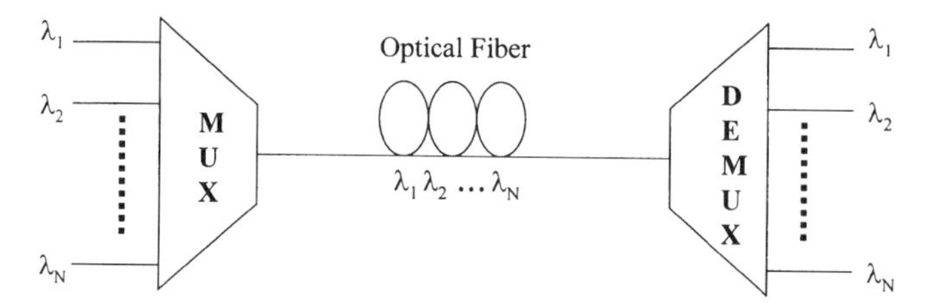

Fig. 12.5. A simple DWDM optical transmission system.

more channels can be multiplexed into a single fiber. This puts DWDM multiplexer/demultiplexer technology at the forefront from the system point of view since ultimately its performance defines how many channels can be utilized in one fiber.

12.2.3.1. *Key Parameters of DWDM Multiplexer/Demultiplexer*

A multiplexer device requires a wavelength-selective mechanism that can be implemented using common technologies, including the interference filter, bulk grating, array waveguide grating (AWG), and Mach–Zehnder (MZ) interferometer. Before comparing different enabling technologies, we first introduce several key optical parameters of the multiplexer, as illustrated in Fig. 12.6.

1. Center frequency: The International Telecommunications Union (ITU) has proposed an allowed channel frequency grid based on 100-GHz

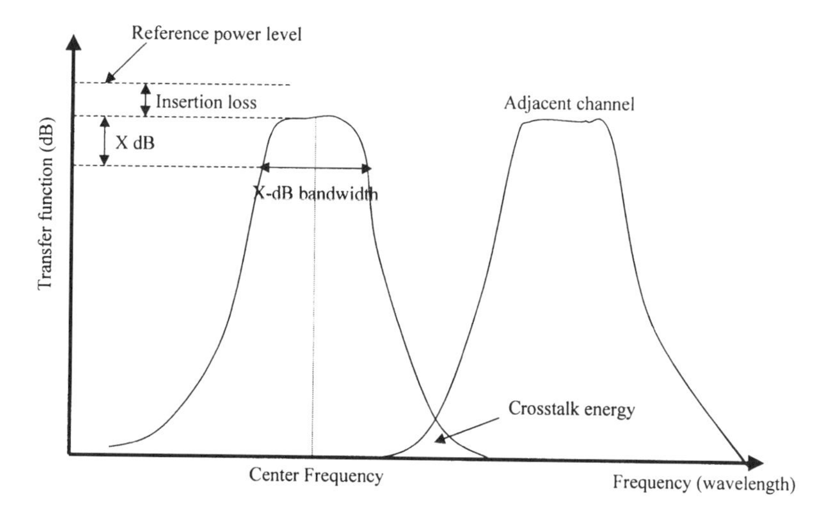

Fig. 12.6. Definition of key parameters of the DWDM multiplexer.

spacing from a reference frequency of 193.1 THz (ITU G.692 Draft). Channel center frequencies are selected from the "ITU Grid."

2. Insertion loss: The insertion loss is the input-to-output power loss of the multiplexer. Lower component insertion loss is better for the system.
3. X-dB Passband (where X can be any number): Passband is the spectrum region around the center wavelength of the multiplexer. For example, 1-dB passband is the bandwidth at insertion loss 1 dB down from the top of the multiplexer spectrum transfer function. Wider and flatter passband makes the system more tolerable to wavelength drift of the laser transmitter.
4. Cross talk: In a DWDM system, there is usually energy leaking from adjacent and nonadjacent channels. This energy leakage is defined as cross talk. If the transfer function of the multiplexer has a sharp cutoff, it can better suppress unwanted cross talk and result in high isolation.

In addition to the above, a high-performance DWDM multiplexer should have small polarization-dependent loss (PDL), low polarization mode dispersion (PMD), and the same performance under hostile environment temperature.

12.2.3.2. *Dielectric Thin-Film Filter*

The dielectric thin-film filter is one of the most popular technologies used to make the DWDM multiplexer. A single-cavity thin-film filter consists of two dielectric stacks separated by a cavity. It behaves as a bandpass filter passing through a particular wavelength (or wavelength band) and rejecting all the other wavelengths. For DWDM applications a multicavity thin-film filter is used since it results in a filter transfer function with a flatter and wider passband as well as a sharper cutoff. The characteristics of single-cavity and multicavity filters are compared in Fig. 12.7.

One configuration of the eight-channel DWDM demultiplexer utilizing the dielectric thin-film filter is depicted in Fig. 12.8. A wide bandpass filter first separates the incoming eight wavelengths into two groups. For each group, three cascaded narrow bandpass filters separate multiwavelength light into four individual wavelengths.

The dielectric thin-film filter has been the dominant DWDM multiplexer/demultiplexer technology because of its high isolation, low PDL and PMD, and acceptable insertion loss. Its performance is extremely stable with regard to temperature variations. However, it is very expensive to manufacture interference filters with less than 100-GHz bandwidth due to extremely low manufacturing yield.

Numerous references, including [14, 15, 16], address the principles and design of dielectric thin-film filters.

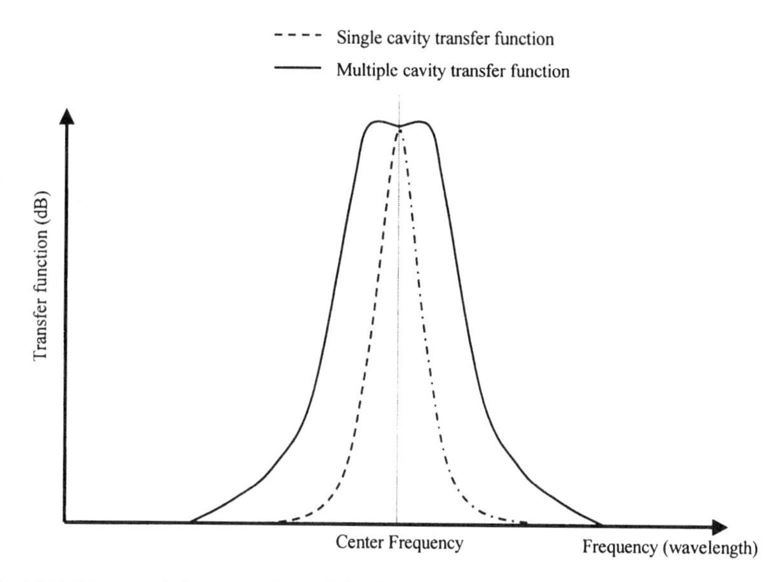

Fig. 12.7. Characteristic comparison of single- and multiple-cavity dielectric thin-film filters.

12.2.3.3. *Diffraction Grating*

As shown in Fig. 12.9(a), a reflective diffraction grating is a mirror with periodically corrugated lines on the surface. The diffraction grating functions to reflect (or transmit in the case of transmission grating) light at an angle proportional to the wavelength of incident light. The general equation for a diffraction grating is

$$d[\sin(\theta_i) - \sin(\theta_d)] = m\lambda, \tag{12.7}$$

where d is the period of the lines on grating surface, λ is the wavelength of the light, θ_i is the incident angle of the light, θ_d is the angle of diffracted light, and m is an integer called the order of the grating. In Fig. 12.9(b), multiwavelength light impinges on the grating at the same angle. Grating separates the wavelengths because each different wavelength leaves the grating at a different angle in order to satisfy Eq. (12.7). This is how diffraction grating works in the DWDM multiplexer/demultiplexer.

The most distinguished property of this technology is that its insertion loss is generally independent of the channel numbers. Hence, it is very attractive for DWDM systems with a large number of channels. However, its performance is very sensitive to polarization effects, and passband flatness is poor, with a round shape rather than a flat top, which requires tighter wavelength control for system operation in the field.

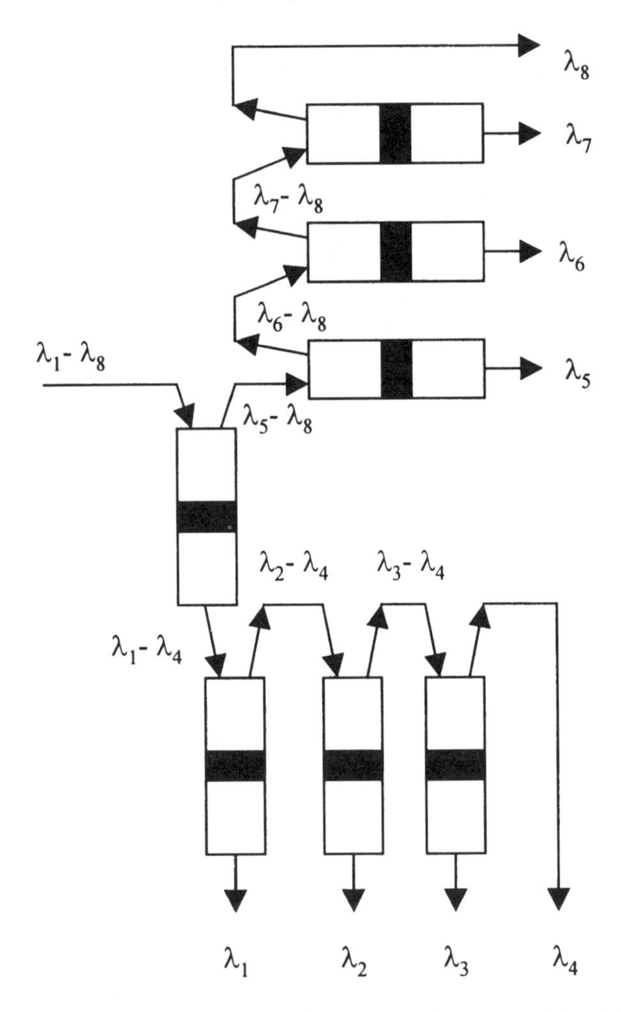

Fig. 12.8. Architecture of an eight-channel DWDM demultiplexer using dielectric thin-film filters.

12.2.3.4. *Array Waveguide Grating*

The array waveguide grating (AWG) is an optical integrated circuits technology fabricated usually on planar silicon substrates. Its operation is illustrated in Fig. 12.10. It consists of two star couplers connected by an array of waveguides that act like a grating. Within the waveguide array there is a constant length difference from one waveguide to the next. Because of this design, the transmission amplitude T_{pq} from the pth input port to the qth

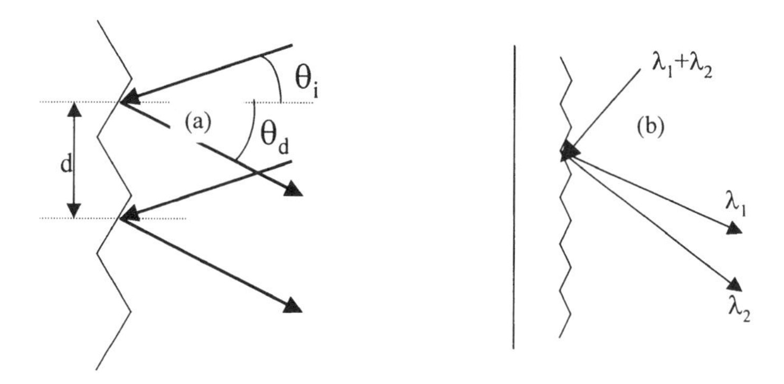

Fig. 12.9. (a) A reflective diffraction grating. (b) Operation principle of the diffraction grating.

output port may be expressed as [17]

$$T_{p,q} \propto \sum_{m=1}^{N} P_m \exp\left[imb\left(\frac{2\pi}{\lambda}\right)(p-q)\right] \tag{12.8}$$

where N is the number of ports of the star coupler, P_m is the power in the mth waveguide, and b is a constant. Clearly T_{pq} is a periodic function when wavelength varies. For use as a demultiplexer there is only one input port at which all wavelengths are present; its operation principles are described as follows. All wavelengths presented at the input ports are coupled into the waveguide grating evenly through the input coupler. The waveguides of the grating are strongly coupled to each other at the output coupler. The differential length difference in the waveguide grating results in a phase difference of the multiwavelength light when it appears in different ports of the output star coupler. The linear length difference and the position of the two

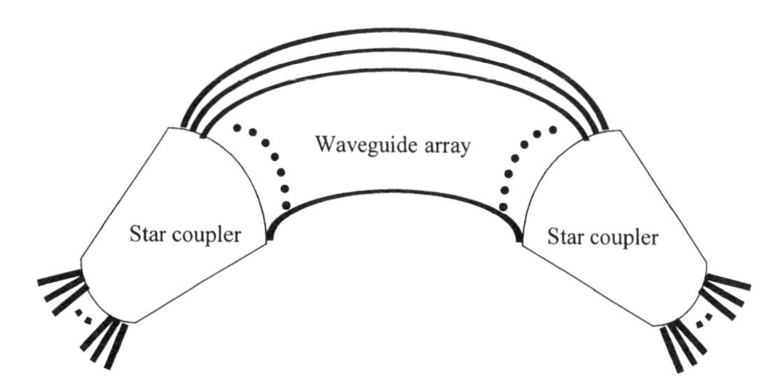

Fig. 12.10. Schematics of an array waveguide grating.

star couplers are designed in such way that at one specific output port only one particular wavelength is constructively interfered. As a result, the multiple wavelengths from the input of the AWG are demultiplexed.

After decades of intensive worldwide research, the AWG–based DWDM device has achieved performance suitable for commercial application. Its main drawback is that its poor temperature coefficient requires an active temperature controller, which adds an additional bundle to network management. Nevertheless, AWG takes advantage of silicon optical bench waveguide technology and can be manufactured in large quantities at one time. It is a cost-effective solution for a single chip handling many channels.

12.2.3.5. *Mach–Zender Interferometer–Based Interleaver*

The Mach–Zender (MZ) interferometer is a basic interference device. It typically consists of two 3 dB couplers interconnected by two optical paths of different lengths, as shown in Fig. 12.11. The first 3 dB coupler evenly splits the input signals into two parts, which experience different phase shift after passing through two different paths. The two lights interfere at the second 3 dB coupler when combined together. Because of the wavelength-dependent phase shift the power transfer function of the device is also wavelength dependent. In matrix form it is simply

$$\begin{bmatrix} T_{01}(\lambda) \\ T_{02}(\lambda) \end{bmatrix} = \begin{bmatrix} \sin^2(\pi \times n \times \Delta L/\lambda) \\ \cos^2(\pi \times n \times \Delta L/\lambda) \end{bmatrix} \tag{12.9}$$

where n is the refraction index of the waveguide material and ΔL is the path difference.

A single input port MZ interferometer with certain value of the path difference ΔL can be used as a 1×2 demultiplexer. Multiwavelength light appears on the input port. When the input wavelength λ_i satisfies the condition $n\Delta L/\lambda_i = m_i/2$ for any positive odd integer m_i, the wavelength λ_i appears on the first output port due to constructive interference. Similarly, the wavelength λ_i, which satisfies the condition $n\Delta L/\lambda_i = m_i/2$ for any positive even integer m_i,

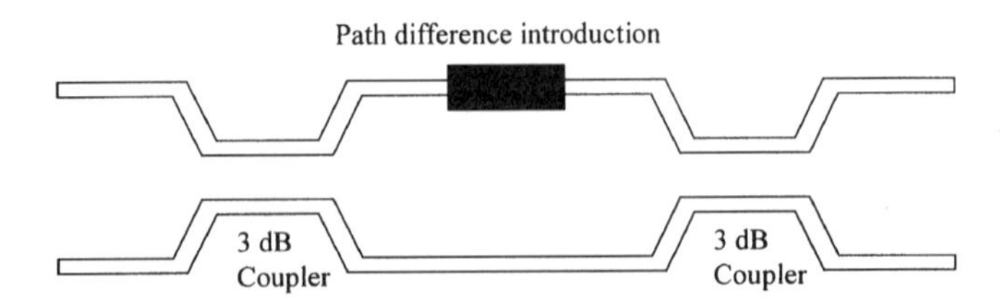

Fig. 12.11. Schematics of an M–Z inteferometer.

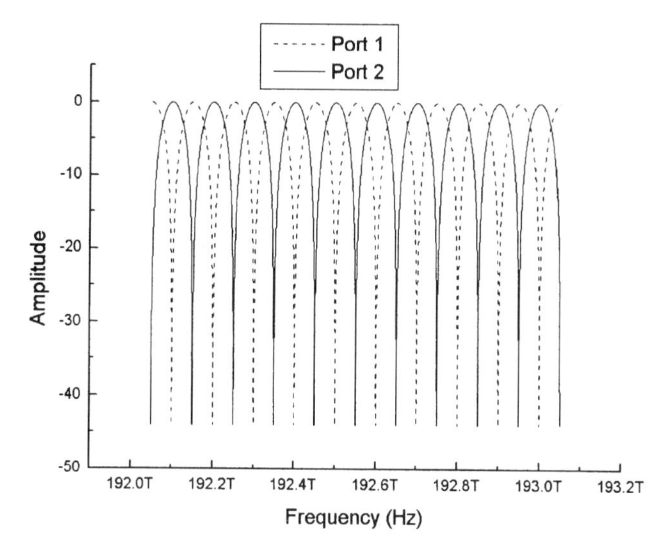

Fig. 12.12. Transfer functions of an M–Z interferometer with non–flat-top shape.

appears on the second output port. The transfer functions of both output ports are plotted in Fig. 12.12. If there are only two input wavelengths, a three-port (one input port and two output ports) MZ interferometer acts as a 1×2 demultiplexer. Theoretically, cascaded $n - 1$ MZ interferometers can be constructed to be a $1 \times n$ demultiplexer. If designed carefully, the MZ interferometer can have very high wavelength resolution. A 0.1-nm channel spacing MZ interferometer–based demultiplexer was demonstrated at SuperCom'99 [18]. The traditional approach to constructing an MZ interferometer usually results in a non–flat top transfer function, as shown in Fig. 12.12. Special designs [19] have been proposed to achieve MZ interferometers with high wavelength resolution and a flat-top transfer function, as shown in Fig. 12.13. Usually, multiple MZ interferometers are concatenated to achieve high isolation among DWDM channels. This design results in a relatively high device insertion loss.

12.2.3.6. *Application Example of DWDM Multiplexing Technologies*

In Table 12.2, the performance of 16-channel demultiplexers based on different technologies is summarized.

From Table 12.2 we see that no single technology is superior in all aspects for all applications. Thin-film filter has been the dominant technology for the past several years in the application of 1.6-nm channel spacing DWDM systems with under 16 channels. AWG is currently a very competitive technol-

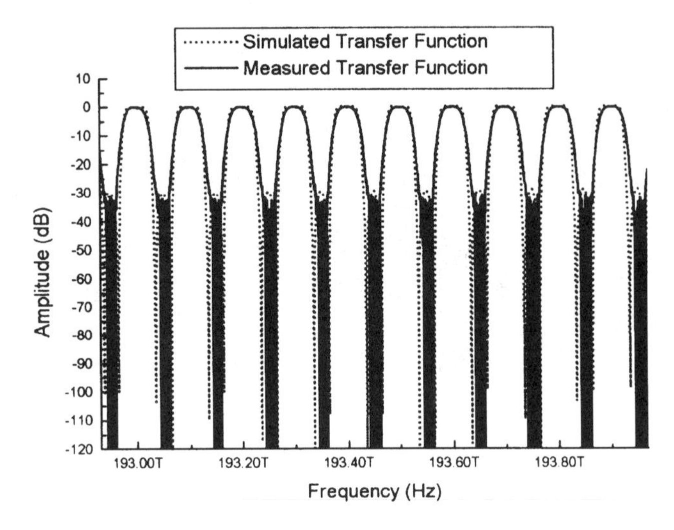

Fig. 12.13. Transfer function of an M–Z interferometer with flat-top shape.

ogy for 0.8 nm channel spacing DWDM systems with more than 16 channels, mainly because of its low cost and improved performance. Despite its high insertion loss, the MZ interferometer is the only commercially available technology now for a DWDM system with 0.4 nm and narrower channel spacing. It is also adaptive to system upgrade.

A design example is presented in Fig. 12.14 to demonstrate how different technologies can be combined together to implement a DWDM demultiplexer with the requirements of 80 channels and 0.4-nm channel spacing. An MZ interferometer–based 1×2 demultiplexer is first used to separate the 80-

Table 12.2

Specification of 16-Channel, 100-GHz Spacing Demultiplexers

Technologies	Thin-film filter	AWG	Bulk grating	Special 1×2 MZI
0.5 dB BW (nm)	0.2	0.2	0.2	0.25
Insertion loss	8 dB	9 dB	6 dB	6 dB
Isolation	>22 dB	>22 dB	>35 dB	>30 dB
PDL	0.1 dB	<0.5 dB	<0.5 dB	0.2 dB
Spacing narrower than 0.4 nm	No	No	No	Yes
Cost	Medium	Low	Low	High

Source: Optical Fiber Conference 1999 and 2000.

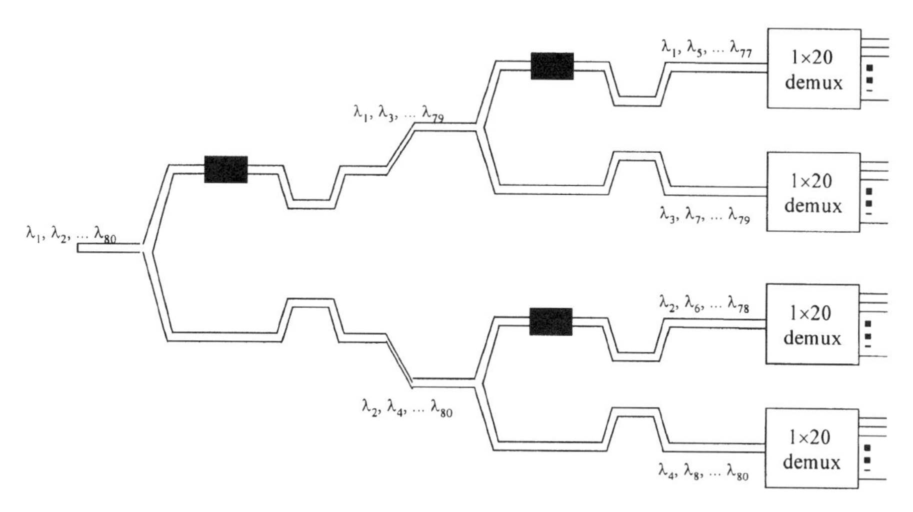

Fig. 12.14. A scalable, upgradable 80-channel demultiplexer combining different technologies.

channel, 0.4-nm channel spacing signal into two streams. Each stream now has 40 channels with 0.8 nm channel spacing. Similarly, the next stage of the MZ interferometer separates signals into four streams and converts the channel spacing to 1.6-nm spacing. Finally, cost-effective AWG or thin-film filter–based 1×20 demultiplexers are used to separate all DWDM channels completely. An additional benefit of using MZ interferometers is that the end users do not have to pay for the entire 80-channel demultiplexer at one time. Instead they may buy and install 1×20 demultiplexer submodules gradually as network traffic grows. This may effectively reduce total system life-cycle cost.

12.2.4. TRANSPONDER

The need for the transponder arises when the DWDM system interfaces with other fiber-optic systems. ITU defined a set of optical frequencies with 100 GHz spacing in the 1530 to 1620 nm optical amplifier bandwidth to be used for DWDM, commonly known as the ITU grid. There are two types of transponders, the transmitter transponder and the receiver transponder. The transmitter transponder takes in non–DWDM–complaint signals, such as a short-reach SONET signal at 1310 nm, and converts them into DWDM–complaint signals at ITU frequency for output. It has a SONET short-reach receiver and a DWDM transmitter, which usually has a DFB LD at ITU frequency, an external modulator, and a wavelength locker. The receiver transponder receives signals off a DWDM link and outputs them at the desired format, such as short-reach SONET at 1310 nm.

A major function of the transmitter transponder is to generate highly precise and stable wavelengths for DWDM transmission. Since a DWDM system transmits many wavelengths carrying a variety of traffic simultaneously down an optical fiber, it is important that these wavelengths do not drift away from their designated value and interfere with traffic carried by other wavelengths. This requires that a transmitter for a DWDM system has very stable wavelength, in addition to a specific wavelength value. At 1550 nm, 100-GHz frequency spacing is about 0.8 nm wavelength spacing. The semiconductor laser diode (LD) wavelength has a temperature shift of about 0.1 nm/°C. An 8° difference in LD temperature will cause the wavelength to move from one ITU grid to the next. In DWDM transmitter design, the LD of choice is the single-frequency multiple–quantum-well (MQW)–distributed feedback (DFB) laser. The DFB is made to the desired ITU frequency/wavelength. A thermal electric cooler (TEC) control loop is applied to keep the LD temperature constant. For a DWDM system with channel spacing $\leqslant$100 GHz, an additional wavelength locker is also used, in combination with a TEC loop, to lock the wavelength to the ITU grid. The most widely used wavelength locker is based on a glass or air-gap Fabry–Perot etalon. Shown in Fig. 12.15, the first-generation wavelength locking operation taps off 1% to 5% of the laser output light, and feeds it both to photodiode A through a highly stable etalon and directly to photodiode B. The etalon acts as a bandpass filter, allowing only ITU grid wavelength to pass through. Deviations in laser wavelength from ITU grid will result in change of optical power detected by photodiode A. The electrical output feedback signal is generated by comparing the signals from the two photodiodes, and is sent to the TEC of the laser to adjust the laser chip temperature, which in turn adjusts the laser wavelength accordingly. The thermal stability ($<$0.04 GHz/°C) of the etalon is a key to this wavelength

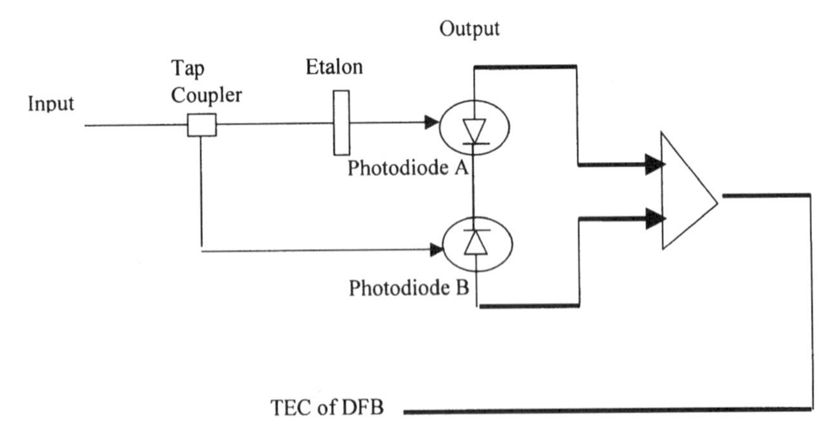

Fig. 12.15. Wavelength locker for DWDM systems.

locking. The second generation of wavelength lockers operate on the same principle, but the etalon and PDs are put inside an LD hermetic package, and in some designs, LD back-facet light is used for locking. The advantages of this technique over the discrete wavelength locker are increased space efficiency and cost reduction.

Most transponders offer the "3R function": regeneration, reshaping, and retiming. Besides wavelength conversion, transponders also process some SONET overhead bytes, such as B1 and J1. Forward error correction (FEC), which helps to combat noise in long-haul transmission, is also done by transponder. Since more and more data traffic tries to take on DWDM express directly, without going through a SONET vehicle, transponders are also designed to interface with signals other than SONET, such as 100 Mb/s Ethernet, gigabit Ethernet, etc. Transponders only offer 2R without retiming when the incoming signals are not SONET.

Two reverse trends are currently taking place in transponder development. One is to add multiplexing functions to an existing transponder. The so-called transmux or muxponder takes in four channels of OC-48 traffic and multiplexes them into one OC-192 DWDM–complaint signal for a DWDM transport system. The other trend is to build a DWDM–complaint transmitter into the optical interfaces of data switches and routers, thus eliminating the need for a transponder altogether.

12.2.5. OPTICAL ADD/DROP MULTIPLEXER

Optical add-drop multiplexer technology substantially reduces the cost of DWDM optical networks. One of the OADM configurations is shown in Fig. 12.16. A multichannel optical signal appears on the input port. A WDM filter is used to drop one of the incoming multiple channels and pass through the rest (the express channels). Another WDM filter is used to add one channel,

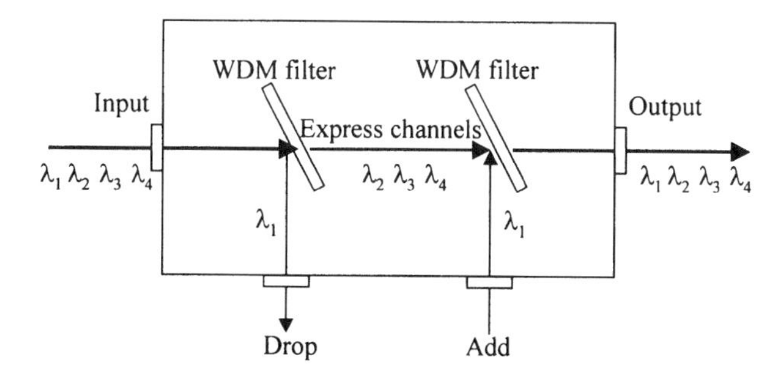

Fig. 12.16. Schematic of a one-channel OADM.

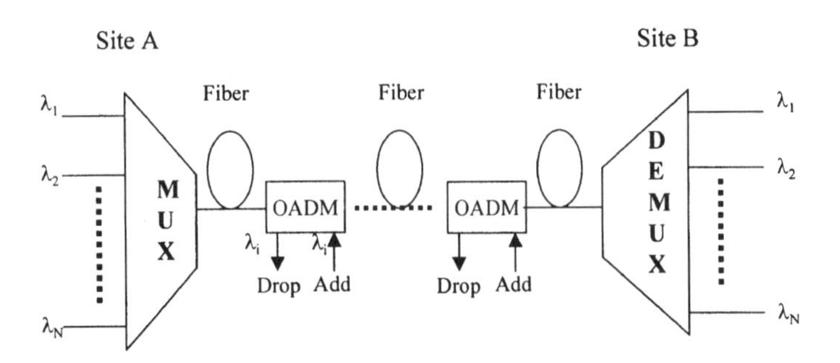

Fig. 12.17. A point-to-point optical link with OADMs.

in a simple form at the same wavelength as the one just dropped, into the express channels. All channels exit at the output port. The application of this technology in a point-to-point network is demonstrated in Fig. 12.17. Multiple channels are transmitted between major locations A and B. In between there are several small network nodes that have some traffic demands. OADMs offer these intermediate nodes access to a portion of the network traffic while maintaining the integrity of the other channels. Without OADM, all channels have to be terminated at the intermediate node even for a small portion of traffic exchange, which results in a much higher cost.

Cascaded OADMs with architecture as shown in Fig. 12.16 can be used to add or drop multiple channels in one intermediate node, but this comes with high overall insertion loss. Since the wavelengths of WDM filters are predetermined, the wavelengths to be added/dropped at each intermediate node have to be carefully preplanned, so this type of OADM is also called fixed OADM (FOADM).

Another type of OADM architecture, as shown in Fig. 12.18, is being aggressively pursued. A pair of DWDM demultiplexer and multiplexers are interconnected by an array of optical switches. The demultiplexer separates

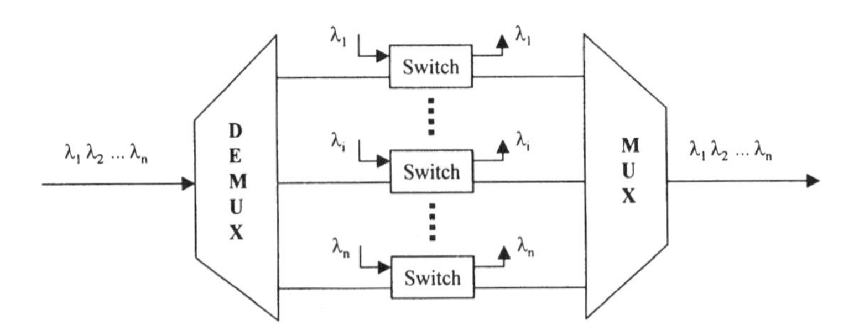

Fig. 12.18. Schematic of a reconfigurable OADM.

incoming wavelengths completely. The optical switch functions in two ways, express or add/drop. For any wavelength λ_i, if the information carried does not need to be exchanged with the local intermediate node the optical switch passes λ_i directly to the multiplexer. In contrast, if information exchange is demanded, the optical switch is controlled to drop λ_i and simultaneously add the same wavelength but with fresh contents. The advantage of this OADM architecture is that it may add/drop any or many of the incoming wavelengths dynamically to accommodate network traffic in real time. It is also called reconfigurable OADM (ROADM).

12.2.6. OPTICAL CROSS-CONNECT

Optical cross-connect (OXC) stands for the optical network element (NE) that provides for incoming optical signals to be switched to any one of the output ports. The difference between OXCs and digital cross-connect DCS systems is the interface rate and switch fabric granularity. DCS interfaces traditionally have been electrical and the matrix granularity that has been less than 50 Mb/s (DS3 or STS-1). OC-3/STM-1 or OC12/STM-4 interfaces are now available for DCS, but matrix granularity remains at less than 155 Mb/s. OXCs, in contrast, interconnect at the optical line rate of DWDM and optical network elements (e.g., OC-48) and have a switch fabric granularity to match (e.g., OC-48). In terms of cost and transparency, OXCs should not contain any O/E/O conversion. In reality, though, the demand for OXCs arose before the advance of optical switch and optical monitoring technology. Commercial OXC products are implemented in mainly three ways: opaque OXC, using O/E/O and electrical switch core; transparent core OXC, using optical switch core and some O/E/O; and totally transparent OXC, using optical switch and no O/E/O. Grooming capability is a key distinction among various OXC products. Grooming is the breaking down of an incoming signal into its constituent parts and the efficient repacking of those parts. Some opaque OXC breaks incoming OC-48 2.5 Gb/s streams into smaller STS-1 52 Mb/s signals. The switch fabrics act at STS-1 level, directing STS-1 streams to the correct outgoing port, and packing STS-1 signals into OC-48 for output. Grooming and switching at lower speed than line rate allows carriers to provide lower speed service to customers directly and automatically. Currently, there are more enterprises interesting in STS-1 service than in OC-48 services. Carriers will have to use expensive digital cross-connect to offer STS-1 service if OXC does not have the necessary granularity.

Three classes of OXCs with optical core have been defined [20, 21] (Telcordia's Optical Cross-Connect Generic Requirements, GR-3009-CORE):

- Fiber switch cross-connect (FXC) switches all of the wavelength channels from one input fiber to an output fiber, in effect acting as an automated fiber patch panel. FXCs are less complex than a wavelength-selective or wavelength-interchanging cross-connect. In parts of the network where protection against fiber cuts is the main concern, FXCs could be a viable solution. They may also make the best use of current proven optical technologies. While FXCs can provide simple provisioning and restoration capabilities, they may not offer the flexibility required to promote new end-to-end wavelength generating services.

- Wavelength selective cross-connect (WSXC) can switch a subset of the wavelength channels from an input fiber to an output fiber. They require demultiplexing of an incoming DWDM signal, and remultiplexing them into its individual constituent wavelengths. This type of cross-connect offers much more flexibility than an FXC, allowing the provisioning of wavelength services, which in turn can support video distribution, distance learning, or a host of other services. A WSXC also offers better flexibility for service restoration; wavelength channels can be protected individually using a mesh, ring, or hybrid protection scheme.
- Wavelength interchanging cross-connect (WIXC), which is a WSXC with the added capacity to translate or change the frequency (or wavelength) of the channel from one frequency to another. This feature reduces the probability of not being able to route a wavelength from an input fiber to an output fiber because of wavelength contention. WIXC offer the greatest flexibility for the restoration and provisioning of services.

Optical switch is the fundamental building block for OXCs with optical core. The key issues for switches are low insertion loss, low cross talk, relatively fast switch time, reliable manufacturing at low cost. Optical switches can be categorized into those based on free-space waves and those based on guided waves. Broadly speaking, free-space optical switches exhibit lower loss than their guided-wave counterparts. Moreover, switches for which the principle of operation relies on the electro-optic effect have faster switching times than their electromechanical counterparts.

The following technologies are being employed for optical switching applications:

- MEMS (microelectromechanical systems). These are arrays of tiny mirrors originally developed for the very large video screens seen at sports events and pop music concerts. It is at the moment the hottest technology for optical switching, attenuation, and wavelength tuning.
- Liquid crystals. Borrowed from laptop-screen technology, electric voltage alters the properties of liquid crystals so that light passing through them is polarized in different ways. Passive optical devices then steer each wavelength of light one way or the other, depending on its polarization.

- "Inkjet". Tiny bubbles act like mirrors, glancing light onto intersecting paths as they traverse microscopic troughs carved in silica. The bubbles are generated using ink-jet printer technology.
- Thermo-optical switches. Light is passed through glass that is heated up or cooled down with electrical coils. The heat alters the refractive index of the glass, bending the light so that it enters one fiber or another.

MEMS is a form of nanotechnology, on the micron scale, that integrates mechanical structures such as mirrors with sensors and electronics [22]. By patterning various layers of polysilicon as they are deposited, one can build structures and etch some part away; the remaining devices are capable of motion. In general, these devices are small, cheap, fast, robust, and can be easily scaled to large numbers and integrated with on-chip electronics using well-established, very large scale integration (VLSI)–complementary metal-oxide–semiconductor (CMOS) foundry processes. MEMS fabrication, though, is much more complicated than integrated circuits (ICs) because of the intricacy of the actuators, sensors, and micromirrors being produced [23]. Electrostatic, magnetic, piezoelectric, and thermal are the principal actuation methods used in MEMS. Piezoresistive, capacitive, and electromagnetic are the different sensing methods used in MEMS. In addition to manufacturing concerns, control software is an important element. Following a period of government-funded basic research in the mid- to late 1980s, commercially produced MEMES have been deployed for the past decade in a number of applications. Early applications included airbag sensors, projection systems, scanners, and microfluidics. Continued technical developments have very recently extended MEMS applications to include optical networking, with devices such as optical switch, variable optical attenuators (VOAs), tunable lasers, tunable filters, dispersion compensation, and EDFA gain equalizers. Commercial MEMS–based all-optical switches basically route photons from an input optical fiber to one of the output fibers by steering light through a collimating lens, reflecting it off a movable mirror, and redirecting the light back into the desired output port. The two basic design approaches are the two-dimensional (2D) digital approach and the three-dimensional (3D) analog approach. In 2D MEMS, micromirrors and fibers are arranged in a planar fashion, and the mirrors can only be in one of the two known positions ("on" or "off") at any given time. An array of mirrors is used to connect N input fibers to N output fibers. This is called N^2 architect, since N^2 number of mirrors are needed. For example, an 8×8 2D switch uses 64 mirrors. A big advantage of the 2D approach is that it requires only simple controls, essentially consisting of very simple transistor to-transistor–logic (TTL) drivers and associated electronic upconverters that provide the required voltage levels at each MEMS micromirror. The 3D analog approach uses the same principle of moving a mirror to redirect light. But in this approach, each mirror has multiple possible positions—at least N positions. It is called $2N$ architect because two arrays of N mirrors are used to connect N input to N output fibers. The 3D approach can scale to thousands of ports with low loss and high uniformity since the distance of light propagation does not increase as the port count grows, whereas in 2D switch, port count increase results in squared increase in light travel distance and increase in the pitch of the micromirrors and the diameter of the light beam, placing tight constraints on collimator performance and mirror alignment tolerance. Such a trade-off can rapidly become unmanageable, leading to very large silicon devices and low yields. Thirty-two ports are currently considered a top-end size for a single-chip solution in 2D MEMS switch. The catch for 3D switch is that a sophisticated analog-driving scheme is needed to ensure that the mirrors are in the correct position at all times. Although MEMS technology can produce $2N$ 3D mirror arrays with impressive stability and repeatability by using a simple open-loop driving scheme, closing the loop with active feedback control is fundamental to achieving the long-term stability required in carrier-class deployment of all optical OXCs. Using a closed-loop control scheme implies that monitoring the beam positions must be implemented in conjunction with computation resources for the active feedback loop and very linear high-voltage drivers.

12.2.7. OPTICAL MONITORING

Optical monitoring provides the base for optical networking. It is crucial not only to offer large capacity for various traffics but also to manage all the traffic streams. Network management (NM) functions include performance monitoring (PM), alarm, provision, and protection/restoration. NM in electrical networks is usually done by monitoring some predefined overhead (OH) bytes or bits in digital signals. For example, SONET/SDH defines path, line, and section overhead bytes. SONET equipment reads these OH bytes for network management. In principle, an optical network should perform the same functions at the optical layer. Unfortunately, optical monitoring technology is still in its infancy stage, way behind optical transmission technology.

The earliest (and still widely used) optical monitoring system uses an asymmetric optical coupler, called a tap coupler, to take a small portion of the traffic-bearing optical signal. After O-to-E conversion, this tapped signal provides the optical power level of the mainstream optical signal to NM, based on which LOS (loss of signal) alarms and protection switching are determined. In some DWDM systems, optical channel or wavelength information is encoded using a low-frequency dither tone to modulate the transmitter laser. This low-frequency content is extracted at the optical monitoring point, so that channel wavelength information is also available, in addition to optical power level.

With increases in the channel count and transmission distance of DWDM systems, demand for optical layer information increases too. Channel monitor has become an attractive product recently because it measures the optical spectrum of the DWDM composite signal, giving valuable information such as individual channel optical signal-to-noise ratio (OSNR) and real-time channel wavelength value. A channel monitor is basically an optical spectrum analyzer (OSA) with a reduced feature set. But unlike OSA as a costly testing instrument, channel monitors are integrated into DWDM system equipment. They must be small in size and cost effective. While OSA is based on tilting a diffraction grating to disperse the spectral content of optical signals, channel monitors can be cataloged into two types: one is based on a photodiode array; the other on a tunable Fabry–Perot (FP) filter. In the first case, the optical signal is dispersed into spatially separated beams per their wavelength/frequency content, by either diffraction grating or fiber Bragg grating. A photodiode (PD) array then detects these single-frequency beams and the full signal spectrum information is obtained when signals from all photodiodes are combined. In the second case, the cavity length of a Fabry–Perot filter, consisting of two reflection facets with an air or glass gap in between, is tuned by a voltage-driven piezoelectric (PZT) actuator. This tunable FP filter acts as a bandpass filter, which allows the DWDM channel to pass through sequentially.

The key issues regarding channel monitors are channel resolution, number of channels, wavelength accuracy, OSNR accuracy, dynamic range, and sensitivity. For channel monitors based on a photodiode array, the number of pixels of PD array posts an intrinsic limit to the number of channels a monitor can resolve. For a channel monitor using 256 PD array, there are only about 3 spectral points to represent each channel when the incoming DWDM signal has 80 channels. To monitor today's DWDM systems with >100 channels, a 521 or bigger PD array is definitely needed. However, the sensitivity of the PD array may suffer as pixel number increases. For channel monitors based on tunable FP filters, PZT as a moving part posts concerns about its long-term reliability.

Channel monitors provide channel power, OSNR, and wavelength information that is valuable for optical layer networking. This information is not adequate for network management to diagnose the health of each transmitted optical channel, though. Chromatic dispersion, polarization mode dispersion (PMD), and nonlinearities and their interplay in transmission fiber all have an impact on signal quality. To this date, bit-error rate (BER) in the electrical domain remains the most trusted measure of the health of optical transmission. There is still quite a long way to go for optical monitoring to overtake BER, thus enabling all-optical networking.

12.3. DESIGN OF OPTICAL NETWORK TRANSPORT

In the previous section we discussed some critical optical network elements such as optical fibers, optical amplifiers, transmitters, and receivers. Now we consider the design of simple optical network transport when these key components are put together. We first will discuss the impacts of fiber dispersion, fiber nonlinearity, and signal-to-noise ratio (SNR); then we will present design examples considering all effects.

12.3.1. OPTICAL FIBER DISPERSION LIMIT

The electric field of a linearly polarized optical signal or pulse can be expressed as

$$\boldsymbol{E}(x, y, z, t) = \boldsymbol{n}[E(x, y, z, t)e^{j[\beta(\omega)z - \omega t]} + c.c.] \tag{12.10}$$

where $\boldsymbol{n}$ is a unit vector, $\beta(\omega)$ is the propagation constant, ω is the angular frequency, and *c.c.* is the acronym of the complex conjugate. $\beta(\omega)$ can be expanded in a Taylor series around the center frequency ω_0 of the light,

$$\beta(\omega) = \sum_{m=0}^{\infty} \frac{1}{m!} \beta_m (\omega - \omega_0)^m, \tag{12.11}$$

where

$$\beta_m = \left. \frac{\partial^m \beta}{\partial \omega^m} \right|_{\omega = \omega_0}.$$

Chromatic dispersion is caused by the dependence of $\beta(\omega)$ on the optical frequency. Optical pulse consists of many frequency elements that travel at different speeds due to the frequency dependent $\beta(\omega)$. As they travel along the optical fiber, the higher-frequency elements of the pulse propagate faster than the lower-frequency ones. Thus, the optical pulse broadens when it reaches its destination. This phenomenon, often referred to as chromatic dispersion, is detrimental to the optical transmission system, as schematically explained by Fig. 12.19. At the origination of the transmission system, binary sequence 101, represented by different power levels of optical pulses, is sent to a trunk of optical fiber. Fiber chromatic dispersion causes the spread of optical pulses. The receiver (not shown in the figure) may see the symbol 0 as 1. As a result, a transmission error occurs.

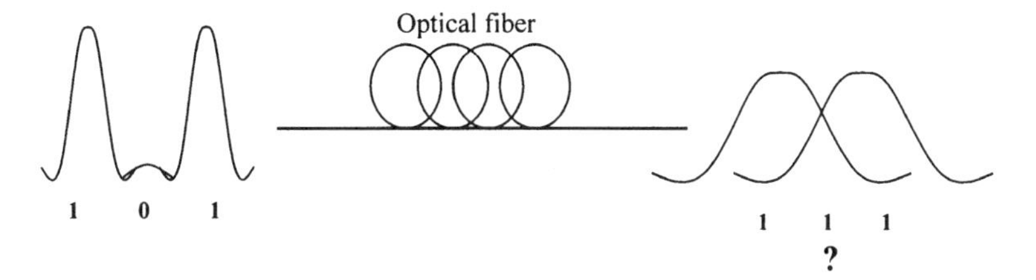

Fig. 12.19. System impairment of chromatic dispersion.

The D parameter, defined as

$$D = \frac{d}{d\lambda}\frac{1}{v_g} = -\frac{2\pi c}{\lambda^2}\beta_2 \tag{12.12}$$

is commonly used to measure chromatic dispersion, where λ is the wavelength and c is the speed of light in vacuum, respectively. For a conventional single mode fiber D is approximately 17 ps/nm-km in the 1.55 μm window. As explained early, chromatic dispersion restricts a maximum optical signal transmission distance. The dispersion-limited transmission distance for a system using external modulated transmitter is approximately $6000/B^2$ kilometers, where B is the bit rate of the signal. It is obvious from the above relationship that the dispersion limit distance drops rapidly as signal speed increases. For example, a linear transmission system with a bit rate of 2.5 Gbits/s can propagate nearly 1000 kilometers without worry about dispersion limitation. However, the limit is only a few kilometer when the bit rate increases to 40 Gbits/s. Hence, in a high–bit rate optical network, dispersion compensation is definitely required.

12.3.2. OPTICAL FIBER NONLINEARITY LIMIT

Nonlinear optical effects are the responses of the dielectric material at the atomic level to the electric fields of an intense light beam. Nonlinear optical effects in fiber generally fall into two categories: inelastic effects such as stimulated Raman scattering (SRS) and stimulated Brillouin scattering (SBS); and elastic effects such as self-phase modulation (SPM), cross-phase modulation (XPM), and four wave mixing (FWM).

12.3.2.1. *Stimulated Light Scattering*

In the case of stimulated light scattering, a photon associated with the incoming optical field, often called a pump, interacts with the dielectric material to create a photon at a lower Stokes frequency and another form of energy called a phonon. The major difference between Raman scattering and Brillouin scattering is that an optical phonon is created in Raman scattering and an acoustic phonon is created in Brillouin scattering.

Although a complete study is very complicated, the initial growth intensity of the Stokes wave I_s in the case of SRS may simply be described by

$$\frac{dI_s}{dz} = -g_R I_p I_s, \tag{12.13}$$

where I_p is the pump intensity and g_R is the Raman gain coefficient ($\approx 1 \times 10^{-13}$ m/W for silica fibers near 1 μm wavelength). A similar relationship holds for SBS with g_R replaced by the Brillouin gain coefficient g_B ($\approx 6 \times 10^{-11}$ m/W for silica fibers near 1 μm). In an optical transmission system the intensity of the scattered light in both SBS and SRS cases grows exponentially at the expense of the transmission signal which acts as a pump. This phenomenon leads to considerable fiber loss. The measure of the power level causing significant SBS and SRS is commonly called threshold power P_{th}, defined as the incident pump power at which half of the power is transferred to the Stokes wave. For SBS, the threshold can be estimated by [24]

$$P_{th} \approx \frac{21 A_{\text{eff}}}{g_B L_{\text{eff}}}, \tag{12.14}$$

where A_{eff} is the effective mode cross section, often referred to as the effective fiber core area; and L_{eff} is the effective interaction length [25]

$$L_{\text{eff}} = \frac{1 - \exp(-\alpha L)}{\alpha}, \tag{12.15}$$

where α is the fiber loss and L is the actual fiber length. For SRS, the threshold power can be estimated by [24]

$$P_{th} \approx \frac{16 A_{\text{eff}}}{g_R L_{\text{eff}}}. \tag{12.16}$$

The threshold power of SBS at conventional single mode fiber can be as low as ~ 1 mW at the most popular 1.55 μm window. Under the same conditions, the threshold power of SRS is more than 500 mW, mainly due to the much smaller Raman gain coefficient.

From the system point of view, it is very important to keep the optical signal of every channel below the SBS and SRS thresholds. The SBS threshold can be increased as the light source line width becomes wider than the Brillouin line width [26]. Another practical approach is to dither the laser source. Slight dither in frequency, for example, tens of MHz, can increase the SBS threshold by more than an order of magnitude. Although the SRS threshold is higher, in DWDM systems it still causes power transfer from the low-wavelength channels to the high ones and therefore leads to SRS cross talk between channels. Large fiber dispersion and power equalization between channels are usually used to alleviate the SRS negative impacts of SRS.

12.3.2.2. *Self-Phase Modulation and Cross-Phase Modulation*

Both self-phase modulation (SPM) and cross-phase modulation (XPM) in optical fiber arise from nonlinear refraction that refers to the intensity dependence of the refractive index. SPM has been well depicted in [35]. The refractive index can be expressed as

$$n(\omega) = n_0(\omega) + n_2 \frac{P}{A_{\text{eff}}}, \tag{12.17}$$

where $n_0(\omega)$ is a linear component and n_2 is the nonlinear refraction index coefficient, which is $\sim 3 \times 10^{-20}\,\text{m}^2/\text{w}$ for silica fibers. This nonlinear refraction results in a nonlinear phase shift to the light traveling in a fiber [27]

$$\Phi_{NL} = \gamma P L_{\text{eff}} \tag{12.18}$$

where γ is the nonlinear coefficient, and

$$\gamma = \frac{2\pi n_2}{\lambda A_{\text{eff}}}, \tag{12.19}$$

where λ is the wavelength of the light. Although the nonlinearity of the silica material itself is very small, the confincd optical power inside the fiber and the long propagation length lead to significant nonlinear phase shift.

SPM is caused by phase shift from the optical field itself. The peak of a pulse induces a higher nonlinear coefficient and travels slower in the fiber than the wings. As a result, in the wavelength domain the leading edge of the pulse acquires shift toward red (longer wavelength) and the trailing edge acquires shift toward blue (shorter wavelength). The signal broadening in the frequency domain is

$$\Delta B = \gamma L_{\text{eff}} \frac{dP}{dt}. \tag{12.20}$$

Practically, SPM should be considered seriously in high-speed systems and may lead to system launch power under a few dBm, depending on fiber types.

The intensity dependence of the refractive index can also cause XPM when two or more channels are transmitted. The nonlinear phase shift for a specific channel arises from the modulation of other channels presented. Similarly, there is spectral broadening

$$\Delta B = 2\gamma L_{\rm eff} \frac{dP}{dt}. \tag{12.21}$$

The factor of 2 indicates that the effect of XPM is twice that of SPM for the same input power. However, in practice XPM does not necessarily impose a strong system penalty if the channel spacing is large enough ($\geqslant$100 GHz) and/or dispersion is well managed.

12.3.2.3. *Four Wave Mixing*

Like SPM and XPM, four wave mixing (FWM) also arises from the intensity dependence of the refractive index of silica. When three optical fields at different wavelengths ω_i, ω_j, and ω_k ($k \neq i, j$) are simultaneously propagating along the fiber, they interact through the third-order electric susceptibility of the silica fiber and generate a fourth optical field

$$\omega_{ijk} = \omega_i \pm \omega_j \pm \omega_k. \tag{12.22}$$

In a WDM system this formula applies to every choice of three channels. Consequently, there are a lot of new frequencies generated. The new frequencies suck the optical power of the original frequencies. Moreover, in the case of equally spaced channels they coincide with the original frequencies, leading to coherent cross talk that degrades system performance severely. The peak power of the new field can be expressed by

$$P_{ijk} \propto P_i P_j P_k \eta, \tag{12.23}$$

where η is called FWM efficiency. Unequal channel spacing has been proven to reduce FWM substantially but it is not a practical approach due to other component and system design considerations. Practically, limited system launch power can reduce FWM generation but may result in other system penalties. At fixed input power, large channel spacing ($\geqslant$100 GHz) and/or high fiber dispersion can alleviate FWM efficiency.

12.3.3. SYSTEM DESIGN EXAMPLES

The design of a practical optical transport system is complicated. Many factors must be considered, such as the detailed channel (wavelength) plan for DWDM systems, appropriate transmitter and receiver pairs, optical fiber types, optical amplifiers, dispersion compensation modules, multiplexers/demultiplexers, and many others. This section briefly discusses system design guidelines. We start from the design of a linear system, considering only the so-called power budget. Then we include the fiber dispersion limit, fiber nonlinearity limit, and noise into system design consideration based on the optical signal-to-noise ratio (OSNR).

12.3.3.1. *Power Budget*

A single-wavelength optical transmission link is shown in Fig. 12.20. The optical transport consists of transmitter, receiver, optical fiber, and connectors among all elements.

The receiver has a minimum optical power threshold under which the system performs poorly. Usually this minimum power threshold is called receiver sensitivity $P_{\min}$. In designing a system, power budget ensures that the power level reaching the receiver exceeds the sensitivity to maintain good system performance. The type of transmitter (wavelength and material) and its launch power P_L are usually specified depending on the desired length of the fiber transmission link. The fiber type and its insertion loss are determined by the system operation wavelength. The power budget criteria requires the condition

$$P_L - L - M - P_{\min} = 0, \tag{12.24}$$

where M is the system margin which is allocated to consider the component aging degradations over their lifetimes and some operation cushion. L is the total loss coming mainly from the insertion loss of optical fiber, plus some connector loss L_{con}. L can be expressed as

$$L = \alpha l + L_{\text{con}}, \tag{12.25}$$

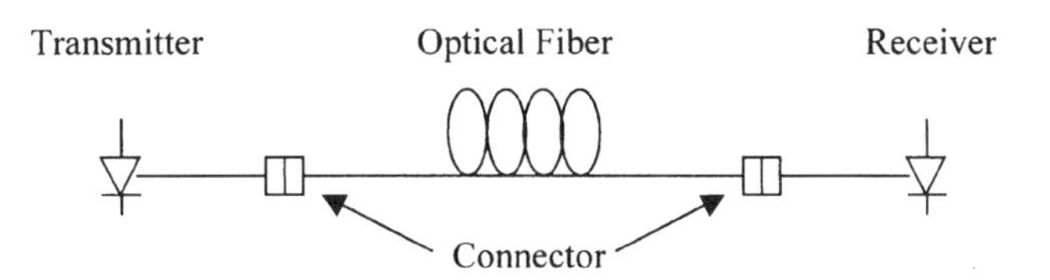

Fig. 12.20. A single-channel, point-to-point optical transmission link.

Table 12.3

Power Budget of an Optical Transmission Link

Transmitter launch power	−8 dB m
Receiver sensitivity	−22 dB m
System margin	6 dB
Optical fiber loss	0.6 dB/km
Connector loss	2 dB
Maximum transmission distance	10 km

where α is the optical insertion loss in dB/km and l is the fiber span length.

In Table 12.3 we list the specifications of a typical set of components, including transmitter, receiver, and optical fiber for a gigabit Ethernet application in the 1.3 μm window. Using Eqs. (12.24) and (12.25) we estimate that the maximum distance of the optical link is 10 km. In Table 12.3, the fiber fusion splice loss is already included in the insertion loss of single mode optical fiber.

12.3.3.2. *Optical Signal-to-Noise Ratio (OSNR)*

The introduction of optical amplifiers and DWDM technologies greatly complicates the power budget approach. The simple math in the last section, based on the fact that the system is insertion loss limited, is no longer valid since the impairments associated with amplifier noise, fiber dispersion, and nonlinearity are the dominant effects in determining maximum link distance. Another methodology, based on the optical signal-to-noise ratio (OSNR), must be adopted. Before addressing this in detail, we examine an important parameter called bit-error ratio (BER).

12.3.3.2.1. *Bit-Error Ratio*

There are many ways to measure the quality of optical transmission systems. The ultimate and most accurate test for any transmission medium is bit-error ratio (BER) performance. This reflects the probability of incorrect identification of a bit by the decision circuit of an optical receiver. In its simplest form BER is a figure of merit, defined as

$$\mathrm{BER} = \frac{B(t)}{N(t)}, \tag{12.26}$$

where $B(t)$ is the number of bits received in error and $N(t)$ is the total number of bits transmitted over time t. A commercial optical transmission system in general requires the BER to vary from 1×10^{-9} to 1×10^{-15}, typically around 1×10^{-12} (a BER of 1×10^{-12} corresponds to, on average, one error per trillion transmitted bits). Since BER is a statistical phenomenon, enough measuring time has to be allocated to ensure accuracy.

12.3.3.2.2. *Optical Signal-to-Noise Ratio*

In a contemporary long-haul DWDM transmission system, as shown in Fig. 12.21, high-power optical amplifiers are periodically used after every certain span of optical fibers, usually 80 or 100 km, to restore the power of the optical signal. Assume that the optical signal from the transmitter is superclean without any noise. The signal first encounters some components, such as the DWDM multiplexer, experiencing insertion loss. The first amplifier not only boosts signal power before it enters into optical fiber, but also generates amplified spontaneous emission (ASE) noise

$$P_{\text{noise}} = 2n_{sp}h\gamma(g - 1)B_0, \tag{12.27}$$

where $h\gamma$ is the photon energy, g is the gain of the amplifier, B_0 is the optical bandwidth in Hertz, and n_{sp} is related to the noise figure of the amplifier. The output of the first amplifier is then launched into the optical fiber. After a certain distance its power is attenuated by the fiber but boosted again by the second amplifier. Now besides the signal, ASE noise generated by the first amplifier is also amplified. Moreover, new ASE noise is generated by the second amplifier and added onto the signal. The process is repeated at every following amplifier site, and the ASE noise accumulates over the entire transmission line. The gain of amplifiers in such a system is usually automatically adjusted to ensure that the output power (signal power plus noise power) of every channel is fixed. Since more and more ASE noise is created and accumulated, signal power is sacrificed to keep total power unchanged. Sometimes the power level of accumulated noise can be very

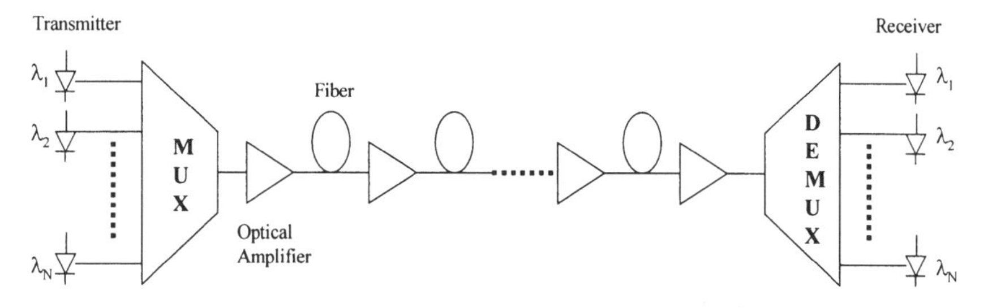

Fig. 12.21. A multichannel amplified optical transmission system.

high. As a result, optical SNR at the output of a chain of N amplifiers is reduced.

We now present some general but useful engineering rules of designing amplified DWDM optical transmission systems.

1. Allocate the minimum OSNR required by the optical receiver to achieve the desired system BER. For example, an OSNR of 21 dB at 0.1-nm measurement bandwidth may be required to achieve 1×10^{-15} BER for a system at 10 Gbits/s transmission rate.
2. Allocate the power penalty from optical fiber nonlinearity. There are many nonlinear phenomena, including SPM, XPM, FWM, SBS, and SRS, that may cause impairments to transmission systems, especially at high speed. Such impairments directly require additional optical power increase in order for the optical receiver to maintain the same system performance integrity. Such a power increase in unit decibels is called the power penalty.
3. Allocate the power penalty from optical fiber dispersion. Similar to nonlinearity, fiber dispersion leads to system performance degradation that adds additional OSNR requirements.
4. Allocate the OSNR margin, considering the potential penalty from the component aging effect, polarization dependent effects, and so on.

To conclude this section, we present a typical 10 Gbits per second system design example in Table 12.4. This example illustrates that system OSNR requirements may be raised substantially to compensate for various impairments.

12.4. APPLICATIONS AND FUTURE DEVELOPMENT OF OPTICAL NETWORKS

12.4.1. LONG-HAUL BACKBONE NETWORKS

Previous sections explained that various technologies, especially optical amplifiers and DWDM, have revolutionized backbone long-distance optical networks. Now we will briefly review the evolution of the backbone optical network from yesterday's short-reach system to tomorrow's ultra–long-haul (ULH), all-optical system. We will also introduce the key enabling technologies for ULH transmission.

12.4.1.1. *Evolution of Backbone Optical Network*

A generic schematic of the backbone optical network before the deployment of WDM and EDFA technologies is presented in Fig. 12.22(a), where TX and

Table 12.4

A Design Example Based on OSNR Allocation

Impairment	OSNR allocation (dB)
Ideal OSNR without impairment	21
Nonlinearity impairment	1
Dispersion impairment	2
System margin	3
Actual requirement	27

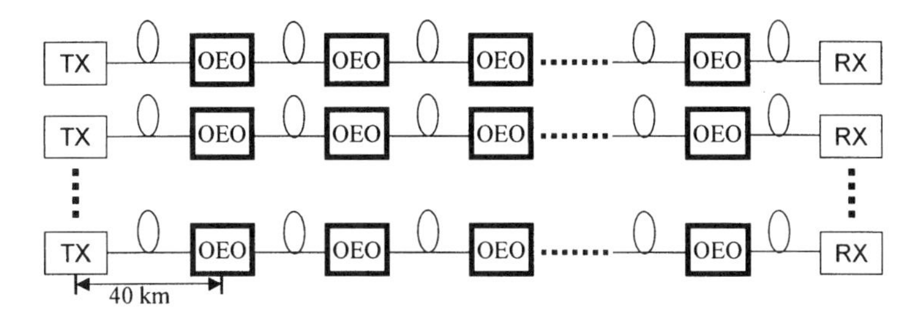

Fig. 12.22(a). Yesterday's backbone optical network.

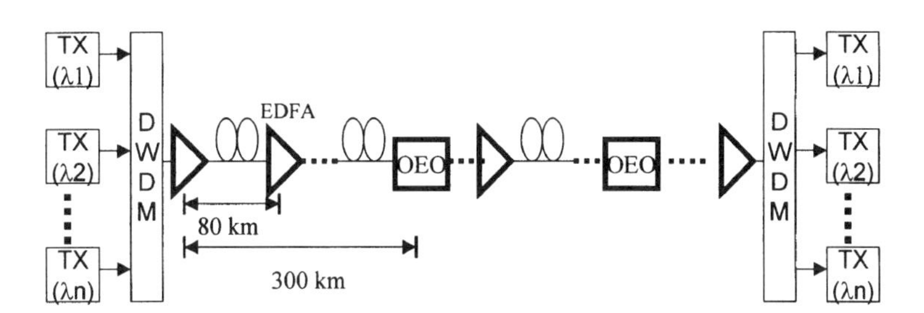

Fig. 12.22(b). Today's backbone optical network.

RX stand for laser transmitter and optical receiver, respectively. In each fiber only one wavelength is transmitted. Due to the attenuation of optical fiber, the optical signal can only propagate about 40 km. Then an optical receiver must be used to convert the signal to an electronic signal, which then modulates a new transmitter to restore the original optical signal. This kind of optical-to-electronic-to-optical (OEO) conversion repeats every other 40 km until the information reaches its final destination.

In today's backbone optical network, as shown in Fig. 12.22(b), EDFA and DWDM technologies are widely applied. DWDM technology is about to

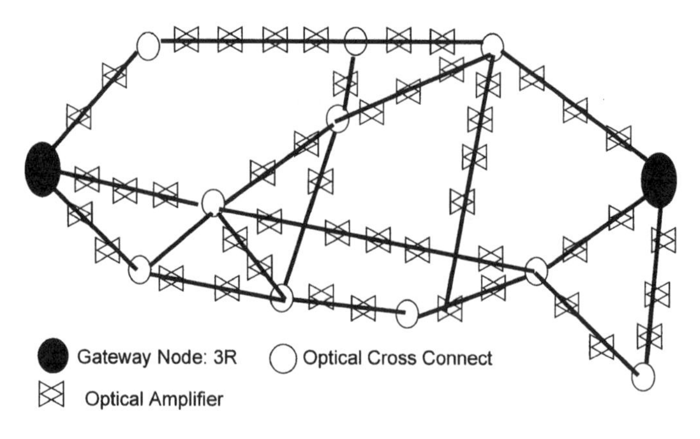

Fig. 12.22(c). Tomorrow's ultra–long-haul backbone network.

combine more than 100 wavelengths carrying life traffic into a single piece of fiber. EDFA, which is a much more cost-effective solution, has substantially reduced the use of OEO devices in today's networks. Nevertheless, the cost of such OEO regenerators is still the largest portion of overall network cost. Because of the economic factor, it is generally agreed that tomorrow's terrestrial backbone network should be an ultra–long-distance (over thousands of kilometers) one between OEO regenerators. Ideally, we should see a network similar to Fig. 12.22(c) without OEO regenerators from coast to coast.

12.4.1.2. *Enabling Technologies for Ultra–Long-Haul Transmission*

In this section several key enabling technologies, including forward error correction, Raman amplification, dynamic gain equalization filters, second-order dispersion compensation, and soliton transmission are briefly discussed.

12.4.1.2.1. Forward Error Correction (FEC)

We have seen previously that in order to achieve required system BER, a minimum OSNR has to be guaranteed. FEC is one of the coding technologies that can improve system BER without additional requirements on OSNR improvement. Figure 12.23 demonstrates how FEC is implemented. The encoder introduces extra bits to the input data on the transmitter site. On the receiver site, the decoder uses the extra bits to restore the input data sequence when the data is corrupted by the various impairments associated with the transmission link. As a result, FEC enables the system to achieve the same BER performance with less channel OSNR compared to a system without

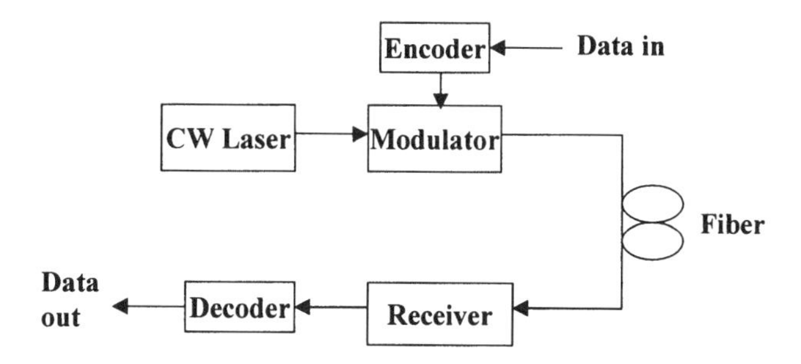

Fig. 12.23. Implementation of FEC in an optical communication system.

FEC. In other words, a system powered by FEC will achieve a longer transmission distance with the same BER.

The extra bits introduced by FEC will increase the bit rate. Practically, a 7% increase in bit rate will comfortably decrease system OSNR requirements by 5 dB or more, which in turn, easily doubles the transmission distance.

12.4.1.2.2. Raman Amplification

In high-capacity DWDM transmission systems, every channel has to keep enough OSNR in order to achieve required BER. Consequently, high-power amplifiers are used to boost the launch power into every fiber span. There are two direct impairments associated with this method, however:

- High-power amplifiers produce more ASE than low-power amplifiers. ASE is the major noise source degrading transmission system performance.
- High launch power causes more severe fiber nonlinearity impairment.

Mixed together with EDFA, Raman amplification becomes a key technology to enable ULH transmission.

Raman scattering was discussed in the previous section. Raman amplification actually takes advantage of Raman scattering. It occurs when high-energy pump photons scatter off the optical phonons of an amplification materials lattice matrix and add energy coherently to the signal photons. When applied to optical networks, pump light is launched directly into the transmission fiber and the transmission fiber serves as the gain medium to the optical channels. Hence, it is also called distributed Raman amplification. The advantages of using distributed Raman amplification are as follows:

- Distributed Raman amplification offers an effective noise figure which is much smaller than EDFA, resulting in higher system OSNR.

- Since the gain medium is distributed, the launch power into the fiber can be reduced substantially to lower the fiber nonlinearity penalty, such as FWM.
- The gain spectrum is flatter than that of EDFA, so the performance is more even for a multichannel DWDM system.

Practically, it is still difficult to manufacture a reliable high-power Raman pump source. It is more likely that the hybrid EDFA and Raman amplification will be deployed in tomorrow's ULH transmission systems.

12.4.1.2.3. *Gain Equalization Filter (GEF)*

The gain spectrum of EDFA is nonuniform, and the nonuniformity worsens in transmission systems with amplifier chains. The channels experiencing less amplification suffer a penalty from reduced OSNR while the channels experiencing more amplification may trigger fiber nonlinearity penalties. A gain equalization filter (GEF) is a device that compensates for the accumulated gain spectrum nonuniformity.

12.4.1.2.4. Second-Order Dispersion Compensation

We can easily find that the D parameter defined in Sec. 12.3.1 is also a function of wavelength. In other words, the dispersion of optical fiber varies with different channels. For tomorrow's higher-capacity ULH network, this second-order dispersion may result in severe impairment in that (a) the DWDM channels will occupy the whole C and L band, so the wavelength swing is at least 40 nm; and (b) at such high wavelength swing, the accumulated dispersion difference from thousands of kilometers of transmission fiber is huge.

As a result, traditional dispersion compensation, which only corrects the D parameter, is not enough for ULH networks because the residual dispersion difference may be fatal to some channels at high bit rates (>10 Gbits/s). Dispersion compensation devices which can correct second-order dispersion become a critical element.

12.4.1.2.5. Soliton Transmission

A soliton is a narrow optical pulse that retains its shape as it travels along a very long distance of optical fiber. For many years researchers all over the world have been studying soliton for high–bit rate, ULH transmission applications. Only recently has soliton emerged as a viable alternative for next-generation ULH optical networks.

In soliton transmission, two impairments to traditional systems; namely, chromatic dispersion of the optical fiber and self-phase modulation (SPM), actually work against each other to generate a stable, nondispersive pulse which can be transmitted over very long distances. Solitons are more robust to fiber polarization mode dispersion (PMD).

12.4.2. METROPOLITAN AND ACCESS NETWORKS

Access networks include networks connecting end users or customer premises equipment (CPE) to a local central office (CO). The metropolitan area network (MAN) is responsible for connecting clusters of access networks to a long-haul backbone network, also known as a wide-area network (WAN). The long-haul market is focused largely on lowest cost per bit, whereas metro networks focus on flexible, low-cost, service-oriented delivery of bandwidth. Because metro networks serve a much larger number of clients, including residential and business customers with varied needs, than long-haul networks, network scalability and software-based automatic provisioning to reduce truck roll for changes in network topology, bit rates, or customers are very crucial.

The overwhelming trend is to simplify network infrastructure by putting more functions in new optical layers. One result of this evolution is that the distinction between metro networks and access networks is blurring. Carriers are building new metro networks to deliver service to CPE in the most direct way possible.

New metro networks can be roughly cataloged as DWDM/router–based networks and SONET–based networks. Figure 12.24 shows an example of a DWDM/router–based network, in which bidirectional DWDMs couplers act as passive optical splitters, dividing the light carried by the two OC-48s into

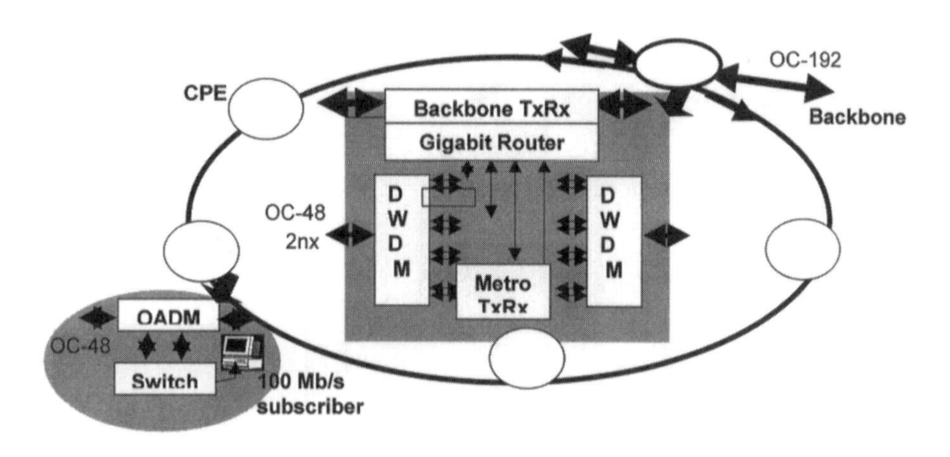

Fig. 12.24. DWDM router–based optical network.

$2n$ wavelengths (n passing clockwise through the ring, n passing counterclockwise). At CPE, an OADM picks up two of the $2n$ wavelengths on the ring—one traveling clockwise, one counterclockwise. The two wavelengths enter a switch. One of the wavelengths is converted directly to bandwidth for use by the building's tenants, at 100 Mb/s per tenant. The other provides an alternate route in case the first link fails due to a break in the fiber or a failure in one of the network devices. If fiber is cut on the way into the building, for instance, the switch will sense that packets are not coming through its primary port. It automatically will shift to the secondary port. Likewise, if a device fails on the customer network, the Internet connection is still guaranteed. This type of DWDM/router–based network is preferred by emerging Internet service providers that are looking to deliver more bandwidth at lower cost to their end users. As voice- and video-over IP matures, this network promises to deliver all services to the end user via an optical IP infrastructure.

SONET–based metro networks focus on providing data intelligence using a voice-optimized SONET platform. The SONET multiservice provisioning platform (MSPP) and passive optical network (PON) [28] are the two most active areas at present. SONET MSPPs use full or slimmed-down versions of SONET with added statistical multiplexing to handle other non–SONET traffic. They are designed to sit in carriers' COs and POPs, and, in some cases, CPEs to switch voice, video, and different types of data traffic. PON products focus on providing a low-cost way for service providers to deliver access capacity. Unlike SONET MSPP, they only work over the last mile, not among COs or POPs. Figure 12.25 shows a typical PON architecture [29]. An optical line terminal (OLT) sitting at a carrier's CO sends traffic downstream to network subscribers and handles its upstream return from subscribers. At the outside plant, passive optical splitters distribute traffic from OLT to CPEs

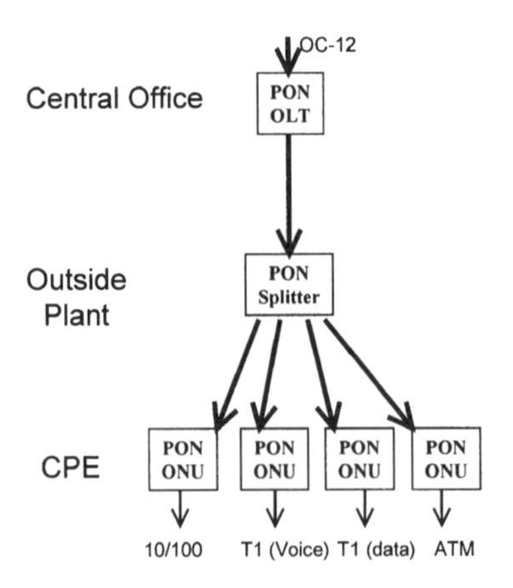

Fig. 12.25. PON architecture.

using star, ring, or tree configurations. Finally, optical network units (ONU) at CPE take in light that is sent from the passive splitters, convert it to specific types of bandwidth such as Ethernet, ATM, and T1 voice and data, and pass it on to routers, PBXs, switches, and other enterprise-networking gear. ONUs also incorporate the lasers that send traffic back to the central office at the command of the OLT. Both SONET MSPP and PON are well suited for incumbent carriers with a need to stretch their SONET infrastructure for more data traffic. And DWDM is supported on both platforms for capacity boosting and for optical-level bandwidth granularity.

The huge traffic demand, coupled with unprecedented capacity and networking capability growth in long-haul back and access networks, is pushing metro networks to grow from <50 km coverage to a much bigger range of 200 to 400 km. DWDM also is becoming a necessity in metro networks, less for its capacity boosting capability, than for the transparency and service flexibility brought forth by wavelength networking. Optical add/drop modules (OADM) are widely deployed in metro networks as an effective optical way to manipulate bandwidth. All this together indicates the beginning of an era of optical networking.

12.4.3. FUTURE DEVELOPMENT

At the beginning of the 21st century, networking with optical wavelengths is still at a very early stage. DWDM transmission is in the middle of evolving into DWDM networking, driven by unprecedented bandwidth demand from widespread use of the Internet. Advances in optical technology for networking have so far been trailing market demand. Breakthroughs in the areas of ultra–long-haul (>3000 km) transmission, optical cross-connect, tunable optical source, optical monitoring, and tunable optical filters will all have a big impact on the progress of optical networking.

Ultra–long-haul optical transmission is revolutionizing the communication infrastructure [30]. It not only cuts down backbone transmission cost and provides transparency by eliminating rate-dependent electrical 3R along the route, but also facilitates optical layer bandwidth manipulation through OADM and OXC. Optical amplification is the base for coast-to-coast ultra–long-haul backbone transmission. Conventional optical amplified systems can only reach <1000 km distance before electrical 3R is used. Transmission of DWDM OC-192 channels over 2000 to 5000-km distances poses major challenges that cannot be met with conventional DWDM technology [31]. Foremost among these problems are dispersion accumulation, impairments due to optical nonlinearities, accumulation of optical noise, and optical amplifier gain nonuniformities. At present, Raman amplification (both distributed and discrete) [32], super FEC, high–spectral efficiency modulation, dispersion slope management, and dynamic gain equalization are the key technologies under study to combat these difficulties. Another effort to realize ultra–long-haul transmission is soliton transmission.

Opaque OXC—optical cross-connect using O/E/O and electrical switch core has been deployed in the field since late 2000. Transparent core OXC uses optical switch core and some O/E/O, and totally transparent OXC uses optical switch and no O/E/O. OXCs based on optical switch core are still striving to make their way to real-world applications. Electrical switch fabrics offer clear advantages. They allow use of proven, off-the-shelf chip technology, and they provide channel monitoring and management capabilities. On the other hand, they bring clear disadvantages as well. Electrical switch fabric requires O/E/O that uses expensive optoelectronic components such as lasers and modulators, and limits the port speed and ultimately scalability of OXC. In contrast, all-optical switches use MEMs and other transparent optical components to redirect lightwaves without regard to the speed of the signal. Thus, they have the benefit of port-speed independence, so that carriers can upgrade their transport equipment without having to replace their switches. However, all-optical switches offer little channel monitoring and line rate bandwidth grooming, making service provisioning and network management quite challenging. Therefore, opaque OXC will be around for quite a while, perhaps forever at the edge of the network. Transparent core OXC is a future-ready interim solution. Totally transparent OXC will dominate the deep core of the network when optical switch as well as optical monitoring technologies mature.

While wavelengths hold the key to commercial optical networking, they are generated by wavelength-stabilized DFB lasers. Wavelength-tunable semiconductor lasers will provide networking flexibility and cost reduction in the following areas:

- Spares. Today, for every discrete wavelength deployed in an 8- or 80-channel DWDM system, 8 or 80 spare transponders must be in inventory in the event of transmitter failure. Tunable semiconductor lasers have been proposed as a means to solve this problem.
- Hot backup. An idle channel containing a tunable laser is used for redundancy to immediately replace any failed transponder.
- Replacement. Using tunable lasers to fully replace the fixed-wavelength lasers in DWDM networks solves inventory management and field-deployment complexity.
- Reconfigurable and/or dynamic OADM. In optical networks using OADM, tunable lasers enable network operators to dynamically allocate

which channels are dropped from the network and which channels are added.

- Optical cross-connect and wavelength routing. Tunable laser transmitters are a way to achieve additional flexibility, reliability, and functionality in an optical network. Tuning the laser from one wavelength to another wavelength can change the path that is taken by an optical signal. A tunable filter at the receiving node is also required.

In the future, nanosecond-speed tunable lasers may make it possible to switch packets in the optical domain. Today, there are four basic technology candidates for tunable lasers:

1. DFB lasers. Limited tuning (5 to 15 nm) is achieved by tuning the temperature of the laser chip.
2. Distributed Bragg Reflector (DBR) lasers. The gratings of DBR lasers are built in the one of the passive sections of the laser cavity, instead of inside the active section as is the case for DFBs. The biggest benefit is that injection current to the grating can result in a much bigger change in the refractive index of the cavity, since carriers in the passive section are not clamped as they are in the active section; this in turn results in wider wavelength tuning through injection current [33].
3. External-cavity diode lasers, which consist of a Fabry–Perot laser diode and an external high-reflectivity diffraction grating. Tuning is achieved through mechanically tilting the grating.
4. Vertical-cavity surface-emitting lasers (VCSELs). VCSELs are compound semiconductor microlaser diodes that emit light vertically perpendicular to their p–n junctions from the surface of a fabricated wafer [34].

Tuning is achieved by moving the top reflection mirror of the laser cavity implemented with MEMS. All of these four tunable laser technologies have their advantages and disadvantages.

Tunable optical filters, which pass any desired wavelength via software-based electrical control, are the next step toward current fixed-wavelength mux/demux used in DWDM networks. Tunable lasers and tunable filters together, they enable the dynamic routing of all wavelengths in an optical network; that is, any wavelength can be routed to any destination via software provisioning.

Optical monitoring must extract all information for transmission, such as power, spectrum, noise, dispersion, and nonlinearity, and for networking, such as destination, routs, priority, and content type, at the optical layer so that the optical network has an OAM&P (operation, administration, maintenance, and provisioning) intelligence close to that of the electrical network. This is a very important area for optical networking with a great deal left to be deployed.

REFERENCES

12.1 C. A. Brackett, June 1990, "Dense Wavelength Division Multiplexing Networks: Principles and Applications," *IEEE J. Select. Areas Commun.*, vol. 8, no. 6, 948.

12.2 N. K. Cheung, K. Nosu, and G. Winzer, Eds., June 1990, "Special Issue on Dense Wavelength Division Multiplexing Techniques for High-Capacity and Multiple-Access Communications Systems," *IEEE J. Select. Areas Commun.*, vol. 8, no. 6, 945.

12.3 P. E. Green, Jr., *Fiber Optic Networks*, Prentice-Hall, Englewood Cliffs, 1993.

12.4 J. R. Jones, October 1982, "A Comparison of Lightwave, Microwave, and Coaxial Transmission Technologies," *IEEE J. Quantum Electronics*, vol. QE-18, no. 10, 1524.

12.5 J. L. Zyskind, C. R. Giles, J. R. Simpson, and D. J. Digiovanni, Jan./Feb. 1992, "Erbium-Doped Amplifiers and the Next Generation of Lightwave Systems," *AT&T Technical Journal*, 53.

12.6 C. R. Giles and T. Li, June 1996, "Optical Amplifiers Transform Long-Distance Lightwave Telecommunications," *Proceedings of the IEEE*, vol. 84, no. 6, 870.

12.7 N. A. Olsson, July 1989, "Lightwave Systems with Optical Amplifiers," *J. Lightwave Technol.*, vol. 7, no. 7, 1071.

12.8 P. E. Green, Jr., June 1996, "Optical Networking Update," *IEEE J. Select. Areas Commun.*, vol. 14, no. 5, 764.

12.9 D. Y. Al-Salameh, et al., Jan.–March 1998, "Optical Networking," *Bell Labs Technical Journal*, 39.

12.10 C. R. Giles and E. Desurvire, Feb. 1991, "Modeling Erbium-Doped Fiber Amplifiers," *J. Lightwave Technol.*, vol. 9, no. 2, 271.

12.11 C. R. Giles and E. Desurvire, Feb. 1991, "Propagation of Signal and Noise in Concatenated Erbium-Doped Fiber-Optical Amplifiers," *J. Lightwave Technol.*, vol. 9, no. 2, 147.

12.12 J.-M. P. Delavaux and J. A. Nagel, May 1995, "Multistage Erbium-Doped Fiber Amplifier Designs," *J. Lightwave Technol.*, vol. 13, no. 5, 703.

12.13 D. Baney, C. Hentschel, and J. Dupre, "Optical Fiber Amplifiers—Measurement of Gain and Noise Figure," Hewlett Packard, Lightwave Symposium, 1993.

12.14 H. A. Macleod, *Thin-Film Optical Filters*, 2nd edition, Macmillan Publishing Company, New York, 1986.

12.15 Z. Knittl, *Optics of Thin Films*, John Wiley, New York, 1976.

12.16 M. Bass, E. Van Stryland, David Williams, and William Wolf, *Handbook of Optics*, 2nd edition, McGraw Hill, London, UK, 1994.

12.17 C. Dragone, 1991, "An $N \times N$ Optical Multiplexer Using a Planar Arrangement of Two Star Couplers," *IEEE Photonic Technology Letters*, vol. 3, 812.

12.18 Simon Cao, Avanex Corporation, personal communication.

12.19 Kuang-Yi Wu and Jian-Yu Liu, Programmable Wavelength Router, U.S. Patent 5,978,116, November 2, 1999.

12.20 P. Perrier, "Optical Cross-Connects—Part 1: Applications and Features," Fiber Optics Online, 08/25/2000.

12.21 P. Perrier, "Optical Cross-Connects—Part 2: Enabling Technologies," Fiber Optics Online, 09/12/2000.

12.22 C. Marxer and N. F. de Rooij, January 1999, "Micro-Optomechanical 2 × 2 Switch for Single Mode Fibers Based on Plasma-Etched Silicon Mirror and Electrostatic Actuation," *J. Lightwave Technol.*, vol. 17, no. 1, 2.

12.23 M. Makihara, et al., January 1999, "Micromechanical Optical Switches Based on Thermocapillary Integrated in Waveguide Substrate," *J. Lightwave Technol.*, vol. 17, no. 1, 14.

12.24 R. G. Smith, 1972, "Optical Power Handling Capability of Low-Loss Optical Fiber as Determined by Stimulated Raman and Brillouin Scattering," *Applied Optics*, vol. 11, 2489–2160.

12.25 E. P. Ippen, in *Laser Applications to Optics and Spectrascopy*, vol. 2, Addison-Wesley, Reading, Mass., 1975, Chapter 6.
12.26 R. H. Stolen, "Nonlinear Properties of Optical Fibers," in *Optical Fiber Telecommunications*, S. E. Miller and A. G. Chynoweth, eds., Academic Press, New York, 1979.
12.27 G. P. Agrawal, Fiber-Optic Communication Systems, 2nd edition, John Wiley & Sons, Inc., New York, 1997
12.28 M. Klimek et al., 2000, "Advances in Passive Optical Networking," *NFOEC'00*, 699.
12.29. Mary Jander, "PONs: Passive Aggression," Lightreading. com. 5/8/2000
12.30 A. A. M. Saleh, 2000, "Transparent Optical Networking in Backbone Networks," *Optical Fiber Conference 2000, ThD7-1*, 62.
12.31 J. L. Zyskind, et al., 2000, "High-Capacity, Ultra–Long-Haul Transmission," *NFOEC'00*, 342.
12.32 N. Takachio et al., "32 × 10 Gb/s Distributed Raman Amplification Transmission with 50-GHz Channel Spacing in the Zero-Dispersion Region over 640 km of 1.55-μm Dispersion-Shifted Fiber," OFC'99, PD9, San Diego, CA, 1999.
12.33 L. Zhang and J. Cartledge, January 1995, "Fast-Wavelength Switching of Three-Section DBR Lasers," *J. Quantum Electr.*, vol. 31, no. 1, 75.
12.34 M. Jiang, et al., "Error-Free 2.5 Gb/s Transmission of a C-Band Tunable 8 dBm Vertical-Cavity Surface-Emitting Laser over 815 km of Conventional Fiber," *NFOEC'00*, 186.
12.35 F. Forghieri, R. W. Tkach, and A. R. Chraplyvy, "Fiber Nonlinearities and Their Impact on Transmission Systems," in *Optical Fiber Telecommunications IIIA*, eds., I. P. Kaminow and T. L. Koch, p. 203–204, Academic Press, 1997.

EXERCISES

12.1 One digital voice channel takes up 64 kb/s. How many voice channels can an OC-192 (9.953.28 Gb/s) system deliver?

12.2 Decibel (dB) is commonly used in communications. It can represent absolute signal power, defined as P (dBm) = 10 log (P(mW)), and relative level, defined as ΔP (dB) = 10 Log (P1 (mW)/P2 (mW)). 0 dBm represents 1 mW and 3 dB represents two times. An optical amplifier has −5 dBm total input from an 80-channel DWDM signal. Suppose all channels have the same power. What is the input power to the amplifier for each channel?

12.3 Noise power in a fiber-optic communication system depends linearly on the measurement bandwidth. The OC-192 receiver requires that the incoming signal has a minimum optical signal-to-noise ratio (OSNR) of 25 dB, for 0.1-nm bandwidth. What is the requirement of OSNR for 0.5-nm bandwidth?

12.4 In DWDM systems, one method is to use an optical bandpass filter (BPF) to pick out a single channel from a composite DWDM signal, and feed it to the receiver. Since the optical receiver reacts to a wide spectrum of light, it falls on the optical BPF to suppress out-of-band noises before they flood the receiver. If the receiver requires that the

BPF suppresses the unwanted channel to 25 dB down the wanted channel, in the case of two-channel DWDM, how much lower should every unwanted channel be relative to the wanted channel if channel counts increase to 80?

12.5 Chromatic dispersion in optical fiber limits the bandwidth–distance product of a system. For 140 km of single mode fiber having 17 ps/nm/km chromatic dispersion, assuming the laser source has 10 nm FWHM (full width at half maxim) and system 20% bandwidth efficiency, what is the maximum dispersion-limited data rate of the system?

12.6 Find the numerical steady-state solutions for amplifier rate equations (12.1) to (12.5), assuming 20 meter Erbium fiber length, 90 mW pump power at 980 nm, and -10 dBm input signal at 1550 nm.

12.7 An optical amplifier produces 2.00 μW amplified spontaneous emission (N_{EDFA}), for a signal wavelength of 1550 nm and a bandwidth of 0.5 nm. What is the corresponding noise figure?

12.8 Draw a configuration of a transmit transponder having a 1310-nm SONET interface, SONET 3R, and FEC, as well as asynchronous 2R capabilities. Describe in detail the functions of each block.

12.9 Refer to Exercise 12.6. Draw the configuration of the corresponding receiver transponder. Describe in detail the functions of each block.

12.10 Modern optical communication systems use the wavelength range 1300–1650 nm in a silica fiber for signal transmission. If the channel spacing is 12.5 GHz, how many channels can be carried by a single fiber?

12.11 Typically, optical signal experiences 0.8 dB insertion loss when it passes through a commercial dielectric thin-film filter, and 0.4 dB insertion loss when it is reflected by the same filter. Estimate the maximum insertion loss of the eight-channel DWDM demultiplexer in Fig. 12.8.

12.12 Derive the transmission expression of a chain of Mach–Zehnder interferometers acting as a filter.

12.13 Demonstrate how the Mach–Zehnder interferometer is used as a demultiplexer and show the design schematic of a 1×4 demultiplexer.

12.14 Design a 32-channel, 50-GHz channel-spacing DWDM demultiplexer based on the knowledge learned from Sec. 12.2.3. There may be multiple solutions; compare the technical merit and economy of different solutions.

12.15 Estimate the threshold power of stimulated Brillouin scattering in conventional single mode fiber.

12.16 Explain why self-phase modulation in optical fiber causes the peak of an optical pulse to travel slower in the fiber than the wings.

12.17 Two wavelengths, at frequency f_1 and f_2, can also interact through optical fiber to generate four wave–mixing sidebands. Derive the expression of the frequencies of the sidebands.

12.18 Given the following parameters for a system working at 0.85 μm window, transmitter launch power -10 dBm and receiver sensitivity -35 dBm, 3.5 dB/km fiber cable loss, estimate the maximum link distance.

Index

Q